MW00451347

SOIL WATER AND AGRONOMIC PRODUCTIVITY

Advances in Soil Science

Series Editors: Rattan Lal and B. A. Stewart

Published Titles

Interacting Processes in Soil Science
R. J. Wagenet, P. Baveye, and B. A. Stewart

Soil Management: Experimental Basis for Sustainability and Environmental Quality
R. Lal and B. A. Stewart

Soil Management and Greenhouse Effect
R. Lal, J. M. Kimble, E. Levine, and B. A. Stewart

Soils and Global Change
R. Lal, J. M. Kimble, E. Levine, and B. A. Stewart

Soil Structure: Its Development and Function
B. A. Stewart and K. H. Hartge

Structure and Organic Matter Storage in Agricultural Soils
M. R. Carter and B. A. Stewart

Methods for Assessment of Soil Degradation
R. Lal, W. H. Blum, C. Valentine, and B. A. Stewart

Soil Processes and the Carbon Cycle
R. Lal, J. M. Kimble, R. F. Follett, and B. A. Stewart

Global Climate Change: Cold Regions Ecosystems
R. Lal, J. M. Kimble, and B. A. Stewart

Assessment Methods for Soil Carbon
R. Lal, J. M. Kimble, R. F. Follett, and B. A. Stewart

Soil Erosion and Carbon Dynamics
E.J. Roose, R. Lal, C. Feller, B. Barthès, and B. A. Stewart

Soil Quality and Biofuel Production
R. Lal and B. A. Stewart

Food Security and Soil Quality
R. Lal and B. A. Stewart

World Soil Resources and Food Security
R. Lal and B. A. Stewart

Soil Water and Agronomic Productivity
R. Lal and B. A. Stewart

Advances in Soil Science

SOIL WATER AND AGRONOMIC PRODUCTIVITY

Edited by

Rattan Lal and B. A. Stewart

CRC Press is an imprint of the
Taylor & Francis Group, an **informa** business

CRC Press
Taylor & Francis Group
6000 Broken Sound Parkway NW, Suite 300
Boca Raton, FL 33487-2742

Printed in the United States of America on acid-free paper
Version Date: 20120523

International Standard Book Number: 978-1-4398-5079-4 (Hardback)

Library of Congress Cataloging-in-Publication Data

Soil water and agronomic productivity / edited by Rattan Lal and B.A. Stewart.
p. cm. -- (Advances in soil science ; v. 19)
Includes bibliographical references and index.
ISBN 978-1-4398-5079-4
1. Soil moisture. 2. Soil moisture conservation. 3. Water in agriculture. 4. Water-supply. I. Lal, R. II. Stewart, B. A. (Bobby Alton), 1932- III. Series: Advances in soil science (Boca Raton, Fla.) ; v. 19.

S594.S683 2012
631.4'32--dc23
2012006491

Visit the Taylor & Francis Web site at
http://www.taylorandfrancis.com

and the CRC Press Web site at
http://www.crcpress.com

Contents

SECTION III Irrigation and Soil Water Management

SECTION IV Agronomic Management of Soil and Crop

SECTION V Policy and Economics

SECTION VI Tools of Watershed Management

SECTION VII Research and Development Priorities

Preface

Only 2.75% ($40.7 \times 10^6 km^3$) of the world's water pools are renewable freshwater resources. Of these, polar ice caps and glaciers (melting rapidly because of climate change) constitute 2% (29.6×10^6 km^3), groundwater constitutes 0.7% ($10.36 \times 10^6 km^3$), and all other water bodies constitute 0.05%. The latter include soil water, permafrost, and wetlands, which constitute 0.0167% ($0.25 \times 10^6 km^3$), and freshwater lakes and rivers, which constitute about 0.033% ($0.048 \times 10^6 km^3$). Thus, renewable freshwater supply is a scarce resource and is unequally distributed among the world regions.

More severe than the scarcity of cropland is the nonavailability of freshwater supply for the rapidly expanding world's population, with competing demands for nonagricultural uses. About 1 billion people, mostly in rural Asia and sub-Saharan Africa (SSA), do not have access to hygienically clean water. By 2050, the annual per capita available freshwater supply will be merely 503 m^3 (the minimum required is $10^3 m^3$/year) for Egypt, 517 m^3 for Ethiopia, 543 m^3 for UAE, 690 m^3 for Iran, 791 m^3 for Burkina Faso, 803 m^3 for Zimbabwe, and 815 m^3 for Afghanistan. There will be 30 densely populated countries (e.g., Pakistan, India, and others in arid and semiarid climates) that will face severe water shortage and recurring drought stress by 2025. Water-related problems will be exacerbated by climate change and the attendant increase in the intensity and frequency of extreme events.

Hunger and related malnutrition affect 1,020 million people around the world. Of these, 230 million live in India and another 220 million in SSA. Low agronomic production is attributed to low crop yields of <1 t/ha under rainfed conditions because of recurring drought stress. Loss of grain production due to water scarcity in developing countries is estimated at 100 Mt in 1995, 300 Mt in 2025, and 425 Mt in 2050. Crop yields can be improved through soil-water conservation and increase in area under supplemental irrigation. Over a 42-year period between 1961 and 2003, cropland area under irrigation increased from 3.5 million hectares (Mha) or 8.6 million acres (Ma) to 7.0 Mha (17.3 Ma) in SSA. Only 5% of the irrigable land area is currently being irrigated in SSA. In comparison, cropland area under irrigation increased between 1961 and 2003 from 30.4 Mha (75.1 Ma) to 54.6 Mha (134.8 Ma) in China and from 24.7 Mha (61.0 Ma) to 55.8 Mha (137.8 Ma) in India.

The need for an efficient use of soil water is also enhanced by the lack of availability of freshwater supply for supplemental irrigation. Global water use for agriculture, as a percentage of the total water use, was 81.4% in 1900, 72.3% in 1950, 68.2% in 1975, and 56.7% in 2000. Global water use for urban purposes (km^3/year) was 20 in 1900, 60 in 1950, 150 in 1975, and 440 in 2000. Similarly, global water use (km^3/year) for industrial purposes was 30 in 1900, 190 in 1950, 630 in 1975, and 1900 in 2000.

Availability of water for irrigation is also constrained by the diversion to fossil fuel production and eutrophication/pollution of water resources. One liter of bioethanol production requires 3500 L of fresh water. Thus, there is a strong and prime need for conserving, recycling, and improving soil-water resources to meet the food demands of the growing world population.

The severity of drought is likely to be exacerbated by the projected climate change because of an increase in the frequency of extreme events. The abrupt climate change may increase the risks of three types of drought. These are (1) meteorological drought caused by the long-term decrease in precipitation, (2) hydrological drought by the long-term decline in surface runoff and severe fall in the groundwater levels, and (3) agronomic drought caused by the reduction in soil moisture availability because of degradation in the structural properties and retention porosity. Severe degradation, especially of soil physical and biological qualities, aggravates agronomic drought through a decrease in the effective rooting depth by accelerated erosion, a decline in the soil organic matter content by decomposition and erosion, a reduction in the magnitude and stability of aggregates, a decline in the water infiltration rate and an increase in the losses by surface runoff, and an increase in soil evaporation. Thus, agronomic yields of upland crops, especially of shallow-rooted seasonals/annuals, are adversely affected by agronomic droughts. A decline in soil fertility and an elemental imbalance, along with an increase in salinization, also impact the vulnerability of crops to drought.

Therefore, soil-water management is crucial to reducing the vulnerability to agronomic drought. Technological innovations to enhance the availability of water for agricultural crops through soil-water management depend on soil- and site-specific conditions. Crop water use can be increased by the management of surface runoff, groundwater, irrigation, and soil water. This volume is devoted to the principles and practices of enhancing water-use efficiency. This 21-chapter volume is thematically divided into seven sections: (1) Water and Agronomic Productivity (two chapters), (2) Water Resources and Agriculture (six chapters), (3) Irrigation Management (four chapters), (4) Agronomic Management of Soil and Crop (six chapters), (5) Policy (one chapter), (6) Tools of Watershed Management (one chapter), and (7) Research and Development Needs (one chapter). World-renowned scientists were invited to contribute chapters to illustrate these seven themes.

The editors thank all the authors for their outstanding contributions to this volume. Thanks are due to the staff of Taylor & Francis for their timely efforts in publishing this volume and also to the staff of the Carbon Management and Sequestration Center, who made valuable contributions. Our special thanks are due to Ms. Theresa L. Colson for her dedication and commitment in handling the editorial production of the chapters and for collaborating with the authors in the review process.

Rattan Lal
Bobby Stewart

Editors

Rattan Lal is a distinguished university professor of soil physics in the College of Food Agriculture and Environment Sciences and director of the Carbon Management and Sequestration Center, Food, Agricultural, and Environmental Sciences/Ohio Agriculture Research and Development Center, The Ohio State University. Before joining Ohio State in 1987, he was a soil physicist for 18 years at the International Institute of Tropical Agriculture, Ibadan, Nigeria. In Africa, Professor Lal conducted long-term experiments on land use; watershed management; soil erosion processes as influenced by rainfall characteristics and soil properties; methods of deforestation; soil-tillage and crop-residue management; cropping systems including cover crops and agroforestry; and mixed/relay cropping methods. He also assessed the impact of soil erosion on crop yield and related erosion-induced changes in soil properties to crop growth and yield. Since joining The Ohio State University in 1987, he has continued research on erosion-induced changes in soil quality and developed a new project on soils and climate change. He has demonstrated that accelerated soil erosion is a major factor affecting the emission of carbon from the soil to the atmosphere. Soil-erosion control and adoption of conservation-effective measures can lead to carbon sequestration and mitigation of the greenhouse effect. His other research interests include soil compaction, conservation tillage, mine soil reclamation, water table management, and sustainable use of soil and water resources of the tropics for enhancing food security. Professor Lal is a fellow of the Soil Science Society of America, American Society of Agronomy, Third World Academy of Sciences, American Association for the Advancement of Sciences, Soil and Waste Conservation Society, and Indian Academy of Agricultural Sciences. He is a recipient of the International Soil Science Award of the Soil Science Society of America, the Hugh Hammond Bennett Award of the Soil and Water Conservation Society, the 2005 Borlaug Award, and the 2009 Swaminathan Award. He also received an honorary degree of Doctor of Science from Punjab Agricultural University, India; from the Norwegian University of Life Sciences, Aas, Norway; and from the Alecu Russo Balti State University, Moldova. He is the past president of the World Association of the Soil and Water Conservation, the International Soil Tillage Research Organization, and the Soil Science Society of America. He is a member of the U.S. National Committee on Soil Science of the National Academy of Sciences (1998–2002, 2007–present). He has served on the panel on Sustainable Agriculture and the Environment in the Humid Tropics of the National Academy of Sciences. He has authored and coauthored about 1500 research papers. He has also written 15 and edited or coedited 48 books.

B.A. Stewart is a distinguished professor of soil science at West Texas A&M University, Canyon, Texas. He is also the director of the Dryland Agriculture Institute and a former director of the USDA Conservation and Production Laboratory at Bushland, Texas; past president of the Soil Science Society of America; and a member of the 1990–1993 Committee on Long-Range Soil and Water Policy, National Research Council, National Academy of Sciences. He is a fellow of the Soil Science Society of America, the American Society of Agronomy, and the Soil and Water Conservation Society; a recipient of the USDA Superior Service Award; a recipient of the Hugh Hammond Bennett Award of the Soil and Water Conservation Society; and an honorary member of the International Union of Soil Sciences in 2008. Dr. Stewart is very supportive of education and research on dryland agriculture. The B.A. and Jane Anne Stewart Dryland Agriculture Scholarship Fund was established at West Texas A&M University in 1994 to provide scholarship to undergraduate and graduate students with a demonstrated interest in dryland agriculture.

Contributors

Francisco J. Arriaga
USDA-ARS National Soil Dynamics Laboratory
Auburn, Alabama

Li Baoguo
Department of Soil and Water Sciences
China Agricultural University
Beijing, China

ElSayed Abdel Bary
Soil Science Department
Zagazig University
Sarqia, Egypt

G. Basch
Institute of Mediterranean Agricultural and Environmental Sciences
University of Evora
Evora, Portugal

R. Louis Baumhardt
USDA-ARS Conservation and Production Research Laboratory
Bushland, Texas

E. Bautista
U.S. Arid-Land Agricultural Research Center
Maricopa, Arizona

Ronald L. Bingner
USDA-ARS National Sedimentation Laboratory
Oxford, Mississippi

J. Bordovsky
Texas AgriLife Research Center
Texas A&M University System
Lubbock, Texas

K.F. Bronson
Irrigation Water Management
U.S. Arid-Land Agricultural Research Center
Maricopa, Arizona

A. Calegari
Soil Science Department
Agronomic Institute of Paraná
Londrina, Brazil

A.J. Clemmens
WEST Consultants, Inc.
Tempe, Arizona

Paul D. Colaizzi
USDA-ARS Conservation and Production Research Laboratory
Bushland, Texas

Seth M. Dabney
USDA-ARS National Sedimentation Laboratory
Oxford, Mississippi

D.R. dos Santos
Soil Science Department
Federal University of Santa Maria
Santa Maria, Brazil

Anjali Dubey
Carbon Management and Sequestration Center
The Ohio State University
Columbus, Ohio

Robert G. Evans
USDA-ARS Northern Plains Agricultural Research Laboratory
Sidney, Montana

Steve R. Evett
USDA-ARS Conservation and Production Research Laboratory
Bushland, Texas

Huang Feng
Department of Soil and Water Sciences
China Agricultural University
Beijing, China

T. Friedrich
Plant Production and Protection Division
Food and Agriculture Organization of the United Nations
Rome, Italy

Huang Gao-bao
Gansu Agricultural University
Gansu, China

Kaushal K. Garg
Resilient Dryland Systems
International Crops Research Institute for the Semi Arid Tropics
Patancheru, India

J. Gastelum
Central Arizona Project
Phoenix, Arizona

P.I. Gubiani
Soil Science Department
Federal University of Santa Maria
Santa Maria, Brazil

N.C. Hansen
Department of Soil and Crop Sciences
Fort Collins, Colorado

Terry A. Howell
USDA-ARS Conservation and Production Research Laboratory
Bushland, Texas

D.J. Hunsaker
U.S. Arid-Land Agricultural Research Center
Maricopa, Arizona

Mostafa Ibrahim
Carbon Management and Sequestration Center
The Ohio State University
Columbus, Ohio

Meharban Singh Kahlon
Department of Soil Science
Punjab Agricultural University
Ludhiana, India

A. Kassam
School of Agriculture
University of Reading
Reading, United Kingdom

Roger A. Kuhnle
USDA-ARS National Sedimentation Laboratory
Oxford, Mississippi

Rattan Lal
Carbon Management and Sequestration Center
The Ohio State University
Columbus, Ohio

Wenzhao Liu
Institute of Soil and Water Conservation
The Chinese Academy of Sciences
Yangling, China

Pritpal Singh Lubana
College of Agricultural Engineering
Punjab Agricultural University
Ludhiana, Punjab, India

P.S. Minhas
Indian Council of Agricultural Research
New Delhi, India

Ephraim Nkonya
International Food Policy Research Institute
Washington, DC

Susan A. O'Shaughnessy
USDA-ARS Conservation and Production Research Laboratory
Bushland, Texas

G.A. Peterson
Department of Soil and Crop Sciences
Fort Collins, Colorado

Xu Qiang
Ningxia University
Yinchuan, China

Raveendra Kumar Rai
Water Resources Division
DHI (India) Water & Environment Private Ltd.
New Delhi, India

J.M. Reichert
Soil Science Department
Federal University of Santa Maria
Santa Maria, Brazil

James R. Rigby
USDA-ARS National Sedimentation Laboratory
Oxford, Mississippi

Claudia Ringler
International Food Policy Research Institute
Washington, DC

Johan Rockström
Stockholm Environment Institute
Stockholm, Sweden

F.L. Santos
Institute of Mediterranean Agricultural and Environmental Sciences
University of Evora
Evora, Portugal

Z. Sheng
Texas AgriLife Research Center
Texas A&M University System
El Paso, Texas

F. Doug Shields
USDA-ARS National Sedimentation Laboratory
Oxford, Mississippi

Anil Kumar Singh
Indian Council of Agricultural Research
New Delhi, India

Vijay P. Singh
Department of Biological and Agricultural Engineering
and
Department of Civil and Environmental Engineering
Texas A&M University
College Station, Texas

B.A. Stewart
Dryland Agriculture Institute
West Texas A&M University
Canyon, Texas

Atef Swelam
Agricultural Engineering Department
Zagazig University
Zagazig, Egypt

Venkatesh Uddameri
Department of Environmental Engineering
Texas A&M University–Kingsville
Kingsville, Texas

Paul W. Unger (retired)
Formerly with: USDA-ARS Conservation and Production Research Laboratory
Bushland, Texas

C. Wang
Department of Agricultural and Applied Economics
Texas Tech University
Lubbock, Texas

Gao Wang-sheng
China Agricultural University
Beijing, China

Suhas P. Wani
Resilient Dryland Systems
International Crops Research Institute for the Semi Arid Tropics
Patancheru, India

Liang Wei-li
Agricultural University of Hebei
Baoding, China

D.G. Westfall
Department of Soil and Crop Sciences
Fort Collins, Colorado

Qingwu Xue
Texas AgriLife Research and Extension Center
Amarillo, Texas

S. Zhao
Department of Agricultural and Applied Economics
Texas Tech University
Lubbock, Texas

Section I

Water and Agronomic Productivity

1 Global Water Balance and Agronomic Production in Relation to Food Security

Raveendra Kumar Rai and Vijay P. Singh

CONTENTS

1.1 WATER BALANCE

The distribution of water on the earth is highly uneven; therefore, sustainable planning and management of water on the earth to ensure environmental protection, economics, and equitability and understanding of water balance are important. For a comprehensive understanding of the hydrological system and its social integration, a knowledge of the hydrologic cycle is vital. The hydrologic cycle is also fundamental to the understanding of the carbon cycle as well as the nitrogen cycle.

1.1.1 Hydrologic Cycle

The hydrologic cycle is the unending and continuous movement of water with no specific start or end point. It is characterized by its variability in space and time. Among the various reservoirs in the cycle, the oceans are the greatest reservoir of water on the earth, covering about three-fourths of the earth's surface. Water from the oceans evaporates into the atmosphere which retains it as vapor. The atmosphere then releases this vapor primarily as precipitation in the form of rain, snow, sleet, or hail. During precipitation, some of the moisture evaporates back to the atmosphere before reaching the ground, some water is intercepted by vegetation, a portion infiltrates the ground, and the remainder flows off the land into lakes, rivers, or back to the oceans. The moisture on and beneath the earth's surface is of particular importance to humans and society.

The water cycle is also intricately intertwined with many other environmental cycles, such as the transport of energy, chemicals, and sediments. About half a million cubic kilometers of water evaporates from the oceans every year, and approximately the same amount falls back as precipitation across the globe, only one-fifth of which falls on land.

Water in the hydrologic cycle can be both a benefit and a hazard, with its extreme variations being particularly dangerous. Societies can thrive in otherwise hostile climates by drawing supplemental water from the ground or diverting it from rivers, but rapid and intense precipitation or snowmelt can also cause devastating floods and contribute to soil erosion. However, more harmful are the extended periods without precipitation causing severe droughts. Such droughts cause hardships today and have contributed to the collapse of civilizations in the past.

1.1.2 Global Water Balance

For global water budgeting, two conventional perspectives are generally used: (a) atmospheric perspective and (b) earth surface perspective. The global water balance from an atmospheric perspective can be expressed as follows:

$$\frac{dS_{atm}}{dt} = E - P, \tag{1.1}$$

where S_{atm} is the total amount of water stored in the entire atmosphere in the form of vapor, liquid, and solid; P and E are the corresponding global fluxes of precipitation and evapotranspiration, respectively. Equation 1.1 does not consider any loss (e.g., molecular diffusion) or gain (e.g., comet material) of water to/from outer space.

When water balance is presented from the earth surface perspective, then the water budget equation is expressed as

$$\frac{dS_{earth}}{dt} = E - P, \tag{1.2}$$

where S_{earth} is the total storage of water in the form of vapor, liquid, and solid over and within the earth's continents and oceans (surfaces and subsurfaces). To depict the processes and mechanisms of the global water cycle, more descriptive forms of Equations 1.1 and 1.2 must be used. A more descriptive form of Equation 1.2 can be expressed as

$$\frac{dS_l}{dt} + \frac{dS_o}{dt} = P_l + P_o - E_l - E_o, \tag{1.3}$$

where the subscripts l and o refer to the land and ocean components, respectively. For the water balance of the land surface, the following equation is used:

$$\frac{dS}{dt} = P - E - R. \tag{1.4}$$

For all global land surfaces, the total runoff term, R, represents the total (i.e., surface and subsurface) flow of water that reaches the oceans and is primarily the global river discharge. However, when water budgeting is considered at a smaller scale (i.e., river basins, catchments, etc.), both surface (i.e., stream/river flow and/or surface flow) and subsurface flows (i.e., groundwater and aquifer flow) would contribute more equally to the runoff terms.

On a global scale, the major reservoirs for water are the ocean, atmosphere, cryosphere (snow and ice), lithosphere (surface and groundwater), and biosphere. Table 1.1 presents the distribution of water on the earth. In the hydrologic cycle, the water is transferred between reservoirs primarily via five fluxes: precipitation, evapotranspiration, sublimation, runoff, and streamflow (Table 1.2). Additionally, there are fluxes that transfer water within a reservoir, such as advection of moisture in the air, percolation in soils, and the so-called thermohaline circulation, which conveys water to and from the ocean's surface and its depth.

1.1.3 Freshwater Balance

Relative to the salt water, the amount of useful fresh water is small. Out of the earth's total amount of water, only 2.4% of fresh water is available in glaciers, groundwater aquifer, lakes, rivers, etc. Unfortunately, these resources are not evenly distributed or easily accessible to everyone. The freshwater distribution on the earth is presented in Table 1.3.

Table 1.3 shows that most of the fresh water is available in the form of ice and permanent snow cover. However, out of 78.51% of glacial fresh water, 69% is found in Antarctic and Arctic regions. The total usable freshwater supply to the ecosystem and humans from river systems, lakes, wetlands, soil moisture, and shallow groundwater is less than 1%, which is only 0.01% of all the earth's water. However, 0.007% of all the earth's water is readily available for human consumption (WHO 1976), which indicates that fresh water on the earth is finite and unevenly distributed as well.

TABLE 1.1
Distribution of Water on Earth

Form of Water	Area Covered ('000 km²)	Volume ('000 km³)	Share of Water Reserves (%)	
			Total Water Reserves	Freshwater Reserves
Oceans	361300	1338000	96.5	—
Groundwater	134800	23400[a]	1.7	—
Fresh groundwater	134800	10530	0.76	30.1
Soil moisture	82000	16.5	0.001	0.05
Glaciers and permanent snow cover	16000	24000	1.74	68.7
Antarctica	14000	22000	1.56	61.7
Greenland	1800	2300	0.17	6.68
Arctic islands	230	83.5	0.006	0.24
Mountainous areas	220	40.6	0.003	0.12
Ground ice zones of permafrost strata	21000	300	0.022	0.86
Water reserves in lakes	2000	180	0.013	—
Fresh water	1240	91	0.007	0.26
Salt water	820	85.4	0.006	—
Marsh water	2700	11.47	0.0008	0.03
Water in rivers	148800	2.12	0.0002	0.006
Biological water	510000	1.12	0.0001	0.003
Atmospheric water	510000	12.9	0.001	0.04
Total water reserves	510000	1390000	100	—
Fresh water	148800	35000	2.35	100

Source: Korzun, V.I. (ed.) *World Water Balance and Water Resources of the Earth*. No. 25 of Studies and Reports in Hydrology. UNESCO, Paris, 1978.

[a] Not including the groundwater reserves in Antarctica.

1.1.4 Freshwater Balance Continentwise

There is large spatial variability in the freshwater resource, which can be seen from the continentwise freshwater resources in the earth (Table 1.4). Based on Table 1.4, it may be stated that the glaciers and ice caps cover about 10% of the world's landmass. These are concentrated in Greenland and Antarctica and contain 70% of the world's fresh water. Unfortunately, most of these resources are located far from human habitation and are not readily accessible for human use. According to the United States Geological Survey (USGS), 96% of the world's frozen fresh water is at the South and North Poles, with the remaining 4% spread over 550,000 km² of glaciers and mountainous ice caps measuring about 180,000 BCM (UNEP 1992; Untersteiner 1975; WGMS 1998, 2002). Groundwater is by far the most abundant and readily available source of fresh water, followed by lakes, reservoirs, rivers, and wetlands. An analysis indicates the following: groundwater represents over 90% of the world's readily available freshwater resource (Boswinkel 2000). About 1.5 billion people depend upon groundwater for their drinking water supply (WRI, UNEP, UNDP, World Bank 1998). The amount of groundwater withdrawn annually is roughly estimated at 600–700 BCM, representing about 20% of global water withdrawals (WMO 1997). A comprehensive picture of the quantity of groundwater withdrawn and consumed annually around the world does not exist. Most freshwater lakes are located at high altitudes, with nearly 50% of the world's lakes located in Canada alone. Many lakes, especially those in arid regions, become salty through evaporation, which concentrates

TABLE 1.2
Estimates of Average Annual Precipitation (P), Evaporation (E), Runoff Rate (P − E), and Runoff Ratio ([P − E]/P)

Region	Surface Area (10^4 km²)	P (mm/year)	E (mm/year)	P − E (mm/year)	(P − E)/P
Europe	10.0	657	375	282	0.43
Asia	44.1	696	420	276	0.40
Africa	29.8	696	582	114	0.16
Australia	8.9	803	534	269	0.33
North America	24.1	645	403	242	0.38
South America	17.9	1564	946	618	0.40
Antarctica	14.1	169	28	141	0.83
All land areas	148.9	746	480	266	0.36
Arctic Ocean	8.5	97	53	44	0.45
Atlantic Ocean	98.0	761	1133	−372	−0.49
Indian Ocean	77.7	1043	1294	−251	−0.24
Pacific Ocean	176.9	1292	1202	90	0.07
All oceans	361.1	1066	1176	−110	−0.10
Globe	510.0	973	973	0	0

Source: Pagano, T.C. and Sorooshian, S., *Encyclopedia of Hydrological Sciences*, John Wiley & Sons, New York, 2005.

TABLE 1.3
Distribution of Freshwater Resources on the Earth

S. No.	Source	Percentage Distribution
1	Glacial ice	78.51
2	Groundwater aquifer	20.64
3	Soil moisture	0.44
4	Lakes	0.38
5	Rivers	0.01
6	Atmospheric moisture	0.01

TABLE 1.4
Continentwise Freshwater Resources by Volume

Continent	Glaciers and Permanent Ice (km³)	Wetlands, Large Lakes, Reservoirs, Rivers (km³)	Groundwater (km³)
Asia	60,984	30,622	7,800,000
Africa	0.2	31,776	5,500,000
Antarctica	30,109,800	—	—
Australia	180	221	1,200,000
Europe	18,216	2,529	600,000
Greenland	2,600,000	—	—
North America	90,000	27,000	4,300,000
South America	900	NA	3,000,000

Source: UNEP. *Glaciers and the Environment.* UNEP/GEMS Environment Library No. 9. p. 8. UNEP, Nairobi, Kenya, 1992.

the inflowing salts. The Caspian Sea, the Dead Sea, and the Great Salt Lake are among the world's major salt lakes. Rivers form a hydrologic mosaic, with an estimated 263 international river basins covering 45.3% (231,059,898 km^2) of the earth's land surface, excluding Antarctica (UNEP, Oregon State University et al., in preparation). The total volume of water in the world's rivers is estimated at 2,115 BCM (Groombridge and Jenkins 1998).

1.1.5 Freshwater Resources of Sample Countries

To determine the freshwater balance of a country, it is necessary to determine the renewable water resources, followed by water quality and estimation of the exploitable water resources. In computing the water resources on a countrywide basis, renewable and nonrenewable water resources need to be considered. Renewable water resources comprise the long-term mean annual flow in rivers and net groundwater availability, whereas nonrenewable water resources comprise the deep aquifer water that has a negligible rate of recharge during the human timescale.

Renewable water resources can also be categorized into natural and actual. Natural renewable water resources are the total amount of a country's water resources (internal and external resources) computed on a yearly basis, both surface water and groundwater, which is generated through the hydrological cycle. Actual renewable water resources, on the other hand, take into consideration the quantity of flow reserved to upstream and downstream countries through formal and informal agreements or treaties and possible reduction of external flow due to upstream water abstraction. Unlike the natural renewable water resources, the actual renewable water resources vary with time and consumption pattern, and therefore these resources must be associated with a specific year.

Besides this, not all natural freshwater (surface water and groundwater) resources are accessible for use. The exploitable water resources (manageable water resources or water development potential) consider such factors as the economic and environmental feasibility of storing floodwater behind dams or extracting groundwater; the physical possibility of catching water that naturally flows out to the sea; and the minimum flow requirements for navigation, environmental services, aquatic life, etc.

The total water resources of sample countries are summarized in Table 1.5, which also includes important water resource indicators, such as dependency ratio, and per capita internal and total renewable water resources of the country. The dependency ratio and per capita water resources can be defined as

$$\text{Dependency ratio} = \frac{\text{ERWR}}{\text{TRWR}} \times 100\%, \tag{1.5}$$

where ERWR and TRWR are the external and total renewable water resources of the country, respectively.

$$\text{Per capita IRWR (in m}^3\text{/year/inhab.)} = \frac{\text{IRWR}}{\text{Total population}}, \tag{1.6}$$

$$\text{Per capita TRWR (in m}^3\text{/year/inhab.)} = \frac{\text{TRWR}}{\text{Total population}}. \tag{1.7}$$

In Equations 1.6 and 1.7, units of IRWR and TRWR are in m^3/year.

Based on internal renewable water resources of sampled countries (IRWR) (Table 1.5), it may be stated that per capita water availabilities are quite less for a few countries and have large spatial variability. The per capita water availability in Pakistan is lowest among the sampled countries (i.e., 1016 lpcd), followed by South Africa (2833 lpcd) and India (3422 lpcd).

TABLE 1.5
Renewable Water Resources of Sample Countries

Parameters	Unit	Afghanistan	Australia	China	Canada	Germany
Total area	km^2	652090	7741220	9561000	9970610	357030
Total pop.	1000 inhab.	21765	19138	1252952	30757	82017
Av. ppt. (1961–1990)	km^3/year	213.4	4136.9	5994.7	5352.2	250
Internal resources: surface	km^3/year	—	440	2711.5	2840	106.3
Internal resources: groundwater	km^3/year	—	72	828.8	370	45.7
Internal resources: overlap	km^3/year	—	20	727.9	360	45
Internal resources: total	km^3/year	55	492	2812.4	2850	107
External resources: natural	km^3/year	10	0	17.2	52	47
External resources: actual	km^3/year	10	0	17.2	52	47
Total resources: natural	km^3/year	65	492	2829.6	2902	154
Total resources: actual	km^3/year	65	492	2829.6	2902	154
Dependency ratio	%	15.4	0	0.6	1.8	30.5
IRWR per capita	m^3/year/inhab.	2527	25708	2245	92662	1305
TRWR per capita	m^3/year/inhab.	2986	25708	2258	94353	1878

Parameters	Unit	India	Italy	New Zealand	Pakistan	South Africa
Total area	km^2	3287260	301340	270530	796100	1221040
Total pop.	1000 inhab.	1008937	57530	3778	141256	43309
Av. ppt. (1961–1990)	km^3/year	3558.8	250.8	468.4	393.3	603.9
Internal resources: surface	km^3/year	1222	170.5	—	47.4	43
Internal resources: groundwater	km^3/year	418.5	43	—	55	4.8
Internal resources: overlap	km^3/year	380	31	—	50	3
Internal resources: total	km^3/year	1260.5	182.5	327	52.4	44.8
External resources: natural	km^3/year	647.2	8.8	0	181.4	5.2
External resources: actual	km^3/year	636.1	8.8	0	170.3	5.2
Total resources: natural	km^3/year	1907.8	191.3	327	233.8	50
Total resources: actual	km^3/year	1896.7	191.3	327	222.7	50
Dependency ratio	%	33.9	4.6	0	76.5	10.4
IRWR per capita	m^3/year/inhab.	1249	3172	86554	371	1034
TRWR per capita	m^3/year/inhab.	1880	3325	86554	1576	1154

Parameters	Unit	Greenland	United Kingdom	United States		
				Alaska	Conterminous	Hawaii
Total area	km^2	341700	242910	—	9629090	—
Total pop.	1000 inhab.	56	59634	627	279583	1212
Av. ppt. (1961–1990)	km^3/year	759	296.3	—	5800.8	32
Internal resources: surface	km^3/year	—	144.2	—	1862	5.2
Internal resources: groundwater	km^3/year	—	9.8	—	1300	13.2
Internal resources: overlap	km^3/year	—	9	—	1162	0
Internal resources: total	km^3/year	603	145	800	2000	18.4
External resources: natural	km^3/year	0	2	180	71	0
External resources: actual	km^3/year	0	2	180	71	0
Total resources: natural	km^3/year	603	147	980	2071	18.4
Total resources: actual	km^3/year	603	147	980	2071	18.4
Dependency ratio	%	0	1.4	18.4	3.4	0
IRWR per capita	m^3/year/inhab.	10767857	2431	1276055	7153	15187
TRWR per capita	m^3/year/inhab.	10767857	2465	1563168	7407	15187

Source: FAO. Review of world water resources by country. Water Reports No. 23, Food and Agriculture Organization, Rome, 2003.

1.2 GLOBAL WARMING AND CLIMATE CHANGE

Climate is the most important driving parameter that causes year-to-year variability in socioeconomic and environmental systems including the availability of water resources. It affects the development and planning of water resources schemes, such as flood prevention and control, drought management, and food and fiber production. Further, any change in climate will increase the uncertainty in water resources planning. Apart from this, changes in the climatic pattern will have profound effects and consequences for natural and agricultural ecosystems and for society as a whole. These changes could even alter the location of the major crop production regions on the earth (Reddy and Hodges 2000). The shift in "normal weather" patterns, with their associated extreme events, will surely change the zones of crop adaptation and cultural practices required for successful crop production. Climate- and weather-induced instability in food and fiber supplies will alter social and economic stability and regional competitiveness (Reddy and Hodges 2000).

In recent years, there has been a considerable concern about global warming and climatic changes. Alteration in climate is governed by a complex system of atmospheric, land surface, and oceanic processes and their interactions. Atmospheric processes also result in an increase in surface-level ultraviolet radiation and changes in temperature and rainfall patterns. Human activities, on the other hand, are responsible for changes in ecosystems due to the increased emission rates of CO_2 and other greenhouse gases. The evidence using state-of-the-art computer models incorporating as much of the theoretical understanding of the earth's weather suggests that global warming is occurring along with shifting patterns of rainfall and incidents of extreme weather events (IPCC 2007a).

1.2.1 Impact on Space–Time Distribution of Rainfall

Global warming will cause an increase in the atmospheric moisture content and thus an increase in the global mean precipitation. The global annual land mean precipitation had a small, but uncertain, upward trend of approximately 1.1 mm per decade (uncertainty ± 1.5 mm) over 1901–2005. During the twentieth century, precipitation generally increased from latitudes 30° to 85°N over land; but notable decreases occurred between latitudes 10°S and 30°N in the last 30–40 years. In western Africa and South Asia, a declining linear trend in rainfall was noticed during 1900–2005 with 7.5% per century (significant statistically at <1% level), whereas much of northwest India showed increased rainfall with more than 20% per century (IPCC 2007a). Figure 1.1 shows the land precipitation changes in the world. It also shows that, excluding Asian and African countries, other parts have increasing rainfall trends.

Changes in the precipitation pattern, evaporation of water from the soil, and transpiration (especially an increase in the extreme precipitation events) are expected to increase the runoff by 2060 in some parts of northern China, East Africa, and India. Runoff is important for replenishment of the water of rivers and lakes and therefore also important for irrigation and maintenance of ecosystem services. Across South Asia (Afghanistan, Bangladesh, Bhutan, India, Maldives, Nepal, Pakistan, and Sri Lanka), large populations depend on semisubsistence agriculture for their livelihoods. Rainfall in the semiarid and subhumid regions of South Asia is highly variable and unreliable, which highly influences agricultural productivity (Lal et al. 2011).

1.2.2 Impact on Snow, Ice, and Glaciers

The greatest asset of altitudes of more than 3500 m is the drinking water reserves that exist in the form of glaciers. These, in addition, play a buffer role in case of drought, releasing their quota of water every year to compensate for water losses in times of drought. With global

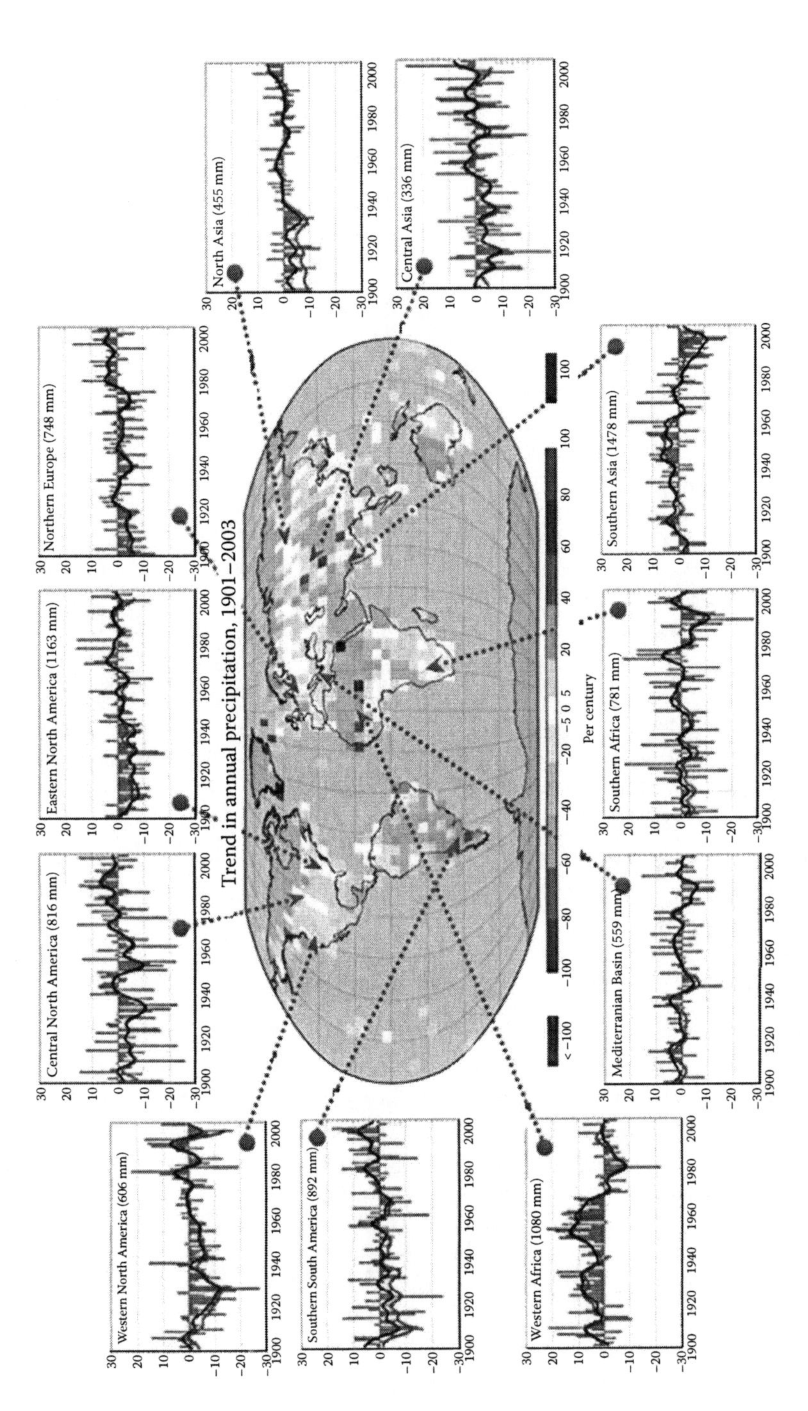

FIGURE 1.1 Global precipitation change. (From IPCC. Climate change 2007: Impacts, adaptation, and vulnerability. Contribution of Working Group II to the Fourth Assessment Report of the Intergovernmental Panel on Climate Change. Cambridge University Press, Cambridge, UK, 2007.)

warming, however, the glaciers will be dramatically affected and consequently so will the communities located at the highest altitudes and the urban complexes located at lower altitudes. The enhanced melting as well as the increased length of the melt season of glaciers lead, at first, to increased river runoff and discharge peaks, while in a longer timeframe, the runoff is expected to decrease. The formation of lakes is occurring as glaciers retreat in several steep mountain ranges, constituting a danger for glacial lake outburst floods.

The IPCC further reported that widespread mass losses from glaciers and reductions in snow cover over recent decades are projected to accelerate throughout the twenty-first century, reducing water availability, hydropower potential, and changing the seasonality of flows in regions supplied by melt water from major mountain ranges (e.g., Hindu-Kush, Himalaya, Andes), where more than one-sixth of the world population currently lives (IPCC 2007a,b).

Central Asia, northern China, and the northern part of South Asia face immense vulnerabilities associated with the retreat of glaciers, at a rate of 10–15 m a year in the Himalayas. Seven of Asia's great river systems will experience an increase in flows over the short term, followed by a decline as glaciers melt. Climate change will be superimposed on wider pressures on water systems.

1.2.3 Impact on Stream Flow

The current stress on water resources from population growth and economic and land-use changes, including urbanization, is expected to be exacerbated by changes in precipitation and temperature due to climate change (lead by changes in runoff and water availability). Due to climate change, the runoff is projected to increase by 10%–40% by mid-century at higher latitudes and in some wet tropical areas, including populous areas in east and southeast Asia, and decrease by 10%–30% over some dry regions at mid-latitudes and dry tropics, due to decreases in rainfall and higher rates of evapotranspiration.

The negative impacts of climate change on freshwater systems outweigh its benefits. Areas in which runoff is projected to decline face a reduction in the value of the services provided by water resources (very high confidence). The beneficial impacts of increased annual runoff in some areas are likely to be tempered by the negative effects of increased precipitation variability and seasonal runoff shifts on water supply, water quality, and flood risk (IPCC 2007a).

1.2.4 Effect on Freshwater Supplies

Water is involved in all components of the climate system (atmosphere, hydrosphere, cryosphere, land surface, and biosphere). Moreover, the hydrological cycle is intimately linked to the changes in atmospheric temperature and radiation balance, affecting it through a number of mechanisms (Bates et al. 2008).

There is a growing concern for water availability due to an increase in temperature and evaporation, a rise in sea level, and variations in rainfall patterns, thus altering the hydrological balance of many ecosystems, as well as a certainty of its negative impacts on animal and plant subsistence and of unleashing conflicts over water resources. Freshwater availability in central, south, east, and southeast Asia, particularly in large river basins, will affect more than a billion people by the 2050s (Parry et al. 2007).

Further, prolonged and repeated droughts can cause loss of crop/animal production, undermining the sustainability of livelihood systems, especially of those based on rainfed agriculture. Figure 1.2 provides a landscape of the vulnerability of freshwater resources around the world. The land precipitation trend showed an increase over the twentieth century between 30°N and 85°N; however, prominent decreases have occurred in the past 40 years (Bates et al. 2008).

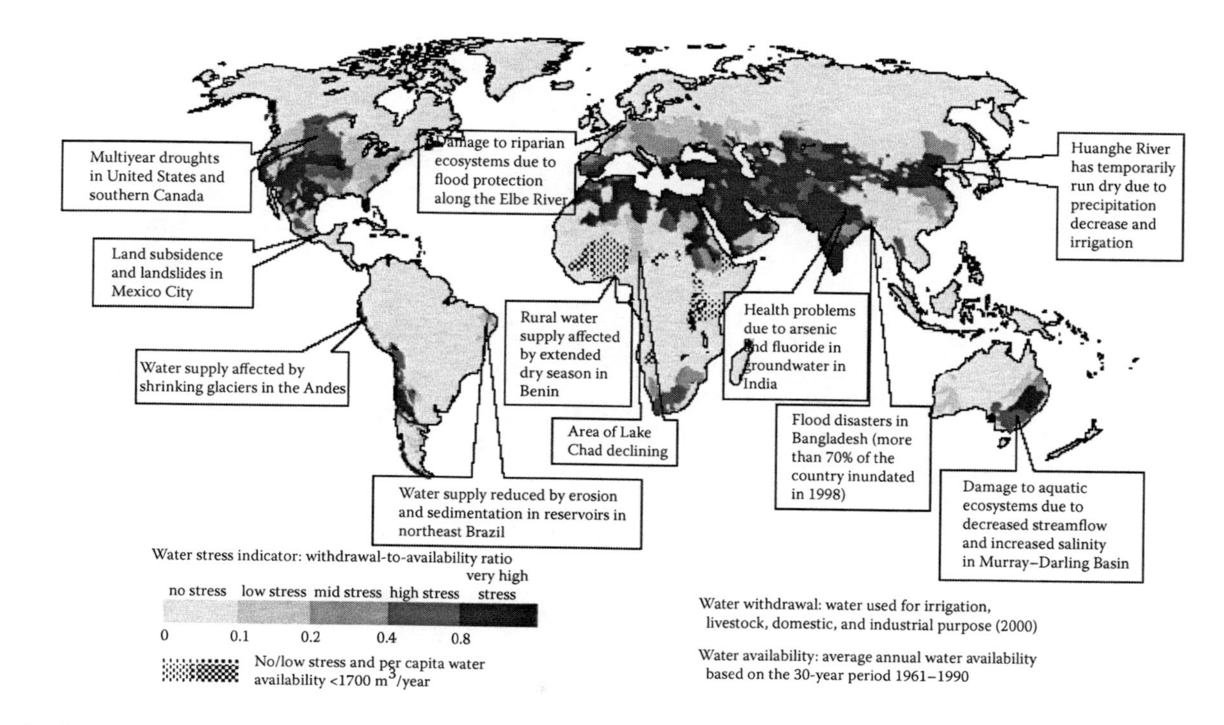

FIGURE 1.2 Example of current vulnerabilities of freshwater resources. (From Bates, B.C., Kundzewicz, Z.W., Wu, S., and Palutikof, J.P. (eds). Climate change and water. Technical Paper of the Intergovernmental Panel on Climate Change, Secretariat IPCC, Geneva, 2008.)

1.2.5 Impact on Hydrological Extremes

Increased temperature and heavy rainfall events due to climate change will result in increased flood frequency and severity. Based on the available evidence, it may be stated that a significant future increase in heavy rainfall events in many regions will take place including some in which the mean rainfall is projected to decrease. The resulting increased flood risk poses challenges to society, physical infrastructure, and water quality. On the other hand, a decrease in the number of rainy days in a hydrologic year will increase the frequency and severity of the drought.

It is likely that up to 20% of the world population will live in areas where river flood potential could increase. Increases in the frequency and severity of floods and droughts are projected to adversely affect sustainable development. Increased temperatures will further affect the physical, chemical, and biological properties of freshwater lakes and rivers, with predominantly adverse impacts on many individual freshwater species, community composition, and water quality. In coastal areas, rising sea levels will exacerbate water resource constraints due to increased salinization of groundwater supplies (IPCC 2007a,b). Effects will be largest in the densely populated and low-lying mega deltas of Asia and Africa, while small islands are especially vulnerable (IPCC 2007a).

Due to climate change and reduced extent of rainfall periods, many semiarid areas (e.g., the Mediterranean Basin, western United States, southern Africa, and northeastern Brazil) will suffer with decreases in water resources. Drought-affected areas are projected to increase in extent, with the potential for adverse impacts on multiple sectors such as agriculture, water supply, energy production, and health. Regionally, large increases in irrigation water demand as a result of climate changes are projected (IPCC 2007a,b).

1.2.6 Impact on Temperature

It has been established that global surface warming took place at a rate of 0.74 ± 0.18°C over the period of 1906–2005 (IPCC 2007a), and it is expected to be more in the next century than what has occurred during the past 10,000 years (IPCC 2007a). Figure 1.3 illustrates the range of historical temperature increases since the 1850s.

Particularly in India, based on the data for the period 1901–2005, it can be shown that the all-India mean annual temperature has been rising at 0.05°C/decade, with the maximum temperature at +0.07°C/decade and the minimum temperature at +0.02°C/decade (Kothawale and Kumar 2005). As a result, the diurnal temperature range shows an increase of 0.05°C/decade. However, in northern India, the average temperature is falling at a rate of −0.38°C and is unlikely to rise in all-India average temperature (i.e., at +0.42°C/century) (Arora et al. 2005).

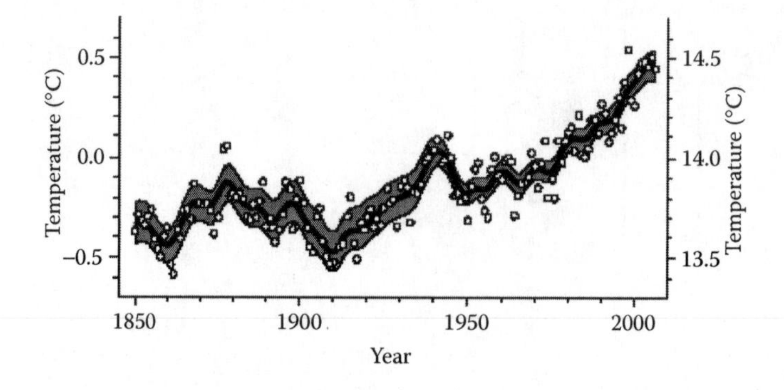

FIGURE 1.3 Global average surface temperature. (From IPCC. Climate change 2007: Impacts, adaptation, and vulnerability. Contribution of Working Group II to the Fourth Assessment Report of the Intergovernmental Panel on Climate Change. Cambridge University Press, Cambridge, UK, 2007.)

1.2.7 Impact on Evapotranspiration

Evapotranspiration is directly related to temperature changes. However, the extent of impact will be spatial. The rate of evapotranspiration affects the crop water requirement and thus freshwater demands for irrigation. In recent years, the demand for irrigation on account of climate changes and drastic land-use pattern changes has exacerbated so much that the dependability on groundwater has increased. As a result, in the Indian subcontinent, nearly 50% of administrative blocks have been declared "dark zones."

1.2.8 Impact on Ecosystems

Based on the IPCC (2007a), it has been stated that the resilience of many ecosystems is likely to be exceeded this century by an unprecedented combination of climate change, associated disturbances (e.g., flooding, drought, wildfire, insects, ocean acidification), and other global change drivers, namely, land-use change, pollution, fragmentation of natural systems, and overexploitation of resources.

Over the course of this century, the net carbon uptake by terrestrial ecosystems is *likely* to peak before mid-century and then weaken or even reverse, thus amplifying climate change (IPCC 2007a).

For increases in the global average temperature exceeding 1.5°C–2.5°C and in the associated atmospheric CO_2 concentrations, major changes have been projected in ecosystem structure and function, species' ecological interactions, and shifts in species' geographical ranges, with largely negative consequences for biodiversity and ecosystem goods and services such as water and food supply (IPCC 2007a).

More specifically, the timing of phenological events such as flowering is often related to environmental variables, such as temperature. Changing environments are, therefore, expected to lead to changes in the life cycle, and these have been recorded for many species of plants (Parmesan and Yohe 2003). These changes have the potential to lead to asynchrony between species, or to change competition between plants. Flowering times in British plants, for example, have changed, leading to annual plants flowering earlier than perennials and insect-pollinated plants flowering earlier than wind-pollinated plants, with potential ecological consequences (Fitter and Fitter 2002). A recently published study has used data recorded by the writer and naturalist Henry David Thoreau to confirm the effects of climate change on the phenology of some species in the area of Concord, Massachusetts (Willis et al. 2008).

An overwhelming majority of studies of regional climate effects on terrestrial species reveal consistent responses to warming trends, including pole-ward and elevation range shifts of flora and fauna. Responses of terrestrial species to warming across the Northern Hemisphere are well documented by changes in the timing of growth stages (i.e., phenological changes), especially the earlier onset of spring events, migration, and lengthening of the growing season (IPCC 2007a). In various regions across the world, some high-altitude and high-latitude ecosystems have already been affected by changes in climate. The IPCC (2007a) report reviewed relevant published studies of biological systems and concluded that 20%–30% of the species assessed may be at risk of extinction from climate change impacts within this century if global mean temperatures exceed 2°C–3°C (3.6°F–5.4°F) relative to preindustrial levels.

These changes can cause adverse or beneficial effects on species. For example, climate change may benefit certain plant or insect species by increasing their ranges. The resulting impacts on ecosystems and humans, however, may be positive or negative, depending on whether these species are invasive (e.g., weeds or mosquitoes) or if they are valuable to humans (e.g., food crops or pollinating insects). The risk of extinction may increase for many species, especially those that are already endangered or at risk due to isolation by geography or human development, low population numbers, or a narrow temperature tolerance range.

Observations of ecosystem impacts are difficult to use in future projections because of the complexities involved in human–nature interactions (e.g., land-use change). Nevertheless, the observed changes are compelling examples of how rising temperatures can affect the natural world and raise questions of how vulnerable populations will adapt to the direct and indirect effects of climate change.

1.2.9 Impact on Agriculture

Climate change will affect rainfall, temperature, and water availability for agriculture in vulnerable areas (IPCC 2007c). For example, drought-affected areas in South Asia will experience losses in agricultural production, undermining efforts to cut rural poverty. According to the Fourth Assessment Report of the IPCC, South Asia is very likely to get warm during this century, putting pressure on some of the prime productive land and reducing agricultural output, biodiversity, and the natural ability of ecosystems to recover. It is expected that climate change impacts will be uneven between countries and regions within Asia: China with 140 million undernourished people should gain 100 million tons in cereal production; India, in turn, with 200 million undernourished people is expected to lose 30 million tons (Bates et al. 2008).

Assuming a 4.4°C increase in temperature and a 2.9% increase in precipitation, the global agricultural output potential is likely to decrease by about 6% or 16% without carbon fertilization (GRID-Arendal and Ahlenius 2007). Climate change has led to projections that by the year 2080, agricultural output potential may be reduced by up to 60% for several African countries, with an average of 16%–27% (Figure 1.4).

At lower latitudes, especially in seasonally dry and tropical regions, crop productivity is projected to decrease for even small local temperature increases (1°C–2°C), which would increase food risk (IPCC 2007a).

This climate–food interaction will affect the poor. In some industrial and industrializing nations, climate change creates new patterns of food production, causing the development of new exporting and importing zones. The access to supplies and the energy to import may become strategic concerns and lead to international conflicts. Besides, popular discontent over livelihood security was

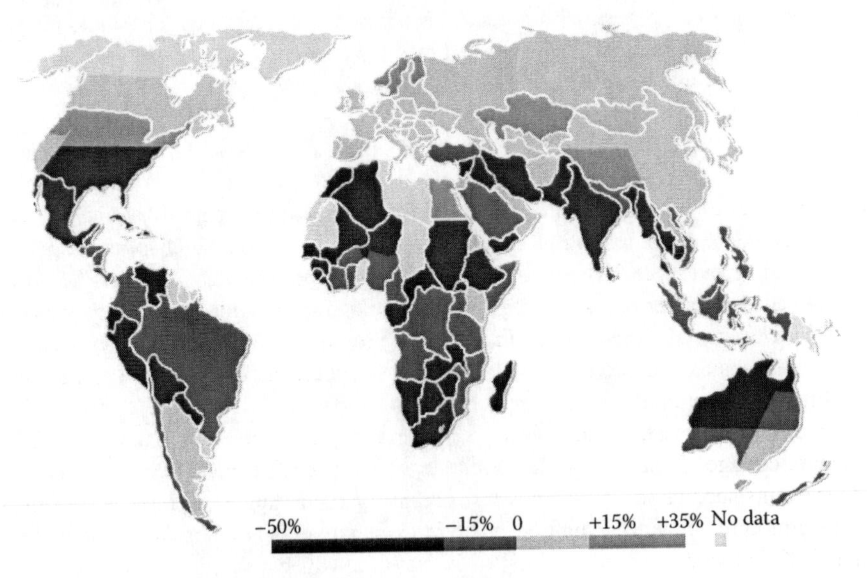

FIGURE 1.4 Projected losses in food grain due to climate change by 2080. (From UNEP. The environmental food crisis: The environment's role in averting future food crises. A UNEP rapid response assessment, United Nations Environment Programmes, Norway, 2009.)

a contributing cause for instability in Africa during the 1980s and 1990s. The same pressures will cause considerable population movements and displacement both within countries and internationally, which, in turn, will increase insecurity in its hard meaning.

1.2.10 Impact on Human Health

Climate change indirectly influences human health on a large scale. According to IPCC (2007a,b) reports, the health status of millions of people is projected to be affected through, for example, increases in malnutrition; increased deaths, diseases, and injuries due to extreme weather events; increased burden of diarrhea diseases; increased frequency of cardiorespiratory diseases due to higher concentration of ground-level ozone in urban areas related to climate change; and the altered spatial distribution of some infectious diseases.

Climate change is projected to bring some benefits in temperate areas, such as fewer deaths from cold exposure, and some mixed effects, such as changes in the range and transmission potential of malaria in Africa (WHO 2008). Overall, it is expected that benefits will be outweighed by negative health effects of rising temperature, especially in developing countries. Critically important will be the factors that directly shape the health of populations, such as education, health care, public health initiatives, and infrastructure and economic development.

1.2.11 Impact on Energy

Energy production and use are sensitive to changes in climate. For example, increasing temperatures will reduce the consumption of energy for heating but increase the energy used for cooling buildings (IPCC 2007b). The implications of climate change for energy supply are less clear than those for energy demand. The net effects of these changes on energy production, energy use, and utility bills will vary by region and by season.

The effects of climate change on energy supply and demand will depend not only on climatic factors but also on patterns of economic growth, land use, population growth and distribution, technological change, and social and cultural trends that shape individual and institutional actions.

There may also be changes in energy consumed for other climate-sensitive processes, such as pumping water for agricultural irrigation. Rising temperatures and associated increases in evaporation may increase the energy needs for irrigation, particularly in dry regions, such as the western United States.

On the other hand, less research has been undertaken on the impact of climate change on energy production. Some of the possible effects may be as follows:

- Hydropower generation is the energy source that is likely to be most directly affected by climate change because it is sensitive to the amount, timing, and geographical pattern of precipitation and temperature. Furthermore, hydropower needs may increasingly conflict with other priorities, such as salmon restoration goals in the Pacific Northwest (IPCC 2007c). However, changes in precipitation are difficult to project at the regional scale, which means that climate change will affect hydropower either positively or negatively, depending on the region.
- Infrastructure for energy production, transmission, and distribution may be affected by climate change. For example, if a warmer climate is characterized by more extreme weather events such as windstorms, ice storms, floods, tornadoes, and hail, the transmission systems of electric utilities may experience a higher rate of failure, with attendant costs (IPCC 2007a).
- Power plant operations can be affected by extreme heat waves. For example, the intake water that is normally used to cool the power plants becomes warm enough during extreme heat events that it compromises power plant operations.

- Some renewable sources of energy can be affected by climate change, although these changes are very difficult to predict. If climate change leads to increased cloudiness, solar energy production can be reduced. Wind energy production would be reduced if wind speeds increase above or fall below the acceptable operating range of the technology. Changes in growing conditions can affect biomass production, transportation, and power plant fuel sources, which are starting to receive more attention (IPCC 2007a).

1.3 GLOBAL POPULATION

The global population is the total population of humans on the planet Earth, which is currently estimated to be 6.92 billion by the U.S. Census Bureau (http://www.census.gov/ipc/www/popclock-world.html). The world population has experienced continuous growth since the end of the Bubonic Plague, the Great Famine, and the Hundred Years War in 1350, with a population of 300 million (IDB 2010). The highest rates of growth—increases above 1.8% per year—were seen briefly during the 1950s and for a longer period during the 1960s and 1970s; the growth rate peaked at 2.2% in 1963 and declined to 1.1% by 2009. Annual births have reduced to 140 million since their peak at 173 million in the late 1990s and are expected to remain constant, while deaths number 57 million per year and are expected to increase to 80 million per year by 2040. Current projections show a continued increase of population (but a steady decline in the population growth rate), with the population expected to reach between 7.5 and 10.5 billion in the year 2050. Based on the projection (UN 2009), the world population growth is shown in Figure 1.5.

In the earth's total population, Asia accounts for over 60% with more than 4 billion people. China and India together have about 37% of the earth's population. Africa follows with 1 billion people, 15% of the world's population. Europe's 733 million people make up 11% of the world's

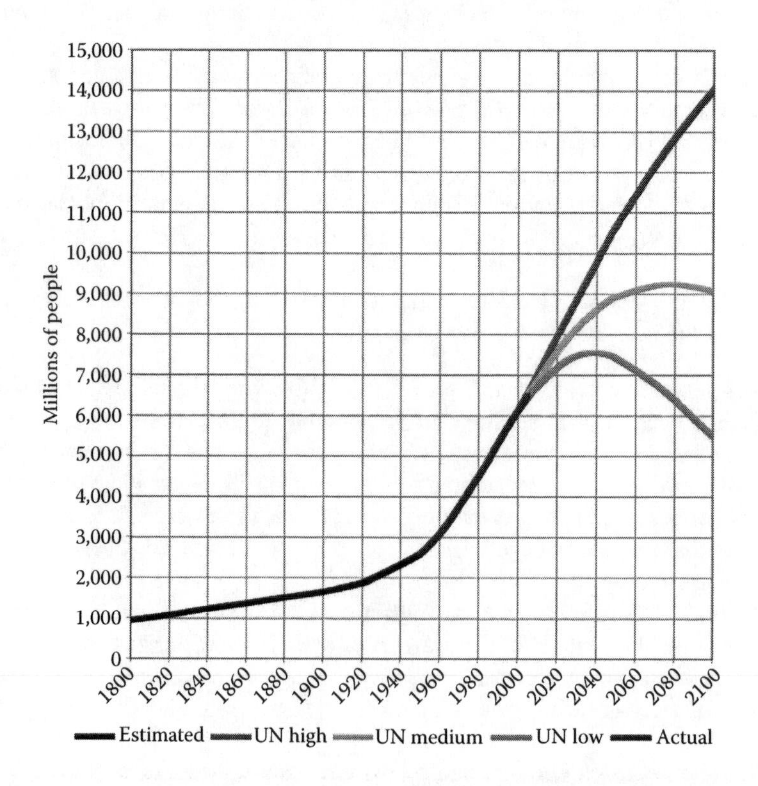

FIGURE 1.5 World population from 1800 to 2100, based on UN (2009) projections and U.S. Census Bureau historical estimates.

population. Latin America and the Caribbean region have a population of 589 million (9%), North America is home to 352 million (5%), and Oceania is home to 35 million (less than 1%) (UN 2009). Based on the latest UN projection, the world's population will be about 9.1 billion by 2050, which is 30% more than the current population of 6.9 billion.

1.3.1 Demographics

The increment in the world's population based on the UN projection (UN 2009) is shown in Figure 1.5, which reveals that the population has increased at a faster rate during 1940–2000 and will be expected to increase up to the years 2040 or 2070. The regional population increase at a temporal scale is shown in Figure 1.6, which shows that Asia and Africa have the largest shares in the population. Figure 1.7 shows the population distribution pattern of the world, whereas the spatial

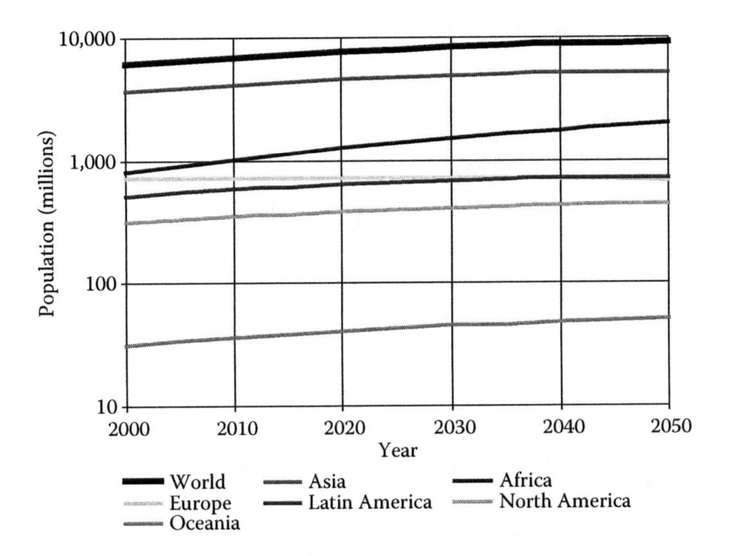

FIGURE 1.6 Population at temporal scale.

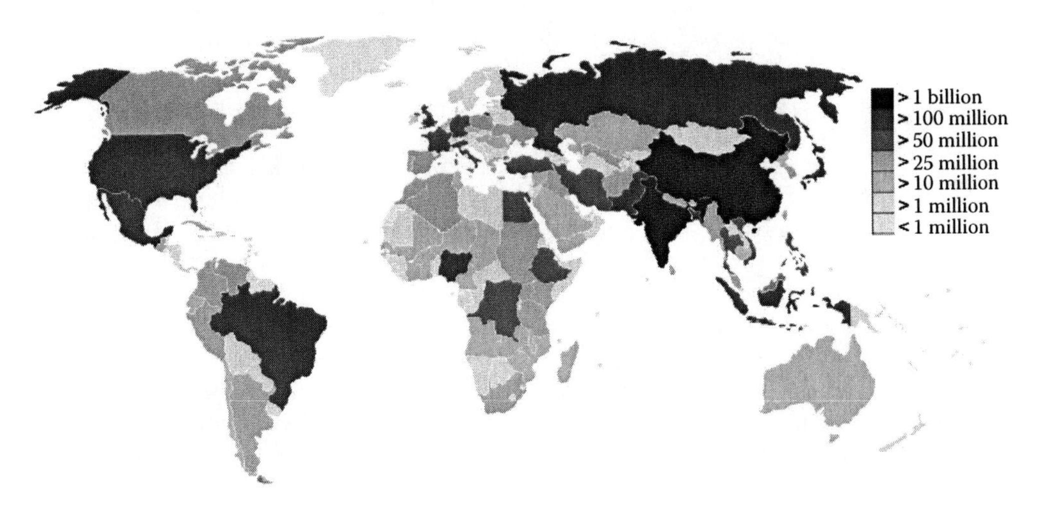

FIGURE 1.7 Spatial distribution of global population. (From UN. World urbanization prospects: The 1994 Revision. United Nations Population Division, Department for Economic and Social Information and Policy Analysis, New York, 1994.)

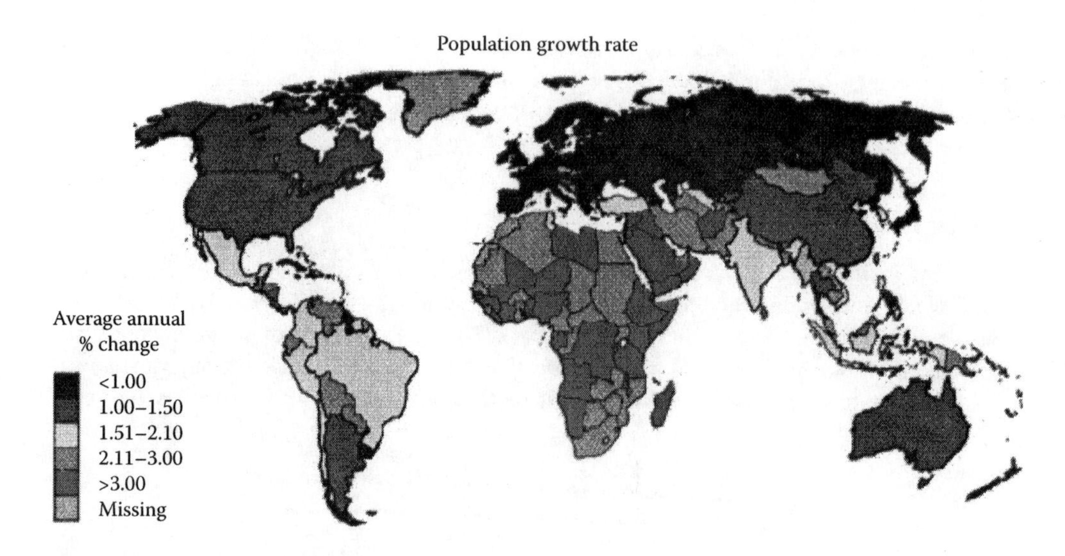

FIGURE 1.8 Spatial variation of population growth rate. (From UN. World urbanization prospects: The 1994 Revision. United Nations Population Division, Department for Economic and Social Information and Policy Analysis, New York, 1994.)

variation in population growth is depicted in Figure 1.8 (UN 1994). The countrywise population and population density are presented in Table 1.6.

1.4 DEMANDS FOR FRESH WATER

Reflecting on the water resources that will be available for future use, Hossain et al. (2011) have separated societally and environmentally important water resources into five major categories: water, food, energy, human health, and ecosystem function. These categories are interrelated, and water is required for each of the remaining four categories. The water that would be available will depend on the demands or stressors placed on the water resources and the uses thereof. This will apply to both the quantity and quality of water resources. Some of the stressors may entail local human population requirements; irrigation, floods, and droughts; biofuel production; weather variability and long-term change; land management practices; waste generation and treatment; animal and insect dynamics; vehicular and industrial emissions; natural landscape change; natural events, such as earthquakes; and others.

Uses of fresh water can be categorized as consumptive and nonconsumptive (sometimes called "renewable"). The use of water is referred to as consumptive when water is not immediately available for another use, although the final product can be directly consumed. For example, irrigation water is consumptive, because the final product is the farm produce associated with losses due to subsurface seepage and evaporation. Water that can be treated and returned as surface water, such as sewage, is generally considered nonconsumptive if that water can be put to additional use. Water use in power generation and industry is generally described using an alternate terminology, focusing on separate measurements of withdrawal and consumption. Withdrawal describes the removal of water from the environment, while consumption describes the conversion of fresh water into some other form, such as atmospheric water vapor or contaminated waste water.

In general, water used by farmers is 75% of the total water used, whereas industrial water use is only of the order of 22%. The water use for domestic and municipal demands is much less and is of the order of 0.10%.

TABLE 1.6
Countrywise Global Population and Domestic Wastewater Generation

S. No.	Country/Regions	Population	Area (km²)	Density (per km²)	Domestic Demand (MCM per Year)	Wastewater Generation (MCM per Year)
	World (land only, excl. Antarctica)	**6,919,484,600**	134,940,000	51	126280.59	101024.48
	World (land only)	**6,919,484,600**	148,940,000	46	126280.59	101024.48
	World (with water)	**6,919,484,600**	510,072,000	14	126280.59	101024.48
1	Macau (China)	541,200	29.2	18,534	9.88	7.90
2	Monaco	33,000	1.95	16,923	0.60	0.48
3	Singapore	5,076,700	710.2	7,148	92.65	74.12
4	Hong Kong (China)	7,003,700	1,104	6,349	127.82	102.25
5	Gibraltar (United Kingdom)	31,000	6.8	4,559	0.57	0.45
6	Vatican City	826	0.44	1,877	0.02	0.01
7	Malta	416,333	316	1,318	7.60	6.08
8	Bermuda (United Kingdom)	65,000	53	1,226	1.19	0.95
9	Bangladesh	162,221,000	143,998	1,127	2960.53	2368.43
10	Saint Maarten (Netherlands)	37,429	34	1,101	0.68	0.55
11	Bahrain	791,000	720	1,099	14.44	11.55
12	Maldives	309,000	298	1,037	5.64	4.51
13	Guernsey	65,726	78	843	1.20	0.96
14	Jersey	92,500	116	797	1.69	1.35
15	Saint-Martin (France)	35,263	53.2	663	0.64	0.51
16	Taiwan	23,069,345	35,980	639	421.02	336.81
17	Mauritius	1,288,000	2,040	631	23.51	18.80
18	Barbados	256,000	430	595	4.67	3.74
19	Aruba (Netherlands)	107,000	193	554	1.95	1.56
20	Palestinian territories (status disputed)	3,050,000	6,020	507	55.66	44.53
21	San Marino	30,800	61	505	0.56	0.45
22	Mayotte (France)	186,452	374	499	3.40	2.72
23	South Korea	48,456,369	99,538	487	884.33	707.46
24	Nauru	10,000	21	476	0.18	0.15
25	Puerto Rico (United States)	3,982,000	8,875	449	72.67	58.14
26	Curaçao (Netherlands)	142,180	444	446	2.59	2.08
27	Lebanon	4,224,000	10,452	404	77.09	61.67
28	Saint-Barthélemy (France)	8,450	21	402	0.15	0.12
29	Tuvalu	10,441	26	402	0.19	0.15
30	Netherlands	16,670,000	41,526	401	304.23	243.38
31	Rwanda	9,998,000	26,338	380	182.46	145.97
32	Israel	7,697,600	20,770	371	140.48	112.38
33	India	1,210,193,422	3,287,240	368	22086.03	17668.82
34	Haiti	10,033,000	27,750	362	183.10	146.48
35	Martinique (France)	402,000	1,128	356	7.34	5.87
36	Belgium	10,827,519	30,528	355	197.60	158.08
37	Marshall Islands	62,000	181	343	1.13	0.91

(continued)

TABLE 1.6 (Continued)
Countrywise Global Population and Domestic Wastewater Generation

S. No.	Country/Regions	Population	Area (km²)	Density (per km²)	Domestic Demand (MCM per Year)	Wastewater Generation (MCM per Year)
38	Japan	127,387,000	377,873	337	2324.81	1859.85
39	American Samoa (United States)	67,000	199	337	1.22	0.98
40	Guam (United States)	178,000	549	324	3.25	2.60
41	Saint Lucia	172,000	539	319	3.14	2.51
42	Virgin Islands (United States)	110,000	347	317	2.01	1.61
43	Réunion (France)	785,139	2,510	313	14.33	11.46
44	Sri Lanka	20,238,000	65,610	308	369.34	295.47
45	Philippines	92,226,600	300,076	307	1683.14	1346.51
46	Comoros	676,000	2,235	302	12.34	9.87
47	Grenada	104,000	344	302	1.90	1.52
48	Burundi	8,303,000	27,834	298	151.53	121.22
49	El Salvador	6,163,000	21,041	293	112.47	89.98
50	Saint Vincent and the Grenadines	109,000	388	281	1.99	1.59
51	Trinidad and Tobago	1,339,000	5,130	261	24.44	19.55
52	Vietnam	85,789,573	331,689	259	1565.66	1252.53
53	United Kingdom	62,041,708	243,610	255	1132.26	905.81
54	Guadeloupe (France)	405,000	1,628	249	7.39	5.91
55	Jamaica	2,719,000	10,991	247	49.62	39.70
56	Germany	81,757,600	357,022	229	1492.08	1193.66
57	Liechtenstein	35,981	160	225	0.66	0.53
58	Cayman Islands (United Kingdom)	56,000	264	212	1.02	0.82
59	Pakistan	176,090,000	803,940	219	3213.64	2570.91
60	Dominican Republic	10,090,000	48,671	207	184.14	147.31
61	Italy	60,200,060	301,318	200	1098.65	878.92
62	North Korea	24,051,706	120,538	200	438.94	351.15
63	Nepal	29,331,000	147,181	199	535.29	428.23
64	Saint Kitts and Nevis	52,000	261	199	0.95	0.76
65	Antigua and Barbuda	88,000	442	199	1.61	1.28
66	Luxembourg	502,207	2,586	194	9.17	7.33
67	Switzerland	7,761,800	41,284	188	141.65	113.32
68	Northern Mariana Islands (United States)	87,000	464	188	1.59	1.27
69	Seychelles	84,000	455	185	1.53	1.23
70	Andorra	86,000	468	184	1.57	1.26
71	Sao Tome and Principe	163,000	964	169	2.97	2.38
72	Kuwait	2,985,000	17,818	168	54.48	43.58
73	Nigeria	154,729,000	923,768	167	2823.80	2259.04
74	Anguilla (United Kingdom)	15,000	91	165	0.27	0.22
75	Federated States of Micronesia	111,000	702	158	2.03	1.62
76	British Virgin Islands (United Kingdom)	23,000	151	152	0.42	0.34
77	Gambia	1,705,000	11,295	151	31.12	24.89

TABLE 1.6 (Continued)
Countrywise Global Population and Domestic Wastewater Generation

S. No.	Country/Regions	Population	Area (km²)	Density (per km²)	Domestic Demand (MCM per Year)	Wastewater Generation (MCM per Year)
78	Isle of Man	80,000	572	140	1.46	1.17
79	Tonga	104,000	747	139	1.90	1.52
80	China	1,344,100,000	9,640,821	139	24529.83	19623.86
81	Kiribati	99,350	726	137	1.81	1.45
82	Uganda	32,710,000	241,038	136	596.96	477.57
83	Transnistria (Moldova)	555,347	4,163	133	10.14	8.11
84	Czech Republic	10,532,770	78,866	134	192.22	153.78
85	Guatemala	14,027,000	108,889	129	255.99	204.79
86	Malawi	15,263,000	118,484	129	278.55	222.84
87	Qatar	1,409,000	11,000	128	25.71	20.57
88	Denmark	5,532,531	43,094	128	100.97	80.77
89	Cape Verde	506,807	4,033	126	9.25	7.40
90	Thailand	64,232,760	513,115	125	1172.25	937.80
91	Poland	38,163,895	312,685	122	696.49	557.19
92	Indonesia	237,556,363	1,904,569	121	4335.40	3468.32
93	Moldova	3,567,500	33,844	105	65.11	52.09
94	Syria	21,906,000	185,180	118	399.78	319.83
95	Togo	6,619,000	56,785	117	120.80	96.64
96	Portugal	10,636,888	92,391	115	194.12	155.30
97	Tokelau (New Zealand)	1,378	12	115	0.03	0.02
98	France (Metropolitan)	62,793,432	551,500	114	1145.98	916.78
99	Slovakia	5,424,057	49,033	111	98.99	79.19
100	Albania	3,195,000	28,748	111	58.31	46.65
101	Armenia	3,230,100	29,800	108	58.95	47.16
102	Hungary	10,013,628	93,032	108	182.75	146.20
103	Dominica	78,940	751	105	1.44	1.15
104	Azerbaijan	8,896,900	86,600	103	162.37	129.89
105	Cuba		109,886	102		
106	Slovenia	2,075,456	20,256	102	37.88	30.30
107	Serbia (excluding Kosovo)	7,800,000	77,474	101	142.35	113.88
108	Ghana	23,837,000	238,533	100	435.03	348.02
109	Austria	8,372,930	83,858	100	152.81	122.24
110	Turkey	77,804,122	783,562	93	1419.93	1135.94
111	Spain	46,087,170	506,030	91	841.09	672.87
112	Romania	21,466,174	238,391	90	391.76	313.41
113	Costa Rica	4,579,000	51,100	90	83.57	66.85
114	Cyprus	801,851	9,251	87	14.63	11.71
115	Malaysia	28,306,700	329,847	86	516.60	413.28
116	Greece	11,306,183	131,957	86	206.34	165.07
117	Cook Islands (New Zealand)	20,000	236	85	0.37	0.29
118	Republic of Macedonia	2,114,550	25,713	82	38.59	30.87
119	Cambodia	14,805,000	181,035	82	270.19	216.15
120	Sierra Leone	5,696,000	71,740	79	103.95	83.16
121	Benin	8,935,000	112,622	79	163.06	130.45

(continued)

TABLE 1.6 (Continued)
Countrywise Global Population and Domestic Wastewater Generation

S. No.	Country/Regions	Population	Area (km²)	Density (per km²)	Domestic Demand (MCM per Year)	Wastewater Generation (MCM per Year)
122	Turks and Caicos Islands (United Kingdom)	33,000	417	79	0.60	0.48
123	Northern Cyprus (status disputed)	264,172	3,355	79	4.82	3.86
124	Croatia	4,443,000	56,538	79	81.08	64.87
125	Ukraine	46,936,000	603,700	78	856.58	685.27
126	Egypt	79,799,922	1,001,449	80	1456.35	1165.08
127	Wallis and Futuna (France)	15,480	200	77	0.28	0.23
128	East Timor	1,134,000	14,874	76	20.70	16.56
129	Myanmar (Burma)	50,020,000	676,578	74	912.87	730.29
130	Bosnia and Herzegovina	3,781,000	51,197	74	69.00	55.20
131	Ethiopia	79,221,000	1,104,300	72	1445.78	1156.63
132	Morocco	32,149,024	446,550	72	586.72	469.38
133	Jordan	6,316,000	89,342	71	115.27	92.21
134	Iraq	30,747,000	438,317	70	561.13	448.91
135	Brunei	400,000	5,765	69	7.30	5.84
136	Kenya	39,802,000	580,367	69	726.39	581.11
137	Swaziland	1,185,000	17,364	68	21.63	17.30
138	Lesotho	2,067,000	30,355	68	37.72	30.18
139	Bulgaria	7,351,234	110,912	66	134.16	107.33
140	Honduras	7,466,000	112,492	66	136.25	109.00
141	Côte d'Ivoire	21,075,000	322,463	65	384.62	307.70
142	Samoa	184,984	2,831	65	3.38	2.70
143	French Polynesia (France)	256,603	4,000	64	4.68	3.75
144	Georgia	4,465,000	69,700	64	81.49	65.19
145	Senegal	12,534,000	196,722	64	228.75	183.00
146	Ireland	4,450,878	70,273	63	81.23	64.98
147	Tunisia	10,327,800	163,610	63	188.48	150.79
148	Uzbekistan	27,488,000	447,400	61	501.66	401.32
149	Montserrat (United Kingdom)	5,900	102	58	0.11	0.09
150	Burkina Faso	15,757,000	274,000	58	287.57	230.05
151	United Arab Emirates	4,599,000	83,600	55	83.93	67.15
152	Mexico	107,550,697	1,958,201	55	1962.80	1570.24
153	Lithuania	3,329,227	65,300	51	60.76	48.61
154	Ecuador	14,681,432	283,561	52	267.94	214.35
155	Tajikistan	6,952,000	143,100	49	126.87	101.50
156	Belarus	9,755,106	207,600	47	178.03	142.42
157	Fiji	849,000	18,274	46	15.49	12.40
158	Tanzania	43,739,000	945,087	46	798.24	638.59
159	Bhutan	2,162,546	47,000	46	39.47	31.57
160	Afghanistan	29,863,010	652,090	46	545.00	436.00
161	Panama	3,454,000	75,517	46	63.04	50.43
162	Iran	74,196,000	1,648,195	45	1354.08	1083.26
163	Montenegro	630,548	14,026	45	11.51	9.21
164	Yemen	23,580,000	527,968	45	430.34	344.27

TABLE 1.6 (Continued)
Countrywise Global Population and Domestic Wastewater Generation

S. No.	Country/Regions	Population	Area (km²)	Density (per km²)	Domestic Demand (MCM per Year)	Wastewater Generation (MCM per Year)
165	Guinea-Bissau	1,611,000	36,125	45	29.40	23.52
166	Nicaragua	5,743,000	130,000	44	104.81	83.85
167	Palau	20,000	459	44	0.37	0.29
168	Eritrea	5,073,000	117,600	43	92.58	74.07
169	Cameroon	19,522,000	475,442	41	356.28	285.02
170	Guinea	10,069,000	245,857	41	183.76	147.01
171	South Africa	49,320,500	1,221,037	40	900.10	720.08
172	Saint Helena (United Kingdom)	4,918	122	40	0.09	0.07
173	Colombia	45,967,392	1,138,914	40	838.90	671.12
174	Djibouti	864,000	23,200	37	15.77	12.61
175	Madagascar	20,653,556	587,041	35	376.93	301.54
176	Faroe Islands (Denmark)	49,006	1,399	35	0.89	0.72
177	Latvia	2,248,961	64,600	35	41.04	32.83
178	Zimbabwe	13,009,530	390,757	33	237.42	189.94
179	United States	311,390,000	9,826,675	32	5682.87	4546.29
180	Liberia	3,476,608	111,369	31	63.45	50.76
181	Venezuela	29,228,598	916,445	32	533.42	426.74
182	Estonia	1,340,021	45,100	30	24.46	19.56
183	Democratic Republic of the Congo	68,692,542	2,344,858	29	1253.64	1002.91
184	Mozambique	22,894,000	801,590	29	417.82	334.25
185	Abkhazia (Georgia; claims independence)	200,000	7,138	28	3.65	2.92
186	Kyrgyzstan	5,482,000	199,900	27	100.05	80.04
187	Laos	6,320,000	236,800	27	115.34	92.27
188	Somaliland (Somalia; claims independence)	3,500,000	137,600	25	63.88	51.10
189	Saint-Pierre and Miquelon (France)	6,125	242	25	0.11	0.09
190	The Bahamas	342,000	13,878	25	6.24	4.99
191	Equatorial Guinea	676,000	28,051	24	12.34	9.87
192	Peru	29,461,933	1,285,216	23	537.68	430.14
193	Brazil	194,718,800	8,514,877	23	3553.62	2842.89
194	Chile	17,241,088	756,096	23	314.65	251.72
195	Sweden	9,366,092	449,964	21	170.93	136.74
196	Uruguay	3,463,197	175,016	20	63.20	50.56
197	Vanuatu	240,000	12,189	20	4.38	3.50
198	Solomon Islands	523,000	28,896	18	9.54	7.64
199	South Ossetia (Georgia; claims independence)	70,000	3,900	18	1.28	1.02
200	Zambia	12,935,000	752,618	17	236.06	188.85
201	New Zealand	4,315,800	270,534	16	78.76	63.01

(continued)

TABLE 1.6 (Continued)
Countrywise Global Population and Domestic Wastewater Generation

S. No.	Country/Regions	Population	Area (km²)	Density (per km²)	Domestic Demand (MCM per Year)	Wastewater Generation (MCM per Year)
202	Finland	5,384,580	338,145	16	98.27	78.61
203	Sudan	39,154,490	2,505,813	16	714.57	571.66
204	Paraguay	6,349,000	406,752	16	115.87	92.70
205	Angola	18,498,000	1,246,700	15	337.59	270.07
206	Algeria	34,895,000	2,381,741	15	636.83	509.47
207	Papua New Guinea	6,732,000	462,840	15	122.86	98.29
208	Argentina	40,091,359	2,780,400	14	731.67	585.33
209	Somalia	9,133,000	637,657	14	166.68	133.34
210	Belize	322,100	23,000	14	5.88	4.70
211	Pitcairn Islands (United Kingdom)	67	5	13	0.00	0.00
212	New Caledonia (France)	244,410	18,575	13	4.46	3.57
213	Norway	4,925,155	385,155	13	89.88	71.91
214	Niger	15,290,000	1,267,000	12	279.04	223.23
215	Saudi Arabia	28,146,658	2,149,690	12	513.68	410.94
216	Mali	14,517,176	1,240,192	12	264.94	211.95
217	Republic of the Congo	3,998,904	342,000	12	72.98	58.38
218	Turkmenistan	5,110,000	488,100	10	93.26	74.61
219	Oman	2,845,000	309,500	9.2	51.92	41.54
220	Bolivia	9,879,000	1,098,581	9	180.29	144.23
221	Chad	11,274,106	1,284,000	8.8	205.75	164.60
222	Russia	141,927,297	17,098,242	8.3	2590.17	2072.14
223	Niue (New Zealand)	2,000	260	7.7	0.04	0.03
224	Central African Republic	4,422,000	622,984	7.1	80.70	64.56
225	Kazakhstan	15,776,492	2,724,900	5.8	287.92	230.34
226	Gabon	1,475,000	267,668	5.5	26.92	21.54
227	Libya	6,420,000	1,759,540	3.6	117.17	93.73
228	Guyana	762,000	214,969	3.5	13.91	11.13
229	Canada	33,740,000	9,984,670	3.4	615.76	492.60
230	Botswana	1,950,000	581,730	3.4	35.59	28.47
231	Mauritania	3,291,000	1,025,520	3.2	60.06	48.05
232	Suriname	520,000	163,820	3.2	9.49	7.59
233	Iceland	318,452	103,000	3.1	5.81	4.65
234	Australia	22,729,504	7,682,300	3	414.81	331.85
235	Namibia	2,171,000	824,292	2.6	39.62	31.70
236	French Guiana (France)	187,056	90,000	2.1	3.41	2.73
237	Western Sahara (status disputed)	513,000	266,000	1.9	9.36	7.49
238	Mongolia	2,671,000	1,564,116	1.7	48.75	39.00
239	Falkland Islands (status disputed)	3,140	12,173	0.26	0.06	0.05
240	Greenland (Denmark)	57,000	2,175,600	0.026	1.04	0.83

Source: List of sovereign states and dependent territories by population density, Wikipedia.

1.4.1 Agriculture

Without water, there is no agriculture. It is estimated that 69% of the world's water is being used for the irrigation sector, with 15%–35% of irrigation withdrawals being unsustainable (WBCSD 2009). Approximately 3000 lpcd of water is required to meet the dietary demands of a person. This is a considerable amount, when compared to that required for drinking, which is between 2 and 5 L. Comparing this figure with the IRWR (i.e., Table 1.5), it may be stated that future dietary water demands will be very difficult to meet in many parts of the world. For example, the per capita IRWRs for Pakistan and South Africa are 1016 and 2833, respectively, and are less than the per capita dietary demand of 3000 lpcd. Therefore, to produce food for the 6.5 billion or so people who inhabit the planet today, the amount of water required is 19.5 BCM per day (i.e., 7117.5 BCM per year), which is equivalent to a canal that is 10 m deep, 100 m wide, and 7.1 million km long. This dietary water demand will increase to 27.3 BCM per day (i.e., 9964.5 BCM per year), when the projected world' population for the year 2050 is to be fed.

1.4.2 Energy

Regardless of the source of energy generation, large quantities of fresh water are needed. Therefore, recently a term—water-energy nexus—has been coined to emphasize their joint consideration. In a Virginia Tech Study, Younos and Hill (2008) have analyzed 11 types of energy sources, including coal, fuel ethanol, natural gas, and oil, and five power-generating methods, including hydroelectric, fossil fuel thermoelectric, and nuclear methods. Based on their calculations on available governmental reports, they estimated the gallons of water required to generate a British Thermal Unit (BTU). The water-use efficiencies of various energy sources are given in Table 1.7. According to

TABLE 1.7
Water-Use Efficiencies of Various Energy Production Technologies

	Low-Range Efficiency		High-Range Efficiency	
Sources	**L/GJ**	**L/kWh**	**L/GJ**	**L/kWh**
Natural gas	13	0.0466	N/A	N/A
Synfuel–coal gasification	47	0.1708	26	112
Tar sands	65	0.2329	38	164
Oil shale	86	0.3105	50	215
Synfuel–Fisher Tropsch	177	0.6366	60	259
Coal	177	0.6366	164	707
Hydrogen	616	2.2202	243	1047
Liquid natural gas	625	2.2513	N/A	N/A
Petroleum/oil–electric sector	5171	18.6311	2,420	10428
Fuel ethanol	10816	38.9702	29,100	125392
Biodiesel	60326	217.3634	75,000	323175
Hydroelectric	86	0.3105	N/A	N/A
Geothermal	560	2.0184	N/A	N/A
Solar thermoelectric	991	3.5710	270	1163
Fossil fuel thermoelectric	4740	17.0786	2,200	9480
Nuclear	10342	37.2623	5,800	24992

Note: 1 gallon = 3.79 U.S. liters; 1 gallon = 4.546 British liters; 1 BTU = 251.9 calories; 1 million BTUs = 292.8 kWh; 1 million BTUs = 1.055 gigajoules (GJ)

the study, the most water-efficient energy sources are natural gas and synthetic fuels produced by coal gasification. The least water-efficient energy sources are fuel ethanol and biodiesel. In terms of power generation, geothermal and hydroelectric energy types use the least amount of water, while nuclear plants use the most (Younos and Hill 2008). For example, electricity production is one of the largest users of water. For example, for a 60-watt incandescent light bulb burning for 12 hours a day for a year in 111 million houses, a power plant would consume about 2.95 BCM of water.

1.4.3 Industry

It is estimated that 22% of the world's water is used for industries. Major industrial users include hydroelectric dams; thermoelectric power plants, which use water for cooling; ore and oil refineries, which use water in chemical processes; and manufacturing plants, which use water as a solvent. Water withdrawal can be very high for certain industries, but consumption is generally much lower than that of agriculture.

1.4.4 Waste Disposal

Once water is abstracted for various uses, it gets polluted and disposed into the environment. Wastewater is generally categorized into domestic and industrial. The distribution of wastewater generation from different sectors in the world is shown in Figure 1.9. It can be estimated that about 80% of water is converted into domestic wastewater and needs adequate treatment for final disposal into the environment. However, treated wastewater, if disposed into natural streams, requires additional water to assimilate the pollutants in the river itself. Based on the population and average domestic water use (at the rate of 50 lpcd), estimated domestic water demands and domestic wastewater generation are given in Table 1.6.

Based on these estimates, the global demand for domestic water is approximately 126.3 BCM per year. Due to this consumption, the likely estimate of wastewater generation from households will be 101 BCM per year. It is assumed that every 100 L of wastewater requires 75 L of fresh water for dilution. If 50% of the generated wastewater is treated and reused, then the remaining 50% of the wastewater requires the dilution of water in the amount of 37.9 BCM/year and will be approximately 62.3 BCM/year when the estimation will be based on the projected population of 2050.

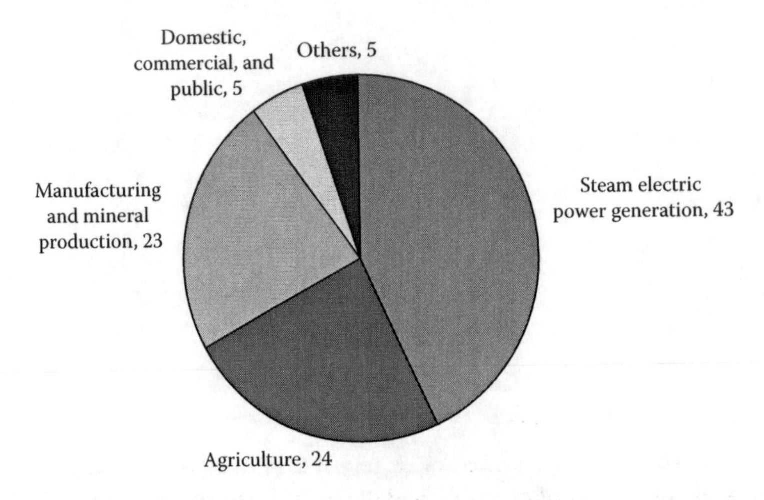

FIGURE 1.9 Percentage distributions of worldwide wastewater generation from different sectors.

1.4.5 Environmental Quality and Management

Water resources and catchment development activities often intercept natural flows in rivers and streams. The interception leads to the impairment of various day-to-day water-based activities and ecosystem functions. To minimize the impairment and maintain the environmental quality of rivers, it is required to allocate a certain percentage of water in the river throughout the year, depending upon the river's natural flow regime, which may be designated as the environmental flow requirement of the river.

Bunn and Arthington (2002) have formulated four basic principles that emphasize the role of flow regime in structuring aquatic life and show the link between flow and ecosystem changes. (i) Flow is a major determinant of physical habitats in rivers and, in turn, is the major determinant of biotic composition. Therefore, river flow modifications eventually lead to changes in the composition and diversity of aquatic communities. (ii) Aquatic species have evolved life history strategies primarily in response to natural flow regimes. Therefore, flow regime alterations can lead to the loss of biodiversity of native species. (iii) Maintenance of natural patterns of longitudinal and lateral connectivity in river floodplains determines the ability of many aquatic species to move between the river and the floodplain or between the main river and its tributaries. The loss of longitudinal and lateral connectivity can lead to the local extinction of species. (iv) The invasion of exotic and introduced species in rivers is facilitated by the alteration of flow regimes. Interbasin water transfer may represent a significant mechanism for the spread of exotic species.

The amount of minimum water to be maintained for a healthy river environment can be decided on the basis of the Tennant method (Tennant 1976). This method establishes stream flow requirements on the basis of a predetermined percentage of mean annual flow (MAF) (i.e., internal renewable surface water resources) and associates aquatic-habitat conditions with different percentages of MAFs. Minimum stream flows for small streams during summer were established by the Tennant method as 40%, 30%, and 10% of MAF, which represent good, fair, and poor habitat conditions, respectively. At 30% MAF, most of the stream substrate is submerged, but at 10% MAF, half or more of the stream substrate can be exposed (Tennant 1976). The 10% MAF value is often used to determine the minimum stream flow requirements in summer. Based on the above discussion, if it is desired to maintain the river system at good habitat condition, then the total volume for environmental water demand in the world may be of the order of 634.5 BCM (i.e., 30% of the total volume of water in the world's rivers, i.e., 2,115 BCM).

1.4.6 Domestic Use

It is estimated that 8% of the world's water is used for domestic purposes. These include drinking water, bathing, cooking, sanitation, and gardening; however, the basic household water requirement is approximately 50 lpcd. Drinking water is water that is of sufficiently high quality that it can be consumed or used without the risk of immediate or long-term harm. Such water is commonly called potable water. In most developed countries, the water supplied to households, commerce, and industry is all of drinking water standard, even though only a very small proportion is actually consumed or used in food preparation. Considering the domestic requirement of 50 lpcd, the annual domestic water use can be estimated to be 118.6 and 166 BCM per year for the current population and the projected population of 2050, respectively.

1.4.7 Availability of Water for Future Use

The amount of water that would be available for future use would depend on the demands and stresses and how these demands are managed, and the overall interaction between the lithosphere, pedosphere, atmosphere, and hydrosphere would be dealt with. Water resources are sensitive to social and environmental variables. Therefore, it is important that a multidisciplinary approach is adopted to develop and manage water resources.

1.5 FOOD SECURITY

The World Food Summit of 1996 defined food security as existing "when all people at all times have access to sufficient, safe, nutritious food to maintain a healthy and active life." Commonly, the concept of food security is defined as including both physical and economic access to food that meets people's dietary needs as well as their food preferences. In many countries, health problems related to dietary access are an ever-increasing threat. In fact, malnutrition and food-borne diarrhea are becoming a double burden. Food security is built on three pillars:

- *Food availability*: sufficient quantities of food available on a consistent basis
- *Food access*: having sufficient resources to obtain appropriate foods for a nutritious diet
- *Food use*: appropriate use based on the knowledge of basic nutrition and care, as well as adequate water and sanitation

Food security is a complex sustainable development issue, linked to health through malnutrition, and also to sustainable economic development, environment, and trade.

1.5.1 Requirements to Feed Population

It is assumed that for self-sufficiency, some 900 m^3 of water per person per year has to be provided. Further to this, the FAO in discussion with the CFWA and the World Water Summit (Falkenmark 1997) made the following approximations: a good nutrition level implies 2700 kcal per person per day, with 2300 kcal plant-based nutrition and 400 kcal animal-based nutrition. The production of the former consumes 1 m^3 per 1000 kcal and the latter, 5 m^3 per 1000 kcal, which altogether amounts to 4.3 m^3 per person per day or 1570 m^3 per person per year.

In an arid climate, all of this would have to be provided by blue water (irrigation water). In a humid climate, all may be provided by green water (soil moisture). In a semiarid climate, 50% may be met by green water and 50% by blue water, leading to 800 m^3 per person per year for food self-sufficiency.

An alternative calculation would be the following: for an acceptable plant-based diet, 250 kg per person per year is required (annual per capita grain use in 1990: India, 200 kg per person; China, 300 kg per person) (Brown and Kane 1994).

To feed the projected population of the world by 2050 (i.e., 9.1 billion) while considering changes in dietary patterns, an additional 70% of agricultural production would be required. This dietary pattern shift is toward the consumption of meat, fish, milk, and high-value protein-based food. In financial terminology, to meet the projected demand for the year 2050, over 80 million U.S. dollars need to be invested in the agriculture sector annually. This is an increase of about 50% than what is currently being invested.

In terms of water requirements, to suffice food production for the projected population, an additional 2200 BCM of consumptive water per year will be required (Swedish Assessment). This corresponds to a 50% increase from the current situation. If this requirement is covered by irrigation only, it would involve more than a doubling of all the water withdrawals from rivers and aquifers today, and that would be absolutely unacceptable in view of the degraded river flow regimes and aquatic ecosystems. However, considering the FAO projecting the average diet in the developing countries for 2030–3000 kcal per person per day, an additional consumptive water use of 4200 BCM/year would be required by 2030, assuming that hunger would be altogether eradicated, whereas by the year 2050, the additional water required would go up to 5200 BCM/year.

Besides the above consumptive water need, a study by IWMI suggests that irrigation might not contribute (blue water) more than some 270 BCM/year by 2015 (520 BCM/year by 2030, 725 BCM/year by 2050). Under such circumstances, a question arises as to how the remaining water need will be managed.

1.5.2 Food Wastage

According to the legal definition of waste by the European Commission (EU), food waste is "any food substance, raw or cooked, which is discarded, or intended or required to be discarded." Food wastage may occur at different stages: planting to harvesting; at harvesting; farm-level storages; food processing, packaging, and cooking; postcooking and feeding. Waste magnitudes at different stages are different, though wastage is highest in processing, packaging, and after cooking.

Food waste can have dramatically varied impacts, depending on the amount produced and how it is dealt with. In some countries, the amount of food waste is negligible, and therefore the impact is quite insignificant. However, in developed countries, especially in the United States and the United Kingdom, where food scrap represents around 19% of the waste dumped in landfills, food waste has enormous environmental impact due to the production of methane, a greenhouse gas. A study by Vaughan (2009) reported that the hunger of 1.5 billion people could be alleviated by eradicating food wastage by British consumers and American retailers, food services, and householders, including arable crops such as wheat, maize, and soybean, to produce wasted meat and dairy products. When this fact is converted into water equivalence, it may be remarked that the irrigation water used by the farmers to grow wasted food would be enough for the domestic water requirement for 9 billion people (i.e., 164.25 BCM per year). Further to this, a fresh estimate from the Ministry of Food Processing published that agricultural food of approximately Rs 580 billion gets wasted every year in India. Therefore, the prevention of food wastage is very important to save water, which requires a change in attitude.

1.5.3 Biotechnology

To meet the food demand for the growing population under climatic changes and limited water availability, intensification of agriculture and reliance on irrigation and chemical inputs have led to environmental degradation. Much of Asia faces problems of salinity, pesticide misuse, and degradation of natural resources. The green revolution technologies were useful in the favorable and irrigated environment, but they had little impact on the millions of small landholders living in rainfed and marginal areas where poverty is concentrated. In addition, declining public investments in the agriculture sector across regions have largely affected food production. These factors have been responsible for the decline in annual agricultural growth from an average of 3.3% during 1977–1986 to about 1.5% during 1987–1996.

It means that to meet food demands for the future population (i.e., an increase of 50% in food production by the year 2030), dependence on traditional technology no longer will be helpful. Under such circumstances, it has been required to encourage biotechnology in agriculture to increase production. The advantages of new techniques of modern biotechnology are (i) speedy plant and animal breeding, (ii) possible solutions to previously intractable problems such as drought tolerance, and (iii) the development of new products such as more nutritious food.

Genetically modified (GM) crops have been commercially cultivated since 1996. Over the past years, their worldwide production has continuously increased. It was reported that by the end of the year 2005, GM crops were grown in 21 countries, and the official global area of GM crops totaled 90 million hectares, which amounts to approximately 5% of the area under cultivation.

The contribution of biotechnology toward increasing yields will be realized by decreasing the losses from diseases and pests while minimizing the use of pesticides. Based on several studies, it is expected that agricultural biotechnology will contribute significantly toward poverty reduction and food security through increased productivity, lower production costs and food prices, and improved nutrition (ADB, 2001). Modern plant breeding may help to achieve productivity by introducing resistance to pests and diseases, reduce pesticide use, improve crop tolerance for abiotic stress, improve the nutritional value of some foods, and enhance the durability of products during harvesting and shipping. Biotechnology may offer cost-effective solutions to vitamin and mineral deficiencies by developing rice varieties that contain vitamin A and minerals. Raising productivity could increase small landholders' income, reduce poverty, increase food access, reduce malnutrition, and improve livelihoods of the poor. Besides the above advantages, the long-term impact of genetically improved foods on human health and the environment is not known; that will require monitoring and further research.

1.5.4 Climatic Extremes: Floods and Droughts

Agriculture is highly sensitive to climate variability and weather extremes, such as droughts, floods, and severe storms. The forces that shape the climate are also critical to farm productivity. Human activities have already changed the atmospheric characteristics such as temperature, rainfall, evapotranspiration, levels of CO_2, and ground-level ozone. The scientific community expects such trends to continue. The increased potential for droughts, floods, and heat waves will pose challenges for farmers. Additionally, due to the enduring climatic changes, available water supply and soil moisture could not make it feasible to continue crop production in certain regions.

The coastal flooding will further reduce the amount of land available for agriculture. Due to climatic extremes (www.who.int/globalchange/en/), the natural disaster impacts have been increasing in different geographical regions (Table 1.8). Based on origin, the distribution of natural disasters (Table 1.9) is shown in Figure 1.10, which shows that natural disasters caused by hydrometeorology are more than the others (Source: International Disaster Database: www.em-dat.net). The trend associated with hydrometeorological disasters is increasing in nature.

TABLE 1.8
Number of Climatic Extremes and People Affected

	1980s			1990s		
Regions	**Events**	**Killed (1000 People)**	**Affected (1000 People)**	**Events**	**Killed (1000 People)**	**Affected (1000 People)**
Africa	243	417	137.8	247	10	104.3
Eastern Europe	66	2	0.1	150	5	12.4
Eastern Mediterranean	94	162	17.8	139	14	36.1
Latin America and Caribbean	265	12	54.1	248	59	30.7
Southeast Asia	242	54	850.5	286	458	427.4
Western Pacific	375	36	273.1	381	48	1199.8
Developed countries	563	10	2.8	577	6	40.8
Total	*1848*	*692*	*1336*	*2078*	*601*	*1851*

Source: United Nations International Strategy for Disaster Reduction, www.unisdr.org/disaster-statistics/.

TABLE 1.9
Distribution of Natural Disasters by Origin

Nature/Origin	1900–1909	1910–1919	1920–1929	1930–1939	1940–1949	1950–1959	1960–1969	1970–1979	1980–1989	1990–1999	2000–2005	Total
Total	73	107	99	112	176	294	588	964	1900	2720	2788	9821
Hydrometeorological	28	72	56	72	120	232	463	776	1498	2034	2135	7486
Geological	40	28	33	37	52	60	88	124	232	325	233	1252
Biological	5	7	10	3	4	2	37	64	170	361	420	1083

Source: Centre for Research on the Epidemiology of Disasters, EM-DAT: The International Disaster Database, www.em-dat.net.

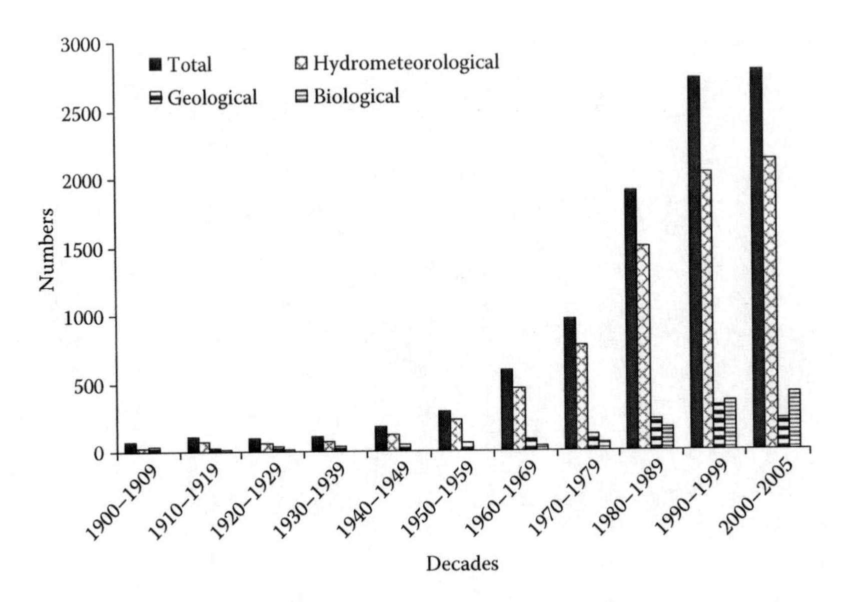

FIGURE 1.10 Distribution of natural disasters by origin. (From Centre for Research on the Epidemiology of Disasters, EM-DAT: The International Disaster Database, www.emdat.be.)

1.5.5 Conflicts and Wars

For centuries, wars and conflicts have been tied to the protection of water resources. With the risk of water shortages around the world becoming more and more of an issue, water has become the fuel of certain conflicts in many regions around the world. Many regions around the world deal with shortages of water, leading to interstate and intercountry disputes. Globally, over 200 water bodies are shared by two or more countries or areas. A few examples are as follows:

i. *The Middle East*: Middle East conflicts are usually tied to religious issues or oil, but water has become a major factor in recent disputes. This water dispute arises from the overuse and political territorial uses resulting from the disagreement in water distribution of the Jordan and the Tigris–Euphrates River Basins. The Jordan River Basin includes parts of Lebanon, Syria, Israel, Jordan, and the West Bank. However, Turkey, Syria, and Iran are the countries having water conflicts due to the Tigris–Euphrates Basin.
ii. *Africa*: In many parts of Africa, the water shortage has become part of daily life. Water conflicts due to sharing of the Nile River water between Uganda, Sudan, and Egypt have been seen but prevented through political will by signing a sharing agreement.
iii. *Asia*: In India and China, water shortages pose both social and economic threats. Throughout India, Pakistan, Nepal, and Bangladesh, water shortages are increasingly triggering conflict. The Indus River Basin and the Ganga River Basin are the major areas of water conflicts between India and Pakistan and between India and Bangladesh, respectively. In India herself, there are interstate disputes due to water sharing of the Yamuna River water and the Cauvery River water.

Generally, it is stated that the third world war will be fought over water.

1.6 WATER AND WATERSHED MANAGEMENT

Starting from the availability of fresh water in terms of total renewable water resources (TRWR), the impact of population growth and climate change, and environmental protection, water has become a critical issue that will lead to future conflicts and wars among various nations and regions. Therefore, by now it is of prime importance to think about how to deal with this issue of future water requirement to meet the various needs. The only way that seems to resolve this issue is watershed management, and the first step will be the inventory of natural resources at the watershed scale.

Watershed management is the process of creating and implementing plans, programs, and projects to sustain and enhance watershed functions that affect the plant, animal, and human communities within a watershed boundary. Watershed management includes the optimal utilization and planning of available natural resources along with the management of water supply, water quality, drainage, storm water runoff, water rights, etc.

For sustainable watershed management, there should be causal linkages between the geophysical science, society, ecology, and economics. The three main components of watershed management are (i) land management, (ii) water management, and (iii) biomass management. The land management is determined by the land characteristics such as terrain, slope, formation, depth, texture, moisture, infiltration rate, and soil capability. Altogether, these factors can be termed as land capabilities. The land management interventions can be categorized as structural, vegetative, production, and protection measures.

Mechanical conservation measures may become necessary in watershed management in the initial stages. Structural measures include interventions such as contour bunds, stone bunds, earthen bunds, graded bunds, compartmental bunds, contour terrace walls, contour trenches, bench terracing, broad based terraces, centripetal terraces, field bunds, channel walls, stream bank stabilization, check dams, etc. Watersheds may contain natural ecosystems such as grasslands, wetlands, mangroves, marshes, and water bodies. All these ecosystems have a specific role in nature. Vegetative measures include vegetative cover, plant cover, mulching, vegetative hedges, grass land management, fencing, agroforestry, etc. The production measures include interventions aimed at increasing the productivity of land, such as mixed cropping, strip cropping, cover cropping, crop rotations, cultivation of shrubs and herbs, contour cultivation conservation tillage, land leveling, use of improved variety of seeds, horticulture, etc. Protective measures, such as landslide control, gully plugging, runoff collection, and so on, can also be adopted. Adoption of all the interventions mentioned above should be done strictly in accordance with the characteristics of the land taken for management.

Water characteristics such as inflows (precipitation, surface water inflow, groundwater inflow), water use (evaporation, evapotranspiration, irrigation, drinking water), outflows (surface water outflow, groundwater outflow), and storage (surface storage, groundwater storage, root zone storage) are the principal factors to be taken care of in sustainable water management. The broad interventions for water management are rainwater harvesting, groundwater recharge, maintenance of water balance, preventing water pollution, economic use of water, preventing various losses, etc.

Rainwater harvesting forms the major component of water management. The rainwater collected can be recharged into the ground. Rooftop water harvesting, diversion of perennial springs and streams into storage structures, farm ponds, and so on, are the methods widely used for rainwater harvesting. Some simple and cost-effective rainwater-harvesting structures are percolation pits/tanks, recharge trenches/rain pits, recharge wells, ferro-cement tanks, farm ponds, V ditch, bench terracing, etc. The economic use of water and avoidance of affluence in the use of water at individual and community levels may be the major concern for water management in the years to come.

Major intervention areas for biomass management include eco-preservation, biomass regeneration, forest management and conservation, plant protection and social forestry, increased productivity of animals, income and employment generation activities, coordination of health and sanitation programs, better living standards for people, eco-friendly lifestyle of people, formation of a learning community, etc.

Besides the above three components of watershed management, public participation and collective action are critical ingredients for sustainable management. Sustainability, equity, and participation are the three basic elements of participatory watershed management. Sustainability involves the conservation and enhancement of the primary productivity of the ecosystem, the main components of which are land, water, and biomass. Equity has to be seen in terms of creating an equitable access to livelihood resources for the watershed community. Participatory watershed management attempts to ensure the sustainability of the ecological, economic, and social exchanges taking place in the watershed territory. This includes natural resource exchange, which is the conventional watershed management, and participatory watershed management additionally considers the economic, political, and cultural exchanges.

1.6.1 Water Allocation and Valuing/Pricing

For sustainable water resources management, a few important statements need to be carefully understood:

i. Dublin Statement on Water and Sustainable Development (ICWE 1992):
"Principle No. 4: Water has an economic value in all its competing uses and should be recognized as an economic good. Within this principle, it is vital to recognize first the basic right of all human beings to have access to clean water and sanitation at an affordable price. Past failure to recognize the economic value of water has led to wasteful and environmentally damaging uses of the resource. Managing water as an economic good is an important way of achieving efficient and equitable use, and of encouraging conservation and protection of water resources."
ii. Agenda 21, Chapter 18 (UNCED 1992):
"Water should be regarded as a finite resource having an economic value with significant social and economic implications regarding the importance of meeting basic needs."
iii. Ministerial Declaration of the 2nd World Water Forum (World Water Council 2000):
"To manage water in a way that reflects its economic, social, environmental and cultural values for all its uses, and to move towards pricing water services to reflect the cost of their provision. This approach should take account of the need for equity and the basic needs of the poor and the vulnerable."
iv. Ministerial Declaration of the 3rd World Water Forum (World Water Council 2003):
"Funds should be raised by adopting cost recovery approaches which suit local climatic, environmental and social conditions and the 'polluter-pays' principle, with due consideration to the poor. All sources of financing, both public and private, national and international, must be mobilized and used in the most efficient and effective way."

The focus on economic efficiency as the primary objective in the development and allocation of water resources is because of its importance as a social objective: efficiency values having viable meaning in resolving conflicts and assessing the opportunity costs of pursuing alternative uses (Young 1996). Although economically efficient allocation of irrigation water is rarely attained in practice, the analysis of economic efficiency provides a useful point of reference for understanding the causes of inefficient allocation and mechanisms for improving the overall economic performance of irrigated production.

Economically efficient allocation of water is desirable to the extent that it maximizes the welfare that society obtains from available water resources. Welfare in this context refers to the economic well-being of society and is determined by the aggregate well-being of its individual citizens. Economically efficient allocation maximizes the value of water across all sectors of the economy. This is achieved through the allocation of water to uses that are of high value to society and away from uses with low value. Efficient allocation occurs in a competitive, freely functioning market when supply is in equilibrium with demand (Tsur and Dinar 1997).

TABLE 1.10
Policy Measures Relevant to Resource Management

Conditions	Public Sector		Private Sector	
Poverty rights	Least-developed country	Developing country	Least-developed country	Developing country
Pricing	Price = Marginal cost Price = Marginal social cost	Price = Marginal social cost	Price = Marginal social cost	Price = Marginal social cost
Quantity trading			Possible emissions and resource quota trading	Emissions and resource quota trading
Command and control	Environmental quality objectives	Environmental quality objectives	Environmental quality objectives	Environmental quality objectives
Investment policy	Cost–benefit analysis	Cost–benefit analysis	Environmental impact assessment	Environmental impact assessment

As indicated above, economic efficiency and equity are important considerations in the allocation of water. Greater efficiency is required in the face of increasing water scarcity, and equity is a concern because of the importance of water to the livelihoods and well-being of rural communities in particular. It is possible to derive a broad classification of policy measures that are relevant to managing resources within the boundaries of a nation. The measures include the redefinition of property rights and investment policies.

One thing to be kept in mind is that many of the large public irrigation schemes that were promoted as part of the green revolution, particularly in Asia, were designed to target poor rural communities and as such were never oriented to maximize economic output, but instead to guarantee production of food staples (Plusquellec 2002).

The proper pricing of inputs (such as raw water) and outputs (such as agricultural irrigation products) can be viewed as a form of property right designation, while command and control measures are also means of defining property rights or modifying existing ones (FAO 2004). Table 1.10 shows the various generic types of policy measures. Once a regime of property rights has been established, the proper pricing of a resource requires that it be priced at least at marginal private costs and preferably at marginal social cost (especially in the longer term and where output prices are below the private production costs). As pricing of water affects the allocation decisions of those with competing wants, then by correctly pricing water, efficient allocation of water is achieved. However, the standard economic efficiency (marginal) cost pricing result is sometimes problematic as regards the specification of production technology. In the water supply sector, inputs to production are often not perfectly divisible. Investments often require large lumps of capital (e.g., for dams and reservoirs). In such cases, marginal cost pricing to achieve economic efficiency requires some form of intervention (Sherman 1989). Table 1.10 lists quantity-based measures as a separate policy option, although they have similar effects on the price-based measures. Finally, investment policy, which is most usually characterized in terms of cost–benefit analysis, is applicable to all public-sector operations (although environmental impact assessments are employed most widely in assessing private-sector environmental impacts).

Water allocation systems differ in the extent to which they address efficiency and equity in objectives. The various systems can be compared according to several criteria (Dinar et al. 1997; Howe et al. 1986; Winpenny 1994):

- *Flexibility in allocation of supplies*: allocation requires flexibility such that supplies can be shifted between uses and sectors, as demand changes, so as to achieve efficiency.

- *Security of tenure for users*: established users require security of tenure if they are to be expected to take the necessary measures to use the resource efficiently. Although this may conflict with flexibility, problems should not arise if sufficient water reserves are available to meet the unexpected demands.
- *Payment of real opportunity costs of water by users*: users should pay the real opportunity costs of their use, so that other demand or external effects are internalized.
- *Predictability of the allocation outcome*: in order to achieve the best allocation and minimize uncertainty, the outcome of the allocation process needs to be predictable.
- *Equity in the allocation process*: users should perceive the allocation process to be equitable.
- *Political and public acceptability*: the allocation should serve the various political and public values and objectives, thereby making it acceptable to the groups in society.
- *Efficacy in achieving desired policy goals*: the form of allocation should change an existing undesirable situation toward one where the desired policy goals are achieved.
- Administrative feasibility and sustainability: the allocation mechanism must be practicable, adaptable, and allow an increasing effect of policy.

Water allocation systems range from government-controlled to market-led systems, and combinations of the two. The prevailing institutional frameworks (including laws, regulations, and organizations) and the water resources infrastructure (Dinar et al., 1997) influence the precise nature of allocation systems. However, they commonly fall into one of only a small number of categories: public allocation, market-based allocation, and user-based allocation.

1.6.2 Water Conservation

Water conservation refers to the process of reducing the usage of water and recycling of wastewater for different purposes such as cleaning, manufacturing, construction, agricultural irrigation, and parks. The goal of water conservation efforts includes sustainability, energy conservation, and habitat conservation. The important components of water conservation are water reuse, water recycling, and water-use efficiency.

1.6.3 Water Recycling and Reuse

Looking into the global water scarcity, wastewater recycling and reuse may be one of the good water conservative options. The motivational factors for wastewater recycle/reuse may include the following:

- Opportunities to augment limited primary water sources
- Prevention of excessive diversion of water from alternative uses, including the natural environment
- Possibilities to manage in situ water sources
- Minimization of infrastructure costs, including total treatment and discharge costs
- Reduction and elimination of discharges of wastewater (treated or untreated) into the receiving environment
- Scope to overcome political, community, and institutional constraints

Reuse of wastewater can be a supplementary source to existing water sources, especially in arid/semiarid climatic regions. Most large-scale reuse schemes are in Israel, South Africa, and arid areas of the United States, where alternative sources of water are limited. This option will be beneficial even in the regions where rainfall is adequate but water shortage is caused due to the spatial and temporal variability. For example, Florida, in the United States, is not a dry area, has limited options for water storage, and suffers from water shortages during dry spells. For this reason, wastewater reuse schemes form an important supplement to the water resource of this region.

1.6.4 Water-Use Efficiency

Water-use efficiency is a tool of water conservation that results in more efficient water use and thus reduces water demand. The value and cost-effectiveness of a water efficiency measure must be evaluated in relation to its effects on the use and cost of other natural resources (e.g., energy and chemicals).

1.7 SUMMARY AND CONCLUSION

The availability of good quality water will vary from one region to another. A multidisciplinary approach will be required to determine accurately the water availability. In this chapter, a holistic approach has been considered to develop a wide and basic understanding of the different issues related to water resources and food security. For water resources, a general baseline in terms of the hydrologic cycle and renewable water resources has been described, followed by the possible impacts on the availability and uses of water due to climate change, population growth, environmental pollution, etc. Based on the consequences, including spatiotemporal variability, it may be stated that the availability of good quality water will be a great challenge in the near future for the livelihood and the environment. Under such circumstances, it will be important to think over each and every drop of water being used and wasted. There should be a proper accounting of water consumption for suppliers and users as well. Based on the aforesaid discussions, the following concluding thoughts can be suggested to meet future demands:

i. Agriculture: It is the highest water consumer, and therefore, it should be optimized on a spatial basis through proper cropping patterns, adequate irrigation methods, adequate crop variety, organic farming, minimizing the conveyance losses, etc. These are the practices at watershed level management. However, for global action, in addition to these, virtual water transfer should be encouraged and brought into the national water and agricultural policy.
ii. Changes in lifestyle and food habits may reduce water consumption. This will require shifting the food habit from nonvegetarian to vegetarian.
iii. Food wastage: Food wastage can save several billions of water per year.
iv. Minimization of losses incurred in domestic water supply.
v. Wastewater management: The disposal of wastewater, either treated or untreated, into rivers must be minimized, because approximately 50%–75% of fresh water is needed for dilution. Wastewater should be treated using such a technology that it can be completely reused for various purposes such as irrigation, construction, and reserved parks.
vi. A sustainable watershed management framework needs to be developed and followed under policy with proper enforcements. It should consider the inventory of all natural resources including land-use capability and water resources availability, livestock, socioeconomics, and environmental health.
vii. For the success of any good water resource or watershed management plan, public participation is important. The first step to involve the public into the plan is by adequate accounting of water consumption and wastewater generation; social and environmental impacts of their water use should be quantified and billed. Misuse of water and the resultant environmental damage should be treated as an offense.
viii. There should be population control in urban centers especially located near river banks to prevent river pollution.
ix. Among the above, the most important constraint is the continuous population growth. How to control it will be a major challenge.

REFERENCES

ADB. 2001. Agricultural biotechnology, poverty reduction and food security. A working paper, Asian Development Bank, May 2001.

Arora, M., N.K. Goel, and P. Singh. 2005. Evaluation of temperature trends over India. *Hydrological Sciences Journal* 50(1): 81–93.

Bates, B.C., Z.W. Kundzewicz, S. Wu, and J.P. Palutikof (eds). 2008. Climate change and water. Technical Paper of the Intergovernmental Panel on Climate Change, Secretariat IPCC, Geneva.

Boswinkel, J.A. 2000. International Groundwater Resources Assessment Centre (IGRAC), Netherlands Institute of Applied Geoscience, Netherlands.

Brown, L.R. and H. Kane. 1994. Full house. Reassessing the Earth's population carrying capacity. The Worldwatch Environment Alert Series, New York/London.

Bunn, S.E. and A.H. Arthigton. 2002. Basic principles and ecological consequences of altered flow regimes for aquatic biodiversity. *Environ. Manag.* 30: 492–507.

Dinar, A., M.W. Rosegrant, and R. Meinzen-Dick. 1997. Water allocation mechanisms: Principles and examples. World Bank Policy Research Working Paper 1779. World Bank, Washington, DC.

Falkenmark, M. 1997. Meeting water requirements of an expanding world population. *Phil. Trans. Royal Soc. Land* 352: 929–936.

FAO. 2003. Review of world water resources by country. Water Reports No. 23, Food and Agriculture Organization, Rome.

FAO. 2004. Water charging in irrigated agriculture: An analysis of internal experience. Water Reports No. 27, Food and Agriculture Organization, Rome.

Fitter, A.H. and R.S. Fitter. 2002. Rapid changes in flowering time in British plants. *Science* 296 (5573): 1689–1691.

GRID-Arendal and H. Ahlenius (ed.) 2007. Environmental knowledge for change. Projected losses in food production due to climate change by 2080, http://maps.grida.no/go/graphic/projected-losses-in-food-production-due-to-climate-change-by-2080 (accessed 3 March 2009).

Groombridge, B. and M. Jenkins 1998. Freshwater biodiversity: A preliminary global assessment. World Conservation Monitoring Centre (UNEP-WCMC), World Conservation Press, Cambridge, UK.

Hossain, F., D. Niyogi, J. Adegoke, G. Kallos, and R. A. Pielke Sr. 2011. Making sense of the water resources that will be available for future use. *Eos* 90: 144–145.

Howe, C.W., D.R. Schurmeier, and W.D. Shaw. 1986. Innovative approaches to water allocation: The potential for water markets. *Water Resour. Res.* 22(4): 439–445.

ICWE. 1992. The Dublin Statement on Water and Sustainable Development. International Conference on Water and the Environment, Dublin, Ireland, January 26–31, 1992.

IDB. 2010 World Population. International Data Base (http://www.census.gov/ipc/www/idb/worldpopinfo.php.)

IPCC. 2007a. Climate change 2007: Impacts, adaptation, and vulnerability. Contribution of Working Group II to the Fourth Assessment Report of the Intergovernmental Panel on Climate Change. Cambridge University Press, Cambridge, UK.

IPCC. 2007b. Climate change 2007: The physical science basis. Contribution of Working Group I to the Fourth Assessment Report of the Intergovernmental Panel on Climate Change. Cambridge University Press, Cambridge, UK.

IPCC. 2007c. Climate change 2007: Fourth Assessment Report. Intergovernmental Panel on Climate Change. Cambridge University Press, Cambridge, UK.

Korzun, V.I. (ed.) 1978. *World Water Balance and Water Resources of the Earth.* No. 25 of Studies and Reports in Hydrology. UNESCO, Paris.

Kothawale, D.R. and K. Rupakumar. 2005. On the recent changes in surface temperature trends over India. *Geophysical Research Letters* 32, L1874. DOI: 1029/2005gl023528.

Lal, R., M.V.K. Sivakumar, S.M.A. Faiz, A.H.M.M. Rahman, and K.R. Islam. 2011. Climate change and food security in South Asia. Springer.

Pagano, T.C. and S. Sorooshian. 2005. Global water cycle (fundamental, theory, mechanisms). *Encyclopedia of Hydrological Sciences.* John Wiley & Sons, New York. DOI: 10.1002/0470848944.hsa179.

Parmesan, C. and G. Yohe. 2003. A globally coherent fingerprint of climate change impacts across natural systems. *Nature* 421 (6918): 37–42.

Parry, M.L., O.F. Canziani, J.P. Palutikof, P.J. van der Linden, and C.E. Hanson (eds). 2007. Climate change 2007: Impacts, adaptation and vulnerability. Contribution of Working Group II to the Fourth Assessment Report of the Intergovernmental Panel on Climate Change, Cambridge University Press, Cambridge, UK, p. 13.

Plusquellec, H. 2002. How design, management and policy affect the performance of irrigation projects. Emerging modernization procedures and design standards. FAO, Bangkok.

Reddy, K.R. and H.F. Hodges. 2000. Climate change and global crop productivity. CAB International Publishing, UK.

Sherman, R. 1989. *The Regulation of Monopoly*. Cambridge University Press, Cambridge, UK.

Tennant, D.L. 1976. Instream flow regimens for fish, wildlife, recreation and related environmental resources. *Fisheries* 1: 6–10.

Tsur, Y. and A. Dinar. 1997. On the relative efficiency of alternative methods for pricing irrigation water and their implementation. *World Bank Economic Review* 11: 243–262.

UN. 1994. World urbanization prospects: The 1994 revision. United Nations Population Division, Department for Economic and Social Information and Policy Analysis, New York.

UN. 2009. World population ageing 2009. United Nations Population Division, Department of Economic and Social Affairs, New York.

UNCED. 1992. The Earth Summit: Agenda 21, Chapter 18: Protection of the quality and supply of freshwater resources—Application of integrated approaches to the development, management and use of water resources. United Nations Conference on Environment and Development, Rio de Janeiro, June 3–14, 1992.

UNDP, UNEP, World Bank, and WRI 2000. *World Resources 2000–2001—People and Ecosystems*. WRI, Washington DC.

UNEP. 1992. Glaciers and the environment. UNEP/GEMS Environment Library No. 9. UNEP, Nairobi, Kenya, p. 8.

UNEP. 2002. *Global Environment Outlook—Past, Present and Future Perspectives*. Earthscan, UK, USA.

UNEP. 2009. The environmental food crisis: The environment's role in averting future food crises. A UNEP rapid response assessment, United Nations Environment Programmes, Norway.

Untersteiner, N. 1975. Sea ice and ice sheets and their role in climatic variations. Appendix 7 in GARP Publication Series No. 16. In *The Physical Basis of Climate and Climate Modelling*. World Meteorological Organisation, Geneva, pp. 206–224.

U.S. Census Bureau. 2011. World POPClock Projection, http://www.census.gov/ipc/www/popclockworld.html.

Vaughan, Adam 2009. Estimation of food waste could lift 1 bn out of hunger, say campaigners, *The Guardian*, http://www.guardian.co.uk/environemt/2009/sep/08/food-waste.

WBCSD. 2009. Water facts & trends, http://www.wbcsd.org/includes/getTarget.asp? type=d&id=MTYyNTA. Retrieved December 3, 2009.

WGMS. 1998. Monitoring strategy, www.geo.unizh.ch/wgms, 1998.

WGMS. 2002. World Glacier Monitoring Service, http://www.kms.dk/fags/ps09wgms.htm, viewed March 2002.

WHO. 1976. Surveillance of drinking-water quality. World Health Organization, Geneva.

WHO. 2008. Climate change and human health. World Health Organization, www.who.int/globalchange/en/.

WHO/UNICEF. 2000. Global water supply and sanitation assessment 2000 report. UNICEF, New York.

Willis, C.G., B. Ruhfel, R.B. Primack, A.J. Miller-Rushing, and C.C. Davis. 2008. Phylogenetic patterns of species loss in Thoreau's woods are driven by climate change. *Proceedings of the National Academy of Sciences of the United States of America* 105(44): 17029–17033.

Winpenny, J. 1994. *Managing Water as an Economic Resource*. Routledge, London.

WMO. 1997. *Comprehensive Assessment of the Freshwater Resources of the World*. World Meteorological Organisation, Geneva, p. 9.

World Water Council. 2000. Ministerial declaration of The Hague on water security in the 21st century. Second World Water Forum on 22nd March 2000.

World Water Council. 2003. Ministerial declaration of the 3rd World Water Forum in Koyota in 2003.

WRI, UNEP, UNDP, and World Bank 1998. *World Resources 1998–1999—A Guide to the Global Environment*. Oxford University Press, New York.

Young, R.A. 1996. Measuring economic benefits for water investments and policies. World Bank Technical Paper No. 338.

Younos, T. and R. Hill. 2008. Water needed to produce various types of energy. *Science Daily*, April 22, 2008.

2 Soil Water and Agronomic Production

Rattan Lal

CONTENTS

2.1 INTRODUCTION

Ever since the evolution of *Homo sapiens*, the human population touched 1 billion by ~1800 AD. The population touched 2 billion by 1925, 3 billion by 1960, 4 billion by 1974, 5 billion by 1987, 6 billion by 1999, and 7 billion by 2011 (UN 2011). The population is projected to touch 8 billion by 2028 and 9 billion by 2054 and may stabilize at 10 billion by 2100 and beyond. Despite the exponential growth in the human population, earth's natural resources are either fixed or dwindling. For example, only 2.5% of the earth's total water reserve is fresh water (Figure 2.1). Freshwater ecosystems cover <1% of the earth's surface (Johnson et al. 2001). There is no alternative to water, it has many competing uses, and it is a scarce commodity. As much as 40% of the world's population and several ecosystems are already vulnerable to water scarcity (Oki and Kanae 2006), and the adverse effects of scarcity will be exacerbated by an increase in the population and its numerous demands (Pfister et al. 2011).

Soil is an important reservoir of fresh water (Table 2.1), and the growth of all terrestrial plants and soil biota depends on its availability and quality. The principal source of soil water is the precipitation and infiltration or transfer into the soil from the soil–atmosphere interphase. Soil water is lost by evaporation, transpiration, and deep seepage or percolation. Soil water storage and its judicious use can reduce the adverse effects of water scarcity, because soil water in the root zone is the essence of all terrestrial life.

Soil water, also called green water (Rockström et al. 2009), is composed of only a fraction of renewable freshwater resources (17×10^3 km^3; Table 2.1). Green water is the soil water held in the unsaturated zone, is formed by precipitation, and is available for plants. Green water is strategically important in relation to international commodity trade (Aldaya et al. 2010). Because of its

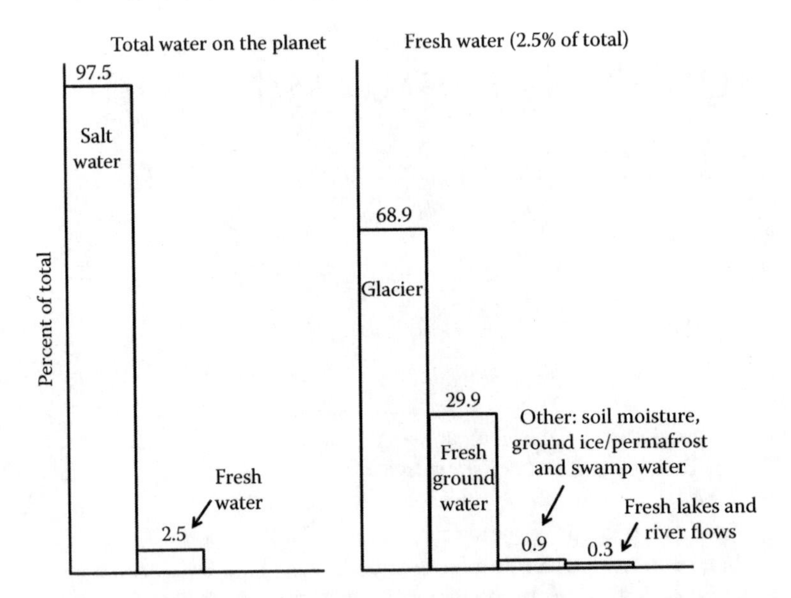

FIGURE 2.1 Distribution of global water. (Adapted from Shiklomanov, I., In *Water in Crisis: A Guide to Word's Fresh Water Resources*, p. 473, Oxford University Press, Oxford, UK, 1993.)

low opportunity cost, the effective and efficient use of green water for crop production has comparatively less environmental externalities than the use of blue water for supplemental irrigation. In this context, therefore, blue water refers to the water in lakes, rivers, and wetlands, which can be withdrawn for irrigation. Thus, irrigated land receives both green and blue water, but rainfed land depends only on green water (Falkenmark and Rockström 2006; Hoff et al.

TABLE 2.1
Terrestrial Water Balance

Parameter		Amount
I	Reservoirs (10^3 km^3)	
	Ocean	1,338,000
	Glacier and snow	24,064
	Permafrost	300
	Lake	175
	Wetland	17
	Soil moisture	17
	Biological water	1.0
	Water vapor overland	3.0
	Water vapor over sea	10.0
II	Fluxes (10^3 km^3/year)	
	Precipitation over ocean	391
	Total terrestrial precipitation	111
	Evaporation over ocean	436.5
	Total terrestrial evapotranspiration	65.5
	Total river flow	45.5
	(i) Surface runoff	15.3
	(ii) Subsurface runoff	30.2

Source: Adapted from Oki, T. and Kanae, S., *Science*, 313, 1068–1072, 2006.

2010). By comparison, gray water is urban and industrial water (polluted/contaminated), which can be reused for irrigation following some treatment. Virtual water is the water traded among nations (regions) through trade of food. It refers to the water required in the production of crops. Thus, the exchange of water through the trade of agricultural produce involves the virtual water trade. The virtual water content of crops grown in the Indo-Gangetic Plains of northern India is estimated at 745–9405 m^3/Mg for wheat and 2502–9562 m^3/Mg for rice (Kumar and Jain 2011). At present, the global virtual water trade is ~1000 km^3/year. Thus, the total water use within a country is a combination of the national water withdrawn/consumed plus the traded virtual water.

Despite a limited supply of renewable freshwater resources, human demand has increased drastically. During the twentieth century, global water withdrawal increased by a factor of 6.8 from 579 km^3/year in 1900 to 3927 km^3/year in 2000 (Gleick 2003; Table 2.2). Irrigation of agricultural lands is the major consumer of freshwater withdrawal, 2504 km^3/year out of 3765 km^3/year in 1995 (~70%, Table 2.2), and the demand is projected to double by 2050 (Tilman et al. 2002). The demand for water will also increase because of the emphasis on biofuels (Melillo et al. 2009). Agriculture uses 1300 m^3 capita/year to produce an adequate diet (Falkenmark and Rockström 2004). Therefore, the sustainable management of soil water (precipitation + irrigation) is critical to enhancing agronomic production and saving water for other competing but essential uses (i.e., industry, domestic, and recreation).

Similar to water, soil resources of good quality are also limited. Both the per capita renewable fresh water and the per capita arable land area are declining with the increase in the population and the demands caused by growing affluence and standards of living. Some 1020 million people face transit and chronic hunger (FAO 2009), and 3.5 billion additional people must be fed within the next five decades (Borlaug 2007). The sustainable management of soils and green water is essential to advancing food security.

Therefore, the objective of this chapter is to describe the state-of-the-world's water resources that are usable for agriculture, to discuss the technological options for sustainable management under an uncertain climate and degrading and dwindling soils, and to outline a technological option for enhancing soil/green water availability for increasing agronomic production to meet the food demands of the increasing population.

TABLE 2.2
World Water Withdrawal and Consumption

Year	Global Water Withdrawal (km^3/year)	Global Water Consumption (km^3/year)	Irrigated Area (10^6 ha)	Global Irrigated Water Withdrawal (km^3/year)	Global Irrigation Water Consumption (km^3/year)
1900	579	331	47.3	513	321
1940	1088	617	76.0	895	586
1950	1382	768	101.0	1080	722
1960	1968	1086	138.8	1481	1005
1970	2526	1341	167.7	1743	1186
1980	3175	1686	209.3	2112	1445
1990	3633	1982	245.2	2425	1691
1995	3765	2074	263.7	2504	1753
2000	3927	2329	276.3	—	—
2010	4323	2501	—	—	—

Source: Modified from Scanlon, B.R., et al., *Water Res.*, 43, 1, 2007.

2.2 AGRONOMIC PRODUCTION AND WATER DEMAND

The increase in agricultural production since 1960 is among the world's greatest success stories. The production of foodgrains increased from 660 million ton (Mt) in 1960 to >2400 Mt in 2010, by a factor of 3.7. Indeed, the growth rate of agronomic/food production surpassed that of the world's population, which more than doubled from 3 billion in 1960 to 7 billion in 2011. In the United States, the grain yields of corn (*Zea mays*) (Figure 2.2), soybean (*Glycine max*) (Figure 2.3), wheat (*Triticum aestivum*) (Figure 2.4), and rice (*Oryza sativa*) (Figure 2.5) increased linearly between 1960 and 2010. A similar increase in the yields of all cereals happened on a global scale (Figure 2.6). Consequently, global food grain production increased drastically over this period (Figure 2.7). This remarkable success, which saved hundreds of millions from starvation, was partly due to the expansion of cropland area equipped with irrigation facilities. Between 1900 and 2000, the agricultural area increased from 813 to 1382 million hectares (Mha) for cropland and from 2118 to 3426 Mha for pastureland. The use of N fertilizer increased from 11.6 Mt in 1950 to 104 Mt in 2008 (Table 2.3). Similar to fertilizer use, the irrigated land area increased from 47.3 Mha in 1900 to 276.3 Mha in 2000 (Table 2.2). Consequently, global irrigated water withdrawal increased by a factor of 4.9 from 513 km^3/year in 1900 to 2504 km^3/year in 1995. Global irrigation water consumption increased by a factor of 5.5 from 321 km^3/year in 1900 to 1753 km^3/year in 1995 (Table 2.2). Because of the increasing demand from industry and urbanization, the supply of renewable fresh water will decrease between 1985 and 2025 (Table 2.4; Rockström et al. 2009). About 40% of the total food production relies on supplemental irrigation (Vico and Porporato 2011), because the green water stored in the effective rooting depth for rainfed/dry farming is inadequate. The potential for expanding irrigation is severely constrained in arid, semiarid, and semihumid regions. Therefore, enhancing soil water storage and its judicious management to increase crop yield per drop are crucial to enhancing agronomic production and achieving global food security.

2.3 REGIONAL WATER BALANCE AND WATER PRODUCTIVITY

The water balances for different biomes are shown in Table 2.5 (Shen and Chen 2010). Runoff (blue water), which is expressed as a percentage of precipitation, is 7% for arid, 9% for semiarid, 20% for

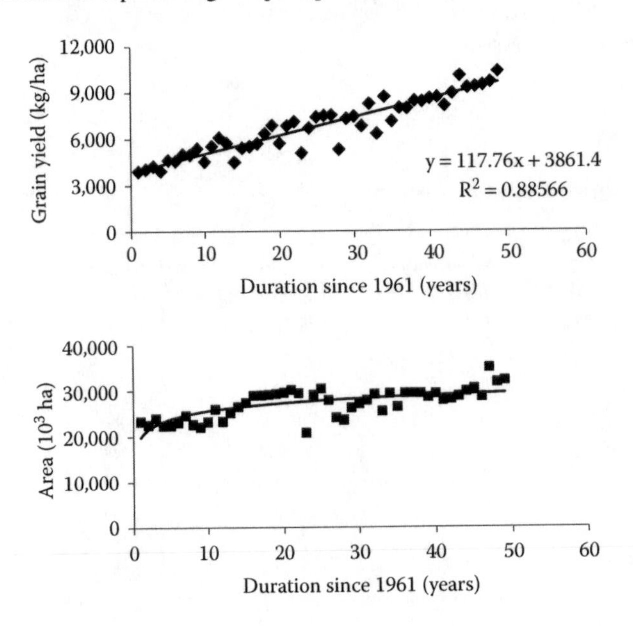

FIGURE 2.2 Change in cultivated area and grain yield of maize in the United States between 1961 and 2010. (Redrawn from USDA. Crop acreage and yields. 2010. http://www.usda.gov/wps/portal/usda/usdahome?parentnav=AGRICULTURE&navid=CROP_PRODUCTION&.)

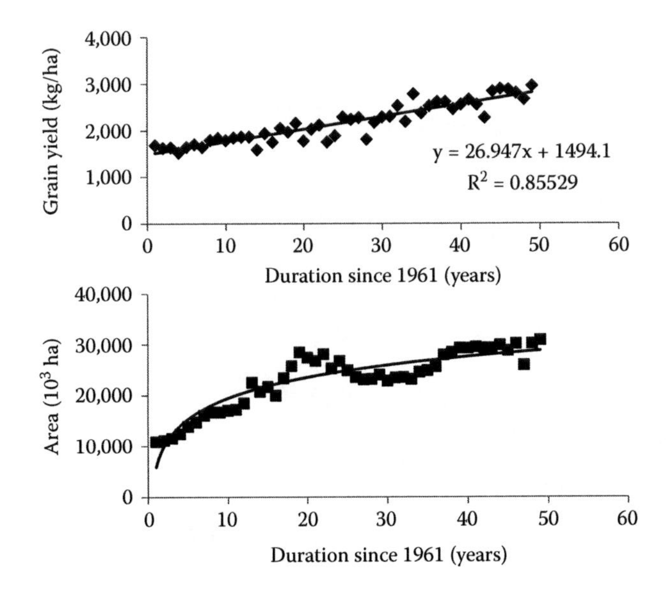

FIGURE 2.3 Change in cultivated area and grain yield of soybean in the United States between 1961 and 2010. (Redrawn from USDA. Crop acreage and yields. 2010. http://www.usda.gov/wps/portal/usda/usdahome?parentnav=AGRICULTURE&navid=CROP_PRODUCTION&.)

semihumid, and 37% for humid regions. Losses due to evaporation are also high in arid and semiarid regions. Thus, the potential for using runoff (blue water) for irrigation is rather low in arid and semiarid regions. There is a severe decline in the groundwater level in the Indo-Gangetic Plains in South Asia (see Chapter 6), China (Cominelli et al. 2009; Fang et al. 2010), and the Ogalalla Aquifer in the United States (see Chapter 4). Even in regions where runoff (blue water) can be used for irrigation, capital investment is prohibitively high for resource-poor, small landholders (Revelle 1976). Thus,

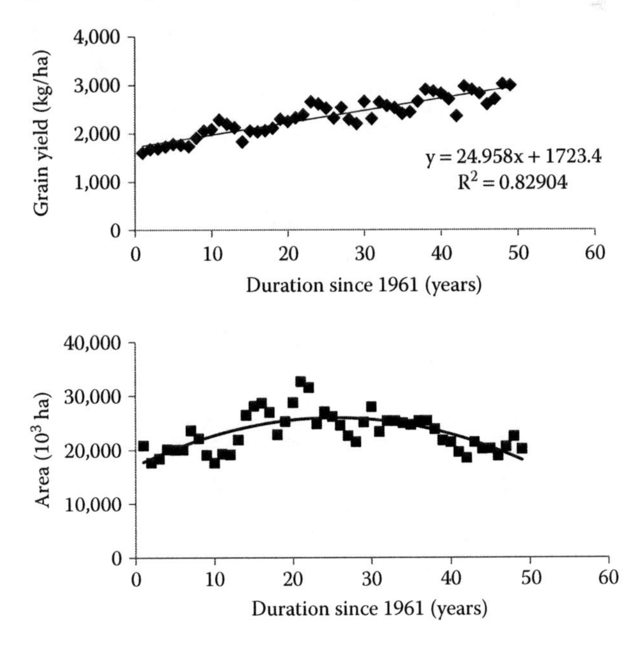

FIGURE 2.4 Change in cultivated area and grain yield of wheat in the United States between 1961 and 2010. (Redrawn from USDA. Crop acreage and yields. 2010. http://www.usda.gov/wps/portal/usda/usdahome?parentnav=AGRICULTURE&navid=CROP_PRODUCTION&.)

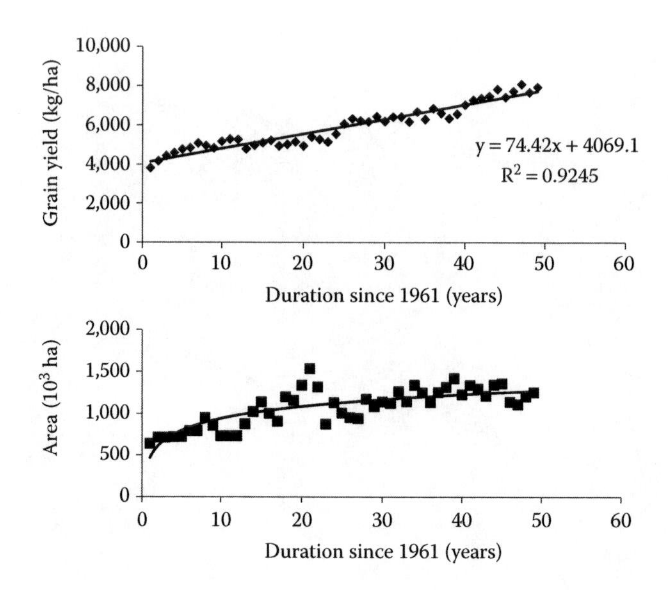

FIGURE 2.5 Change in cultivated area and grain yield of rice in the United States between 1961 and 2010. (Redrawn from USDA. Crop acreage and yields. 2010. http://www.usda.gov/wps/portal/usda/usdahome?parentnav=AGRICULTURE&navid=CROP_PRODUCTION&.)

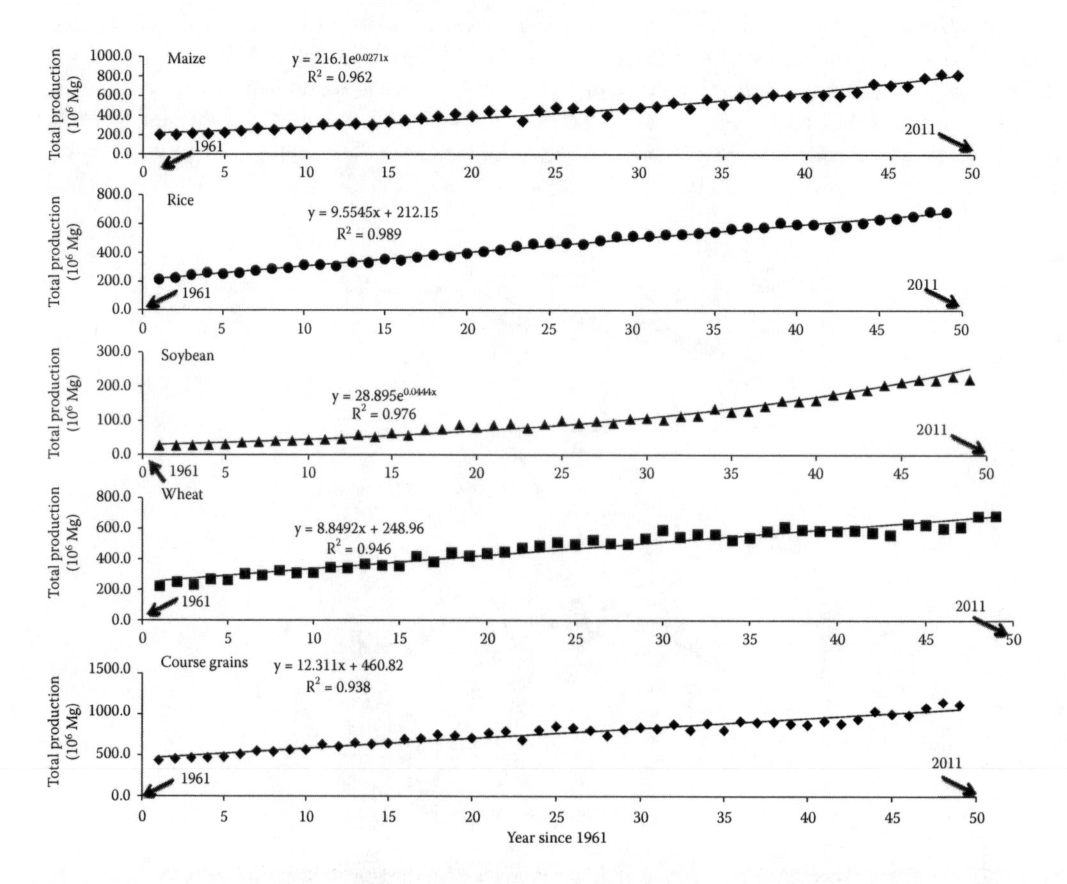

FIGURE 2.6 Changes in global grain production of principal grain crops. (Redrawn from FAOSTAT. Crop production. 2010. http://faostat.fao.org/site/567/desktopDefault.aspx?PageID=567#ancor.)

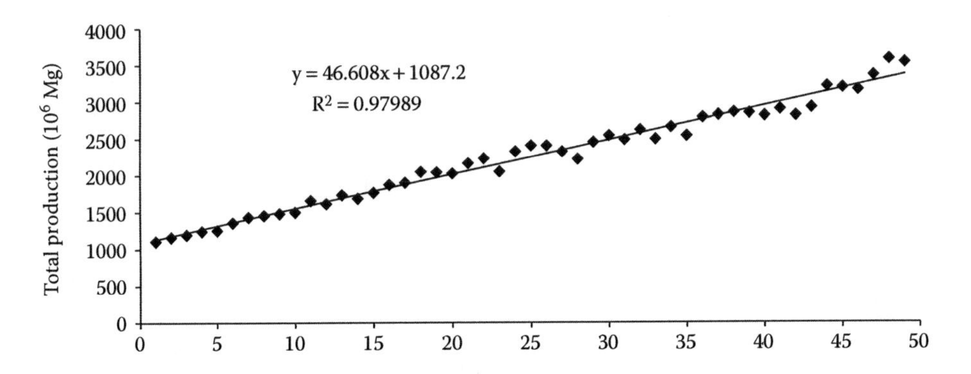

FIGURE 2.7 Change in total grain production in the world. (Redrawn from FAO. FAOSTAT. 2011. http://faostat.fao.org/site/567/desktopdefault.aspx?PageID=567#ancor. Accessed August 1, 2011.)

the rate of future expansion of irrigation will be slower (Plusquellec 2002). Therefore, the strategy is to produce more from less and more crop per drop of renewable fresh water consumed in agricultural lands. This strategy has also been termed "greening the global water system" (Hoff et al. 2010).

The effective use of water for agricultural production is measured by water-use efficiency (WUE) and water productivity (WP). The term, "water-use efficiency," used by irrigation specialists, assesses the effectiveness of the water delivery to crops as affected by losses during delivery (i.e., seepage and evaporation). Others have defined WUE as the ratio of the amount of agricultural output to the input or flux of water used in its production (Moore et al. 2011). However, a wide range of variables are used as the denominators in computing the ratio (Perry et al. 2009), including precipitation, irrigation, precipitation plus irrigation, evapotranspiration, and transpiration. Similarly, a wide range of variables are used as the numerators, including photosynthetic rates, shoot biomass, grain yield, income, and profit. (Peden et al. 2007).

In this chapter, however, the term WP is used. Yet, WP also includes the benefits and costs of water use (Molden et al. 2010). There are two types of WP: (i) physical WP is the ratio of

TABLE 2.3
Global Land Use

Year	Cropland (10^6 ha)	Pastureland (10^6 ha)	Nitrogen Fertilizer Use (10^6 Mg)	Total Fertilizer Use (10^6 Mg)
1700	265	524	—	—
1750	321	697	—	—
1800	402	942	—	—
1850	537	1955	—	—
1900	813	2118	—	—
1950	1230	2930	—	—
1960	1361	3208	11.6	31.1
1970	1405	3276	31.8	69.3
1980	1444	3357	60.8	116.7
1990	1478	3450	77.2	137.8
2000	1382	3426	—	135.2
2008	1380	3357	103.8	—

Source: Modified from Scanlon, B.R., et al., *Water Res.*, 43, 1, 2007; FAO. FAOSTAT. 2011. http://faostat.fao.org/site/567/desktopdefault.aspx?PageID=567#ancor. Accessed August 1, 2011; IFDC. Global and regional data on fertilizer production and consumption 1961/62–2002/3. Muscle Shoals, AL, 2004.

TABLE 2.4
Present and Future Demands of Water Resources for Irrigation

	Population		Water Supply (km^3/year)	
Region	**1985**	**2025**	**1985**	**2025**
Africa	543	1,440	4,520	4,100
Asia	2,930	4,800	13,700	13,300
Australia/Oceania	22	33	714	692
Europe	667	682	2,770	2,790
North America	395	601	5,890	5,870
South America	267	454	11,700	10,400
World	4,830	8,010	39,300	37,100

Source: Adapted from Vörösmarty, C.J., et al., *Science*, 289, 284, 2000.

agricultural production to the amount of water consumed and (ii) economic WP is the magnitude of economic return per unit of water used. Therefore, the strategy is to enhance WP through the use of innovative options. Important among these innovative irrigation practices are regulated deficit irrigation, limited irrigation, and controlled alternative partial root zone drying (PRD) (Fang et al. 2010). These irrigation practices can save 15%–35% of water and increase WP by 10%–30% (Kang et al. 2002). Wastewater reuse is also an ecologically sustainable and hygienically safe approach (Neubert 2009; Chapter 11).

2.4 MANAGING GREEN WATER IN RAINFED AGRICULTURE

The grain yields in developing countries for rainfed (dryland) farming are about half of those from irrigated croplands (1.5 vs. 3.1 Mg/ha) (Rosegrant et al. 2009). The low productivity of rainfed agriculture is largely attributed to recurring drought stress during the growing season. The large yield gap between the actual and attainable yields (Table 2.6) can be narrowed by improving the use efficiency of green water or the precipitate stored in the soil. The adverse impacts of the extreme interannual variability in the rainfall, large uncertainty, and erratic distribution can be partly addressed by enhancing storage and improving the WP (2.8). The water requirements also depend on the crop type, growth duration, rooting depth, canopy cover, leaf characteristics, etc. Thus, there are large differences in

TABLE 2.5
Comparative Water Balances for Different Biomes

	Components of the Hydrologic Cycle (mm/year)				
Biome	**Precipitation (P)**	**Runoff (R)**	**Evapotranspiration (ET)**	**ET:P**	**R:P**
Arid	111	8	103	0.93	0.07
Semiarid	385	35	350	0.91	0.09
Semihumid	726	145	581	0.80	0.20
Humid	1142	426	716	0.63	0.37
Tropical	2580	1480	1100	0.43	0.57
Hyper humid	1093	688	405	0.37	0.63
Global	825	340	489	0.59	0.41

Source: Adapted from Shen, Y. and Chen, Y., *Hydrol. Process.*, 24, 129–135, 2010.

TABLE 2.6
Actual Yields of Rainfed Grain Crops as a Percentage of Attainable Yield

Country	% of Attainable Yield	Yield Improvement Factor by RMPs
Botswana	27.6	3.6
Burkina Faso	23.7	4.2
Ethiopia	31.6	3.2
India	42.1	2.4
Iran	18.4	5.4
Iraq	17.1	5.8
Jordan	18.4	5.4
Kenya	28.9	3.5
Morocco	25.0	4.0
Niger	26.3	3.4
Pakistan	10.5	9.5
Syria	18.4	5.4
Tanzania	23.7	4.2
Thailand	52.6	1.9
Uganda	23.7	4.2
Vietnam	63.1	1.6
Yemen	10.5	9.5
Zambia	31.6	3.2
Zimbabwe	31.6	3.2

Source: Recalculated from Rockström, J., et al., *Agr. Water Manage.*, 97, 543, 2010.

consumptive water use, depending on the land use, the crop type and its management, the soil quality and its management, and the evaporative demand of the atmosphere. Drought stress on rainfed croplands is likely to be exacerbated with the projected climate change, and the decline in productivity may not be compensated by the CO_2 fertilization effect (Lobell et al. 2008; The Royal Society 2009). Yet, there are several options for improving crop production in dry environments (Chaves and Davies 2010). Exploiting genetic variations and selecting drought-tolerant/avoiding germplasms are important strategies (Edmeades et al. 1999; Richards et al. 2010; Saint Pierre et al. 2010). The use of agronomic practices for conserving soil water (i.e., mulch farming, conservation tillage, nutrient management, and time of planting) is also strategically important, especially to realize the genetic potential of improved cultivars. The basic principles and practices of enhancing green water storage in rainfed agriculture are outlined in Figure 2.8.

2.5 MANAGING WATER IN IRRIGATED AGRICULTURE

Only 16% of the land that is equipped for irrigation yields 40% of the global production (Fereres and Connor 2004; Muralidharan and Knapp 2009). Globally, 7100 km^3/year are used for food production, of which 5500 km^3/year are used in rainfed agriculture and 1600 km^3/year are used in irrigated agriculture (de Fraiture et al. 2007). By 2050, the total water demand for agriculture is expected to increase to 8,500–11,000 km^3/year (Rockström et al. 2010), depending on the development and adoption of new water saving technology. The competing demand for water for industry and urban uses and the climate change are likely to reduce the availability of water for irrigation. Thus, enhancing WUE and WP by decreasing losses, especially of soil water, is more important now than

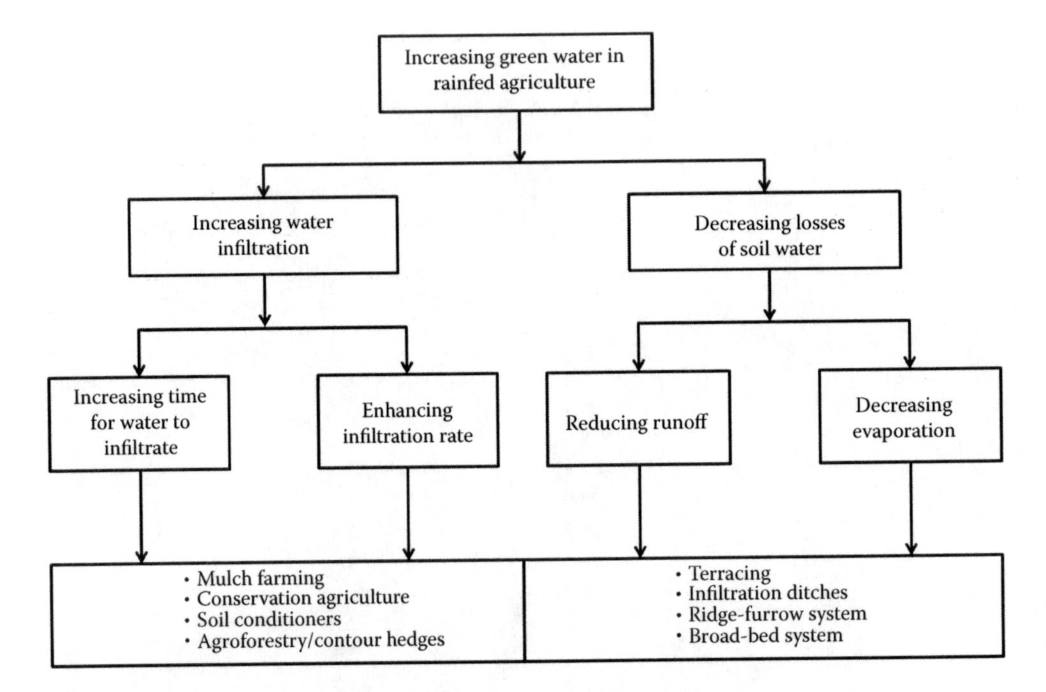

FIGURE 2.8 Principles and practices of enhancing green water in rainfed agriculture.

ever before. The water footprint of irrigated crops must be reduced, so that the severe decline in the groundwater level in aquifers can be curtailed. The goal is to produce more crop per drop of water by managing the climate, water, soil, and crop. Rather than indiscriminate and excessive use of water, a holistic approach is needed to integrate the management of all natural resources. In addition, it is important for land managers to realize that irrigation water management is a dynamic process throughout the growing season (Muralidharan and Knapp 2009) and involves making critical choices with regard to the time to irrigate, the amount to irrigate, the method to irrigate, the fraction of soil area to irrigate, etc. Optimizing the use of irrigation water is essential to saving the dwindling water resources. The concept of PRD (Stoll et al. 2000) is proposed on the basis of root-to-shoot signaling (Dodd 2009). Repeated cycles of wetting and drying by PRD enhance the availability of plant nutrients. The use of root growth hormones and enhanced nodulation can also increase the agronomic yield in water-deficit environments (Diaz-Zorita and Fernández-Canigia 2009; Belimov et al. 2009). Water and plant nutrients (fertigation) must be delivered directly to the plant roots by drip subirrigation to minimize losses. Condensation irrigation, delivering water to the plant roots as vapors (as is the case in desert plants from the subsoil to the roots in the surface layer at night), is another option that must be explored. Strategies to enhance the WUE in irrigated agriculture are outlined in Figure 2.9.

2.6 ENHANCING WATER PRODUCTIVITY IN RAINFED AGRICULTURE

Rainfed agriculture accounts for 60% of the total agronomic production, but it has a low WP. Thus, improving WP can enhance global food production. There are several options for increasing the green water in rainfed agriculture. The basic principles are (i) increase water infiltration; (ii) store any runoff for recycling; (iii) decrease losses by evaporation and uptake by weeds; (iv) increase root penetration in the subsoil; (v) create a favorable balance of essential plant nutrients, especially of P; (vi) grow drought avoidance/adaptable species and varieties; (vii) adopt cropping/farming systems that produce a minimum assured agronomic yield in a bad season rather than those that produce the maximum yield in a good season; (viii) invest in soil/land restoration measures (i.e., terraces and

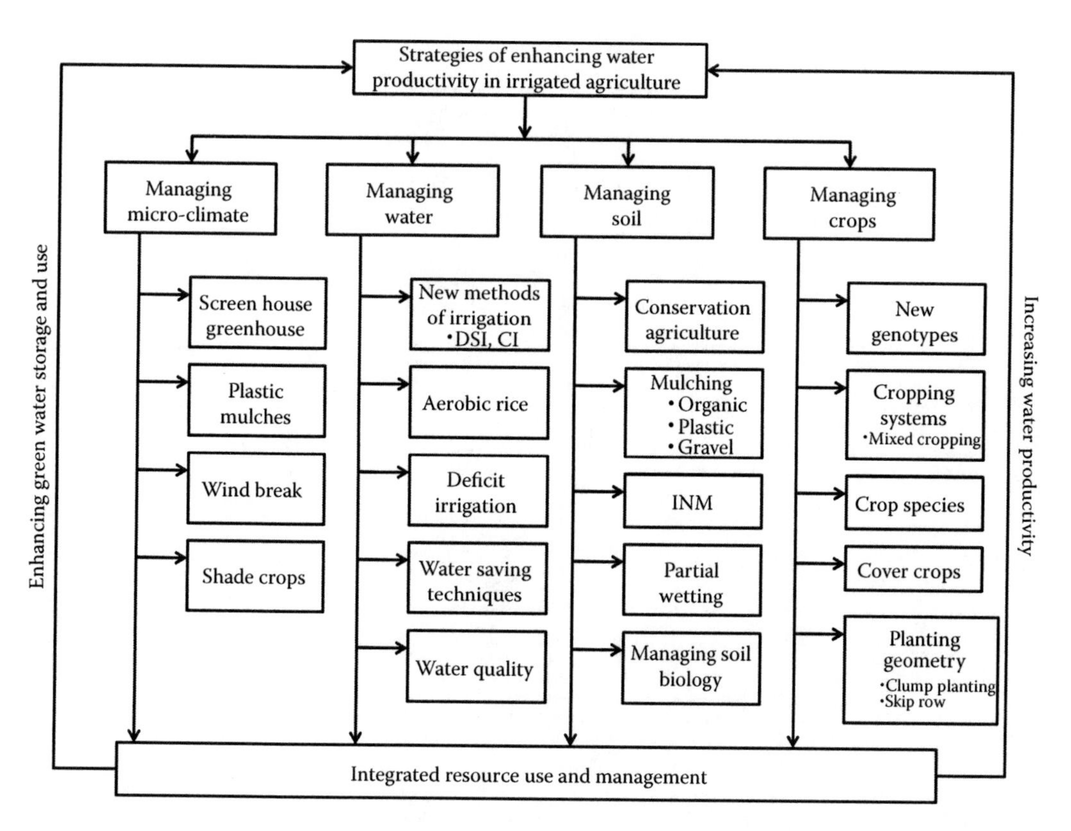

FIGURE 2.9 Strategies of enhancing water-use efficiency in irrigated agriculture.

shelter belts); (ix) develop and use weather forecasting technology to facilitate the planning of farm operations; and (x) use precision or soil-specific farming technology using legume-based cropping systems to reduce losses of C and N (Drinkwater et al. 1998) and to improve soil fertility. Similarly, growing crops and varieties with better root systems is a useful strategy (Eissenstat et al. 2000) to reduce the risks in a harsh environment. The root system is important to drought resistance/tolerances (Passioura 1983). Based on these principles, the technological options for enhancing WUE in rainfed agriculture are listed in Figure 2.10.

2.7 RESTORING DEGRADED SOILS TO ENHANCE WATER PRODUCTIVITY

Alleviating soil-related constraints (i.e., nutrient deficiency, shallow rooting depth, and excessive soil erosion) is relevant to improving WP in both rainfed and irrigated agricultures. An adequate N availability, at the critical stage, is crucial to improving WP (Hatfield et al. 2001; Fang et al. 2010). In addition to N, the application of P can also improve WP (Liang 1996). A judicious use of fertilizer enhanced the WP in the irrigated cropland in the North China Plain from 0.23 to 0.90 kg/m^3 for cereal crops (Xu and Zhao 2001; Fang et al. 2010). The land application of organic wastes (Oldare et al. 2011) can improve the soil quality and increase the WP. The goal is to reduce water consumption without reducing the yields through agronomic management. This is achievable by (i) reducing water delivery losses, (ii) improving soil water availability to crop roots, and (iii) increasing the WP (Stanhill 1986). The agronomic practices that are effective in conserving soil water (green water), which also alter the soil surface energy balance, include mulch farming, plastic sheeting, and conservation tillage (Hatfield et al. 2001). The adoption of conservation-effective cropping systems, which reduce the soil erosion risks, is also relevant to improving the WP of soils prone to accelerated erosion. For highly erodible

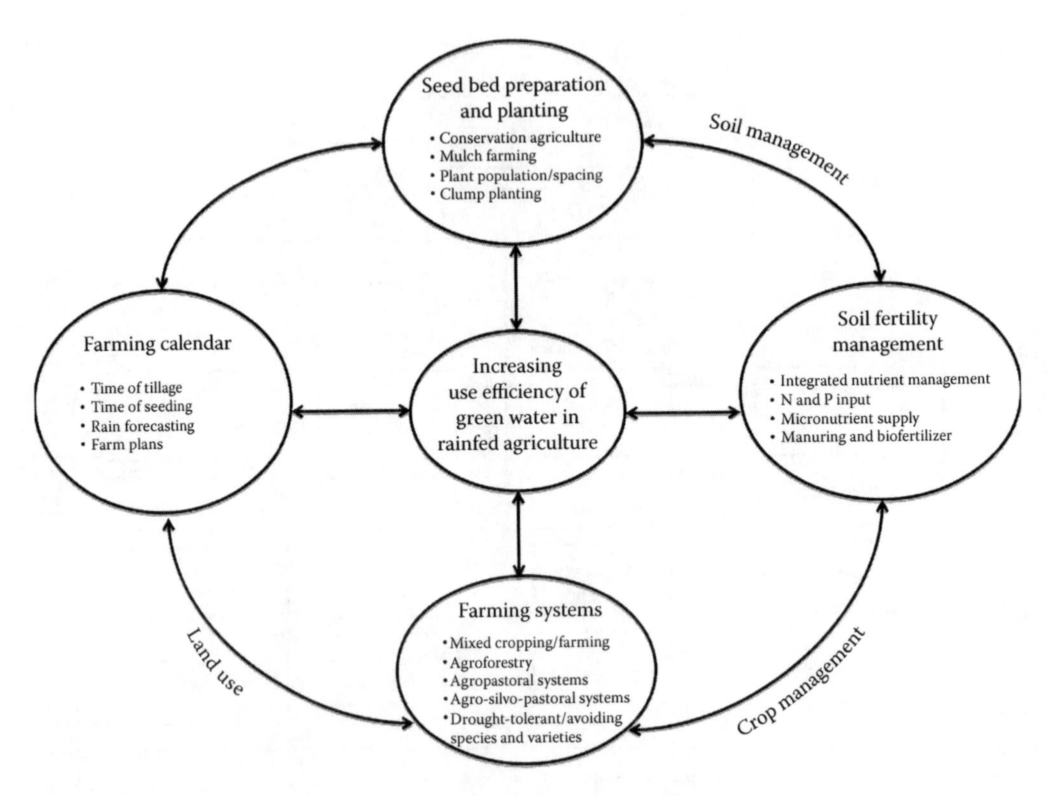

FIGURE 2.10 Some recommended management practices for improving water-use efficiency in rainfed agriculture.

soils of the semiarid Loess Plateau of China, Jun et al. (2010) observed that a forage–food–crop rotation is effective in reducing soil erosion and improving WP. The establishment of soil conservation practices in the semiarid Loess Plateau of China also improved the groundwater recharge (Gates et al. 2010).

The hydrological cycle, the C cycle, the nutrient cycle, and land management are intricately linked. Thus, a prudent approach is to enhance soil C sequestration by restoring the degraded soil and improving the WP (Bossio et al. 2008). Addressing soil degradation to improve WP is also a rational strategy. Meeting the food demands of the projected population of 9.2 billion by 2050 would require 30%–40% more water for agriculture with an optimistic scenario (de Fraiture et al. 2007), and 70%–90% more if the WP is not improved through soil restoration and the adoption of the recommended management practices (RMPs) (Bossio et al. 2008). With the realization that every land-use decision is also a water-use decision (Bossio et al. 2008), the adoption of sustainable agricultural techniques based on soil-specific RMPs can enhance WP. The data in Table 2.7, based on a survey of 144 projects, show that adopting RMPs improved WP by 16%–29% in irrigated agriculture and 70%–108% in rainfed agriculture (Pretty et al. 2006). Thus, adopting an RMP for agronomic improvement is essential to enhancing the WP.

2.8 BIOFUEL AND WATER DEMANDS

The conversion of biomass into modern biofuel is gaining momentum because of the climatic and environmental impacts of fossil fuel combustion and rising energy prices. Some examples of modern biofuels include bioethanol from corn grains and sugarcane (*Saccharum officinarum*) juice, and biodiesel from soybean and rape seeds (*Brassica napus*). The first-generation bioethanol is sugar-based or starch-based, the second generation is cellulose-based, the third generation is

TABLE 2.7
Increase in Water Productivity by Adoption of Sustainable Technologies Based on Recommended Management Practices in 144 Projects

Water Regime		Crops	Mean Water Productivity (kg/m³)		
			Without RMP	With RMP	Increase (%)
I	Rainfed				
		Cereals	0.47	0.80	70.2
		Legumes	0.43	0.87	102.3
		Roots and tubers	2.79	5.79	107.5
II	Irrigation				
		Rice	1.03	1.19	5.5
		Cotton	0.17	0.22	29.4

Source: Pretty, J., et al., *Environ. Sci. Technol.*, 40, 1114, 2006.

from the biomass of algae and cyanobacteria, and the fourth generation is hydrogen from the biomass. Consequently, there is a wide range of biofuel feedstock sources (Hattori and Morita 2010). Yet, the production of biofuel can put additional strain on water resources, especially in China and India (de Fraiture et al. 2008). Currently, biofuels account for about 100 km³ of water use. However, relative water use is more for the production of feedstock on irrigated land, estimated at 44 km³ out of a total of 2630 km³. It takes approximately 2500 L of crop evapotranspiration and 820 L of irrigation water withdrawn to produce 1 L of biofuel (de Fraiture et al. 2008). Stone et al. (2010) estimated that the cubic meter of water that is required per milligram of ethanol is 580 for sugarcane, 2580 for corn grain, 1980 for switchgrass (*Panicum virgatum*), 9460 for grain sorghum (*Sorghum bicolor*), and 931 for sweet sorghum (Table 2.8). To meet the U.S. billion-ton vision, the current water requirement is 8.64×10^9 m³, which will increase to 5.01×10^{10} m³ by 2030 (Stone et al. 2010).

TABLE 2.8
Water Requirements for Biofuel

Crop		Crop Water Requirements for Biofuel	
		(m³ Water/Mg Fuel)	(m³ Water/GJ)
I	Ethanol		
	Corn (grain)	2380	97
	Sugarcane	580	22
	Corn (stover)	2465	92
	Corn (stover + grain)	1093	41
	Switchgrass	1980	74
	Grain sorghum	9460	354
	Sweet sorghum	931	35
II	Biodiesel		
	Soybean	9791	259
	Canola	4323	130

Source: Stone, K.C., et al., *Bioresour. Technol.*, 101, 2014, 2010.

2.9 AEROBIC RICE

Flooded rice is a water-intensive crop and has a low WP compared with upland rice. Rice requires two to three times more water to produce 1 kg of grains than other cereals. The intensive cultivation of flooded rice paddies in the rice–wheat (*T. aestivum*) system of the Indo-Gangetic Plains has depleted the groundwater reserves in northwest India (Rodell et al. 2009; Shan et al. 2006; Kerr 2009). Thus, there is a strong need to develop water-saving techniques for rice cultivation with a high WP. The goal is to produce similar yields but with a lower water consumption and a slight or no productivity loss. The strategy is to develop aerobic rice through genetic improvement (Farooq et al. 2009) and to identify agronomic practices involving rice cultivation without prolonged periods of inundation. There are improved systems of irrigation with considerable savings in water use (Kang and Zhang 2004). There are also physiological and ecological bases for water saving, including agronomic management involving seedbed preparation, the method of sowing, the date of sowing, nutrient management, weed control, and water management. The latter includes the irrigation method, the frequency of irrigation, and the amount of irrigation. Aerobic rice, growing rice in a method similar to other upland cereals (i.e., wheat), saves water and has a high WP (Bouwman et al. 2006, 2007; Bouwman and Tuong 2001). In an aerobic system, rice is grown in well-drained and unpuddled soil without flooding. A principal agronomic constraint in aerobic rice is weed control. Thus, the yield of aerobic rice can be 20%–30% lower than the total of flooded rice, but the WP is 1.6–1.9 times higher, resulting in a water saving of 35%–57% (Farooq et al. 2009). The WP of aerobic rice is more than that by alternative wetting and drying irrigation and mid-season drainage.

2.10 SOIL QUALITY AND WATER PRODUCTIVITY

Enhancing and sustaining soil quality are essential to improving WP. The soil organic matter (SOM) is an important parameter that affects the physical, chemical, biological, and ecological components of soil quality. Increasing the SOM concentration improves the soil's physical quality in terms of aggregation and aggregate strength, water infiltration rate, runoff and erosion, available water-holding capacity, and soil temperature; its chemical quality in terms of ion-exchange and cation-exchange capacities, nutrient storage and availability, nutrient use efficiency, and buffering against pH; its biological quality in terms of biodiversity, biological habitat and reservoir for the gene pool, gaseous exchange between the soil and the atmosphere, and C sequestration; and its ecological quality in terms of elemental cycling, ecosystem C budget, denaturing and filtering of pollutants, net primary and ecosystem productivity, etc. Increasing the SOM concentration implies creating a positive C (and N, P, S) budget(s) through adopting the following practices: conservation agriculture, mulching, cover cropping, manuring and biofertilizers, agroforestry, and other complex cropping/farming systems; the establishment of deep-rooted species/varieties, controlled grazing and improved pasture, precision farming, the use of biochar, etc. Principal challenges to increasing the SOM concentration are the lack of or the low rate of adoption of RMPs (listed above) and an abruptly changing climate. There are numerous uncertainties with regard to the response of the terrestrial (Friend 2010) and soil C pools (Norby et al. 2004; Heimann and Reichstein 2008) to changes in climate (i.e., temperature, precipitation, extreme events, soil moisture storage, and elemental cycling). Uncertainties are also attributed to the complexities created by the CO_2 fertilization effect, plant respiration and nutrient response (Leakey et al. 2009a,b), partitioning of biomass (Pendall et al. 2004), increased risks of soil erosion (Lal 2010), and alternation in decomposition because of any changes in the C:N ratio (Torbert et al. 2000; Norby et al. 2001).

2.11 DIETARY PREFERENCE AND WATER DEMAND

A meat-based diet has a large water and carbon footprint. The data in Table 2.9 show several orders of magnitude difference in the water requirement for plant-based and animal-based food. Food sources can be grouped on the basis of its water requirements into different categories, ranging from

TABLE 2.9
Liters of Water Needed to Produce 1 kg of Food

Water Requirement	Produce	Liters of Water (kg)	Relative Water Need
Low	Lettuce	125	1
	Tomatoes	183	1.5
	Cabbage	200	1.6
	Cucumber	233	1.9
	Potatoes	250	2.0
Medium	Oranges	459	3.7
	Apples	692	5.5
	Bananas	850	6.8
	Corn	892	7.1
High	Peaches/nectarines	1,183	9.5
	Wheat bread	1,284	10.3
	Wheat	1,350	10.8
	Mangos	1,584	12.7
	Avocados	1,834	14.7
	Tofu	1,867	14.9
	Groundnuts	3,068	24.5
	Rice	3,360	26.9
Very high	Chicken	3,900	31.2
	Olives	4,352	34.8
	Pork	4,800	38.4
	Cheese	6,000	48.0
Extremely high	Beef	16,000	128.0

Source: Modified and recalculated from Hoekstra, A.J. The water footprint of food. Twente Water Center, University of Twente, the Netherlands, 2011.

low to extremely high (Table 2.9). The preference for an animal-based diet, with a water requirement ranging from very high to extremely high, can exacerbate the already severe problem of water scarcity in many populous countries (i.e., China and India). Dietary choices must be discussed objectively at various levels of society and must be included in the school curricula.

2.12 STRATEGIES OF WATER MANAGEMENT

In arid and semiarid regions, already faced with the severe problems of water stress, the loss of blue water must be minimized by storing it for future use within the watershed. Because of the high evaporative demands both now and in the future, storage in aboveground impoundments is prone to high losses by evaporation. Thus, recharging the aquifers and creating belowground storage are preferred strategies. Gray water must be recycled and used for enhancing the production of the third generation of biofuels (i.e., algae and cyanobacteria) and for promoting urban agriculture, both of which are a high priority to meet the growing demands of an increasing and urbanizing human population.

2.13 CONCLUSION

A renewable freshwater supply is a finite resource. Similar to soil, water is also prone to misuse, contamination/pollution, and eutrophication. As much as 70% of total water withdrawal has been used for agriculture, mostly irrigation. However, there are numerous competing and essential uses (i.e., domestic, industrial, recreational, and aquaculture). While equipping some arable land in sub-Saharan Africa with irrigation facilities, the WP of existing irrigated land (i.e., China, India, Pakistan, Egypt, and Iran) must be improved, and new and innovative irrigation methods must be adopted. Because of its strong interaction, WP can be enhanced by improving the soil quality and increasing the efficient use of N and other nutrients. The choice of biofuel feedstock must be based on water (land and nutrient) requirements. School curricula must teach about water (and C or energy) footprints of plant-based vs. animal-based diets. The relative proportion of green water must be increased by improving water infiltration and enhancing soil water storage. Losses of blue water (surface runoff, interflow, river discharge) must be minimized by improving the aquifer recharge and enhancing the subsurface storage. Every drop must be recycled and used for multiple purposes. Win-win scenarios must be identified.

REFERENCES

Aldaya, M.M., J.A. Allan, and A.Y. Hoekstra. 2010. Strategic importance of green water in international crop trade. *Ecological Economics* 69: 887–894.

Belimov, A.A., I.C. Dodd, N. Hontzeas, et al. 2009. Rhizosphere bacteria containing l-aminocyclopropane-1-carboxylate deaminase increase yield of plants grown in drying soil via both local and systemic hormone signalling. *New Phytologist* 181: 413–423.

Borlaug, N. 2007. Feeding a hungry world. *Science* 318: 359.

Bossio, D., A. Noble, D. Molden, et al. 2008. Land degradation and water productivity in agricultural landscapes. In *Conserving Land, Protecting Water. Comprehensive Assessment of Water Management in Agriculture*, eds. D. Bossio and K. Geheb, Series 6, pp. 20–32. CABI, Wallingford, UK.

Bouwman, B.A.M. and T.P. Tuong. 2001. Field water management to save water and increase its productivity in irrigated lowland rice. *Agricultural Water Management* 49: 11–30.

Bouwman, B.A.M., X.G. Yang, H. Wang, et al. 2006. Performance of aerobic rice varieties under irrigated conditions in North China. *Field Crops Research* 97: 53–65.

Bouwman, B.A.M., R.M. Lampayan, and T.P. Tuong. 2007. Water management in irrigated rice—Coping with water scarcity. International Rice Research Institute, Los Baños, the Philippines.

Chaves, M. and B. Davies. 2010. Drought effects and water use efficiency: Improving crop production in dry environments. *Functional Plant Biology* 37: 3–6.

Cominelli, E., M. Galbiati, and C. Tonelli, et al. 2009. Water: The invisible problem. *EMBO Reports* 10: 671–676.

de Fraiture, C., D. Wichels, J. Rockström, et al. 2007. Looking ahead to 2050: Scenarios of alternative investment approaches. In *Water for Food, Water for Life: A Comprehensive Assessment of Water Management in Agriculture*, ed. D. Molden, pp. 90–145. Earthscan, London and IWMI, Colombo.

de Fraiture, C., M. Giordano, and Y. Liao. 2008. Biofuels and implications for agricultural water use: Blue impacts of green energy. *Water Policy 10 Supplement* 1: 67–81.

Diaz-Zorita, M. and M.V. Fernández-Canigia. 2009. Field performance of a liquid formulation of *Azospirillum basilense* on dryland wheat productivity. *European Journal of Soil Biology* 45: 3–11.

Dodd, I.C. 2009. Rhizosphere manipulations to maximize 'crop per drop' during deficit irrigation. *Journal of Experimental Botany* 60: 2454–2459.

Drinkwater, L.E., P. Wagoner, and M. Sarrantonio. 1998. Legume-based cropping system have reduced carbon and nitrogen losses. *Nature* 396: 262–265.

Edmeades, G.O., J. Bolaños, S.C. Chapman, et al. 1999. Selection improves drought tolerance in tropical maize populations: I. Gains in biomass, grain yield, and harvest index. *Crop Science* 39: 1306–1315.

Eissenstat, D.M., C.E. Wells, R.D. Yanai, and J.L. Whitbeck. 2000. Building roots in a changing environment: Implications for root longevity. *New Phytologist* 147: 33–42.

Falkenmark, M. and J. Rockström. 2004. *Balancing Water for Humans and Nature: The New Approach in Ecohydrology*. Earthscan, London.

Falkenmark, M. and J. Rockström. 2006. The new blue and green water paradigm: Breaking new ground for water resources planning and management. *Journal of Water Resources Planning and Management* 51: 185–198.

Fang, Q.X., L. Ma, T.R. Green, et al. 2010. Water resources and water use efficiency in the North China Plain: Current status and agronomic management options. *Agriculture Water Management* 97: 1102–1116.

FAO. 2009. 1.02 billion people hungry. FAO News Room. http://www.fao.org/news/story/on/item/20568/icode/en (accessed August 1, 2011).

FAO. 2011. FAOSTAT. http://faostat.fao.org/site/567/desktopdefault.aspx?PageID=567#ancor (accessed August 1, 2011).

FAOSTAT. 2010. Crop production. http://faostat.fao.org/site/567/desktopDefault.aspx?PageID=567#ancor

Farooq, M., Kobayashi, N., Wahid, A., et al. 2009. Strategies for producing more rice with less water. In *Advances in Agronomy*, ed. D.L. Spark, Vol. 101, pp. 351–387. Academic Press, London.

Fereres, E. and D.J. Connor. 2004. Sustainable water management in agriculture. In *Challenges of the New Water Policies for XXI Century*, eds. E. Cabrera and R. Cobacho. A.A. Balkema, pp. 157–170. The Netherlands.

Friend, A.D. 2010. Terrestrial plant production and climate change. *J. Exp. Bot.* doi: 10.1093/hxb/erq019.

Gates, J.B., B.R. Canlon, X. Mu, et al. 2010. Impacts of soil conservation on groundwater recharge in the semi-arid Loess Plateau, China. *Hydrogeology Journal* 19: 865–875.

Gleick, P.H. 2003. Global fresh water resources: Soft-path solutions for the 21st century. *Science* 302: 1524–1528.

Hatfield, J.L., J.T. Sauer, and J.H. Prueger. 2001. Managing soils to achieve greater water use efficiency: A review. *Agronomy Journal* 93: 271–280.

Hattori, T. and S. Morita. 2010. Energy crops for sustainable bioethanol production: Which, where and how? *Plant Production Science* 13: 221–234.

Heimann, M. and M. Reichstein. 2008. Terrestrial ecosystem carbon dynamics and climate feedbacks. *Nature* 451: 289–292.

Hoekstra, A.J. 2011. The water footprint of food. Twente Water Center, University Of Twente, the Netherlands.

Hoff, H., Falkenmark, M., Gerten, D., et al. 2010. Greening the global water system. *Journal of Hydrology* 384: 177–186.

IFDC. 2004. Global and regional data on fertilizer production and consumption 1961/62–2002/3. Muscle Shoals, AL.

Johnson, N., C. Revenga, and J. Echeverria. 2001. Managing water for people and nature. *Science* 292: 1071–1074.

Jun, F., S. Mingan, W. QuanJiu, et al. 2010. Toward sustainable soil and water resources use in China's highly erodible semi-arid Loess Plateau. *Geoderma* 155: 93–100.

Kang, S. and J. Zhang. 2004. Controlled alternate partial root-zone irrigation: Its physiological consequences and impact on water use efficiency. *Journal of Experimental Botany* 55: 2437–2446.

Kang, S.Z., L. Zhang, Y.L. Liang, et al. 2002. Effects of limited irrigation on yield and water use efficiency of winter wheat in the Loess Plateau of China. *Agriculture Water Management* 555: 203–216.

Kerr, R.A. 2009. Northern India's ground water is going, going, going ... *Science* 325: 798.

Kumar, V. and S.K. Jain. 2011. Export and import of virtual water from different states of Indian through food grain trade. *Hydrology Research* 42(2/3): 229–238.

Lal, R. 2010. Beyond Copenhagen: Mitigating climate change and achieving food security through soil carbon sequestration. *Food Security* 2: 169–177.

Leakey, A.D.B., E.A. Ainsworth, C.J. Bernacchi, A. Rogers, S.P. Long, and D.R. Ort. 2009a. Elevated CO_2 effects on plant carbon, nitrogen, and water relations: Six important lesions from FACE. *Journal of Experimental Botany* 60: 2859–2876.

Leakey, A.D.B., E.A. Ainsworth, C.J. Bernacchi, A. Rogers, S.P. Long, and D.R. Ort. 2009b. Genomic basis for simulated respiration by plants growing under elevated carbon dioxide. *PNAS* 106: 3597–3602.

Liang, Y.L. 1996. The adjustment of soil water and nitrogen phosphorus nutrition on root system growth of wheat and water use. *Acta Ecologica Sinica* 16: 258–264.

Lobell, D.B., M.B. Burke, C. Tebaldi, et al. 2008. Prioritizing climate change adaptation needs for food security in 2030. *Science* 319: 607–610.

Melillo, J.M., J.M. Reilly, D.W. Kicklighter, et al. 2009. Indirect emissions from biofuels: How important? *Science* 326:1397–1399.

Molden, D., T. Oweis, P. Steduto, et al. 2010. Improving agricultural water productivity: Between optimism and caution. *Agricultural Water Management* 97: 528–535.

Moore, A.D., M.J. Robertson, and R. Routley. 2011. Evaluation of the water use efficiency of alternative farm practices at a range of spatial and temporal scales: A conceptual framework and a modeling approach. *Agricultural Systems* 104: 162–174.

Muralidharan, D. and K.C. Knapp. 2009. Spatial dynamics of water management in irrigated agriculture. *Water Resources Research*. doi:10.1029/2007WR006756.

Neubert, S. 2009. Wastewater reuse: How "integrated" and sustainable is the strategy? *Water Policy* 11: 37–53.

Norby, R.J., M. Francesca Cotrufo, P. Ineson, and E.G. O'Neill. 2001. Elevated CO_2, litter chemistry, and decomposition: A synthesis. *Oecologia* 127: 153–165.

Norby, R.J., J. Ledford, C.D. Rilley, E. Miller, and E.G. Neill. 2004. Fine-root production dominate response of a deciduous forest to atmospheric CO_2 enrichment. *PNAS* 101: 9689–9693.

Oki, T. and S. Kanae. 2006. Global hydrological cycles and world water resources. *Science* 313: 1068–1072.

Oldare, M., V. Arthurson, M. Pell, et al. 2011. Land application of organic waste—Effects on the soil ecosystem. *Applied Energy* 88: 2210–2218.

Passioura, J.B. 1983. Roots and drought resistance. *Agricultural Water Management* 7: 265–280.

Pendall, E., S. Bridgham, P.J. Hanson, et al. 2004. Below-ground process responses to elevated CO_2 and temperature: A discussion of observation, measurement methods and model. *New Phytologist* 162: 311–322.

Peden, D., G. Tadesse, and A. Misra. 2007. Water and livestock for human development. In *Water for Food, Water for Life: A Comprehensive Assessment of Water Management in Agriculture*, ed. D. Molden, 485–514. London: International Water Management Institute, Colombo and Earthscan.

Perry, C., P. Steduto, and R.G. Allen. 2009. Increasing productivity in irrigated agriculture: Agronomic constraints and hydrological realities. *Agricultural Water Management* 96: 1517–1524.

Pfister, S., P. Bayer, A. Koehler, et al. 2011. Environmental impacts of water use in global crop production: Hotspots and trade-offs with land use. *Environmental Science and Technology* 45: 5761–5768.

Plusquellec, H. 2002. Is the daunting challenge of irrigation achievable? *Irrigation and Drainage* 51: 185–198.

Pretty, J., A. Noble, D. Bossio, et al. 2006. Resource-conserving agriculture increases yields in developing countries. *Environmental Science and Technology* 40: 1114–1119.

Revelle, R. 1976. The resources available for agriculture. *Scientific American* 235: 164–179.

Revelle, R. 1982. Carbon dioxide and world climate. *Scientific American* 247: 35–43.

Richards, R.A., G.J. Rebetzke, M. Watt, et al. 2010. Breeding for improved water productivity in temperate cereals: Phenotyping, quantitative trait loci, markers and the selection environment. *Functional Plant Biology* 37: 85–97.

Rockström, J., M. Falkenmark, L. Karlberg, et al. 2009. Future water availability for global food production: The potential of green water to build resilience to global change. *Water Research*. doi:10.1029/2007WR006767.

Rockström, J., L. Karlberg, S.P. Wani, et al. 2010. Managing water in rainfed agriculture—The need for a paradigm shift. *Agricultural Water Management* 97: 543–550.

Rodell, M., I. Velicogna, and J. Famiglietti. 2009. Satellite-based estimates of groundwater depletion in India. *Nature* 460: 999–1002.

Rosegrant, M.W., C. Ringler, and T. Zhu. 2009. Water for agriculture: Maintaining food security under growing scarcity. *Annual Review of Environmental Research* 34: 205–222.

Saint Pierre, C., R. Trethowan, and M. Reynolds. 2010. Stem solidness and its relationship to water-soluble carbohydrates: Association with wheat yield under water deficit. *Functional Plant Biology* 37: 166–174.

Scanlon, B.R., A. Jolly, M. Sophocleous, et al. 2007. Global impacts of conversion from natural to agricultural ecosystems on water resources: Quantity versus quality. *Water Research* 43: 1–18. doi: 10.1029/2006WR005486.207.

Shan, L., X.P. Deng, and S.Q. Zhang. 2006. Advances in biological water saving research: Challenge and perspectives. *Bulletin of National Natural Science Foundation of China*. 20: 66–71.

Shen, Y. and Y. Chen. 2010. Global perspective on hydrology, water balance, and water resources management in arid basins. *Hydro Process* 24: 129–135.

Shiklomanov, I. 1993. World fresh water resources. In *Water in Crisis: A Guide to Word's Fresh Water Resources*, ed. P. H. Gleik, p. 473, Oxford University Press, Oxford, UK.

Stanhill, G. 1986. Water-use efficiency. *Advances in Agronomy* 39: 53–85.

Stoll, M., B.R. Loveys, and P.R. Dry. 2000. Hormonal changes induced by partial rootzone drying of irrigated grapevine. *Journal of Experimental Botany* 51: 1627–1634.

Stone, K.C., P.G. Hunt, K.B. Cantrell, et al. 2010. The potential impacts of biomass feedstock production on water resource availability. *Bioresource Technology* 101: 2014–2025.

The Royal Society. 2009. Reaping the benefits: Science and the sustainable intensification of global agriculture. RS Policy document 11/09.

Tilman, D., K.G. Cassman, P.A. Matson, et al. 2002. Agricultural sustainability and intensive production practices. *Nature* 418: 671–677.

Torbert, H.A., S.A. Prior, H.A. Rogers, and C.W. Wood. 2000. Review of elevated atmospheric CO_2 effect on agroecosystems: Residue decomposition processes and soil C storage. *Plant and Soil* 224: 59–73.

UN Population Division. 2011. Revision of World Population Prospects. UN Department of Economics and Social Affairs, New York.

United Nations Statistics Division. 2010. United Nations Statistics Division Statistical Database. http://esa.un.org/unpd/wpp/index.htm (accessed August 1, 2011).

USDA. 2010. Crop acreage and yields. http://www.usda.gov/wps/portal/usda/usdahome?parentnav=AGRICULTURE&navid=CROP PRODUCTION&

USDA. 2011. USDA-FAS Current World Production. Markets and Trade Reports. http://www.fas.usda.gov/report.asp (accessed August 1, 2011).

Vico, G. and A. Porporato. 2011. From rainfed agriculture to stress avoidance irrigation: I. a generalized irrigation scheme with stochastic soil moisture. *Advances in Water Resources* 34: 263–271.

Vörösmarty, C.J., P. Green, J. Salisbury, et al. 2000. Global water resources: Vulnerability from climate change and population growth. *Science* 289: 284–288.

Xu, F.A. and B.Z. Zhao. 2001. Development of crop yield and water use efficiency in Fenqiu county, China. *Acta Pedologica Sinica* 38: 491–497.

Section II

Water Resources and Agriculture

3 Changes in Precipitation during the Twentieth Century across a Latitude Gradient in the United States

Anjali Dubey and Rattan Lal

CONTENTS

3.1 INTRODUCTION

An increase in the atmospheric concentration of greenhouse gases (GHGs) is causing global warming and abrupt climate change (IPCC 2007). The effects of climate change are multiple and vary across the planet. An important concern is the change in precipitation across the biomes. An increase in anthropogenic radiative forcing may increase the amount of atmospheric water vapor, which may destabilize the atmosphere and alter the precipitation regime (Kunkel 2003).

It is generally perceived that dry areas are reportedly getting drier and wet areas are getting wetter (Dore 2005). For example, Karl and Knight (1998) reported a 10% increase in the annual precipitation across the United States, while Groisman and Easterling (1994) reported only a 4% increase in the annual precipitation in the continental United States within the last century. During the early part of the twentieth century, however, precipitation either decreased or did not change significantly. Yet, it has increased significantly in the latter half of the twentieth century (Bradley et al. 1987; Diaz and Quayle 1980; Klugman 1983). The total increase in precipitation is attributed to the increase in its frequency and intensity (Karl and Knight 1998). A general trend of short-duration extreme precipitation indicates an increase in the frequency of events between 1931 and 1996, especially in the southwestern United States, across the southern Great Plains, and into the southern Great Lakes region and the Northeast (Kunkel et al. 1999). By contrast, a downward trend has been observed in the northwestern United States and Florida (Kunkel et al. 1999). These trends are in accord with the conclusions of the IPCC (1996) that the hydrological cycle is likely to become more intense in warmer climates, which may lead to an increase in heavy rain events.

Establishing long-term trends in precipitation across a large area can be difficult because of several inconsistencies in the instruments used over such a long period of time (Lettenmaier et al. 1994). Major factors that can affect precipitation include the mean latitude, longitude, elevation, distance from the coast, and slope aspect (Keim et al. 2005). The point precipitation measurements are also affected by the gauge undercatch bias, which is often larger in winter than in summer (Groisman and Legates 1994).

The changing climate has had a strong impact on increasing the magnitude of flood damage in the United States in the latter part of the twentieth century (Pielke and Downton 2000). Economic losses caused by floods in the United States are second only to those caused by hurricanes among all natural hazards, averaging $3–$4 billion annually (Changnon and Hewings 2001). An increase in the frequency of flooding in response to climate change is related to an alteration in the precipitation regimes in the United States (Karl and Knight 1998). The changing patterns of precipitation also affect the production of food crops, which is one of the major global concerns in the twenty-first century (Dore 2005). Excessive precipitation can damage crops, and the costs of crop losses and agricultural damage are expected to increase significantly with climate change (McCarthy 2001; Reilly et al. 2003; Rosenzweig et al. 2002). Between 1990 and 2010, floods in the Midwest, North Dakota, Red River, and Mississippi caused massive damage to crops, resulting in the loss of billions of dollars, and also delayed planting (Rosenzweig et al. 2002). An increase in precipitation also increases soil wetness and the risks of anaerobiosis, which makes plants more prone to diseases and insect infestation (Ashraf and Habib-ur-Rehman 1999). Farming operations, planting, and harvesting are also delayed because of the inability to operate machinery due to excessive rains and poor trafficability. Rosenzweig et al. (2002) observed that excessive soil moisture due to an increase in precipitation may double the maize production losses by 2030 in the United States. Loss in yields by the inundation of crops in the Midwest can also inflate food prices in the United States (Clemmitt 2008).

Nationally and internationally, water availability and water quality are relevant issues. Green water constitutes the rainwater that percolates into the soil, is held in retention pores, and is available for plant roots to absorb. By contrast, blue water is the water present in rivers, lakes, and aquifers (Rockström et al. 2009). By 2050, 59% of the global population will have a scarcity of blue water and 36% will have a shortage of both green and blue water (Rockström et al. 2009). Large biomes, which are responsible for the major ecosystem services, depend on approximately 90% of green water or terrestrial vapor flow to the atmosphere (Rockström and Gordon 2001). It is important to identify the technology that will provide efficient ways to manage green water resources, thereby lowering the risks of agricultural droughts and promoting global food security (Rockström et al. 2009).

It is in this context that the ecological and economic implications of changes in the hydrological cycle caused by climate change are extremely important and relevant. Thus, this chapter focuses on establishing any trends in precipitation across several states in the United States. The overall goal is to establish any fingerprints of climate change on the amount and distribution of annual and seasonal precipitation in diverse regions of the United States.

3.2 DATA AND METHODOLOGY

The total annual precipitation for stations in 11 states—Alaska, Minnesota, Ohio, Tennessee, Alabama, Hawaii, Oklahoma, Florida, Arizona, New Mexico, and Texas—was analyzed for 80–100 years. These sites were selected to represent diverse climates and ecoregions, such as boreal forests, tropical rainforests, temperate continental, humid subtropical, and semiarid climates. The selection of the stations was based on the accuracy of the data, the length of the period, the availability, and on the least missing data, since the historical climate records have been archived for these stations. The precipitation trends were analyzed both on an annual basis and for the growing season. In this chapter, the growing season is defined as the frost-free period (Table 3.1).

TABLE 3.1
Locations and Descriptions of Stations

Station	Latitude and Longitude	Height (Meters above Sea Level)	Data Coverage	MAP (cm)	Köppen Climate Classification
Waseca, MN	44°04′N, 93°32′W	351.4	1915–2010	78.97	Humid continental
Jackson, TN	35°37′N, 88°51′W	121.9	1903–2010	129.78	Humid subtropical
Fairhope, AL	30°33′N, 87°53′W	7.0	1920–2010	165.94	Humid subtropical
Moore Haven Lock, FL	26°50′N, 81°05′W	10.7	1922–2010	122.43	Tropical monsoon
Lihue, HI	21°59′N, 159°20′W	30.5	1950–2010	103.42	Tropical rainforest
University Experiment Station, AK	64°51′N, 147°52′W	144.8	1916–2010	30.61	Continental subarctic (boreal)
Coshocton, OH	40°22′N, 81°47′W	347.5	1956–2010	94.21	Hot summer continental
Wooster, OH	40°47′N, 81°55′W	310.9	1897–2010	93.62	Hot summer continental
Bellefontaine, OH	40°21′N, 83°46′W	361.2	1895–2010	92.06	Hot summer continental
Bowling Green, OH	41°23′N, 83°37′W	205.7	1894–2010	82.56	Hot summer continental
Circleville, OH	39°37′N, 82°57′W	205.1	1896–2010	99.32	Hot summer continental
Anahuac, TX	29°47′N, 94°38′W	7.3	1910–2010	133	Humid subtropical

(continued)

TABLE 3.1 (Continued)
Locations and Descriptions of Stations

Station	Latitude and Longitude	Height (Meters above Sea Level)	Data Coverage	MAP (cm)	Köppen Climate Classification
Whitney Dam, TX	31°52′N, 97°23′W	175	1950–2010	85.95	Humid subtropical
San Jon, NM	35°07′N, 103°20′W	1289.3	1910–2010	41.94	Semiarid
Caballo Dam, NM	32°54′N, 107°19′W	1277.1	1937–2010	24.25	Semiarid
Roosevelt, AZ	33°40′N, 111°09′W	672.1	1906–2010	140.77	Semiarid
Bartlett Dam, AZ	33°49′N, 111°39′W	502.9	1940–2010	34.79	Semiarid
Stillwater, OK	36°07′N, 97°06′W	272.8	1893–2010	85	Humid subtropical

Source: National Climatic Data Center.
Note: MAP: mean annual precipitation.

The sites selected for the study were located in 11 states across a north–south gradient in the United States: Waseca (Minnesota), Jackson (Tennessee), Fairhope (Alabama), Moore Haven Lock (Florida), Lihue (Hawaii), University Experiment Station (Alaska), Stillwater (Oklahoma), Roosevelt and Bartlett Dam (Arizona), San Jon and Caballo Dam (New Mexico), Anahuac and Whitney Dam (Texas), and five stations in Ohio: Coshocton, Wooster, Circleville, Bowling Green, and Bellefontaine (Figures 3.1 and 3.2). Data representing the precipitation measurements were collected from the National Climatic Data Center (www.ncdc.noaa.gov). These data were analyzed to establish trends in the annual and seasonal precipitation for all of the stations. Regression analysis was performed using Microsoft Excel (2007) and StatTools 5.7. A confidence interval of 90% was used while calculating the p values of the models. The growing season is the time between the last frost and the first frost of the year (Burkhead 1972). The period between May and October was used to categorize the growing season data for all states.

FIGURE 3.1 The locations of the stations used in the study.

FIGURE 3.2 The locations of the stations in Ohio used in the study.

3.3 RESULTS

3.3.1 University Experiment Station, Alaska

This site was selected for the study because Alaska is situated in the northwest extremity of the North American continent and represents an arctic biome. Thus, it is important to analyze and compare the data from this site with those in the continental United States. No major significant changes were observed on average in the amount of annual precipitation or that during the growing season. The annual precipitation showed a nonsignificant decreasing trend at the rate of 0.003 cm/year, while the seasonal precipitation increased at 0.01 cm/year (Figure 3.3). The p values of 0.9 and 0.6 of the annual and the growing season trends, respectively, are not statistically significant.

3.3.2 Waseca, Minnesota

The annual precipitation in Waseca has increased over the last 95 years at an average rate of 0.28 cm/year (Figure 3.4). The trend during the growing season is similar to that of the annual precipitation trend, but the rate of increase is smaller. Thus, precipitation during the growing season increased at an average rate of 0.12 cm/year (Figure 3.4), which is almost half the rate of increase on the annual basis. The p values for both the annual and the growing season trends are statistically significant (Figure 3.4).

3.3.3 Ohio

Several stations were selected within Ohio to establish any north–south gradients in precipitation.

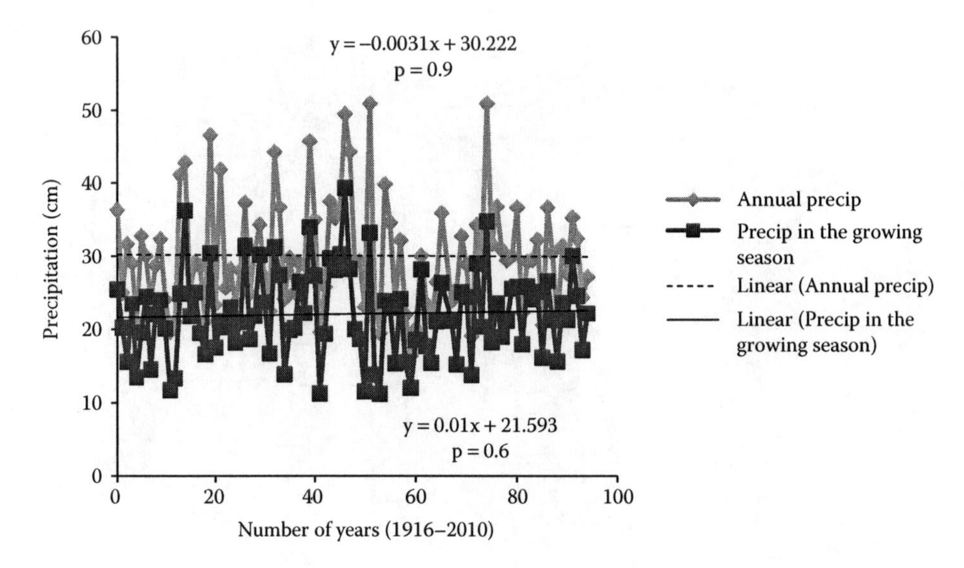

FIGURE 3.3 The trends in precipitation in University Experiment Station, AK, from 1916 to 2010.

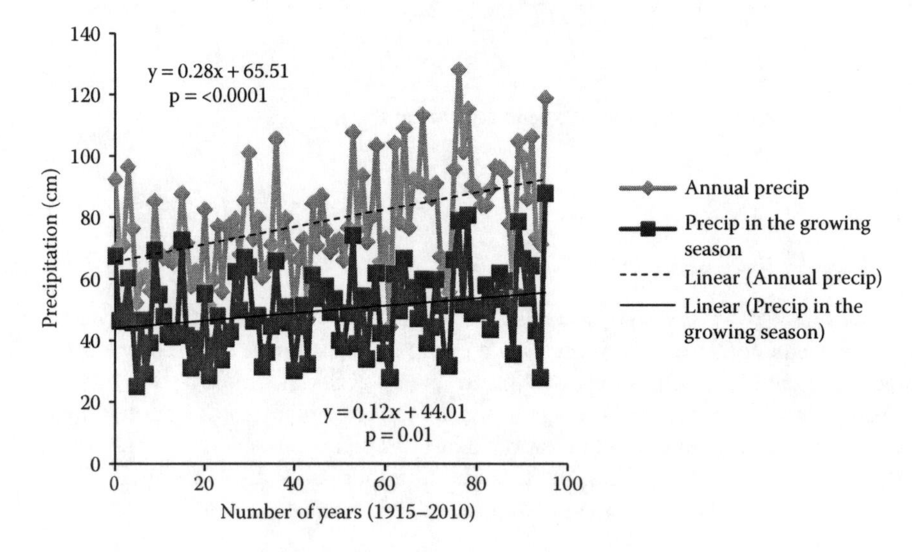

FIGURE 3.4 The trends in precipitation in Waseca, MN, from 1915 to 2010.

3.3.3.1 Coshocton

Coshocton is located in east central Ohio (Figure 3.5). Even though the National Climatic Data Center has data dating back to only 1956 for this site, there is a clear trend in the precipitation records. The annual precipitation increased at an approximate rate of 0.25 cm/year (Figure 3.5). The p value of 0.06 is statistically significant. The rate of increase in precipitation during the growing season is ~0.14 cm/year (Figure 3.5). However, the p value for this model is 0.2, which is not statistically significant.

3.3.3.2 Wooster

Wooster is located in northeastern Ohio (Figure 3.2) and the trend in precipitation is similar to that of Coshocton. The annual precipitation increased over the last 113 years (Figure 3.6). The rate of increase

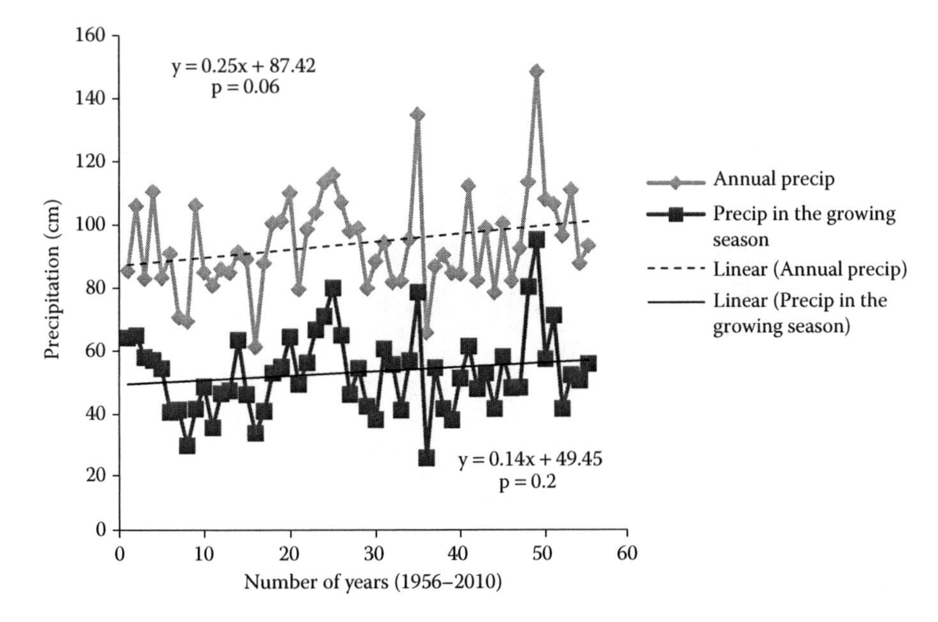

FIGURE 3.5 The trends in precipitation in Coshocton, OH, from 1956 to 2010.

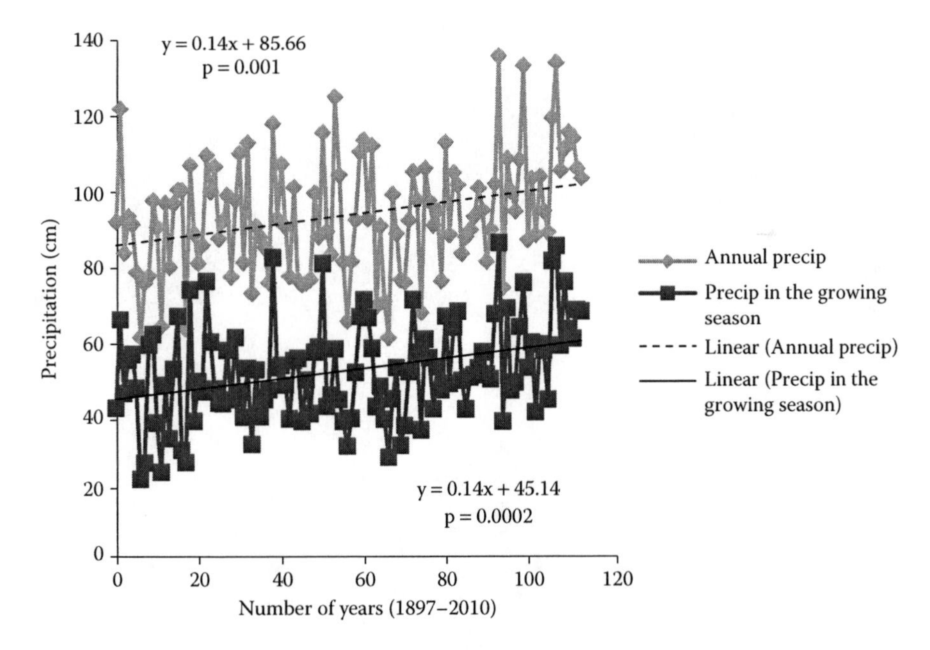

FIGURE 3.6 The trends in precipitation in Wooster, OH, from 1897 to 2010.

in the annual precipitation is ~0.14 cm/year, which is the same as the rate of increase during the growing season (Figure 3.6). The *p* values for both the trends were statistically significant (Figure 3.6).

3.3.3.3 Bellefontaine

Bellefontaine is located in west central Ohio (Figure 3.2). The precipitation records date back to 1895. An analysis of the data indicates some trends over the 115 years. The annual precipitation has increased at an average rate of ~0.1 cm/year (Figure 3.7), and it has a significant *p* value of 0.02. The precipitation during the growing season increased by ~0.05 cm/year (Figure 3.7), but the *p* value for the same is not

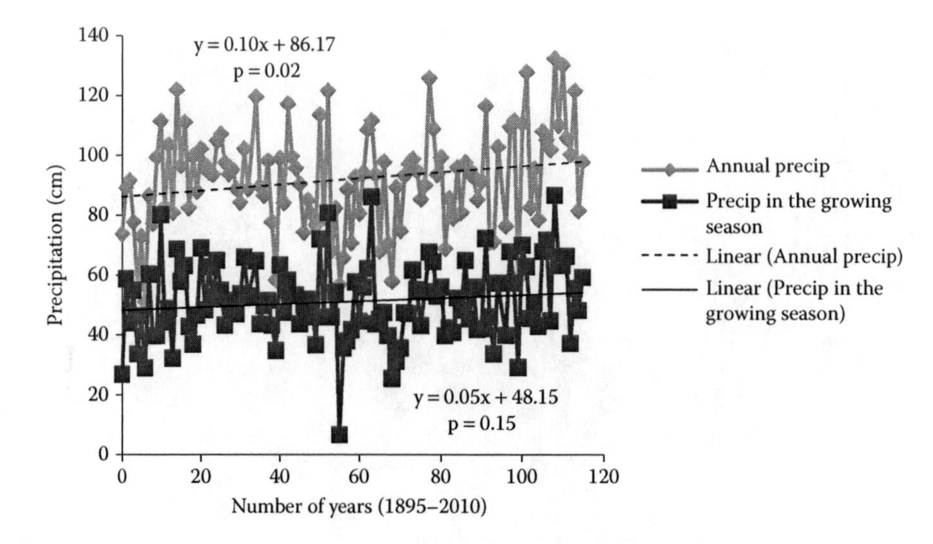

FIGURE 3.7 The trends in precipitation in Bellefontaine, OH, from 1895 to 2010.

statistically significant. These trends are not as prominent as for the sites in east central and northeastern Ohio.

3.3.3.4 Bowling Green

Bowling Green is located in northwest Ohio (Figure 3.2). Similar to the Bellefontaine site, only minor changes are observed in precipitation in Bowling Green between 1894 and 2010 (Figure 3.8). The rates of increase in precipitation are ~0.03 and ~0.04 cm/year annually and during the growing season, respectively. Both the trends have nonsignificant p values.

3.3.3.5 Circleville

Circleville is located in south central Ohio (Figure 3.2). For the period between 1896 and 2010, there are only minor changes in the precipitation patterns and the p values for the trends are nonsignificant.

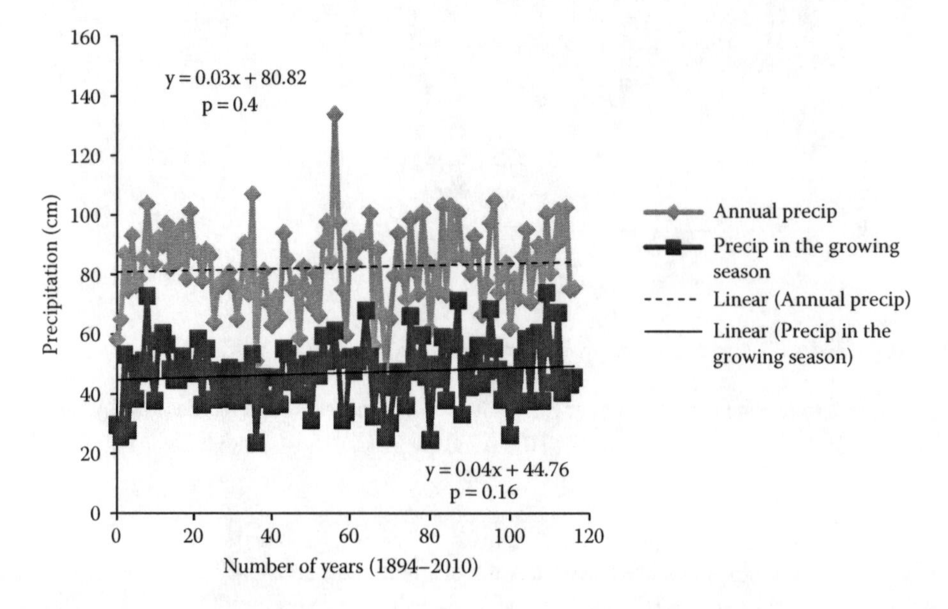

FIGURE 3.8 The trends in precipitation in Bowling Green, OH, from 1894 to 2010.

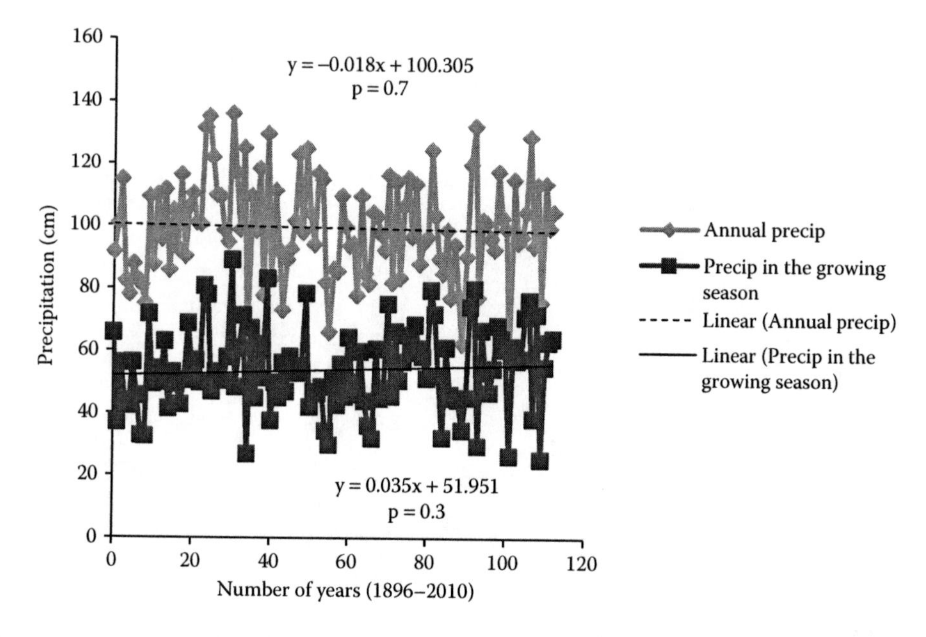

FIGURE 3.9 The trends in precipitation in Circleville, OH, from 1896 to 2010.

There has been a minor decrease in the annual precipitation at an average rate of ~0.018 cm/year (Figure 3.9). By contrast, precipitation has increased during the growing season at an average rate of 0.03 cm/year (Figure 3.9).

3.3.4 Jackson, Tennessee

This station is located far south of Waseca, Minnesota (Figure 3.1), but the precipitation trend over the last century is rather similar. The annual precipitation has increased between 1903 and 2010, at an average rate of 0.2 cm/year (Figure 3.10), while the rate of increase during the growing season is ~0.14 cm/year (Figure 3.10). The *p* values for both the models are significant.

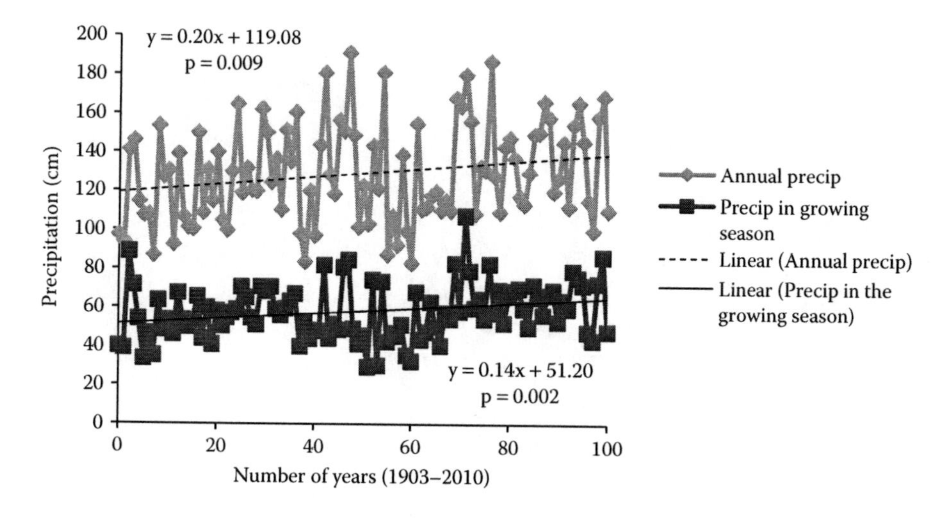

FIGURE 3.10 The trends in precipitation in Jackson, TN, from 1903 to 2010.

3.3.5 Arizona

The precipitation patterns were analyzed for two stations in central Arizona.

3.3.5.1 Bartlett Dam

Bartlett Dam is located in Maricopa County in central Arizona (Figure 3.1). There is no prominent trend in precipitation during the 61-year period between 1940 and 2010 and the *p* values are nonsignificant. The data in Figure 3.11a show a slight increase in the annual precipitation at a rate of ~0.12 cm/year. Precipitation during the growing season does not show any significant trend (Figure 3.11a). It is interesting to note that in the last two decades, however, precipitation has decreased significantly (*p* value = 0.03) at a rate of ~1 and 0.5 cm/year annually and during the growing season, respectively (Figure 3.11b).

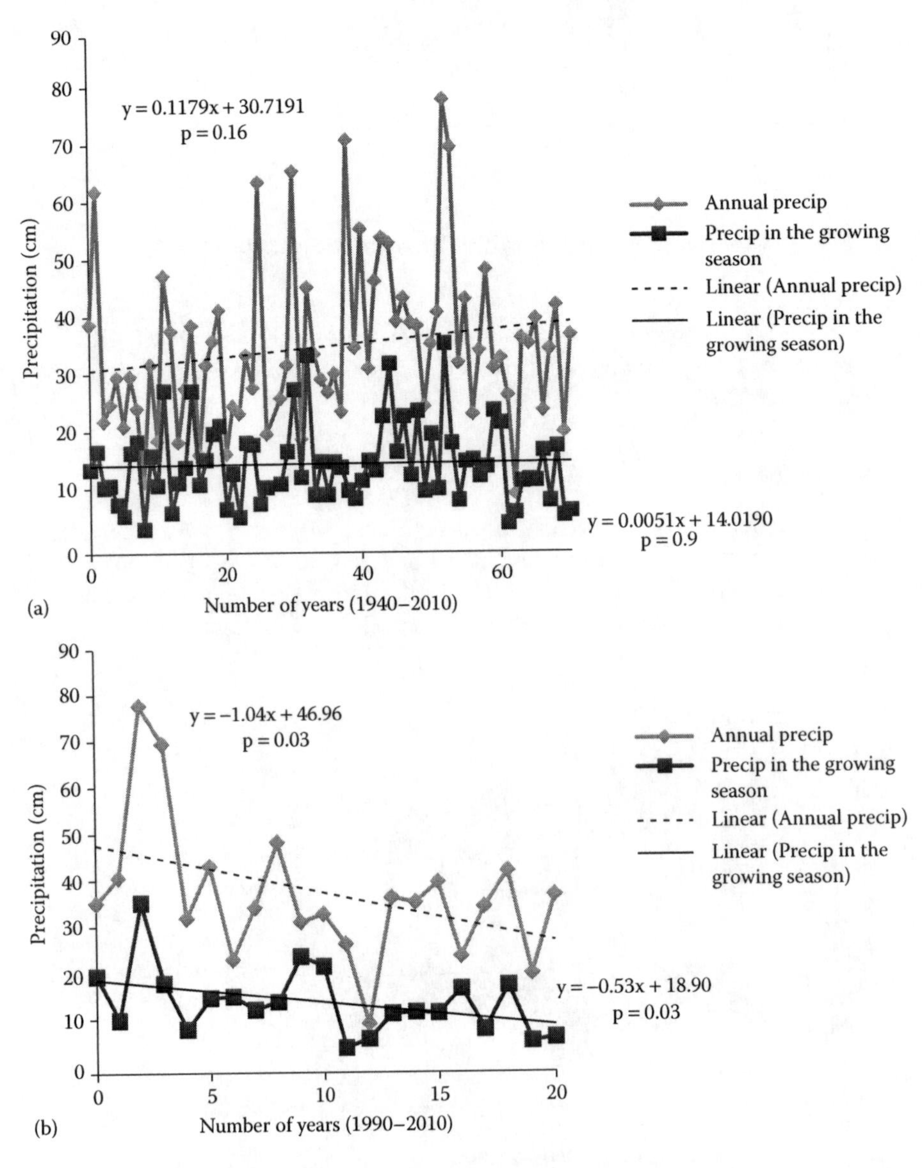

FIGURE 3.11 The trends in precipitation in Bartlett Dam, AZ: (a) from 1940 to 2010 and (b) from 1990 to 2010.

3.3.5.2 Roosevelt

Roosevelt is also located in central Arizona (Figure 3.1). The precipitation trends look similar to those in Bartlett Dam, Arizona, but are not statistically significant. The average precipitation has decreased annually as well as during the growing season within 105 years. Between 1906 and 2010, precipitation has decreased at an average rate of ~0.13 and 0.07 cm/year annually and during the growing season, respectively (Figure 3.12a). There has been a sharper decrease in the last two decades. The annual precipitation decreased significantly at a rate of ~1.54 cm/year, while the seasonal precipitation decreased at a rate of ~0.3 cm/year (Figure 3.12b).

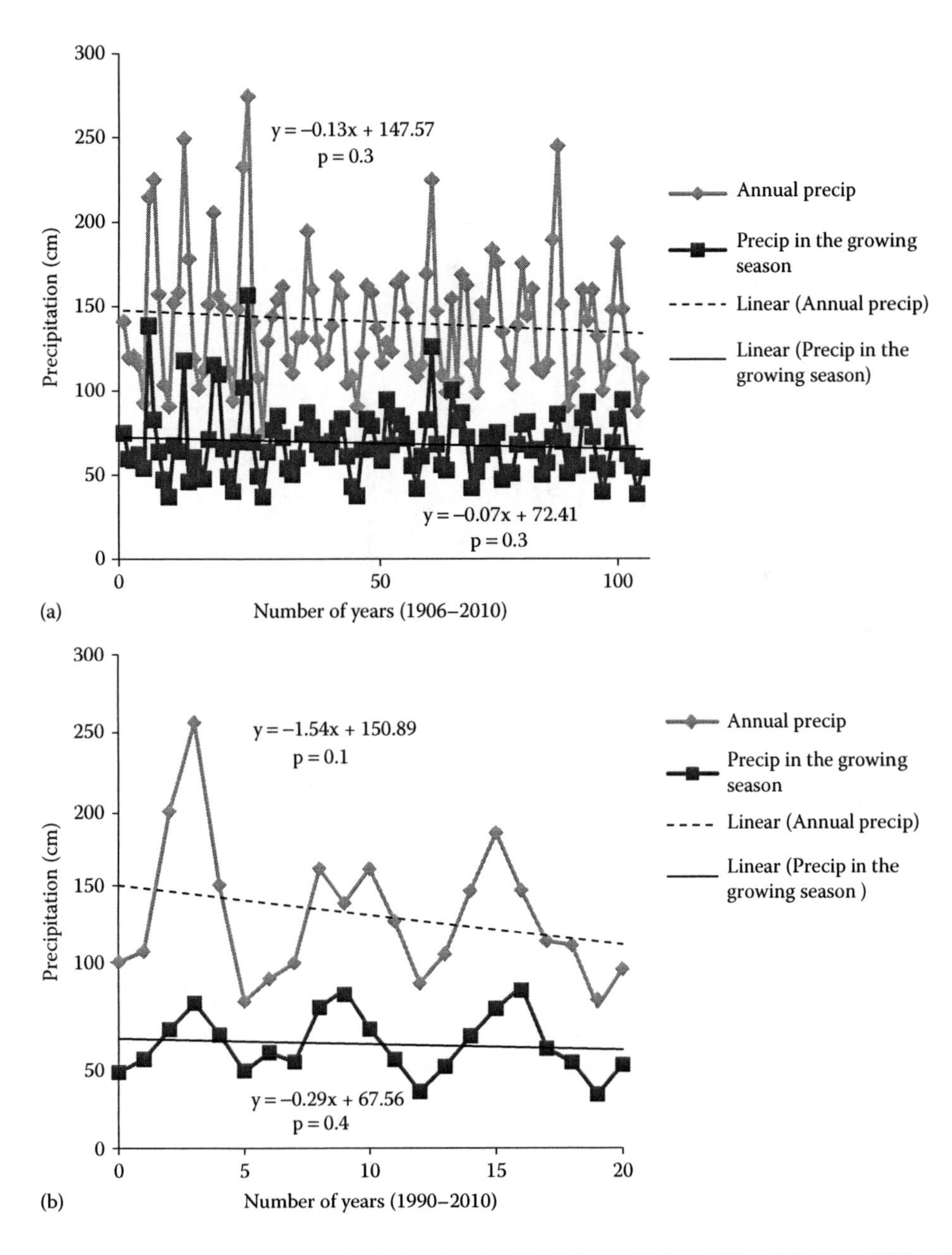

FIGURE 3.12 The trends in precipitation in Roosevelt, AZ: (a) from 1906 to 2010 and (b) from 1990 to 2010.

3.3.6 New Mexico

Two sites from different parts of New Mexico were studied.

3.3.6.1 San Jon

San Jon is located in northeastern New Mexico. The data in Figure 3.13a show that during the last 101 years, the annual precipitation has increased at ~0.05 cm/year. There is no clear trend in the growing season precipitation between 1910 and 2010 (Figure 3.13a). The *p* values for the 101-year period are not statistically significant. However, similar to Arizona, a very sharp decrease in the precipitation trend is observed in New Mexico since 1990. Precipitation has decreased at an

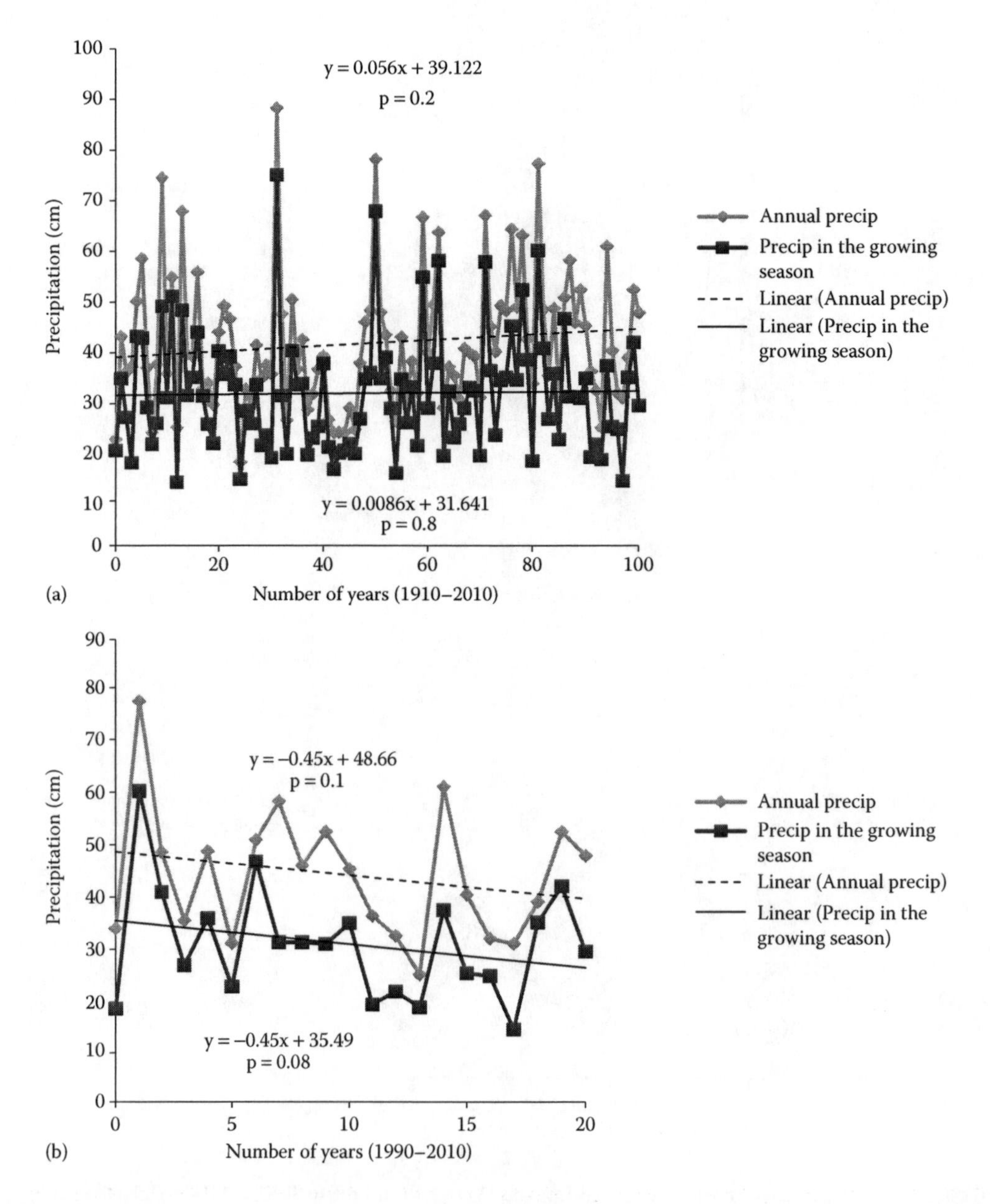

FIGURE 3.13 The trends in precipitation in San Jon, NM: (a) from 1910 to 2010 and (b) from 1990 to 2010.

average rate of ~0.45 cm/year between 1990 and 2010, both annually and during the growing season (Figure 3.13b) with significant *p* values.

3.3.6.2 Caballo Dam

Caballo Dam is located in southwestern New Mexico (Figure 3.1). The precipitation data were analyzed for a 71-year period between 1937 and 2010. An increasing trend was observed in the long-term data. The trend is statistically significant with the respective *p* values for the annual trend and the growing season trend—0.05 and 0.06. The rate of increase in the annual precipitation is ~0.12 cm/year (Figure 3.14a). However, the rate of increase in precipitation during the growing season is lower at ~0.8 cm/year (Figure 3.14a). Similar to the trends in other southern stations, the Caballo Dam site also shows a sharp decreasing trend in precipitation during the last two decades

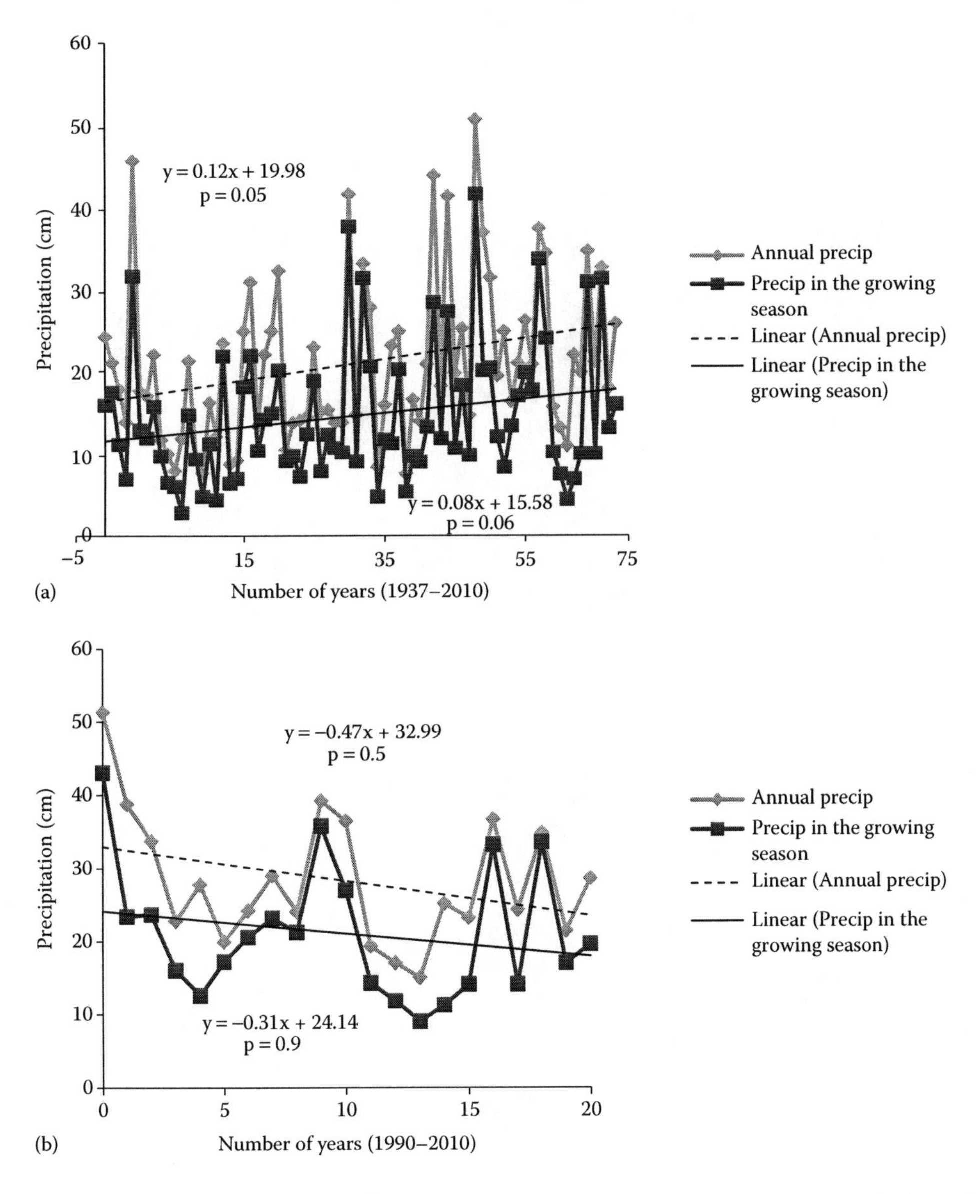

FIGURE 3.14 The trends in precipitation in Caballo Dam, NM: (a) from 1937 to 2010 and (b) from 1990 to 2010.

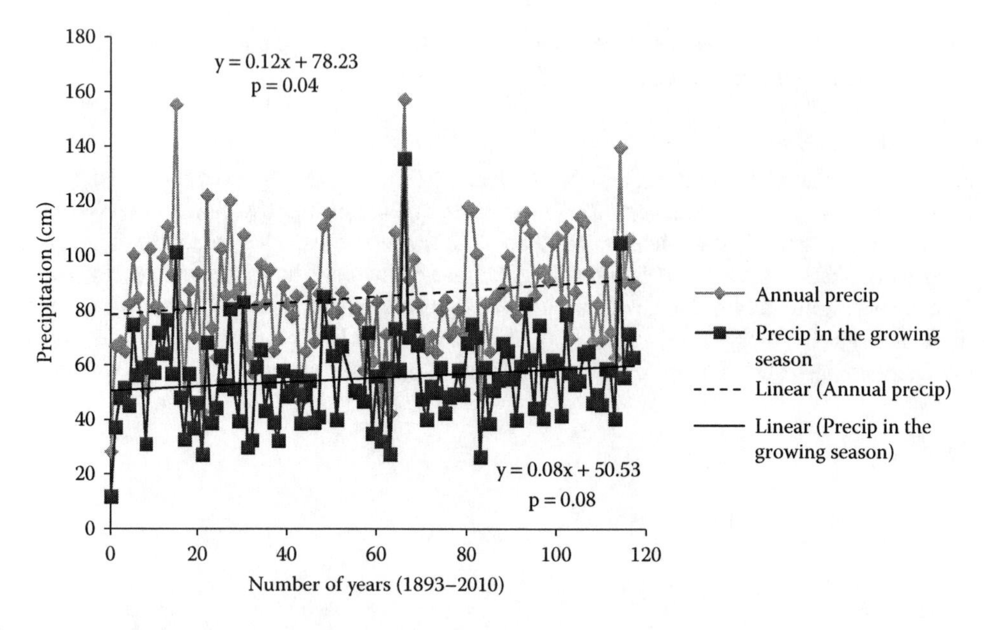

FIGURE 3.15 The trends in precipitation in Stillwater, OK, from 1893 to 2010.

(Figure 3.14b). The rate of decrease in precipitation between 1990 and 2010 is ~0.5 and 0.31 cm/year for the annual and seasonal precipitation, respectively (Figure 3.14b). However, the trends in the last 20 years are not statistically significant.

3.3.7 Stillwater, Oklahoma

Stillwater is located in north central Oklahoma (Figure 3.1). The precipitation data were studied from 1893 to 2010. The annual precipitation has increased at a rate of ~0.1112 cm/year, with an especially high increase in the last two decades (Figure 3.15). Precipitation during the growing season has also increased at a rate of ~0.08 cm/year (Figure 3.15). Both the results are statistically significant.

3.3.8 Texas

Several parts of Texas experienced a severe drought during 2011. Thus, two stations were selected to analyze the long-term and short-term trends in precipitation.

3.3.8.1 Anahuac

Anahuac is located in eastern Texas. The precipitation records date back to 1910. A slight increasing trend in precipitation is observed over 101 years. Both the annual precipitation and that during the growing season have increased slightly at an average rate of ~0.1 cm/year between 1910 and 2010 (Figure 3.16a). However, during the last two decades, the annual precipitation has decreased at a rate of 0.95 cm/year, but the seasonal precipitation does not show any specific trend (Figure 3.16b). None of the results at this site is statistically significant.

3.3.8.2 Whitney Dam

This site is located in central Texas. The trends in precipitation are similar to the station in eastern Texas. There is a small increasing trend in precipitation within the last 61 years. The rate of increase

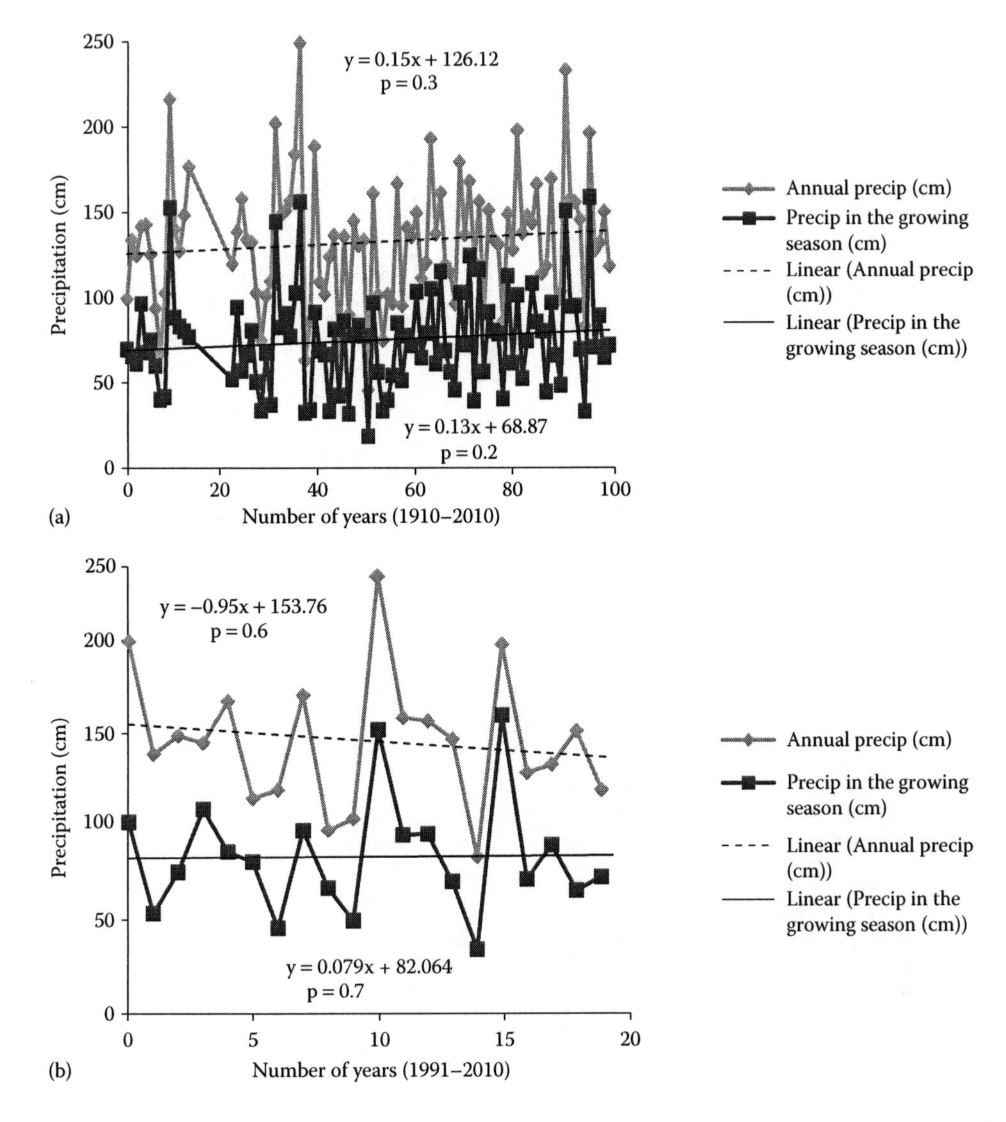

FIGURE 3.16 The trends in precipitation in Anahuac, TX: (a) from 1910 to 2010 and (b) from 1991 to 2010.

in the annual precipitation is ~0.18 cm/year (Figure 3.17). Precipitation during the growing season does not show any definite statistically significant trend (Figure 3.17).

3.3.9 Fairhope, Alabama

The data in Figure 3.18 show an overall increase in the annual precipitation in Fairhope, Alabama, even though the rate of increase is not as high as that in Waseca, Minnesota. Similarly, the rate of increase in precipitation during the growing season is much smaller than the rate of increase in the annual precipitation. Between 1920 and 2010, precipitation increased at a rate of ~0.2 cm/year (Figure 3.18). The increase is statistically significant. However, the rate of increase during the growing season is merely ~0.0809 cm/year and is nonsignificant (Figure 3.18).

3.3.10 Moore Haven Lock, Florida

Rather than an increasing trend, the annual precipitation in Moore Haven, Florida, and other southern sites in the United States indicates statistically significant declining trends. For example, the

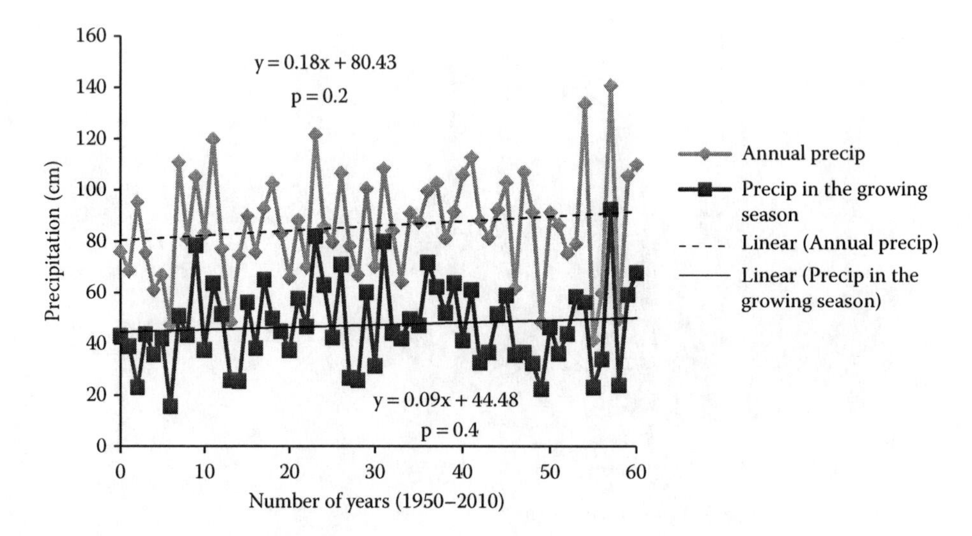

FIGURE 3.17 The trends in precipitation in Whitney Dam, TX, from 1950 to 2010.

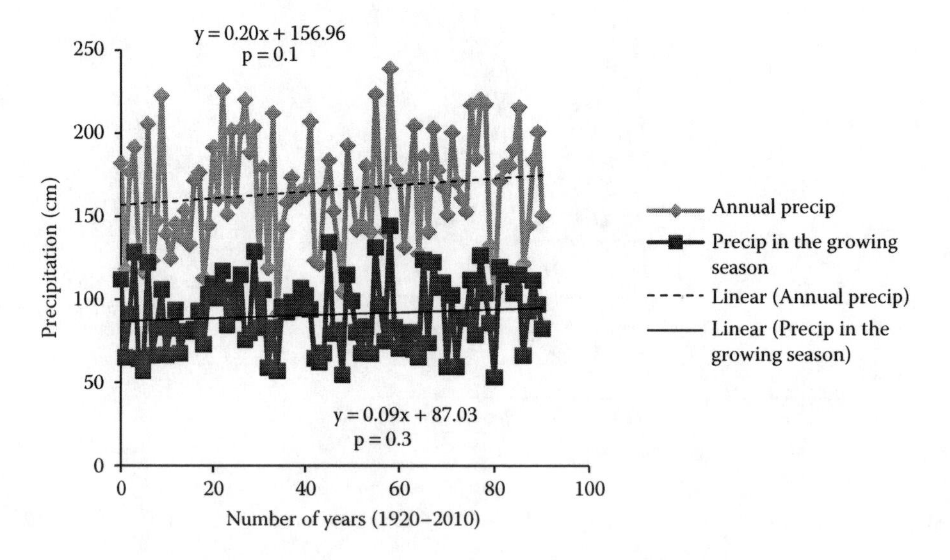

FIGURE 3.18 The trends in precipitation in Fairhope, AL, from 1920 to 2010.

annual precipitation has decreased at Moore Haven Lock in Florida over the 88-year period between 1922 and 2010. Between 1922 and 2010, the annual precipitation decreased at a rate of ~0.14 cm/year (Figure 3.19). Furthermore, the rate of decline is even more during the growing season, at an average rate of 0.2 cm/year (Figure 3.19).

3.3.11 Lihue, Hawaii

Hawaii was included in this study because it is located southwest of the continental United States and represents the tropical rainforest biome. It has a warm tropical climate, which differs from all other sites included in this study. However, the data availability for Lihue, Hawaii, is limited to 60 years, from 1950 to 2010. Yet, there is a statistically significant declining trend in the annual precipitation during the last 60 years. The rate of decrease in the annual precipitation is approximately 0.4 cm/year

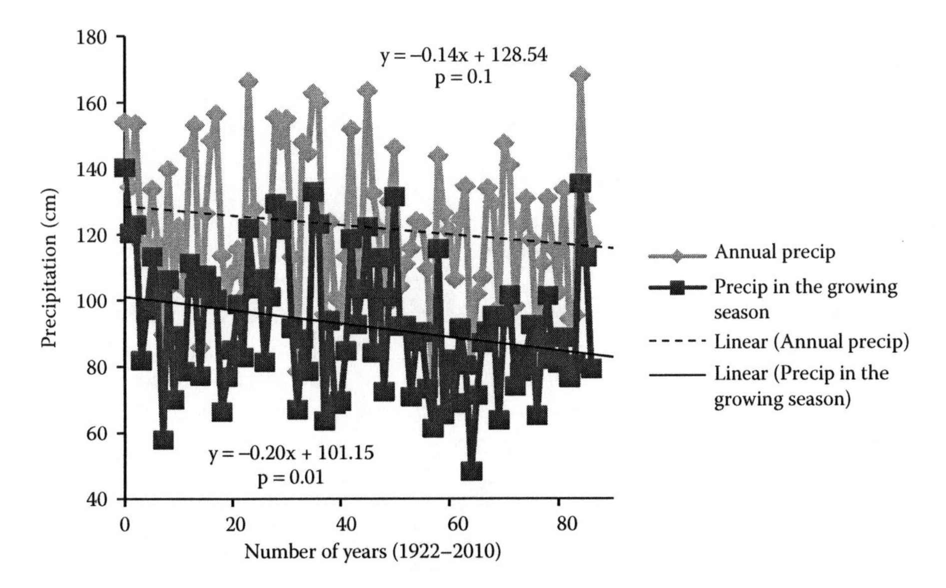

FIGURE 3.19 The trends in precipitation in Moore Haven Lock, FL, from 1922 to 2010.

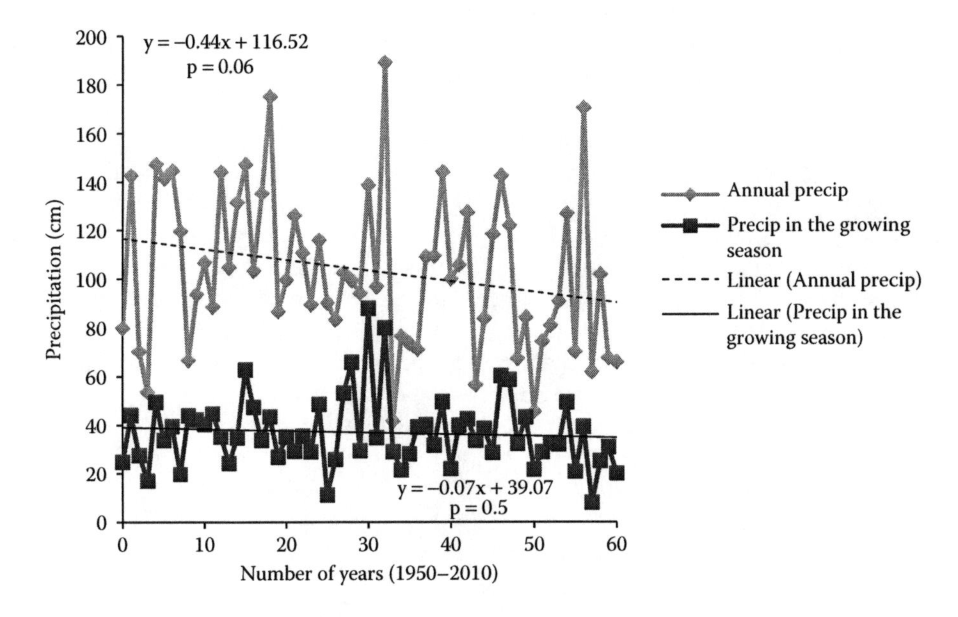

FIGURE 3.20 The trends in precipitation in Lihue, HI, from 1950 to 2010.

(Figure 3.20). However, the rate of decrease in precipitation during the growing season is relatively small and statistically nonsignificant at 0.07 cm/year (Figure 3.20).

3.4 DISCUSSION

This study shows no significant precipitation trend in the Alaskan site. However, an increasing trend is observed for sites located in Minnesota, Tennessee, Alabama, and Oklahoma. By contrast, a decreasing trend is observed for sites in Florida and Hawaii. In Ohio, three sites show an increasing trend while two sites do not show any specific trend. In Texas, New Mexico, and Arizona, an overall increasing trend is observed during the last century. However, in the past two decades, the precipitation has decreased, resulting in droughts.

Stafford et al. (2000) reported an increase in the total precipitation for three of the four seasons throughout most of Alaska in a 50-year period (1948–1998), while a slight decreasing trend was observed during the summer months. The present study is based on the 95-year precipitation records from the University Experiment Station located in Fairbanks. However, neither the annual nor the growing season precipitation shows any specific trends. There is a slight but nonsignificant decrease in the annual precipitation and a nonsignificant increase in the seasonal precipitation.

In contrast to the data of the present study for Minnesota, Baker (1962) reported a declining trend in both the seasonal and the annual precipitation between 1900 and 1958 in Minnesota for five stations studied across the state. Yet, in the current study, the Waseca site located in southern Minnesota indicates an increasing trend in the long-term precipitation between 1915 and 2010 for both the annual precipitation and the growing season precipitation.

In the present study, the sites in east central and northeastern Ohio indicate an increasing trend in precipitation, while the increase is not significant in the sites located in northwestern, south central, and west central Ohio (Bellefontaine, Circleville, and Bowling Green). Harstine (1991) studied the precipitation trends in Ohio between 1931 and 1980 and found precipitation to be highest in southern and eastern Ohio and decreasing in northwestern Ohio.

The present study indicates a declining trend in the precipitation in Florida and in the Caribbean region. The precipitation data from land-based sites indicate a similar declining trend in some regions (Coleman 1982; Kunkel et al. 1999; Neelin et al. 2006). The data from Moore Haven Lock located in southern Florida also indicate a declining trend. The annual precipitation decreased between 1922 and 2010, and the rate of decrease in precipitation during the growing season is even higher.

In the Hawaiian Islands, the records from stations along the windward coastal side of the islands have shown a general decreasing trend in precipitation. By contrast, an increasing trend is reported along the southeast and northwestern sides of the Hawaiian Islands (Doty 1982; Woodcock and Jones 1970). The precipitation data from 1951 to 2000 have been analyzed from more than 100 stations across four major islands in Hawaii. A declining trend in precipitation has been reported since 1980 (Diaz et al. 2005). In the present study, data from Lihue also show a similarly decreasing trend in rainfall from 1950 to 2010. However, the rate of decrease is less pronounced for the growing season precipitation than for the annual precipitation.

For east central Texas, Harmel et al. (2003) reported that precipitation increased over a 61-year period between 1939 and 1999. It was also reported that precipitation in most sites in southern and central Texas increased historically between 1895 and 2006 (Nielsen-Gammon 2011). In the present study, both the Anahuac and Whitney Dam sites located in eastern and western Texas, respectively, also indicate an increasing trend. However, during the last 20 years, precipitation has a definite decreasing trend.

An increasing trend in precipitation has been reported in most of the Great Plains areas, including the stations located in south central, southwestern, northern, and eastern Oklahoma between 1981 and 2001 (Garbrecht et al. 2004). The data in the present study from Stillwater, located in north central Oklahoma, are in agreement with the hypothesis by Garbrecht et al. and also indicate an increasing trend in precipitation during the twentieth century.

It has been reported that dry areas in the United States are getting drier and wet areas are getting wetter (Dore 2005). The data analysis in the present study indicates that sites in the northern and midwestern states (Minnesota and Ohio) show an increasing trend in precipitation. The rate of increase is larger in the northern (Minnesota and Ohio) than in the southern states of Tennessee and Alabama. The trend in the annual precipitation is negative for some sites in the southwestern and southeastern United States—Hawaii, Arizona, and Florida. However, the annual precipitation is seemingly increasing in the south central United States—Texas and Oklahoma. Precipitation in Alaska, located in the extreme northwestern United States, does not show any definite trend. In New Mexico and Arizona, the precipitation in the last two decades has been decreasing.

Several states in the southern United States have experienced severe drought during 2011 (CNN 2011). The rainfall received during 2011 has been extremely low in Texas, Florida, New Mexico, and Oklahoma. The present study shows a sharp decreasing trend in precipitation during the last two decades in Florida and New Mexico, corroborating the drought experienced in 2011. However, no decreasing trend is observed for the sites located in Texas and Oklahoma. These observations are in agreement with the published literature (Garbrecht et al. 2004; Harmel et al. 2003; Nielsen-Gammon 2011). Several other factors may be responsible for severe droughts, such as an increase in temperature, evaporation, and the changing pattern of the oceanic currents (Fawcett et al. 2011; Seager et al. 2009). The rising demand for water resources due to the increasing population is another factor responsible for the deficit (Manuel 2008). The results and findings of the present study can be used to develop guidelines for adapting agricultural, industrial, ecological, and residential water management strategies in order to make the most efficient use of natural precipitation, since surface water and groundwater (blue water) supplies are also dependent on it. With the increasing population and further increasing water demand in the midst of climate change, sustainable water resource management can play a crucial role in agricultural and industrial economies.

3.5 CONCLUSION

The data presented support the following conclusions:

1. Precipitation trends are increasing in some regions across the United States, such as Minnesota, Tennessee, Ohio, Oklahoma, and Alabama, while decreasing in other regions, such as Florida and Hawaii, based on their geographic location, climate, and other factors (Table 3.2).
2. In New Mexico, Arizona, and Texas, there is an increasing trend in the long-term precipitation. However, in the last 20 years, precipitation has been decreasing.

TABLE 3.2
Summary of *p* Values and Significance for Different Stations

Station	Trend	*p* Values	Significant
Waseca, MN (1915–2010)	Annual	<0.0001	✓
	Growing season	0.01	✓
Jackson, TN (1903–2010)	Annual	0.009	✓
	Growing season	0.002	✓
Fairhope, AL (1920–2010)	Annual	0.1	✓
	Growing season	0.3	×
Moore Haven Lock, FL (1922–2010)	Annual	0.1	✓
	Growing season	0.01	✓
Lihue, HI (1950–2010)	Annual	0.06	✓
	Growing season	0.5	×
University Exp Stn, AK (1916–2010)	Annual	0.9	×
	Growing season	0.6	×
Coshocton, OH (1956–2010)	Annual	0.06	✓
	Growing season	0.2	×
Wooster, OH (1897–2010)	Annual	0.001	✓
	Growing season	0.0002	✓
Bellefontaine, OH (1895–2010)	Annual	0.02	✓
	Growing season	0.15	×

(*continued*)

TABLE 3.2 (Continued)
Summary of *p* Values and Significance for Different Stations

Station	Trend	*p* Values	Significant
Bowling Green, OH (1894–2010)	Annual	0.4	×
	Growing season	0.16	×
Circleville, OH (1896–2010)	Annual	0.7	×
	Growing season	0.3	×
Anahuac, TX (1910–2010)	Annual	0.3	×
	Growing season	0.2	×
Whitney Dam, TX (1950–2010)	Annual	0.2	×
	Growing season	0.4	×
San Jon, NM (1910–2010)	Annual	0.2	×
	Growing season	0.8	×
San Jon, NM (1990–2010)	Annual	0.1	✓
	Growing season	0.08	✓
Caballo Dam, NM (1937–2010)	Annual	0.05	✓
	Growing season	0.06	✓
Roosevelt, AZ (1906–2010)	Annual	0.3	×
	Growing season	0.3	×
Roosevelt, AZ (1990–2010)	Annual	0.1	✓
	Growing season	0.4	×
Bartlett Dam, AZ (1940–2010)	Annual	0.16	×
	Growing season	0.9	×
Bartlett Dam, AZ (1990–2010)	Annual	0.03	✓
	Growing season	0.03	✓
Stillwater, OK (1893–2010)	Annual	0.04	✓
	Growing season	0.08	✓

3. In Ohio, the data from the northeastern and the east central regions indicate a larger increase in precipitation over the twentieth century as compared with the western and the central parts.
4. In some cases, the increase in precipitation is more pronounced on an annual basis rather than during the growing season. In others, the trends are almost the same. Therefore, there can be a possible scenario where even though the total precipitation is increasing annually, the rate of increase during the growing season is rather small. A change in precipitation during the growing season is important for agricultural production.

ACKNOWLEDGMENT

All the primary data for the study were obtained from the National Climatic Data Center (www.ncdc.noaa.gov) and is gratefully acknowledged.

REFERENCES

Ashraf, M. and K. Habib-ur-Rehman. 1999. Interactive effects of nitrate and long-term waterlogging on growth, water relations, and gaseous exchange properties of maize (*Zea mays* L.). *Plant Science* 144(1): 35–43.

Baker, D.G. 1962. Seasonal temperature and precipitation trends at five Minnesota stations. *Monthly Weather Review* 90(7): 283–286.

Bradley, R.S., H.F. Diaz, J.K. Eischeid, P.D. Jones, P.M. Kelly, and C.M. Goodess. 1987. Precipitation fluctuations over Northern Hemisphere land areas since the mid-19th century. *Science* 237(4811): 171.

Burkhead, C.E. 1972. Usual planting and harvesting dates. U.S. Department of Agriculture, Statistical Reporting Service.

Changnon, S.A. and G.J.D. Hewings. 2001. Losses from weather extremes in the United States. *Natural Hazards Review* 2: 113.

Clemmitt, M. 2008. Global food crisis. *CQ researcher* 18(24): 555–574.

CNN. 2011. Record percentage of United States experiences exceptional drought. http://www.cnn.com/2011/US/08/01/us.record.drought/index.html?iref=allsearch (accessed August 8, 2011).

Coleman, J.M. 1982. Recent seasonal rainfall and temperature relationships in peninsular Florida. *Quaternary Research* 18(2): 144–151.

Diaz, H.F. and R.G. Quayle. 1980. The climate of the United States since 1895: Spatial and temporal changes. *Monthly Weather Review* 108: 249.

Diaz, H.F., P.S. Chu, and J.K. Eischeid. 2005. Rainfall changes in Hawaii during the last century. *16th Conference on Climate Variability and Change*, San Diego, CA.

Dore, M.H.I. 2005. Climate change and changes in global precipitation patterns: What do we know? *Environment International* 31(8): 1167–1181.

Doty, R.D. 1982. Annual precipitation on the island of Hawaii between 1890 and 1977. *Pacific Science* 36(4): 421–425.

Fawcett, P.J., J.P. Werne, R.S. Anderson, J.M. Heikoop, E.T. Brown, M.A. Berke, and S.J. Smith. 2011. Extended megadroughts in the southwestern United States during Pleistocene interglacials. *Nature* 470(7335): 518–521. DOI:10.1038/nature09839.

Garbrecht, J., M. Van Liew, and G.O. Brown. 2004. Trends in precipitation, streamflow, and evapotranspiration in the Great Plains of the United States. *Journal of Hydrologic Engineering* 9(5): 360–367.

Groisman, P.Y. and D.R. Easterling. 1994. Variability and trends of total precipitation and snowfall over the United States and Canada. *Journal of Climate* 7(1): 184–205.

Groisman, P.Y. and D.R. Legates. 1994. The accuracy of United States precipitation data. *Bulletin of the American Meteorological Society (United States)* 75(2): 215–227.

Harmel, R.D., K.W. King, C.W. Richardson, and J.R. Williams. 2003. Long-term precipitation analyses for the Central Texas Blackland Prairie. *Transactions of the ASAE* 46(5): 1381–1388.

Harstine, L. 1991. Hydrologic atlas for Ohio: Average annual precipitation, temperature, streamflow and water loss for 50 year period, 1931–1980. Water Inventory Report No. 28. Ohio Department of Natural Resources, Division of Water, Ground Water Resources Section. http://www.dnr.state.oh.us/water/pubs/hydatlas/default/tabid/4187/Default.aspx (accessed July 15, 2011).

IPCC. 1996. Climate change 1995: The science of climate change. Contribution of Working Group I to the Second Assessment Report of the Intergovernmental Panel in Climate Change.

IPCC. 2007. Climate change 2007: The physical science basis. Contribution of Working Group I to the Fourth Assessment Report of the Intergovernmental Panel on Climate Change.

Karl, T.R. and R.W. Knight. 1998. Secular trends of precipitation amount, frequency, and intensity in the United States. *Bulletin of the American Meteorological Society* 79(2): 231–241.

Keim, B.D., M.R. Fischer, and A.M. Wilson. 2005. Are there spurious precipitation trends in the United States Climate Division database. *Geophysical Research Letters* 32(4): L04702.

Klugman, M.R. 1983. Evidence of climatic change in United States seasonal precipitation data, 1948–1976. *Journal of Climate and Applied Meteorology* 22(8): 1367–1376.

Köppen Climate Classification. http://snow.cals.uidaho.edu/clim_map/koppen_usa_map.htm (accessed August 14, 2011).

Kunkel, K.E. 2003. North American trends in extreme precipitation. *Natural Hazards* 29(2): 291–305.

Kunkel, K.E., K. Andsager, and D.R. Easterling. 1999. Long-term trends in extreme precipitation events over the conterminous United States and Canada. *Journal of Climate* 12(8): 2515–2527.

Lettenmaier, D.P., E.F. Wood, and J.R. Wallis. 1994. Hydro-climatological trends in the continental United States, 1948–1988. *Journal of Climate* 7(4): 586–607.

Manuel, J. 2008. Drought in the southeast: Lessons for water management. *Environmental Health Perspectives* 116(4): A168–A171.

McCarthy, J.J. 2001. Climate change 2001: Impacts, adaptation, and vulnerability. Contribution of Working Group II to the Third Assessment Report of the Intergovernmental Panel on Climate Change. Cambridge University Press.

Neelin, J.D., M. Münnich, H. Su, J.E. Meyerson, and C.E. Holloway. 2006. Tropical drying trends in global warming models and observations. *Proceedings of the National Academy of Sciences of the United States of America* 103(16): 21349–21354.

Nielsen-Gammon, J.W. 2011. The changing climate of Texas. In *The Impact of Global Warming on Texas,* eds. J. Schmandt, G.R. North, and J. Clarkson, pp. 39–68. University of Texas Press, Austin, TX.

Pielke Jr, R.A. and M.W. Downton. 2000. Precipitation and damaging floods: Trends in the United States, 1932–1997. *Journal of Climate* 13(20): 3625–3637.

Reilly, J., F. Tubiello, B. McCarl, D. Abler, R. Darwin, K. Fuglie, and S. Hollinger. 2003. U.S. agriculture and climate change: New results. *Climatic Change* 57(1): 43–67.

Rockström, J. and L. Gordon. 2001. Assessment of green water flows to sustain major biomes of the world: Implications for future ecohydrological landscape management. *Physics and Chemistry of the Earth, Part B: Hydrology, Oceans and Atmosphere* 26(11–12): 843–851.

Rockström, J., M. Falkenmark, L. Karlberg, H. Hoff, S. Rost, and D. Gerten. 2009. Future water availability for global food production: The potential of green water for increasing resilience to global change. *Water Resources Research* 45, W00A12, doi:10.1029/2007WR006767.

Rosenzweig, C., F.N. Tubiello, R. Goldberg, E. Mills, and J. Bloomfield. 2002. Increased crop damage in the U.S. from excess precipitation under climate change. *Global Environmental Change* 12(3): 197–202.

Seager, R., A. Tzanova, and J. Nakamura. 2009. Drought in the southeastern United States: Causes, variability over the last millennium, and the potential for future hydroclimate change. *Journal of Climate* 22(19): 5021–5045.

Stafford, J.M., G. Wendler, and J. Curtis. 2000. Temperature and precipitation of Alaska: 50 year trend analysis. *Theoretical and Applied Climatology* 67(1): 33–44.

Woodcock, A.H. and R.H. Jones. 1970. Rainfall trends in Hawaii. *Journal of Applied Meteorology* 9: 690–696.

4 Desired Future Conditions for Groundwater Availability in the High Plains Aquifer System

Z. Sheng, C. Wang, J. Gastelum, S. Zhao, and J. Bordovsky

CONTENTS

4.1 INTRODUCTION

The High Plains Aquifer, also known as the Ogallala Aquifer, has been used extensively for agricultural production in the Great Plains. It covers part of eight states, extending from west Texas to South Dakota (USGS 2011a). The High Plains Aquifer provides water for agricultural production and urban uses. In 2000, a total of 24.2 billion cubic meters of groundwater (or at a rate of 66.2 million cubic meters per day, m^3/day) was pumped from the High Plains Aquifer (Maupin and Barber 2005), accounting for 30% of all groundwater used for irrigation in the United States (Guru and Horne 2000). Almost 97% of the pumped groundwater was used for irrigation; the remaining 3% was used for public, industrial, domestic, and other uses (Maupin and Barber 2005). In 2002, there were a total of 6.1 million ha of irrigated land in the High Plains area (USDA Census of Agriculture 2002; USGS 2011b), accounting for approximately one-fifth of all cropland in the United States. Alfalfa, corn, cotton, sorghum, soybeans, peanuts, and wheat are major crops in the High Plains region. These crops provide Midwest cattle operations with enormous amounts of feed and account for 40% of the feedlot beef output here in the United States (Guru and Horne 2000). The use of groundwater from the High Plains Aquifer has transformed this area into an important agricultural region that sustains more than one-fourth of the nation's agricultural production. Therefore, the current availability and future sustainable supply of high-quality groundwater are central to the overall health of the High Plains agricultural economy, the growth of its cities and rural communities, and the well-being of the ecosystem.

Historically, groundwater pumping since predevelopment (about the 1950s) has resulted in the depletion of groundwater in most of the aquifer area because pumping far exceeded its capacity to recharge and to capture the discharge. In addition, pumping has also caused depletion of surface water where the river is hydraulically connected with the shallow aquifer. Moreover, groundwater has also been contaminated due to long-term agricultural applications and other sources.

To secure water supplies for future agricultural production and community development, best management practice (BMP) strategies, including the adoption of high-efficiency irrigation technology, the Conservation Reserve Program of Natural Resources Conservation Service, and other programs have been implemented to slow down the depletion of groundwater storage and prevent contamination. Many projects have been conducted and are under way to gain a better understanding of the hydrological processes and develop better BMP strategies to extend the life of the aquifer and to sustain the economic development in the region. One of the recent attempts in preserving aquifer storage for future generations is to manage the utilization of the groundwater resources with the desired future conditions (DFCs). This chapter summarizes the current status of aquifers, impacts of historic pumping, and DFCs. Current DFCs aim at slowing down the depletion of aquifer storage, primarily based on water quantity and not water quality. In this chapter, we also examine factors that have influenced historic/current groundwater pumping practices and will have impacts on future groundwater availability and recommend a holistic analysis tool to link a hydrological model with an econometric model for the assessment of water users' behavior in response to groundwater depletion and socioeconomic factors.

4.2 PHYSICAL AND CULTURAL CHARACTERISTICS OF THE HIGH PLAINS AQUIFER

The High Plains Aquifer, the largest fresh aquifer system in the United States, covers 450,000 km^2 in parts of eight states—Colorado, Kansas, Nebraska, New Mexico, Oklahoma, South Dakota, Texas, and Wyoming (Figure 4.1; USGS 2011a). The aquifer is further divided into three regional subdivisions—Northern High Plains (NHP), Central High Plains (CHP), and Southern High Plains (SHP); there is little groundwater flow in the aquifer between the regional subdivisions (Figure 4.1; Weeks et al. 1988). The High Plains Aquifer extends from south of 32° to almost 44° north latitude and from east of 96°30′ to almost 106° west longitude. The land surface elevation of the High Plains area

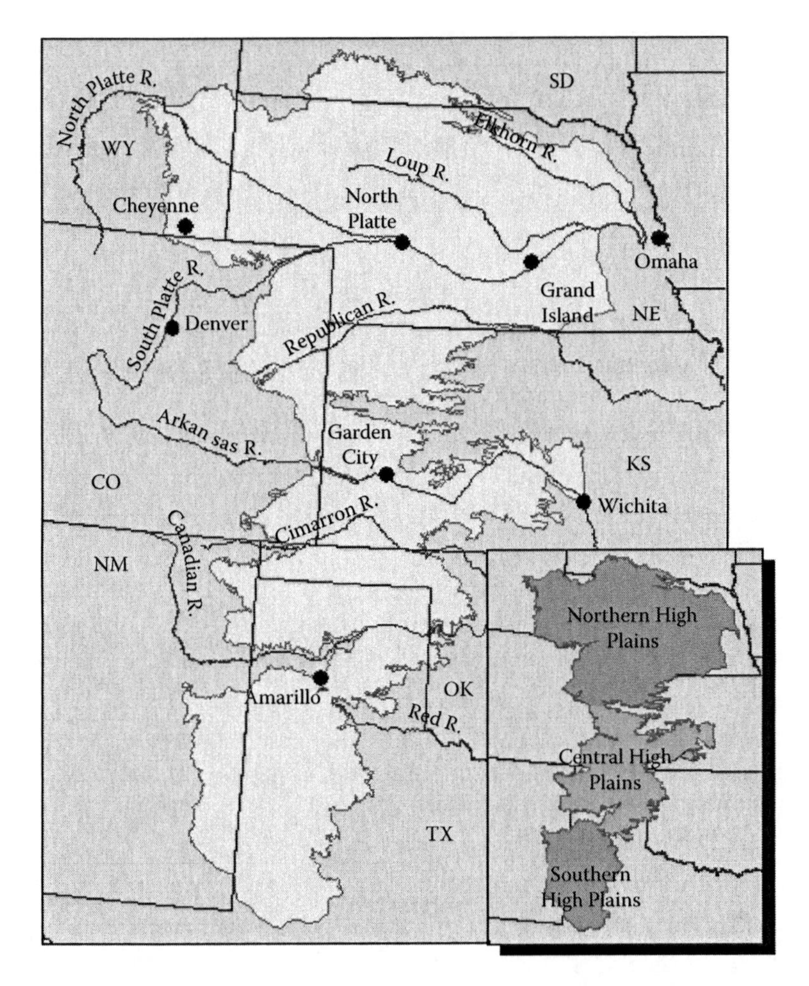

FIGURE 4.1 The High Plains Aquifer boundary and its subdivisions. (From USGS, High Plains Regional Ground Water Study, 2011, http://co.water.usgs.gov/nawqa/hpgw/images/figure1.jpg. Accessed September 30, 2011.)

ranges from about 2377 m above the National Geodetic Vertical Datum of 1929 (NGVD 29) on the western boundary to 354 m above NGVD 29 on the eastern boundary (USGS 2011a).

4.2.1 Climate

The High Plains area is characterized by a middle-latitude dry continental climate with abundant sunshine, moderate precipitation, frequent winds, low humidity, and a high rate of evaporation. Mean annual temperature ranges from about 6.1°C in the north to 17.2°C in the south. Mean annual precipitation in the region varies from 305 mm in the west to 838 mm in the east (Figure 4.2). Pan evaporation ranges from about 1524 mm/year in northern Colorado to about 2667 mm/year in southeastern New Mexico. During most years, in much of the area, irrigation is required for economic yields of typical crops—alfalfa, corn, cotton, sorghum, soybeans, peanuts, and wheat (USGS 2011b).

4.2.2 Hydrogeological Units

The High Plains Aquifer consists of all or parts of several geologic units of the Quaternary and Tertiary age, as shown in Table 4.1 (Weeks and Gutentag 1984). The Brule Formation of

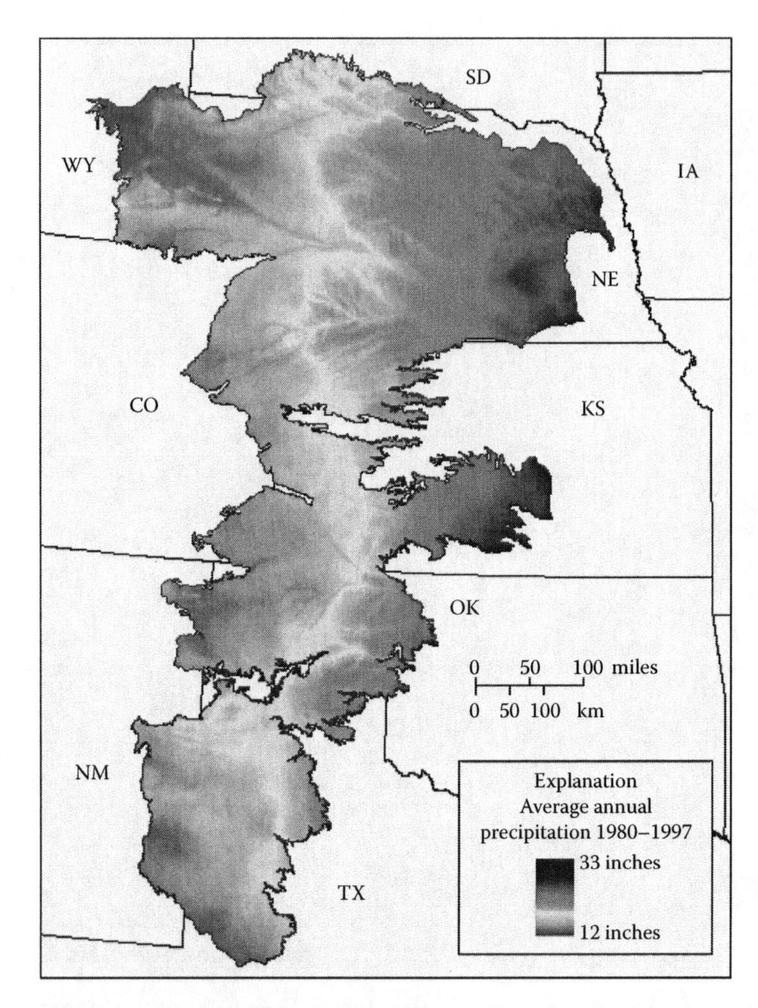

FIGURE 4.2 Precipitation distributions in the High Plains area. (From USGS, High Plains Water-Level Monitoring Study, 2011, http://ne.water.usgs.gov/ogw/hpwlms/physsett.html. Accessed September 30, 2011.)

Oligocene age is the oldest geologic unit included in the aquifer. The Brule Formation is the upper unit of the White River Group and is primarily massive siltstone with beds and channel deposits of sandstone. It includes lenticular beds of volcanic ash, clay, and fine sand. The Brule underlies much of western Nebraska and is included in the aquifer only where it has been fractured or where the formation contains solution openings. Such secondary porosity and permeability are developed only where the Brule crops out or is near the land surface (Figure 4.3; USGS 2011b).

The Arikaree Group of Miocene and Oligocene age overlies the Brule Formation and consists primarily of massive, very fine to fine-grained sandstone. The Arikaree Group crops out in western Nebraska and pinches out to the south and east, as does the White River Group, which includes the Brule Formation. The maximum thickness of the Arikaree is about 305 m in western Nebraska (Weeks and Gutentag 1984; USGS 2011b).

The Ogallala Formation of Miocene age is the principal geologic unit in the High Plains Aquifer and extends to the land surface throughout most of the aquifer area (Table 4.1). The Ogallala Formation consists of unconsolidated gravel, sand, silt, and clay. It also includes caliche, a hard deposit of calcium carbonate that precipitated when part of the groundwater that

TABLE 4.1
The Stratigraphic Columns of the High Plains Aquifer

System	Series	Geologic Unit	Thickness [meters (feet)]	Physical Character
Quaternary	Pleistocene and Holocene	Valley-fill deposits	0–18 (0–60)	Stream-laid deposits of gravel, sand, silt, and clay associated with the most recent cycle of erosion and deposition along present streams. Forms part of High Plains Aquifer when it is hydraulically connected to underlying Quaternary and Tertiary deposits
		Dune sand	0–91 (0–300)	Fine to medium sand with small amounts of clay, silt, and coarse sand formed into hills and ridges by the wind. Forms part of High Plains Aquifer when saturated
		Loess	0–76 (0–250)	Silt with lesser amounts of very fine sand and clay deposited as windblown dust
	Pleistocene	Unconsolidated alluvial deposits	0–168 (0–550)	Stream-laid deposits of gravel, sand, silt, and clay locally cemented by calcium carbonate into caliche or mortar beds. Forms part of High Plains Aquifer when it is hydraulically connected laterally or vertically to Tertiary deposits
Tertiary	Miocene	Ogallala formation	0–213 (0–700)	Poorly sorted clay, silt, sand, and gravel generally unconsolidated; forms caliche layers or mortar beds when cemented by calcium carbonate. Includes units equivalent to the locally used terms "Ash Hollow," "Kimball," "Sidney Gravel," and "Valentine" Members or Formations assigned to the Ogallala Formation or "Group," and Delmore and Laverne Formations. Ogallala comprises a large part of High Plains Aquifer when saturated
		Arikaree Group	0–305 (0–1000)	Predominantly massive, very fine to fine-grained sandstone with localized beds of volcanic ash, silty sand, siltstone, claystone, sandy clay, limestone, marl, and mortar beds. Includes units assigned to the Hemingford Group of Lugn (1938), Marsland Formation, Rosebud Formation used in South Dakota by Harksen and Macdonald (1969), and Sheep Creek Formation. Also includes units equivalent to Gering Formation, Harrison Sandstone, and Monroe Creek Sandstone. Forms part of the High Plains Aquifer
	Oligocene	White River Group	0–213 (0–700)	Upper unit, Brule Formation, predominantly massive siltstone containing sarrastone beds and channel deposits of sandstone with localized lenticular beds of volcanic ash, claystone, and fine sand. The Brule Formation is considered a part of the High Plains Aquifer only where it contains saturated sandstones or interconnected fractures. Lower unit, Chadron Formation, mainly consists of varicolored, bentonitic, loosely to moderately cemented clay and silt that contains channel deposits of sandstone and conglomerate

Source: Modified from Weeks, J.B. and Gutentag, E.D., In *Ogallala Aquifer Symposium II*, ed. G.A. Whetstone. Texas Tech University, Water Resources Center, Lubbock, TX, 1984.

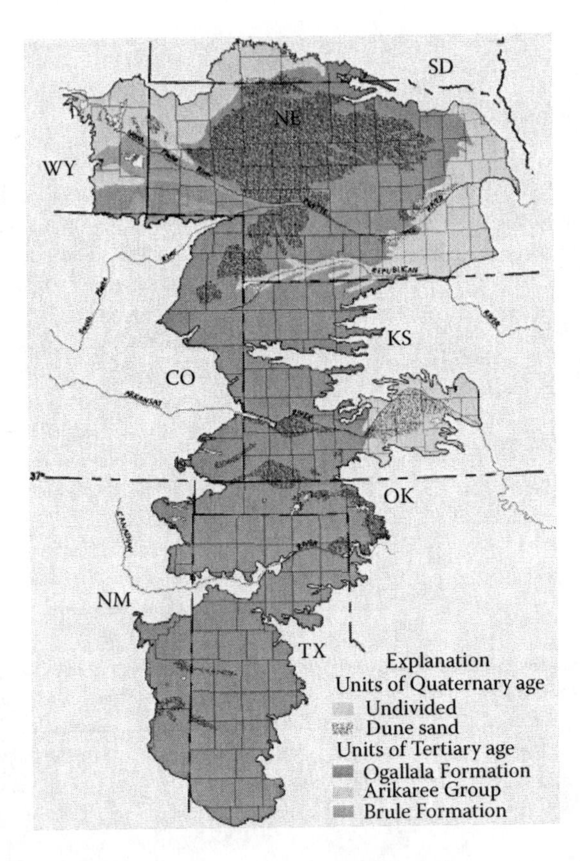

FIGURE 4.3 Hydrologic units of the High Plains Aquifer. (From USGS, High Plains Water-Level Monitoring Study, 2011, http://ne.water.usgs.gov/ogw/hpwlms/physsett.html. Accessed September 30, 2011.)

moved through the formation evaporated. During late Tertiary time, the Ogallala Formation was deposited by an extensive eastward-flowing system of braided streams that drained the eastern slopes of the Rocky Mountains over about 347,000 km^2 in eastern Colorado, Kansas, Nebraska, New Mexico, Oklahoma, South Dakota, Texas, and Wyoming (Weeks and Gutentag 1984; USGS 2011b).

Unconsolidated deposits of Quaternary age overlie the Ogallala Formation. These Quaternary deposits consist of gravel, sand, silt, and clay, much of which is reworked material that was derived from the Ogallala Formation. In places where these unconsolidated deposits are saturated, such as in southeastern Nebraska and south-central Kansas, they compose part of the High Plains Aquifer (Table 4.1). Deposits of loess, deposited as windblown material and consisting mostly of silt with small quantities of very fine-grained sand and clay, overlie the Ogallala Formation or the unconsolidated Quaternary sediments in some locations. Where the loess is thick, it forms the upper confining unit of the High Plains Aquifer. The dune sands of Quaternary age are most extensive in west-central Nebraska, where they cover about 51,800 km^2 and attain a maximum thickness of about 91 m. Saturated dune sands are also part of the High Plains Aquifer south of the Arkansas River in southwest and south-central Kansas. The dune sands are highly porous and, therefore, quickly absorb rainfall that recharges the High Plains Aquifer. Valley-fill deposits along the channels of streams, such as the Platte and the Arkansas Rivers, are also considered to be part of the High Plains Aquifer when they are hydraulically connected with the underlying Quaternary and Tertiary deposits (Weeks and Gutentag 1984). The High Plains Aquifer is underlain by rocks that range in age from Tertiary to Permian.

4.2.3 Population, Land Use, and Agricultural Production

The population of the High Plains area grew more than double from 0.9 million people in 1900 to 2.3 million people in 2000, of which 40% live in the cities with populations greater than 20,000 (U.S. Census Bureau 2000). Many of these cities are located near the major rivers or near the aquifer boundary. Based on USGS (1992) National Land-Cover Data, in the High Plains area 55.6% of land is rangeland, 40.9% agricultural, and 3.5% a combination of wetlands, forest, urban, water, and barren lands. The agricultural lands include 53% row crops (such as sorghum, corn, and cotton), 33% small grains (mostly wheat), and 14% pasture, alfalfa, and fallow lands; and 28% of the agricultural land is irrigated. Crops produced in the High Plains area account for a substantial percentage of the total crop production for the United States. In 1997, the High Plains area yielded 19% of the wheat, 19% of the cotton, 15% of the corn, and 3% of the sorghum of the nation. In addition, the High Plains area also produced about 18% of the total cattle production and an increasing percentage of total swine production in the United States (USDA Census of Agriculture 1998).

4.3 CURRENT STATUS OF HYDROLOGICAL CONDITION AND IMPACTS OF GROUNDWATER DEVELOPMENT

A good knowledge of current hydrological conditions and impacts of historic groundwater development is a key to securing future water supplies for the growing water demands in the region. Large-scale groundwater development in the aquifer has caused great water level drops in SHP, CHP, and in the southern part of the NHP Aquifer. Agricultural recharge has also increased the concentrations of dissolved solids, nitrates, pesticides, and other pollutants, especially in the shallow groundwater. In the area where surface water and groundwater are hydraulically connected, groundwater pumping has also caused depletion of surface water.

4.3.1 Depth to Water

The depth to water in a particular area is the difference between the altitude of land surface and the altitude of the groundwater surface. The generalized depth to water in the High Plains Aquifer in 2000 is shown in Figure 4.4 (McMahon et al. 2007). In most places, the water levels shown are lower than those that existed before widespread irrigation withdrawals began. The depth to water in the High Plains Aquifer is less than 30 m in about one-half of the area of the aquifer and less than 61 m in most of Nebraska and Kansas. The depth to water generally is less near the Platte and the Arkansas Rivers than in areas farther from the rivers, because the rivers are hydraulically connected to the aquifer through the stream valley aquifers that parallel the rivers. The groundwater surface is between 61 and 91 m below the land surface in parts of western and southwestern Nebraska and in parts of southwestern Kansas. The depth to water is as much as 122 m in a small area in southwestern Kansas where development of the aquifer began earlier than in most parts of Kansas; consequently, water-level declines are greater (USGS 2011b; McMahon et al. 2007).

4.3.2 Saturated Thickness and Well Yield

The saturated thickness of an aquifer is the vertical distance between the groundwater surface and the base of the aquifer and is one of the factors that determine the quantity of water that can be pumped from a well. Other factors that affect well yield include well construction and the hydraulic properties of the aquifer. The characteristics of the High Plains Aquifer are listed in Table 4.2.

The saturated thickness of the High Plains Aquifer in 1980 ranged from 0 (where the sediments that compose the aquifer were unsaturated) to about 305 m (Weeks and Gutentag 1984). The greatest saturated thickness is in north-central Nebraska, where the aquifer consists of the Ogallala Formation and overlying dune sands. Locally in southwestern Kansas, dissolution of salt in the

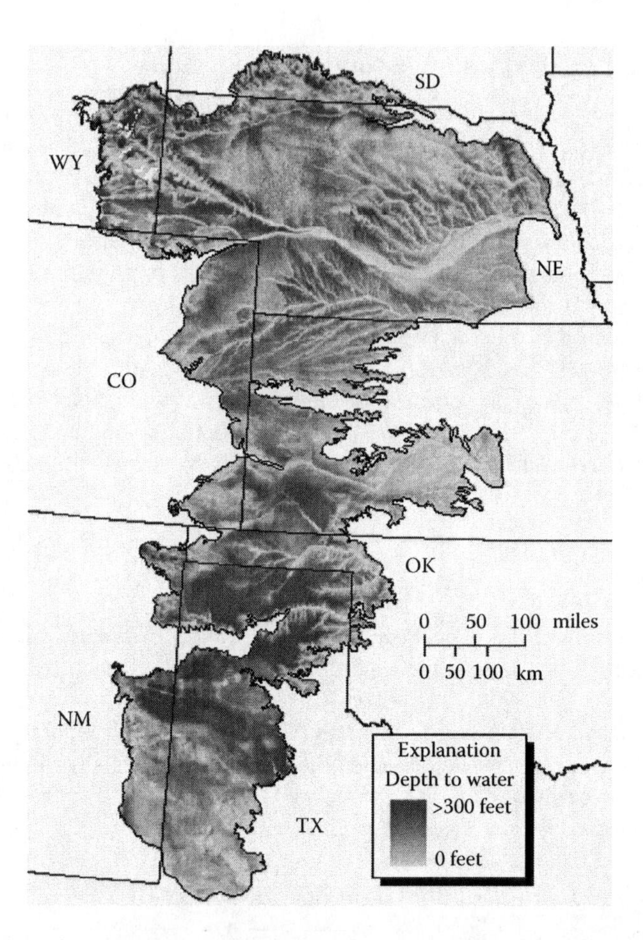

FIGURE 4.4 Depth to water in 2000 in the High Plains Aquifer. (From McMahon, P.B., Dennehy, K.F., Bruce, B.W., et al., U.S. Geological Survey, Professional Paper 1749, Reston, VA, 2007; USGS, High Plains Water-Level Monitoring Study, 2011, http://ne.water.usgs.gov/ogw/hpwlms/physsett.html.)

Permian bedrock that underlies the High Plains Aquifer has resulted in collapse features that were filled with younger sediments. These anomalously thick accumulations of sediments coincide with thick sequences of saturated aquifer materials. The average saturated thickness of the entire aquifer in 2009 was about 57 m. In Nebraska, the average saturated thickness was 104 m, but in Kansas, it was only about 27 m (Table 4.2; McGuire 2011).

Changes in the saturated thickness of the High Plains Aquifer have resulted from groundwater development. Saturated thickness has decreased in most places, but in two areas in south-central Nebraska, recharge to the aquifer from surface-water irrigation, combined with the downward leakage of water from canals and reservoirs, has increased the saturated thickness. In large areas in southwestern Kansas, large-scale irrigation development has decreased the saturated thickness of the aquifer by more than 25%. Decreases of more than 10% in saturated thickness result in a decrease in well yields and an increase in pumping costs because of the increased depth at which the pump must be set to lift the water (Weeks and Gutentag 1984).

The quantity of drainable water in the aquifer can be estimated by multiplying the volume of saturated material by the average specific yield of 0.15 (Weeks and Gutentag 1984; McGuire 2011). According to an estimate by McGuire (2011), there were about 3583 billion m^3 of drainable water stored in the High Plains Aquifer in 2009. About 67% of the drainable water in storage in the entire aquifer is in Nebraska and 11% in Texas. The remaining 22% drainable water is in Colorado, Kansas, New Mexico, South Dakota, Oklahoma, and Wyoming (Table 4.1). Not all drainable water

TABLE 4.2
Characteristics of the High Plains Aquifer

Characteristics	Total	CO	KS	NE	NM	OK	SD	TX	WY
Area underlain by aquifer (km^2)	450,790	38,591	78,995	164,854	24,476	19,037	12,303	91,816	20,720
% of total aquifer area	100	8.6	17.5	36.6	5.4	4.2	2.7	20.4	4.6
% of each state underlain by aquifer	—	14	38	83	8	11	7	13	8
Ave. area weighted saturated thickness for predevelopment (m)	60.9	25.4	33.8	104.2	18.5	43.1	63.3	43.8	55.5
Avg. area weighted saturated thickness in 2009 (m)	56.7	21.3	26.9	104.0	13.9	39.3	63.3	32.6	55.4
Volume of drainable water in storage in 2009 (BCM)	3,583	120	324	2,412	53	120	66	409	79

Source: Modified from Weeks, J.B. and Gutentag, E.D., In *Ogallala Aquifer Symposium II*, ed. G.A. Whetstone. Texas Tech University, Water Resources Center, Lubbock, TX, 1984; USGS, High Plains Water-Level Monitoring Study, 2011, http://ne.water.usgs.gov/ogw/hpwlms/physsett.html. Accessed September 30, 2011; Updated using the data from McGuire, V.L., U.S. Geological Survey Scientific Investigations Report 2011-5089, 13 pp, 2011, http://pubs.usgs.gov/sir/2011/5089/. Accessed August 31, 2011.

in storage within the aquifer can be recovered for use. The quantity of water that can be recovered from the aquifer varies with location, depending on the lithology, saturated thickness, hydraulic conductivity, specific yield of the aquifer at that location, and well construction. Water has been almost completely removed from about 29% of the formerly saturated aquifer material in Texas, whereas the quantity of material dewatered in Nebraska is negligible (Table 4.1).

The greatest yields of water generally are obtained from wells that are completed in coarse-grained aquifer material in places where the saturated thickness of the High Plains Aquifer is great. The potential yield of wells is greater than 47.3 liters per second (L/sec) in most of Nebraska and large parts of Kansas. A well capable of producing 47.3 L/sec can irrigate 51 ha and effectively supply one center-pivot irrigation system. Well yields vary from one formation to the other. Yields from the Brule Formation typically are less than 18.9 L/sec. Wells completed in the Arikaree Group generally do not yield large quantities of water but might yield as much as 22 L/sec in western Nebraska, where the saturated thickness is about 61 m. Well yields from the Brule Formation and the Arikaree Group are greatest where secondary porosity, such as fractures or solution openings, has been developed in the rocks. Well yields from the Ogallala Formation are 63 L/sec from 30 m of saturated sand and gravel in many parts of Kansas and Nebraska but are only 6.3 L/sec from 6 m of saturated sand and gravel in western Kansas (Weeks and Gutentag 1984).

4.3.3 Groundwater Flow

Water in the High Plains Aquifer generally is under unconfined conditions. Locally, water levels in wells completed in some parts of the aquifer may rise slightly above the regional groundwater surface because of artesian pressure created by local confining beds. The altitude and configuration of the groundwater surface of the High Plains Aquifer are shown in Figure 4.5. The configuration and slope of the groundwater surface are similar to the configuration and slope of the land surface. Water in the aquifer generally moves from west to east, or perpendicular to the contours and in the direction of the arrows shown in Figure 4.5. Water moves in response to the slope of

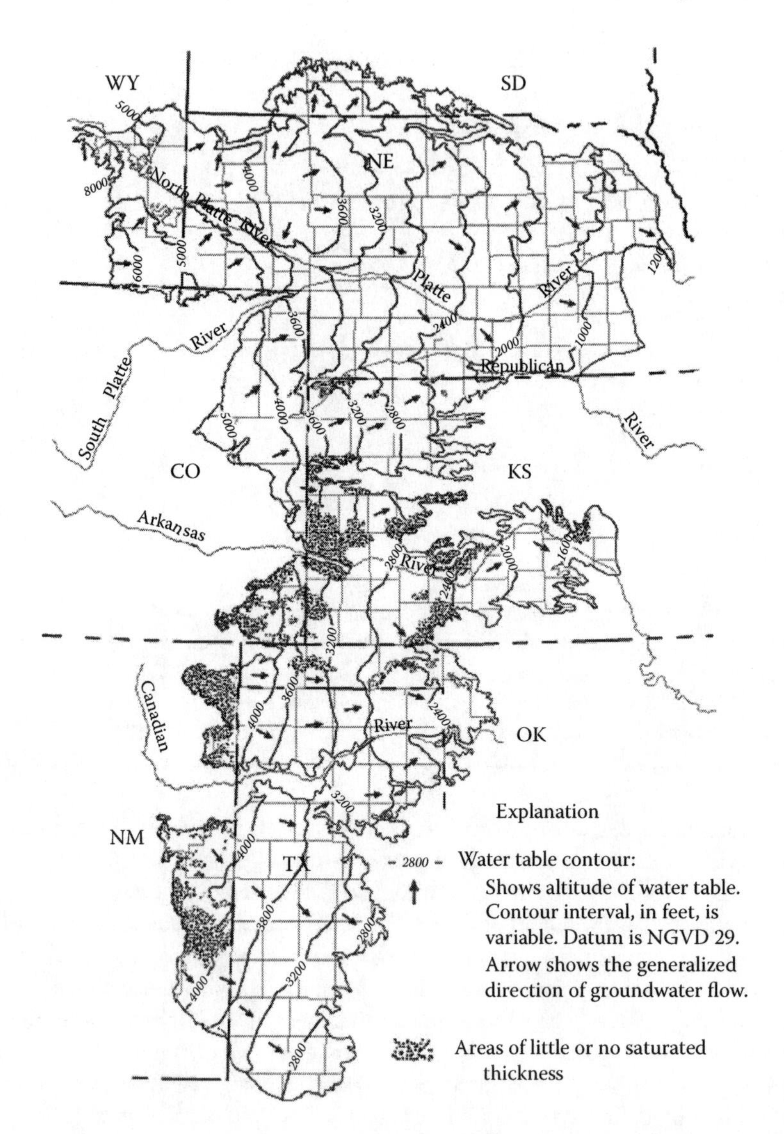

FIGURE 4.5 Predevelopment water level contours and direction of groundwater flow. (From Gutentag, E.D., Heimes, F.J., Krothe, N.C., et al., U.S. Geological Survey Professional Paper 1400-B, Reston, VA, 1984; USGS, High Plains Water-Level Monitoring Study, 2011b, http://ne.water.usgs.gov/ogw/hpwlms/physsett.html.)

groundwater surface, which typically averages between 1.89 and 2.84 m/km. On the basis of this average slope and aquifer hydraulic properties, the velocity of water that moves through the aquifer is estimated to average about 0.3 m/day. The spacing of the groundwater level contours is affected by different hydrologic conditions. For example, where contours are widely separated, such as in western Nebraska, the slope of the groundwater surface is gentler than where the contours are more closely spaced. Widespread recharge to the aquifer by infiltration of precipitation through dune sands occurs in western Nebraska, and, thus, the slope of the groundwater surface there is relatively gentle and water will move much slower than in other areas (USGS 2011b).

Where the groundwater level contours cross streams, the configuration of the contours indicates the relation of the water in the aquifer to the water in the stream. For example, where the contours from 975 to 1220 m cross the North Platte River in western Nebraska, the contours bend upstream (Figure 4.5). This upstream flexure indicates that water moves from the aquifer to the stream, and the North Platte River is a gaining stream in this area. By contrast, where the 610 m contour crosses the Platte River in west-central Nebraska, a slight downstream bend in the contour indicates that water is moving from the stream to the aquifer; the Platte River is a losing stream in this area, and the water from the river recharges the aquifer. In southwestern Kansas, the Bear Creek and the Crooked Creek Fault Zones (Figure 4.5) have displaced the High Plains Aquifer and little or no saturated thickness of the aquifer exists on the upthrown side of the faults. In these areas, the groundwater level contours end abruptly at the faults (Figure 4.5; USGS 2011b).

4.3.4 Recharge and Discharge

In an undisturbed groundwater flow system, the amount of water that moves into an aquifer (recharge) is equal to the amount of water that moves out of the aquifer (discharge) over a specific time period, and the flow system is in equilibrium. Prior to development in the High Plains Aquifer, the groundwater level of the aquifer and the quantity of water stored in the aquifer vary little in response to changes in precipitation, stream flow, and the amount and types of vegetation. A groundwater flow system is no longer in equilibrium when the long-term discharge is not equal to the long-term recharge. The altitude of the groundwater surface rises when the recharge rate exceeds the discharge rate and declines when the discharge rate exceeds the recharge rate. Withdrawal of large quantities of groundwater by wells and redistribution of surface water in ditches and canals, all for irrigation purposes, have changed the natural recharge and discharge of the High Plains Aquifer (Weeks and Gutentag 1984).

Recharge to the High Plains Aquifer is primarily by the infiltration of precipitation and locally by the infiltration from streams and canals. Some surface water that is used for crop irrigation also percolates downward and recharges the aquifer. A small quantity of water from the underlying bedrock moves upward and mixes with water in the High Plains Aquifer; this water is also considered to be recharge. The aquifer is recharged at total rates that range between 1.27 and 152 mm/year in Nebraska and Kansas. The rates of recharge are highly variable and range from about 0.3% to 20% of the average annual precipitation in the dry and wet parts of these states. The greatest rates of recharge by precipitation are in areas where dune sand or other highly permeable material is at the land surface. Recharge by infiltration of stream flow usually is greatest when stream flow is high and, thus, provides a large difference between stream and aquifer water levels (Weeks and Gutentag 1984). According to a comprehensive review of published research on the connection between playas and the High Plains Aquifer, most studies indicate that recharge rates are significantly higher beneath playas than in the surrounding nonplaya environment (Gurdak and Roe 2009). Recharge rates, however, do vary from playa to playa. Characteristics that affect recharge include playa size and depth, size of a playa's drainage area, depth of the clay that lines the basin, depth of sediment that has accumulated in the basin, subsurface sediment between the playa and the aquifer, and depth of the groundwater surface.

Natural discharge from the High Plains occurs at springs, seeps, streams, and as evapotranspiration by plants. Where the groundwater surface is near the land surface, groundwater can evaporate directly. Transpiration rates are greatest along stream valleys where deep-rooted salt cedar, willows, cottonwoods, and sedges grow. Where the High Plains Aquifer is locally underlain by permeable bedrock and the groundwater surface in the aquifer is higher than that in the bedrock, small amounts of water move downward from the aquifer into the bedrock (Weeks and Gutentag 1984).

Large quantities of water are withdrawn from the aquifer by wells, and in some areas large amounts of water discharge from the aquifer to streams. For example, a study was done during 1975 to determine how much of the flow of the Platte River in Nebraska was derived from groundwater. The annual gain in stream flow within Nebraska was about 3.7 billion m^3, most of which was groundwater discharge from the High Plains Aquifer to the river. Large quantities of surface water are diverted from the Platte River and used for irrigation; thus, the amount of groundwater discharge to the river probably was significantly greater than the measured gain in stream flow.

Most of the discharge from the High Plains Aquifer is by withdrawals from wells, and practically all of the water withdrawn is used for irrigation purposes (Weeks and Gutentag 1984).

4.3.5 Groundwater Quality

The chemical quality of water in the High Plains Aquifer is affected by many factors. These factors include the chemical composition and solubility of aquifer materials, the increase in dissolved-solids concentrations in groundwater in areas where the water discharges by evapotranspiration, and the chemical composition of water that recharges the aquifer. Groundwater generally contains smaller concentrations of dissolved minerals near recharge areas where the residence time of the water in the aquifer has been short, and, therefore, dissolution of aquifer minerals has been less. The water generally is more mineralized near discharge areas because residence time has been longer and more dissolution of minerals has taken place.

The dissolved-solids concentration in groundwater is a general indicator of the chemical quality of the water. Dissolved-solids concentrations in groundwater from the High Plains Aquifer are less than 500 milligrams per liter (mg/L) in most of Kansas and Nebraska but locally exceed 1000 mg/L in both states. The limit of dissolved solids recommended by the U.S. Environmental Protection Agency for drinking water is 500 mg/L. Most crops can tolerate water in which the dissolved-solids concentration is 500 mg/L or less. In places with well-drained soils, many types of crops can tolerate water with a dissolved-solids concentration between 500 and 1500 mg/L. In southwestern and south-central Kansas, the High Plains Aquifer overlies Permian bedrock that contains bedded salt. Where circulating groundwater has dissolved some of this salt and the mineralized water has subsequently moved upward into the High Plains Aquifer, the dissolved-solids concentration of the water in the High Plains Aquifer is greatly increased. Also, dissolved-solids concentrations are generally greater near streams where water from the High Plains Aquifer discharges. Groundwater near the streams is shallow enough to be transpired by plants or to be evaporated directly from the soil. Concentrations of dissolved solids in the groundwater are increased by the evapotranspiration process. Rates of transpiration are greatest where deep-rooted phreatophytes, such as sedges, cottonwood, willows, and salt cedar, grow.

Excessive concentrations of sodium in water adversely affect plant growth and soil properties, and constitute salinity and sodium hazards that may limit irrigation development. Sodium that has been concentrated in the soil by evapotranspiration and ion exchange decreases the tillability and permeability of soil. Areas of high or very high sodium hazard occur in parts of Kansas. Sodium hazard is evaluated by the sodium adsorption ratio, which relates the concentration of sodium to calcium plus magnesium; if this ratio is high, then the sodium can destroy any clay in the soil and thus affect soil structure. Concentrations are less than 25 mg/L in most of Nebraska and northern Kansas. Concentrations are greatest in southwestern Kansas where evapotranspiration rates are high and in south-central Kansas where the High Plains Aquifer overlies Permian bedrock that contains saline water derived from the partial dissolution of salt beds. Sodium concentrations are high along the Platte and the Republican Rivers where the evapotranspiration rates also are high. Salinity and sodium hazards are generally low in Nebraska where the High Plains Aquifer primarily consists of sand and gravel, which contain few sodium-bearing minerals (Gurdak et al. 2009).

Excessive fluoride concentrations are a widespread problem in water from the High Plains Aquifer. Some of the fluoride is derived from the dissolution of fluoride-bearing minerals in parts of the aquifer that contain sand and gravel, such as the Ogallala Formation. Extremely large concentrations (2–8 mg/L) of fluoride are reported where the aquifer contains volcanic ash deposits or where it is underlain by rocks of Cretaceous age. Large concentrations of fluoride in drinking water cause staining of teeth, but fluoride is not a concern in irrigation water (Gurdak et al. 2009).

The generally shallow depth of the groundwater surface in the High Plains Aquifer makes water in the aquifer susceptible to contamination. Application of fertilizers and organic pesticides to cropland has greatly increased since the 1960s, thereby increasing the amount of potential contaminants present in the soil. Increased concentrations of sodium, alkalinity, nitrate, and triazine (a herbicide) have been found in water that underlies small areas of irrigated croplands in Nebraska and Kansas. Of the 132 wells sampled during 1984–1985 in Nebraska, 43 had measurable concentrations (>0.04 micrograms per liter, μg/L) of the herbicide atrazine. Increased concentrations of 2,4-dichlorophen-oxyacetic acid (2,4-d, a pesticide) were found in water that underlies rangeland in a small part of the Great Bend area of the Arkansas River in Kansas (Gurdak et al. 2009).

4.3.6 Impact of Groundwater Development

4.3.6.1 Water Level Drop: Depletion of the Aquifer

Decline of the groundwater level started after the implementation of irrigation development in the High Plains Aquifer. The acreage irrigated with groundwater increased from 0.85 million ha in 1949, to 2.47 million ha in 1959, to 3.64 million ha in 1969, and to 5.54 million ha in 1980 (Gutentag et al. 1984). For the period 1980–2005, irrigated acreage had minor fluctuations ranging from 5.14 million ha (2002) to 6.27 million (2005) (McGuire et al. 2003).

In 2000, almost 97% of the total withdrawals (66.2 million m^3/day or an annual total of 24.2 billion m^3) from the High Plains Aquifer were used for irrigation; the remaining 3% were used for public, industrial, domestic, and other uses. Nebraska, Texas, and Kansas accounted for 88% of the total withdrawals, which was almost entirely used for irrigation (Maupin and Barber 2005). SHP started experiencing water level declines by 1940, the CHP by 1950, and the NHP by 1960. From predevelopment until 1980, the highest water level drops in Texas were around 61 m, and the maximum declines in CHP of southern Kansas exceeded 31 m but were registered only in two small areas in that region. Water level declines of more than 15 m were detected in different small sections of CHP of southwestern Kansas, the north central panhandle of Texas, the central panhandle of Texas, and the central panhandle of Oklahoma (McGrath and Dugan 1993). During the period of 1980–1999, considering water levels from 4818 wells in the High Plains Aquifer, the water level variation ranged from 10 m rise to 20 m decline (McGuire 2001). The average area-weighted water level decline was 1 m as compared with 3 m experienced during the period from predevelopment to 1980. South of the Canadian River in New Mexico and Texas and an area southwest of Kansas had water level declines of more than 18 m, the highest in the High Plains Aquifer (McGuire 2001).

From predevelopment until 2009, water level declines of more than 3, 7.6, and 15.2 m were present, respectively, in 26%, 18%, and 11% of the aquifer. Only 2% of the aquifer registered water level increases of more than 3 m. From predevelopment to 2009, the aquifer has had a water storage decline of around 338 billion m^3. As a result, the total drainable water storage in the aquifer was approximately 3583 billion m^3 (McGuire 2011). Figure 4.6 depicts groundwater level changes from predevelopment to 2009. This figure shows that the highest water level declines occurred in Texas and the southwest portion of Kansas. Water level rising occurred in Nebraska perhaps as a result of irrigation using surface water as well seepage from the irrigation network. The three states exclusively located in the NHP, Nebraska, South Dakota, and Wyoming, have had virtually no change in average saturated thickness since predevelopment. The larger saturated thickness, larger recharge rates, smaller consumptive irrigation requirements, and later development of groundwater irrigation have contributed to less depletion of groundwater storage in the region. The acreage-weighted average saturated thickness in Colorado, Kansas, and Oklahoma, which covers most of the CHP area, was less than 30 m or about 85% of the predevelopment saturated thickness. The average saturated aquifer thickness in New Mexico and Texas in the SHP was less than 29 m or less than 80% of the predevelopment saturated thickness. The larger declines in saturated thickness in the SHP than the other High Plains Aquifer areas can be attributed to earlier groundwater development and less recharge (Dugan et al. 1994).

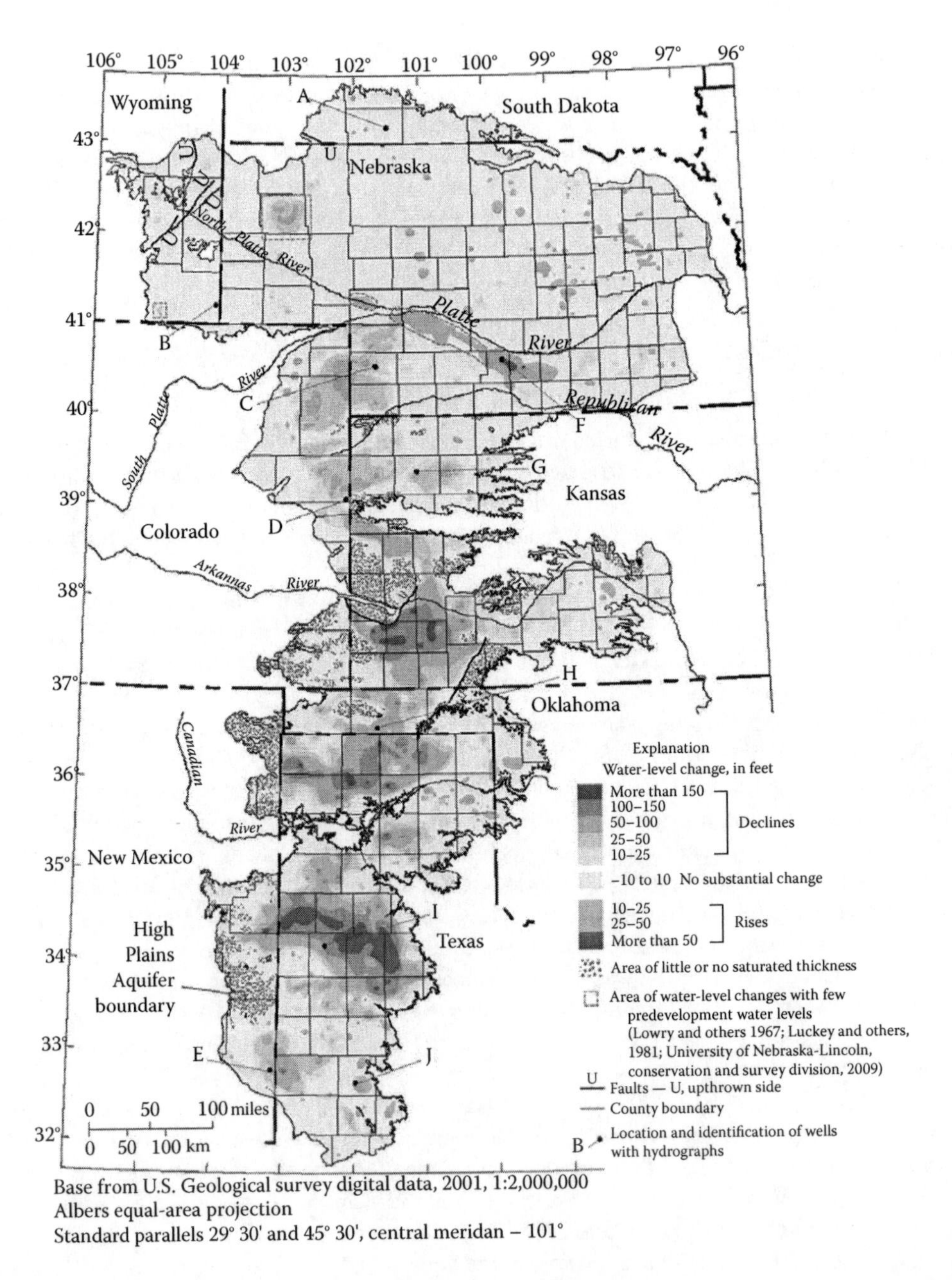

FIGURE 4.6 Water level changes from predevelopment to 2009. (From McGuire, V.L., U.S. Geological Survey Scientific Investigations Report 2011-5089, 13 pp, 2011, http://pubs.usgs.gov/sir/2011/5089/. Accessed August 31, 2011.)

4.3.6.2 Deterioration of Water Quality and Soil Salinity

Groundwater sustainability of an aquifer depends on both water quantity and water quality. In the High Plains Aquifer, deep groundwater has been recharged during the past several thousand years, while shallow groundwater, near the groundwater surface and underneath irrigated land, has been recharged during the last 50 years. The water of the High Plains Aquifer has been affected by irrigation farming activities. Agricultural recharge has increased the concentrations, mainly in the shallow groundwater, of dissolved solids, nitrate, pesticides, and other pollutants.

The transport of chemicals from land surface to water could take decades to centuries; however, natural depressions could reduce the transport time of pollutants to the groundwater surface (Gurdak et al. 2009).

For the period 1999–2004, the U.S. Geological Survey's National Water-Quality Assessment Program (NAWQA) offered the first systematic effort to assess water quality of the High Plains Aquifer in three subdivisions, NHP, CHP, and SHP, respectively (Gurdak et al. 2009). According to the study, based on the maximum contaminant level (MCL) established by the federal national drinking standards, the Ogallala Formation of the SHP has the poorest domestic well water quality, while the central Ogallala Formation in the CHP and NHP had the better water quality. Most of the exceedance values in the MCL of national standards were for arsenic, dissolved solids, fluoride, iron, manganese, and nitrate. Arsenic values exceeded primary federal standard values (MCL = 10 μg/L) in at least one water sample of each of the hydrogeologic units except for the CHP. The average value of arsenic concentration in the aquifer was close to 4.91 μg/L, and the highest percentage, around 59%, of arsenic concentration above the MCL was on the SHP (McMahon et al. 2007).

The median nitrate concentration values remained very similar, around 2 mg/L as nitrogen, for the period 1930–1970, and increased to almost 3 mg/L for the period 1980–1990 (Litke 2001; McMahon et al. 2007). For the period 1999–2004, the average nitrate concentration values for the three regions were close to 2.8 mg/L, with a maximum value of 20.3 mg/L in the CHP region (Ogallala formation). Atrazine, deethylatrazine, and chloroform were the most commonly detected pesticides and volatile organic compound, respectively; however, none of them exceeded primary drinking water standards (McMahon et al. 2007).

Elevated salinity in irrigation water is greatest in the southern part of the High Plains Aquifer. The total dissolved-solids concentrations of 3% of samples exceeded 2000 mg/L, which means severe salinity restrictions should be applied if the water is used for irrigation. No salinity restrictions were exceeded for samples from the northern part of the High Plains Aquifer. About half of the samples from the central and southern parts of the aquifer could be characterized as having slight to moderate salinity with restrictions on the use of irrigation water (McMahon et al. 2007).

The USDA (1988) Soil Conservation Service identified the High Plains region as an area of high potential for water and soil salinity problems based on spatial scales of river basins or watersheds and salt indicators, which include Total Dissolved Solids (TDS) and chloride in waters and salts in surface soils and geologic formations. Soil salinity is more widespread in the northern Plains than anywhere else in the United States. Salinity adversely affects crop growth in the region's most northern states. Saline conditions in the root zone severely affect nearly 10% of the northern Plains landscapes (USDA 1996). The areas of the saline-affected soils were based on the presence of a horizon with greater than 4 dS/m within 50 cm of the land surface. The electrical conductivity estimates of the soils followed the saturated paste method of the Soil Survey Staff (1995).

4.3.6.3 Effects on Surface Water and Ecosystem

Groundwater depletion can drastically affect rivers and streams, such as turning once perennial rivers into ephemeral ones (Stromberg et al. 1996). Groundwater pumping in the High Plains Aquifer has resulted in a decrease in base flow or discharge of the aquifer into the streams, especially in Kansas and Nebraska. During extreme droughts such as the ones in the 1950s, heavy groundwater pumping could even turn a gaining stream into a losing stream. This sudden decrease threatened the riparian ecosystem. Usually, the first species to be affected by the drop in groundwater levels are the water-sensitive plants, like some wetland and riparian plants. Such plants as juvenile *Salix gooddingii*, juvenile *Populusfremontii*, and juvenile *Fraxinusvelutina* only have a range of about 1.8 m change in depth to groundwater. This means that these plants will not be able to survive if the groundwater level drops more than 1.9 m from the surface (Stromberg et al. 1996). Younger plants are also generally more affected by the lack of groundwater than the older plants (Stromberg et al. 1996). Currently, in some part of the High Plains Aquifer, groundwater has dropped more than 45 m from predevelopment conditions (McGuire 2011). Hence, in areas around the High Plains Aquifer, such riparian ecosystems have already been destroyed. Similar conditions have been observed throughout the western United States. Since these species have such a narrow range of tolerance for water table changes, any decrease in groundwater surface can result in a decrease in the biodiversity of the ecosystem. In addition, as groundwater levels drop and native species begin to die, new exotic, nonnative species will begin to invade the ecosystem. These invasive species have a higher tolerance for a decrease in groundwater and can better adapt to this environment. Consequently, an invasion by exotic species greatly exacerbates the decrease of biodiversity in an ecosystem (Stromberg et al. 1996).

4.4 GROUNDWATER BUDGET, GROUNDWATER AVAILABILITY, AND DFCs

To determine the DFC, we need to know the current groundwater storage and groundwater budget. However, not all the groundwater stored in the aquifer will become available due to technological, environmental, economic, and legal constraints.

4.4.1 Groundwater Budget and Availability

Assuming that the groundwater budget in an aquifer system is in long-term equilibrium, under predevelopment conditions, inflow (recharge) into the aquifer system is approximately equal to outflow (discharge). The quantity of water stored in the aquifer system is constant or varies about some average condition in response to annual or longer-term climatic variations (Alley et al. 1999). The groundwater budget can be described in terms of rates (or volumes over a specified period of time) as (Alley et al. 1999; Sheng 2011)

$$\sum Q_i^{in} - \sum Q_i^{out} = \Delta S, \tag{4.1}$$

where Q_i^{in} and Q_i^{out} are, respectively, inflow and outflow rates of the groundwater system and ΔS is the change in storage, reflecting annual or longer-term climatic variations. Under natural conditions, inflow (recharge) includes (1) areal recharge from precipitations through an unsaturated zone to the groundwater surface (Q_r); (2) recharge from losing streams, lakes, and wetlands (Q_s); and (3) subsurface (groundwater) inflow or interbasin exchange (Q_{fin}): while outflow (discharge) includes (1) discharge to streams, lake, wetlands, and springs (Q_d); (2) evapotranspiration (Q_e); and (3) subsurface outflow (Q_{fout}) (ASCE 1987; Alley et al. 1999; Sheng 2011).

Such equilibrium can be broken by pumping and artificial recharge using imported water, treated wastewater, or storm water. If only considering the effects of pumping, the groundwater budget can then be written as

$$\sum Q_i^P = \sum Q_i^{in} - \sum Q_i^{out} - \Delta S. \tag{4.2}$$

As a dynamic system, pumpage Q_i^P will be achieved by increasing inflow Q^{in}, for example, by increased seepage, decreasing outflow Q_i^{out}, decreasing storage S, or a combination of all three.

The inflows can be further expanded as

$$\sum Q_i^{in} = Q_r + Q_s + Q_{fin}, \tag{4.3a}$$

$$\sum Q_i^{out} = Q_d + Q_e + Q_{fout}. \tag{4.3b}$$

All the inflow and outflow terms on the right sides of the above equations are the sum of each component over the whole basin during a specific time period of interest. For example, Q_r is the recharge from precipitation through infiltration and deep percolation. It is controlled by land cover, hydrological properties of geological materials, intensity, and duration of precipitation. Irrigation (crops, consumptive uses, irrigation scheduling) and urbanization (pavement) affect soil moisture, infiltration, deep percolation, and eventually influence recharge to the aquifer. Native vegetation covers have been replaced by invasive plants, in turn affecting recharge, especially mountain front recharge. Climate variations could have altered precipitation patterns, resulting in variable recharge. Other terms also vary to different degrees in response to human activities and climate variability. Therefore, groundwater budget terms inherit a great deal of uncertainty due to the heterogeneity of materials and lack of historical data. It should be noted that the above equations only consider water quantity; they do not consider the effects of water quality on water availability.

Once pumping starts, the aquifer system will adjust itself and try to reach a new equilibrium after some time, called the aquifer system response time, which is defined as the time taken for the water level and storage changes throughout the aquifer system to become negligible after an increase or decrease in supply withdrawal (Walton 2011) or the time taken to full capture (Bredehoeft and Durbin 2009). The aquifer response time can range from days to centuries or more (Bredehoeft et al. 1982; Sophocleous 2000; Alley et al. 2002; Bredehoeft and Kendy 2008), depending on many factors, such as aquifer system dimensions, aquifer transmissivity and storativity, confining layer storage, confining layer leakage, aquifer system boundary conditions, and well location and penetration.

Even though the groundwater budget can be calculated or estimated with acceptable accuracy, water availability or sustainability has proved to be an elusive and multifaceted concept to define in a precise manner and with universal applicability. *Water availability* is not a simple function of the quantity and quality of water in an aquifer system but is also constrained by the physical structures, laws, regulations, and socioeconomic factors that control its demand and use (Alley and Leake 2004). Therefore, determining *groundwater availability* means more than calculating the volume of groundwater underlying a particular area or within an aquifer. One must also consider that some of the water may not be economically recoverable or of poor quality as well as the fact that groundwater is connected to the rest of the hydrologic system. Groundwater withdrawals can and usually do affect the amount (and quality) of surface water. For example, depletion of a small part of the total volume of groundwater in storage (sometimes only a few percent) can have substantial and undesirable effects on the availability of surface water, which becomes the limiting factor in the development of the groundwater resource (Alley 2007). The Texas Water Development Board (TWDB) defines groundwater availability as the effective recharge plus the amount of water that can be recovered annually from storage over a specified planning period without causing irreversible harm such as land-surface subsidence or water-quality deterioration (Muller and Price 1979; Mace et al. 2006).

A number of other terms have been used to describe groundwater availability. One of them is *groundwater sustainability,* which was defined by Alley et al. (1999) as the "development and use of groundwater in a manner that can be maintained for an indefinite time without causing unacceptable environmental, economic, or social consequences." The term *safe yield* has been used and amended to quantify sustainable groundwater development (Lee 1915; Meinzer 1920, 1923; Williams and Lohman 1949; Bear and Levin 1967; Bear 1979; Domenico and Schwartz 1990; Fetter 1994; Sophocleous 1997, 2000). The term *groundwater mining* or *groundwater overdraft* usually refers to a prolonged and progressive decrease in the amount of water stored in a groundwater system (Reilly et al. 2008), for example, in heavily pumped aquifers in arid and semiarid regions as the study area, especially in Texas and Kansas. The relative contributions of changes in storage, changes in recharge, and changes in discharge evolve with time. The initial response to withdrawal of water is changes in storage. If the system can come to a new equilibrium, the changes in storage will stop and inflows will again balance outflows, for example, the balanced groundwater budget in Nebraska.

Groundwater availability can range from nothing to all of the drainable water from an aquifer. Therefore, any method for determining groundwater availability should recognize such characteristics. Quantifying groundwater availability requires an intersection of policy and science: policy to define socioeconomic and environmental goals and science to estimate the actual amount of water that can be produced based on the socioeconomic goals (Mace et al. 2001).

4.4.2 Desired Future Conditions

The concept of desired conditions (DCs) as the social, economic, and ecological attributes that management strives to attain is well established (IEMTF 1995). DCs and desired future conditions (DFCs), pioneered by the U.S. Forest Service as part of its planning process in the 1970s and 1980s, have evolved over time as a result of criticisms and different applications (Leslie et al. 1996; Sutter et al. 2001). In a science context, the term tends to imply that DCs can be expressed specifically and measurably; however, in the planning process of most resource management agencies, it often implies a more broad description from which more specific objectives are tiered. In the application of planning processes, the term generally implies a timescale that is relatively long-term (e.g., >15 years), for example, 50 years for water resources planning in Texas. In other applications, however, such a timescale may or may not be applied (Bennetts and Bingham 2007). Planning processes generally include a hierarchy of goals and objectives, ranging from a broad vision or mission statement down to specific objectives or targets within a timescales ranging from long (e.g., into perpetuity) to short (e.g., annual or less). Two additional elements in the planning process are goals expressed in terms of desired resource conditions and goals expressed in terms of management strategies or activities intended to achieve those desired resource conditions.

In the 1970s, the noticeable depletion of the Ogallala Aquifer was the beginning of the public policy concern about the future condition of groundwater resources. The greatest levels of depletion have occurred in northern Texas and west-central Kansas. Certain portions of these areas have presented water level declines of more than 45.7 m, with decreases, especially in the Texas portion, of more than 50% in the saturated thickness of the aquifer. A great portion of the aquifer is beneath the Nebraska Sandhills, which has not been considerably exploited because irrigation farming is not economically feasible (McGuire and Fischer 1999; Peterson et al. 2003; McGuire 2009).

The Ogallala Aquifer is the largest aquifer in North America. The aquifer is shared by eight states, which makes its management complicated. These states have different interests and laws and regulations. Texas groundwater law is based on the "rule of capture," which some interpret as a "use it or lose it" rule. Under this rule, land owners are entitled to the water underneath their land (Kromm and White 1992; Verchick 1999). Nebraska and Oklahoma apply the general rules of the correlative rights doctrine (groundwater shared by the owners overlying the aquifer, extracted water applicable under the overlying land, and groundwater rights are usufructuary). The rest of the other states utilize the prior appropriation doctrine, which in a similar fashion to that of surface water protects water users on the basis of seniority in time (Gardner et al. 1997).

The states have been implementing regulations to manage groundwater. For instance, Kansas, New Mexico, and Colorado established policies to deny new water pumping permits if it is demonstrated that water availability in surrounding wells could be affected (Peterson et al. 2003). In 1972, Kansas authorized the creation of groundwater management districts (GMDs) to have more direct control over groundwater management. In 1991, the northwest Kansas GMD 4 implemented the "zero depletion" policy, which limited pumping to not exceed the aquifer natural recharge (Sophocleous 2000).

In September 2005, the Texas legislature passed House Bill 1763, which brought significant changes to groundwater resources management in Texas. Among the most important changes are the following: besides defining their groundwater availability, GMDs will be in charge of developing DFCs; groundwater conservation districts (GCDs) are tasked to define groundwater availability for regional water planning processes; and groundwater districts will issue permits based on groundwater availability (Mace et al. 2006). DFCs serve as management goals and provide answers as to how the aquifer would look in the future (Sheng et al. 2011).

In 2008, the North Plains GCD in Texas proposed to establish an Ogallala Aquifer DFC for Dallam, Hartley, Moore, and Sherman Counties; and Hansford, Hutchinson, Ochiltree, and Lipscomb Counties in Groundwater Management Area 1 (GMA-1): at least 40% of the aquifer storage to be remaining in 50 years for the area of Dallam, Hartley, Moore, and Sherman Counties and at least 50% of the aquifer storage to be remaining in 50 years for the area of Hansford, Hutchinson, Ochiltree, and Lipscomb Counties within the district. The district proposed two DFCs because the uses or conditions for the aquifer within the district differ substantially from one geographic area to another (North Plains Groundwater Conservation District 2008). Then, managed available groundwater will be determined by the TWDB based on the DFCs. The impacts of DFCs on other districts will be evaluated.

Mittelstet et al. (2011) (cited in President's message of April 2011, AWRA Water Blog<awramedia.org/mainblog/page7/>) compared two approaches to the administration of groundwater law on a hydrologic model of the North Canadian River, an alluvial aquifer in northwestern Oklahoma. Oklahoma limits pumping rates to retain 50% aquifer saturated thickness after 20 years of groundwater use. The Texas Panhandle GDC's rules limit pumping to a rate that consumes no more than 50% of saturated thickness in 50 years, with reevaluation and readjustment of permits every 5 years, using a hydrologic model (MODFLOW). Their results show that Oklahoma's approach initially would limit groundwater extraction more than the Texas GCD approach, but the Texas GCD approach would be more protective in the long run. Both the Oklahoma and Texas Panhandle GCD approaches would deplete alluvial base flow at approximately 10% development. Results also suggested that periodic reviews of permits could protect aquifer storage and river base flow.

In sum, DFCs set goals for retaining fresh groundwater storage in the aquifer within a time frame (50 years) based on current hydrological conditions and legal framework. To achieve those DFC goals, GMDs/state agencies also develop management strategies or activities for implementation. Some DFCs target the sustainable development of natural resources, such as the "zero depletion" policy in Kansas, others do not. Determination of DFC relies on multiple factors, including physical constraints of natural resources (aquifer and groundwater), and technology (irrigation methods), and social–economical, legal, and institutional constraints.

4.4.3 Constraints

How much groundwater is available for use depends upon how these changes in inflow and outflow affect the surrounding environment (physical and technical constraints) and what the public defines as undesirable effects on the environment (socioeconomic and institutional constraints). In determining the effects of pumping and the amount of water available for use, it is critical to recognize that not all the water pumped is necessarily consumed. For example, not all the water pumped for irrigation is consumed by crop evapotranspiration. Part of the irrigated water returns to the groundwater system as infiltration (irrigation return flow). Most other uses of groundwater are similar in that some of the water pumped is not consumed but is returned to the system (Alley et al. 1999). Thus, it is important to differentiate between the amount of water pumped and the amount of water consumed when we estimate future water availability, develop management strategies, and set up DFC goals.

4.4.3.1 Physical Constraints

Two key elements should be considered in the estimation of the amount of groundwater that is available for use. First, the use of groundwater and surface water must be evaluated together as a system. This evaluation includes all of the water available from changes in groundwater recharge, from changes in groundwater discharge, and from changes in storage for different levels of water consumption as shown in Equation 4.2. Second, because any use of groundwater changes the subsurface and surface environment (i.e., the water must come from somewhere), the trade-off between groundwater use and changes to the environment should be determined and a threshold should be set at which the level of change becomes undesirable. In conjunction with a system-wide analysis of the groundwater and surface-water resources this threshold can then be used to determine appropriate limits for consumptive use (Alley et al. 1999).

System-wide hydrologic analyses typically use numerical models to aid in estimating water availability and the effects of extracting water on the groundwater and surface-water system. If constructed correctly, the computer models attempt to represent the complex relations among the inflows, outflows, changes in storage, movement of water in the system, and possibly other important features using mathematical equations. As a mathematical representation of the system, the models can be used to estimate the response of the system to various development options and provide insight into appropriate management strategies. However, a computer model is a simplified representation of the actual system, and the judgment of water-management professionals is required to evaluate model simulation results and plan appropriate actions (Alley et al. 1999).

Extensive pumping of groundwater for irrigation has led to groundwater-level declines in excess of 46 m in parts of the High Plains Aquifer in Kansas, New Mexico, Oklahoma, and Texas. These large water-level declines have led to reductions in the saturated thickness of the aquifer exceeding 50% of the predevelopment saturated thickness in some areas. Lower groundwater levels cause increases in pumping lifts. Decreases in the saturated thickness result in declining well yields. Surface-water irrigation has resulted in water-level rises in some parts of the aquifer system, such as along the Platte River in Nebraska.

A long delay between pumping and its effects on natural discharge from the High Plains Aquifer is caused by the large distance between many of the pumping wells and the location of the springs and seeps that discharge from the groundwater system. The SHP is perhaps the best known example of significant, long-term nonequilibrium for a regional groundwater system in the United States. That is, water levels continue to decline without reaching a new balance (equilibrium) between recharge to and discharge from the groundwater system. When the groundwater level declines, less water will be available for irrigation and the cost for pumping will increase as pumping lift increases. As a result, dryland acreage tends to increase. Figure 4.7 shows that from 1972 to 2000 the dryland acreage share in 16 counties in the Southern Texas High Plains increased as the groundwater levels dropped. In summary, in the SHP and CHP areas, DFCs will mostly be limited by saturated thickness and depth to water, while in the northern part of the NHP, DFCs will mostly likely be limited by water-quality concerns instead of water quantity within the planning horizon.

4.4.3.2 Economic Constraints

Agricultural economics can be a limiting factor to set up DFC goals. Most economic factors, such as crop prices, ad energy, land, fertilizer, labor, and equipment costs, are independent of water resources, but the cost of pumping is closely related to groundwater availability, controlled by depth to water (pump lifting) and well yield. As the groundwater level declines (increasing pumping lift), energy consumption (gas/electricity) is proportional to pumping lift. Depth to water and well yield could become important constraints to future agricultural economics and development. Advanced irrigation technology requires high capital investment for irrigated agricultural production, while at the same time there are clearly economic benefits from the improved technology, such as reduced labor costs and more crop revenue per unit of water applied (Upendram and Peterson 2007). What are the incentives for water users to conserve water? As suggested by Upendram and Peterson (2007), the public funds could be used more effectively directed to programs that ensure a reduction in overall water depletion, such as the purchase and retirement of water rights. Therefore, DFC can

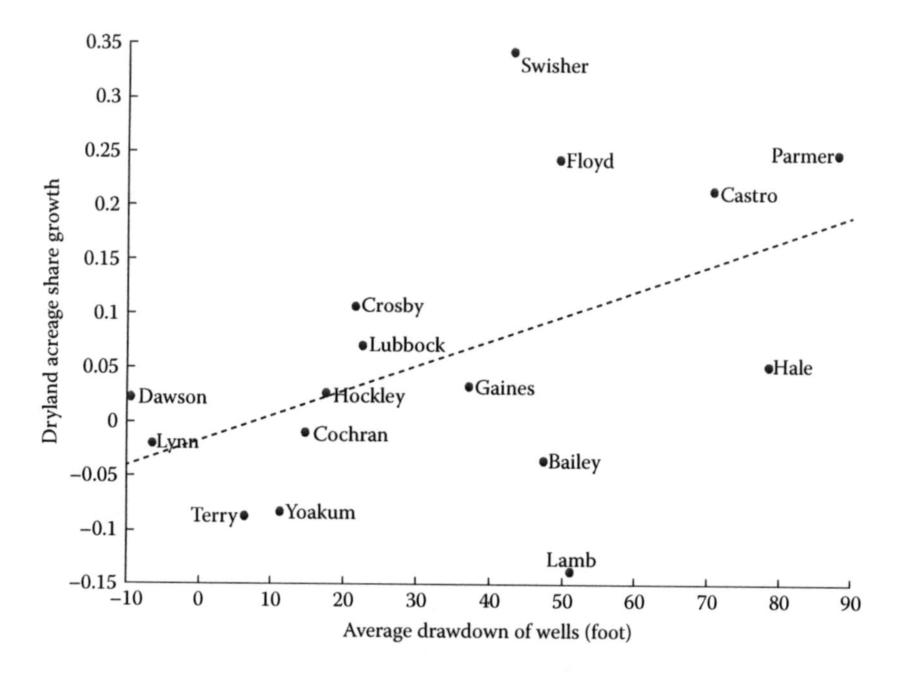

FIGURE 4.7 Dryland acreage expands as groundwater level declines.

also be constrained by water users' behavior in terms of economic benefits and availability of public funds for water conservation.

4.4.3.3 Effects of Irrigation Technology

Advances in irrigation technology have resulted in water-use efficiency and high crop production per unit of water applied. Irrigation efficiencies vary for different types of systems, depending upon application rates, crops, soil type, and field slope, but are typically in the range of 40%–60% for surface or gravity flow systems, 75%–85% for improved surface or gravity systems, 75%–85% for sprinkler systems, and 90%–98% for low energy precise application (LEPA) systems used with furrow dikes (Sloggett and Dickason 1986). Different water conservation techniques to maximize water resources are summarized in Table 4.3. Figure 4.8 shows changes in crop production in the Texas High Plains from 1950 to 2010, which reflect the effects of multiple factors, including advances in irrigation technology, groundwater available for irrigation, costs for pumping, fertilizer, labor, prices of crops, drought, and others. Further study is on the way to identify relationship among the farmers' decision on crop selection, irrigation technology, and groundwater availability (see Section 4.4.4 for details). Conventionally, people anticipated that the overall consumptive uses/depletion of natural groundwater storage would be reduced by the adoption of new irrigation technologies or conservation measures. However, recent research findings of Upendram and Peterson (2007) indicated that an improvement in irrigation efficiency by adapting new irrigation technology can result in either more or less water consumption, with the direction of impact depending on the changes in irrigated acreage and crop choice following the efficiency improvement. In contrast to widely held beliefs, Ward and Pulido-Velazquez's (2008a) study results in the Upper Rio Grande Basin indicated that water conservation subsides are unlikely to reduce water use under conditions that occur in many river basins. Adoption of more efficient technologies could reduce valuable return flows and limit aquifer recharge; therefore, conservation programs that target reduced water diversions or applications do not necessarily guarantee water savings. To achieve DFC goals, by reducing depletion by adopting more efficient irrigation technologies, we should design and

TABLE 4.3
Maximize Total Water Resources—Rainfall and Irrigation

Reduce crop constraints

- Drought-tolerant crops (grain sorghum vs. corn)
- Deep rooted crops
- Rotation of crops—spread water demand

Prevent rainfall runoff and soil surface evaporation

- Land forming techniques
  - Leveling
  - Contour farming
  - Terracing
- Furrow diking
- Crop residue—reduced or no-till

Soil water accounting

- Soil profile full of water prior to peak crop demand
- Hydrologic frequency analysis—rainfall probability relative to crop demand

Irrigation

- Accommodate critical crop growth periods
- Ratio of irrigated to dryland becomes a function of early rainfall
- Staggered planting dates to stagger water demand
- Variety selection for drought stress
- Application systems
  - Eliminate nonevapotranspiration losses
    - Water distribution losses, runoff, deep percolation
  - Furrow systems—short runs
  - Spray—sprinkler systems: large droplet size, applicator close to soil
  - Low energy precision application systems
  - Subsurface drip irrigation

Source: Modified from Lyle, W.M. and Bordovsky, J.P., Paper 85-2602, ASAE, St. Joseph, MI, 1985.

implement technical, accounting, and institutional measures that accurately track and economically reward reduced water depletion (Ward and Pulido-Velazquez 2008a).

4.4.3.4 Legal and Institutional Constraints

In the High Plains, state laws that regulate the development of groundwater vary from one to another. Oklahoma and Texas are the only two states that recognize private ownership of groundwater, while other six state laws dedicate water to the people of the state. Some set specific depletion caps, while others only require that water must go to "beneficial uses." In all states, a well registration or permitting system exists. In several states, there is a limit on the minimum spacing between wells.

In Colorado, one must obtain a permit to use groundwater, which is considered property rights and can be bought and sold independently of the land upon which the well rests. The Colorado Groundwater Commission is in charge of issuing and changing water permits. It divided the groundwater in Colorado into two districts: the NHP Designated Basin and the SHP Designated Basin. Designated groundwater is defined by the Colorado legislature as "groundwater which in its natural course would not be available to and required for the fulfillment of decreed surface rights, or groundwater in areas not adjacent to a continually flowing stream wherein groundwater

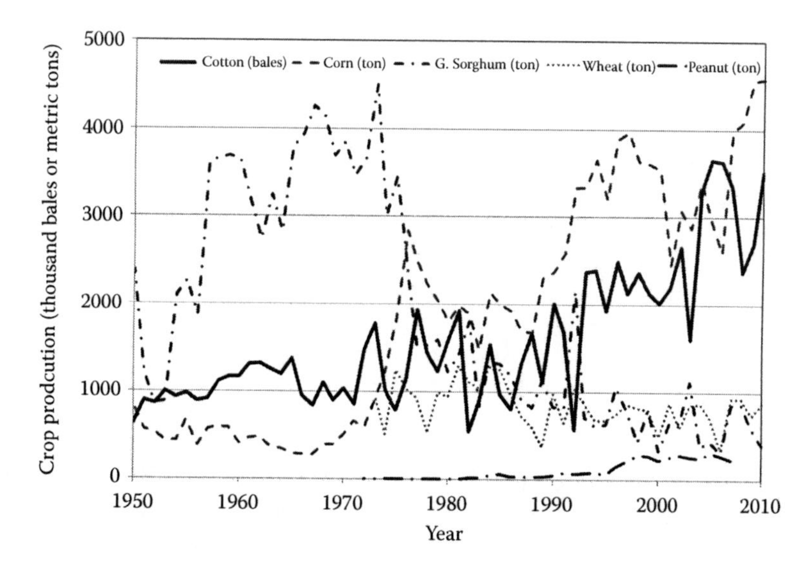

FIGURE 4.8 Irrigated crop production form the Texas High Plains (District 1-N and 1-S). (Data from http://quickstats.nass.usda.gov/, http://www.nass.usda.gov/Statistics_by_State/Texas/Publications/Historic_Estimates/he_corn.pdf, http://www.nass.usda.gov/Statistics_by_State/Texas/Publications/Historic_Estimates/he_cotton.pdf, http://www.nass.usda.gov/Statistics_by_State/Texas/Publications/Historic_Estimates/he_sorghum.pdf.)

withdrawals have constituted the principal water usage for at least fifteen years" (Colorado Statutes, Title 37 Article 90). These districts can put additional rules regarding groundwater management into place. Since 1990, permits have been issued with the policy that no permit will be issued for a new well if it is projected that the well will deplete the aquifer within a 4.8 km (3 mile) radius by 40% within 100 years (McGuire et al. 2003).

In Kansas, anyone with a permit or a vested right can use groundwater for nondomestic purposes. Landowners must "use the water for beneficial purposes" (McGuire et al. 2003). Five GMDs are in charge of regulating nondomestic water use and they can propose regulations that do not conflict with state law. Of these governing bodies, some follow the idea of "safe yield" and others follow the idea of "allowable depletion." "Safe yield" is the concept that total groundwater withdrawal in the district must be a certain percentage of the aquifer recharge in that radius. "Allowable depletion" is the concept that total groundwater withdrawal in the district must not deplete the aquifer in that radius by more than a specific amount in a specific time (Sophocleous 1998; McGuire et al. 2003; Kansas Department of Agriculture 2009).

In Nebraska, groundwater is a public resource. One has the right to use "a reasonable amount of the groundwater under their land for beneficial use on that land" (McGuire et al. 2003). The Nebraska Department of Natural Resources (NDNR) is in charge of groundwater withdrawals. In the early 1970s, the legislature created 23 Natural Resource Districts (NRDs), which have the authority to manage groundwater use. Each NRD is required to have a groundwater management plan approved by the NDNR, which outlines what the NRD will do to manage depletion and quality concerns in its area. In areas where there is no NDNR governance, residents follow the Nebraska correlative rights doctrine that states that residents must share when groundwater supplies are limiting (McGuire et al. 2003).

In New Mexico, water rights in six groundwater basins are based on the prior appropriation governed by the state engineer (New Mexico Office of the State Engineer 2005). In general, the state engineer approves most permits for domestic use. For nondomestic use, however, the state engineer will approve permits only if these four criteria are met: "(1) no objections are filed,

(2) unappropriated water exists in the basin, (3) no infringement on the water rights of prior appropriators occurs, and (4) it is not detrimental to the public welfare or the water conservation goals of the State" (McGuire et al. 2003). In addition, water rights are bought and sold independently of the land upon which the well sits. As long as the state engineer approves, water rights can be sold for out-of-state use (McGuire et al. 2003). The state engineer determines whether unappropriated water exists in the basins by monitoring water-level declines within a 23–65 km^2 area, depending on aquifer properties, around the site of a proposed new appropriation. If the annual water-level decline exceeds 0.76 m, the state engineer will not approve the permit because the rate of decline is considered excessive (Templer 1992; Ashley and Smith 1999).

In Oklahoma groundwater management policy is based on the reasonable use doctrine. The groundwater is owned by the landowners whose properties overlie the aquifer. The Oklahoma Water Resources Board (OWRB) grants licenses for withdrawing water from the aquifer. The amount of water withdrawn is determined by the OWRB and varies from region to region. This amount is determined by the maximum amount of water that can be withdrawn and still secure the availability of water at least 20 years from the time of the license; hence, the specific amount of water is uniquely calculated for each region. This amount is adjusted yearly when each license holder must check in with the OWRB and report how much water they pumped and set limits for the amount of water allowed for the next year (OWRB 2002; McGuire et al. 2003).

In South Dakota, water as a public resource is managed by the South Dakota Department of Environment and Natural Resources (SDDENR). The SDDENR works with the Water Management Board (WMB), a group of citizens, to manage groundwater resources. However, water can be withdrawn once a permit has been approved. The state also sets a specified amount of water that can be withdrawn. Generally, the amount of water withdrawn does not exceed the natural recharge rate. However, most of the land overlying the High Plains Aquifer in South Dakota lies within the boundary of either the Pine Ridge or Rosebud Indian Reservations, and the tribes assert the right to control the development and use of groundwater resources within reservation boundaries (South Dakota, Codified Laws, Title 46; South Dakota Administrative Rules, Article 74:02). All permits, other than irrigation permits, can be transferred to out-of-state (McGuire et al. 2003).

In Texas, groundwater is regarded as a public resource as it is still in the aquifer. However, once the groundwater is withdrawn, it is specific to the landowner. Groundwater can be sold to other locations, including out-of-state. No landowner can purposefully withdraw groundwater for "malicious reasons or to willfully waste the water." The GCD can enact some limitations on the groundwater withdrawal rate, the total amount withdrawn, and the well spacing (State of Texas Statutes, Chapters 35 and 36). Fourteen GCDs regulate to a varying extent more than 81% of the area that overlies the High Plains Aquifer in Texas (TWDB 2010). No permits are needed for wells that withdraw less than 94.6 m^3/day (or 25,000 gallons/day) or for domestic/livestock wells (McGuire et al. 2003; TWDB 2010).

In Wyoming, groundwater management policy is based on prior appropriation. Wyoming law mandates that all natural waters be the property of the state. Permits are given out to users for beneficial use. There is no specific limit on the amount of water that can be withdrawn. If the user would like to change any aspect of the permit, the user must first appeal to the state. No permits are needed if the out-of-state transfer is less than 1.23 million m^3/year (McGuire et al. 2003; Wyoming State Engineer's Office 2005).

4.4.4 The Linkage between Aquifer Hydrology and Water User Behavior

Evaluating alternative DFC-based management plans requires accurate predictions of future aquifer conditions. One of the many challenges facing groundwater policy analysts is to model the complex interactions between water-user behavior and aquifer conditions. This section presents a critical review of the state of the research in this area and points out a new direction for future work.

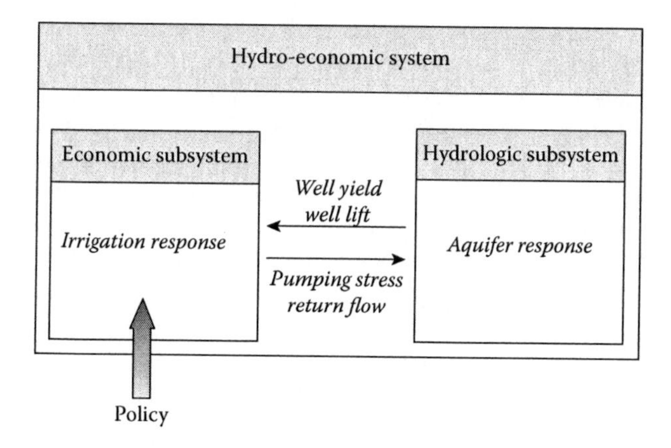

FIGURE 4.9 A conceptual framework for water policy evaluation.

4.4.4.1 A Conceptual Framework

A general characteristic of human exploitation of natural resources is that the behavior of the resource users and resource stock are dynamically interdependent. While resource exploitation perturbs the physical equilibrium of the resource stock, the change in that stock's properties, such as quality, quantity, and availability, can conversely affect the resource user's cost–benefit equation, resulting in a change in utilization of the resource. In the case of groundwater use, pumping alters the groundwater level and well yield, which in turn changes the water user's future pumping costs and water demand.

To be more specific, consider the evaluation of alternative management plans to achieve a predetermined set of DFCs. Figure 4.9 illustrates the various channels thorough which a management plan influences the aquifer system. Suppose that the management plan is aimed at reducing the pumping of water for irrigation by way of introducing incentives or regulations. The resulting alterations in the pumping for irrigation will change the speed at which the groundwater surface declines. This will change the pumping lift and well yield facing the irrigator, which in turn will affect future water-use decisions. Failing to account for this dynamic effect can lead to biased estimations of future aquifer conditions.

The extant hydrological models for simulating future aquifer conditions do not explicitly account for this dynamic process. Rather, future groundwater demands are frequently estimated separately and then incorporated into the hydrological model as an independent external force. The future conditions simulated in this manner are subject to prediction errors, which can lead to misinformed decisions. These limitations are illustrated below with the groundwater policy analysis framework presently adopted in the state of Texas.

4.4.4.2 The Hydrology-Centered Modeling Approach

The TWDB has developed sophisticated hydrological models to facilitate water management and planning. In particular, a MODFLOW-based aquifer model, a groundwater availability model (GAM) (Blandford et al. 2008), has been developed for the southernmost part of the Ogallala Aquifer underlying portions of Texas and New Mexico. GAM has served as the basis for the regional water planning groups in the Texas High Plains to develop DFCs and evaluate alternative water management plans.

The TWDB has developed separate estimation methods for municipal, industry and mining, steam electric power, agricultural irrigation, and livestock (TWDB 2011). Here we focus only on agricultural irrigation water use because it accounts for over 95% of the water use in the Texas Ogallala Aquifer. In GAM future irrigation water demands are estimated in two steps. Historic water demands are first estimated, and then projected into the planning period to generate the required future water demands. Estimation of historic irrigation water demands is based on satellite imagery data and irrigation surveys conducted by the National Agriculture Statistics Service (NASS 2007). The acreage planted for each crop under irrigation is estimated for each county using remote sensors or satellite sensors. Because the water applications in each crop depend on the irrigation technologies used, the crop acreage data are combined with the NASS irrigation survey data to generate county-level water use. Future water demands are then obtained using a simple linear extrapolation method, with minor adjustments made in accordance with the discrepancy between previous predictions and actual observations (TWDB 2002, 2007).

The approach described above evidently oversimplifies the water user's decision-making process. Indeed, it lacks a model representing the water user's decision-making process under a constantly changing environment dictated by a variety of economic, hydrological, and policy factors. Although updating the extrapolative model periodically does help improve the predictive power of the model in the short run, it is inadequate for long-run policy analysis—such as water planning over a 50 year horizon to achieve the 50/50 groundwater management goal.

4.4.4.3 The Economic Analysis of Groundwater Policy

Most economic analyses of groundwater policy have been cast in a rational choice model, in which the water users are assumed to maximize an economic objective subject to a set of economic, hydrologic, and institutional constraints. Solving the optimization model produces a set of decisions to be made by the water user, including the quantity of water demanded. Using this approach, a large body of literature has developed to evaluate groundwater policy in the Texas portion of the Ogallala Aquifer (Arabiyat 1998; Feng 1992; Johnson 2003; Wheeler 2005). Parallel research has been conducted for other parts of the United States (e.g., Peterson and Ding 2005; Scheierling et al. 2006; Ward and Pulido-Velázquez 2008b).

This approach renders the analyst a great deal of modeling flexibility, permitting detailed information to be analyzed in complex, large-scale optimization models. However, it also forces the analyst to assume—sometimes in an arbitrary fashion—producer behavior, rather than discover it empirically from past data. Consequently, conclusions often are drawn on a case-by-case basis. Arbitrary behavioral assumptions in these hydroeconomic models limit their prediction power.

Meanwhile, these optimization models tend to oversimplify the evolution of the aquifer system. The complex aquifer system is frequently represented by a few equations with variables defined at the county level, leaving unexploited a great deal of hydrological information available.

4.4.4.4 Toward a Holistic Approach

The two types of analyses reviewed above have each contributed in their own ways to our ability to understand the aquifer system's future conditions, leaving substantial gaps that can be closed only by combining their respective strengths to form a unified framework for policy analysis. The best approach for assessing groundwater management plans should take explicit account of the dynamic interactions of the hydrologic and economic systems. And such a holistic approach should be taken not only in estimation or calibration of the model, but also in use of the model to predict future conditions, on which we elaborate below.

The usefulness of a model depends ultimately on its ability to explain the historic data and predict the future behavior of the system under concern. A holistic approach for model development can improve the performances of the economic and hydrologic models simultaneously. Consider the economic model first. To estimate groundwater use requires a model explaining the cropping and irrigation technology adoption patterns. A large body of econometric analyses has been conducted to understand the producer's irrigation technology and crop choice, and the estimation method is well established (Baerenklau and Knapp 2007; Caswell and Zilberman 1985; Green et al. 1996; Lichtenberg 1989; Plantinga et al. 1999; Shah et al. 1995; Wu and Brorsen 1995; Wu and Segerson 1995). Most of these models, however, treat resource conditions in an oversimplified manner. In groundwater-related studies, the typical approach is to construct a few variables summarizing the hydrologic conditions facing the water user and include them as explanatory variables in the regression. The most frequently used proxies are well depth and fuel cost (e.g., Caswell et al. 1990; Negri and Brooks 1990). This approach in essence takes hydrologic conditions as exogenous variables independent of economic activity, ignoring the complex dynamic process described above. Hydrological models can generate a wealth of detailed hydrologic data, providing an opportunity to improve these econometric models. While most of the land use models are developed at the county level, for example, hydrological models can generate hydrological data at a much more refined level. Incorporating such data into the econometric model is bound to improve estimation precision.

Hydrological models, such as the GAM described above, can also be improved if more accurate water-use data are used for model calibration. Because the econometric model takes full account of various forces of influence on water-use decisions, it is bound to provide a more accurate estimation of groundwater demand than the methods presently adopted by hydrologists. Incorporating the econometrically estimated groundwater use data into hydrological models, therefore, will improve these models' predictive power.

The procedures described above build on existing models and therefore are easy to implement. They essentially require a data-sharing protocol between the hydrologist and the economist. A more challenging yet more interesting approach is to completely integrate econometric estimation and hydrologic model calibration. That is, they are executed in a unified computer program so that the interchange of the hydrologic and economic data described above can be automated and iterated. This approach will maximize the performance of the two models but require a much close collaboration between the economist and the hydrologist.

Once the hydrologic and economic models are developed, they can be used to simulate future groundwater conditions to examine whether a proposed water management plan can achieve DFCs. The integrated approach can improve the simulation results. Unlike in the simple extrapolative model mentioned above, the econometrically estimated water-use model allows water-use patterns to change in response to a variety of economic and environmental conditions. The policy analyst, then, will be able to run the simulation under a large number of contingencies, allowing a robust test of water management plans.

4.5 SUMMARY

The High Plains Aquifer is a vital water resource of great importance to the economy of the High Plains region. It supplies water for the region's crop, livestock, and meat processing industries. Because the High Plains Aquifer receives limited and slow recharge, especially in the SHP and CHP areas, the High Plains economy is dependent on a finite water resource.

Historical overdraft of groundwater has caused large water level drops, especially in the SHP and CHP. In Texas, a decline of 71 m from the predevelopment water level was observed in 2009. Heavy groundwater pumping and intensive agricultural production have also caused deterioration of groundwater quality, which in turn further reduced groundwater availability, as the high salinity causes damage to crops and soil structures. Heavy groundwater pumping has also caused depletion of streams adjacent to the shallow aquifer, as groundwater pumping alters the hydraulic connection between the aquifer and the stream.

To assure future water supplies, different measures and policies have been implemented. One concept is to set up DFCs for the High Plains Aquifer. Though the term DFC may mean different things for different people, the planning process generally reflects a hierarchy of goals and objectives, ranging from a broad vision or mission statement down to specific objectives or targets within a timescale. Two additional elements include goals expressed in terms of desired resource conditions and goals expressed in terms of management strategies or activities intended to achieve those desired resource conditions. DFCs set by Texas, Kansas, and Oklahoma intend to slow down or curtail the ongoing depletion of groundwater in some highly stressed areas. Some DFCs tend to create more sustainable development, while others may not. DFCs are affected by multiple factors, including physical constraints of the aquifer, irrigation technology, economic, and legal/institutional constraints. To develop effective DFC goals for future groundwater development, we need an integrated model taking account of the hydrological, economic, and institutional forces that influences groundwater availability. In this chapter, we propose such a framework linking the existing hydrological model to a new econometric model to assess alternative groundwater management plans for achieving DFC goals.

ACKNOWLEDGMENT

The work presented in this chapter was supported in part by the U.S. Department of Agriculture-Cooperative State Research, Education, and Extension Service, under agreement of 2009-65102-05841. We would also like to thank Professor Rattan Lal for his encouragement and constructive review comments and Ms. Theresa L. Colson for her editorial comments.

ABBREVIATIONS

BMPs	Best management practices
CHP	Central High Plains
DFC	Desired future condition
GAM	Groundwater availability model
GCD	Groundwater conservation district
GMA	Groundwater Management Area
GMDs	Groundwater management districts
MCL	Maximum contaminant level
NAWQA	National Water-Quality of Assessment Program
NDNR	Nebraska Department of Natural Resources
NGVD	National Geodetic Vertical Datum
NHP	Northern High Plains
NRD	Natural Resource District
OWRB	Oklahoma Water Resources Board
SDDENR	South Dakota Department of Environment and Natural Resources
SHP	Southern High Plains
TWDB	Texas Water Development Board
USGS	U.S. Geological Survey
WMB	Water Management Board

REFERENCES

Alley, W.M. 2007. Another water budget myth: The significance of recoverable ground water in storage. *Ground Water* 45:251.

Alley, W.M. and S.A. Leake. 2004. The journey from safe yield to sustainability. *Ground Water* 42:12–16.

Alley, W.M., T.E. Reilly, and O.L. Franke. 1999. Sustainability of ground-water resources. U.S. Department of the Interior, U.S. Geological Survey.

Alley, W.M., R.W. Healy, J.W. LaBaugh, et al. 2002. Flow and storage in groundwater systems. *Science* 296:1985.

American Society of Civil Engineers (ASCE). 1987. *Ground Water Management,* 3rd edition. ASCE Manual and Reports on Engineering Practice No. 40, 281 pp, New York.

Arabiyat, S.T. 1998. Agricultural sustainability in the Texas High Plains: The role of advanced irrigation technology and biotechnology. M.S. Thesis, Texas Tech University, Lubbock, TX.

Ashley, J.S. and Z.A. Smith. 1999. *Groundwater Management in the West.* 310 pp, Lincoln, NE: University of Nebraska Press.

Baerenklau, A.K. and K.C. Knapp. 2007. Dynamics of agricultural technology adoption: Age structure, reversibility, and uncertainty. *Am J Agric Econ* 89(1):190–201.

Bear, J. 1979. *Hydraulics of Groundwater*. New York: McGraw-Hill.

Bear, J. and O. Levin. 1967. The optimal yield of an aquifer. *Int Assoc Sci Hydrol Bull* 72:401–412.

Bennetts, R.E. and B.B. Bingham. 2007. Comparing current and desired conditions of resource values for evaluating management performance: A cautionary note on an otherwise useful concept. *The George Wright Forum* 24(2):108–116.

Blandford, T.N., D.J. Blazer, K.C. Calhoun, et al. 2008. Groundwater availability model of the Edwards-Trinity (High Plains) Aquifer in Texas and New Mexico. Austin, TX: Texas Water Development Board.

Bredehoeft, J. and T. Durbin. 2009. Ground water development—The time to full capture problem. *Ground Water* 47:506–514.

Bredehoeft, J. and E. Kendy. 2008. Strategies for offsetting seasonal impacts of pumping on a nearby stream. *Ground Water* 46:23–29.

Bredehoeft, J.D., S.S. Papadopulos, and H. Cooper Jr. 1982. Groundwater: The water budget myth. In *Scientific Basis of Water Resource Management*. pp. 51–57, Washington, DC: National Academy Press.

Caswell, M. and D. Zilberman. 1985. The choices of irrigation technologies in California. *Am J Agric Econ* 67(2):224–234.

Caswell, M., E. Lichtenberg, and D. Zilberman. 1990. The effects of pricing policies on water conservation and drainage. *Am J Agric Econ* 72(4):883–890.

Domenico, P.A. and F.W. Schwartz. 1990. *Physical and Chemical Hydrology*. New York: John Willey and Sons.

Dugan, J.T., T. McGrath, and R.B. Zelt. 1994. Water-level changes in the High Plains Aquifer—Predevelopment to 1992. U.S. Geological Survey Water-Resources Investigations Report 94-4027, Reston, VA.

Feng, Y. 1992. Optimal intertemporal allocation of groundwater for irrigation in the Texas High Plains. Ph.D. Diss.,Texas Tech University, Lubbock, TX.

Fetter Jr, C.W. 1994. *Applied Hydrogeology*, 4th edition. Upper Saddle River, NJ: Prentice-Hall.

Gardner, R., M.R. Moore, and J.M. Walker. 1997. Governing a groundwater commons: A strategic and laboratory analysis of western water law. *Econ Inq* 35(2):218–234.

Green, G., D. Sunding, D. Zilberman, et al. 1996. Explaining irrigation technology choices: A microparameter approach. *Am J Agric Econ* 78(4):1064–1072.

Guru, M.V. and J.E. Horne. 2000. *The Ogallala Aquifer*. Poteau, OK: Kerr Center for Sustainable Agriculture.

Gurdak, J.J. and C.D. Roe. 2009. Recharge rates and chemistry beneath playas of the High Plains Aquifer—A literature review and synthesis. Reston, VA: U.S. Geological Survey Circular 1333.

Gurdak, J.J., P.B. McMahon, K. Dennehy, et al. 2009. Water quality in the High Plains Aquifer, Colorado, Kansas, Nebraska, New Mexico, Oklahoma, South Dakota, Texas, and Wyoming, 1999–2004. Reston, VA: Circular 1337.

Gutentag, E.D., F.J. Heimes, N.C. Krothe, et al. 1984. Geohydrology of the High Plains Aquifer in parts of Colorado, Kansas, Nebraska, New Mexico, Oklahoma, South Dakota, Texas, and Wyoming, U.S. Geological Survey Professional Paper 1400-B, Reston, VA.

Interagency Ecosystem Management Task Force (IEMTF). 1995. *The Ecosystem Approach: Healthy Ecosystems and Sustainable Economies*. Washington, DC: White House Office of Environmental Policy.

Johnson, W.J. 2003. Water conservation policy alternatives for the southern portion of the Ogallala Aquifer. Ph.D. Diss.,Texas Tech University, Lubbock, TX.

Kansas Department of Agriculture. 2009. Kansas Water Appropriation Act. http://www.ksda.gov/includes/statute_regulations/mainportal/KSWaterAppropriationAct82a_701.pdf (accessed August 31, 2011).

Kromm, D.E. and S.E. White. 1992. *Groundwater Exploitation in the High Plains*. Lawrence, KS: University Press of Kansas.

Lee, C.H. 1915. The determination of safe yields of underground reservoirs of the closed basin type. *Trans Am Soc Civ Eng* 79:148–218.

Leslie, M., G.K. Meffe, J.L. Hardesty, et al. 1996. *Conserving Biodiversity on Military Lands: A Handbook for Natural Resources Managers*. Arlington, VA: The Nature Conservancy.

Lichtenberg, E. 1989. Land quality, irrigation development, and cropping patterns in the northern High Plains. *Am J Agric Econ* 71(1):187–194.

Litke, D.W. 2001. Historical water-quality data for the High Plains regional groundwater study area in Colorado, Kansas, Nebraska, New Mexico, Oklahoma, South Dakota, Texas, and Wyoming. 1930–1998, U.S. Geological Survey Water-Resources Investigations Report 00-4254.

Lugn, A. L. 1938. The Nebraska state geological survey and the "Valentine problem". *Am. Jour. Sci.*, 36: 220–227.

Lyle, W.M. and J.P. Bordovsky. 1985. Water conservation techniques and equipment for limited supplied. Paper 85-2602. St. Joseph, MI: ASAE.

Mace, R.E., W.F. Mullican III, and T.S.-C. Way. 2001. Estimating groundwater availability in Texas. In Proceedings of the 1st Annual Texas Rural Water Association and Texas Water Conservation Association Water Law Seminar. *Water Allocation in Texas: The Legal Issues*, Austin, TX, January 25–26, 2001, Section 1, 16, http://www.twdb.state.tx.us/gam/gam_documents/gw_avail.pdf (accessed August 31, 2011).

Mace, R.E., R. Petrossian, R. Bradley, et al. 2006. A streetcar named desired future condition: The new groundwater availability for Texas (revised). 8th Annual Changing Face of Water Rights in Texas.

Maupin, M.A. and N.L. Barber. 2005. Estimated withdrawals from principal aquifers in the United States, 2000. Reston, VA: U.S. Geological Survey Circular 1279.

McGrath, T.J. and J.T. Dugan. 1993. Water-level changes in the High Plains Aquifer—Predevelopment to 1991. U.S. Geological Survey Water-Resources Investigations Report 93-4088.

McGuire, V.L. 2001. Water-level changes in the High Plains Aquifer, 1980 to 1999. Report USGS FS-029-01. Lincoln, NE: U.S. Geological Survey.

McGuire, V.L. 2009. Water-level changes in the High Plains Aquifer, predevelopment to 2007, 2005–2006, and 2006–2007. U.S. Geological Survey Scientific Investigations Report 2009-5019.

McGuire, V.L. 2011. Water-level changes in the High Plains Aquifer, predevelopment to 2009, 2007–2008, and 2008–2009, and change in water in storage, predevelopment to 2009. U.S. Geological Survey Scientific Investigations Report 2011-5089, 13 pp, http://pubs.usgs.gov/sir/2011/5089/ (accessed August 31, 2011).

McGuire, V.L. and B.C. Fischer. 1999. Water-level changes, 1980 to 1997, and saturated thickness, 1996–1997, in the High Plains Aquifer. U.S. Geological Survey Fact Sheet 124-99.

McGuire, V.L., M.R. Johnson, R.L. Schieffer, et al. 2003. Water in storage and approaches to ground-water management, High Plains Aquifer, 2000. Reston, VA: U.S. Geological Survey Circular 1243.

McMahon, P.B., K.F. Dennehy, B.W. Bruce, et al. 2007. Water-quality assessment of the High Plains Aquifer, 1999–2004, U.S. Geological Survey, Professional Paper 1749, Reston, VA.

Meinzer, O.E. 1920. Quantitative methods of estimating ground-water supplies. *Bull Geol Soc Am* 31:329.

Meinzer, O.E. 1923. Outline of ground-water hydrology. U.S. Geological Survey Water Supply Paper 494, 5 pp.

Mittelstet, A.R., M.D. Smolen, G.A. Fox, and C.A. Damian. 2011. Comparison of aquifer sustainability under groundwater administrations in Oklahoma and Texas. *J Am Water Resour Assoc* 47(2):424–431. DOI: 10.1111/j.1752-1688.2011.00524.x.

Muller, D.A. and R.D. Price. 1979. Ground-water availability in Texas, estimates and projections through 2030, 77. Texas Department of Water Resources Report 238.

National Agriculture Statistics Service (NASS). 2007. The Census of Agriculture, http://www.agcensus.usda.gov/ (accessed 15 August, 2011).

Negri, D.H. and D.H. Brooks. 1990. Determinants of irrigation technology choice. *West J Agric Econ* 15(2):213–223.

New Mexico Office of the State Engineer. 2005. Underground water basins in New Mexico, http://www.ose.state.nm.us/PDF/Maps/underground_water.pdf (accessed August 31, 2011).

North Plains Groundwater Conservation District. 2008. Proposed desired future conditions for the Ogallala Aquifer, http://www.npwd.org/DFC/DFC%20discussion%20%20_3_.pdf (accessed August 31, 2011).

Oklahoma Water Resources Board (OWRB). 2002. 2001 Annual Report 18. Norman, OK: Oklahoma University Printing Services.

Peterson, J.M. and Y. Ding. 2005. Economic adjustments to groundwater depletion in the High Plains: Do water-saving irrigation systems save water? *Am J Agric Econ* 87:147–159.

Peterson, J.M., T.L. Marsh, and J.R. Williams. 2003. Conserving the Ogallala Aquifer: Efficiency, equity, and moral motives. *Choices* 1:15–18.

Plantinga, A.J., T. Mauldin, and D.J. Miller. 1999. An econometric analysis of the costs of sequestering carbon in forests. *Am J Agric Econ* 81(4):812–824.

Reilly, T.E., K. Dennehy, W. Alley, et al. 2008. Ground-water availability in the United States: Reston, VA: U.S. Geological Survey Circular 1323, http://pubs. usgs. gov/circ/1323 (accessed August 31, 2011).

Scheierling, S.M., R.A. Young, and G.E. Cardon. 2006. Public subsidies for water-conserving irrigation investments: Hydrologic, agronomic, and economic assessment. *Water Resour Res* 42. DOI: 10.1029/2004WR003809.

Shah, F.A., D. Zilberman, and U. Chakravorty. 1995. Technology adoption in the presence of an exhaustible resource: The case of groundwater extraction. *Am J Agric Econ* 77(2):291–299.

Sheng, Z. 2011. Impacts of groundwater pumping and climate variability on groundwater availability in the Rio Grande Basin, proceedings on sustainability on the border: Water, climate, and social change in a fragile landscape, May 16–18, 2011, The University of Texas at El Paso.

Sheng, Z., J.R. Gastélum, C. Wang, et al. 2011. Historic pumping and future desired conditions of the stressed aquifers within a groundwater management area in Texas. Land Grant and Sea Grant National Water Conference, January 31–February 1, 2011.

Sloggett, G. and C. Dickason. 1986. Ground-water mining in the United States. Washington, DC: U.S. Department of Agriculture, Econ. Res. Serv. AER No. 555.

Soil Survey Staff. 1995. Soil survey laboratory information manual. Soil Survey Investigations Report 45. Lincoln, NE: USDA Natural Resources Conservation Service, National Soil Survey Center, Soil Survey Laboratory.

Sophocleous, M.A. 1997. Managing water resources systems: Why "safe yield" is not sustainable. *Ground Water* 35:561–561.

Sophocleous, M.A. 1998. Water resources of Kansas: A comprehensive outline. In *Perspectives on Sustainable Development of Water Resources in Kansas*, ed. M.A. Sophocleous, 239 pp. Lawrence, KS: Kansas Geological Survey Bulletin.

Sophocleous, M.A. 2000. From safe yield to sustainable development of water resources—The Kansas experience. *J Hydrol* 235(1–2):27–43.

Stromberg, J., R. Tiller, and B. Richter. 1996. Effects of groundwater decline on riparian vegetation of semiarid regions: The San Pedro, Arizona. *Ecol Appl* 6(1):113–131.

Sutter, R.D., J.J. Bachant, D.R. Gordon, et al. 2001. An assessment of the desired future conditions for focal conservation targets on Eglin Air Force Base. Report to Natural Resources Division, Eglin Air Force Base, Niceville, FL. Gainesville, FL: The Nature Conservancy.

Templer, O.W. 1992. The legal context of groundwater use. In *Groundwater Exploitation in the High Plains*, eds. D.E. Kromm and S.E. White, 240 pp. Lawrence, KS: University of Kansas Press.

TWDB. 2002. Water for Texas-2002. Texas Water Development Board, January 2002, http://www.twdb.state.tx.us/publications/reports/State_Water_Plan/2002/WaterforTexas2002.pdf (accessed August 15, 2011).

TWDB. 2007. Water for Texas-2007. Texas Water Development Board, January 2007, http://www.twdb.state.tx.us/publications/reports/State_Water_Plan/2007/2007StateWaterPlan/vol%201_FINAL%20113006.pdf (accessed August 15, 2011).

TWDB. 2010. Groundwater conservation districts, http://www.twdb.state.tx.us/mapping/maps/pdf/gcd_only_24x24.pdf (accessed August 15, 2011).

TWDB. 2011. Water demand projections methodology, http://www.twdb.state.tx.us/wrpi/data/proj/docu.asp (accessed August 15, 2011).

United States Census Bureau. 2000. United States Census 2000, http://www.census.gov/main/www/cen2000.html (accessed September 30, 2011).

Upendram, S. and J.M. Peterson. 2007. Irrigation technology and water conservation in the High Plains Aquifer region. *J Contemp Water Res Educ* 137:40–46.

USDA. 1988. Water quality education and technical assistance plan. USDA–Soil Conservation Service and USDA–Extension Service Report.

USDA. 1996. America's Northern Plains: An overview and assessment of natural resources. U.S. Department of Agriculture, Natural Resources Conservation Service, 16 pp. Jamestown, ND: Northern Prairie Wildlife Research Center Online, http://www.npwrc.usgs.gov/resource/habitat/amnorpln/index.htm (accessed August 31, 2011).

USDA Census of Agriculture. 1998. 1997 Census of Agriculture, http://www.agcensus.usda.gov/Publications/1997/index.asp (accessed October 10, 2011).

USDA Census of Agriculture. 2002. 2002 Census of Agriculture, http://www.agcensus.usda.gov/Publications/2002/index.asp (accessed October 10, 2011).

USGS. 1992. 1992 National Land Cover Data, http://www.epa.gov/mrlc/nlcd.html (accessed September 30, 2011).

USGS. 2011a. High Plains regional ground water study, http://co.water.usgs.gov/nawqa/hpgw/images/figure1.jpg (accessed September 30, 2011).

USGS. 2011b. High Plains water-level monitoring study, http://ne.water.usgs.gov/ogw/hpwlms/physsett.html (accessed September 30, 2011).

Verchick, R.R.M. 1999. Dust bowl blues: Saving and sharing the Ogallala Aquifer. *J Environ Law Litigat* 14:13.

Walton, W.C. 2011. Aquifer system response time and groundwater supply management. *Ground Water* 49(2):126–127.

Ward, F.A. and M. Pulido-Velazquez. 2008a. Water conservation in irrigation can increase water use. *Proc Natl Acad Sci* 105(47):18215–18220.

Ward, F.A. and M. Pulido-Velázquez. 2008b. Efficiency, equity, and sustainability in a water quantity–quality optimization model in the Rio Grande Basin. *Ecol Econ* 66:23–37.

Weeks, J.B. and E.D. Gutentag. 1984. The High Plains regional aquifer—Geohydrology. In *Ogallala Aquifer Symposium II*, ed. G.A. Whetstone. Lubbock, TX: Texas Tech University, Water Resources Center: 6–25.

Weeks, J.B., E.D. Gutentag, F.J. Heimes, et al. 1988. Summary of the High Plains regional aquifer-system analysis in parts of Colorado, Kansas, Nebraska, New Mexico, Oklahoma, South Dakota, Texas, and Wyoming. Reston, VA: U.S. Geological Survey Professional Paper 1400-A.

Wheeler, E. 2005. Water policy alternatives for the southern Ogallala Aquifer: Economic and hydrologic implications. M.S. Thesis, Texas Tech University, Lubbock, TX.

Williams, C.C. and S.W. Lohman. 1949. Geology and ground-water resources of a part of south-central Kansas, with special references to the Wichita municipal water supply, with analyses by Robert H. Hess and others. *Kans Geol Surv Bull* 79:455.

Wu, J. and B.W. Brorsen. 1995. The impacts of government programs and land characteristics on cropping patterns. *Can J Agric Econ* 43(1):87–104.

Wu, J. and K. Segerson. 1995. The impacts of policies and land characteristics on potential groundwater pollution in Wisconsin. *Am J Agric Econ* 77(4):1033–1047.

Wyoming State Engineer's Office. 2005. Regulating and administering Wyoming's water resources, http://seo.state.wy.us/about.aspx (accessed August 15, 2011).

5 Competition between Environmental, Urban, and Rural Groundwater Demands and the Impacts on Agriculture in Edwards Aquifer Area, Texas

Venkatesh Uddameri and Vijay P. Singh

CONTENTS

5.1 INTRODUCTION

The Edwards Aquifer is a prolific limestone formation in central Texas and includes the Edwards and associated limestones. The aquifer spans across 12 counties (Figure 5.1) and serves as the main water supply to nearly 2 million residents. The major cities sustained by the aquifer include the city of San Antonio and the metropolitan areas of San Marcos and south Austin. South-central Texas is one of the fastest-growing areas in the nation. Between 2000 and 2010, the population in the region increased by approximately 20% (U.S. Census 2010). Austin and San Antonio were ranked first and second, respectively, among the large cities on Forbes' best places for jobs list in 2010 (Fisher 2011). The Edwards Aquifer is a shared resource that not only drives economic growth, but also sustains the unique ecological flora and fauna of the Texas hill country. The aquifer is home to several endangered and threatened species, such as the Fountain Darter and the Texas Blind Salamander (Fridell 2008). The south-central part of Texas is also the prime agricultural area in the state that depends on the Edwards Aquifer resources. The major crops grown in the area include vegetables, hay, cotton, and cereal crops, such as barley, corn, sorghum, and wheat, which are important to meet the food and fiber requirements of the state and the nation. Because it is a water-scarce region, the competition for water between agricultural, urban, and ecological interests is intense; therefore, the Edwards Aquifer resources are highly regulated and managed using a complex set of policies and protocols (EAA 2011).

Water is a major natural constraint that limits growth in semiarid regions, such as Texas. The Edwards Aquifer region lies at the cusp of two major physiographic regions—the Low Western

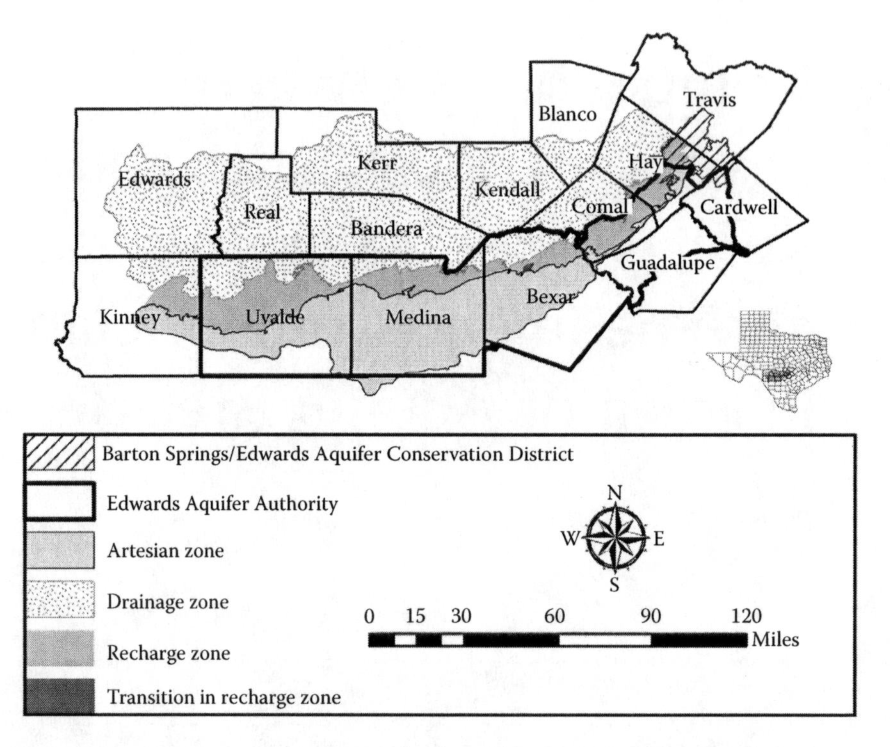

FIGURE 5.1 Edwards Aquifer area and jurisdictional boundaries of the Edwards Aquifer Authority and the Barton Springs/Edwards Aquifer Conservation District.

Plains and the Gulf Coastal Plains—and crosses through the Edwards Plateau and south-central climatic regions of the state (Bomar 1995). The climate in the region varies from semiarid to subtropical, humid zones (Larkin and Bomar 1983). As such, the rainfall is highly erratic and is characterized by prolonged droughts or years of extremely high rainfall. These problematic climatic patterns, coupled with the increasing need for continuous and sustained water supplies to meet the urban demands and ecological requirements, pose a unique threat to the viability of agriculture in the region. The transfer of water from the agricultural sector to meet the municipal and industrial demands is becoming increasingly common in Texas and is likely to continue in the future (TWDB 2007). According to the south-central region water planning group's projections, the municipal demands in the Edwards Aquifer region are likely to increase by 47% in the year 2060 from the current needs of approximately 370,000 AFY. On the other hand, the water demand for irrigation is projected to decrease by 27% from the current usage of 380,000 AFY and some of this will be used to offset the growing municipal needs (TWDB 2011). These shifts in water uses will increase the risk of food security in the region because agricultural products have to be imported in greater quantities. There is a renewed interest in using agricultural produce for biofuel production. Many crops produced in the Edwards Aquifer region, such as wheat, corn, and sorghum, have lower carbon emissions when compared with gasoline (McCarl 2006). Therefore, the possibility of using crops in the region to produce biofuels further exacerbates the potential risks to the self-reliance and food security of the region.

Understanding the nature and the extent of the risk to the agricultural activities in the Edwards Aquifer region and the steps that can be taken to mitigate the impacts, particularly in an altered climate regime of the future, is an important aspect of topical interest to land and water planners in the region and the nation. The overall goal of this chapter is to undertake a historical evaluation of rural to urban land use changes along with an assessment of the available water supplies and the salient hydrologic factors and regulatory frameworks controlling their use, to understand the agricultural shifts in the region. This understanding of the historical trends is used to evaluate the potential risks to agriculture, particularly in the context of an altered climate in the future.

5.2 EDWARDS AQUIFER: HYDROGEOLOGICAL SETTING

The Edwards Aquifer is composed of regionally extensive carbonate rocks and is one of the most permeable aquifers in the world. The Edwards Aquifer is divided into two segments, namely, the San Antonio segment, which extends from a groundwater divide near the city of Brackettville in Kinney County to a groundwater divide near the city of Kyle in Hays County (Figure 5.1). The San Antonio segment includes all or parts of Uvalde, Medina, Bexar, Comal, and Hays Counties. The San Antonio segment is approximately 180 mi long and 5–40 mi wide and discharges into the Comal Springs and San Marcos Springs (Hamilton et al. 2008). The northern portions of the aquifer extend to the outcrop regions of the Edwards rocks, while the "bad water line" (i.e., 10,000 mg/L of total dissolved solids or the fresh water–saline water interface) is assumed as the southern boundary (Lindgren et al. 2004).

The Barton Springs segment extends from Kyle to south Austin in Travis County (Figure 5.1), covering an area of approximately 130 mi^2 and discharging into the Barton Springs and Cold Springs in the vicinity of Austin, Texas (Barrett and Charbeneau1997). The Barton Springs segment is bounded by a hydraulic divide on the south along Onion Creek and by the Colorado River in the north. The Mount Bonnell fault is taken as the western boundary and the aquifer is considered to end at the bad water line along the east (Scanlon et al. 2003).

Four depositional provinces were formed during the Lower Cretaceous era—the central Texas platform on the Edwards Plateau, the Maverick Basin, the Devils River trend, and the San Marcos platform (Maclay 1995). Three of these provinces (Maverick Basin, Devils River trend, and San Marcos platform) occupy much of the present-day Edwards Aquifer. The Edwards Aquifer is underlain by the Trinity Aquifer, which is composed of sediments from the Glen Rose limestone formation deposited during the early portions of the Lower Cretaceous era and, for the most part, is overlain by Anacacho limestone, Austin chalk, Eagle Ford group, Buda limestone, and Del Rio clay deposits from the Upper Cretaceous era (Figure 5.2). The recharge zone of the Edwards Aquifer occurs along the northern boundary where it outcrops uncomfortably (Puente 1978).

Abbott (1975) has described the geochemical evolution of the Edwards Aquifer. Because it is a karst system, the aquifer is characterized by a highly heterogeneous and anisotropic hydraulic conductivity distribution. Hovorka et al. (1998, 2004), Mace and Hovorka (2000), and Halihan et al. (2000) present information pertaining to the matrix, fracture, and conduit permeabilities and suggest the locations of major conduits. The conduit development is more pronounced in the deeper portions of the aquifer, which tend to be saturated, than the shallower portions, which may only retain water for relatively short periods. The geochemistry of the aquifer sediments was initially composed of aragonite, calcite, dolomite, and gypsum. These minerals have been replaced by calcite within the aquifer (Hovorka et al. 2004). The reader is referred to Lindgren et al. (2004) and the above references for additional details pertaining to the hydrogeologic conceptualization of the Edwards Aquifer. In addition, literature reviews pertaining to the Edwards Aquifer have been compiled by Menard (1995) and Esquilin (2004) and provide useful hydrogeochemical information.

The aquifer thickness ranges from about 450 ft near the recharge zone in Bexar, Comal, and Hays Counties to about 1100 ft in Kinney County. The flow of water in the aquifer is highly heterogeneous and anisotropic and, to a large extent, depends on the extent of karstification. Generally speaking, the flow in the aquifer occurs under unsaturated conditions in caves by turbulent flow in solution-enlarged conduits (Gale 1984). Worthington (2001) reports an average velocity along the conduits of about 6000 ft/day. The hydraulic conductivity and transmissivity obtained using the pump tests reflect the combined effects of the matrix, fracture, and conduit permeability, and each varies by several orders of magnitude. The transmissivity in the recharge area is less than 430,000 ft^2/day and that in the confined portions of the freshwater zone is estimated to range from 430,000 to 2,200,000 ft^2/day. The transmissivity near the fresh water–saline water boundary is extremely

	Maverick Basin	Devils River trend	San Marcos platform	Hydrogeologic unit
Upper Cretaceous	Anacacho limestone and Austin chalk			Upper confining unit
	Eagle Ford group (small)			
	Buda limestone (small)			
	Del Rio group (very small)			
Lower Cretaceous	Solomon Peak Formation	Devils River limestone	Georgetown Formation (small)	Edwards Aquifer
	West Nueces and McKnight Formations		Edwards group	
	Glen Rose Formation	Glen Rose Formation	Glen Rose Formation	Glen Rose Formation

FIGURE 5.2 Hydrostratigraphic structure of the Edwards Aquifer Formation. (Modified from Lindgren, R.J., Dutton, A.R., Hovorka, S.D. et al., Conceptualization and simulation of the Edwards Aquifer, San Antonio region, TX. Reston, VA: U.S. Geological Survey, 2004; Maclay, R.W. Geology and hydrology of the Edwards Aquifer in the San Antonio area, Texas. Reston, VA: U.S. Geological Survey, 1995.)

low and on the order of 100 ft^2/day (Maclay and Land 1988). The specific yield of the aquifer ranges from 0.05 to 0.2 in the San Antonio segment and from 0.005 to 0.06 in the Barton Springs segment (Maclay and Land 1988; Senger and Kreitler 1984). As can be seen, the aquifer is capable of accommodating and transmitting significant amounts of water, but has limited storage capabilities when compared with clastic aquifer formations.

5.3 RECHARGE AND DISCHARGE

The recharge to the Edwards Aquifer occurs in the north and northeastern parts of the Balcones fault zone, where streams lose flow. The recharge from the losing creeks is considered to be a major recharge mechanism in the Barton Springs segment of the aquifer as well (Barrett and Charbeneau 1997). Due to direct precipitation on the aquifer, the diffuse recharge is estimated to be about 15% of the total annual recharge (Scanlon et al. 2003). The regional groundwater flow in the San Antonio segment consists of a flow from the unconfined formation through the confined formation to eventual discharge into the San Marcos Springs and Comal Springs. The regional flow in the Barton Springs Formation is from Onion Creek toward Barton Springs. In addition to the recharge, the Edwards Aquifer also receives some water from the Trinity Formation. The underflow from the Trinity Formation is estimated to be about 2% of the total recharge in the San Antonio segment and about 9% in the Barton Springs segment (Mace et al. 2000). In addition to the recharge, the regional flow is affected by anthropogenic pumping and discharges to the springs in the aquifer, which is discussed next.

The discharges from the Edwards Aquifer are due to springflows and withdrawals by humans. The springflows in the San Antonio segment, particularly in the Comal Springs and San Marcos Springs, are depicted in Figure 5.3. As can be seen, the springflows vary considerably and were lowest during the 1950s, when central Texas experienced severe and prolonged droughts. Well yields of greater than 1000 gal/min have been noted in the confined portions of the Edwards Aquifer, and

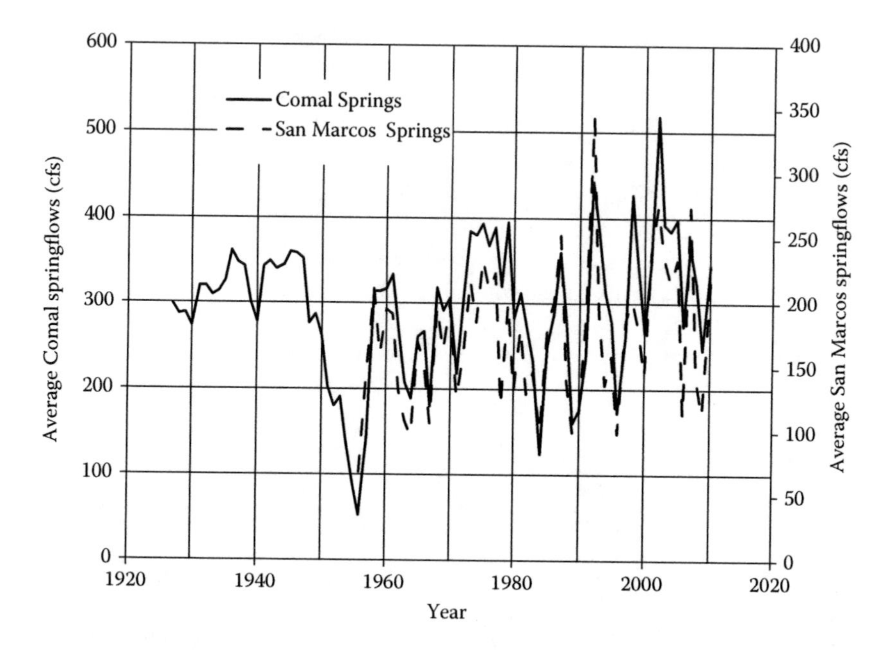

FIGURE 5.3 Annual average Edwards Aquifer San Antonio pool discharges into Comal Springs and San Marcos Springs. (Adapted from EAA. 2011. http://edwardsaquifer.org/display_policies_rules_s.php?pg=groundwater_management_plan. Accessed October 2011.)

yields of several hundred gallons can be obtained from the unconfined portions. More than 60% of the groundwater use in Uvalde County and more than 80% in Medina County are for irrigation purposes. Groundwater withdrawals for municipal uses are another important source of discharge from the Edwards Aquifer. The withdrawals by the city of San Antonio are depicted in Figure 5.4 and are seen to steadily increase, as the population of Bexar County increased from nearly 293,000

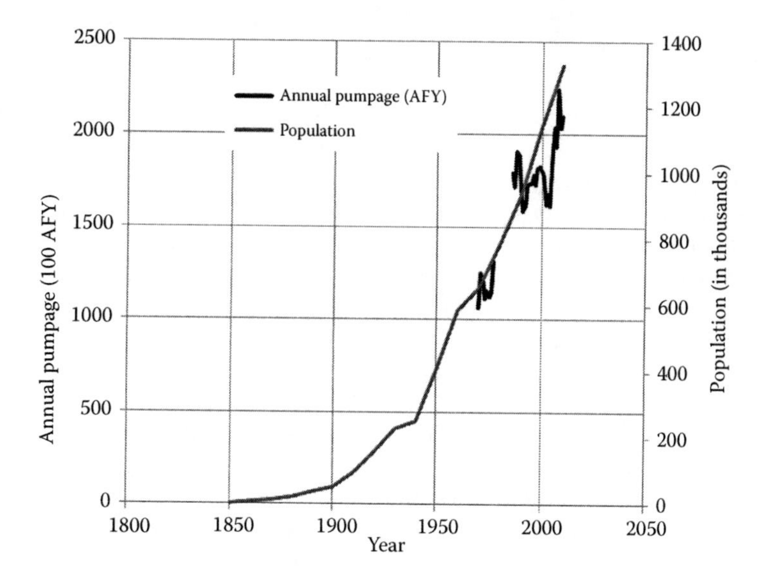

FIGURE 5.4 Population and annual groundwater withdrawals by the city of San Antonio. (Adapted from U.S. Census. United States Census Bureau quickfacts. 2010. http://www.census.gov/popfinder/. Accessed October 2011; EAA. 2011. http://edwardsaquifer.org/display_policies_rules_s.php?pg=groundwater_management_plan. Accessed October 2011.)

in 1939 to over 1.7 million in 2010. Municipal water withdrawals account for more than 85% of the total groundwater extractions in Bexar, Hays, Kinney, and Travis Counties, which are predominantly urban. The withdrawals in the San Antonio segment are concentrated around the city of San Antonio (Bexar County) and the city of Uvalde.

The increased groundwater withdrawals from the Edwards Aquifer have led to conflicts between municipal interests and rural and environmental groups. These conflicts have paved the way for creating a complex regulatory framework to manage the aquifer and to ensure that both the economic demands and the ecological needs are met in a sustainable manner. The conflicts between the municipal and environmental groups have also had a profound impact on the agricultural activities in the region, along with the conflicts between agricultural and environmental groups. The conflict over the Edwards Aquifer waters and the creation of the current aquifer management framework are discussed next.

5.4 REGULATORY FRAMEWORK FOR MANAGEMENT OF THE EDWARDS AQUIFER

Historically, the groundwater resources in Texas have been considered private property and the landmark "rule-of-capture" decision in the *Houston & Texas Central Railroad Co. v. East* by the Texas Supreme Court in 1904 has been the basis of the regulation (or lack thereof) of groundwater resources (Potter 2004). As per the "rule of capture," a property owner has the right to produce groundwater underneath his/her property as long as it is for beneficial use, even if it limits the ability of the neighbors to withdraw and use the water. This law has aptly been dubbed as the law of the largest pump because it has resulted in a tragedy of the commons situation and an inefficient use of groundwater resources. The recognition of the overexploitation of the aquifer resources under the rule of capture prompted the state of Texas to develop legislation that allowed for the creation of underground water or groundwater conservation districts (UGWCDs or GCDs). While the Edwards Underground Water District (EUWD) was created by the 56th Legislature of Texas in 1959 to protect and preserve the Edwards Aquifer, it did not have significant regulatory authority. In 1989, Uvalde and Medina Counties pulled out of the EUWD because of disagreements over pumping limits and established single county underground water districts. In 1991, the Sierra Club joined the Guadalupe-Blanco River Authority (GBRA) and others to file a suit in the U.S. District Court alleging that the Secretary of the Interior and the U.S. Fish and Wildlife Service (USFWS) failed to protect the endangered species dependent on the aquifer. In February 1993, Judge Lucius Bunton ruled in favor of the plaintiffs, determining that unabated pumping will lead to takings (of water rights) from the endangered and threatened species.

The Edwards Aquifer Authority (EAA) was created by the 73rd Legislature in response to the judge's ruling with the regulatory authority to regulate groundwater withdrawals in the aquifer such that sufficient environmental flows are maintained in the Comal Springs and the San Marcos Springs. The permitted withdrawal (excluding exempt use and those by federal agencies) from the aquifer within the jurisdictional boundaries of the Edwards Aquifer is capped at 572,000 AFY (EAA 2011). As the Edwards Aquifer responds quickly to pumping and drought events, the EAA uses a critical period management plan (CPMP) to reduce withdrawals from the aquifer based on the observed springflows and water levels in the aquifer. The flows in the San Marcos Springs and the Comal Springs as well as the water levels in the J-17 index well in Bexar County are used to regulate the San Antonio pool of the Edwards Aquifer. By the same token, the J-27 index well in Uvalde County is used as an indicator to specify critical management periods in the Uvalde pool of the aquifer. As can be seen from Table 5.1, as much as a 40% reduction from the originally permitted amount may become necessary under prolonged drought situations. In a similar fashion, the Barton Springs Edwards Aquifer Conservation District (BSEACD) uses the Barton Springs discharge and water levels in the Lovelady index well to regulate pumping in the aquifer.

TABLE 5.1
Pumping Reduction Policies of the Edwards Aquifer Authority for the San Antonio Pool

	Critical Period			
Trigger	**Stage I**	**Stage II**	**Stage III**	**Stage IV**
Index well J-17 (MSL)	<660	<650	<640	<639
San Marcos springflow (cfs)	<96	<80	N/A	N/A
Comal springflow (cfs)	<225	<200	<150	<100
Withdrawal reduction	20%	30%	35%	40%

Source: EAA. Edwards Aquifer Authority. 2011. http://edwardsaquifer.org/display_policies_rules_s.php?pg=groundwater_management_plan. Accessed October 2011.

5.5 AGRICULTURE IN THE EDWARDS AQUIFER AREA

Historically, agriculture and ranching have been an important water-use group in the Edwards Aquifer region. Irrigated farming has been used to grow crops, such as cotton, corn, sorghum, nuts, and vegetables. However, as can be seen from Table 5.2, the majority of the farmland is used as pastures, and cattle rearing and ranching activities are predominant in the area. Generally speaking, agricultural activities are more prominent in the western sections of the study area and generally decrease in the eastward direction because of the urban sprawl of the Austin–San Antonio metropolitan corridor and also because of poor quality water.

Water needs for agricultural practices have also been a source of controversy in the Edwards Aquifer region. Historically, the farming interests in the region have fought hard to retain the rule of

TABLE 5.2
Distribution of Farmland and Average Farm Owner Age in Edwards Aquifer Study Area

	Distribution of Farmland for Various Activities (%)			Change in Total Farmland Since 2002 (%)	Average Farm Owner Age
County	**Pasture**	**Cropland**	**Other**		
Bandera	76.43	6.5	17.07	−10.75	62
Bexar	61.75	29.44	8.81	−3.53	60.2
Comal	64.86	19.47	15.67	−5.48	59.8
Edwards	95.06	0.00	4.94	2.00	58.4
Hays	65.99	16.67	17.34	2.33	60
Kendall	77.68	9.95	12.37	−16.65	59.6
Kerr	82.67	6.56	10.77	4.65	60.9
Kinney	91.11	0.00	8.89	8.35	60.1
Medina	63.08	23.2	13.72	−2.04	59.2
Real	81.08	7.01	11.91	−7.31	62.7
Travis	57.42	28.9	13.68	−7.13	59.6
Uvalde	79.11	13.3	7.59	−12.82	60.4

Source: USDA. The agricultural census of 2007. 2011. http://www.agcensus.usda.gov/Publications/2007/Full_Report/Volume_1,_Chapter_2_County_Level/Texas/index.asp. Accessed October 2011.

capture. In 1989, Uvalde County and Medina County pulled out of the original EUWD and formed single county districts over disputes arising from establishing production limits and drought management plans (Votteler 1998). Today, these districts continue to locally regulate the groundwater resources under the broader umbrella of the EAA. The region has also seen conflicts between agricultural and environmental groups, which have had an impact on the agricultural activities. In 1995, the Sierra Club sued the U.S. Department of Agriculture (USDA) under the Endangered Species Act, alleging that the USDA was allowing agricultural activities to harm the species in the aquifer. In 1996, in response to this lawsuit, the USDA was ordered to develop a species conservation plan. An investigation by Schaible et al. (1999), however, indicated that the USDA farm programs do not apparently play a substantial role in the total debate of sharing waters.

While the establishment of the EAA and the subsequent regulation of the Edwards Aquifer are viewed as a careful compromise between the environmental, urban, and agricultural interest groups, several provisions are considered particularly favorable for agricultural interests. In particular, there are provisions that require the EAA to treat agricultural water uses on a par with municipal and environmental needs. The agriculture share is guaranteed to be proportional to the historical share and the farmers are guaranteed a minimum of 2 AFY/acre of land used historically for farming. While the EAA allows for water transfers and leases between different water-user groups, transfers from agricultural permits are limited such that at least 1 AFY/acre must be retained for agricultural use. Agricultural users are only charged $2.00/acre-feet of water, whereas municipal and industrial permit holders are currently charged $39.00/acre-feet of water (EAA 2011). Furthermore, farmers who have planted crops prior to the implementation of the Stage I critical management plan (drought contingency measure) are allowed to continue irrigating to properly close out their crops (EAA 2011). These provisions and rules are intended to address the concerns related to the preservation of irrigated agriculture and the mitigation of economic and social consequences in areas where few profitable alternatives exist.

The land cover within the Edwards Aquifer region in the years 1992, 2001, and 2006 is schematically depicted in Figure 5.5. As can be seen, the urban area has more than doubled during the 1990s. The agricultural area has decreased from over 10% of the study area to about 3.5% over a

Land cover (%)	2006	2001	1992
Water and wetlands	0.8	0.8	0.5
Urban	8.1	7.7	3.6
Barren	0.2	0.1	0.7
Forest	31.8	32.5	44.8
Shrubland	46.0	45.9	25.0
Herbaceous	9.7	9.8	14.9
Pasture/Hay	0.6	0.6	4.5
Cultivated land	2.8	2.7	6.1

FIGURE 5.5 Land cover changes in the Edwards Aquifer area during 1992–1996. Darker regions represent urban areas, while the lighter regions represent agricultural and rangelands.

15-year period. The total area under farms during the years 2002 and 2007 was analyzed to evaluate whether special provisions aimed at preserving agricultural practices in the region were serving their intended purpose, using agricultural census data for these 2 years. The total acreage under agriculture increased in Kerr and Kendall Counties in the north and Uvalde and Edwards Counties in the west. However, the land under agriculture decreased in the remaining eight counties. The reduction in agricultural farmland was particularly significant in Hays and Travis Counties, which lie close to the Austin city limits, as well as in Real and Medina Counties, which are adjacent to the city of San Antonio. Overall, the region saw a loss of over 130,000 acres in the 5-year evaluation period between 2002 and 2007.

The estimated historical use of the Edwards Aquifer has been compiled by the U.S. Geological Survey (USGS) and the EAA. These data provide another indirect approach to evaluate the conflicts between urban and rural interests and temporal shifts between the two water-use groups. The water-use data presented in Figure 5.6 demonstrate how both agricultural and municipal demands continued to rise in the period between 1950 and 1980. In the last few decades, the municipal demand has clearly slowed down but continues to see a slow increase. On the other hand, the agricultural water use has seen a remarkable reduction, particularly after 1995 (i.e., after the creation of the EAA). The average agricultural water use during the period 1975–1995 was approximately 125,000 AFY; however, during the 1996–2007 period, the average agricultural use dropped to nearly 100,000 AFY. On the other hand, the municipal use averages have been 245,000 AFY (during 1975–1995) and 250,000 AFY (1996–2007). The drop in the agricultural water use along with the changes in the land cover clearly point toward diminished agricultural activities, and a vast majority of the agricultural land is now being used as pasture (USDA 2011).

The regulatory mandates of the EAA also include the establishment of groundwater rights and the development of instruments that facilitate water sales and leases. The transfer of water from agriculture to municipal uses is favored in the economic literature as it generally moves water from low-valued use to a higher-valued use and increases economic efficiency (e.g., Titenberg 2005). The theoretical investigation of McCarl et al. (1999) indicates that the emergence of a water market is likely under tight pumping regulations in the Edwards Aquifer region. In particular,

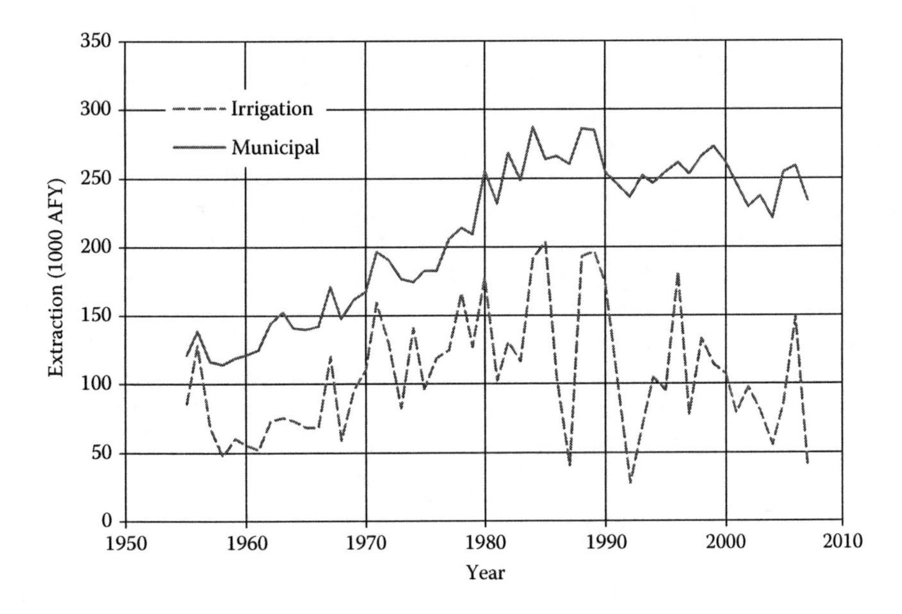

FIGURE 5.6 Historical agriculture and municipal pumping in the Edwards Aquifer area. (Adapted from EAA. 2011. http://edwardsaquifer.org/display_policies_rules_s.php?pg=groundwater_management_plan. Accessed October 2011.)

agricultural guarantees, such as proportional pumping and at least a 2 AFY/acre allocation, are seen to be necessary to preserve agricultural welfare, but also point toward the development of a viable water market where nonagricultural users can obtain agricultural water by making fair compensations. The empirical investigation by Kaiser and Phillips (1999) indicates that many farmers in the Edwards Aquifer region generally have expressed a willingness to engage in such transactions under a suitable compensatory framework. The reallocation of water between the agricultural and the municipal sectors partially explains the decrease in the agricultural activities and dry-year transfers from low-valued crops have been noted in the area (Keplinger and McCarl 2000). Anecdotal evidence indicates that social issues and local concerns could impose huge transaction costs on such transfers and, as such, this is not the only factor affecting the decreasing agricultural activities.

The shift from growing water-intensive crops, such as corn, to ranching-oriented operations (pasture) is an important factor that affects the reduction in agricultural water use. The EAA also requires that farmers exhibit at least 60% irrigation efficiency or implement a groundwater conservation management plan. This strategy has played a significant part in reducing groundwater use for irrigation, which, in turn, has been used to augment other needs (IEA 2002). According to one estimate, approximately 11% of conserved water has been transferred to municipal uses in Uvalde County (IEA 2002).

5.6 FUTURE OF AGRICULTURE IN THE EDWARDS AQUIFER REGION

The sustainability of the fast-paced urbanization of the San Antonio–Austin metropolitan corridor critically hinges on the availability of water resources. The State of Texas Water Plan indicates that the region could experience shortfalls in supplies by the year 2030 (TWDB 2007). The predictions of the general circulation models (GCM) indicate that the problematic climate of south Texas will be further exacerbated with an increase in the annual temperature of about 4°C by the turn of the century (IPCC 2007). While an ensemble of GCMs predict no significant changes in the annual precipitation, the region is projected to experience prolonged droughts and greater intensity storms, which result in greater runoff and limited storage opportunities (Norwine et al. 2007; IPCC 2007). Predictive modeling appears to indicate that climate change is likely to have a significant impact on the soil water availability in the region (Uddameri and Parvathinathan 2007) and, as such, it will increase the need for irrigation. It is projected that the economic losses to the region due to climate change could be almost $10.0 million/year (Chen 2001).

Water use for agricultural and municipal uses is strongly correlated in the Edwards Aquifer region, as it appears from Table 5.3. A county-by-county assessment of the groundwater production data indicates that the competition between urban and agricultural uses has increased over the last few years, both in the urban and rural areas. This competition becomes more intense during the periods of drought as the total amount of available water is insufficient to meet all demands and some sacrifices have to be made. An increasing population in the urban areas will increase the purchasing power of the municipalities to obtain water from irrigators, and farmers may find it profitable to sell their water rights to municipalities rather than engage in agricultural activities. The dry-year water-use reduction incentives are shown to be a major policy instrument to reduce conflicts between urban, rural, and environmental interests (Schaible et al. 1999; IEA 2002). Crop substitution, conversion to dryland farming, and the use of drought-tolerant varieties are likely to increase in the future.

The sustainability of agriculture in the Edwards Aquifer region greatly hinges on the ability of the urban areas to find alternative supplies of water. Clearly, a reduction in the uptake of water by one user group benefits the other user groups. The city of San Antonio has in place a strong municipal water conservation program. The city has a pricing structure that explicitly funds a conservation program and has a water consumption rate goal of 110–133 gpcd, which is one of the best in the

TABLE 5.3
Correlation between Competing Groundwater Uses in the Edwards Aquifer Area

	Irrigation	Municipal	Livestock	Industrial	Spring
Irrigation	1.00				
Municipal	0.49	1.00			
Livestock	0.24	–0.02	1.00		
Industrial	–0.14	0.23	–0.63	1.00	
Springs	–0.44	0.22	–0.09	0.09	1.00

Source: Based on data from EAA. Edwards Aquifer Authority. 2011. http://edwardsaquifer.org/display_policies_rules_s.php?pg=groundwater_management_plan. Accessed October 2011.

nation (NWF 2010). The city has also invested in the development of the Twin Oaks Aquifer storage and recovery (ASR) system in southern Bexar County to store available surplus water from the Edwards Aquifer in the Carrizo Aquifer to meet the city's demands during times of low supplies. The utility of this facility in augmenting San Antonio's water demands during low-flow periods was demonstrated during the droughts experienced in 2006 and 2009. In addition, San Antonio is actively evaluating and pursuing other demand management and supply augmentation strategies, such as the use of treated wastewater for watering golf courses, ocean water desalination, and recharge enhancement of the Edwards Aquifer using check dams. The city's goal of reducing its reliance on the Edwards Aquifer not only helps environmental interests, but also has clear benefits for the agricultural sector in the region. However, certain long-term strategies being pursued by the San Antonio Water System (SAWS), such as the leasing of water in the neighboring Carrizo Aquifer as well as the LCRA/SAWS project in the Gulf Coast Aquifer, have the potential of removing water from the agricultural sector albeit not in the Edwards Aquifer region. Thus, the competition and contention in the Edwards Aquifer region can have negative spillover effects on agriculture in the neighboring areas as well.

On a much broader scale, there is a diminishing trend in the number of students pursuing agricultural studies in the United States (Harlin and Weeks 2001). Furthermore, younger farmers are noted to be more economically diversified and are more apt to simultaneously pursue both agricultural and nonagricultural interests (Caldwell 2011). The data pertaining to the average farmer age and changes in agricultural land are summarized in Table 5.2 and appear to point toward that trend in the study area as well. A detailed evaluation of the U.S. Agricultural Census indicates that the peri-urban counties in the study area have a greater percentage of younger farmers and it is very likely that they will be more inclined to curtail farming than older farmers who may not have other skill sets to diversify. Thus, it is critical that the socioeconomic importance of agriculture to the region be recognized and imparted to younger audiences in the region if agricultural activities are to be sustained at present-day levels.

5.7 CONCLUSION

The overall goal of this study is to undertake a historical assessment of water conflicts in the Edwards Aquifer region in an effort to understand how the evolving groundwater management policies and actions by competing municipal and environmental groups have impacted the agricultural activities in the region. The competing demands on scarce water resources arising from a growing population and a strong interest in protecting endangered species unique to the Edwards Aquifer place constraints on sustaining agriculture in the region. The current regulatory framework in place, however, provides several incentives, such as equal protection of agricultural water rights, subsidized water costs, and provisions to continue irrigation for crop closeouts during droughts. Nonetheless, there has been a net reduction of about 130,000 acres of total farmland between the years 2002 and 2007. These farmland losses are more pronounced in the periurban areas surrounding Austin and San Antonio. The current regulatory framework allows for partial water transfers between agricultural and nonagricultural uses, and there is a general willingness noted among farmers to market water under sufficient economic gains. It is likely that the projected drier, warmer climate for the region may increase the transfer of water from the agriculture to the nonagricultural sector. The city of San Antonio has an aggressive plan to minimize its reliance on the Edwards Aquifer, which will eventually benefit agricultural interests in the region, but may do so at the cost of curtailing agriculture in the surrounding areas of the state. For long-term agricultural sustainability, it is important that the critical role of agriculture in shaping the region's economy is highlighted and is used to encourage younger minds to pursue agriculture-oriented education.

ABBREVIATIONS

EAA	Edwards Aquifer Authority
EUWD	Edwards Underground Water District
FESA	Federal Endangered Species Act
GBRA	Guadalupe-Blanco River Authority
GCDs, UGWCDs	Groundwater conservation districts
GCM	General circulation models
SAWS	San Antonio Water System

REFERENCES

Abbott, P.L. 1975. Calcitization of Edwards Group dolomites in the Balcones fault zone aquifer, south central Texas. *Geology* 2: 359–362.

Barrett, M.E. and R.J. Charbeneau. 1997. A parsimonious model for simulating flow in a karst aquifer. *Journal of Hydrology* 196: 47–65.

Bomar, G.W. 1995. *Texas Weather*, 2nd edn, revised. Austin, TX: University of Texas Press.

Caldwell, J. 2011. Young farmers see a diverse future. http://www.agriculture.com/news/business/young-farmers-see-a-diverse-future_5 ar19008 (accessed October 2011).

Chen, C.C. 2001. Effects of climate change on a water dependent regional economy. *Climate Change* 49: 397–409.

EAA. 2011. Edwards Aquifer Authority. http://edwardsaquifer.org/display_policies_rules_s.php?pg=groundwater_management_plan (accessed October 2011).

Esquilin, R. 2004. Edwards aquifer bibliography through 2003. San Antonio, TX: Edwards Aquifer Authority.

Fisher, D. 2011. America's fast growing cities. Forbes. http://www.forbes.com/2011/05/20/fastest-growing-cities.html (accessed October 2011).

Fridell, J. 2008. *Protecting Earth's Water Supply*. Minneapolis, MN: Lerner.

Gale, S.J. 1984. The hydraulics of conduit flow in carbonate aquifers. *Journal of Hydrology* 70: 309–327.

Halihan, T., R.E. Mace, and J.M. Sharp. 2000. Flow in the San Antonio segment of the Edwards Aquifer: Matrix, fractures or conduits? In *Groundwater Flow and Contaminant Transport in Carbonate Aquifers*, eds. R.E. Mace and J.M. Sharp, pp. 129–146. Brookfield: VT, A.A. Balkema.

Hamilton, J.M., S. Johnson, R. Esquilin, et al. 2008. Edwards Aquifer Authority hydrologic data report 2008. San Antonio, TX: Edwards Aquifer Authority.

Harlin, J. and B. Weeks. 2001. A comparison of traditional and non-traditional students' reasons for enrolling in an agricultural education course. *Journal of Southern Agricultural Education Research* 51: 150–160.

Hovorka, S.D., T. Phu, J.P. Nicot, and A. Lindley. 2004. Refining the conceptual model for flow in the Edwards Aquifer—Characterizing the role of fractures and conduits in the Balcones fault zone segment. Austin, TX: Bureau of Economic Geology.

IEA. 2002. The economic impact of lease payments for irrigation water rights in Uvalde County. Antonio, TX: The Institute for Economic Development at the University of Texas at San Antonio.

IPCC. 2007. Summary for policymakers. Climate change 2007: The physical science basis. Contribution of Working Group I to the Fourth Assessment Report of the Intergovernmental Panel on Climate Change. http://www.ipcc.ch/pdf/assessment-report/ar4/wg1/ar4-wg1-spm.pdf (accessed October 2011).

Kaiser, D. and L.M. Phillips. 1999. Dividing the waters: Water marketing as a conflict resolution strategy in the Edwards Aquifer region. *Natural Resources Journal* 38: 411–444.

Keplinger, K.O. and B. McCarl. 2000. An evaluation of the 1997 Edwards Aquifer irrigation suspension. *Journal of the American Water Research Assocociation* 36: 889–901.

Larkin, T.J. and G.W. Bomar. 1983. *Climatic Atlas of Texas*. Austin, TX: Texas Department of Water Resources.

Lindgren, R.J., A.R. Dutton, S.D. Hovorka, et al. 2004. Conceptualization and simulation of the Edwards Aquifer, San Antonio region, TX. Reston, VA: U.S. Geological Survey.

Mace, R.E. and S.D. Hovorka. 2000. Estimating porosity and permeability in a karstic aquifer using core plugs, well tests, and outcrop measurements. In *Groundwater Flow and Contaminant Transport in Carbonate Aquifers*, eds. I.D. Sasowski and C.M. Wicks, pp. 93–111, Brookfield, VT: A.A. Balkema.

Mace, R.E., A.H. Chowdhury, R. Anaya, and S.C. Way. 2000. Groundwater availability of the Trinity Aquifer, Hill Country area, Texas—Numerical simulations through 2050. Austin, TX: Texas Water Development Board.

Maclay, R.W. 1995. Geology and hydrology of the Edwards Aquifer in the San Antonio area, Texas. Reston, VA: U.S. Geological Survey.

Maclay, R.W. and L.F. Land. 1988. Simulation of flow in the Edwards Aquifer, San Antonio Region, TX and refinements of storage and flow concepts. Reston, VA: U.S. Geological Survey.

McCarl, B., C.R. Dillon, K.O. Keplinger, and R L. Williams. 1999. Limiting pumping from the Edwards Aquifer: An economic investigation of proposals, water markets, spring flow guarantees. *Water Resources Research* 35: 1257–1268.

McCarl, B.A. 2006. Biofuels: Will we see more? An economists view. Texas Renewables '06 Conference. Austin, TX: Texas Renewable Energy Industries Association.

Menard, J.A. 1995. Bibliography of the Edwards Aquifer through 1993. Reston, VA: U.S. Geological Survey.

Norwine, J., R. Harriss, J. Yu, C. Tebaldi, and R. Bingham. 2007. The problematic climate of South Texas 1900–2100. In *The Changing Climate of South Texas*, eds. J. Norwine and K. John, pp. 15–56. Kingsville, TX: Center for Research on Environmental Science and Technology—Research on Environmental Sustainability in Semi-Arid Coastal Areas.

NWF. 2010. Seven ways in which Texas cities can conserve water and what 19 cities around the state are—and are not—doing to make it happen. Joint Publication. National Wildlife Federation and Lone Star chapter of Sierra Club, TX.

Potter III, H.G. 2004. History and evolution of the rule of capture. Austin, TX: Texas Water Development Board.

Puente, C. 1978. Method of estimating natural recharge to the Edwards Aquifer in the San Antonio area, Texas. Reston, VA: U.S. Geological Survey.

Scanlon, B.R., R.E. Mace, M.E. Barrett, and B. Smith. 2003. Can we simulate regional groundwater flow in a karst system using equivalent porous media model? Case study, Barton Springs Edwards Aquifer. *Journal of Hydrology* 276: 137–158.

Schaible, G., B.A. McCarl, and R.D. Lacewell. 1999. The Edwards Aquifer water resource conflict: USDA farm program resource-use incentives? *Water Resources Research* 35: 3171–3183.

Senger, R.K. and C.W. Kreitler. 1984. Hydrogeology of the Edwards Aquifer, Austin area, Central Texas. Austin: Bureau of Economic Geology.

Titenberg, T. 2005. *Environmental and Natural Resources Economics*. Englewood, NJ: Pearson Learning.

TWDB. 2007. *Water for Texas*, Vol. I. Austin, TX: Texas Water Development Board.

TWDB. 2011. State of Texas water plan (Draft). Austin, TX: Texas Water Development Board.

Uddameri, V. and G. Parvathinathan. 2007. Climate change impacts on water resources in South Texas. In *The Changing Climate of South Texas*, eds. J. Norwine and K. John, pp. 109–126. Kingsville, TX: Center for Research on Environmental Science and Technology—Research on Environmental Sustainability in Semi-Arid Coastal Areas.

U.S. Census. 2010. United States Census Bureau quickfacts. http://www.census.gov/popfinder/ (accessed October 2011).

USDA. 2011. The agricultural census of 2007. http://www.agcensus.usda.gov/Publications/2007/Full_Report/Volume_1,_Chapter_2_County_Level/Texas/index.asp (accessed October 2011).

Votteler, T.H. 1998. The little fish that roared: The endangered species act, state groundwater law and private property rights collide over the Texas Edwards Aquifer. *Environmental Law* Winter Issue: 845–905.

Worthington, S.R.H. 2001. Depth of conduit flow in unconfined carbonate aquifers. *Geology* 29: 335–338.

6 Sustaining Groundwater Use in South Asia

Meharban Singh Kahlon, Rattan Lal, and Pritpal Singh Lubana

CONTENTS

6.1 INTRODUCTION

Sustainable groundwater (GW) management is important to revitalize the green revolution (GR) in South Asia (SA). Together, the SA nations have over 2800 km^3 or billion cubic meters (BCM) of renewable freshwater resources (Gleick 1998). The GW overdraft and unsound management strategies now threaten the sustainability of the GR in the SA region. India is the largest GW user in the world. The annual replenishable GW resources of India are 433 BCM compared with the net annual GW availability of 399 BCM and the overall GW development of 58%. The renewable water resources of India are about 4% of the global availability (IWRS 1998). More than 60% of India's irrigated agriculture depends on GW, which was only 29% in 1950–1951 (WPB 2002). Similarly, in Pakistan, Bangladesh, and Nepal, GW is used in over 75% of the irrigated areas. At present, GW irrigation has surpassed surface (canal) irrigation as the primary source of food production and income generation in many countries of SA. GW depletion is a big problem in India and Pakistan (Postel 1999), whereas GW pollution is a major issue in Bangladesh and Nepal (Khan 1994). Due to the excessive mining of GW, the water table (WT) has been depleting by as much as 30 cm/year in northwestern India. GW pumping with electricity and diesel pump sets also accounts for an estimated 16–25 Tg of carbon (C) emission in the atmosphere. Likewise, there is also the problem of water logging and rises in the water table in some parts of India and Pakistan caused by seepage from water sources (e.g., unlined canals), excessive irrigation by flooding, and lack of drainage to safely remove the excess water. The key question for policy makers, planners, and researchers is how to exploit the resource without exhausting its supply and without damaging the environment. This chapter outlines the concepts and approaches for sustainable GW use and describes the factors affecting its availability without adversely affecting the hydrogeological balance, crop production, and the environment. This chapter also discusses a range of techniques for the sustainable management of GW, including the need for rainwater harvesting (RWH) and artificial recharge, and it outlines some relevant policy interventions at local and regional levels.

GW is a primary source of fresh water in many parts of the world and it constitutes about 89% of the total freshwater resources of the planet (Gleick 1993; Shah et al. 2007; Menon 2007). The growth of GW resources in selected SA countries is closely linked to the dependence on extensive irrigation in SA (Figure 6.1; Shah and Mukherjee 2001; Roy and Shah 2002; Shah 2007; Scott and Sharma 2009). GW irrigation has played a significant role, along with surface (canal) irrigation, in creating the GR and enhancing the livelihoods of hundreds of millions of people. In India, food production increased from 50.8 million Mg (M Mg) in 1950–1951 to 215.0 M Mg in 2005–2006, partly because of the increase in the irrigated area under surface and GW irrigation (Figure 6.2). Indian agriculture primarily depends on

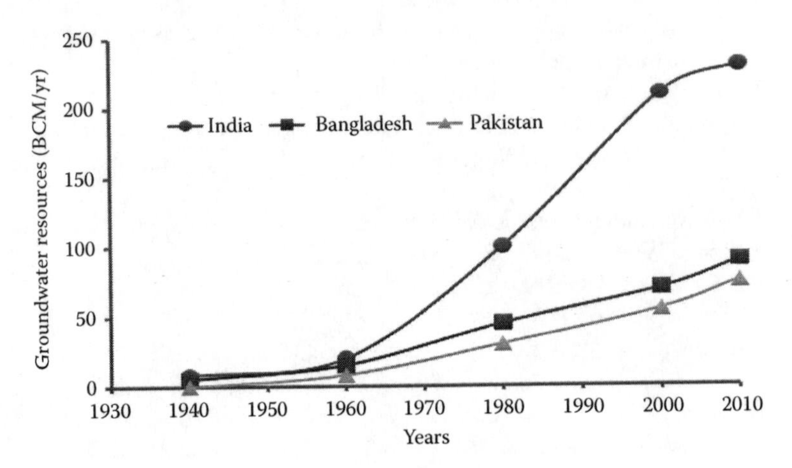

FIGURE 6.1 The growth of groundwater in South Asia. (Modified from Shah, T., *The Agricultural Groundwater Revolution: Opportunities and Threat to Development*, CABI Publishing, Wallingford, UK, 2007.)

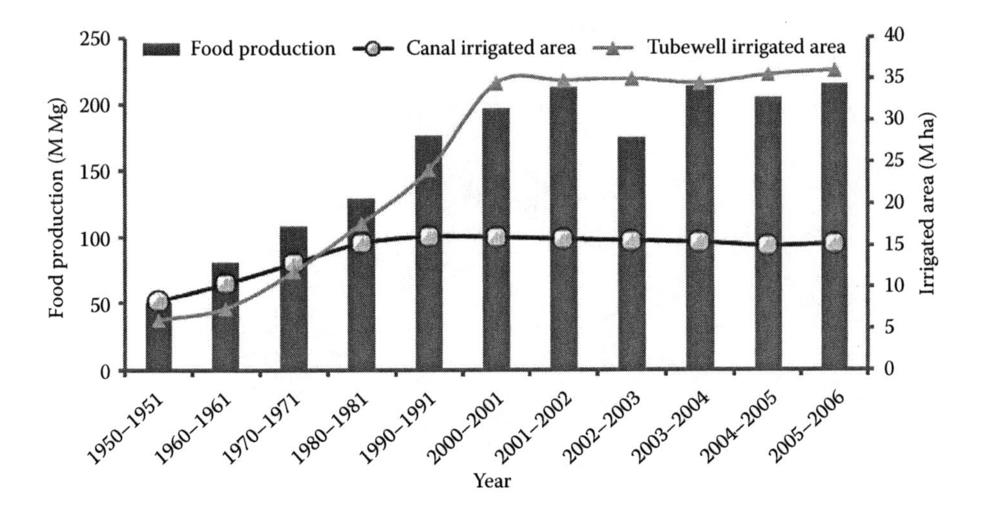

FIGURE 6.2 The increase in food production with an increase in the irrigation facilities in India. (Modified from GOI, Ministry of Agriculture, as given in Vaidyanathan, 1999: 61; Central Water Commission, Water and Related Statistics 1994, p. 81, 1995/96: Fertilizer Association of India, Fertilizer Statistics 1998/99; Planning Commission, GOI, Eleventh 5 Year Plan, Vol. III, 2008.)

GW obtained by large numbers of pump sets (Shah 2009a). Between 1970 and 1994, the area under GW irrigation increased by 105%, compared with an increase of only 28% of the area under surface water irrigation in India (Shah 2002). The number of mechanized wells and tube wells in India increased from less than 1 million in 1960 to 19 million in 2000, registering the maximum pump sets in the world. In Pakistan's Punjab, the number of wells and/or tube wells increased from a few thousand in 1960 to 0.5 million in 2000. In Bangladesh, which hardly had any GW irrigation until 1960, the area under irrigated cropland by tube wells increased from 4% in 1972 to 70% in 1999 (Mainuddin 2002). In India, Bangladesh, and Nepal, the modal pump size is 6.5 hp and the average duration of operation is 400–500 h/year (Shah 1993). GW depletion is primarily caused by sustained pumping throughout SA. The mean GW level in the region is decreasing at a rate of 4.0 ± 1.0 cm/year (an equivalent water height of 17.7 ± 4.5 BCM/year), as observed over the Indian states of Rajasthan, Punjab, and Haryana (including Delhi) (Rodell et al. 2009). This rate of GW depletion is equivalent to a net loss of 109 BCM of water, which is double the capacity of India's largest surface water reservoir. In the current scenario, water has become the single most important constraint to increasing food production (Seckler et al. 1998a,b; Dennehy et al. 2002; Chatterjee and Purchit 2009). In humid areas, rainfall distribution is such that there is no need for additional water application. In arid and semiarid areas, where crop failures occur due to water deficiency, farmers resort to supplemental irrigation. Because the SA region has a predominantly arid and semiarid climate, the rainfall is not adequate to recharge the GW, and excessive mining of GW for irrigation causes WT depletion. Only 58% of India's GW is recharged every year (www.indusenviro.com Regulatory News Update No. 39 August, 2009).

In the Indian subcontinent, GW use increased from 10–20 BCM in 1950 to 240–260 BCM in 2000 (Shah 2005). Farming has turned into a race to keep food production ahead of the population, which unfortunately requires the extraction of more and more natural resources. High-yielding varieties (HYV) in combination with fertilizer and pesticides produce two to three times more than the traditional varieties (Lappe et al. 1998), but they also require three times more water. In terms of water use, therefore, they are less than half as productive (Shiva 1991). Rice (*Oryza sativa* L.), which is the second dominant crop of the SA region, further adds to the decline in GW in the region because it demands six times more water than wheat (*Triticum aestivum* L). Another reason for the depletion of GW is the increasing demand from the domestic, industrial, and hydroelectric sectors. In Pakistan's Punjab, the WT is dropping at 3–9 m/year. The growing shortages of GW are leading

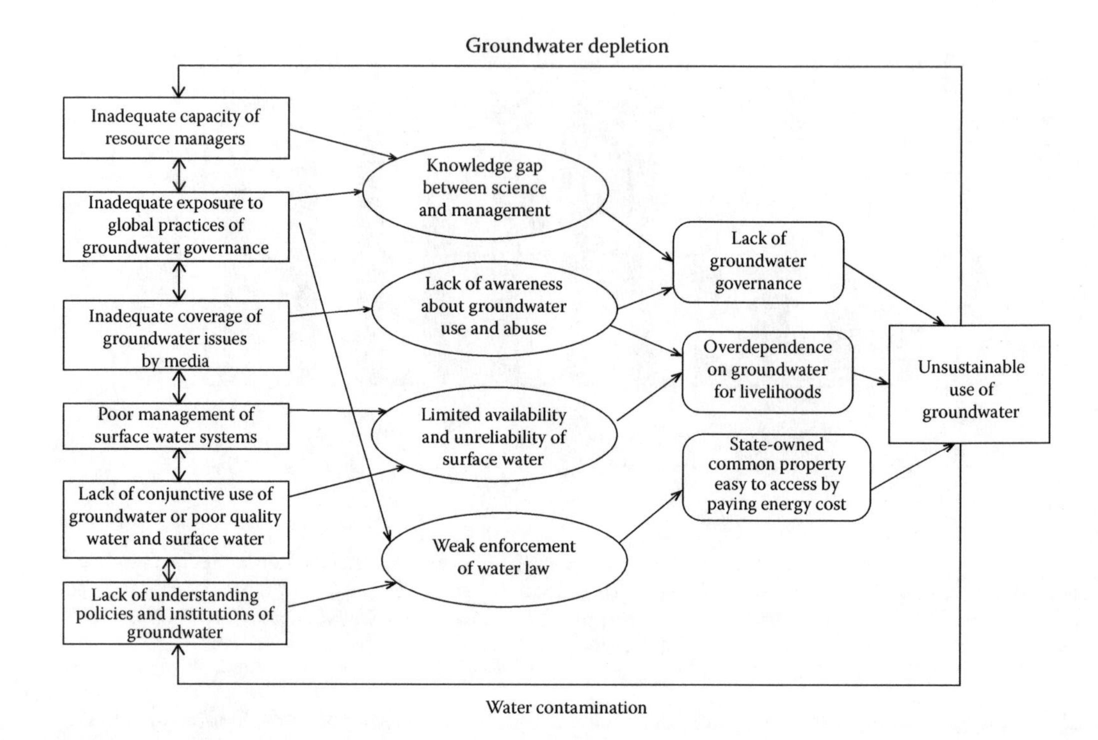

FIGURE 6.3 The factors responsible for the unsustainable use of groundwater. (Adapted from Challenge Program on Water and Food [CPWF] project report, IWMI, No. 42, 2009.)

to social and ethnic conflicts. Some of the negative effects of GW depletion include the drying up of wells, the reduction of water in streams and lakes, the deterioration of the water quality, increased pumping and electricity costs, and land subsidence.

There are several reasons for overdrafting (Figure 6.3), but the electrical supply and pricing are the primary reasons for GW irrigation pumping, particularly in India (Scott and Shah 2004). Freshwater withdrawal for various uses is the highest at ~380 BCM/year in India, 155 BCM in Pakistan, 22 BCM in Bangladesh, and 6 BCM in Sri Lanka (Subramanian 2000). By 2040, India as a whole will come under the group of water-stressed countries if measures are not taken soon to ensure sustainable GW usage. It will cause a reduction in agricultural output and shortages of potable water, leading to extensive socioeconomic stresses. Khair et al. (2010) suggested that resource management should be based on information gathering and resource planning by establishing appropriate systems for resource monitoring on a regular basis; demand side management through the registration of users through permits and a licensing system; appropriate laws and a regulatory mechanism; a pricing system for efficient water use; the promotion of conjunctive water use; supply side management through RWH and surface water use for increasing recharge; and improving water conservation and undertaking GW management at river basin level. Molden (2007) suggested that water can be efficiently managed by changing the way we think about water and agriculture, by improving access to agricultural water and its use, by managing agriculture to enhance the ecosystem service, and by increasing the productivity of water and upgrading the rainfed systems.

6.2 SIGNIFICANCE OF GW

Water is essential for all living beings—plants, animals, and people. GW is fresh water retained in the pores of soil and rocks. It is also the water that flows within aquifers below the WT. It is a unique resource, widely available, providing security against droughts and yet it is closely linked to the surface water resources and the hydrological cycle. Its reliable supply, uniform quality, temperature, relative turbidity, low pollution concentration, minimal evaporation losses, and low cost of development are the critical attributes that make GW a precious resource. GW is a finite but renewable resource that is intrinsically linked to surface water and other natural resources (Custodio et al. 2004). It is generally a more reliable freshwater resource than surface water and can be readily developed to meet human needs and agricultural demands.

6.2.1 Economic Perspectives of GW

During the 1980s, the contribution of GW irrigation to India's gross domestic product (GDP) was around 10% (Daines and Pawar 1987). Extrapolating to the present, the size of the GW irrigation economy of India would be \$50–\$55 billion during the 2000s. India, Pakistan, and Bangladesh have active markets in pump irrigation services in which tubewell owners sell GW irrigation to their neighbors at a price that exceeds their marginal cost of pumping. Giordano and Villholth (2007) reported that this price offers a market valuation of GW use in irrigation. In India, for instance, a large number of farmers paid their neighboring borewell owners \$0.04 per m^3 for GW irrigation in 2003; applying this price to the annual GW use of 200 BCM yields \$8 billion as the annual economic value of the GW used in Indian agriculture (Table 6.1). For the Indian subcontinent as a whole, the corresponding estimate is around \$10 billion (Shah 2007). In many parts of water-scarce India, water buyers commonly enter into pump irrigation contracts, offering as much as one-third of their crop share to the irrigation service provider. In water-abundant areas, by contrast,

TABLE 6.1
Groundwater Resources, Economy, and Contribution to Irrigation in South Asia

	India	Pakistan	Bangladesh	Nepal
Annual GW use (BCM)[a]	189–204	54.5	31.2	0.37
Percentage withdrawal to renewable GW[b]	45.3	109.1	52.4	—
Percent of national share of global withdrawals[b]	28.9	9.1	1.7	—
Number of wells (million)[a]	21.0	0.5	0.8	0.06
Average output/well (m^3/h)[a]	25–27	100	30	30
Average hours of operation/ (well year)[a]	360	1,090	1,300	205
Percent of population dependent on GW	55–60	60–65	64	—
Percent of GW as part of water resource used	53	34	69	12
Total irrigation water use (million m^3/year)	460,000	150,600	12,600	28,700
Price of pumping irrigation (\$/h)[a]	1–1.1	2	1.5	1.5
Value of GW used per year in billion dollars[a]	7.6–8.3	1.1	1.6	0.02

[a] Shah, T., *The Agricultural Groundwater Revolution: Opportunities and Threat to Development*, CABI Publishing, Wallingford, UK, 2007.
[b] FAO, AQUASTAT (http://www.fao.org/nr/water/aquastat/main/index.stm; 40, 41).

TABLE 6.2
Contribution of Surface Water (SW) and Groundwater (GW) to Irrigation and Total Agricultural Output of India

Year/Indicator (At 1990 Dollar/Rupee Exchange Rate)	1970–1973	1990–1993	Change (%)
Avg. agri. productivity ($/ha)	261.4	470.3	79.9
Contribution of SW ($/ha)	41.3	62.6	51.6
Contribution of GW ($/ha)	13.3	74	456.4
Contribution of SW (million $)	4,680	7,005	49.7
Contribution of GW (million $)	1,320	7,297	452.8
Contribution of SW as a percentage of the total agricultural output	15.5	13.9	−1.6% points
Contribution of GW as a percentage of the total agricultural output	4.4	14.5	+10.1% points
Total agricultural output/year (million $)	28,282	49,891	76.4

Source: DebRoy, A. and Shah, T., *Intensive Use of Groundwater: Challenges and Opportunities*, Swets & Zetlinger, The Netherlands, 2003.

the purchased pump irrigation cost generally amounts to 15%–18% of the gross value of the output it supports (Giordano and Villholth 2007).

6.2.2 Land Productivity and Food and Social Security

The contribution of the GW irrigated area to the total agricultural output of India is ten times more than that of the surface irrigated area (Table 6.2). Evidence in India suggests that a crop yield per cubic meter of water applied on GW irrigated farms tends to be three times higher than that applied on surface water irrigated farms (Dhawan 1989). Land productivity is also observed to be more under tubewell irrigation than under surface (canal) irrigation (Table 6.3). When irrigation is applied through GW, it adds a lot of nutrients, particularly sulfur (S) and

TABLE 6.3
Land Productivity per Net Irrigated Hectare by Sources of Irrigation (Mg/ha in Food Grains Energy Equivalent Units)

State	Wells	Canal Irrigation	Tanks
Andhra Pradesh	5.7 (67.6)[a]	3.4	2.0
Tamil Nadu	6.5 (150)[a]	2.6	2.3
Punjab	5.5 (71.9)[a]	3.2	—
Haryana	5.7 (137.5)[a]	2.4	—
Madhya Pradesh	2.8 (40)[a]	2.0	1.5
Karnataka	4.2 (20)[a]	3.5	2.3

Source: The World Bank, *Groundwater Regulation and Management in India*, The World Bank, vol II, p.7, Washington, DC and Allied Publishers, New Delhi, India, 1991.

[a] Figures in parentheses in the second column are the percentage by which productivity in well-irrigated areas is higher than in canal irrigated areas.

TABLE 6.4
Growth of Wells (Thousands) in India

Year	Dug Wells	Private and Public Tube Wells	Total
1951	3,860	5	3,865
1980	7,786	2,165	9,951
1985	8,742	3,405	12,147
1990	9,407	4,817	14,224
1992	10,120	5,446	15,566
1997	10,501	6,833	17,334

Source: Planning Commission report, India 2007: Groundwater Management and Ownership.

potassium (K), to enrich the soil. Since the 1990s, food production has hardly been affected by a single drought (Sharma and Mehta 2002), despite a string of two to three drought years, because of GW availability during a stress or drought period. Thus, GW development has been a major factor in India's food security and has been a key determinant in transforming Bangladesh from an endemic rice importer into a net rice exporter (Palmer-Jones 1999). Wells have also created greater spatial, social, ethnical, and interpersonal equity in the access to irrigation, especially when compared with large public canal irrigation systems that have created islands of agrarian prosperity (Giordano and Villholth 2007). Whereas the amount of GW contributing to agriculture is less than that of surface water on the global scale, its unique advantages (i.e., reliability, accessibility, on-demand availability, less capital investment, and high productivity) outweigh those of access to surface water. In India, Pakistan, Bangladesh, and Nepal, the GW contribution to the irrigated areas exceeds that of surface water (DebRoy and Shah 2003; Shah et al. 2006). It is also observed that in northwestern India, despite massive investments in canal irrigation, the bulk of the irrigation is delivered by wells and tube wells. From 1950 onward, there has been a continuous increase in the number of wells dug and the number of private and public tube wells (Table 6.4). No wonder, then, that in developing regions of SA, GW development has become the central element of livelihood creation programs for the poor (Kahnert and Levine 1993; Shah 1993; Rosegrant 1997; Calow et al. 1997).

6.3 GW RESOURCES IN SOUTH ASIAN REGION

6.3.1 Water Resources in India

The GW resources of India constitute one of its vital assets. Kumar (2006) assessed that the total annual replenishable GW resources of India are 432 BCM (contributed by rainfall and other sources in the proportion of 67% and 33%, respectively). Other sources include canal seepage, return flow from irrigation, seepage from water bodies, and artificial recharge due to water conservation structures. Together, the annual utilizable surface water and GW are estimated at 690 BCM (Kumar et al. 2005). India's GW resources would be about 300 M ha-m or about 10 times the annual precipitation (Shah 2005). India uses an estimated 230 BCM of GW every year, of which 212 BCM (92%) is used for irrigation purposes and 18 BCM (8%) is used for domestic and industrial purposes (Chatterjee and Purchit 2009). According to the World Bank's report, GW supports >80% of the rural and urban water supplies in India (www.waterworld.com/./waterworld/world./2010//Groundwater-supplies-depleting-at-alarming-rate-in-India.html). The southwest monsoon, the most prevalent contributor of rainfall in India, recharges about 73% of the country's annual replenishable GW during the summer. The irrigation potential was only 22.6 M ha in 1950–1951, but it has now reached about 100 Mha. The projections for the future population and food requirements of the country indicate that the population of India may stabilize at around 1.6–1.7 billion by 2050 AD, which would require about 450 M Mg of food grains annually at the desired level of food consumption. It is necessary to equip irrigation on at least 130 Mha for food crops alone and on 160 Mha for all crops to meet the demands of the country in 2050 AD and ensure food security.

In India, as much as 195 BCM of water is available out of a total annual precipitation of about 4000 BCM. This water is harnessed as both surface water and GW. Surface utilization is mainly through dams constructed across rivers served by larger catchments and tanks served by smaller catchments. The GW is utilized mainly through both open wells and tube wells. The canal irrigated area has increased from 8.3 to 18 Mha during 1950–1951 and 1999–2000 (Table 6.5). Likewise, the well irrigated area has increased from 6 to 34 Mha during the 50 years that ended in 2010. During the period from 1950 to 1999, the area under irrigation with tanks decreased from 3.6 to 2.7 Mha. Tanks are mostly concentrated in regions where other sources of irrigation are either less or are completely absent. The annual potential GW recharge augmentation from the canal irrigation system is about 89.5 BCM (Kumar et al. 2005; CWP 2005). The available GW resource for irrigation is 360 BCM, of which the utilizable quantity is 325 BCM (90%). Precipitation and snowmelts are the only sources of fresh water that are stored in aboveground reservoirs or in GW aquifers below the surface. The total average annual flow per year for Indian rivers is estimated at 1953 BCM (Rama Krishna and Shiva Kumar 2008).

6.3.2 GW Development in Pakistan

Pakistan has a long history of using GW for irrigated agriculture. Until the 1960s, GW extraction was done by means of open wells with a rope and bucket, Persian wheels, karezes (a form of subterranean aqueduct), reciprocating pumps, and hand pumps. Large-scale extraction and use of GW for irrigated agriculture in Pakistan started during the 1960s with the launch of Salinity Control and Reclamation Projects (SCARPs). Under this program, 16,700 tube wells (supplying an area of 2.6 Mha) with an average discharge of 0.09 m^3/sec were installed to lower the WT, to create favorable crop growth conditions in the root zone, and to reduce the risk of soil salinization (Bhutta and Smedema 2007). The pumped GW was discharged into the existing public canal system to increase the irrigation supplies (Qureshi et al. 2008). The demonstration of SCARP tube wells was followed by an explosive development of private tube wells with an average discharge of 0.03 m^3/sec. The provision of subsidized electricity by the government and the introduction of locally made diesel engines provided an impetus for a dramatic increase in the number of private tube wells. Currently, about 0.8 million small capacity private tube wells are operational in Pakistan, of which more than 90% are used for agriculture (Qureshi et al. 2008). The geographic distribution of electric and diesel irrigation pumps for Pakistan and other SA countries is presented in Table. 6.6. Investments in the installation of private tube wells are of the order of Rs. 25 billion (US$400 million), whereas the

TABLE 6.5
Areas and Percentage Shares of Different Irrigation Sources in India

Sources of Irrigation	1950–1951 Area (Mha)	Percentage	1999–2000 Area (Mha)	Percentage
Canals	8.3	40	18	31.5
Wells and tube wells	6	29	33.6	58.7
Tanks	3.6	17	2.7	4.7
Other sources	3	14	2.9	5.1
Total	20.9	100	57.2	100

Source: Central Statistical Organization (CSO), Statistical Abstract, India, 2002.

TABLE 6.6
Geographic Distribution of Electric and Diesel Irrigation Pumps (%) in South Asia

Country	Diesel (%)	Electric (%)
Pakistan[a]	89.6	10.4
Bangladesh[b]	96.7	3.3
Eastern India[c]	84.0	16.0
Western India[c]	19.4	80.6

[a] Pakistan Agricultural Machinery Census 2004.
[b] Bangladesh: Mandal, 2006.
[c] Third Minor Irrigation Census 2000–2001.

annual benefits in the form of agricultural production are about Rs. 150 billion (US$2.5 billion) (Shah et al. 2003a). GW is currently providing more than 50% of the total crop water requirements with the flexibility of its availability as and when it is needed (Shah 2007). Punjab Province has taken the lead in the development of private tube wells (PWP 2001; GOP 2000). As a result, GW extraction increased from 9 to 45 BCM between 1965 and 2002 (Bhutta 2002; World Bank 2007).

6.3.3 Surface and GW Resources of Bangladesh

Bangladesh is endowed with plenty of surface water and GW resources. The average annual internal renewable water resources of Bangladesh are 105 BCM (WRI 2000). Bangladesh is located within the flood plains of three great rivers: the Ganges, Brahmaputra, and Meghna. The combined discharge of the three main rivers is among the highest in the world. Peak discharges are of the order of 1×10^{-4}, 7×10^{-5}, and 1.8×10^{-4} BCM/sec for Brahmaputra, Ganges, and Meghna, respectively. On average, almost 1106 BCM of water crosses the borders of Bangladesh annually, of which 85% crosses between June and October. About 54% (599 BCM) is contributed by the Brahmaputra, 31% (344 BCM) by the Ganges, and nearly 15% (163 BCM) by the tributaries of the Meghna and other minor rivers (www.fao.org/nr/water/aquastat/.../bangladesh/index.stm).

The availability of GW resources in Bangladesh is determined by the properties of the aquifers and the volume of annual recharge. The key factors that determine GW availability include the capacity of the country's aquifers to store water and its characteristics, which govern economic withdrawal of GW for irrigation, domestic, and industrial needs. The sources of recharge are rainfall, flooding, and stream flow in rivers. The quaternary alluvium of Bangladesh constitutes a huge aquifer with reasonably good transmission and storage properties. Heavy rainfall and inundation during the monsoon season help the aquifers to be substantially recharged annually. The total internal renewable water resources of the country are estimated at 105 BCM, with a negligible overlap (the volume of water resources common to both surface and groundwater). This includes 84 BCM of surface water produced internally as stream flows from rainfall and 21 BCM of GW resources produced within the country. In 2008, the total water withdrawal was estimated at about 35.9 BCM of which about 31.5 BCM (88%) was used for agriculture, 3.6 BCM (10%) for municipalities, and 0.8 BCM (2%) for industries. Approximately 28.5 BCM, or 79% of the total water withdrawal, comes from GW and 7.4 BCM, or 21%, from surface water. Arsenic (As) contamination in Bangladesh reached a drastic level during the last decade due to the overexploitation of GW and inefficient water resources management (Safiuddin and Karim 2003).

6.3.4 GW Resources in Nepal

Nepal is a landlocked SA country. The average annual rainfall in the capital city Kathmandu is about 1300 mm (Jacobson 1996). Most of the rainfall occurs during the summer monsoon months

(June–September). These rains can cause severe flooding, especially in the Terai plains. GW is abundant in the aquifers of the Terai and Kathmandu valleys. About 50% of the water used in the city of Kathmandu is derived from GW (BGS 2001). GW availability is more limited in the populated hill regions because of the lower permeability of the indurated and crystalline rock types. Despite abundant rainfall, agricultural development is restricted by the limited development of irrigation. In the Kathmandu valley (area of 0.05 Mha), GW is extracted from two main aquifers within the thick alluvial sediment sequence. A shallow unconfined aquifer occurs at around 0–10 m depth and a deep confined aquifer occurs at around 310–370 m (Khadka 1993). The exploitation of these aquifers, especially the shallow aquifer, has increased rapidly in recent years as a result of the increasing urbanization of the region. Recent abstraction of GW from the deep aquifer has led to a decrease in the WT level by 15–20 m since the mid-1980s (Khadka 1993). The annual recharge in the Terai region of Nepal is 6–16 BCM, and the annual extraction is only 1 BCM, whereas in the Kathmandu valley, the annual recharge is only 5.5×10^{-4} BCM and the abstraction for domestic and industrial use is 1.8×10^{-3} BCM, creating an annual deficit of 1.3×10^{-3} BCM (verification-unit.org/./nepal./5-SPKhan_and_PS%20Tater_hydrogeologyandgroundwaterResoourcesofNepal.pdf). Extensive GW development for irrigation in the Terai plains started in the 1960s. Today, about 0.02 Mha are under GW irrigation by some 60,000 shallow tube wells and 1,000 deep tube wells, with a total GW abstraction of 700 m^3. About 200,000 shallow hand pump and tube wells now serve most of the 11 million population of the Terai plains.

The resource availability and development are not always in tune with each other (Figure 6.4); for example, in northwestern India, the dependence on pure tubewell irrigation does not match its resource availability, and there is much scope for further development in eastern India, the rest of Bangladesh, and Nepal's Terai region (Shah et al. 2006).

6.4 MAJOR CAUSES OF GW DEPLETION

In India, irrigation water and energy prices are highly subsidized by the government. GW for small-scale irrigation is also free to all farmers who can privately invest in tube wells and open wells (Singh 2001). These kinds of policies are the major reasons for the excessive use and the depletion

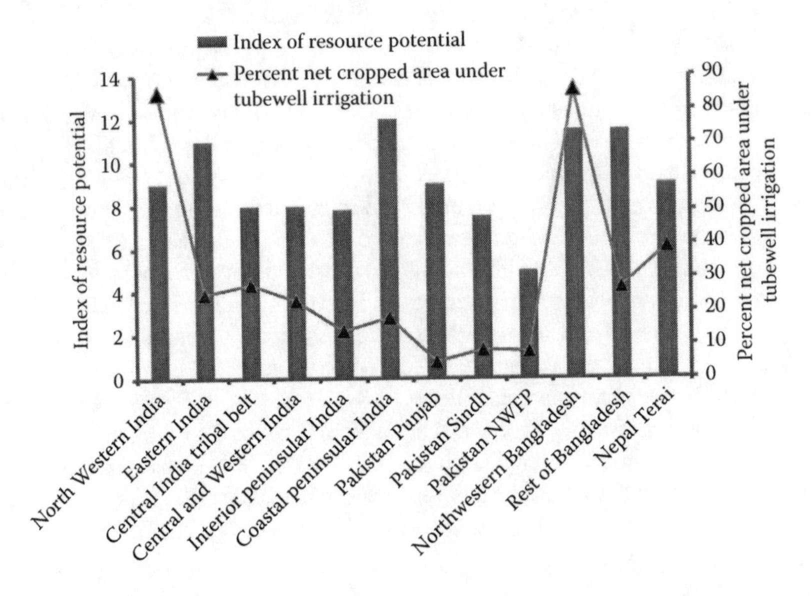

FIGURE 6.4 The mismatch between groundwater availability (resource potential) and development. (Modified from Shah, T., Singh, O.P., and Mukherji, A., *Hydrogeol. J.*, 14, 286–309, 2006.)

of GW reserves (World Bank 1993). With increasing investment in small-scale irrigation, the depletion of GW in many dry land villages is occurring at alarming rates. The basic incentive structures that induce overexploitation of GW are related to a lack of clearly defined and secure property rights that encourage cooperation (Shiferaw et al. 2003). This may eventually lead to overpumping and the exhaustion of the aquifer. The enormous development of bore wells also threatens aquifers and causes GW depletion, particularly in densely populated areas (Narain 1998). The GW policies are not in accord with the ground realities. There is less emphasis on the scarcity, depletion, and quality of GW. The enforcement of laws and administrative regulations are weak. Water rights are linked to land ownership. This virtually makes many aquifers open access resources to land owners (Shiferaw et al. 2003). The excessive and indiscriminate exploitation of GW has created a declining WT situation in the state. The problem is most critical in central Punjab. The average rate of decline over the last few years has been 55 cm/year. The worst affected districts are Moga, Sangrur, Nawanshahar, Ludhiana, and Jalandhar. This has resulted in extra power consumption, it affects the socioeconomic conditions of the small farmers, it destroys the ecological balance, and it adversely affects the sustainable agricultural production and economy of the state (Aggarwal et al. 2009). The major reason for this is a change in cropping pattern, for example, the introduction of HYV, urbanization, a decrease in the natural recharge, an increase in the population, industrialization, and more dependency on GW.

6.5 CONSEQUENCES OF GW OVEREXPLOITATION

6.5.1 Depletion of Aquifers and Crop Productivity

The most severe consequence of excessive GW pumping is the depletion of aquifers (Figure 6.5) and the subsequent lowering of the WT depth below which the ground is saturated with water, as observed in India's central Punjab, where the average WT depth decreased at a rate of 94 cm/year for the year 2004–2005 (Figure 6.6). The latest study of GW balance estimates shows that out of 138 development blocks, 103 fall in the "overexploited" category with a GW extraction of more than 100% of annual replenishment, 5 are classified as critical (stage of development 90%–100%), 4 are "semicritical" with a stage of development between 70% and 90%, and 25 are in the "safe" category with GW drafts of less than 70% of annual recharge. Block Nihalsinghwala in Moga District, for example, has the

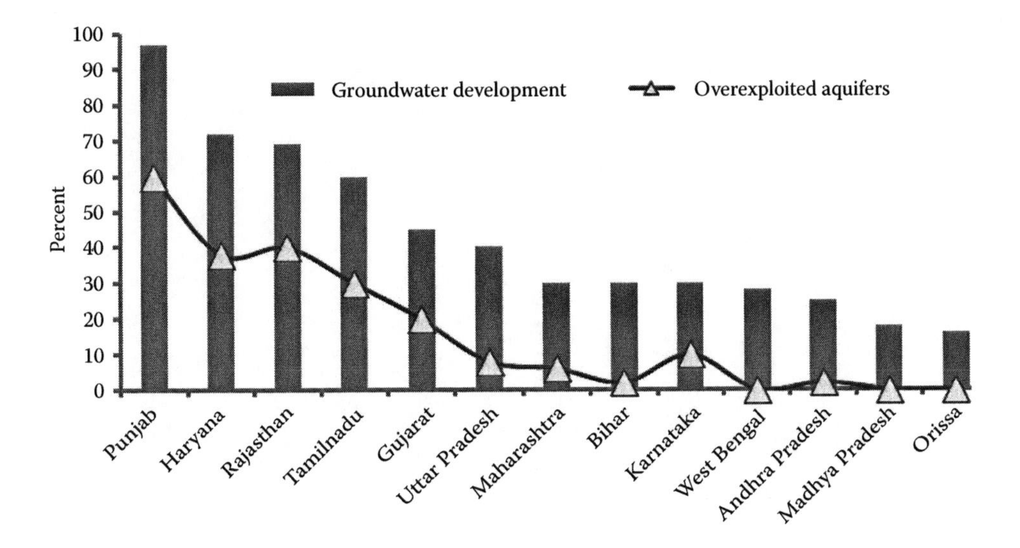

FIGURE 6.5 Groundwater aquifers in India are being depleted. (Modified from World Bank, *Revitalizing Punjab's Agriculture*, New Delhi, India, 2003.)

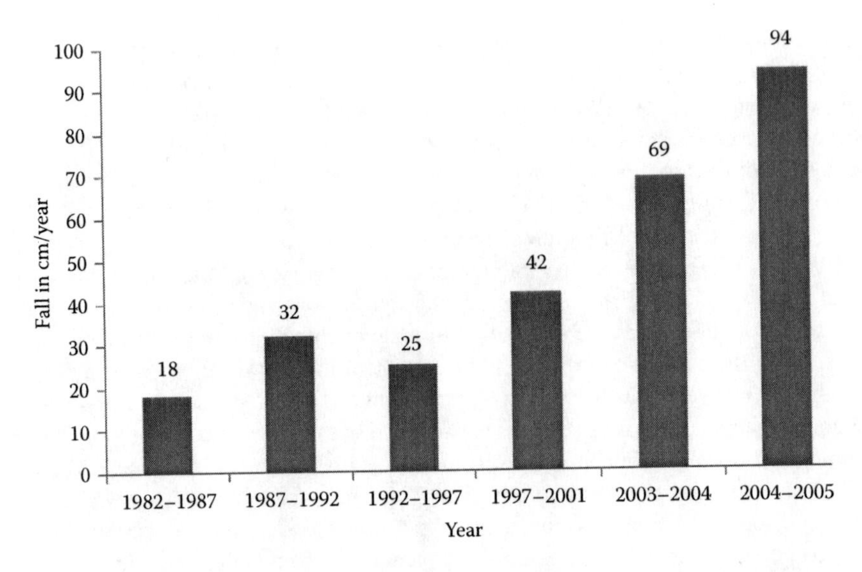

FIGURE 6.6 Average fall in the groundwater level in central Punjab, India. (Modified from Sidhu, R. and Vatta, K., Toward sustainable water use in India: Case studies from Punjab. New York, Columbia Water Center, 2010, water.columbia.edu/sitefiles/file/Uploads/Sidhu11-2010.pdf)

maximum stage of development of more than 200%. Safe blocks lie in southwest Punjab, where the GW is of poor quality, and in the Kandi area, where extraction is restricted due to deeper GW aquifers. Hira and Khera (2000) observed that in India's central Punjab, the percentage area with a WT depth of >10 m increased from 3% in 1973 to 53% and that with a WT depth of <5 m decreased from 45% to 5% for the same period because of the excessive mining of GW (Figures 6.7 and 6.8). The total number of tube wells in a small state like India's Punjab has increased from 192,000 to 935,000 over the 36 years between 1970 and 2006 (Table 6.7). The rate of depletion was less in the southwest zone as compared with the central zone due to the fast development of the canal network and better conjunctive use of surface water and GW (Table 6.8). About 75% of the state of Punjab is in the overexploited category, and only 18% is in the safe category (Figure 6.9). Thus, the WT in 90% of the Indian state of

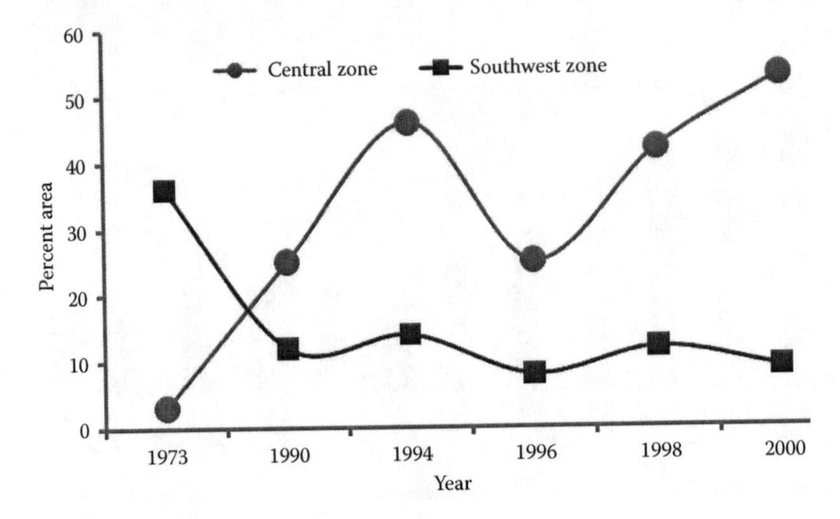

FIGURE 6.7 The increase in percent area with a water table depth of more than 10 m in the central zone of Punjab, India. (Modified from Hira, G.S. and Khera, K.L., Research Bulletin, Department of Soils, Punjab Agricultural University, Ludhiana, India, 2000.)

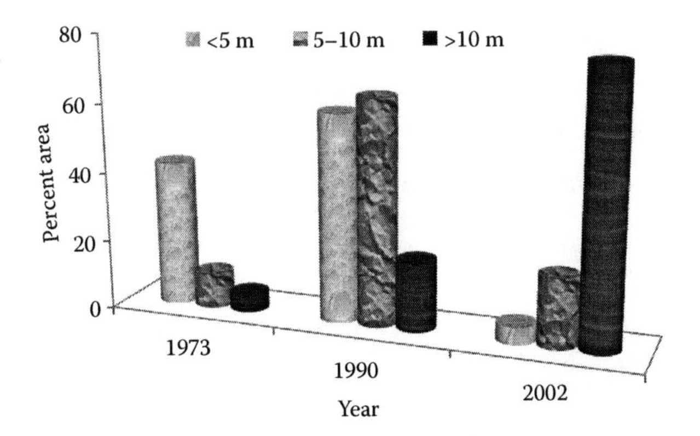

FIGURE 6.8 Percentage of area under different water table depths in central Punjab, India. (Modified from Hira, G.S. and Khera, K.L., Research Bulletin, Department of Soils, Punjab Agricultural University, Ludhiana, India, 2000.)

Punjab has decreased drastically since 1970. The WT was below 50 m in 2010, compared with 6–8 m in the 1970s. The key culprits are rice cultivation, the procurement price for rice, cheap labor, and the freedom to dig a tube well to get GW. The state government exacerbated the problem by supplying virtually free electricity, which resulted in intensive rice farming in Punjab. Paddies and wheat consume 66% of the total irrigation water (Hira et al. 2004). Recently, the Punjab State Farmers Commission has estimated that the aquifer level declined by about 4.7 m between 2001 and 2006 against 0.6 m between 1980 and 1986. In part of the state, the aquifer has declined over 10 m since the 1980s, which has many economic and ecological implications.

GW depletion is threatening the sustainability of the cropping system in the region. More electricity is required to withdraw water from lower depths (Figure 6.10), which increases the energy costs. GW depletion has directly impacted on the drying up of wells, the increased cost of pumping and well infrastructure, land subsidence (Galloway et al. 2001), salt-water intrusion, the changes in surface albedo, and related climate change (Ponce et al. 1997). The unsustainable use of GW will significantly impact a host of hydrological, ecological, and other natural resources and services,

TABLE 6.7
Number of Tube Wells (Million) in Punjab, India

Year	Diesel	Electric	Total
1970–1971	1.01	0.91	1.92
1980–1981	3.20	2.80	6.00
1990–1991	2.00	6.00	8.00
2000–2001	2.00	7.88	9.88
2001–2002	2.85	8.21	11.06
2002–2003	2.88	8.45	11.33
2003–2004	2.88	8.56	11.44
2004–2005	2.88	8.8	11.68
2005–2006	2.88	9.05	11.93
2006–2007	2.80	9.52	12.32

Source: Statistical Abstracts of Punjab. The Economic Advisor to Govt. of Punjab and Director of Agriculture, Punjab, 2008.

TABLE 6.8
Areas under Different Water Depths in the Central and Southwest Zones in Punjab, India: 1973–2000 (Percentage Area)

	Central Zone (Amritsar)			South-West Zone (Faridkot)		
Year	**<5 m**	**5–10 m**	**>10 m**	**<5 m**	**5–10 m**	**>10 m**
1973	39	58	3	39	25	36
1990	9	66	25	39	49	12
1994	6	48	46	30	56	14
1996	6	69	25	45	47	8
1998	6	49	42	45	43	12
2000	6	41	53	41	50	9

Source: Hira, G.S. and Khera, K.L., Research Bulletin, Department of Soils, Punjab Agricultural University, Ludhiana, India, 2000.

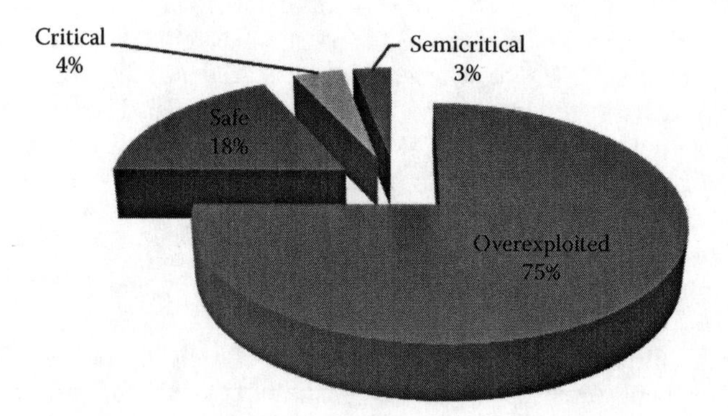

FIGURE 6.9 Extent of groundwater exploitation in Punjab. (From Jain, A.K. and Kumar, R., Water management issues—Punjab, 2007.)

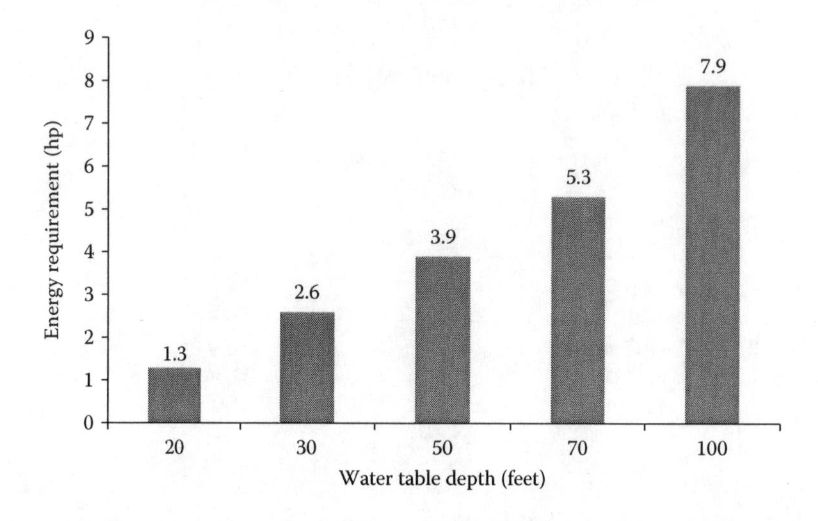

FIGURE 6.10 The increase in energy requirement with the increase in tubewell depth in Punjab, India. (Modified from Hira, G.S., Workshop papers presented in groundwater use in northwest India. Centre for Advancement of Sustainable Agriculture, New Delhi, India, 2004.)

including freshwater bodies, and aquatic, riparian, transitional, and terrestrial ecosystems. The GW depletion causes huge environmental, social, and economic costs. A few years ago, David Seckler, the then director general of IWMI, wrote alarmingly that a quarter of India's food harvest is at risk if the country fails to manage the GW properly. Today, many people think that Seckler might well have underestimated the situation, and that if India does not take charge of its GW, the agricultural economy may crash. The government initiatives to popularize deep bore wells to get more water have led to a drastic decline of the WT (Menon 2007). At the same time, there is neither interest nor any initiative in favor of recharging the GW. As the WT declines, poor farmers find it difficult to meet the huge energy requirement for deepening wells, which ultimately leads to a decline in food production.

6.5.2 Deterioration of GW Quality and Soil Salinization

Another adverse effect of the falling WT is the deterioration of the GW quality. The quality of the GW in the Indus Plains varies widely, both spatially and with depth, and is related to the pattern of GW movement in the aquifer (Qureshi et al. 2008). There are large numbers of saline GW pockets in the canal command areas of the Punjab and Sindh Provinces in Pakistan and the southwestern districts of India's Punjab. The GW is of poor quality in 23% of the area in Pakistan's Punjab Province and 78% in Sindh Province (Haider 2000). In the lower parts of the Indus Plains, the area of fresh GW is confined to a narrow strip along the Indus River. Similar situations exist in central areas of Punjab Province, where a layer of fresh GW floats over the saline water. Due to excessive pumping of this thin, fresh GW layer, the downward gradients are increasing, thereby inducing salt-water intrusion into the fresh GW areas. As a result of saline GW intrusion, about 200 public tube wells installed in the fresh GW zone of Pakistan's Punjab and Sindh Provinces had to be abandoned due to an increase in the GW salinity (EC: 10–12 dS/m). The salt-affected soils associated with the use of poor-quality GW for irrigation have become an important ecological disaster in India and Pakistan (WAPDA 2007). The potential to leach out the salt is very limited due to highly saline soils at shallow depths and highly saline GW at deeper depths (Bhutta and Smedema 2007). At present, 7 Mha of agricultural soils are affected by salt in India. Minhas (1996) observed that 32%–84% of the GW surveyed in different Indian states is rated either saline or alkaline. Because of the continental monsoonal climate, the basic principles of saline water management need some adaptation, for example, providing for a leaching requirement is not appropriate when the growing season for postmonsoon winter crops starts with a surface-leached soil profile, because it would increase the salt load. To minimize the salinity hazards in soils with a high WT, the salts are usually leached down and the waterlogging problems are alleviated by the installation of a subsurface drainage system. The use of GW by crops is also related to its depth and salt content (Chaudhary et al. 1974; Minhas et al. 1989, 1994).

6.5.3 Socioeconomic and Environmental Impacts

The declining WT and the soil degradation as a result of poor-quality GW use for irrigation have seriously affected the social and economic status of farmers in SA countries. The drying of wells has increased the burden of providing a livelihood for mankind. Women have to walk long distances to bring fresh drinking water from natural streams because the GW is very deep; sometimes, the GW is hazardous to health as it may contain heavy metals like As. Soil degradation caused by the use of brackish water has reduced the production potential of major crops by as much as 25%, valued at an estimated loss of US$250 million/year (Haider et al. 1999). The GW overdraft has also led to seawater intrusion in the coastal areas of the Indus Basin, and is threatening the wetlands ecosystems. Important aquatic resources, mangrove forests, and coastal areas need to be protected. Mangrove forests cover 130,000 ha; they are an important source of firewood and provide a natural breeding ground for shrimps and other wildlife.

6.6 EXTENT OF GW DEPLETION IN SA

6.6.1 India

Almost 54% of the GW blocks in the six states of India (i.e., Gujarat, Haryana, Maharashtra, Punjab, Rajasthan, and Tamil Nadu) are nearly depleted (*Economic Times* 2010). In fact, if the number of overexploited blocks continues to grow at the present rate of 5.5% per year, by 2018 roughly 36% of India's GW blocks will face serious problems. By 2030, 60% of the aquifers will be in a critical condition. Overexploitation is more prevalent in northwestern, western, and peninsular India (Shah 2002; Chatterjee and Purchit 2009). The status of GW development is comparatively high in northwestern India, where the stage of development is more than 100%, implying that in this region the average annual GW consumption is more than the average annual recharge. These states depleted, on average, 17.7 BCM of water annually, more than the government's estimate of 13.2 BCM in the same period. The GW depletion in the region was equivalent to a net irreplaceable loss of 109 BCM, or nearly 20% of Indians' annual water consumption of 634 BCM (Kerr 2009). Irrigation accounts for over 90% of water consumption in India, as is the case in many SA countries (Rosegrant et al. 2002; FAO 2003). Intensive agriculture leads to the depletion of GW resources, the buildup of salinity, waterlogging, the formation of hardpan (subsoil compaction), soil–nutrient imbalance, and an increased incidence of pests (Pingali and Rosegrant 2001). Some studies indicate that up to 50% of wells (especially open wells), once in use in many parts of India, have completely dried up. Agricultural economies built on the basis of GW irrigation may eventually collapse as nonsustainable water use leads to a depletion of the resource (Shiferaw et al. 2003). In India, the withdrawal of water for nonagricultural uses (i.e., domestic and industrial) accounted for 8% in the mid-1990s and is expected to increase to 14% by 2025 (Rosegrant et al. 2002). Although there are numerous strategies for the sustainable management of the GW resources, they often fail to create any positive impact on sustainable use for a number of reasons. Important among these is the absence of any pricing mechanism or strict regulation, indiscriminate GW exploitation, its wasteful utilization, and the land disposal of urban and industrial wastes (Menon 2007).

6.6.2 Pakistan

The availability of inexpensive drilling technologies allows even poor farmers to access GW to increase their crop production and improve livelihoods in Pakistan. The unreliability of the surface water supplies has turned more and more farmers to using GW without full awareness of the hazard represented by its quality. The trend of a continuous decline in the WT has been observed in many areas of Pakistan (Qureshi et al. 2009), illustrating the serious imbalance between abstraction and recharge. In the fresh GW areas of the Punjab and Baluchistan Provinces, the falling WT is a major issue. The excessive lowering of the WT has made pumping more expensive. As a result, many wells have gone out of production, yet the WT continues to decline and salinity continues to increase. In many areas of Balochistan, the WT is dropping at a rate of 2–3 m/year. Over 80% of the GW exploitation in Pakistan takes place through small capacity, private tube wells (Qureshi et al. 2010). These shallow tube wells (up to 6 m in depth) were initially installed by the farmers to capture the seepage from unlined canals to supplement their irrigation supplies for meeting the crop water requirements. Therefore, their installation and operational costs were very low and farmers were enjoying their benefits without much financial burden. However, the WT receded to depths that were inaccessible in 5% and 15% of the irrigated areas of the Punjab and Balochistan provinces, respectively. Under the business as usual scenario, this area is expected to increase to 15% in Punjab and 20% in Baluchistan by 2010 (PPSGDP 2000). With the receding WT depths (>15 m), farmers are left with no choice but to drill deeper wells. This transformation has increased the installation and operational costs. The construction cost of a deep electric tube well (>20 m) is US$5000 compared with US$1000 for a shallow tube well (<6 m). The present cost of pumping GW from a shallow tube well is US$4.2/1000 m^3 as compared with US$12/1000 m^3 from a deep tube well. Of course, all these costs are affected by the increasing energy prices. Beyond 20 m depth, turbine and submersible pumps are needed to extract the GW. The average cost of installation of such a pump is about US$10,000 in Balochistan and the maintenance of these deep tube wells is generally beyond the capacity of poor farmers. Under these conditions, access to GW has been restricted to large and rich farmers who can still afford this price. Thus, the sufferings of poor small farmers are further compounded. In Balochistan, the installation of more than 20,000 deep private tube wells over the last 10 years has negatively affected the traditional production systems. The declining WT has resulted in the desertification of the lands and the drying up of high-value fruit orchards. Qureshi et al. (2003) reported that more than 70% of the farmers in the Punjab Province depend directly or indirectly on GW to meet their crop demands. Over 80% of GW exploitation in Pakistan is in the private sector and there is no restriction or control on its abstraction. Nor is there a mechanism for allocating GW rights or for regulating its use. Anybody can install a tube well anywhere on his/her land and extract any amount of water at any time without consideration of the detrimental effects of this action on the resources and on others.

6.6.3 Bangladesh and Other South Asian Countries

In Bangladesh, fluctuations in the WT are drying up shallow wells, particularly during summer. This creates major difficulties for villagers in obtaining drinking-water supplies (Sadeque 1996), which has also had major inequity effects. While wealthy farmers can afford to deepen their existing wells or install new ones, resource-poor, small landholders cannot afford the cost of chasing the WT. There are also severe adverse environmental impacts. The dry-season flows at Farakha Barrage near the Bangladesh border could decline by about 75% if historical GW development patterns continue (Ilich 1996). Declines in the dry-season flow are a point of contention between India and Bangladesh. For Bangladesh, these flows are critical for irrigation, drinking-water supply, and for sustaining the mangrove areas along the coast. In Bangladesh's capital city Dhaka, the WT has dropped from 26.6 m in 1996 to 61.2 m in 2007, due to a lack of recharging, and this has put the sprawling metropolis at great risk. In Nepal, GW has declined by 15–20 m in the Kathmandu Valley due to excessive extraction and the use of GW for industrial and commercial purposes (www.ngoforum.net/index.php? option=com_content&task).

6.7 SUSTAINABLE MANAGEMENT OF GW IN SA

Food, water, and environmental security are the principal global issues of the twenty-first century. Although the stagnation and decline in agricultural production can be due to political and social reasons, the degradation of soil and water resources and the lack of appropriate technology to address the basic issues of resource mobilization and management are the primary factors responsible for low agricultural productivity (Lal 2000). Water scarcity and poor water quality are major concerns in SA countries, which mainly depend on agriculture for the livelihood of the people. Understanding the physics and hydrology of GW movement and measurement, the sociology and mindset of GW users, the economics of the water and agriculture sector, and the politics of the laws and the institutions are necessary for sustainable GW management (Shah 2007). Conjunctive use of the region's surface water and GW resources would help in minimizing the problems of waterlogging and GW mining (Shah 2000). Different approaches for the sustainable use of GW in SA are discussed next.

6.7.1 GW Management in India

India's GW governance pentagram is presented in Figure 6.11. The data on GW levels in India are neither widely published nor made available outside government organizations. Extraction and recharge estimates are also unreliable. As a result, discussions on GW depletion are always based on unreliable data (Menon 2007). The "boom and bust" in the GW-based agricultural economies has

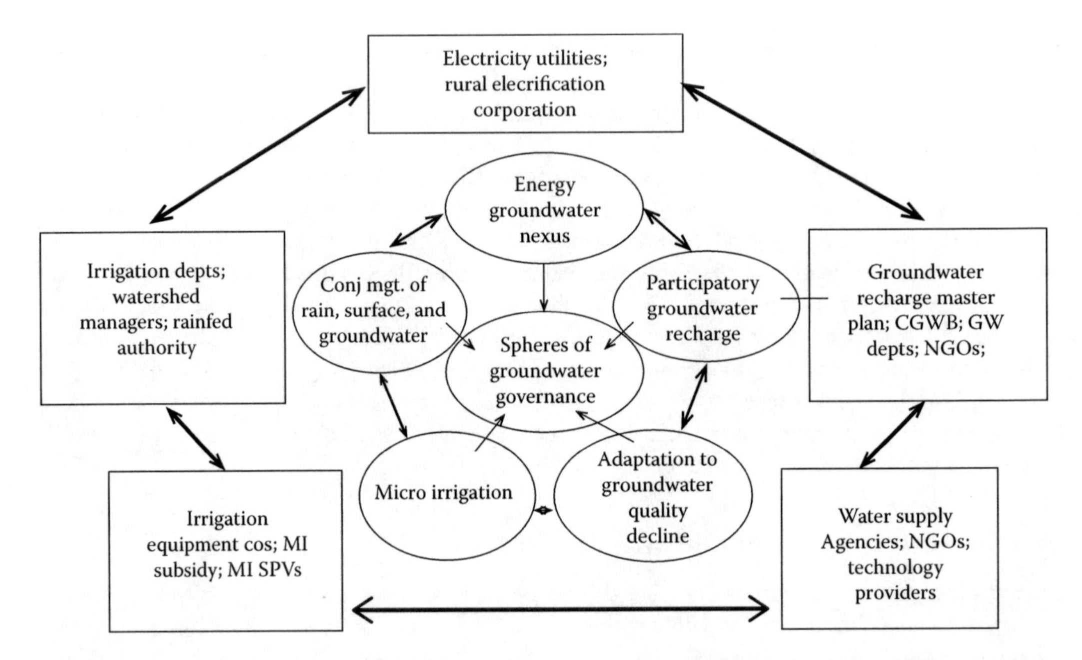

FIGURE 6.11 India's GW governance pentagram. (Adapted from Shah, T., *Environ. Res. Lett.* 4, 035005, 2009a.)

been widely observed in India (Shah 2002; Rao et al. 2003). Economic incentives and water charges are needed to regulate GW abstraction and to shift cropping patterns toward water-saving options and to crops with high net water productivity. Programs of GW management must account for the current and future water requirements and the ecological, economical, and social costs of the overexploitation of the GW resources (World Bank 1999; Custodio et al. 2004).

6.7.1.1 Various Approaches for Efficient GW Management

In India, the different strategies on sustainable GW management clearly show that successful management needs an interdisciplinary and holistic approach, involving all stakeholders, technocrats, policy makers, and farmers. Further, hydrogeological conditions, local specific environmental issues, and indigenous methods of water conservation and usage must also be objectively assessed. There should be an effective policy framework that considers all the multidimensional aspects of the issues of water scarcity and overexploitation (Menon 2007). Such a framework must also include improving the public water supply, using energy pricing and supply to manage agricultural GW draft, increasing RWH and GW recharge, transferring surface water in lieu of GW pumping, increasing the economic growth, reducing the dependence on agriculture, and formalizing the water sector (Shah 2009b). An accurate assessment of the available surface water and GW resources, considering the anthropogenic changes, is needed for the planning, designing, and operationalizing of the water resources projects as well as for watershed management. There should be a periodic reassessment of the surface water and GW potential on a scientific basis, taking into consideration the quality of the water available and the economic viability of its extraction (MWR 2009). In critically overexploited areas, bore well drilling must be regulated until the WT attains the desired level. The sustainability of GW utilization must be assessed from an interdisciplinary perspective because hydrology, ecology, geomorphology, climatology, pedology, agronomy, and economics play important roles (Ponce 2006). These interdisciplinary aspects of GW utilization have brought into question the concept of safe yield, defined as the maintenance of a long-term balance between the amount of withdrawal and the amount of recharge (Sophocleous 2000). Thus, the issue of GW sustainability is critical to overall development (Alley and Leake 2004). As a dynamic and replenishable resource, the GW has to be estimated primarily based on the component of annual recharge. The later is subjected to development by means of suitable structures whose choice depends on the hydrogeological and climatic conditions. The exploitation of GW resources must be regulated so as not to exceed the recharging possibilities, as well as to ensure social equity. The detrimental environmental consequences of overexploitation of GW need to be effectively prevented by the central and state governments (Jain et al. 2007).

6.7.1.2 Institutional Capacity Building

Another means of enhancing the use efficiency of GW is through institutional capacity building. This involves empowering and equipping people and organizations with the appropriate tools and sustainable resources to address their problems (Reddy and Vijaya Kumari 2007). Clearly, a joint management approach combining government administration with active people participation is a promising solution.

6.7.1.3 Proper Development of GW Resources and Interbasin Transfers

GW management not only requires the proper assessment of the available resources and an understanding of the interconnection between the surface water and GW system, but also actions for proper resource allocation and prevention of the adverse effects of the uncontrolled development of GW resources. In many parts of the country, deep aquifers are not fully utilized or developed, leading to underutilization of the available GW resources (Table 6.9). It is estimated that the basin transfer of water from surplus to deficit basins, by linking rivers in the northern region to those in the eastern region, may increase the irrigation potential. Such an undertaking, however, must be done on a win-win basis for all states concerned (MWR 2006). Policy makers often pay attention to the regions where GW development has the greatest potential and neglect the regions with low or hidden potential. In India, the eastern and northeastern regions are yet to develop their GW resources (Menon 2007). Yet, there is wide scope for the development of GW in these areas, which often faces floods during the rainy season. Ironically, some regions of India are suffering from flood damage and other regions are prone to acute and perpetual water shortages. Karnataka, Tamil Nadu, Rajasthan, Gujarat, Andhra Pradesh, and Maharashtra are the worst drought-prone states. Uttar Pradesh, Bihar, West Bengal, Orissa, and Assam face severe flooding. The interbasin transfer of water in India is a long-term option to partly overcome the spatial and temporal imbalance of the availability and demand of water resources (Kumar et al. 2005). Flood plains are good reservoirs of GW. Thus, the sustainable management of flood plain aquifers offers excellent scope for its development and the additional requirement of water. The development of GW in the Yamuna flood plain area in Delhi is an example of the scientific management of resources (Shekhar and Prasad 2009). The waterlogged

TABLE 6.9
Availability of Surplus/Unutilized Groundwater in the Eastern States of India

State	Net Annual Groundwater Availability (BCM)	Annual Groundwater Draft (BCM)	Annual Groundwater Available as Surplus/Unutilized	
			BCM	(%)
Assam	24.9	5.5	19.5	78
Bihar	27.4	10.8	16.7	61
Chhattisgarh	13.7	2.8	10.9	80
Jharkhand	5.3	1.1	4.2	79
Orissa	21.0	3.9	17.2	82
West Bengal	27.5	11.7	15.8	58
Punjab	21.4	31.2	−9.7	−45

Source: National Rainfed Authority Area, Food security, water and energy nexus in India, 2008. www.corecentre.co.in/Database/Docs/DocFiles/food_security.pdf.

areas in canal commands offer scope for GW development by lowering the WT by up to 6 m or more. One of the effective strategies for sustainable GW management is to use surface water in one area, utilizing the recharge by developing GW in areas adjacent to the canal commands.

6.7.1.4 Artificial Recharge and RWH

There is a need to recharge aquifers and conserve rainwater through the installation of RWH structures (Sivanappan 2006). In urban areas, rainwater can be harvested from rooftops and open spaces. RWH reduces the possibility of flooding, and also decreases the community's dependence on GW for domestic uses. Apart from bridging the demand–supply gap, recharging improves the quality of the GW, raises the WT in wells and bore wells, and prevents the flooding and choking of drains. One can also save energy to pump GW as the WT rises due to recharging of the wells. In many states in India, RWH is being undertaken on a massive scale. Substantial benefits of RWH exist in urban areas as water demand has already outstripped the supply in most cities. It is estimated that, annually, about 36.4 BCM of surplus surface runoff can be recharged to augment the GW (Romani 2005). In rural areas, techniques are demonstrating the feasibility of artificial recharge by modifying the natural movement of the water through suitable civil structures (e.g., percolation tanks, check dams, stream bunds, and gully plugs). The government should take initiatives, such as offering soft loans and passing legislation to install devices in all new government and private buildings to promote the RWH system. Moreover, publicity and training are a must to make people aware of the rainwater quality and the harvesting system. The collected rainwater could be used for washing, gardening, and toilet flushing without prior treatment. Areas that suffer from a severe shortage of water could use the RWH system for domestic water uses (potable water). Harvested water can be treated by chlorination, solar or UV radiation, or heating prior to human consumption. The states of Karnataka, Andhra Pradesh and Tamilnadu have over 200,000 tanks, and a strategy that has been widely recommended is to transform these tanks into recharge tanks by filling them up with canal water (Kulandaivelu and Jayachandran 1990; Reddy et al. 1990). In the western region of India, local NGOs and communities have spontaneously created a massive well-recharge movement based on the principle of "water on your roof, stays in your compound; water on your field stays on your field; and water on your village, stays in your village." People have modified some 300,000 wells—open and bore—to divert rainwater to them; they have also constructed thousands of ponds, check dams, and other RWH and recharge structures on the self-help principle to keep the rainwater from gushing into the Arabian Sea (Shah 2000). Percolation of a portion of the rainfall, through the vadose zone, is the principal source of natural recharge to the aquifer systems in India. The recharge rates range from 4.1% to 19.7% of the local average seasonal rainfall (Rangarajan and Athavale 2000). Annually, some 1150 BCM (out of a total annual rainfall of 4000 BCM) of India's rainwater runs off into the seas (INCID 1999; Kumar 2006). It is necessary to hold this water by reducing the velocity of the runoff and providing time for recharge and enhancing the GW availability. Check dams, village ponds, rooftop water harvesting, and interlinking different rivers are some options to increase the GW recharge. The sustainable use of GW should begin by tapping primarily deep percolation and secondarily shallow percolation. The latter should be exploited only if its effects on the base flow of the neighboring streams and water bodies are minimal. Detailed hydrological and hydrogeological studies are required to determine the percolation amounts. In the Indian watersheds, the public and watershed communities pay the investment costs needed to recharge the GW. A number of check dams are built in each watershed at selected locations along the watercourse to retain the stream flow and increase the infiltration into the ground (Kerr et al. 2002). In the coastal desert of North Chile, a fog collection project has been able to provide an average of 11,000 L/day of water to a community of 330 people (Schemenauer and Cereceda 1991).

6.7.1.5 Effective Water Pricing on GW

Subsidized energy prices, open access externalities, low pumping costs, and free access to water jointly encourage GW overpumping and depletion; therefore, water charges may be considered a suitable policy option to promote water saving and to counter depletion (Shiferaw et al. 2003). In India, irrigation water is charged (partly) only for public sources (i.e., canal and tank irrigation systems). There is considerable diversity in the system of levying irrigation charges across states (Sampath 1992). The rates are often levied on the basis of the area irrigated, differentiated by crop and season, but may be uniform throughout the state (FAO 2003). Generally, in the presence of pumping subsidies, the rates are small and small-scale private tube- and open well irrigation systems are exempted from payment. Regardless of its scarcity, water is a free resource to all smallholder farmers who are able to invest in the infrastructure to tap existing aquifers. With the increasing scarcity, local informal markets have developed in some areas, where water-deficient farmers rent water seasonally from water-surplus farmers (Shah 1993; Meinzen-Dick 1998; Saleth 1998; Shah et al. 2003b). The water rates may vary by season and the type of crop grown. As much as 25% of the output is paid for irrigation water. In some cases, payments may occur based on the hours of irrigation. The removal of price and energy subsidies for pumping GW and the proper costing of the public irrigation services are critical policy instruments for managing the demand (Shiferaw et al. 2003; Scott and Shah 2004). For smallholder agriculture, the actual levels of GW extraction and use are not metered and hence cannot be directly observed. Therefore, farm-level direct volumetric charges are infeasible. Many farmers consider free and unrestricted access to GW a fundamental and ancestral right. Despite the efforts of community watershed management projects to regulate water use, this entrenched perception of private rights has contributed to the lack of cooperation and to the dramatic increase in unregulated small-scale irrigation. The incentive-based approach is likely to perform better than the regulatory approach, while also providing higher and sustainable economic benefits from water conservation to small farmers (Schaible 2000; Tiwari and Dinar 2002). Two incentive-based approaches are (1) charges based on hours of irrigation, and (2) charges per unit of output. The first is directly related to the volume of water used. The second is related to the productivity of water. The water charges should be large enough to shift the cropping patterns to high-value and water-efficient crops and induce efficient and judicious use of the GW. Since farmers do not pay (or pay a minimal flat rate) energy tariffs, the water charges can be considered to include the energy prices for pumping. Water-deficit farmers in SA buy water seasonally from adjacent farmers through various informal arrangements (Saleth 1998; Meinzen-Dick 1998). As in land contracts, the transactions may vary from in-kind labor contracts to upfront cash payments. The fixed share approach rather than a fixed quantity or cash rental is preferred perhaps due to its risk-sharing benefits. Unlike the area-based approach, it also allows flexibility in crop choice and permits actual payments to vary according to the crop grown. When water is free to irrigating farmers, it leads to the shifting of cropping patterns to more intensive crops (e.g., rice), which should not be encouraged in water-scarce areas. For the semiarid production systems discussed here, irrigation seems to be an attractive option until the water charges reach about Rs. 25–30 per h ($0.55) (Shiferaw et al. 2003). Alternatively, the output share charges could be increased up to 20%. If the water charges are reinvested in improving the availability of water, small farmers will directly benefit from them. Even while subsidizing electricity, many state governments have begun restricting the power supply to agriculture to cut their losses. Shah (2007) reported that IWMI studies have shown that with the judicious management of the power supply to agriculture, the energy–irrigation nexus can be a powerful tool for GW demand management in livelihood-supporting socioecological conditions to create tradable property rights in GW. Even as governments evolve the GW regulations and their enforcement mechanisms, more practical strategies for GW governance need to be evolved in five spheres, as outlined in Figure 6.11. Synergizing the working of agencies in these spheres offers the best chance to bring a modicum of order and method to the region's water-scavenging irrigation economy (Shah 2009a).

6.7.2 GW Resource Management in Pakistan

In addition to institutional reforms, supply and demand management approaches also need to be followed in Pakistan (Qureshi et al. 2010). For demand management, farmers must adopt improved irrigation and agronomic practices. Current land-use patterns must be reviewed to replace water-intensive crops with water-efficient crops. For supply management, the use of wastewater and saline drainage effluents for agriculture can be promoted. In rainfed areas, more efforts should be focused on RWH. The effective coordination between organizations responsible for the management of GW resources also needs to be enhanced. Van Steenbergen (2002) reported the benefits of local GW management through the use of social norms in Baluchistan, Pakistan. The irrigation system brought water through karezes (a form of subterranean aqueduct) from the subsurface flow of the river and from the infiltrated runoff from the surrounding hills. Typically, no well may be dug within 5 km of an existing subterranean aqueduct.

6.7.2.1 Use of Alternate Water Resources

Some parts of Pakistan suffer from the problem of brackish water, and studies show that this water can be used for irrigating a range of crops grown in different soil types under diverse environmental conditions (Sharma and Rao 1998; Qadir et al. 2001). Brackish GW has been successfully used to irrigate wheat, cotton, pearl millet, sugar beet, etc. Yield reductions of up to 15%–20% have been observed when compared with freshwater irrigation. In the deep GW areas, the excessive accumulation of salts is adequately managed during the summer monsoon rains. In shallow GW areas, properly managed drainage systems are mandatory for the successful use of brackish water for irrigation without causing soil salinization.

6.7.2.2 Promoting RWH

The rainfed areas of Pakistan contribute about 10% of the total agricultural production. However, the production levels of the rainfed areas are low (i.e., 1–1.5 Mg/ha). Yet, the production potential of these areas can be doubled by providing one or two supplemental irrigations at the critical growth stages of the crop (Oweis and Hachum 2001). Wherever farmers are already using supplemental irrigation, it is being done through GW extraction. However, farmers should be encouraged to adopt RWH and watershed management strategies to both improve the productivity of the rainfed systems and reduce the demands on the GW. While many fragmented efforts have been made in the past, there is a need to develop a comprehensive policy at the government level to ensure continuity in this regard. Furthermore, farmers must be educated to optimize the crop yields by using less water rather than maximize it through excessive GW irrigation.

6.7.2.3 Managing Conjunctive Use of Surface Water and GW

In most of the canal command areas in Pakistan, the conjunctive use of surface water and GW is equally practiced in the head and tail ends of the canal system. The canal water delivered to the head-end farmers is generally 32% and 11% more than to the farmers at the tail end and middle end, respectively (Haider et al. 1999). The unmanaged conjunctive use of surface water and GW at the head ends of the canals causes the WT to rise, resulting in waterlogging, whereas at the tail ends, salinity problems are increasing due to the excessive use of bad quality GW for irrigation. Therefore, planned conjunctive use is needed, whereby the upstream farmers make better use of the surface supplies in the canals, which are more reliable for them. For this purpose, the canal department needs to regulate the canal flows to match the crop water requirements. The inferior quality water can be mixed with the canal water in a proportion acceptable for irrigation (Hussain et al. 2010). Farmers also need to be educated on the proper mixing ratios of the surface water and GW resources in order to keep the salinity of the irrigation water within permissible limits and to avoid the risk of secondary salinization.

6.7.3 Managing GW Pollution in Bangladesh

A comprehensive water distribution system and a national water resources management policy need to be implemented to limit the indiscriminate extraction of GW in Bangladesh. An efficient monitoring system should also be established to ensure the supply of potable water throughout the country, and to prevent further As contamination in drinking water. For the overall improvement of the present As disaster in Bangladesh, it is highly recommended that every donor project of As mitigation be legally required to ensure community participation in water resources management. In any water supply and As mitigation project, water resources management with the involvement of the local community would play an important role in the remediation of GW As contamination. The rapid population growth and the lack of overall coordination have increased the demand for water and created ethnic conflicts. The widening gap between the demand and the supply of water is causing severe socioeconomic and environmental problems in Bangladesh. The organizations concerned with water resources management have failed to attain the required level of efficiency. Their failure in water resources management has played a major role in creating the As contamination in Bangladesh and elsewhere in SA. The As contamination of Bangladesh's GW with its genesis and toxic effects on humans has been widely reported (Dhar et al. 1997; DCH 1997; Karim et al. 1997; Nickson et al. 1998; Ahmed 2000; Karim 2000). Several studies indicate that the GW is severely contaminated with As above the maximum permissible limit of 0.05 mg/L for drinking water. About 400 measurements conducted in Bangladesh in 1996 (Smith et al. 2000) indicate that As concentrations in about half of the measurements were above the maximum permissible limit. In 1998, the British Geological Survey (BGS) collected 2022 tubewell water samples from 41 As-affected districts (Smith et al. 2000). Laboratory tests revealed that 35% of these water samples had As concentrations of >0.05 mg/L. In order to mitigate the As disaster, it is essential to treat As-contaminated water and water sources or to avoid using the As-contaminated water. The removal of As from water is possible by ultraviolet radiation, oxidation, chemical precipitation, and filtration (Safiuddin and Karim 2001). Further, As-contaminated soils and shallow GW aquifers could also be cleaned by flushing out the As contaminants. The avoidance of As is also feasible by the use of surface water, rainwater, and alternative GW sources.

6.8 RISING WATER LEVEL AND WATERLOGGING IN SA

Similar to the problem of GW depletion in many parts of SA, there is also the problem of a rise in the WT in parts of India and Pakistan. However, it is difficult to evaluate the extent of the problem based on the available data. In India, the Ministry of Agriculture estimated that the total area affected by waterlogging as a result of both GW rise and poorly controlled irrigation was 8.5 Mha in 1990 (Vaidyanathan 1994). By contrast, the Central Water Commission for 1990, which considered only the areas affected by a GW rise, estimated the waterlogged areas at 1.6 Mha (Vaidyanathan 1994). Regardless of the actual extent, waterlogging represents a major challenge in surface water and GW management. This challenge cannot be addressed in the absence of an integrated approach that incorporates surface water imports and use in conjunction with that of the GW. Large areas of Pakistan face similar problems. The rising WT in the command of surface irrigation systems has fundamental implications on the sustainability of social objectives in the GW-dependent regions. Irrigation-induced salinity and waterlogging reduce crop yields in Pakistan by 30% (FAO 1997). In India, the problem is serious enough to threaten the growth of the agricultural economy (Joshi et al. 1995).

The adverse impact of waterlogging and salinization on farmers and regional economies can be insidious. Initially, the introduction of irrigation often causes a dynamic transformation of the regional and household economies. Farmers introduce HYV of crops and are able to grow valuable market/cash crops. Wealth is created. However, as the WT rises, the "bubble economy" based on unsustainable water management practices deflates. Once salinized, the land and the unsaturated zone of the soil are difficult and expensive to reclaim. Ultimately, many farm families (and regional economies) may be worse off than before the introduction of irrigation unless sustainable and affordable methods of remediation are used. In India's Punjab, the problem of a rise in the WT and a salt buildup has been caused in the southwestern districts (Bajwa and Josan 1989) because of seepage from unlined canals and poor drainage facilities. Moreover, the GW of the region is brackish, for example, saline-sodic water (Minhas and Gupta 1992), and farmers mostly use canal water for irrigation. The seepage from the unlined canals has raised the WT in the entire region.

6.9 CLIMATE CHANGE IMPACTS ON GW IN SA

Global climate change may also profoundly affect hydrologic systems worldwide. Glacial melting and increasing ocean temperatures lead to a rise in the sea level. The changes in temperatures and rainfall influence the growth rates and the leaf size of plants, which affect the GW recharge (Kundzewicz and Doll 2007). Changing river flows in response to the changing mean precipitation, the rising sea levels, and the changing temperatures will influence the natural recharge rates (http://www.gwclim.org/ presentations/plenary/kundzewicz.pdf). In the SA region, particularly India, the frequency and severity of floods and droughts are expected to increase, while higher temperatures will reduce the winter snow pack and hasten the spring snowmelt from mountainous areas (Kumar et al. 2005). Climate change may increase uncertainties in the supply and management of water resources. Temperature increases affect the hydrologic cycle by directly increasing the evaporation of the available surface water and vegetation transpiration. Consequently, these changes can influence the precipitation amounts, the timings and intensity of rains, and indirectly impact the flux and storage of water in surface and subsurface reservoirs (i.e., lakes, soil moisture, and GW). In addition, there may be other associated impacts, such as seawater intrusion, water quality deterioration, and potable water shortages.

GW resources are related to climate change through direct interaction with surface water resources, such as lakes and rivers, and indirectly through the recharge process. The direct effect of climate change on GW resources depends on the change in the volume and distribution of the GW recharge. Therefore, quantifying the impact of climate change on the GW resources requires not only the reliable forecasting of changes in the major climatic variables, but also an accurate estimation of GW recharge (www.angelfire.com/nh/cpkumar/publication/CC_RDS.pdf). Unchecked, GW depletion can exacerbate the impacts of these changes; conversely, controlled management of GW depletion can contribute to their mitigation. Assuming that the volume of GW depleted during the past 100 years is much greater than what can be accounted for by nontransient increases in the volumes of water stored in the soil, natural channels and lakes, or the atmosphere, then the ultimate sink for the "missing" GW is the oceans. Worldwide, the magnitude of GW depletion from storage may be so large as to constitute a measurable contributor to the rise in sea level. For example, the total volume depleted from the High Plains Aquifer equates to about 0.75 mm, or about 0.5%, of the observed sea-level rise during the twentieth century (Konikow and Kendy 2005).

Climate change will act as a force multiplier; it will enhance the GW criticality for drought-proofing agriculture and simultaneously exacerbate the threat to the resource. From a climate change point of view, India's GW hotspots are western and peninsular India. These regions are critical to climate change mitigation as well as adaptation (Shah 2009a). A long-term temperature rise will increase the need to store water for distribution over a longer dry season. In some areas, an integrated solution can be achieved by artificially recharging the excess runoff, when available. Thus, depleted aquifers can be transformed into underground "reservoirs" to supplement the flood-buffering and drought-buffering capacity of the existing surface water reservoirs. Under a changing climate scenario, the annual variability in the monsoon's precipitation levels will increase to more intense floods and droughts. Thus, climate change in the future is expected to have implications on the river flows in SA (especially India), which will directly affect the GW status.

The developing countries of temperate and tropical Asia are already highly vulnerable to extreme climate events, such as floods, droughts, and cyclones. Climate change and variability would exacerbate these vulnerabilities. The GW management approaches for effectively dealing with climate change will have to be different from those that have been used in the past. Climate change due to global warming is increasing the complexities and vulnerability of food security in irrigated and rainfed agriculture, both in developing and developed countries (Brown and Hansen 2008; FAO 2008).

6.10 CARBON FOOTPRINTS OF SA'S GW ECONOMY

GW pumping with electricity and diesel accounts for an estimated 16–25 Tg of C emissions (Shah 2009a), which is 4%–6% of India's total C emissions. Total electricity consumption in GW irrigation is 87 billion kWh. GW pumping in India results in the emission of 14.38 Tg of C (i.e., 11.09 Tg by electric pumps and 3.29 Tg by diesel pump sets). The IFPRI (2009) estimates that the use of coal-based electricity as well as diesel fuel to pump the GW for irrigation are major sources of CO_2 emission. The IFPRI estimates that the C emission from GW irrigation is higher at 16 Tg, roughly 4% of India's total C emissions, and reducing the subsidies on the electricity used to pump GW for irrigation would lower the CO_2 emission by 14% (IFPRI 2009). Two interesting aspects of the C footprint of India's GW economy are as follows. (1) Lifting 1000 m^3/m of water using electricity emits 5.5 times more C than using diesel, and diesel pumps are concentrated in eastern India with rich alluvial aquifers. (2) The C emission from GW irrigation is highly sensitive to the dynamic head over which GW is lifted because, for one, a higher head leads to higher energy use and higher C emissions, and second, beyond a depth of 10–15 m, the diesel pumps become extremely inefficient, forcing irrigators to switch to electricity, which has a larger C footprint anyway (Shah 2009a). Most of India's diesel pumps are concentrated in eastern India and electric pumps are concentrated in western and peninsular India. Deep tube wells have a high C footprint (IFPRI 2009; Nelson et al. 2009). Climate change and GW discussions are at a very early stage in India. However, preliminary studies show massive scope for reducing the C footprint of India's GW economy. Using data from Haryana and Andhra Pradesh, Shukla et al. (2003) reported that every 1 m decline in pumping the WT increases greenhouse gas (GHG) emissions by 4.4% in Haryana and 6% in Andhra Pradesh. The most important determinant of the C footprint of India's pump irrigation economy is the dynamic head over which farmers lift water to irrigate the crops. The larger the head, the higher the energy consumption and the more likely that electrified deep tube wells are used for pumping GW, multiplying the C footprint of GW pumping. The C emissions to lift 1000 m^3 of water to 1 m are 0.665 kg C with diesel-fueled pumps and 3.873 kg C with electric pumps (Nelson et al. 2009). Deep wells powered by electricity are the largest single source of CO_2 emissions. They accounted for 65% of the total in 2000 and are projected to account for 87% in 2050. The 38 million mt of CO_2 emitted from deep well pumping accounted for more than 5% of India's total GHG emissions from all sectors of the economy in 2000 (World Resources Institute 2009). As expected, both higher transmission losses and deeper wells result in more CO_2 emissions. The increase in transmission losses raises CO_2 emissions by about 11%. Deeper wells increase emissions by 33%. Pump efficiency has the most dramatic effect on our estimates of carbon emissions. If pumps are only 20% efficient instead of the 30% assumption of the baseline, carbon emissions increase by 50% over the baseline (Nelson et al. 2009).

6.11 CONCLUSION

The sustainable management of GW offers opportunities for alleviating hunger and enhancing the livelihood of the SA population. However, there is a dire need to use and restore this vital resource through judicious management. A big part of the solution is massive initiatives to augment GW recharge in regions suffering from depletion throughout SA. However, this strategy cannot work without the appropriate demand-side interventions. The removal of price and energy subsidies for pumping GW and the proper costing of public irrigation services are critical policy instruments for managing this vital resource. There is a need for proper legislation to make RWH mandatory for public and private buildings, especially in urban areas. There should be awareness creation about the significance of recharging GW among the public through various media. There is a need to change the mindset by shifting water management skills from resource development to resource planning. All users of water should conduct a water audit to see how water can be saved and used. Wastewater can also be effectively recycled and reused by facilitating ecological activities. There is a strong need to monitor the database related to the GW properties and variables necessary to detect significant trends, using modern tools to generate relevant and accurate data. Because of the changing scenarios, a regular reassessment of the GW resources is essential, and this requires strengthening of the available database. Therefore, the judicious management of GW is an important strategy. Encouraging and implementing artificial recharge, conservation, water-saving irrigation, growing less-water-requiring crops, the conjunctive use of surface water and GW, fresh and brackish water, treatment and reuse of wastewater, and land-use planning and land zoning as per the availability of water, and taking appropriate measures to avoid pollution are all important. A proper capacity buildup and raising public awareness by better communication, coordination, and collaboration with water managers, planners, decision makers, scientists, and water users will help in better GW management. There is a need to establish a legal and regulatory framework regarding the development and use of GW by revising the policies on subsidized and free power in the agricultural sector. These problems must be addressed in view of the increasing risks of water pollution and contamination (i.e., As) and the changing climate of the SA region.

ABBREVIATIONS

BCM	Billion cubic meters
BGS	British Geological Survey
CPWF	Challenge Program on Water and Food
DCH	Dhaka Community Hospital
FAO	Food and Agricultural Organization
GOP	Government of Pakistan
GR	Green revolution
GDP	Gross domestic product
GW	Groundwater
HYV	High yielding varieties
hp	Horse power
IWRS	Indian Water Resources Society
INCID	Indian National Committee on Irrigation and Drainage
IFPRI	International Food Policy Research Institute
MWR	Ministry of Water Resources
PWP	Pakistan Water Partnership
PPSGDP	Punjab Private Sector Groundwater Development Project
RWH	Rainwater harvesting
SA	South Asia
WAPDA	Water and Power Development Authority
WPB	Water policy briefing
WT	Water table

REFERENCES

Aggarwal, R., Kaushal, M.P., Kaur, S., and Singh, B. 2009. Water resource management for sustainable agriculture in Punjab, India. *Water Sci. Technol.* 60: 2905–2911.

Ahmed, K.M. 2000. Groundwater arsenic contamination in Bangladesh: An overview. In: Bhattacharya, P. and Welch, A.H. (eds), *Arsenic in Groundwater of Sedimentary Aquifers.* 31st International Geological Congress, Rio de Janeiro, Brazil, August 2–3, 2000, pp. 3–5.

Alley, W.M. and Leake, S.A. 2004. The journey from safe yield to sustainability. *Ground Water* 42: 12–16.

Bajwa, M.S. and Josan, A.S. 1989. Effect of alternating sodic and non-sodic irrigations on build-up of sodium in soil and crop yield in northern India. *Exp. Agric.* 25: 199–205.

Bhutta, M.N. 2002. Sustainable management of groundwater in the Indus Basin. Paper presented at Second South Asia Water Forum, Pakistan Water Partnership, Islamabad, Pakistan, December 14–16, 2002.

Bhutta, M.N. and Smedema, L.K. 2007. One hundred years of waterlogging and salinity control in the Indus Valley, Pakistan: A historical review. *Irrig. Drain.* 56: 581–590.

British Geological Survey (BGS). 2001. Groundwater quality: Nepal. Natural Environment Research Council (NERC).

Brown, C. and Hansen, J.W. 2008. Agricultural water management and climate risk. IRI Tech. Rep. No. 08-01. International Research Institute for Climate and Society, Palisades, NY, 19 pp.

Calow, R.C., Robins, N.S., Macdonald, A.M., Macdonald, D.M.J., Gibbs, B.R., Orpen, W.R.G., Mtembezeka, P., Andrews, A.J., and Appiah, S.O. 1997. Groundwater management in drought-prone areas of Africa. *Water Resour. Dev.* 13: 241–261.

Centre for Water Policy (CWP). 2005. Some critical issues on groundwater in India. June 2005 report.

Chatterjee, R. and Purchit, R.R. 2009. Estimation of replenishable groundwater resources of India and their status of utilization. *Curr. Sci.* 96: 1581–1791.

Chaudhary, T.N., Bhatnagar, V.K., and Prihar, S.S. 1974. Growth response of crop to depth and salinity of water and soil submergence. I. Wheat. *Agron. J.* 66: 32–35.

Custodio, E., Kretsingerb, V., and Ramon Llamas, M. 2004. Intensive development of groundwater: Concept, facts and suggestions. *Water Pol.* 7: 157–162.

Daines, S.R. and Pawar, J.R. 1987. Economic returns to irrigation in India. Report prepared by SRD Research Group Inc., USA, Agency for International Development Mission to India, New Delhi.

DebRoy, A. and Shah, T. 2003. Socio-ecology of groundwater irrigation in India. In Llamas, R. and Custodio, E. (eds), *Intensive Use of Groundwater: Challenges and Opportunities*. Swets & Zetlinger, The Netherlands, pp. 307–336.

Dennehy, K.F., Litke, D.W., and McMahon, P.B. 2002. Approaches to groundwater resources management. The High Plains Aquifer, USA: Groundwater development and sustainability. *Geol. Soc. London* (Special Publications) 193: 99–119.

Dhaka Community Hospital (DCH). 1997. Arsenic pollution in groundwater of Bangladesh. Monthly Newsletter, June, Dhaka, Bangladesh.

Dhar, R.K., Biswas, B.K., Samanta, G., et al. 1997. Groundwater arsenic calamity in Bangladesh. *Curr. Sci.* 73: 48–59.

Dhawan, B.D. 1989. *Studies in Irrigation and Water Management*. Commonwealth Publishers, New Delhi, India.

Economic Times. 2010. Groundwater depleting at alarming rate: Report, March 6.

Food and Agricultural Organization (FAO). 1997. Irrigation in the Near East region in figures. Water Reports No. 9. FAO, Rome.

Food and Agricultural Organization (FAO). 2003. FAO Aquastat. http://www. fao.org/ag/agl/aglw/aquastat/main/index.stm.

Food and Agricultural Organization (FAO). 2008. Climate change and food security. A framework document. FAO, Rome.

Galloway, D., Jones, D.R., and Ingebritsen, S.E. 2001. Land subsidence in the United States. Geological Survey Circular 1182, Denver, CO, 175 pp.

Giordano, M. and Villholth, K. 2007. *The Agricultural Groundwater Revolution: Opportunities and Threats to Development. Vol. 3: Comprehensive Assessment*. CABI Publishing, Wallingford, UK, 419 pp.

Gleick, P.H. 1993. *Water in Crisis: A Guide to the World's Freshwater Resources*. Oxford University Press, New York.

Gleick, P.H. 1998. *The World's Water*. Island Press, Washington, DC, 198 pp.

Govt. of Pakistan (GOP). 2000. Agricultural statistics of Pakistan. Ministry of Food, Agriculture and Livestock, Economics Division, and Govt. of Pakistan, Islamabad.

Haider, G. 2000. Proceedings of the International Conference on Regional Groundwater Management, October 9–11, Islambad, Pakistan.

Haider, G., Prathapar, S.A., Afzal, M., and Qureshi, A.S. 1999. Water for environment in Pakistan. Global Water Partnership Workshop, Islamabad, Pakistan.

Hira, G.S. 2004. Status of groundwater resources in Punjab and management strategies. Workshop papers presented in groundwater use in northwest India. Centre for Advancement of Sustainable Agriculture, New Delhi, India.

Hira, G.S. and Khera, K.L. 2000. Water resource management in Punjab under rice–wheat production system. Research Bulletin, Department of Soils, Punjab Agricultural University, Ludhiana, India.

Hira, G.S., Jalota, S.K., and Arora, V.K. 2004. Efficient management of water resources for sustainable cropping in Punjab. Research Bulletin, Department of Soils, Punjab Agricultural University, Ludhiana, India.

Hussain, G., Alquwaizany, A., and Al-Zarah, A. 2010. Guidelines for irrigation water quality and water management in the kingdom of Saudi Arabia. *J. Appl. Sci.* 10(2): 79–96.

Ilich, N. 1996. Ganges River Basin water allocation modeling study. Calgary. Consultant report to the World Bank.

Indian National Committee on Irrigation and Drainage (INCID). 1999. Water for food and rural development 2025. Paper presented at the PODIUM workshop, December 15–16, 1999. Central Water Commission, New Delhi, India.

Indian Water Resources Society (IWRS). 1998. Five decades of water resources development in India. IWRA report, New Delhi, India.

International Food Policy Research Institute (IFPRI). 2009. Tapping Indian's agricultural mitigation potential. Rural climate exchange. http://cgiarclimatechange.wordpress.com/.../the-promise-of-indias-agriculture-for-climate-change-mitigation.

Jacobson, G. 1996. Urban groundwater database. AGSO report, Australia. http://www.clw.csiro.au/UGD/DB/Kathmandu/Kathmandu.html.

Jain, S.K., Agarwal, P.K., and Singh, V.P. 2007. Concepts of water governance for India. In Singh, V.P. (ed.), *Hydrology and Water Resources of India*. Springer, The Netherlands, 1155–1190.

Joshi, P.K., Tyagi, N.K., and Svendsen, M. 1995. Measuring crop damage due to soil salinity. In Gulati, M.S.A. (ed.), *Strategic Change in Indian Irrigation*. Macmillan, New Delhi, India.

Kahnert, F. and Levine, G. 1993. *Groundwater Irrigation and the Rural Poor: Options for Development in the Gangetic Basin*. World Bank, Washington, DC.

Karim, M.M. 2000. Arsenic in groundwater and health problems in Bangladesh. *Water Resour.* 34(1): 304–310.

Karim, M.M., Komori, Y., and Alam, M. 1997. Subsurface arsenic occurrence and depth of contamination in Bangladesh. *J. Environ. Chem.* 7: 783–792.

Kerr, R.A. 2009. North India. Groundwater is going going going…. *Science* 325(5942): 798.

Kerr, J., Pangare, G., and Pangare, V.L. 2002. Watershed development projects in India. Research Report No. 127. IFPRI, Washington, DC.

Khadka, M.S. 1993. The groundwater quality situation in alluvial aquifers of the Kathmandu valley, Nepal. *AGSO J. Aust. Geol. Geophys.* 14: 207–211.

Khair, M.S., Richard, J., Culas, J., and Hafeez, M. 2010. The causes of groundwater decline in upland Balochistan region of Pakistan: Implication for water management policies. Paper presented at the 39th Australian Conference of Economists (ACE 2010), Sydney, NSW, September 27–29, 2010.

Khan, H.R. 1994. *Management of Groundwater Resources for Irrigation in Bangladesh*. FAO, Rome.

Konikow, L.F. and Kendy, E. 2005. Groundwater depletion: A global problem. *Hydrogeol. J.* 13: 317–320.

Kulandaivelu, R. and Jayachandran, K. 1990. Groundwater utilization and recharge through percolation ponds—A case study. In *Percolation Ponds: Pre-workshop*. Center for Water Resources, Madras, India, pp. 21–25.

Kumar, C.P. 2006. Management of groundwater in salt water ingress coastal aquifers. In Ghosh, N.C. and Sharma, K.D. (eds), *Groundwater Modeling and Management*. Capital Publishing Company, New Delhi, India, pp. 540–560.

Kumar, R., Singh, R.D., and Sharma, K.D. 2005. Water resources of India. *Curr. Sci.* 89(5): 10.

Kundzewicz, Z.W. and Doll, P. 2007. Will groundwater ease freshwater stress under climate change? International Conference on Groundwater and Climate in Africa, Kampala, Uganda, June 2008.

Lal, R. 2000. Rationale for watershed as a basis for sustainable management of soil and water resources. In Lal, R. (ed.), *Integrated Watershed Management in the Global Eco-System*. CRC Press, Boca Raton, FL, pp. 2–16.

Lappe, F.M., Collins, J., and Rosset, P. 1998. *World Hunger: Twelve Myths*. Grove Press, New York, 2–16.

Mainuddin, M. 2002. Groundwater irrigation in Bangladesh: 'Tool for Poverty Alleviation' or 'Cause of Mass Poisoning'? Proceedings of the Symposium on Intensive Use of Groundwater: Challenges and Opportunities, Valencia, Spain, December 10–14, 2002.

Mandal, M.A.S. 2006. Groundwater irrigation issues and research experience in Bangladesh. International workshop on groundwater governance in Asia, Indian Institute of Technology, Roorkee, India.

Meinzen-Dick, R.S. 1998. Groundwater markets in Pakistan: Institutional development and productivity impacts. In Easter, W., Rosegrant, M., and Dinar, A. (eds), *Markets for Water: Potential and Performance*. Kluwer Academic Publishers, Boston, MA, pp. 207–222.

Menon, S.V. 2007. Ground water management: Need for sustainable approach. http://mpra.ub.uni-muenchen.de/6078/ MPRA Paper No. 6078, posted 04. December 4, 2007. www.wikipedia.org./wiki/groundwater

Minhas, P.S. 1996. Saline water management for irrigation in India. *Agric. Water Manag.* 30: 1–24.

Minhas, P.S. and Gupta, R.K. 1992. *Quality of Irrigation Water: Assessment and Management*. Indian Council of Agricultural Research, Publications & Information Division, Pusa, New Delhi, India.

Minhas, P.S., Sharma, D.R., and Khosla, B.K. 1989. Response of sorghum to the use of saline waters. *J. Indian Soc. Soil Sci.* 37: 140–146.

Minhas, P.S., Singh, Y.P., Tomar, O.S., and Gupta, R.K. 1994. Saline irrigation and its schedules for *Acacia nilotica* and *Dalhergia sissoo* on a highly calcareous soil. In Singh, P., Pathak, P.S., and Roy, M.M. (eds), *Agroforestry Systems for Degraded Lands*. Oxford and IBH, New Delhi, India, pp. 357–364.

Ministry of Water Resources (MWR). 2006. Report of subcommittee on more crop and income per drop of water. Advisory council on artificial recharge of groundwater. GOI, New Delhi, India.

Ministry of Water Resources (MWR). 2009. Report of the groundwater resource estimation committee. Groundwater resource estimation methodology. GOI, New Delhi, India.

Molden, D. 2007. *Water for Food, Water for Life: A Comprehensive Assessment of Water Management in Agriculture*. International Water Management Institute, Colombo and Earthscan, London, UK.

Narain, V. 1998. Towards a new groundwater institution for India. *Water Pol.* 1: 357–365.

Nelson, G.C., Robertson, R., Msangi, S., Zhu, T., Liao, X., and Jawajar, P. 2009. Greenhouse gas mitigation. Issues for Indian agriculture. IFPRI discussion paper 00900.

Nickson, R.T., McArthur, J.M., Burgess, W.G., Ahmed, K.M., Ravenscroft, P., and Rahman, M. 1998. Arsenic poisoning of Bangladesh groundwater. *Nature* 395(6700): 338.

Oweis, T. and Hachum, A. 2001. Reducing peak supplemental irrigation demand by extending sowing dates. *Agric. Water Manage.* 50: 109–123.

Pakistan Water Partnership (PWP). 2001. The framework for action for achieving the Pakistan water vision 2025. WAPDA-Lahore, Pakistan.

Palmer-Jones, R.W. 1999. Slowdown in agricultural growth in Bangladesh. In Rogaly, B., Harriss-White, B., and Bose, S. (eds), *Sonar Bangla? Agricultural Growth and Agrarian Change in West Bengal and Bangladesh*. Sage Publications, New Delhi, India.

Pingali, P.L. and Rosegrant, M. 2001. Intensive food systems in Asia: Can the degradation problems be reversed? In Lee, D.R. and Barrett, C.B. (eds), *Tradeoffs or Synergies? Agricultural Intensification, Economic Development and the Environment*. CABI Publishing, Wallingford, UK, pp. 383–397.

Ponce, V.M. 2006. Groundwater utilization and sustainability. http://groundwater.sdsu.edu.

Ponce, V.M., Lohani, A.K., and Huston, P.T. 1997. Surface albedo and water resources: Hydroclimatological impact of human activities. *J. Hydrol. Eng. ASCE* 2(4): 197–203. http://ponce.sdsu.edu/albedo197.html.

Postel, S. 1999. *Pillar of Sand: Can the Irrigation Miracle Last?* W.W. Norton, New York.

Punjab Private Sector Groundwater Development Project (PPSGDP). 2000. Legal and regulatory framework for Punjab Province. PPSGDP. Technical Report No. 45, Lahore, Pakistan.

Qadir, M., Ghafoor, A., and Murtaze, G. 2001. Use of saline-sodic waters through phytoremediation of calcarous saline sodic soils. *Agric. Water Manag.* 50: 197–210.

Qureshi, A.S., Shah, T., and Akhtar, M. 2003. The groundwater economy of Pakistan. IWMI Regional Office, Lahore, Pakistan, vol. 3, 24pp. (IWMI working paper 64/Pakistan country series no. 19.)

Qureshi, A.S., McCornick, P.G., Qadir, M., and Aslam, Z. 2008. Managing salinity and waterlogging in the Indus Basin of Pakistan. *Agric. Water Manag.* 95: 1–10.

Qureshi, A.S., McCornick, P.G., Sarwar, A., and Sharma, B.R. 2009. Challenges and prospects of sustainable groundwater management in the Indus Basin, Pakistan. *Water Resour. Manag.* DOI: 10.1007/s11269-009-9513-3.

Qureshi, A.S., Gill, M.A., and Sarwar, A. 2010. Sustainable groundwater management in Pakistan: Challenges and opportunities. *Irrig. Drain.* 59: 107–116.

Rama Krishna, I.S. and Shiva Kumar, S. 2008. Perennial water resources of Eastern Ghats of India—An overview. *EPTRI–ENVIS Newslett.* (ISSN: 0974-2336) 14(1): 7–12.

Rangarajan, R. and Athavale, R.N. 2000. Annual replenishable ground water potential of India—An estimate based on injected tritium studies. *J. Hydrol.* 234: 38–53.

Rao, K.P.C., Bantilan, M.C.S., Rao, Y.M., and Chopde, V.K. 2003. What does village-level evidence suggest for research and development priorities? Policy Brief No. 4. International Crops Research Institute for the Semi-Arid Tropics, Patancheru, India.

Reddy, M.D. and Vijaya Kumari, R. 2007. Water management in Andhra Pradesh, India. Paper presented in Indo–U.S. workshop on innovative E-technologies for distance education and extension/outreach for efficient water management, ICRISAT, Patancheru/Hyderabad, India, March 5–9, 2007.

Reddy, N., Rao, K.S.V., and Prakasam, P. 1990. Impact of percolation ponds on groundwater regime in hard rock areas of Andhra Pradesh. In *Percolation Ponds: Pre-workshop*. Center for Water Resources, Madras, India, pp. 137–144.

Rodell, M., Velicogna, I., and Famiglietti, J.S. 2009. Satellite-based estimates of groundwater depletion in India. *Nature* 460: 999–1002.

Romani, S. 2005. Groundwater management: A key for sustainability. Prof. C. Karunakaran Lecture Series No. 6. Akshara Offset Printers, Thiruvananthapuram, India.

Rosegrant, M. 1997. Water resources in the 21st century: Challenges and implications for action. Food, agriculture and the environment discussion paper 20. IFPRI, Washington, DC.

Rosegrant, M.W., Cai, X., and Cline, S. 2002. World water and food to 2025: Dealing with scarcity. International Food Policy Research Institute, Washington, DC.

Roy, A.D. and Shah, T. 2002. The socio-ecology of groundwater in India. International Water Management Institute, *Water Policy Briefing*, vol. 4, pp. 1–6.

Sadeque, S.Z. 1996. Nature's bounty or scarce commodity—Competition and consensus over ground water use in rural Bangladesh. Annual Conference of the International Association for the Study of Common Property, University of California, Berkeley, CA.

Safiuddin, M. and Karim, M.M. 2001. Groundwater arsenic contamination in Bangladesh: Causes, effects and remediation. Proceedings of the 1st IEB International Conference and 7th Annual Paper Meet on Civil Engineering, Chittagong, Bangladesh, November 2–3, 2001. Institution of Engineers, Bangladesh, pp. 220–230.

Safiuddin, M. and Karim, M.M. 2003. Water resources management in the remediation of groundwater arsenic contamination in Bangladesh. In Murphy, T. and Guo, J. (eds), *Aquatic Arsenic Toxicity and Treatment*. Backhuys Publishers, Leiden, The Netherlands, pp. 1–17.

Saleth, M.R. 1998. Water markets in India: Economic and institutional aspects. In Easter, W., Rosegrant M., and Dinar, A. (eds), *Markets for Water: Potential and Performance*. Kluwer Academic Publishers, Boston, MA, pp. 187–205.

Sampath, R.K. 1992. Issues in irrigation pricing in developing countries. *World Dev*. 20: 967–977.

Schaible, G.D. 2000. Economic and conservation tradeoffs of regulatory vs. incentive based water policy in the Pacific North West. *J. Water Resour. Dev.* 16: 221–238.

Schemenauer, R.S. and Cereceda, P. 1991. Fog water collection in arid coastal locations. *Ambio* 20: 303–308.

Scott, C.A. and Shah, T. 2004. Groundwater overdraft reduction through agricultural energy policy: Insights from India and Mexico. *Int. J. Water Resour. Dev.* 20: 149–164.

Scott, C.A. and Sharma, B. 2009. Energy supply and expansion of groundwater irrigation in Indus–Ganges basin. *Int. J. River Basin Manag.* 7: 1–6.

Seckler, D., Molden, D., and Barker, R. 1998a. Water scarcity in the twenty-first century. IWMI Water Brief 2. International Water Management Institute, Colombo, Sri Lanka.

Seckler, D., Amarasinghe, U., Molden, D., De Silva, R., and Barker, R. 1998b. World water demand and supply 1990 to 2025: Scenarios and issues. Research Report No. 19. International Water Management Institute, Colombo, Sri Lanka.

Shah, T. 1993. *Groundwater Markets and Irrigation Development. Political Economy and Practical Policy*. Oxford University Press, New Delhi, India.

Shah, T. 2000. Mobilizing social energy against environmental challenge: Understanding the groundwater recharge movement in western India. *Nat. Resour. Forum* 24: 197–209.

Shah, T. 2002. The socio-ecology of groundwater in India. Water Policy Briefing, Issue No. 4. International Water Management Institute, Colombo, Sri Lanka.

Shah, T. 2005. Groundwater and human development: Challenges and opportunities in livelihoods and environment. *Water Sci. Technol.* 51: 27–37.

Shah, T. 2007. The groundwater economy of South-Asia: An assessment of size, significance and socio-ecological impacts. In Giordano, M. and Villholth, K.G. (eds), *The Agricultural Groundwater Revolution: Opportunities and Threat to Development*. CABI Publishing, Wallingford, UK, pp. 7–36.

Shah, T. 2009a. Climate change and groundwater: India's opportunities for mitigation and adaptation. *Environ. Res. Lett.* 4 (July–September): 035005.

Shah, T. 2009b. *Taming the Anarchy? Groundwater Governance in South Asia*. RFF, Washington, DC.

Shah, T. and Mukherjee, A. 2001. The socio-ecology of groundwater in Asia. IWMI–Tata working paper. International Water Management Institute, Anand, India.

Shah, T., Molden, D., Sakthivadivel, R., and Seckler, D. 2000. *The Global Groundwater Situation: Overview of Opportunities and Challenges*. International Water Management Institute, Colombo, Sri Lanka.

Shah, T., DebRoy, A., Qureshi, A.S., and Wang, J. 2003a. Sustaining Asia's groundwater boom: An overview of issues and evidence. *Nat. Resour. Forum* 27: 130–140.

Shah, T., Scott, C., Kishore, A., and Sharma, A. 2003b. Energy–irrigation nexus in South Asia: Approaches to agrarian prosperity with viable power industry. Research Report No. 70. International Water Management Institute, Colombo, Sri Lanka.

Shah, T., Singh, O.P., and Mukherji, A. 2006. Some aspects of South Asia's groundwater irrigation economy: Analyses from a survey in India, Pakistan, Nepal Terai and Bangladesh. *Hydrogeol. J.* 14: 286–309.

Shah, T., Burke, J., Villholth, K., Angelica, M., and Custodio, E. 2007a. Groundwater: A global assessment of scale and significance. In Molden, D. (ed.), *Water for Food, Water for Life: A Comprehensive Assessment of Water Management in Agriculture*. International Water Management Institute, Colombo and Earthscan, London, UK, pp. 395–419.

Sharma, D.P. and Rao, K.V.G.K. 1998. Strategy for long-term use of saline drainage water in semi-arid regions. *Soil Till. Res.* 47: 287–295.

Sharma, S.K. and Mehta, M. 2002. Groundwater development scenario: Management issues and options in India. Paper presented at the IWMI–ICAR–Colombo Plan Sponsored Policy Dialogue on 'Forward-Thinking Policies for Groundwater Management: Energy, Water Resources, and Economic Approaches' at India International Center, New Delhi, India, September 2–6, 2002.

Shekhar, S. and Prasad, R.K. 2009. The groundwater in the Yamuna flood plain of Delhi (India) and the management options. *Hydrogeol. J.* 17: 1557–1560.

Shiferaw, B.A., Wani, S.P., and Nageswara Rao, G.D. 2003. Irrigation investments and groundwater depletion in Indian semi-arid villages: The effect of alternative water pricing regimes. Working paper series no. 17. International Crops Research Institute for the Semi-Arid Tropics, Patancheru 502 324, Andhra Pradesh, India. An Open Access Journal published by ICRISAT, 24 pp.

Shiva, V. 1991. The green revolution in Punjab. *Ecologist* 21(2): 57–60.

Shukla, P.R., Nair, R., Kapshe, M., Garg, A., Balasubramaniam, S., Menon, D., and Sharma, K. 2003. Development and climate: An assessment for India. Unpublished report submitted to UCCEE, Denmark. Indian Institute of Management, Ahmedabad, India.

Singh, R.B. 2001. Impact of land-use change on groundwater in the Punjab–Haryana plains, India. Impact of human activity on groundwater dynamics. Proceedings of a symposium held during the Sixth IAHS Scientific Assembly at Maastricht, The Netherlands, July 2001. IAHS Publ. No. 269.

Sivanappan, R.K. 2006. Rainwater harvesting, conservation and management strategies for urban and rural sectors. Paper presented in National Seminar on Rainwater Harvesting and Water Management, Nagpur, India, November 11–12, 2006.

Smith, A.H., Lingas, E.O., and Rahman, M. 2000. Contamination of drinking-water by arsenic in Bangladesh: A public health emergency. *Bull. World Health Organ.* 78(9): 1093–1103.

Sophocleous, M. 2000. From safe yield to sustainable development of water resources—The Kansas experience. *J. Hydrol.* 235: 27–43.

Subramanian, V. 2000. *Water: Quantity–Quality Perspectives in South Asia.* Kingston International Publishers, Surrey, UK, 256 pp.

Tiwari, D. and Dinar, A. 2002. Balancing future food demand and water supply: The role of economic incentives in irrigated agriculture. *Q. J. Int. Agric.* 41: 77–97.

Vaidyanathan, A. 1994. *Second India Studies Revisited: Food, Agriculture and Water.* Institute of Development Studies, Madras, India.

Van Steenbergen, F. 2002. Local groundwater regulation. Water Praxis Document No. 14. Land and Water Product Management Group. Arcadis Euroconsult, Arnhem, The Netherlands.

Water Policy Briefing (WPB). 2002. IWMI–Tata water policy programme. Issue 4. June 2002.

Water and Power Development Authority (WAPDA). 2007. Waterlogging, salinity and drainage situation. SCARP Monitoring Organization, WAPDA, Lahore, Pakistan.

World Bank. 1991. Water resources management. A World Bank policy paper. The World Bank, Washington, DC.

World Bank. 1999. *Groundwater Regulation and Management in India.* The World Bank, Washington, DC and Allied Publishers, New Delhi, India.

World Bank. 2003. Sustainable groundwater management: Concepts and tools. GW-MATE (Groundwater Management Advisory Team). http://www.world bank.ogr/gwmate.

World Bank. 2007. Punjab groundwater policy—Mission report. WB-SA-PK-Punjab. GW mission report. http://www.worldbank.org/gwmate (accessed June 2007).

World Resources Institute (WRI). 2000. *Water Resources 2000–2001: People and Ecosystem: The Fraying Web of Life.* WRI, Washington, DC.

7 Water Resources and Agronomic Productivity in the West Asia and North Africa Region

Mostafa Ibrahim, Rattan Lal, ElSayed Abdel Bary, and Atef Swelam

CONTENTS

7.1 INTRODUCTION

The West Asia and North Africa (WANA) region comprises a large part of the Middle East and spans over two continents (Africa and Asia) ranging from 14° to 38° north and 12° west to 59° east (Figure 7.1). Geographically, it spans from the Atlantic Ocean in the west to the Arabian/Persian Gulf in the east and the Mediterranean Sea in the north to the Arabian Sea and Indian Ocean in the south (Figure 7.1). The region includes 22 different countries: Algeria, Bahrain, Cyprus, Egypt, Eritrea, Ethiopia, Iran, Iraq, Jordan, Kuwait, Lebanon, Libya, Morocco, Oman, Qatar, Saudi Arabia, Sudan, Syria, Tunisia, Turkey, the United Arab Emirates (UAE), and Yemen. The region's land area covers 1584 million hectare (Mha) or 10.6% of the world land area (Table 7.1). The cultivated area

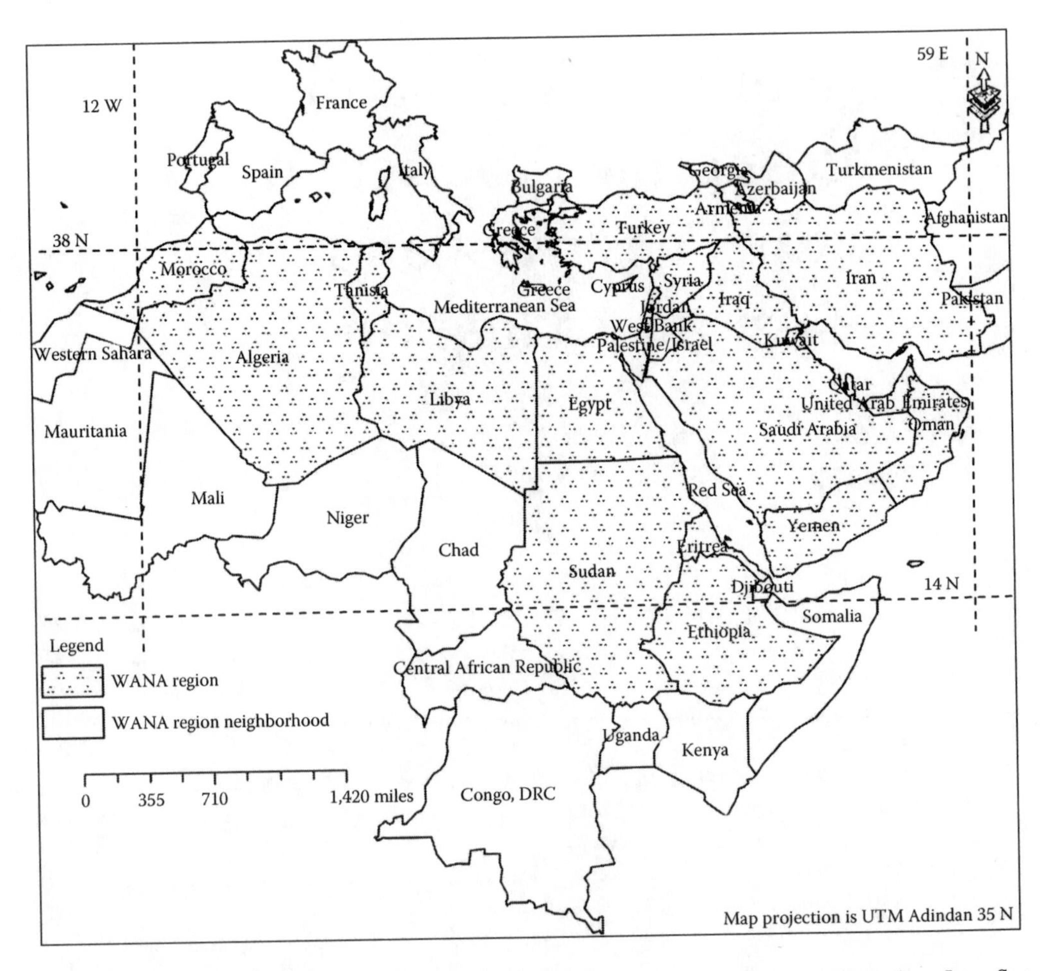

FIGURE 7.1 The West Asia and North Africa (WANA) region and its neighborhood. (Data from Iowa State University, CRP 551 Course. Created by M. Ibrahim, 2011.)

covers 124.83 Mha or 7.9% of the region's total land area. A wide variation in the total and cultivated areas occurs at the individual country level. For example, Bahrain and Cyprus have the smallest land areas of 0.08 and 0.93 Mha, respectively. By contrast, Algeria and Sudan have the largest land areas of 238.17 and 250.58 Mha, respectively (Table 7.1).

The WANA territory is geographically categorized into five regions: the North Africa region (Algeria, Libya, Morocco, and Tunisia), the Nile Valley/Red Sea region (Egypt, Eritrea, Ethiopia, and Sudan), the West Asia region (Cyprus, Iraq, Jordan, Lebanon, and Syria), the Highlands region (Iran and Turkey), and the Arabian Peninsula region (Bahrain, Kuwait, Oman, Qatar, Saudi Arabia, the UAE, and Yemen).

Most of the WANA region countries are characterized by their arid and semiarid climate. However, the large area is composed of diverse climatic regimes. Four broad climatic regions can be distinguished in this large area. The arid region comprises 46.6% of the total land area with a mean annual precipitation (MAP) of 100 mm; the semiarid region comprises 21.9% of the total land area with a MAP of 100–300 mm; the semimoist region comprises 19.7% of the total land area with a MAP of 300–500 mm; and the moist region comprises 12% of the total land area with a MAP of more than 500 mm. The mean annual temperature (MAT) of the region ranges from 11.4°C in Turkey to 29.2°C in Sudan and Yemen (Table 7.1).

The cultivated land area is currently constant, but it could decrease either quantitatively or qualitatively through desertification by land misuse, soil mismanagement, and land use change (urbanization). Similarly, water resources, especially fresh water, which is composed of surface water and groundwater,

TABLE 7.1
General Facts about Countries of the Region

	Location[a]		Total Area[b]	Cultivated Area			
Country	Long.	Lat.	(×10⁶ ha)	(×10⁶ ha)	(%)	MAP (mm)	MAT[c] (°C)
Algeria	3°00′E	28°00′N	238.17	8.42	3.54	89	16.8
Bahrain	50°33′E	26°00′N	0.08	0.00	5.50	83	26.5
Cyprus	33°00′E	35°00′N	0.93	0.11	12.28	498	18.9
Egypt	30°00′E	27°00′N	100.15	3.54	3.54	51	21.0
Eritrea	39°00′E	15°00′N	11.76	0.67	5.71	384	16.0
Ethiopia	38°00′E	08°00′N	110.43	14.51	13.14	848	16.6
Iran	53°00′E	32°00′N	174.52	18.77	10.76	228	16.7
Iraq	44°00′E	33°00′N	43.83	5.45	12.43	216	22.7
Jordan	36°00′E	31°00′N	8.88	0.23	2.60	111	17.3
Kuwait	45°45′E	29°30′N	1.78	0.02	0.84	121	25.6
Lebanon	35°50′E	33°50′N	1.05	0.29	27.24	661	20.6
Libya	17°00′E	25°00′N	175.95	2.05	1.17	56	20.4
Morocco	5°00′W	32°00′N	44.66	8.98	20.11	346	17.7
Oman	57°00′E	21°00′N	30.95	0.09	0.30	125	28.6
Qatar	51°15′E	25°30′N	1.16	0.02	1.38	74	27.1
Saudi Arabia	45°00′E	25°00′N	214.97	3.68	1.71	59	25.4
Sudan	30°00′E	15°00′N	250.58	20.91	8.34	416	29.2
Syria	38°00′E	35°00′N	18.52	5.67	30.59	252	17.0
Tunisia	9°00′E	34°00′N	16.36	5.04	30.81	207	17.7
Turkey	35°00′E	39°00′N	78.36	24.51	31.27	593	11.4
UAE	54°00′E	24°00′N	8.36	0.27	3.17	78	27.1
Yemen	48°00′E	15°00′N	52.80	1.61	3.04	167	29.2
Total	—	—	1584.22	124.83	7.88	—	—

Source: World Climate. 2011. http://www.worldclimate.com. Accessed September 15, 2011; TravelBlog. 2011. http://www.travelblog.org/World/ag-geog.html. Accessed September 21, 2011; Food and Agriculture Organization. Aquastat database query. 2011. http://www.fao.org/nr/water/aquastat/data/query/index.html?lang=en. Accessed September 17, 2011.

[a] Location was obtained from TravelBlog (2011). Longitude and latitude represent the central point of each country.
[b] Total and cultivated area information and mean annual precipitation (MAP) were obtained from FAO (2011).
[c] Mean annual temperature was obtained from World Climate (2011).

are almost constant, but may decrease as a result of anthropogenic activities and/or climate change, which may alter the temperature and rainfall. By contrast, the total population of the region is rapidly increasing. Consequently, the demand for food is also increasing. The total population during the middle of the twentieth century was not an issue either in terms of food security or for other demands because both the water and land resources were adequate. Since 1950, however, the region has witnessed a drastic increase in the total population. Yet, the water and land resources are limited. Consequently, food security and water resources are among the major constraints. This imbalance between people and natural resources is receiving the attention of international organizations such as the United Nations (UN), the Food and Agriculture Organization (FAO), the World Bank, national/regional organizations such as the Arab Organization for Agricultural Development (AOAD), and the International Center for Agricultural Research in the Dry Areas (ICARDA). Food security is emerging as a major issue in the region.

This chapter focuses on water resources, both conventional and nonconventional, in view of the increasing demographic pressure, with a focus on soil water. The availability of water resources is addressed in terms of the water withdrawals of the agricultural, industrial, and municipal sectors. It also addresses per capita food availability, in general, and food security in Egypt as a specific case study.

7.2 THE REGION

7.2.1 POPULATION

In the middle of the twentieth century, the total population of the region was 131.4 million inhabitants, which was 5.2% of the world's total population at that time (Table 7.2 and Figure 7.2). By 1960, this population had increased to 167.7 million at an average annual growth rate of 2.8%. Between 1960 and 1990, the annual rate of population growth in the region increased to 3.2% compared with a 2% growth rate in the world's population. Thus, the total population of the WANA region reached 378.3 million inhabitants in 1990 or 7.13% of the total world population (Table 7.2 and Figure 7.2). Shortly after the 1990 census, the governments of the region realized the problems associated with a high population (e.g., housing shortage, rising food prices, and lack of health facilities). Thus, several policy interventions were implemented to slow down the population growth. Consequently, by 2009, the growth rate had decreased from 3.2% to 2.2% and continues to decline (Figure 7.2). Therefore, the projected population will be 775 million (9.3% of the world's total population) by 2030, with an average growth rate of 1.4%, and 925 million (9.9% of the world's total population) by 2050, with an average annual growth rate of 0.8% (Table 7.2 and Figure 7.2).

TABLE 7.2
Total Population of the WANA Region between 1950 and 2050

	Total Population (Million Inhabitants)										
Country	**1950**	**1960**	**1970**	**1980**	**1990**	**2000**	**2010**	**2020**	**2030**	**2040**	**2050**
Algeria	8.8	10.8	13.8	18.8	25.3	30.5	35.5	40.2	43.5	45.5	46.5
Bahrain	0.1	0.2	0.2	0.4	0.5	0.6	1.3	1.5	1.7	1.8	1.8
Cyprus	0.5	0.6	0.6	0.7	0.8	0.9	1.1	1.2	1.3	1.3	1.3
Egypt	21.5	27.9	35.9	45.0	56.8	67.7	81.1	94.8	106.5	116.2	123.5
Eritrea	1.1	1.4	1.8	2.5	3.2	3.7	5.3	6.8	8.4	10.0	11.6
Ethiopia	18.4	22.6	29.0	35.4	48.3	65.6	83.0	101.0	118.5	133.5	145.2
Iran	17.4	22.0	28.7	38.6	54.9	65.3	74.0	81.0	84.4	85.9	85.3
Iraq	5.7	7.4	10.0	13.7	17.4	23.9	31.7	42.7	55.3	69.0	83.4
Jordan	0.5	0.9	1.7	2.3	3.4	4.8	6.2	7.4	8.4	9.3	9.9
Kuwait	0.2	0.3	0.8	1.4	2.1	1.9	2.7	3.4	4.0	4.6	5.2
Lebanon	1.4	1.9	2.5	2.8	3.0	3.7	4.2	4.5	4.7	4.8	4.7
Libya	1.0	1.4	2.0	3.1	4.3	5.2	6.4	7.1	7.8	8.4	8.8
Morocco	9.0	11.6	15.3	19.6	24.8	28.8	32.0	35.1	37.5	38.8	39.2
Oman	0.9	1.1	1.1	1.5	2.1	3.2	4.0	5.3	6.8	8.2	9.7
Qatar	0.0	0.1	0.1	0.2	0.5	0.6	1.8	2.2	2.4	2.5	2.6
Saudi Arabia	3.1	4.0	5.8	9.8	16.1	20.1	27.5	33.5	38.5	42.2	44.9
Sudan	9.2	11.6	14.8	20.1	26.5	34.2	43.6	54.9	66.9	79.1	91.0
Syria	3.4	4.6	6.4	8.9	12.3	16.0	20.4	24.1	27.9	30.9	33.1
Tunisia	3.5	4.2	5.1	6.5	8.2	9.5	10.5	11.5	12.2	12.5	12.7
Turkey	21.2	28.2	35.5	44.1	54.1	63.6	72.8	80.8	86.7	90.3	91.6
UAE	0.1	0.1	0.2	1.0	1.8	3.0	7.5	9.2	10.5	11.5	12.2
Yemen	4.3	5.1	6.2	8.0	12.0	17.7	24.1	32.2	41.3	51.3	61.6
Total	131.4	167.7	217.3	284.2	378.3	470.6	576.3	680.5	775.0	857.5	925.6
World	2532.2	3038.4	3696.2	4453.0	5306.4	6122.8	6895.9	7656.5	8321.4	8874.0	9306.1

Source: United Nations. Department of Economic and Social Affairs, Population Division, Population Estimates and Projections Section. World population prospects, the 2010 revision. 2010. http://esa.un.org/unpd/wpp/unpp/panel_population.htm. Accessed September 25, 2011.

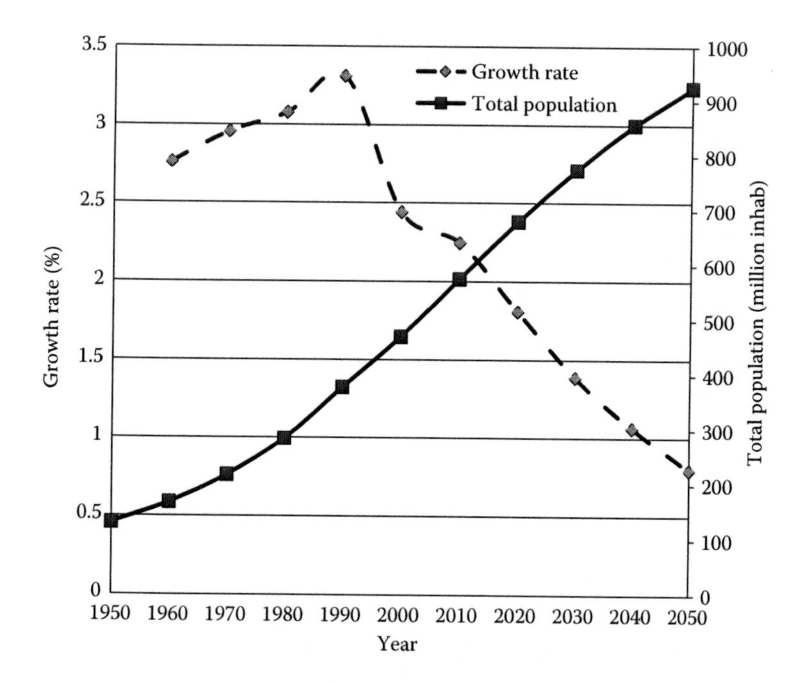

FIGURE 7.2 The total population and the average annual growth percentage in the WANA region between 1950 and 2050. (From United Nations. Department of Economic and Social Affairs, Population Division, Population Estimates and Projections Section. World population prospects, the 2010 revision. 2010. http://esa.un.org/unpd/wpp/unpp/panel_population.htm. Accessed September 25, 2011.)

7.2.2 Water Resources

Water is an important constituent of biophysical, social, and economic resources. It also plays a significant and strategic role in the relationships among the countries that share surface water and groundwater. Although water is an important resource, it must be used as a finite commodity and must be accessible to all inhabitants. Yet, water is a scarce commodity and the region is prone to drought and water scarcity. The scarcity of water results from numerous causes, such as the low MAP and the lack of investment in improving water availability. Water resources are finite and constant, but the demand for water is accelerating in all sectors. Mismanagement of what is available is a serious issue; consumers of water lack awareness and are poorly informed; the database is sketchy and unreliable; shared rivers and tributaries are disputed and are the cause of political unrest; and surface water and groundwater are prone to pollution and eutrophication by the untreated disposal of agricultural, industrial, and municipal effluents in rivers.

Although the total population of the WANA countries was composed of around 8.4% of the world's total population in 2010, the total available renewable water resources were only 1.6% of the world's total surface water (FAO 2011a). In addition to rainfall, water resources in the region originate from conventional and nonconventional sources. Conventional sources are rivers and groundwater. Nonconventional sources are the reused water that comes from agricultural and industrial drainage and the desalinated water that is produced from either seawater or brackish groundwater.

A large part of the WANA is located in the semiarid and arid biomes. Yet, the region receives about 3862.1 km^3 of rain every year (FAO 2011a). The regional distribution of the water received as rainfall is as follows: 498.9 km^3 (12.9% of the total rainfall of the region) in the North Africa region, 2074.6 km^3 (53.7%) in the Nile Valley/Red Sea region, 162.7 km^3 (4.2%) in the West Asia region, 862.6 km^3 (22.3%) in the Highlands region, and 263.3 km^3 (6.8%) in the Arabian Peninsula region. Because of the close link between rainfall and the groundwater and river flow, rainfall ultimately reaches rivers and touches the groundwater level or evaporates (70%–80% of the rainfall evaporates) as components of the hydrologic cycle. Consequently, rainwater is not given the strategic emphasis that it deserves vis-à-vis other water resources. A large fraction of it must be stored as soil water for direct use by crops and pasture.

The total available quantity of water resources—conventional and nonconventional—in the WANA is 857.5 km^3 (Table 7.3). The majority of the water, 836.3 km^3 (>97%), comes from conventional sources (surface water and groundwater), which is distributed among the regions as follows: 49.4 km^3 (5.8%) in the North Africa region, 273.5 km^3 (31.9%) in the Nile Valley/Red Sea region, 105.5 km^3 (12.3%) in the West Asia region, 397.2 km^3 (46.3%) in the Highlands region, and 10.7 km^3 (1.25%) in the Arabian Peninsula region. Surface water represents the largest water resource in the region and is 76.7% of the total water resources (Table 7.3). Of the major rivers in the region, the Nile, the Tigris, and the Euphrates are the largest and account for 84, 48, and 29 km^3/year, respectively (ESCWA 2007). Groundwater, the other constituent of the conventional water sources, can be found in renewable aquifers, such as the coastal and shallow aquifers, and

TABLE 7.3
Total Water Resources of the WANA Region between 2005 and 2009

	Conventional (km^3/year)		Nonconventional (10^6 m^3/year)		
Country	Surface Water	Groundwater	Reused	Desalinated	Total (km^3/year)
Algeria	10.15	1.52	400.0	209.98	12.28
Bahrain	0.00	0.11	17.50	285.80	0.41
Cyprus	0.56	0.41	7.00	33.50	1.01
Egypt	56.0	1.30	4,790	259.88	62.35
Eritrea	6.20	0.50	0.00	0.00	6.70
Ethiopia	120.0	20.0	0.00	0.00	140.0
Iran	106.3	49.3	130.0	200.00	155.93
Iraq	74.33	3.28	1,500	113.15	79.22
Jordan	0.65	0.51	61.00	82.86	1.30
Kuwait	0.00	0.02	0.15	872.35	0.89
Lebanon	3.80	3.20	21.5	10.22	7.03
Libya	0.20	0.50	110.0	210.00	1.02
Morocco	22.00	10.00	350.0	1.20	32.35
Oman	1.05	1.30	21.50	350.40	2.72
Qatar	0.00	0.06	33.00	747.89	0.84
Saudi Arabia	2.20	2.20	131.0	3,868.27	8.40
Sudan	62.50	7.00	0.00	16.06	69.52
Syria	12.63	6.17	1,965	4.75	20.77
Tunisia	3.40	1.60	6.00	8.70	5.01
Turkey	173.8	67.80	1,000	0.50	242.6
UAE	0.15	0.12	108.0	3,191.2	3.57
Yemen	2.00	1.50	52.00	21.17	3.57
Total	657.92	178.4	10,703.65	10,487.88	857.51

Source: Food and Agriculture Organization. Aquastat database query. 2011. http://www.fao.org/nr/water/aquastat/data/query/index.html?lang=en. Accessed September 17, 2011; ESCWA Water Development Report 3. State of water resources in the ESCWA region. United Nations, New York; AOAD 2009.

nonrenewable aquifers, such as the deep aquifers (Attia 2004). The renewable groundwater source is estimated at 178.4 km^3, which is distributed among the WANA countries as follows: 13.6 km^3 (7.6%) in the North Africa region, 28.8 km^3 (16.1%) in the Nile Valley/Red Sea region, 13.6 km^3 (7.6%) in West Asia region, 117 km^3 (65.7%) in the Highlands region, and 5.3 km^3 (3.0%) in the Arabian Peninsula region (Table 7.3).

The groundwater resource is crucially important in the WANA countries because a large part of the total land area is desert, which relies on groundwater only. However, the consumption of groundwater is higher than the recharge, leading to a depletion of the groundwater quality while increasing its salinity (Tolba 2001). A comparison of the conventional water resources between the Highlands (Iran and Turkey) and the rest of the WANA indicates that the Highlands region solely has 46.3% of the total conventional water resources, but covers only 16% of the total area of the region. The remainder of the WANA region (84% of the total area) has only 53.7% of the water resources. This comparison indicates the problem of drought in the majority of the region, especially in the Arabian Peninsula. Consequently, the governments of the region have had to harness other, nonconventional water resources.

Nonconventional water resources comprise reused and desalinated water. Reused water originates from agricultural and industrial drainage water as well as from treated wastewater from municipalities. The total reused water was estimated at 10.7 km^3/year in 2008, of which 45% was used in Egypt (AOAD 2009a). For irrigation, agricultural drainage water is more economical compared with desalinated water. In 1928, Egypt started using agricultural drainage water as an irrigation water source by mixing it with fresh water coming from the Damietta branch of the Nile River. However, in 1970, this use became official across the country by the mixing of agricultural drainage water with fresh water in the main and branch canals (Abdel-Dayem 1997). Additionally, wastewater and industrial drainage water have been used for irrigation after some treatment; however, there are some environmental concerns about the use of these two water resources.

Desalinated water can be produced from brackish aquifers or seawater and is primarily produced for potable uses. Despite the high cost, a desalinated water resource is important for countries that do not have rivers (i.e., the Arabian Peninsula region and Libya). Countries of the Gulf Cooperation Council (GCC)—Bahrain, Kuwait, Oman, Qatar, Saudi Arabia, and the UAE—use the largest quantity of desalinated water worldwide, estimated at 9.32 km^3/year, 88% of the total desalinated water of the region (ESCWA 2009). Saudi Arabia produces 24 million m^3/day of desalinated water or 50% of the world's total supply (Lee 2010). Also, the UAE, Kuwait, and Qatar rank second, sixth, and seventh in the world for producing desalinated water, respectively (Murad 2010; ESCWA 2009). These countries are bordered by the sources from which they can produce desalinated water, such as the Mediterranean Sea, the Red Sea, and the Arabian/Persian Gulf.

In 1990, the amount of available renewable freshwater resources (surface water and groundwater) for the whole region was adequate for the water demand when the total population was relatively small, and the per capita available water was above the water secure limit of 1000 m^3/year (Figure 7.3). While the data pertain to the entire region, only 50% of the countries had an adequate water supply. The number of countries prone to water shortage is increasing over time. For example, 70% of the countries had a scarcity of water in 2010, and most of the region will face water scarcity (812.4 m^3/year/capita) by 2050, except Iran and Turkey, which will have an adequate water supply (Figure 7.3 and Table 7.4).

7.2.3 Water Use

The available water resources are consumed for agricultural, municipal, and industrial uses. The total withdrawal of water between 2000 and 2007 was 392 km^3/year. The agricultural sector consumes the largest amount of the available water resources: 334.8 km^3/year (84.6% of the total water

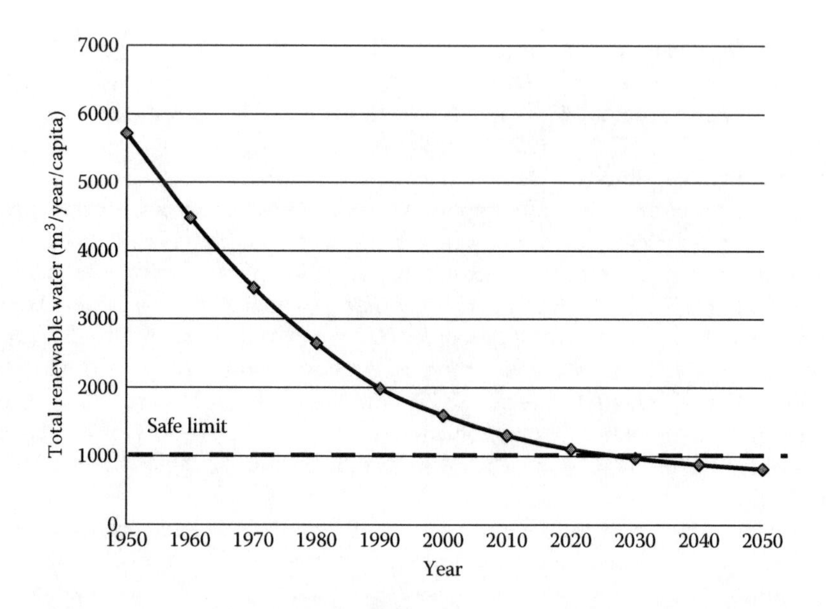

FIGURE 7.3 The total renewable fresh water per capita from 1950 to 2050 in the WANA region. (From United Nations. Department of Economic and Social Affairs, Population Division, Population Estimates and Projections Section. World population prospects, the 2010 revision. 2010. http://esa.un.org/unpd/wpp/unpp/panel_population.htm. Accessed September 25, 2011; Food and Agriculture Organization. Aquastat database query. 2011. http://www.fao.org/nr/water/aquastat/data/query/index.html?lang=en. Accessed September 17, 2011.)

used) (Table 7.5). The agricultural sector includes cultivated lands, animal husbandry, livestock farms, and fish farms or aquaculture. In fact, irrigated cultivated lands consume most of the water used by the agricultural sector. Surface irrigation is widely practiced in Algeria, Egypt, Eritrea, Ethiopia, Iran, Iraq, Lebanon, Morocco, Sudan, Syria, Turkey, and Yemen. The Gulf countries and Libya rely primarily on groundwater for irrigation. The second largest consumer is the municipal sector, which uses 34.3 km^3/year (8.8% of the total used water). The industrial sector uses only 22.85 km^3/year (5.8%). With the increase in the total population and rising standards of living, the demand on water will increase in the coming decades. Thus, there is a strong need to improve the water-use efficiency (WUE).

Members of the Economic and Social Commission for Western Asia (ESCWA) have projected that the municipal water use by the region will be 23.98 km^3 in 2025 compared with 15.08 km^3 in 2005, an increase of 8.9 km^3 (ESCWA 2007). Similarly, the cultivated land areas will increase by 2025 and will need more water. In fact, many of the WANA region's countries have implemented different scenarios to increase WUE. Surface irrigation is being replaced by sprinkler and localized irrigation in many countries. Total water resources are being increased by increasing the nonconventional resources, such as reused and desalinated water.

7.2.4 Future Plans to Increase Water Resources in the WANA Region

Water resources within the WANA region differ widely among countries. For example, Iran and Turkey have a surplus, and all others have a scarcity of water resources. By strengthening the relationships within and among countries, some projects could be implemented to increase the water resources in the region. The following suggestions may involve domestic and international projects. (1) Increasing irrigation efficiency by replacing surface (flooding) irrigation with new systems, such as sprinkler and drip irrigation; and by managing the time periods when irrigation takes place, for example, choosing night time or early morning to irrigate, in order to minimize losses by evaporation. (2) Using low-quality water when irrigating forests and sporting facilities and saving

TABLE 7.4
Water Withdrawal per Sector in the WANA Region between 2000 and 2007

Country	Water Use (km³/year)				Water Withdrawal (m³/capita/year)	
	Agricultural	Municipal	Industrial	Total	Total	Municipal
Algeria	3.94	1.58	0.95	6.47	185.39	45.27
Bahrain	0.16	0.18	0.02	0.36	450.0	225.0
Cyprus	0.15	0.06	0.00	0.21	252.9	69.09
Egypt	59.0	5.30	4.00	68.3	822.89	63.86
Eritrea	0.55	0.03	0.00	0.58	121.7	6.49
Ethiopia	5.20	0.81	0.05	6.06	85.48	10.30
Iran	86.0	6.20	1.10	93.3	128.8	85.60
Iraq	52.0	4.30	9.70	66.0	2095.24	136.51
Jordan	0.61	0.29	0.04	0.94	156.67	48.50
Kuwait	0.49	0.45	0.02	0.96	343.93	160.0
Lebanon	0.78	0.38	0.15	1.31	311.9	90.48
Libya	3.58	0.61	0.13	4.32	675.94	95.31
Morocco	11.01	1.63	0.48	13.12	409.84	50.88
Oman	1.17	0.13	0.02	1.32	471.79	47.86
Qatar	0.26	0.17	0.01	0.44	317.14	124.29
Saudi Arabia	20.83	2.13	0.71	23.67	931.89	83.86
Sudan	36.07	1.14	0.30	37.51	886.83	27.02
Syria	14.67	1.48	0.62	16.77	794.31	69.91
Tunisia	2.17	0.37	0.11	2.65	253.85	35.10
Turkey	29.6	6.20	4.30	40.1	549.30	84.93
UAE	3.31	0.62	0.07	4.00	869.13	134.13
Yemen	3.24	0.27	0.07	3.58	151.06	11.23
Total	334.79	34.33	22.85	391.97	—	—

Source: Food and Agriculture Organization. Aquastat database query. 2011. http://www.fao.org/nr/water/aquastat/data/query/index.html?lang=en. Accessed September 17, 2011.

high-quality water for other uses. (3) Increasing awareness about the scarcity of water among the public, in particular among farmers. (4) Implementing the project of connecting the Congo River with the Nile River, which will increase the water quantity of the Nile, thereby increasing the allocations for Sudan and Egypt. (5) Transporting ice blocks from the southern pole to the Arabian Peninsula region. The latter strategy was suggested by one of the princes of Saudi Arabia in the middle of the 1970s when the Arabian Peninsula experienced a severe drought (Alabbasy 2007). (6) Transporting fresh water from Pakistan to the Arabian Peninsula region using gigantic ships via the Arabian/Persian Gulf or through constructing a pipeline connecting the Mangua River and the UEA with a flow capacity of 520×10^3 m³/day. (7) Establishing a pipeline between Iran and Qatar (the Faith Project) to provide fresh water to the Arabian Peninsula region as a strategic project to enhance the relationships between Iran and the Gulf countries. This project was discussed between Qatar and Iran in 1991 during a visit to Tehran by Muhammad Al Khalifa Ben Than, the sovereign of Qatar. (8) Transporting fresh water from the Karun River in Iran to Kuwait via a pipeline with a flow capacity of 0.8 million m³/day. (9) Implementing the Peace Pipes project connecting Turkey and the Arabian Peninsula region with a flow capacity of 6×10^6 m³/day. This project was proposed in 1987 by Torkot Awzal, the prime minister of Turkey, to sell surplus water to the Arabian Peninsula region.

TABLE 7.5
Trend of Renewable Water Resources per Capita in the WANA Region

Country	Per Capita Renewable Water Resources (m^3/year/person)									
	1960	1970	1980	1990	2000	2010	2020	2030	2040	2050
Algeria	1,080.6	848.7	620.4	461.3	382.3	329.1	290.4	268.4	256.5	250.9
Bahrain	725.0	552.4	322.2	236.7	181.3	92.1	76.8	70.3	65.9	64.4
Cyprus	1,361.3	1,270.4	1,137.1	1,016.9	827.2	706.5	640.4	599.5	580.4	579.1
Egypt	2,053.8	1,595.2	1,274.8	1,008.1	847.1	706.4	604.4	538.1	493.0	464.2
Eritrea	4,424.2	3,410.9	2,551.6	1,994.9	1,717.6	1,199.1	919.9	750.5	627.9	544.6
Ethiopia	5,409.5	4,212.9	3,443.8	2,524.2	1,860.4	1,470.8	1,207.4	1,029.4	914.1	840.3
Iran	6,250.3	4,797.3	3,564.3	2,505.9	2,104.3	1,858.8	1,696.6	1,628.4	1,600.8	1,611.1
Iraq	10,243.9	7,544.9	5,502.2	4,352.3	3,168.5	2,387.1	1,771.3	1,368.1	1,096.5	906.9
Jordan	1,041.1	561.1	407.4	273.9	194.0	151.4	127.1	111.3	100.9	94.8
Kuwait	76.90	26.7	14.5	9.6	10.3	7.3	5.9	5.0	4.3	3.9
Lebanon	2,356.0	1,829.3	1,607.1	1,525.4	1,203.2	1,063.8	995.6	957.5	947.4	961.5
Libya	444.5	301.5	196.1	138.6	114.7	94.3	84.8	77.1	71.8	68.4
Morocco	2,493.6	1,894.2	1,481.9	1,170.3	1,007.3	907.7	826.7	773.3	747.2	739.8
Oman	1,308.4	1,238.9	927.2	673.1	437.5	346.5	263.2	207.1	170.1	143.9
Qatar	1,200.0	545.5	272.7	127.7	101.7	34.1	27.3	25.3	23.8	23.0
Saudi Arabia	594.10	415.9	244.9	148.7	119.7	87.4	71.6	62.4	56.9	53.4
Sudan	5,579.6	4,366.9	3,213.8	2,434.9	1,886.5	1,481.1	1,174.4	964.7	815.8	709.1
Syria	3,676.2	2,637.4	1,885.5	1,363.6	1,050.7	823.1	697.7	603.0	543.3	508.3
Tunisia	1,090.1	896.7	712.1	559.6	486.3	438.9	399.3	376.7	367.1	363.6
Turkey	7,585.0	6,023.0	4,842.9	3,946.1	3,357.1	2,936.0	2,645.1	2,464.7	2,365.4	2,331.5
UAE	1,666.7	652.2	147.1	82.9	49.5	19.9	16.4	14.3	13.0	12.4
Yemen	410.20	341.5	264.2	175.7	118.5	87.3	65.2	50.8	41.0	34.1

Source: Food and Agriculture Organization. Aquastat database query. 2011. http://www.fao.org/nr/water/aquastat/data/query/index.html?lang=en. Accessed September 17, 2011; United Nations. Department of Economic and Social Affairs, Population Division, Population Estimates and Projections Section. World population prospects, the 2010 revision. 2010. http://esa.un.org/unpd/wpp/unpp/panel_population.htm. Accessed September 25, 2011.

7.2.5 Arable Land Area

Deserts comprise a large area of the WANA region in both Africa and Asia, limiting the total area of arable land, which increased from 93.3 Mha in 1962 to 112 Mha (7% of the region's total area) in 2008 (Figure 7.4 and Table 7.6). There is an imbalance between the rate of population growth and the increase in the arable land area. For example, the arable land area increased at an annual rate of <0.01%, while the population increased at an annual rate of 2.2%. Consequently, the per capita arable land area decreased from 0.60 ha in 1962 to 0.17 ha in 2008. Yet, the per capita arable land area in 2008 varied widely among countries: 0.002 ha in Bahrain to 0.501 ha in Sudan. The differences in the per capita arable land area are attributed to the differences in the total population and the existing arable land in each country.

7.2.6 Food Security in the WANA Region

The concept of food security has been evolving since the 1990s. There are more than 200 definitions of food security (Smith et al. 1993). The World Food Summit of 1996 and the FAO (2002) have defined it as "food security exists when all people, at all times, have physical, social, and economic access to sufficient, safe, and nutritious food that meets their dietary needs and food preferences for an active and healthy life." It implies the availability of food, the purchasing power of consumers according to their wages, and food prices (Schmidhuber and Tubiello 2007). In addition, it is important to differentiate between food security at the national level and at the household or community level because of the income disparity that affects access to food.

The WANA countries are suffering from a shortage of all food commodities, especially cereals, despite governments' efforts to overcome such shortages. In the past, most of the WANA countries focused on the concept of self-sufficiency in terms of food security without taking care of the food quality. However, they started to consider food quality in the first decade of the twenty-first century. Food security refers to the availability of many different commodities, such as cereals, legumes,

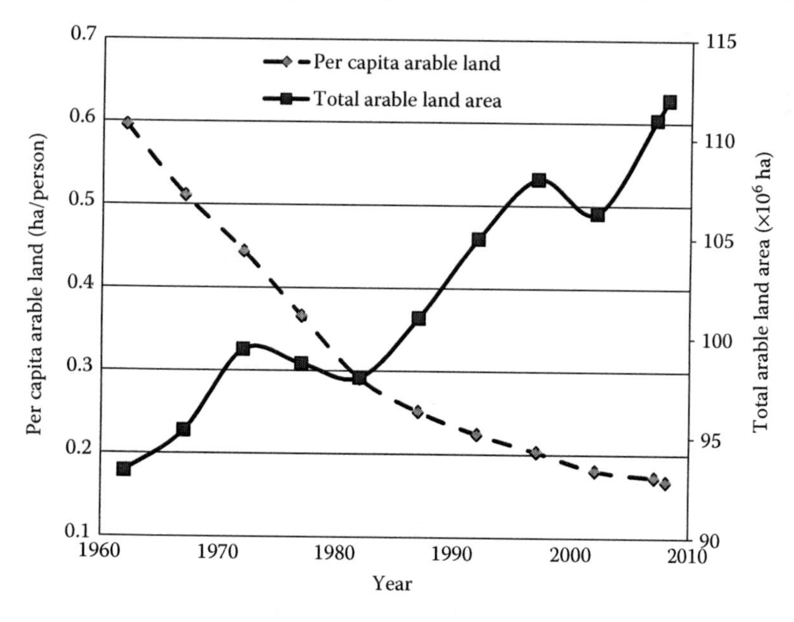

FIGURE 7.4 The temporal changes in and per capita of the arable land area in the WANA region. (From Food and Agriculture Organization. Aquastat database query. 2011. http://www.fao.org/nr/water/aquastat/data/query/index.html?lang=en. Accessed September 17, 2011; United Nations. Department of Economic and Social Affairs, Population Division, Population Estimates and Projections Section. World population prospects, the 2010 revision. 2010. http://esa.un.org/unpd/wpp/unpp/panel_population.htm. Accessed September 25, 2011.

TABLE 7.6
Total and per Capita Arable Land Area in the WANA Region

	Total Arable Area (1000 ha)						Per Capita Arable Area (ha)					
Country	**1962**	**1972**	**1982**	**1992**	**2002**	**2008**	**1962**	**1972**	**1982**	**1992**	**2002**	**2008**
Algeria	6,300	6,510	6,881	7,562	7,547	7,489	0.562	0.446	0.342	0.285	0.240	0.218
Bahrain	1	1	2	2	2	1.4	0.006	0.004	0.005	0.004	0.003	0.002
Cyprus	100	100	103	108	100	82.4	0.170	0.170	0.150	0.140	0.110	0.070
Egypt	2,433	2,725	2,305	2,519	2,936	2,773	0.083	0.073	0.049	0.042	0.040	0.034
Eritrea	391	391	391	391	562	670	0.280	0.220	0.160	0.120	0.150	0.130
Ethiopia	9,900	9,900	9,900	9,900	9,936	13,606	0.440	0.340	0.280	0.200	0.150	0.160
Iran	14,990	16,227	14,141	16,969	16,029	17,037	0.680	0.570	0.370	0.310	0.250	0.230
Iraq	4,650	5,000	5,250	5,770	5,600	5,200	0.588	0.458	0.353	0.302	0.214	0.173
Jordan	272	287	296.1	259.7	195	149.5	0.283	0.162	0.123	0.071	0.038	0.024
Kuwait	1	1	2	4	12	11.4	0.003	0.001	0.001	0.002	0.005	0.004
Lebanon	180	240	208	180	130	144	0.090	0.093	0.074	0.057	0.033	0.034
Libya	1,710	1,730	1,765	1,815	1,815	1,750	1.178	0.797	0.522	0.399	0.326	0.278
Morocco	6,590	7,131	7,767	8,934	8,402	8,055	0.537	0.443	0.376	0.348	0.285	0.255
Oman	20	23	23	35	38	55	0.034	0.029	0.017	0.018	0.015	0.020
Qatar	1	1	6	12	13	13	0.019	0.007	0.022	0.024	0.019	0.010
Saudi Arabia	1,160	1,450	1,990	3,650	3,600	3,446	0.268	0.232	0.184	0.213	0.164	0.137
Sudan	10,875	11,861	12,450	12,900	16,519	20,698	0.886	0.744	0.568	0.453	0.454	0.501
Syria	6,200	5,711	5,288	4,766	4,593	4,699	1.262	0.838	0.547	0.353	0.263	0.221
Tunisia	3,100	3,230	3,160	2,908	2,771	2,835	0.708	0.606	0.465	0.341	0.288	0.279
Turkey	23,131	25,573	24,199	24,514	23,994	21,555	0.820	0.720	0.550	0.450	0.380	0.300
UAE	5	7	18	38	75	65	0.046	0.022	0.015	0.018	0.021	0.014
Yemen	1,273	1,345	1,368	1,378	1,415	1,279	0.234	0.203	0.151	0.102	0.073	0.056
Total	44,771	47,253	48,779	52,733	55,663	58,663	—	—	—	—	—	—

Source: Food and Agriculture Organization. Aquastat database query. 2011. http://www.fao.org/nr/water/aquastat/data/query/index.html?lang=en. Accessed September 17, 2011.

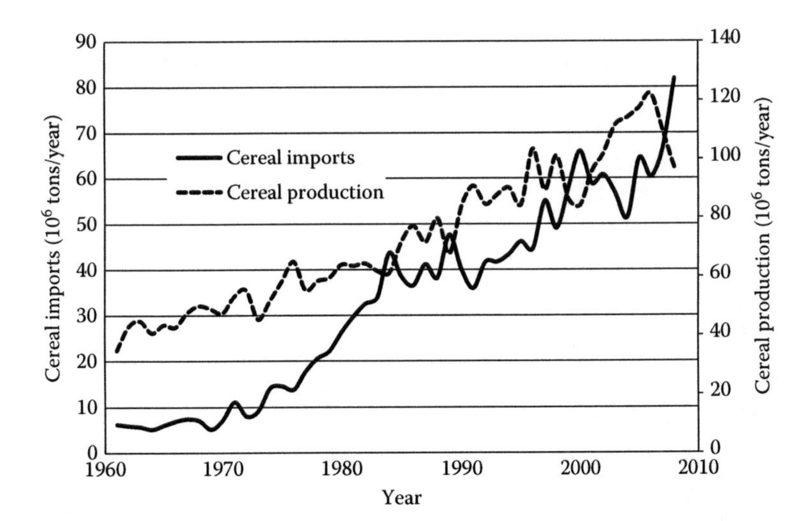

FIGURE 7.5 The changes in cereal production and imports in the WANA region. (From Food and Agriculture Organization. FAOSTAT. 2011. http://faostat.fao.org/site/567/DesktopDefault.aspx?PageID=567#ancor. Accessed September 20, 2011.)

carbohydrates, proteins, and fish. However, this chapter focuses only on cereals, because of their importance to the diet in the region. Cereals are the most common commodity used in the region because farmers prefer to grow cereals to feed themselves and their domestic animals. According to the small area of arable land compared with the total area and total population in the region, cereal production has never been sufficient. Consequently, the region has been importing significant amounts of cereals (Figure 7.5). For example, imports of cereals were 6.3, 36, and 81 million Mg/year in 1961, 1991, and 2008, respectively (Figure 7.6). Wheat is the most important cereal crop and its production in the Arab countries comprised 45% of the total cereal production between 2005 and 2008 (AOAD 2009b). It is the main source of protein and carbohydrate and is accessible to most of the rural communities. Wheat imports have been increasing since the 1960s; the region imported 3.6, 19.0, and 41.5 million Mg in 1961, 1991, and 2008, respectively (Figure 7.6). The increase in cereal imports is attributed to several factors such as the decline in cereal yield as a consequence of

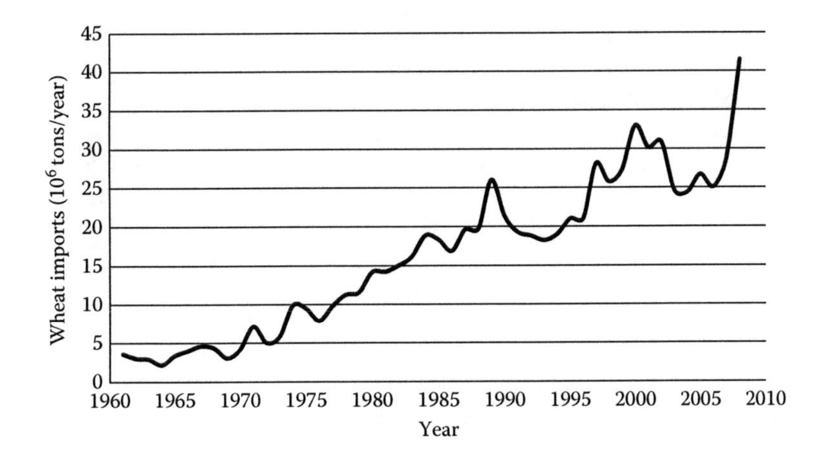

FIGURE 7.6 The history of wheat imports in the WANA region. (From Food and Agriculture Organization. FAOSTAT. 2011. http://faostat.fao.org/site/567/DesktopDefault.aspx?PageID=567#ancor. Accessed September 20, 2011.)

soil degradation, land use change, the increase in urbanization exacerbating labor and power shortages, the migration of the rural population to urban centers, and the contemporary high expenses of agricultural inputs reducing the profit margin.

Yet, the WANA region can be self-sufficient in wheat production. It is anticipated that cereal production can be increased significantly in the WANA region, particularly in the Arab countries, because of recent policy interventions to meet public demands. In addition, political relationships among the countries in the region are improving, thereby facilitating investment in human and natural resources.

7.3 EGYPT: A CASE STUDY

Egypt is located at the northeastern corner of Africa and covers a total area of ~100 Mha (one million square kilometers); the majority of it is desert land. It is bordered by the Mediterranean Sea in the north; Palestine, Israel, and the Red Sea in the east; Libya in the west; and Sudan in the south. The Nile River divides Egypt into two major sections: east and west. The cultivated area is less than 5% of the total area of the country (Attia 2004), and a large fraction of it exists within the Nile Delta and along the Nile River. Furthermore, 95% of the cultivated area is irrigated. Even the small rainfed area along the Mediterranean coast needs supplementary irrigation.

The climate of Egypt is hot and dry in summer and cold and mild in winter. The temperature ranges from 8°C to 18°C in winter and 21°C to 36°C in summer with an extreme temperature of 45°C in the southern part of Egypt and in the Western Desert. The MAT is 21°C and the MAP is 51 mm/year (Table 7.1). Most of the rainfall is received in the northern part of Egypt, especially along the Mediterranean coast.

7.3.1 POPULATION

The population of Egypt has increased substantially during the past six decades and is expected to increase during the next several decades. Most of the population is concentrated on 4% of the total land area of Egypt along the Nile Valley and within the Delta region (Attia 2004). The total population tripled between 1950 and 1990 from 21.5 million to 56.8 million, at an average growth rate of 2.7%/year. However, a large population was not an issue prior to 1990 because the water and land resources were adequate. Several million Egyptians were working abroad, especiallyin Iraq, Libya, and the Gulf countries, earning foreign exchange and increasing the national income.

The war in Iraq and Kuwait forced millions of Egyptians to return, overstressing the civil structures and facilities. Therefore, the Egyptian government initiated several programs to limit the rate of population growth. Consequently, the average rate of population growth decreased from 2.65% to 1.9%/year between 1990 and 2008 (Figure 7.7). The growth rate is still declining and will continue to decline for several decades because of the increase in health awareness among women, high living expenses, the modern lifestyles, and advances in the media. Despite these measures, the population is projected to increase from 81.1 million in 2010 to 123.5 million in 2050 (Figure 7.7).

7.3.2 WATER RESOURCES

With the exception of the northern region, rainfall in Egypt is rather low. Some intermittent torrents occur in the peninsula of Sinai and Upper Egypt; however, their water goes to either streams and rivers or groundwater. The total water resources are either conventional or nonconventional. Conventional resources are confined to the Egyptian allocation of the withdrawal of water from the Nile River, the groundwater along the Nile and its delta, the groundwater in the northern and eastern coasts (shallow groundwater), and the groundwater in the eastern and western deserts (mostly deep and nonrenewable groundwater). Nonconventional water resources include the reused water coming from agricultural and industrial drainage, treated wastewater, and desalinated water produced either

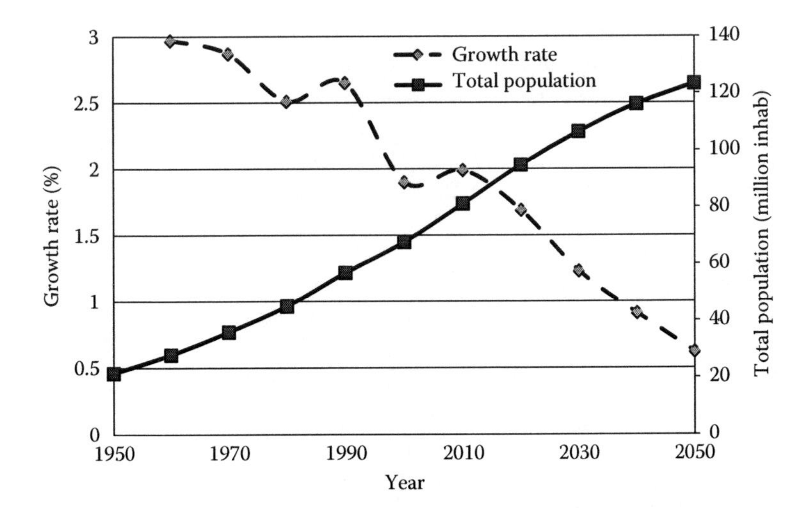

FIGURE 7.7 The temporal changes in the total population and the average annual growth rate in Egypt between 1950 and 2050. (From Food and Agriculture Organization. Aquastat database query. 2011. http://www.fao.org/nr/water/aquastat/data/query/index.html?lang=en. Accessed September 17, 2011; United Nations. Department of Economic and Social Affairs, Population Division, Population Estimates and Projections Section. World population prospects, the 2010 revision. 2010. http://esa.un.org/unpd/wpp/unpp/panel_population.htm. Accessed September 25, 2011.)

from seawater or brackish groundwater (Allam and Allam 2007). According to the treaty between Egypt and the other countries in the Nile Basin, Egypt's water allocation from the Nile River is 56 km^3/year, accounting for 98% of the total renewable fresh water (Abu Zeid 2003).

Since the western and eastern deserts comprise the majority of the total area, groundwater (renewable and nonrenewable) is extremely important for agriculture, municipal, and industry in the deserts (El-Fellaly and Saleh 2004). Moreover, river water is not accessible in large areas within the delta. Consequently, farmers have to dig wells in order to obtain fresh water, which originates from percolation from the irrigation canals. A large number of Egyptian villages do not have a potable water network; consequently, residents drill small pumps, which are manually driven, in order to obtain fresh water for potable uses. The total renewable groundwater provides 1.3 km^3/year (Table 7.3).

The Nile water provides Egypt with 56 km^3/year, which is insufficient for water demands. Several Egyptian governments have tried to identify other water resources, especially nonconventional water resources. Recycling the agricultural drainage water has been practiced for a long time. In 1928, drainage water from the Serw station was used for irrigation. Thus, a supporting station was constructed in 1930 besides the Serw station. Shortly after, the government learned that the quality of the agricultural drainage water was similar to that of the Nile water. Therefore, the drainage water from the supporting station was pumped to the Damietta branch of the Nile. This strategy was officially adopted in the 1970s (Abdel-Dayem 1997). The policy calls for the pumping of agricultural drainage water from the main and branch drains and mixing it with fresh water in the main and branch canals. However, many farmers use the agriculture drainage water directly from the drains because it may be closer to their farms compared with the irrigation canals. In addition, many villages drain their wastewater into agricultural drains, from which large areas are irrigated. Treated wastewater is also used for irrigation. Yet, large areas are unofficially irrigated using untreated wastewater. The official reused water resources—agricultural and industrial drainage and treated wastewater—provide 4.79 km^3/year. In the coming decades, more land will be drained and more wastewater treatment stations will be constructed, which will lead to an increase in the use of reused water, 11 km^3/year by 2017 (El-Fellaly and Saleh 2004).

Desalination is used wherever conventional water resources are inadequate (e.g., the coastal areas of the Mediterranean Sea and the Red Sea). Because of its high cost, desalinated water is primarily used by the municipalities. However, many studies are now being conducted to identify more economic sources of power, such as solar and wind, to produce large quantities of desalinated water. The desalinated water resource amounts to 0.26 km^3/year. Thus, the total quantity of water from all sources between 2005 and 2009 was 62.4 km^3/year (Table 7.3).

7.3.3 Water Use

Water demand in Egypt is increasing because of the increase in the population, living standards, economic activities, and cultivated land, especially with the land reclamation plans implemented by sequential Egyptian governments. The total water withdrawal between 2000 and 2008 was 68.3 km^3/year (Table 7.4), which is distributed among three sectors: agricultural, municipal, and industrial sectors. The agricultural sector is considered the core sector of the Egyptian economy. It contributes ~20% of the gross domestic product (GDP) and uses 31% of the total labor force (Attia 2004; AOAD 2009a). The agricultural sector consumes 59 km^3/year (86.4% of the total water demand) for irrigation and other agricultural activities (Table 7.4). With double cropping or two crops per year, intensive agriculture has doubled the water demand. Rice and sugarcane are grown over large areas, and they are the largest consumers of irrigation water due to their high water requirements. However, the government plans to gradually decrease the area under these crops in order to save water for other crops. The loss of water by evapotranspiration from the entire surface water and irrigation network is estimated at 3 km^3/year.

The second largest user and consumer of water is the municipal sector. This sector consumes 5.3 km^3/year (7.8% of the total water demand) (Table 7.4), primarily from the Nile River, groundwater, and a small portion from desalination, supplying water to 216 cities and 4525 villages, in addition to navigation and tourism activities (Attia 2004). Most of the municipal water is returned to the drains and approximately 1 km^3/year is actually consumed. According to Egyptian government statistics between 2000 and 2004, all cities and 42% of villages have been provided with piped fresh water; 52% of villages are partially provided and only 6% are not provided with fresh water. However, the efficiency of the delivery network to the municipal sector is as low as 50% and even less in some places (Attia 2004). In fact, there is no reliable estimate of the water use for the industrial sector. However, the approximate estimate is 4 km^3/year (Table 7.4).

In Tables 7.3 and 7.4, a comparison of the total water resources and the total water withdrawals indicates that the total water withdrawals exceeded the water resources by 5.95 km^3/year. The deficit may even increase in the future because of the increase in the population, expansion of the cultivated areas, modern lifestyles, and increase in the industrial sector.

Finite water resources in Egypt also affect the per capita water availability. The latter was more than 1000 m^3/year prior to 1987, but declined to 703 m^3/capita/year in 2008 (Table 7.5 and Figure 7.8). The per capita available water supply may decline to less than 500 m^3/capita/year in 2050 (UN 2010; FAO 2011a). Moreover, the quality of the surface water or the groundwater is degraded by pollution from industrial and wastewater drainage in streams and the infiltration of agricultural drainage into groundwater contaminated by fertilizers (especially nitrates) and pesticides.

7.3.4 Arable Land Area

As Egypt is an agrarian nation, the arable land area has been a major issue since the beginning of the twentieth century. The arable land area comprises a small portion of the total land area of Egypt, 2.43% in 1962 and 2.8% in 2008 (Tables 7.1 and 7.6). The arable land area increased between 1962 and 1972, decreased between 1977 and 1987, and increased again between 1992 and 2008 (Figure 7.9 and Table 7.6). In 2008, the total cultivated land area was 3.5 Mha, and 90% of this was irrigated. The old land, which is located in the Nile Valley and Delta, covers 2.3 Mha. The new land,

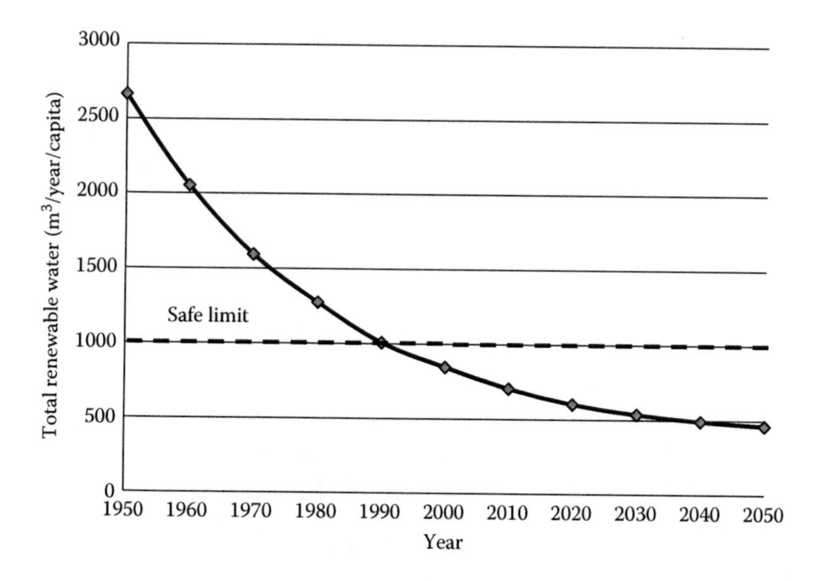

FIGURE 7.8 The trend of total renewable per capita water from 1950 to 2050 in Egypt. (From Food and Agriculture Organization. Aquastat database query. 2011. http://www.fao.org/nr/water/aquastat/data/query/index.html?lang=en. Accessed September 17, 2011; United Nations. Department of Economic and Social Affairs, Population Division, Population Estimates and Projections Section. World population prospects, the 2010 revision. 2010. http://esa.un.org/unpd/wpp/unpp/panel_population.htm. Accessed September 25, 2011.)

which is located on the east and west sides of the Delta and in the eastern and western deserts, covers 1.0 Mha. Oases include 40,000 ha, and the rainfed area located in the north coast covers 0.17 Mha (FAO 2005). The total area irrigated by surface water was 2.84 Mha, which comprised 89% of the total irrigated area. The principal irrigation systems are composed of a mixture of traditional and modern techniques. Flooding irrigation (surface and furrow) is practiced on 89% of the cultivated

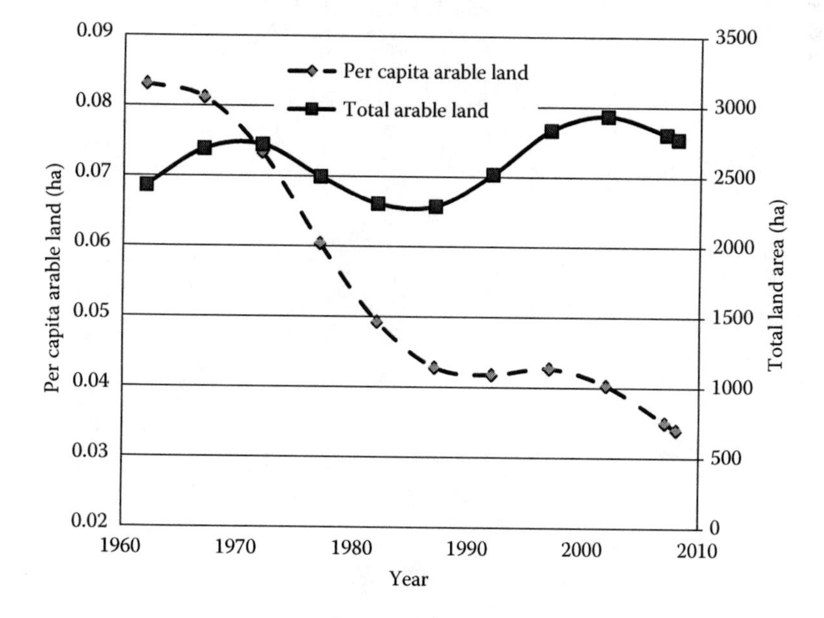

FIGURE 7.9 The trend of the total and per capita arable land area in Egypt between 1962 and 2008. (From Food and Agriculture Organization. Aquastat database query. 2011. http://www.fao.org/nr/water/aquastat/data/query/index.html?lang=en. Accessed September 17, 2011.)

area, especially along the Nile River and within the Nile Delta (Figure 7.10). Thus, water losses due to evaporation are high. Traditional irrigation methods were composed of manually powered tools, such as shadoof and tambour (Figure 7.11), and animal-driven saqias (Figure 7.11). Gradually, these traditional techniques have been replaced by mobile pump sets (Figure 7.12). Furthermore, modern irrigation systems, such as sprinkler and drip irrigation, are also being used (Figure 7.13). These modern systems are practiced in either greenhouse agriculture (Figure 7.13) or new reclaimed soils because they are in large areas and have a low-water-holding capacity.

7.3.5 Future Plans to Increase Water Resources in Egypt

In 1993, the Egyptian government implemented a policy to increase the water resources and meet the water demands by 2017, when the projected total water use will be 67 km^3 for the agricultural sector, 3 km^3 losses by evaporation, 7 km^3 for the municipal sector, and 10 km^3 for the industrial sector, with a total use of 87 km^3. The expected increase in the water demand can be met through: (1) increasing the Egyptian quota of the Nile River to 57 km^3 through implementing the Jonglei canal project; (2) using 7.5 km^3 from the Nile Aquifer; (3) increasing the reused water supply to 10.4 km^3; (4) utilizing 5.3 km^3 of the groundwater and surface water in Sinai; (5) decreasing the loss due to evapotranspiration by cleaning the vegetation growing on the surface of the Nile and from all exposed water bodies (i.e., streams, irrigation, and drainage canals) and improving the irrigation systems; (6) changing the cropping patterns by decreasing the areas grown to for rice (*Oryza sativa*) and sugarcane (*Saccharum officinarum*) and increasing that for sugar beet (*Beta vulgaris*); and (7) creating an awareness about the judicious use of water among the Egyptian people, especially farmers.

FIGURE 7.10 Flooding irrigation: (a) surface, (b, c) furrow, and (d) improved furrow.

FIGURE 7.11 Manually powered and animal-driven irrigation tools: (a) shadoof, (b) tambour, and (c, d) saqias.

7.3.6 Food Security in Egypt

Egypt has been experiencing food scarcity for a long time. Shortages are occurring in all kinds of food, even in the least expensive commodity, bread. In 1977, the Egyptian people protested against the government because of the shortage of bread. Cereal production increased from 5 million Mg in 1961 to 23 million Mg in 2008 (FAO 2011b). However, the production was never sufficient to meet

FIGURE 7.12 (a, b) Mobile pump irrigation machines.

FIGURE 7.13 Modern irrigation systems: (a–c) different sprinkler irrigation designs used in new reclaimed soils; (d, e) drip irrigation used in new reclaimed soils; and (f) drip irrigation used in greenhouse agriculture.

the food demand. Consequently, sequential Egyptian governments have been importing cereals; the imports were 1.4 million Mg in 1961 and 12.3 million Mg in 2008 (FAO 2011b). Furthermore, Egypt is considered the largest wheat importer in the world, importing 10 million Mg in 2010 (FAO 2011c). The food security problem in Egypt has several causes, including international and domestic factors.

Since 2000, many countries around the world have started to produce biofuels to replace oil, which consume large portions of the world's cereal production. Consequently, food prices have risen dramatically, especially the price of agricultural commodities that are scarce in the global market (Weber and Harris 2008). In addition, climate change has affected rainfall patterns in many different areas across the globe, which has changed the agrarian landscape. Also, the increase in the cost of transportation has affected food prices in Egypt. The local causes can be attributed to the small arable land area and the increase in the total population. Also, urbanization and encroachment on agricultural land have diverted high-quality soils for urban and industrial uses. The mismanagement of cropping patterns is another cause, for example, large areas of arable land are owned by wealthy farmers who grow nonstrategic cash crops, such as cantaloupe (*Cucumis melo*) and grapes (*Vitis vinifera*), for export, instead of wheat (*Triticum aestivum*) and corn (*Zea mays*). Even a strategic crop such as rice is exported for more profit, despite the need of the Egyptian communities. Ironically, the government has encouraged the production of cash crops. On the contrary, the prices of fertilizers have increased and those of farm produce have decreased. Consequently, a large number of farmers have shifted cropping patterns. They tend to grow legumes, forages, and vegetables rather than wheat, rice, and corn.

7.3.7 Plant Residue Management

Small landholders are resource-poor. They use everything from the farm to meet their daily demands. The residues of different crops are used for competing purposes. For example, the residues of corn and cotton (*Gossypium barbadense*) are used for cooking and baking; the residues of legumes and wheat are used as fodder; and the residue of rice is used to manufacture bricks or to make dough as a mortar between the bricks and as a filler of the cracks in their construction. However, during the first

FIGURE 7.14 Black clouds resulting from burning rice hay in Egypt. (From Alshahid Centre for Research and Media Studies, UK.)

decade of the twenty-first century, the government encouraged farmers to burn rice hay, corn, and cotton stalks in the fields. The purpose of the burning was to dispose of plant residues and to enrich the soil fertility with the remaining ash; however, overwhelming environmental issues emerged. For example, during the season of burning the rice hay, gigantic black clouds engulfed many cities, in particular Cairo (Figure 7.14). Thus, the Egyptian government has banned the burning of residues. Furthermore, some farmers use corn and cotton residues in the subsurface drainage of their farms; they dig ditches and fill them with these residues up to half the depth of the ditches and then fill the remaining depth with soil. In this process, they dispose of residues that are needed to enhance the soil organic matter (SOM) and to improve soil quality.

7.4 SOIL WATER MANAGEMENT

The economic importance of water is implied in its name "the blue gold." The agricultural sector is the largest consumer, consuming ~70% of the world's total water consumption, which may increase in the future (Green and Deurer 2010). The water consumed in agricultural production is called virtual water (Allan 1993, 1998). Virtual water has three components: blue, green, and gray. Blue water refers to surface water and groundwater. Green water refers to rainfall water and soil water. The latter might originate from rainfall or be added as irrigation water from surface water or groundwater. When water drains from soil contaminated by chemicals (e.g., fertilizers and pesticides) or has been disposed of as municipal or industrial effluents, it is called gray water (Chapagain and Hoekstra 2004). It is important to manage the processes of changing blue water and gray water (water drainage) to green water (irrigation process) to increase WUE. Using modern irrigation systems and irrigating at night and early in the morning increase the WUE of an irrigation system and decrease the water losses by evaporation. The other challenge is to preserve the soil water from being lost through evaporation, percolation, and uptake by weeds.

The soil's physical, hydrological, and biological properties strongly impact the hydrologic balance and can be managed to preserve the green water against losses by leaching and evaporation. The strategy is to enhance the soil water-holding capacity (SWHC) and decrease losses by evaporation. Soil texture is strongly correlated with SWHC; the finer the texture, the higher is the SWHC (Figure 7.15). Since deserts dominate the WANA countries, most cropland soils are coarse-textured and have a low SWHC. Thus, enhancing the SOM by using manure, guano, green manure, and plant residues can increase the SWHC. The SOM has a high-water-holding capacity and can hold 10–1000 times more water than soil minerals (USDA 2003). Therefore, increasing the SOM increases the SWHC and decreases the water losses by leaching and evaporation. Moreover, the SOM enhances

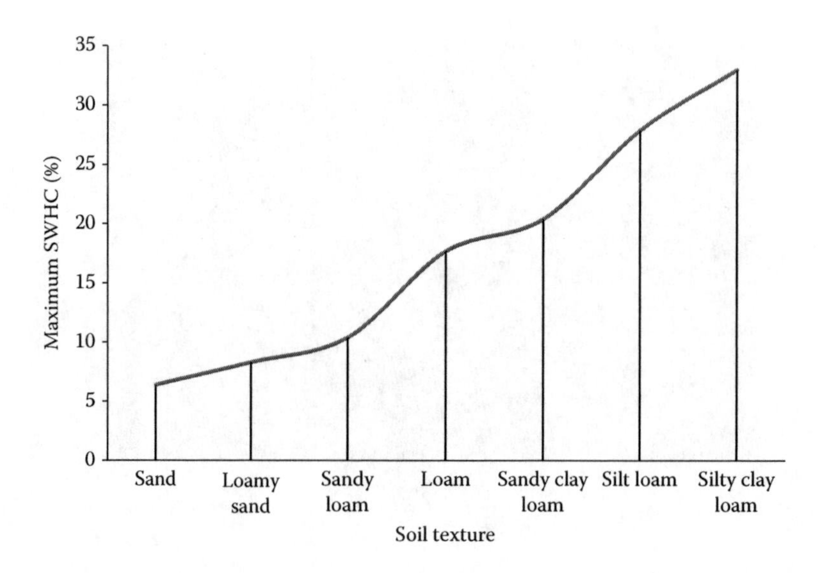

FIGURER 7.15 Different soil textures hold variable amounts of water. (Modified from Lane, D.R., Coffin, D.P., and Lauenroth, W.K., *J. Veg. Sci.*, 9, 239–250, 1998.)

the soil aggregation, which decreases the permeability of sandy soils (Khaleel et al. 1981; Metzger and Yaron 1987). In addition, synthetic and natural soil conditioners increase the SWHC when applied to coarse-textured soils. For example, adding biopolymers (e.g., agar, alginate, guar gum, bacterial polymer, cellulose, psyllium, and bacterial polymer) to sandy loam soils improves the maximum SWHC (Patil et al. 2011). Similarly, applying hydrogel to sandy soils in an arid region with intense evaporation increases the SWHC, delays the permanent wilting point, and reduces the irrigation requirements of different crops (Taylor and Halfacre 1986; Taban and Naeini2006).

In comparison with a bare soil surface, a covered soil has less evaporation. A high temperature and a low rainfall in the WANA region make the soil water susceptible to loss by evaporation. Different scenarios have been implemented to preserve the soil water, such as covering the soil surface with plant residue, which acts as a partial isolator between the soil surface and the atmosphere and reduces water losses by evaporation (Karlen et al. 2009). The practice of no-tillage and conservation tillage leaves more plant residues on the soil surface, which increases the SOM, preserves the soil water, and increases the crop yields (Mrabet 2002; Thomas et al. 2006; Anyanzwa et al. 2010). Therefore, mulching with plant residues is beneficial for the environment and preserves the soil water; thus, it must not be burnt. Similarly, plastic mulch can be used to cover the soil surface, especially sandy soils, to reduce the water losses by evaporation, increase the SOM, and increase the crop yields (Li et al. 1999).

7.5 CONCLUSION

Most of the WANA countries are located in arid and semiarid regions with a low rainfall and a high temperature. Deserts occupy large areas in the region, and the water availability is not sufficient to meet the demands. Surface water comprises 78.8% of the total renewable water resources in the region. Yet, many countries (e.g., Arabian Peninsula) rely entirely on groundwater. Finite conventional water resources and the continuous increase in the total population are the principal causes of water shortage. Nonconventional water resources (i.e., reusing agricultural drainage water, treated industrial and wastewater drainage, and desalinated water) are being explored to increase the water resources. Water desalination is common in the Arabian Peninsula. Saudi Arabia, the UAE, and Qatar are among the world's largest producers of desalinated water. The economic techniques of desalinating water are based on solar and wind power. Green water can be increased by soil water conservation scenarios, such as increasing the SOM content, covering the soil surface with plant residues or plastic mulch, and conservation tillage methods. Also, surface irrigation, which is practiced extensively, can be replaced by modern irrigation systems (sprinkler, drip, etc.). Only 3.7% of the region's total area is arable. Consequently, food production is not sufficient, and most countries import different kinds of food commodities to meet their food demands. Cereals, in particular wheat, are the food staple. Egypt has experienced water scarcity since 1992 as a result of a high total population and limited conventional water resources. Egypt is the largest wheat importer in the world; 10 million Mg of wheat was imported in 2010. Thus, the sustainable management of soil water (green water) is crucial to improving agronomic production.

ABBREVIATIONS

SOM Soil organic matter
WANA West Asia and North Africa

REFERENCES

Abdel-Dayem, M.S. 1997. Drainage water reuse: Conservation, environmental and land reclamation challenges. The Fourth World Water Congress of IWRA; A special session on water management under scarcity conditions. The Egyptian experience, Montreal, Canada.

Abu Zeid, M. 2003. Adopted measures to face major challenges in the Egyptian water sector. Country report presented at the 3rd World Water Forum, Japan.

Alabbasy, R. 2007. Water crisis in the Arabian Gulf region and the suggested solutions for it. In Arabic, Donia Al-Raai electronic newspaper. http://pulpit.alwatanvoice.com/articles/2007/07/02/95154.html (accessed September 6, 2011).

Allam, M.N. and G.I. Allam. 2007. Water resources in Egypt: Future challenges and opportunities. *IWRA, Water Int.* 32: 205–218.

Allan, J.A. 1993. Fortunately there are substitutes for water otherwise our hydro-political futures would be impossible. In *Priorities for Water Resources Allocation and Management.* ODA, London, pp. 13–26.

Allan, J.A. 1998. Virtual water: A strategic resource global solutions to regional deficits. *Ground Water* 36: 545–546.

Anyanzwa, H., J.R. Okalebo, C.O. Othieno, A. Bationo, B.S. Waswa, and J. Kihara. 2010. Effects of conservation tillage, crop residue and cropping systems on changes in soil organic matter and maize-legume production: A case study in Teso District. *Nutr. Cycl. Agroecosys.* 88: 39–47.

Arab Organization for Agricultural Development (AOAD). 2009a. *The Annual Book of Arabic Agricultural Statistics*, Vol. 29. Arab Organization for Agricultural Development, Khartoum, Sudan.

Arab Organization for Agricultural Development (AOAD). 2009b. Situations of food security in the Arab countries. Arab Organization for Agricultural Development, Khartoum, Sudan. Report in Arabic.

Attia, B.B. 2004. Water as a human right: The understanding of water in the Arab countries of the Middle East—A four country analysis. Global Issue Papers. No. 11. Heinrich Boll Foundation, Berlin, Germany.

Chapagain, A.K. and A.Y. Hoekstra. 2004. Water footprints of nations. Volume 1: Main report. UNESCO-IHE, Delft, the Netherlands.

Economic and Social Commission for Western Asia Staff (ESCWA). 2007. ESCWA water development report 2. State of water resources in the ESCWA region. United Nations, New York.

Economic and Social Commission for Western Asia Staff (ESCWA). 2009. ESCWA water development report 3. State of water resources in the ESCWA region. United Nations, New York.

El-Fellaly, S.H. and E.M. Saleh. 2004. Egypt experience with regard to water demand management in agriculture. Eighth International Water Technology Conference, Alexandria, Egypt.

Food and Agriculture Organization (FAO). 2002. The state of food insecurity in the world 2001. Rome, Italy.

Food and Agriculture Organization (FAO). 2005. Fertilizer use by crop in Egypt. Land and plant nutrition management service. Land and water development division. Rome, Italy.

Food and Agriculture Organization (FAO). 2011a. Aquastat database query. http://www.fao.org/nr/water/aquastat/data/query/index.html?lang=en (accessed September 17, 2011).

Food and Agriculture Organization (FAO). 2011b. FAOSTAT. http://faostat.fao.org/site/567/DesktopDefault.aspx?PageI 67#ancor (accessed September 20, 2011).

Food and Agriculture Organization (FAO). 2011c. Wheat imports requirements expected to decline slightly in 2011/12 (July, June) marketing year. http://www.fao.org/giews/countrybrief/country.jsp?code=EGY (accessed September 19, 2011).

Green, S. and M. Deurer. 2010. Green, blue and grey waters: Minimizing the footprint using soil physics. Production footprints, Plant & Food Research, Palmerston, New Zealand.

Karlen, D., R. Lal, R.F. Follett, J.M. Kimble, J.L. Hatfield, J.M. Miranowski, C.A. Cambardella, A. Manale, R.P. Anex, and C.W. Rice. 2009. Crop residues: The rest of the story. *Environ. Sci. Technol.* 43: 8011–8015.

Khaleel, R., K.R. Reddy, and M.R. Overcash. 1981. Changes in soil physical properties due to organic waste applications: A review. *J. Environ. Qual.* 10: 133–141.

Lane, D.R., D.P. Coffin, and W.K. Lauenroth. 1998. Effects of soil texture and precipitation on above-ground net primary productivity and vegetation structure across the central grassland region of the United States. *J. Veg. Sci.* 9: 239–250.

Lee, E. 2010. Saudi Arabia and desalinization. Harvard international review. Harvard University. http://hir.harvard.edu/pressing-change/saudi-arabia-and-desalination-0 (accessed September 22, 2011).

Li, F., A. Guo, and H. Wei. 1999. Effects of clear plastic film mulch on yield of spring wheat. *Field Crop. Res.* 63: 79–86.

Metzger, L. and B. Yaron. 1987. Influence of sludge organic matter on soil physical properties. *Adv. Soil Sci.* 7: 141–163.

Mrabet, R. 2002. Stratification of soil aggregation and organic matter under conservation tillage systems in Africa. *Soil Till. Res.* 66: 119–128.

Murad, A.A. 2010. An overview of conventional and non-conventional water resources in arid region: Assessment and constrains of the United Arab Emirates (UAE). *J. Water Resour. Protect.* 2: 181–190.

Patil, S.V., B.K. Salunke, C.D. Patil, and R.B. Salunkhe. 2011. Studies on amendment of different biopolymers in sandy loam and their effect on germination, seedling growth of *Gossypium herbaceum* L. *Appl. Biochem. Biotechnol.* 163: 780–791.

Schmidhuber, J. and F.N. Tubiello. 2007. Global food security under climate change. *Proc. Natl. Acad. Sci.* 104: 19703–19708.

Smith, M., J. Pointing, and S. Maxwell. 1993. Household food security: Concepts and definitions—An annotated bibliography. Development Bibliography. No. 8. Institute of Development Studies, University of Sussex, Brighton, UK.

Taban, M. and S.A.R. Naeini. 2006. Effect of aquasorb and organic compost amendments on soil water retention and evaporation with different evaporation potentials and soil textures. *Commun. Soil Sci. Plant Anal.* 37: 2031–2055.

Taylor, K.C. and R.G. Halfacre. 1986. The effect of hydrophilic polymer on media water retention and nutrient availability to *Ligustrum lucidum*. *Hort. Sci.* 21: 1159–1161.

Thomas, G.A., R.C. Dalal, and J. Standley. 2006. No-till effects on organic matter, pH, cation exchange capacity and nutrient distribution in a Luvisol in semi-arid subtropics. *Soil Till. Res.* 94: 295–304.

Tolba, M. 2001. Serious challenges are facing the sustainable environment in the Arab world. The future of the environmental work in the Arab world. Proceeding (in Arabic). United Arab Emirates.

TravelBlog. 2011. http://www.travelblog.org/World/ag-geog.html (accessed September 21, 2011).

United Nations (UN). 2010. World population prospects, the 2010 revision. Department of Economic and Social Affairs, Population Division, Population Estimates and Projections Section. http://esa.un.org/unpd/wpp/unpp/panel_population.htm (accessed September 25, 2011).

United States Department of Agriculture (USDA). 2003. Managing soil organic matter: The key to air and water quality. Soil quality technical note. 5. http://soils.usda.gov/sqi/concepts/soil_organic_matter/files/sq_tn_5.pdf (accessed September 18, 2011).

Weber, P. and J. Harris. 2008. Egypt and food security. *Alahram* weekly newspaper; Egyptian press. http://weekly.ahram.org.eg/2008/919/sc6.htm (accessed September 16. 2011).

8 Water Management for Crop Production in Arid Lands

A.J. Clemmens, K.F. Bronson, D.J. Hunsaker, and E. Bautista

CONTENTS

8.1 INTRODUCTION

Food security is a pressing issue for the international community (Clothier et al. 2010). Rockström et al. (2010) discuss the need for changes in the management of water in rainfed systems to address food security. They highlight the need for more effective management of water at river-basin scales. Molle et al. (2010) discuss the impact that human water uses have made on the environment of rivers. Turral et al. (2010) discuss the need for greater investment in irrigation schemes, not new schemes but better management of existing schemes. To summarize their work, water is in short supply; human uses of water harm the environment; without better management of water, we will be faced with both inadequate food and further environmental degradation. Making effective changes in water management is a slow process. As Turral et al. (2010) point out, there is no silver bullet. It is a large-scale problem that has to be solved locally—for each farm, project, and watershed. Too often, watershed-scale solutions are needed, yet farmers make individual decisions based on their individual constraints. These are often in conflict.

Irrigation has a significant impact on world food supplies. In the United States, irrigated cropland produced roughly 53% of the market value of crops harvested on 17% of the harvested cropland, while fully irrigated farms produced roughly 40% of the value on 9% of the land. This increased value is the result of both improved crop yield and quality and the use of irrigation on high-value crops (Clemmens et al. 2008; National Agricultural Statistics Service 2002). Worldwide, roughly 40% of the world's food supply comes from irrigation on less than 20% of its land (Turral et al. 2010; FAO 2003).

In this chapter, we focus on water management in irrigation. We start with a discussion on the effects of water supply and irrigation uniformity on crop productivity. While the initial focus is on individual fields, larger-scale water supply constraints can also impose additional limitations on productivity. Nitrogen management is an important constraint to crop productivity. Understanding the interactive effects of nitrogen and water on crop production is important for improved productivity. Remote sensing (RS) is an underutilized technology that shows much promise for improving water and nitrogen management. Current RS images/practices are often too infrequent, have too low a resolution, are not timely, or are too expensive. Plant growth modeling can be used to project future trends, and thus it can aid in management.

8.2 WATER MANAGEMENT AND ITS INFLUENCE ON CROP PRODUCTIVITY

8.2.1 Uniformity and Efficiency

At the field level, the importance of irrigation uniformity and its influence on crop production and application efficiency (AE) are well known. The low-quarter distribution uniformity is a common method for describing irrigation uniformity. It is the average amount in the quarter area receiving the least amount of water (not necessarily contiguous areas) divided by the average amount of water received. For a normal distribution of values, this can be approximated by (Clemmens and Solomon 1997)

$$DU_{lq} = 1 - 1.27\ CV, \tag{8.1}$$

where CV is the coefficient of variation of the water received (typically infiltrated depths), or the standard deviation divided by the mean. For $CV = 0.2$, $DU_{lq} = 0.75$. For $CV = 0.1$, $DU_{lq} = 0.87$.

The adequacy of the low quarter (AD_{lq}) is a measure of whether or not the irrigation was adequate. It is simply the maximum of the low-quarter amount or the required amount, divided by the required amount. It has an upper limit of 1.0. If the low-quarter value is less than that required, AD_{lq} will be less than 1.0. As an adequacy measure, it is consistent with the low-quarter criteria for determining the amount of water to apply, which is often used in practice. However, even if the low-quarter value equals the required amount, some part of the field may be underirrigated.

The storage fraction (SF) is another measure of adequacy, where SF is defined as the average amount of useful water received (e.g., stored in soil). This fraction is determined by integrating the larger of the actual amount of water received or the required amount, over the field area, and dividing by the amount required.

In Figure 8.1 with $CV = 0.2$, the low-quarter requirement results in roughly 10% of the field receiving a deficit, on average 87.3% of the requirement. The resulting SF would be $10\% \times 87.3/100 + 90\% \times 100/100 = 98.7/100 = 0.987$. At $CV = 0.1$, $SF = 0.995$ when the low-quarter depth just meets the requirement.

The distribution uniformity can be used to develop an estimate of the potential AE from (Burt et al. 1997)

$$PAE_{lq} = DU_{lq}(1 - RO), \tag{8.2}$$

where RO is the fraction of applied water that runs off (i.e., it does not contribute to the distribution of water on the field). This can also be used to estimate the amount of water to apply during an irrigation event. The actual AE differs from the potential efficiency both because more water is applied than is needed to meet the low-quarter amount and because, even if the low-quarter amount is adequate, some portion of the field may still be underirrigated. The actual AE can be estimated using Equation 8.3:

$$AE = SF\frac{\text{Required depth}}{\text{Average depth applied}}. \tag{8.3}$$

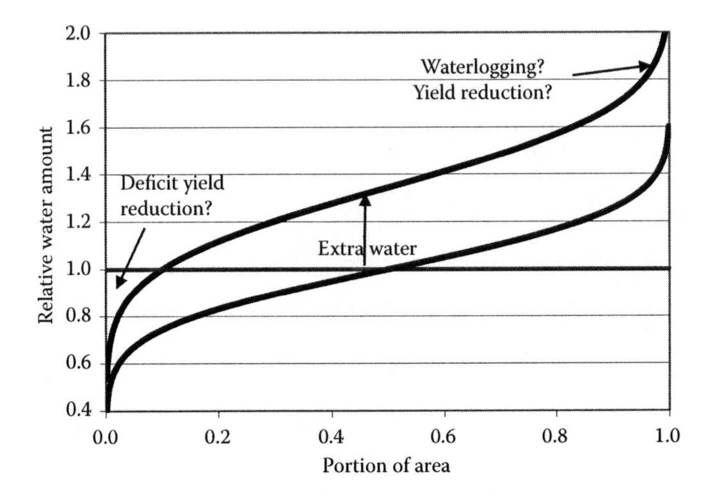

FIGURE 8.1 Adding extra water results in less of the field receiving too little water. CV = 20%.

Consider the example shown in Figure 8.1. If the net amount of water applied during an irrigation (amount applied less runoff) is the same as that required (e.g., to fill the soil water deficit), then because the distribution of water is never perfect, half of the field will get too much water while the other half will not get enough. We show the water distribution as a normal distribution. While many individual factors that influence the water distribution may differ from a normal distribution, when multiple factors are combined, the overall water distribution often resembles a normal distribution (Clemmens and Solomon 1997). For the current discussion, we ignore the runoff. The normal response of a farmer to this distribution of water is to apply more water so that a larger fraction of the field has an adequate amount. In Figure 8.1, with CV = 0.2 (20%), we would have to add 134% of the required amount to provide an average in the low quarter equal to the required amount (relative water amount equal to 1.0). The irrigator would have to add roughly 170% of the required amount to provide 98% of the field with adequate water. Because of this trade-off in extra water versus the amount in deficit, as well as other practical considerations, satisfying the average of the low quarter of the field has been a practical guideline in the United States for half a century.

An alternative to adding extra water is to improve the uniformity of water. By reducing the coefficient of variation from 0.2 to 0.1, less water has to be applied to satisfy the low-quarter criteria, where only 15% extra water is needed (CV = 0.1), as opposed to 34% (CV = 0.2), as shown in Figure 8.2. Here also notice that when the uniformity is improved, the amount of deficit in the underirrigated area is less and the potential for waterlogging and salinization is reduced. The application of these principles to irrigation management is discussed in the work done by Clemmens (1991). An often unexpected result of improving the irrigation uniformity is that both the water consumption and the yield increase because less of the field is underirrigated. *Improving irrigation uniformity is the key to providing both improved yield and reduced water application.*

At this point, we have applied these concepts to a single irrigation event. When one irrigates many times, one might expect to be able to compensate for the variation in the amount during subsequent irrigations. Unfortunately, the variation in the infiltrated (or supplied in the case of pressurized irrigation) depths over a field tends to be systematic. The same area that receives a deficit during one irrigation event usually has a deficit in subsequent irrigation events. There is some randomness, but this is generally minor compared to the systematic effects. However, one can compensate for adding too little water during one irrigation event by irrigating sooner the next time, provided that such flexibility is available in the water supply. Often, the excess water from overirrigation cannot be utilized, except perhaps through drainage water recovery or groundwater pumping.

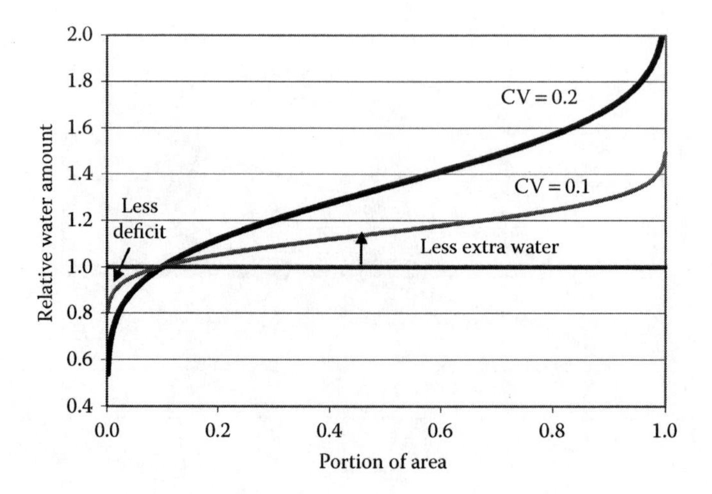

FIGURE 8.2 Improving the uniformity results in less extra water required and less deficit in the area receiving an inadequate supply.

Burt et al. (1997) make an important distinction between the AE, which is applied for a given irrigation event, and the irrigation efficiency, which is the accumulated effect over the irrigation season. Further, the AE is based on the addition of irrigation water to soil water storage, while the irrigation efficiency is based on the crop consumption of irrigation water over time. Figure 8.3 shows an example of measured application and irrigation efficiencies on a surface-irrigated field (Rice et al. 2001). Note that individual AE values vary over the season. The real demand for water is generally changing, while the application amount remains more or less the same. Note also that the irrigation efficiency is initially quite low, because losses of irrigation water occur before water is consumed; however, it gradually builds over the season. Reuse of tail-water runoff would have significantly increased the irrigation efficiency.

Burt et al. (1999) report practical, attainable values of potential AE for various irrigation systems. These vary from less than 50% to roughly 90%. Practical experience suggests that values above 90% are hard to attain for any type of irrigation system. In essence, this limitation is the

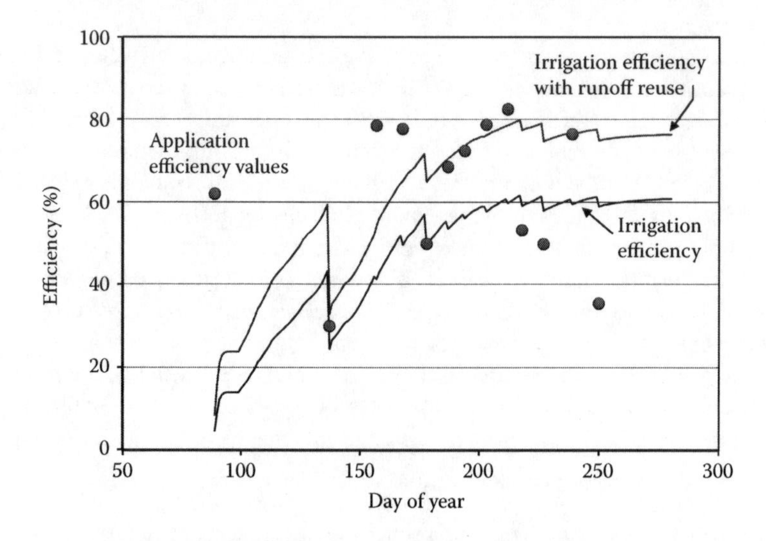

FIGURE 8.3 The application and irrigation efficiencies for a sloping furrow cotton field, 1994. (From Rice, R.C., et al., *Trans. ASAE*, 44, 555, 2001.)

result of the practical limit on the distribution uniformity. DU_{lq} values above 0.9 (or 90%) are hard to attain consistently over time. Even though microirrigation systems can be designed for a $DU_{lq} > 0.9$, field evaluations show the average DU_{lq} values in the range of 0.7–0.85, or not much different from other irrigation methods (Kennedy 1994). The range of values for practical, attainable application efficiencies varies with the sophistication of the irrigation method and the degree of management effort, including maintenance. The attainable PAE_{lq} values for surface methods range from 40% to 90% (higher values imply runoff recovery), for sprinkler systems 60%–90%, and for microirrigation from 80% to above 90%, with reductions for poor management and maintenance.

8.2.2 Influence of Uniformity on Yield

The systematic nature of irrigation system nonuniformity has important implications for yield. If the same areas receive a deficit after each irrigation event, this amount of deficit may be directly related to the yield loss for many crops. The same concept applies to areas of excess water, where the same areas receive an excess amount during each irrigation event. This is further exacerbated by the topography and physical conditions that cause areas with high water tables. Because of these systematic effects, it is possible to have areas of both consistent waterlogging and deficit irrigation in the same field. Because of these systematic patterns, we can use the DU_{lq} concept and apply it to determine the yield reductions based on the irrigation uniformity and the water supply amount. The SF then becomes an indicator of yield reduction due to the deficit. A similar yield loss associated with waterlogging could be developed based on the amount of excess (e.g., above some upper threshold).

Doorenbos and Kassam (1979) suggest that for many crops, the yield is directly related to water consumption (i.e., linear). However, the yield per unit of water consumed for a given crop is dependent on both the crop variety and the climate. For that reason, they express the yield per unit of water consumed relative to the maximum yield and the associated maximum water consumption. Solomon (1983) examined various functions to describe the yield as a function of available water. These included both the rising portion where the yield increases as the available water increases and also the decrease in the yield when the water was in excess. For simplicity, we can approximate these relationships with a trapezoid, as shown in Figure 8.4. When applied to our concern for irrigation uniformity, clearly the amount applied does not exactly match Solomon's meaning of water available, which itself is not always clear. Excess water infiltrated into one soil may percolate through and not have a negative influence on the crop growth, while in another soil, this same amount of water might cause waterlogging. Thus, the declining limb for water excess must be considered much more site specific than the limb for water deficit. Both are influenced by the crop variety and the climate.

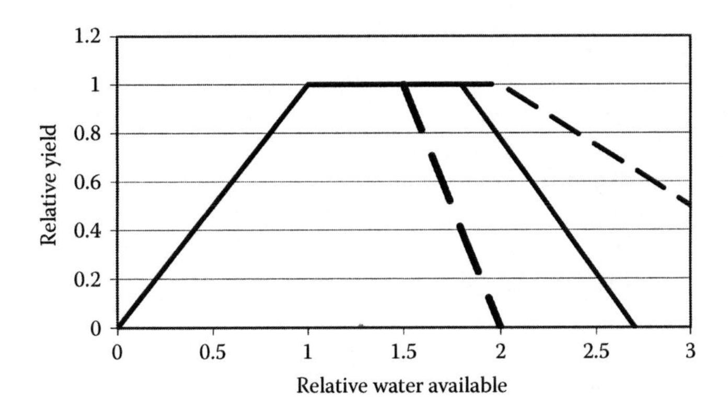

FIGURE 8.4 The influence of water availability on yield.

The relative yield for a field can be obtained by integrating the relative yield function with the relative irrigation water distribution:

$$Y_f = \int_A Y dA = \int_A Y[D(A)] dA, \tag{8.4}$$

where Y_f is the relative yield for the field, Y[D] is the relative yield function (i.e., the yield curve in Figure 8.3), and D(A) is the depth of water (e.g., the water distribution curve in Figure 8.2).

Suarez (2011) suggests that underirrigation does not create soil salinity problems. With low water application, the plant reduces its transpiration such that the soil salinity is maintained at an acceptable level. Soil salinity problems occur because of overirrigation and a lack of adequate drainage. Where sufficient drainage occurs, salinity-related yield reductions are generally insignificant. Tile drainage lines and surface drains are effective mechanisms for providing adequate drainage. In some hydrologic settings, high water tables at one location may be caused by overirrigation at a different location within an irrigation project. In such a setting, those who overirrigate may have little incentive to reduce their water application, since those farmers are not negatively affected. In such settings, both the irrigation water supply and the drainage need to be managed at a project scale.

Depending on the method of nitrogen application, overirrigation can cause nitrogen to be leached from the soil. This can have a negative influence on the yield and can cause water quality problems in receiving waters. The yield–water relationship in Figure 8.4 assumes an adequate nitrogen application. This issue is addressed in a later section.

Irrigation timing also has an influence on the yield. Studies have shown that frequent irrigation does not have a negative influence on the yield as long as the application depth matches the irrigation water requirement. Too long a wait for irrigation can reduce the yield, but generally through reduced transpiration. If irrigation scheduling is based on the low-quarter depth, for those areas that are underirrigated (relative to the average need for water), each irrigation event will simply grow smaller plants (and thus perhaps not needing much water). A more significant concern is too delayed irrigation during critical growth stages. Here, the yields can decline more than would be indicated by the reduction in water use. This is an area of concern where farmers do not have control over their water supply, such as large government-run irrigation projects.

8.3 NITROGEN MANAGEMENT FOR IRRIGATED CROPPING SYSTEMS IN ARID LANDS

Following water, nitrogen (N) fertilizer is the main constraint to crop production (Morrow and Krieg 1990). The canal infrastructure of the irrigation water in western states such as Arizona means basin, flood, and furrow irrigation are still the predominant choices of irrigation methods. Navarro et al. (1997) in Arizona and Booker et al. (2007) and Bronson et al. (2007) in Texas reported that recovery efficiency for ground-based N applications in furrow-irrigated cotton ranged from only 15% to 34%. The N recovery efficiency in furrow-irrigated grain sorghum was similar in west Texas (Booker et al. 2007) but was as high as 60% of the applied N in furrow-irrigated forage sorghum (Tamang et al. 2011).

Fertigation with liquid urea ammonium nitrate (UAN) fertilizer is commonly practiced in the western United States. In this practice, the liquid N fertilizer is typically dribbled into the irrigation canal and then transported to the field through the surface irrigation water stream. However, ammonia-N losses may be high with this type of fertigation (Mikkelsen 2009). Furthermore, the nonuniformity of the irrigation would result in highly variable N fertilizer applications down the furrow, and NO_3 leaching would occur in areas of the field that receive excess water during fertigations (Silvertooth et al. 1992; Jaynes et al. 1992). Knife applications of UAN in 7.5 cm deep bands along the furrow should improve the uniformity of N versus fertigation with UAN. However, a comparison of these two N methods has not been adequately investigated.

In the western United States, a preplant soil test to a depth of 30 cm is a common starting point for N management of irrigated row crops (Zelinski 1985; Zhang et al. 1998; Booker et al. 2007; Bronson et al. 2001, 2009). An estimated yield goal and a N use efficiency factor are used together with the preplant soil NO_3 test to calculate a seasonal N fertilizer rate. The efficiency of the application and the uptake of the N fertilizer vary depending on the irrigation system and the number of applications (splits) (Bronson 2008). The recovery efficiency of the UAN fertilizer that is injected or fertigated on a near daily time step in the subsurface drip-irrigated cotton (*Gossypium hirsutum*) can be as high as 75% (Yabaji et al. 2009; Bronson et al. 2011). When liquid UAN was knifed into cotton beds under a center pivot, the recovery efficiency of N was 40% (Bronson 2008).

Yield goals are an important component of N fertilizer needs and recommendation algorithm. In the semiarid and arid regions of the United States, the cotton yield goals are highly dependent on water availability, for instance, whether the crop will be deficit irrigated or fully irrigated. In deficit-irrigated areas such as west Texas, the yields vary with the irrigation system in the following order: subsurface drip > center pivots > furrow irrigation (Bronson 2008). The N requirements, therefore, are strongly linked to the irrigation mode and the water availability. In Arizona's high-yielding level-basin irrigation systems, the cotton lint yield goals can be 2000 kg/ha, similar to the drip-irrigated cotton in Texas (Bronson 2008). In Texas, the furrow, deficit-irrigated cotton only yields about 1000 kg/ha. Bronson et al. (2001) reported the increasing N fertilizer response and the N requirements as the irrigation level increased from 25% to 75% ET replacement for cotton. Internal (i.e., plant) N requirements vary from 18 to 34 kg/(N bale) in Arizona (Silvertooth et al. 2011).

Petiole NO_3 sampling and analysis is a commonly used approach to monitor the in-season cotton plant N status in the western United States. In-season N fertilizer is ground applied or fertigated, depending on whether the petiole NO_3 levels are deficient, optimum, or adequate (Table 8.1). However, petiole sampling is laborious, and laboratory turnaround is time-consuming. Additionally, petiole NO_3 analysis can be highly variable (Bronson et al. 2001). Canopy reflectance using a spectroradiometer, on the other hand, is a rapid, nondestructive method to assess the in-season cotton N status (Chua et al. 2003; Bronson et al. 2003). Canopy reflectance-based N management in subsurface drip systems in Texas resulted in reduced N fertilizer use, without hurting the lint yields (Bronson et al. 2011). In that research, the N fertilizer was initially applied at half the rate of a regional soil test–based recommendation. When the normalized difference vegetative index (NDVI, a common RS vegetative index) in the reflectance treatment fell below the NDVI of the soil test/adequately fertilized plot, N fertigation was increased. A less expensive but more laborious alternative to canopy reflectance is the chlorophyll meter. The chlorophyll meter measures the greenness of the leaf by red and near-infrared transmittance. Readings are highly correlated to leaf N and petiole NO_3 concentrations (Bronson et al. 2001, 2003). Similar to canopy reflectance, a "sufficiency index" approach was used, where chlorophyll meter readings were referenced or ratioed

TABLE 8.1
Guidelines for Optimum Petiole NO_3-N Concentrations in American Pima and Upland Cotton in Arizona

	Early Squaring	Early Flowering ppm NO_3-N	Early Bolls
Pima cotton	10,000+	8,000+	4,000+
Upland cotton	18,000+	14,000+	8,000+

Source: Adapted from Silvertooth, J.C., et al., *Better Crops with Plant Food*, 95, 21, 2011.

to readings in the adequately fertilized, soil test–based N management plot. Again, modest savings in seasonal N fertilizer use were realized in the chlorophyll meter-based N management plots, without a lint yield reduction.

8.4 IRRIGATION SCHEDULING FOR IMPROVED CROP PRODUCTIVITY

Irrigation scheduling techniques provide agriculture with sound methodologies for managing the timing and the amount of water applications to crops. In arid environments, irrigation scheduling is an essential management practice for maintaining healthy plants and high crop productivity. The primary purpose is to replenish the soil water lost by crop evapotranspiration (ET) that is not supplied by precipitation. Maximizing the crop yields and the water savings can be achieved by scheduling the irrigation depths to closely match the amount of water needed to replenish the soil water lost by ET (English et al. 2002). However, the decision process for choosing the depth of the irrigation water and the time to apply it involves the consideration of several factors in addition to ET, including the particular irrigation method used, the soil type, and the agronomic and economic goals of the farmer (Evett et al. 2012). The general requirements to successful irrigation scheduling, as summarized by Pereira (1999) and Pereira et al. (2002), include the knowledge of crop water requirements and how the crop yield responds to water; understanding the constraints imposed by the specific irrigation method, for example, uniformity of application; knowing the water supply system capabilities, for example, reliability and flexibility; and understanding the financial and economic implications of the particular irrigation practice.

Research has established several methods for determining irrigation scheduling. Martin (2009) summarizes the irrigation scheduling techniques that are commonly used in arid agriculture. The focus of this chapter will be on the widely used soil water balance accounting method (Allen et al. 1998). This method, which keeps track of daily water inputs and outputs for the crop root zone, enables an irrigation farm manager to monitor a field's daily soil water depletion (i.e., the water deficit) in the root zone, which can be used to plan the next irrigation. The soil water balance to determine the daily soil water depletion can be written as (Equation 8.5)

$$D_{r,i} = D_{r,i-1} + ET_i - I_i - R_i + DP_i, \tag{8.5}$$

where the subscripts i and i − 1 represent the current day and the previous day, respectively; D_r is the root zone soil water depletion (mm); ET is the crop evapotranspiration (mm); and I, R, and DP are the depth of irrigation applied (mm), the rainfall (mm), and the deep percolation (mm), respectively.

8.4.1 Crop Coefficients and ET Prediction

Accurate and reliable estimations of crop ET are key inputs for determining appropriate irrigation schedules. The most widely used ET estimation method for irrigation scheduling is the crop coefficient (K_c) and the reference evapotranspiration (ET_o) paradigm (Jensen and Allen 2000; Allen and Pereira 2009). A crop coefficient relates the actual ET of a crop at a given stage of development to the ET_o reference, calculated from meteorological data, that is, $ET = K_c\ ET_o$. The empirically derived K_c values are crop specific and vary during the season to reflect the changes in ET due to crop growth. If sufficient weather data are available, as from automated meteorological stations, the FAO Penman–Monteith formula is recommended for computing the daily ET_o. These data are then combined with the K_c curves provided in FAO Irrigation and Drainage Paper No. 56 (FAO-56) (Allen et al. 1998). The American Society of Civil Engineers (ASCE) adopted the standardized ET_o equations based on the FAO-56 model (Allen et al. 2005). The standardized equations provide a common ET_o model that facilitates the sharing and transferring of K_c information from one location to another. Subsequently, the K_c paradigm with the standardized ET_o equations is now used by U.S. weather station manufacturers and by state and federal-run weather networks that provide Internet access to these data.

Allen et al. (1998) suggest that the $K_c \cdot ET_o$ paradigm can be more effective and accurate on a daily basis when the K_c is separated into two coefficients: one for the basal or transpiration component of the crop ET (i.e., K_{cb}) and one for the wet soil evaporation component (K_e), where $K_c = K_{cb} + K_e$. This method, called the dual crop coefficient approach, was initially developed by Wright (1982). Computations for the dual K_c are provided in FAO-56. Crop ET estimation using the dual K_c approach is especially valuable for improving irrigation scheduling for crops grown in the arid regions of the United States, where high water-use and soil evaporation rates from irrigated crops compete for scarce and increasingly costly water supplies. This irrigation scheduling management is especially important early in the growing season, when crop canopies are small and soil evaporation dominates.

The crop coefficients presented in the literature are generally developed through field experiments using lysimeters or soil water balance techniques and are predominantly developed for crops grown under optimum agronomic conditions (Allen et al. 1998). Therefore, the K_c values are only useful approximations of the actual ET and the water requirements for a given crop, since the actual crop ET in the field can vary from the optimum K_c-based ET for a number of reasons, including crop variety differences, planting density, climatic factors, nutrient status, irrigation management, salinity, and other conditions (Hunsaker et al. 2003). Often these factors reduce the actual water demand below that expected for a crop grown under optimal conditions. Therefore, using crop coefficients for irrigation scheduling can lead to overirrigation of the crops, which can be a serious concern at the district and regional levels of water management, especially in water-short arid and semiarid areas of the world (Santos et al. 2008). Effective irrigation scheduling is also hindered by the occurrences of spatially variable ET fluxes within fields or watersheds, created, for example, by nonuniform water application or precipitation and variable soil water properties. Despite improvements and renewed interest in crop coefficient applications, quantifying ET for large and spatially diverse systems with standard K_c approaches remains difficult. However, RS, which can observe variable vegetation densities and surface thermal conditions, provides information that can be used to quantify the spatial dynamics of ET for a wide range of pertinent management scales.

8.4.2 Crop Coefficients Derived from Remotely Sensed Vegetation Index

Multispectral vegetation index (VI) methods can replace (or supplement) tabulated crop coefficients with a VI that reflects the actual growth stage of the crop at the time of measurement. Starting more than 20 years ago, Neale et al. (1989) developed VI-based crop coefficients for corn. A useful VI for this approach is the NDVI, which is based on canopy irradiance in the red and near-infrared bands and which can be remotely sensed (Glenn et al. 2010). More recent research has shown that observations of multispectral VIs can provide real-time surrogates of crop coefficients for a number of different crops (Bausch 1995; Neale et al. 2005; Hunsaker et al. 2005, 2007; Trout et al. 2008). Moreover, the use of RS to infer the spatial distribution of K_c across the landscape can improve the ability of the standard weather-based ET methods to more accurately estimate the spatial water use within an irrigated field (Hunsaker et al. 2007) at the farm-scale level (Johnson and Scholasch 2005) and at the local, regional, and global scales for natural ecosystems (Glenn et al. 2010). Figure 8.5, derived from field data for wheat (Hunsaker et al. 2007), demonstrates the effectiveness of using an NDVI-based crop coefficient versus a standard FAO-56 approach. For a wheat crop planted at a higher than normal density, the NDVI-based K_{cb} is able to track the measured K_{cb} throughout the season. Conversely, the standard K_{cb} curve for wheat is shown to be a poor estimator of the actual K_{cb} for this plant density, where K_{cb} is underestimated early in the season and overestimated later in the season.

For irrigation management, the VI-based crop coefficient approach has a strong practical appeal due to the long-standing familiarity and the widespread use of crop coefficient methods and their

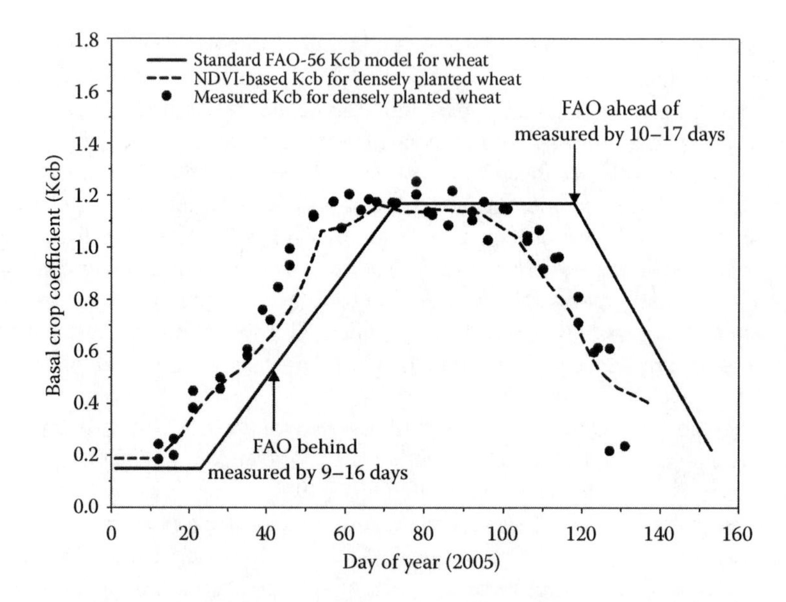

FIGURE 8.5 Comparisons of seasonal trends for the standard FAO-56 basal crop coefficient (K_{cb}), the NDVI-based K_{cb}, and the measured K_{cb} for wheat (From Hunsaker, D.J., et al., *Trans. ASABE*, 50, 2017, 2007.)

relative operational simplicity. Nevertheless, the implementation of the approach could be hindered by its reliance on the empirical relationships between VIs and crop coefficients, by problems associated with the transferability of RS-based crop coefficients from one region to the next, and by the timeliness and the cost-effectiveness of the necessary imagery (Gowda et al. 2008). RS estimates of VI may not be available on a continuous basis. Thorp et al. (2010) recommend the use of crop growth models to infer conditions in between the VI measurements. Root-water extraction is an important aspect of these models. Most use some empirical method to estimate where within the root zone the water is extracted. Next, we propose a simple model based on potential energy.

8.4.3 Deficit Irrigation

Deficit irrigation is commonly used in many irrigated regions where the available water supplies or the irrigation system capacities are limited (Howell et al. 2007). With deficit irrigation, the total irrigation water application is less than that needed to meet the full crop ET requirements for the cropping cycle (Fereres and Soriano 2007). Consequently, some plant water stress occurs and the growth and the yield are usually decreased below their potentials (Bronson et al. 2006). However, a primary goal of deficit irrigation is that water-use efficiency (WUE) can be increased; in some scenarios, it may be more profitable than full irrigation (English 1990). Because of nonuniform water application, it is not possible to provide a uniform deficit over the field. Some areas will have more deficit than other areas. Fereres and Soriano (2007) suggest that the deficit irrigation level should be between 60% and 100% of full ET. In practice, a general approach for determining crop ET under deficit irrigation is by including the water stress coefficient (K_s) within the FAO-56 dual procedures, where $ET = (K_{cb} \cdot K_s + K_e) \cdot ET_o$. In FAO-56, the water stress coefficient on a given day is calculated based on the current level of the root zone soil water deficit and the current evaporative demand (Allen et al. 1998).

8.4.4 Soil Water Extraction

The old rule of thumb for the reduction in soil water from plant ET is that the water is extracted in the following proportions: 40%, 30%, 20%, and 10% from the uppermost to the lowest quarters of the root zone. Nimah and Hanks (1973) developed a model for the removal of soil water through ET based on the soil water suction and the root resistance to flow. This model forms the basis for many soil water balance models in use today. Here, we suggest a conceptually simpler model to aid in understanding the process.

When water is ponded at the surface such that the soil below is saturated, the pressure head at all depths is the elevation of the soil surface (hydrostatic pressure). As the water drains from the soil, the pressure head reduces. If hydrostatic pressure is assumed (no significant transient behavior), the pressure head in the soil above the water table is uniform, until the water table (where the pressure head is the same as the water table elevation) drops to the point where the soil at the surface is at field capacity (i.e., 1/3–1/10 bar depending on the soil). This is shown in Figure 8.6, which shows the pressure needed to raise the soil water to the soil surface. At the far left, the soil is saturated at the surface, so no additional pressure is needed. As the water table drops, the pressure head required is shown as a vertical line from the surface to the water table (saturation). As drainage increases, the pressure head moves from left to right, but always at a constant pressure. The distance between saturation and field capacity is essentially the height of the capillary fringe. Therefore, once the water table drops below the height of the capillary fringe, the pressure head above the capillary fringe remains at the pressure associated with field capacity. Thus, drainage can be modeled as vertical lines from left to right: first between the soil surface and the water table and then between the line for field capacity and the water table. The actual moisture content for any location within the soil depends on the relationship between the pressure and the moisture content.

The above-mentioned drainage model simply assumes hydrostatic pressure. In this model, we also assume that the plant extracts water according to the pressure. To minimize energy, the plant will extract water from the soil with the highest pressure head. For a soil at field capacity, the highest pressure head is at the soil surface. However, as the water is extracted from the soil layer, the pressure head for the soil just below will essentially be the same. Thus, the plant should be able to extract water from all the soil that is at the same pressure head. This is shown by the vertical lines above the field capacity line. The plant will extract water equally from all depths for which the pressure head is equal. The amount of water that can be extracted from the soil between any two vertical lines can be determined by integrating the water content change caused by the soil pressure change, which is the pressure head plus the soil depth. Soil evaporation creates a disturbance to this simple pressure model. Models are available to estimate soil evaporation (Allen et al. 1998). An example of the change in the pressure caused by soil evaporation is shown in Figure 8.7.

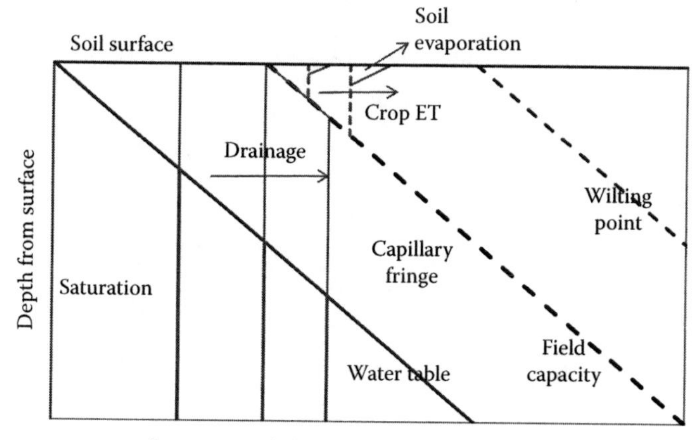

FIGURE 8.6 The constant pressure model for determining drainage and crop water extraction.

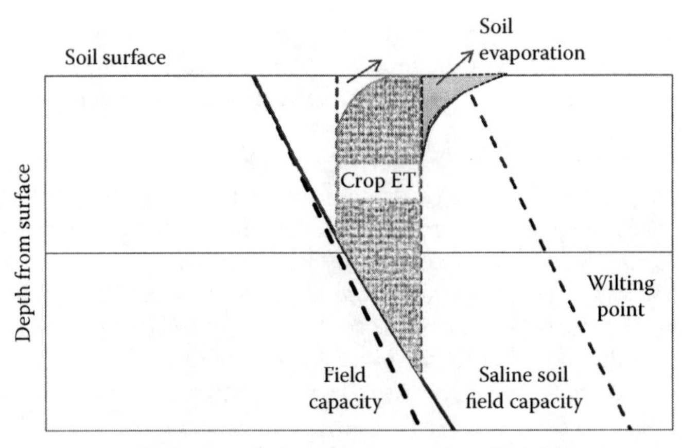

FIGURE 8.7 Crop water use from the constant pressure model.

This model assumes that the roots will distribute themselves such that they can extract this water uniformly. If the soil water changes are consistent over the growing season, this is probably a reasonable assumption.

Figure 8.7 also shows the influence of soil salinity on the ability of plants to extract water. If the soil salinity were uniform, then the extra suction caused by the salt would increase the pressure needed to raise the water to the soil surface. Since the soil salinity tends to be greater at a greater depth, this is shown as a curved line, with essentially no effect at the surface and an increasing effect with depth (Lety and Feng 2007). While prior theories based on steady-state solutions suggest that decreases in the leaching fraction would cause greater salinity, particularly at lower depths, Lety et al. (2011) and Suarez (2011) show that dynamic modeling of soil salinity suggests that much less leaching is required. Further, crop ET would decrease to maintain reasonable salinity levels. By observing the pressure diagram in Figure 8.7, we see that the plant must work harder and harder since the pressure required to extract the water increases. Further, the increased salinity makes it so that the plant extracts less water at a given pressure. We know that drought-tolerant plants will begin to extract less water as the pressure required to extract the water increases, supposedly to preserve more water for future use. Suarez's observations would suggest this also happens with salinity.

8.5 WATER PRODUCTIVITY AT FIELD AND WATERSHED SCALES

Water productivity is the amount of production (output) per unit of water (input). Production/output can be in terms of either the mass of product or the economic value. Water input can be defined in terms of water consumed or water supplied. The farmer and water purveyor would most likely be interested in the economic value per unit of water applied. The farmer needs to stay in business, and the water purveyor needs to justify his/her use of the water supply. An agronomist might be interested in the amount of mass produced per unit of water consumed by ET. This determines the overall efficiency of the crop production process. Society should be interested in the economic value per unit of water consumed, where the water consumed includes the water used for crop ET and other water lost and not available for reuse elsewhere. This is the perspective of getting the best economic value for the water consumed. From the perspective of global food security, the interest might be in the mass of production per unit of water consumed. As we can see, water productivity is viewed quite differently, depending on our perspective.

From a water-rights perspective, we need to understand how irrigation systems interact with the hydrologic system. Irrigation replaces rainwater when rain is not sufficient for effective crop production. Once irrigation water is applied to a field, it essentially becomes part of the hydrologic

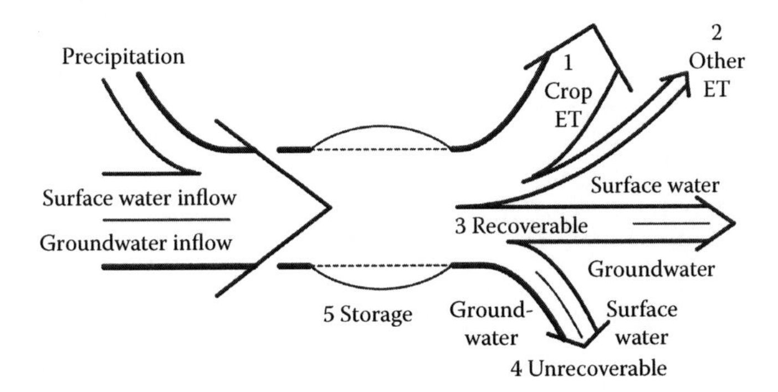

FIGURE 8.8 Predominant water inflows and outflows for irrigation systems where components 1–5 are (1) water consumed by the crop within the area for beneficial purposes; (2) water consumed within the area under consideration but not beneficially; (3) water that leaves the boundaries of the area under consideration but is recovered and reused; (4) water that leaves the boundaries of the area under consideration but is either not recoverable or not reusable; and (5) water that is in storage within the boundaries.

process. Most goes into soil water storage for eventual crop ET, but part of it may percolate below the root zone or run off the field to a natural watercourse. Where irrigation is supplied by canals, spills often return to a stream or river. Figure 8.8 shows a hydrologic water balance for an irrigated area. The water supply can include rainwater, irrigation water from a canal, or irrigation water from groundwater. Outflows include crop ET, ET from other sources, water that is recoverable for use downstream, and water that is not recoverable. ET from other sources can include evaporation from canals and reservoirs, evaporation from weeds, evaporation from vegetation that grows where water runs off fields and spills from canals, and so on. Most of the water that leaves an irrigated area is reusable downstream. However, in some cases, the water goes to a saline sink or is otherwise degraded in quality such that it is of little or no value. The water consumed includes water lost to the atmosphere (evaporation and ET) and water that is unrecoverable.

To determine the benefit of irrigation water use, we would prefer to use the increase in production over rainfed or dryland agriculture, but such estimates may be difficult to obtain. Separating rainwater consumption from irrigation water consumption can be difficult. Unless an individual rainfall event fills the soil water profile, it is difficult to know how much rainwater was available for crop water use, unless soil water is measured. While this is done routinely for research studies, few farmers make these measurements. Thus, determining the water productivity for an individual field may be difficult. For an irrigation project with well-defined hydrologic boundaries, determining a water balance can sometimes be more reliable. A good project water balance example is given in Clemmens (2008).

In the previous sections, irrigation uniformity and scheduling were discussed for an individual field. The assumption in these discussions is that the farmer has control of the water supply. Lamacq et al. (1996) suggest that farm constraints, such as canal capacities, crop mix, and farm labor, can reduce the ability of farmers to apply modern irrigation scheduling and can reduce the overall potential AE. The issue is that the needed timing of irrigation water, field by field, does not always match what is practical. Farmers tend to irrigate fields in a given sequence regardless of the real demand for water. This can result in overirrigation of one field while another is under water stress. The issue is not a distribution uniformity issue, but one of imperfect timing of irrigation events.

For some large irrigation schemes, the farmer's ability to control his/her water supply is limited. Often, the timing of water is determined by the irrigation project personnel and the flow rate may be too low. Depending on the location within the distribution network, farmers may receive a very different service and large differences in the amount of water supplied. This creates significant inequity among farmers. Those at the tail end of canals (tailenders) historically have a poorer service.

This inequity in water supply can also be viewed as a problem with irrigation uniformity. Some farmers will receive excess water, some not enough. Clemmens and Molden (2007) propose that the uniformity for an irrigation project can be determined from combining the uniformity of the field water distribution and the uniformity of the delivery volume. This essentially determines the distribution of water to individual small field units for the project as a whole. They suggest that the coefficient of variation for the application of water within an irrigation project ($CV_{project}$) can be calculated using the following equation:

$$CV_{project} = \sqrt{CV_{in\text{-}field}^2 + CV_{delivery\text{-}volume}^2}, \tag{8.6}$$

where $CV_{in\text{-}field}$ is the coefficient of variation for an irrigation event on an individual field, as used in Equation 8.1, and $CV_{delivery\text{-}volume}$ is the variation in the volume of water supplied to users relative to the crop need. One could also add the variation due to poor delivery timing, but this may be less straightforward. With Equation 8.6, we can develop an estimate, for the project as a whole, of irrigation uniformity (Equation 8.1), potential AE (Equation 8.2), and actual AE (Equation 8.3). If the yield function is known, we can also estimate the influence of the distribution of water on production for the project as a whole.

Consider the yield function shown in Figure 8.4. If we use the solid curve for the influence of large irrigation amounts, the yield function is

$$\begin{aligned} &Y = D \quad \text{for } D \le 1 \\ &Y = 1 \quad \text{for } 1 \le D \le 1.8 \\ &Y = 1 - (D - 1.8)/0.9 \quad \text{for } 1.8 \le D \le 2.7 \\ &Y = 0 \quad \text{for } D \ge 2.7. \end{aligned} \tag{8.7}$$

In order to utilize this function, we have to know the relative amount of irrigation water supplied (RIS), relative to the requirement. If we have $CV_{in\text{-}field} = 0.2$ and $CV_{delivery\text{-}volume} = 0.1$, then from Equation 8.6, $CV_{project} = 0.224$, resulting in $DU_{lq} = 0.72$. If we use this value of CV with a normal distribution of values and apply the yield function as given in Equation 8.7, we can determine the distribution of the yield over the project area, as shown in Figure 8.9. The area under the respective curves gives the relative yield, as defined by Equation 8.4. The curves in Figure 8.9 assume that

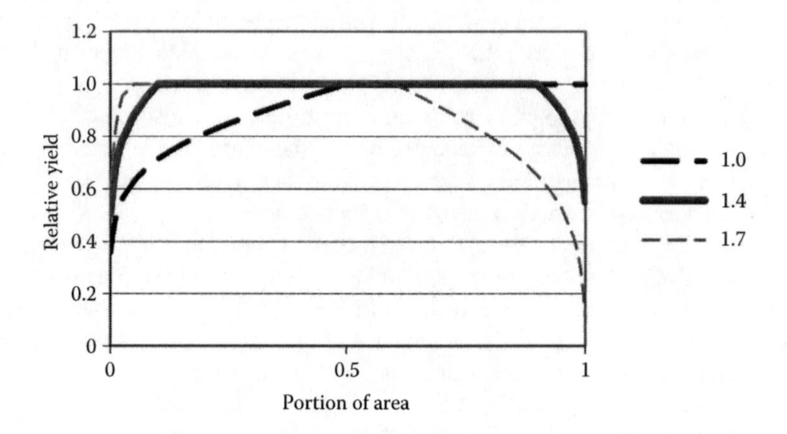

FIGURE 8.9 The yield for the combined distribution with $CV_c = 0.223$ and the yield decline described by the middle curve in Figure 8.4 with RIS values of 1.4, 1.0, and 1.7.

there is no runoff or spills. Where excess water results in runoff, the reductions in the yield due to excess water would be less.

Note that, just as for an individual field, if the amount of water supplied to the project equals the amount required (RIS = 1), that portion of the project where the fields get an inadequate supply would have lower yields. For this uniformity and this yield function, RIS = 1 results in no yield decreases due to excess water application. To match the depth required to the low-quarter depth would give RIS = 1.4. With this amount of supplied water, we see reductions in the yield due to both underirrigation and overirrigation. For this example, this amount of water supply provides roughly the maximum total yield over the area considered. If we apply additional water, RIS = 1.7 in Figure 8.9, we see that we have only a small increase in the yield in the area receiving deficits and a large reduction in the yield for the area receiving excess water. In this case, supplying extra water reduces the overall project yield.

While RIS = 1.4 provided roughly the maximum yield in our example, it may not result in the best overall productivity. If the unused water is reusable downstream, then it will be close to the maximum productivity. Exact calculations depend on the amount of water used in the overirrigated area of the project. Generally, only a portion of the unconsumed water is reusable downstream. The relative amount depends on where the project is in the hydrologic system (Clemmens et al. 2008). If none of the unconsumed water is reused downstream, then based on Equation 8.7, any scenario that has no application depths above the water requirement will have the same productivity (Y = D). This would not be economical for farmers, and, fortunately, this is not common for large irrigation projects.

8.6 CONCLUSION

Maintaining world food production with limited water supplies will require significant improvements in both water and nitrogen management. Irrigation systems need high irrigation uniformity, which requires either pressurized irrigation or modern (high-efficiency) surface irrigation. Poor uniformity, at field and project scales, results in low production, waste of fertilizers, poor water quality, salinity, and so on.

The appropriate timing of irrigation events is also important for maintaining high productivity. Where the water supply is not limited, an improvement in the scheduling with RS and other technologies offers an opportunity to increase crop production and productivity. For some crops, deficit irrigation can provide increases in crop quality (and income) with less water consumed.

For large-scale irrigation systems, large-scale nonuniformity (i.e., related to equity) has a significant impact on production and water productivity. Improvements in the management and operation of these large-scale systems are important for increasing world food production.

ABBREVIATIONS

ET	Crop evapotranspiration
NDVI	Normalized difference vegetation index
VI	Multispectral vegetation index

REFERENCES

Allen, R.G. and L.S. Pereira. 2009. Estimating crop coefficients from fraction of ground cover and height. *Irrigation Science* 28(1): 17–34.

Allen, R.G., L.S. Pereira, D. Raes, and M. Smith. 1998. Crop evapotranspiration—Guidelines for computing crop water requirements. FAO Irrigation and Drainage Paper 56. Food and Agriculture Organization of the United Nations, Rome, Italy.

Allen, R.G., I.A. Walter, R.L. Elliott, T.A. Howell, D. Itenfisu, M.E. Jensen, and R.L. Snyder. 2005. *The ASCE Standardized Reference Evapotranspiration Equation.* American Society of Civil Engineering, Reston, VA, 59 pp.

Bausch, W.C. 1995. Remote sensing of crop coefficients for improving the irrigation scheduling of corn. *Agricultural Water Management* 27(1): 55–68.

Booker, J.D., K.F. Bronson, C.L. Trostle, J.W. Keeling, and A. Malapati. 2007. Nitrogen and phosphorus fertilizer and residual response in cotton-sorghum and cotton-cotton sequences. *Agronomy Journal* 99: 607–613.

Bronson, K.F. 2008. Nitrogen use efficiency varies with irrigation system. *Better Crops with Plant Food* 92(4): 20–22.

Bronson, K.F., A.B. Onken, J.W. Keeling, J.D. Booker, and H.A. Torbert. 2001. Nitrogen response in cotton as affected by tillage system and irrigation level. *Soil Science Society of America Journal* 65: 1153–1163.

Bronson, K.F., T.T. Chua, J.D. Booker, J.W. Keeling, and R.J. Lascano. 2003. In-season nitrogen status sensing in irrigated cotton: II. Leaf nitrogen and biomass. *Soil Science Society of America Journal.* 67: 1439–1448.

Bronson, K.F., J.D. Booker, J.P. Bordovsky, J.W. Keeling, T.A. Wheeler, R.K. Boman, M.N. Parajulee, E. Segarra, and R.L. Nichols. 2006. Site-specific irrigation and nitrogen management for cotton production in the southern High Plains. *Agronomy Journal* 98: 212–219.

Bronson, K.F., J.C. Silvertooth, and A. Malapati. 2007. Nitrogen fertilizer recovery efficiency of cotton for different irrigation systems. 2007 Proceedings Beltwide Cotton Conferences. [CD-ROM]. National Cotton Council of America, Memphis, TN.

Bronson, K.F., A. Malapati, J.D. Booker, B.R. Scanlon, W.H. Hudnall, and A.M. Schubert. 2009. Residual soil nitrate in irrigated southern High Plains cotton fields and Ogallala groundwater nitrate. *Journal of Soil Water Conservation* 64: 98–104.

Bronson, K.F., A. Malapati, J.W. Nusz, P. Lama, P.C. Scharf, E.M. Barnes, and R.L. Nichols. 2011. Canopy reflectance-based nitrogen management strategies for subsurface drip irrigated cotton in the Texas High Plains. *Agronomy Journal* 103: 422–430.

Burt, C.M., A.J. Clemmens, T.S. Strelkoff, K.H. Solomon, R.D. Bliesner, L.A. Hardy, R.A. Howell, and D.E. Eisenhauer. 1997. Irrigation performance measures—Efficiency and uniformity. *Journal of Irrigation and Drainage Engineering* 123(6): 423–442.

Burt, C.M., A.J. Clemmens, R. Bliesner, J.L. Merriam, and L. Hardy. 1999. *Selection of Irrigation Methods for Agriculture.* ASCE On-Farm Irrigation Committee Report, ASCE, Reston, VA, 129 pp.

Chua, T.T., K.F. Bronson, J.D. Booker, J.W. Keeling, A.R. Mosier, J.P. Bordovsky, R.J. Lascano, C.J. Green, and E. Segarra. 2003. In-season nitrogen status sensing in irrigated cotton: I. Yield and nitrogen-15 recovery. *Soil Science Society of America Journal* 67: 1428–1438.

Clemmens, A.J. 1991. Irrigation uniformity relationships for irrigation system management. *Journal of Irrigation and Drainage Engineering* 117(5): 582–699.

Clemmens, A.J. 2008. Accuracy of project-wide water uses from a water balance: A case study from southern California. *Irrigation and Drainage Systems* 22: 287–309.

Clemmens, A.J. and D.J. Molden. 2007. Water uses and productivity of irrigation systems. *Irrigation Science* 25(3): 247–261.

Clemmens, A.J. and K.H. Solomon. 1997. Estimation of global distribution uniformity. *Journal of Irrigation and Drainage Engineering* 123(6): 454–461.

Clemmens, A.J., R.G. Allen, and C.M. Burt. 2008. Technical concepts related to conservation of irrigation and rainwater in agricultural systems. *Water Resources Research* 44: W00E03.

Clothier, B., W. Dierickx, J.D. Oster, C.J. Perry, and D. Wichelns. 2010. Investing in water for food, ecosystems, and livelihoods. *Agricultural Water Management* 97: 493–494.

Doorenbos, J. and Kassam, A.H. 1979. *Yield Response to Water.* FAO Irrigation and Drainage Paper 33. FAO, Rome. 193 p.

English, M.J. 1990. Deficit irrigation I. Analytical framework. *Journal of Irrigation and Drainage Engineering* 116: 399–412.

English, M.J., K.H. Solomon, and G.J. Hoffman. 2002. A paradigm shift in irrigation management. *Journal of Irrigation and Drainage Engineering* 128(5): 267–277.

Evett, S.R., P.D. Colaizzi, S.A. O'Shaughnessy, D.J. Hunsaker, and R.G. Evans. 2011. Irrigation management. In Njoku, E.G. (ed.), *Encyclopedia of Remote Sensing*, Springer Science+Business Media, LLC, New York.

FAO. 2003. World Agriculture Towards 2015/2030: An FAO Perspective. FAO/Earthscan, Rome/London.

Fereres, E. and M.A. Soriano. 2007. Deficit irrigation for reducing agricultural water use. *Journal of Experimental Botany* 58(2): 147–159.

Glenn, E.P., P.L. Nagler, and A.R. Huete. 2010. Vegetation index methods for estimating evapotranspiration by remote sensing. *Surveys in Geophysics* 31: 531–555.

Gowda, P.H., J.L. Chavez, P.D. Colaizzi, S.R. Evett, T.A. Howell, and J.A. Tolk. 2008. ET mapping for agricultural water management: Present status and challenges. *Irrigation Science* 26(3): 223–237.

Howell, T.A., J.A. Tolk, S.R. Evett, K.S. Copeland, and D.A. Dusek. 2007. Evapotranspiration of deficit irrigated sorghum and winter wheat. In Clemmens, A.J. and Anderson, S.S. (eds), Proceedings of *2007 USCID 4th International Conference on Irrigation and Drainage*, pp. 223–239. Sacramento, CA, October 3–6. USCID: Denver, CO.

Hunsaker, D.J., E.M. Barnes, T.R. Clarke, G.J. Fitzgerald, and P.J. Pinter Jr. 2005. Cotton irrigation scheduling using remotely sensed and FAO-56 basal crop coefficients. *Transactions of the ASAE* 48(4): 1395–1407.

Hunsaker, D.J., P.J. Pinter Jr, E.M. Barnes, and B.A. Kimball. 2003. Estimating cotton evapotranspiration crop coefficients with a multispectral vegetation index. *Irrigation Science* 22(2): 95–104.

Hunsaker, D.J., G.J. Fitzgerald, A.N. French, T.R. Clarke, M.J. Ottman, and P.J. Pinter Jr. 2007. Wheat irrigation management using multispectral crop coefficients. I. Crop evapotranspiration prediction. *Transactions of the ASABE* 50(6): 2017–2033.

Jaynes, D.B., R.C. Rice, and D.J. Hunksaker. 1992. Solute transport during chemigation of a level basin. *Transactions of the ASAE* 35: 1809–1815.

Jensen, M.E. and R.G. Allen. 2000. Evolution of practical ET estimating methods. In Evans, R.G., Benham, B.L., and Trooien, T.P. (eds), Proceedings of the *4th National Irrigation Symposium*, pp. 52–65. Phoenix, AZ, November 14–16. ASAE: St. Joseph, Mich.

Johnson, L. and T. Scholasch. 2005. Remote sensing of shaded area in vineyards. *Hort Technology* 15(4): 859–863.

Kennedy, D.N. 1994. California Water Plan Update. California Department of Water Resources, Sacramento, CA, Vol. 1, Bulletin 160-93, p. 166.

Lamacq, S., Y. Le Gal Pierre, E. Bautista, and A.J. Clemmens. 1996. Farmer Irrigation scheduling: A case study in Arizona. pp. 97–102. In Camp, C.R., Sadler, E.J., and Yoder, R.E. (eds), *Proceedings of International Conference on Evapotranspiration and Irrigation Scheduling ASAE*, San Antonio, TX, November 3–6, 1996.

Lety, J. and G.L. Feng. 2007. Dynamic versus steady-state approaches to evaluate irrigation management of saline water. *Agricultural Water Management* 91: 1–10.

Lety, J., G.B. Hoffman, J.W. Hopmans, S.R. Grattan, D. Suarez, D.L. Corwin, J.D. Oster, L. Wu, and C. Amrheina. 2011. Evaluation of soil salinity leaching requirement guidelines. *Agricultural Water Management*. 98: 502–506.

Martin, E.C. 2009. Methods for measuring for irrigation scheduling-when. Arizona Water Series No. 30. Cooperative Extension, College of Agriculture & Life Sciences, The University of Arizona: Tucson, AZ.

Mikkelsen, R. 2009. Ammonia emissions from agricultural operations: Fertilizer. *Better Crops with Plant Food* 93: 9–11.

Molle, F., P. Wester, and P. Hirsch. 2010. River basin closure: Processes, implications and responses. *Agricultural Water Management* 97: 569–577.

Morrow, M.R. and D.R. Krieg. 1990. Cotton management strategies for a short growing season environment: Water–nitrogen considerations. *Agronomy Journal* 82: 52–56.

National Agricultural Statistics Service. 2002, Census of Agriculture 2002, National Agricultural Statistics Service, USDA, Washington, DC. See http://www.nass.usda.gov/Census_of_Agriculture/index.asp, http://www.nass.usda.gov/census/census02/volume1/us/index1.htm, and http://www.agcensus.usda.gov/Publications/2002/index.asp.

Navarro, J.C., J.C. Silvertooth, and A. Galadima. 1997. Fertilizer nitrogen recovery in irrigated upland cotton. A College of Agriculture Report. Series P-108, pp. 402–407, University of Arizona, Tucson, AZ.

Neale, C.M.U., W.C. Bausch, and D.F. Heerman. 1989. Development of reflectance-based crop coefficients for corn. *Transactions of the ASAE* 32(6): 1891–1899.

Neale, C.M.U., H. Jayanthi, and J.L. Wright. 2005. Irrigation water management using high resolution airborne remote sensing. *Irrigation and Drainage Systems* 19(3–4): 321–336.

Nimah, M.N. and R.J. Hanks. 1973. Model for estimating soil water, plant, and atmospheric interactions: I. Description and Sensitivity. *Soil Science Society of America Proceedings* 37: 522–527.

Pereira, L.S. 1999. High performance through combined improvements in irrigation methods and scheduling: A discussion. *Agricultural Water Management* 40(1): 153–169.

Pereira, L.S., T. Oweis, and A. Zairi. 2002. Irrigation management under water scarcity. *Agricultural Water Management* 57(1): 175–206.

Rice, R.C., D.J. Hunsaker, F.J. Adamsen, and A.J. Clemmens. 2001. Irrigation and nitrate movement evaluation in conventional and alternate-furrow irrigated cotton. *Transactions of the ASAE* 44(3): 555–568.

Rockström, J., L. Karlberg, S.P. Wani, J. Barron, N. Hatibu, T. Oweis, A. Bruggeman, J. Farahani, and Z. Zhu Qiang. 2010. Managing water in rainfed agriculture—The need for a paradigm shift. *Agricultural Water Management* 97: 543–550.

Santos, C., I.J. Lorite, M. Tasumi, R.G. Allen, and E. Fereres. 2008. Integrating satellite-based evapotranspiration with simulation models for irrigation management at the scheme level. *Irrigation Science* 26: 277–288.

Silvertooth, J.C., J.E. Watson, J.E. Malcuitt, and T.A. Doerge. 1992. Bromide and nitrate movement in an irrigated cotton production system. *Soil Science Society of America Journal* 56: 548–555.

Silvertooth, J.C., K.F. Bronson, E.R. Norton, and R. Mikkelsen. 2011. Nitrogen utilization by western U.S. cotton. *Better Crops with Plant Food* 95: 21–23.

Solomon, K. 1983. Irrigation uniformity and yield theory. PhD Dissertation, Utah State U., Logan, UT, 271 p.

Suarez, D. 2011. Soil salinization and management options for sustainable crop production. In M. Pessarakli (ed.), *Handbook of Crop and Plant Stresses*, 3rd edn, Chapter 3. CRC Press, Boca Raton.

Tamang, P.L., K.F. Bronson, A. Malapati, and R. Schwartz. 2011. Nitrogen fertilizer requirements for ethanol production from sweet and photoperiod sensitive sorghums in the southern High Plains. *Agronomy Journal* 103: 431–440.

Thorp, K., D.J. Hunsaker, and A.N. French. 2010. Assimilating leaf area index estimates from remote sensing into the simulations of a cropping systems model. *Transactions of the ASAE*. 53(1): 251–262.

Trout, T.J., L.F. Johnson, and J. Gartung. 2008. Remote sensing of canopy cover in horticultural crops. *HortScience* 43(2): 333–337.

Turral, H., M. Svendsen, and J.M. Faures. 2010. Investing in irrigation: Reviewing the past and looking to the future. *Agricultural Water Management* 97: 551–560.

Wright, J.L. 1982. New evapotranspiration crop coefficients. *Journal of the Irrigation and Drainage Division ASCE* 108(1): 57–74.

Yabaji, R., J.W. Nusz, K.F. Bronson, A. Malapati, J.D. Booker, R.L. Nichols, and T.L. Thompson. 2009. Nitrogen management for subsurface drip irrigated cotton: Ammonium thiosufalte, timing, and canopy reflectance. *Soil Science Society of America Journal* 73: 589–597.

Zelinski, L.J. 1985. Development of a soil nitrogen test for cotton. In *Proceedings of Beltwide Cotton Production Research Conference*, National Cotton Council of America, Memphis, TN.

Zhang, H., G. Johnson, B. Raun, N. Basta, and J. Hattey. 1998. OSU soil test interpretations. Oklahoma Cooperative Extension Service Fact Sheet No. 2225. Oklahoma State University, Stillwater, OK.

Section III

Irrigation and Soil Water Management

9 Site-Specific Irrigation Management

Precision Agriculture for Improved Water-Use Efficiency

Susan A. O'Shaughnessy, Robert G. Evans, Steven R. Evett, Paul D. Colaizzi, and Terry A. Howell

CONTENTS

9.1 INTRODUCTION

Irrigation has had a far-reaching effect on human civilization over the last 6000 years, not only for its provision of sustenance, but also because of its influence on the integration of various elements from soil science, agronomy, hydraulics, and hydrology (Cuenca 1989), shaping institutions, cultures, politics, and regulatory policies (National Research Council 1996), and stabilizing rural areas (Playán and Mateos 2006). Irrigated agriculture is a vital component of agriculture and supplies many of the fruits, vegetables, and cereal foods consumed by humans; the grains fed to animals that are used as human food; and the feed to sustain animals that are used for work in many parts of the world (Howell 2001). Irrigation spawned an evolution of technology, ranging from the design and construction of dams and vast water distribution systems to the design of centrifugal pumps, gated pipe, siphon tubes, drip and sprinkler irrigation equipment, sensors and communications for improved irrigation management, and farming equipment to maximize the net returns from irrigated cropping systems. In recent years, irrigated agriculture has become more efficient due to better irrigation scheduling by farmers and to the conversion of furrow or flood irrigation to pressurized systems.

Although irrigation practices have improved, irrigated agriculture still faces compounding and complex challenges. These challenges include meeting the demand for increased global food production, which is expected to double by 2050 (UNESCO 2009); cultivating a greater amount of water-intensive crops to meet changing dietary preferences (Chen et al. 2009; Khan et al. 2009); addressing the declining soil resources (Snapp et al. 2010), the growing competition for water from the municipal and industrial sectors, the escalating energy prices, the greater variability in interannual and intra-annual climate (Thomas 2008), and a call to reduce environmental degradation brought about by poor irrigation practices. The obvious clash between meeting global nutrition and quality-of-life issues will most certainly require major changes in agricultural systems in the next few decades. These pressures will probably force more significant changes in the management and operation of irrigated agriculture in the next 50 years than in the previous 5000 years.

As the problems impacting production agriculture become more complicated, the solutions must become more creative and combine new and existing technologies in original ways with a focus on improving the efficiency of land and water use. Today, we must expect better design and better matching technologies that are appropriate to the managerial capacity of on-farm systems (e.g., limited well capacity and effective water delivery); incorporate irrigation strategies (such as deficit and limited irrigation schemes); and integrate systems of appropriate financial and economical design to ensure proper operation, maintenance (Turral et al. 2009), and producer profitability. This, in part, refers to system management through automation and a parallel adoption of precision technologies for site-specific delivery of water to optimize irrigation management.

9.2 SITE-SPECIFIC IRRIGATION MANAGEMENT AND ITS ROLE IN IMPROVING WATER-USE EFFICIENCY

Water-use efficiency (WUE), the ratio of the crop yield to the amount of seasonal water used, has been widely used to describe irrigation effectiveness in terms of crop yield. But it is a cause for concern when we consider that the high global demand for agricultural yields is coupled with a limited supply of quality water and a diminishing amount of arable land. Improving WUE is a persuasive summons to combine the best agricultural technologies. The Equation 9.1 offered by Howell et al. (1990) presents an explicit expression on how agronomic and engineering mechanisms affect WUE:

$$\mathrm{WUE} = \frac{(\mathrm{HI} \times \mathrm{DM})}{\left\{\mathrm{T}(1-\mathrm{WC})\left[1+\dfrac{\mathrm{E}}{(\mathrm{P}+\mathrm{I}+\mathrm{SW}-\mathrm{D}-\mathrm{Q}-\mathrm{E})}\right]\right\}}, \tag{9.1}$$

where HI is the harvest index, DM is dry matter (g/m^2), T is transpiration (mm), WC is the standard water content used to express the economic yield, E is soil water evaporation (mm), P is precipitation (mm), I is irrigation (mm), SW is soil water depletion from the root zone (mm), D is deep percolation below the root zone (mm), and Q is surface runoff (mm).

Site-specific irrigation management (SSIM) is an irrigation technology that allows for the control of where to place a specific amount of water when the crop requires it; SSIM could be used to control WUE by managing the soil water movement at the root zone and reducing the surface runoff through efficient irrigation scheduling.

Optimal WUE also becomes an issue whenever water supplies are limited due to drought conditions, inadequate well capacities, or water regulations that drive the adoption of deficit irrigation

(DI) strategies (English 2010). DI, applying less than the full crop evapotranspiration (ET) requirements, is a plausible solution to economically sustain the production of many annual and perennial crops. SSIM could also help farmers who must implement DI schemes. Managed DI can be an efficient strategy to apply less water as long as the method allows for sufficient yield or profitability (Zhang 2003). While the adoption of severe water-deficit strategies would likely lead to significant yield losses, managed DI strategies whereby the majority of the irrigation water is applied during critical growth stages (e.g., reproductive stages) have been shown to sustain yields at levels that are economically feasible while conserving water on many crops (Payero et al. 2009). Outside these critical periods (e.g., vegetative stages), irrigation may be limited or unnecessary if sufficient rainfall occurs (Geerts and Raes 2009). The implementation of DI implies an appropriate knowledge of actual crop evapotranspiration (ET_c), crop responses to the timing and the severity of the water deficits, and the economic impacts of yield reduction strategies (Pereira et al. 2002). Employing SSIM methods with the intent to schedule DIs in combination with other precision agriculture (PA) technologies, such as remote crop monitoring and decision support systems, could provide significant water savings with limited crop yield or quality reductions (Zhang 2003; Fereres and Soriano 2007; Sadler et al. 2007; Evans and King 2010).

9.3 ADDRESSING VARIABILITY WITH SSIM

Agricultural fields are inherently spatially variable with respect to topographic relief, soil chemical (e.g., pH, salinity, and organic matter) and physical properties (e.g., texture and electrical conductivity [EC]) (Sadler et al. 1998, 2000; Zhang et al. 2002), fertility differences, and biotic effects. PA technologies evolved to address the various aspects of spatial and temporal variabilities that occur. Spatial variability can include natural genetic variation within a variety, plant structure and stand density (e.g., leaf area index [LAI] m^2/m^{-2}), row spacing and orientation, crop height, pests, fertility, drought, and field history (e.g., carryover effects from previous herbicide applications). A variation in the soil topography and properties can play an important role in the infiltration, runoff, and ponding in low areas of a field such that varying water application depths may be needed in different areas to compensate for inconsistencies (Sadler et al. 2000, 2002). The variability in the soil properties have been strongly correlated with the yields of some crops, including corn (DeJonge et al. 2007; Ko et al. 2009; Nyiraneza et al. 2009), wheat (Vrindts et al. 2003; Ko et al. 2009), soybean (Cox et al. 2003; Irmak et al. 2002), sorghum (Tolk and Howell 2008), cotton (Ping et al. 2008), sugarcane (Johnson and Richard 2005), and grapes (Bramley and Hamilton 2004; Ramos 2006).

The second type of variability is temporal variability, which could be caused by variations across or within production years due to climatic or seasonal events, such as rainfall, frost, or pestilence. Temporal variability causes plants to require varying amounts of water distributed throughout the season. Short-term variability can range from a few days to a few weeks, whereas long-term variability may be for months or years. In either case, temporal variability can be influenced by crop phenology (age and growth stage), soil compaction, soil fertility, weather, pests, nonuniformity of irrigation and fertilizer applications, runoff, and external causes such as herbicide drift, and climatic variability. An automated irrigation system that controls irrigation to respond to the spatial and temporal variabilities and mitigate the economic risk would be a useful farming tool.

9.4 COMPONENTS OF SSIM

Pressurized irrigation systems, such as drip irrigation and self-propelled sprinkler irrigation, can be adapted to address the spatial and temporal within-field variabilities. Additional components are essential hardware such as valves and manifolds, control systems for water delivery and movement control, defined management zones, geographical positioning systems, sensors for monitoring the crop and/or the soil water status, decision support algorithms to trigger or withhold irrigations and estimate the amount of water to deliver, and remote communication technology (Figure 9.1).

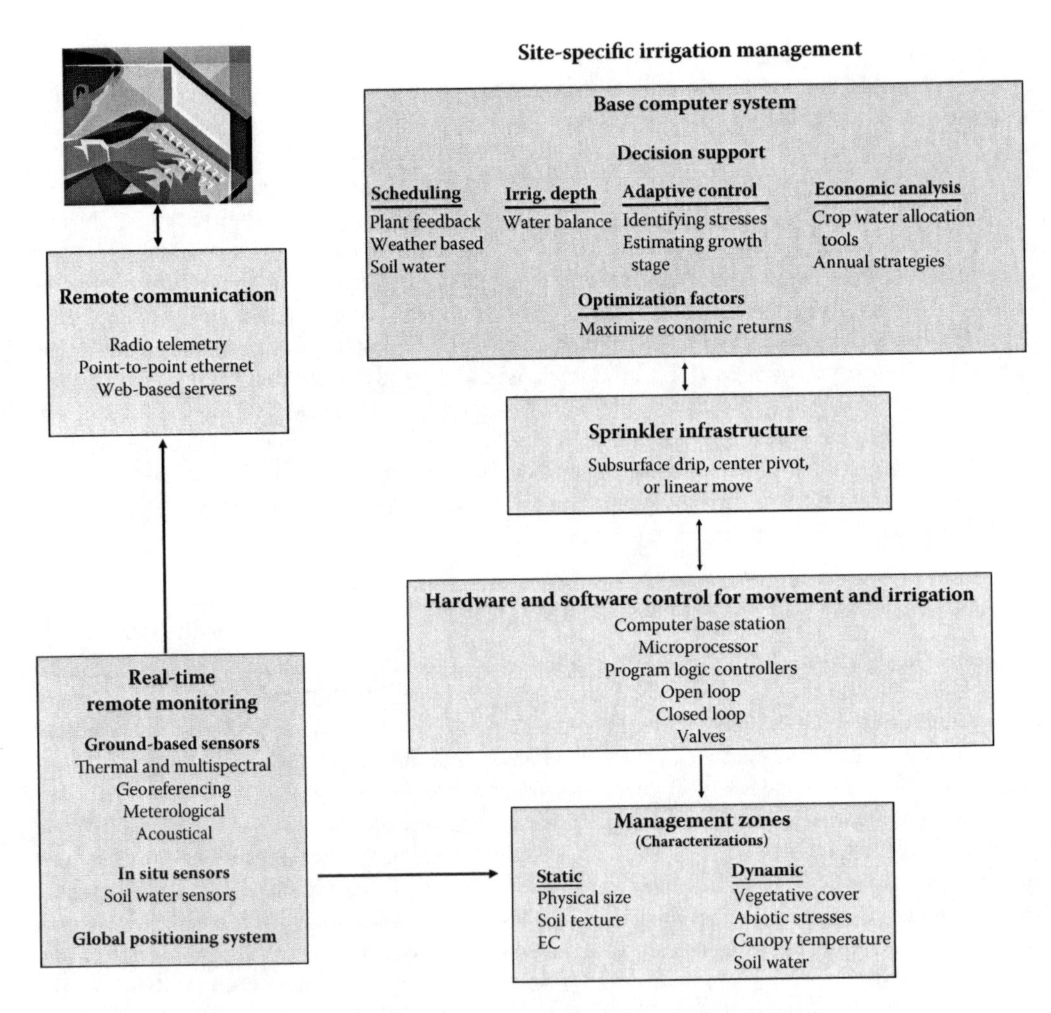

FIGURE 9.1 A block diagram showing the key components for site-specific irrigation management.

From this definition, it is clear that SSIM is a method that depends on the synergy of mechanized hardware functionality and sensor feedback governed by sound science, engineering models, and algorithms; for example, a real-time supervisory control and data acquisition (SCADA) system to achieve optimal and sustainable agricultural production (King et al. 2000). Each of these components will be discussed to give the reader background information.

9.4.1 Pressurized Systems for Conveyance and Application of Water

In the United States, the percentage of irrigated land that is irrigated by sprinkler and drip irrigation systems is approximately 63%, about half again greater than the land area irrigated by gravity systems (NASS 2009) (Figure 9.2). Although the majority of the existing pressurized systems will need moderate investment to achieve SSIM, their basic infrastructure lends itself to retrofitting.

9.4.1.1 Subsurface and Surface Drip Irrigation

While this chapter is primarily directed toward the self-propelled sprinkler irrigation, it is important to mention the site-specific capabilities offered by drip irrigation methods. With drip irrigation (surface or subsurface), growers can routinely achieve application efficiency and WUE exceeding 85% (Ayars et al. 1999). Its adoption can provide a potential solution to the problem of low water and

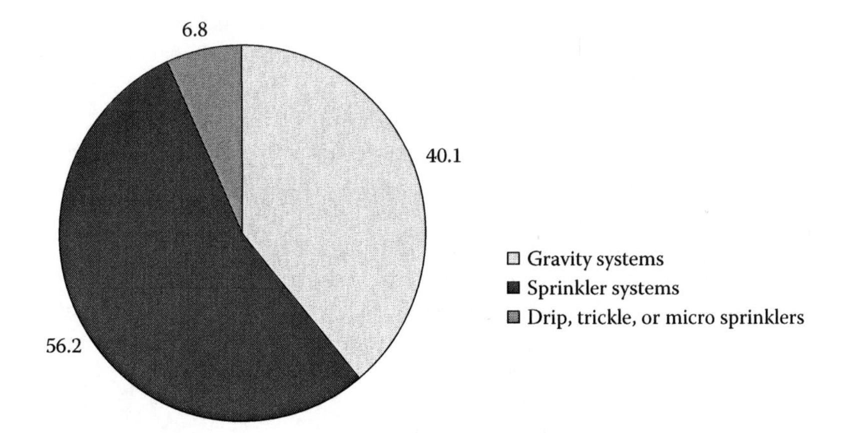

FIGURE 9.2 Percentage of land irrigated by the water distribution method. (Data from NASS, 2008 Farm and ranch irrigation survey. USDA, 2009.)

nutrient use efficiencies due to better control of the applied water, resulting in less water and nutrient losses through deep percolation and runoff, and reduced soil water evaporation. Water deficits can affect yield by impacting crop growth, development, and carbon assimilation. The response of maize, wheat, and sunflower to DI applied via subsurface drip irrigation (SDI) demonstrated that the harvest index (percentage of dry mass partitioned to grain) remained constant until the biomass was reduced to less than 60% of the maximum biomass production (Fereres and Soriano 2007). The timing of the irrigation relative to the anthesis stage was critical, with preanthesis stress showing higher levels of water productivity than postanthesis stress. The DI of sorghum at 25% and 50% of ET using SDI resulted in a higher grain weight than did mid elevation spray and low energy precision application (LEPA) irrigation methods (Schneider et al. 2001; Colaizzi et al. 2004). Lamm and Trooien (2003) demonstrated that the careful management of SDI systems in Colby, Kansas, reduced the net irrigation needs of corn by roughly 25% without reducing the yields due to the minimization of evaporation and deep percolation losses. Some examples are summarized in Table 9.1.

TABLE 9.1
Examples of Impact of Subsurface Drip Irrigation on Water-Use Efficiency

Crop	Water Applied	Water Savings	Improvement	Location
Alfalfa[a]	1174 mm	136 mm (compared with furrow irrigation)	Yield increases of 26%–35%	Imperial Valley, CA
Corn[b]	293 mm	25% (relative to long-term net irrigation requirement)	Reduced in-season drainage, surface runoff, and soil evaporation	Central Plains
Grain sorghum[c]	25% and 50% of crop ET	None—same levels applied as LEPA, LESA, and MESA	Greater yield and greater WUE via reduced evaporation	Texas High Plains
Cotton[d]	185 mm	None—same levels applied as LEPA	Greater lint yield and WUE	Texas High Plains

[a] Ayars, J.E., Phene, C.J., Hutmacher, R.B., Davis, K.R., Schoneman, R.A., Vail, S.S., and Mead, R.M., *Agric. Water Manag.*, 42, 1–27, 1999.

[b] Lamm, F.R. and Trooien, T.P., *Irrig. Sci.*, 22, 195–200, 2003.

[c] Colaizzi, P.D., Schneider, A.D., Howell, T.A., and Evett, S.R., *Trans. ASAE*, 47, 1477–1492, 2004.

[d] Bordovsky, J.P., Lyle, W.M., and Segarra, E., *Texas J. Agric. Nat. Resour.*, 13, 67–73, 2000.

Other advantages of SDI include reducing the evaporation losses and minimizing the leaching losses of N when compared with common surface irrigation methods, by applying water and nutrients to the most active part of the root zone. In some cases, SDI has led to better yields, depending on the crop, soil type, and climate (Colaizzi et al. 2004). One final advantage of SDI is its ability to irrigate with wastewater while minimizing human contact (Choi et al. 2004; Thompson et al. 2009).

A subsurface drip functions as an SSIM tool to deliver water to predetermined management zones when augmented with a distributed irrigation control system, for example, multiple controllers, each controlling specific solenoid valves, and a decision support system that provides some type of feedback control, such as soil water sensing (e.g., Miranda et al. 2005). Site-specific irrigation with fixed systems is mainly used on high-value fruit and vegetable crops, but may also be suitable for row crops, such as corn, cotton, and sorghum, and in horticulture fields, vineyards, orchards, landscapes, nurseries, and greenhouses.

9.4.1.2 Self-Propelled Sprinkler Irrigation Systems

Self-propelled (center pivot and linear move sprinkler) irrigation systems are currently used on more than 47% of all U.S. irrigated cropland (NASS 2009), and the area covered by these machines is expected to increase in the United States and internationally. Center pivot and linear move irrigation systems provide a natural platform on which to develop SSIM technologies due to their current use and high degree of automation (King and Kincaid 2004). Many of the newer models are outfitted with control panels and have remote communication capabilities. These highly mechanized systems offer advantages over furrow irrigation, including good uniformity over large areas at a reasonable cost and substantial labor savings. In addition, the high frequency of the irrigations under these machines potentially reduces the temporal variability of the field soil water content (Sadler et al. 2000). Moving systems are primarily used on large areas of field crops, such as corn, small grains, cotton, and forages. It is fairly certain that these machines will play a major role in advancing future irrigation technologies and strategies around the world, especially on lower-valued food, bioenergy, and feed crops.

Site-specific sprinkler irrigation systems are also well suited for chemigation (Evans and Han 1994; Evans et al. 1995; Sumner et al. 1997; Duke et al. 1998, 2000; Sadler et al. 2002; Palacin et al. 2005; Farahani et al. 2006) and when applied with the appropriate amount of water, they will reduce the potential for runoff and the movement of nutrients below the plant root zone (King et al. 1995, 2009; Sadler et al. 2000, 2005; Evans and King 2010). Site-specific sprinkler application technologies can be used to treat a whole field or smaller management zones within a field with simple on/off sprinkler controls in single-span-wide treatment areas (Evans et al. 2000; King and Wall 2005). These systems can be used to manage irrigation in well-defined areas where the cost of a full precision irrigation control system is not justified, such as rock outcrops, or roads, houses, and waterways (Sadler et al. 2000). The development of automated site-specific irrigation systems would potentially allow the producers to maximize irrigation efficiency while simultaneously minimizing the negative effects on their productivity (Kim et al. 2009).

9.4.2 Essential Hardware

There are different combinations of variable rate hardware to accomplish site-specific irrigation. One design includes outfitting the moving sprinkler with individually controlled manifolds, each capable of delivering discrete flow rates in various combinations of different-sized sprinkler heads to achieve a series of incremental application rates (Evans et al. 2000). Another variable rate technique uses a double sprinkler arrangement at each active outlet along the system lateral, where one sprinkler is sized for one-third of the design flow rate and the second sprinkler for two-thirds of the design flow rate for the outlet. A stepwise variable flow rate is achieved by controlling the operation of each sprinkler using solenoid-activated diaphragm valves on each sprinkler (McCann et al. 1997; King et al. 1997; King and Wall 2005). A similar variable rate application technique was used in

FIGURE 9.3 Dual sprinkler options (LESA and MESA) mounted on a single linear move system. (Photograph from Robert Evans, USDA-ARS, Sidney, Montana, 2010.)

Florence, South Carolina (Sadler et al. 1996; Omary et al. 1997; Camp et al. 1998), and in Garden City, Kansas (Klocke et al. 2003), using three sprinkler heads. The South Carolina machine sized sprinklers at one-seventh, two-sevenths, and four-sevenths of the maximum application depth are used in combination to achieve the desired depth of application.

Pulse modulation for time-proportional control is a third and the most common paradigm for variable rate hardware, where the prescribed irrigation amount in a management zone is accomplished by cycling the sprinklers on and off at selected intervals over short periods of time, ranging from 60 to 250 sec (Duke et al. 1992; Fraisse et al. 1992; Evans et al. 1996, 2010a; Evans and Harting 1999; Perry et al. 2003, 2004; Dukes and Perry 2006; Pierce et al. 2006; Han et al. 2009; Chávez et al. 2010a,b,c). Cycling the valves "on" and "off" does not cause degradation of sprinkler uniformity, as concluded by Perry et al. (2004). Evans et al. (2010b) developed a unique site-specific irrigation system that combined variable rate watering capabilities with MESA and LESA application methods by augmenting a linear move system (Figure 9.3).

Finally, there is the technique of addressing the variable flow by altering the aperture of a nozzle with an inserted pin to achieve different flow rates. King et al. (1997, 2009) and King and Kincaid (2004) adjusted the nozzle apertures to achieve two different flow rates and then cycled the pin placement for time-proportional control of water applications. Whichever design is used, the systems must account for interactions between individual sprinkler valves and the start and stop movements of each tower, in the case of moving sprinklers. It is also important to note that when other combinations of farming and irrigation practices are used in conjunction with hardware augmentation, SSIM is enhanced. Examples include planting in concentric rows using global positioning system (GPS) tractor steering, irrigating every other furrow usingdrag socks for LEPA (Lyle and Bordovsky 1983) to help reduce the evaporation of applied irrigation, and creating furrow dikes or small reservoirs between crop rows to control the runoff (Schneider and Howell 2000).

9.4.3 Control Systems

Control systems for SSIM are hardware and software with specific functionality for (1) sequencing valves to either adjust the flow rates or pulse water through manifolds or individual drop hoses and (2) managing the sprinkler system movement (Fraisse et al. 1995; Evans et al. 2000). These control systems can be classified as open-loop and closed-loop controls depending on how the process is managed and adjusted for optimal performance.

Open-loop systems rely on specified relationships (e.g., state equations and models) between the inputs and the outputs to control the outcome. These systems are sometimes called feedforward systems because there is no real-time feedback mechanism to evaluate the quality of the outputs or to compensate for unexpected influences on the operation of the system. For example, most of the current irrigation decision support programs (often called scientific irrigation scheduling) are basically open-loop systems where irrigation timing and amounts may not equally benefit all areas of a field. The timing and the duration of the water applications may be based on algorithms predicting the performance of an irrigation system based on historical and predicted climatic and soil water conditions. Feedback to the process is usually made by spot measurements (e.g., soil water) and climate data after the operation has been completed and adjustments have been made for the following irrigation event. Use of real-time plant health and soil water status measurements is absent in this process. Open-loop irrigation control systems can also be accomplished with simple timers (e.g., preset time-on and time-off) and little or no feedback, although rain sensors are sometimes used to eliminate or shorten unnecessary irrigations (Cardenas-Lailhacar and Dukes 2008). In general, open-loop types of control systems are not adequate for controlling SSIM sprinklers on large fields.

On the other hand, there are several alternative approaches for closed-loop systems, which appear to be suitable for SSIM sprinklers. The closed-loop systems measure the output of the process (feedback), resulting in periodic adjustments to the controlled parameters during the process in order to minimize the differences between the inputs and the measured outputs. These types of systems are often referred to as adaptive control systems and have the flexibility to change the control parameters to adjust for changing conditions in space and time, depending on the feedback mechanisms and limitations (Smith et al. 2009).

9.4.4 Management Zones

The effective management of spatial variability requires the development and mapping of discrete, georeferenced management zones and the provision of within-field digital elevation maps. Management zones are areas within a field that are relatively homogeneous with regard to at least one characteristic that can be variably treated using irrigation. The basis for such zones can be soil texture or fertility (Watkins et al. 1998, 2002; Pierce and Nowak 1999; Fraisse et al. 2001; Ferguson et al. 2003; Han et al. 2003), soil compaction and/or structure (Upadhyaya et al. 1994; Chung et al. 2006; Andrade-Sánchez et al. 2001, 2002, 2008), apparent soil EC (Sudduth and Hummel 1993; Sudduth et al. 1995, 2000; Newman and Hummel 1999; Drummond et al. 2000; Farahani et al. 2005; Corwin and Lesch 2005a,b; Jabro et al. 2006), topography, microclimate, harvested yields (Jaynes et al. 2005; Zhang et al. 2002), pest pressures, and plant responses. The management zone maps should specify the amount of water to be applied to each defined management area within the field for that irrigation event (Evans et al. 2000; Nijbroek et al. 2003). These management maps tend to be fixed over the season, especially for the management of variable rate irrigation (Hedley et al. 2009); however, ideally, maps of the management zones should be frequently updated based on the real-time, spatially distributed data of actual field conditions. Recent work by Lopez-Lozano et al. (2010) made joint use of satellite imagery (Quickbird*) to derive and update LAI maps overlain onto digital soil maps derived from soil sampling to define site-specific management units.

If frequent updating cannot be supported, various modeling techniques can be used to predict the variability across fields. Modeling is critical in cases of single point measurements, which are generally inadequate for optimal management over a growing season because they can vary independently over time. Similarly, where finely spaced data are unavailable, but data have been obtained

* The mention of trade names of commercial products in this chapter is solely for the purpose of providing specific information and does not imply a recommendation or endorsement by the U.S. Department of Agriculture.

on a coarser grid, the spatial resolution may be increased by interpolating (e.g., kriging and distance weighting) the available data and assigning estimated values to each unknown zone (Han et al. 2003). Hedley and Yule (2009) developed models to predict the daily soil water status from EC data, to control a variable rate irrigation system with the goal of improving WUE. The model was used in three simulated case studies on irrigated dairy pastures, a 24 ha potato field, and a 22 ha maize field, using center pivot irrigation on all sites. Soil management zones were used to define the variable rate irrigation, and the soil EC survey points were kriged along with the use of a spherical semivariogram model to produce a map of soil EC. The amount of plant available water held in each EC-defined soil zone on any one day was calculated using a water balance model.

Peters and Evett (2007) provide a second example in which daily temperature curves were predicted from one-time-of-day temperatures taken over a field by a moving pivot. The temperatures were standardized against the maximum daily air temperature values to provide a seasonal average for locations within each management zone. The ability to dynamically define management zones will help ensure that water is applied only where, when, and in the amount needed at the right times for economically and agronomically efficient production (Corwin and Lesch 2005a,b).

9.4.5 Sensor Systems

A number of different sensors are available for identifying the location and monitoring the crops, soil water, and meteorological data. Sensors are critical to SSIM as their measurements provide the information that is used to spatially characterize the health and water status of crops, to estimate the amount of available soil water, and to approximate crop ET.

GPS is a satellite-based radio navigation system that provides a reliable method for determining a location and a time reference (universal coordinated time; UTC) with a GPS receiver. Existing moving irrigation systems can be retrofitted with a low-cost GPS unit to report a moving sprinkler field position (Peters and Evett 2005). Commercially available moving sprinkler systems now come with the option of a GPS receiver that typically employs the wide area augmentation system (WAAS) for data correction. The site-specific irrigation systems utilized in research detail their use of GPS units to calculate field position and provide accurate spatial water delivery (Peters and Evett 2008; Han et al. 2009; Kim and Evans 2009).

Other sensors, such as satellites, provide information on the patterns of water use, land use, irrigated crop type, crop yield, daily ET, seasonal ET, crop stress, and salinity over large field and regional scales. Information on the land surface can be obtained at a wide range of spatial (5–5000 m) and temporal resolutions (0.5–24 days), in the visible, near infrared, or thermal range (Bastiaanssen et al. 2000). Multispectral data from satellites might be used to enhance water and energy conservation by helping to determine the exact causes of the nonuniform appearance (and yield) of the crop (Gowda et al. 2008; Hornbuckle et al. 2009). Advanced pattern recognition software and other tools for satellite multispectral or other remotely sensed data can be used to map LAI and detect many problems in agricultural fields, such as the presence of weeds (Tellaeche et al. 2008; Aitkenhead et al. 2003), disease, and lack of nutrients (Bausch and Diker 2001; Clay et al. 2006).

Greater temporal monitoring of within-field variability can be acquired from aerial imagery using aircraft (Tilling et al. 2007) or an unmanned aerial vehicle (Berni et al. 2009) as a platform to fix the equipment and obtain digital, multispectral, or thermal images (Falkenberg et al. 2007; DeTar and Penner 2007; Ben-Gal et al. 2009). Currently, a combination of satellite and aerial imagery and GIS mapping services are being integrated into the SSIM (Pierce et al. 2010) of the vineyards in the Napa Valley region.

Crop monitoring can also be achieved by ground-based remote sensing, which includes photometric sensors (bands in the visible range), thermal imaging sensors (typically placed on a lift), and infrared and near infrared (band pass greater than 780 nm) sensors. Remote sensing using infrared thermometers (IRTs) in close proximity to the field was investigated early on by Aston and van Bavel (1972) to remotely detect soil water depletion and by Idso et al. (1981) and Jackson et al.

(1981) to characterize crop water stress, and Clawson and Blad (1982) used these sensors for irrigation scheduling on corn. In the case of moving irrigation systems, it is ideal to use the system lateral as a platform for remote sensing; this has been achieved using IRTs (Upchurch et al. 1998; Peters and Evett 2004, 2008). Using location-specific and time-specific data, Peters and Evett (2004) also derived a method to produce field canopy temperature maps from IRTs mounted on a moving irrigation lateral.

Ground-based remote sensors using photodiode filters in the visible range offer a method of disease detection with an enhanced spatial resolution compared with aircraft or satellite systems (West et al. 2003). Geographic and spectral data were shown to be useful for detecting the onset and progression of plant disease (Steddom et al. 2003, 2005) and its effects on crop yield, WUE (Price et al. 2010; Workneh et al. 2009), and insect infestation (Mirik et al. 2007; Yang et al. 2009). The detection of these diseased patches can potentially improve WUE if the response is to withhold irrigation when the disease is detected early in the season and the yield potential is forecasted to be less than profitable. Reflectance-based crop coefficients (K_c) have also been used for irrigation scheduling of corn (Neale et al. 1989; Bausch 1993, 1994) and cotton (Hunsaker et al. 2003, 2005). The crop coefficients are predicted as a function of spectral reflectance and are crop-specific.

9.4.6 Decision Support Algorithms

Climatic variability, disease, and pest invasion of a field, as well as on-farm management decisions, often contribute to within-field variability on both spatial and temporal scales. Examples of on-farm management decisions include tillage, fertilization practices, and pesticide spray programs, seeding rates, cultivar and variety selection, herbicide carryover effects, long-term crop rotations, and irrigation applications (Pierce and Nowak 1999; Zhang et al. 2002; Brase 2006). In addition, management is also affected by irrigation water quality, soil variability, crop rotation, water application patterns, soil texture, topography, and irrigation equipment limits. Thus, it is important to be able to characterize both the spatial and temporal variabilities throughout the growing season and respond accordingly. Within every decision support program structure, the irrigator predefines the criteria and guidelines to be used by the software structure and simulation models of the irrigation system and plant growth in making basic decisions to be implemented by a microprocessor-based control system.

A simulated adaptive control system that is capable of evaluating real-time data and seeks to optimize irrigation management is VARIwise (McCarthy et al. 2010). This decision support system is a general-purpose adaptive control simulation algorithm for the center pivot irrigation of cotton that is constructed to accept a range of inputs, including spatially variable soil properties, nonuniform irrigation application strategies, weather profiles, and crop varieties. Both the plant growth and irrigation systems models are used along with real-time feedback from field sensors. Although the reported results are of simulations only, its methodology appears solid and can be modified for other crops.

Decision support systems can also provide the grower with instructions for chemigation (e.g., nitrogen fertilizer) and alerts (e.g., insects and diseases) using established nutrition and pest models and real-time environmental data. In short, decision support provides more management flexibility by implementing short-term, routine commands to direct irrigation schedules and other basic operations, which frees the irrigator to concentrate on managing other unforeseen and more immediate issues to minimize risk and reduce costs.

Irrigation management algorithms will need to provide support options toward maximizing the net benefits while taking into account the resource limitations, sustainability, and economic outcomes. Such strategies may focus on minimizing costs or maximizing profits and may include decisions on whether to improve the lower-performing zones toward mean production levels or to maximize productivity in the superior yielding zones. Crespo et al. (2010) used the Pareto dominance rules to enable the discrimination of multiple criteria into groups and provide a choice of optimal simulated irrigation strategies. Classifying optimum irrigation strategies will require further detailed models of the relationships between applied water, crop production, economics, and irrigation efficiency (English et al. 2002).

9.4.7 Irrigation Scheduling Strategies

It is not atypical for farmers to overirrigate their crops, exacerbating the water scarcity issues and reducing yields. The reasons for overwatering vary and include the risks of underwatering, the lack of appropriate data to prevent overirrigation, and the lack of time to adequately monitor multiple fields. Overirrigation can lead to waterlogging, a condition that promotes hyphal growth, spore dispersal, infection, and disease development (Price et al. 2010), and can negatively impact the yield response in some crops such as cotton (O'Shaughnessy and Evett 2010a; Kock et al. 1990). Site-specific irrigation requires a scientific method of determining when to deliver a prescribed amount of water to a particular location in the field when crops require it. Soil water content, plant feedback, and weather monitoring are used to schedule irrigations, while the soil water-holding capacity and the crop type are used to determine how much water should be applied during irrigation (Han et al. 1996).

Farmers can use a number of resources to schedule irrigations. The dominant paradigm in the United States is to estimate crop evapotranspiration (mm), $ET_c = K_c \times ET_0$, using data from a weather station to calculate a reference evapotranspiration (ET_0 in millimeters), which is multiplied by a crop coefficient (K_c) specific for a region and for a crop and its growth stage (Evett 2012). In some cases, crop ET estimates have been modeled along with soil water content, for example, ISAREG, which adjusts the ET_c when soil water depletion exceeds a stress depletion fraction (Pereira et al. 2009).

Other algorithms have used plant feedback information to schedule irrigations successfully. For example, the time temperature threshold (TTT) algorithm, patented as the Biologically Identified Optimal Temperature Interactive Console (BIOTIC; Upchurch et al. 1998), enables automatic irrigation scheduling and control of plant WUE for corn in drip-irrigated plots and soybean and cotton in LEPA-irrigated plots (Evett et al. 1996, 2002; Peters and Evett 2008; O'Shaughnessy and Evett 2010b). The results from automatic irrigation scheduling demonstrated that there was not a significant difference between the yields from manually and automatically controlled irrigation methods for corn and soybean (Evett et al. 1996; Peters and Evett 2008) and, at times, automatic irrigation scheduling resulted in improved WUE and irrigation WUE (O'Shaughnessy and Evett 2010b; Table 9.2). Evett et al. (2002) also showed that the WUE levels can be controlled using differing time thresholds when applying the TTT method of irrigation scheduling.

TABLE 9.2
Impact on Water-Use Efficiency Using Automatic Irrigation Scheduling with Moving Sprinkler Systems

Pressurized Irrigation System	Crop	Water Applied (mm)	Water Savings	Improvement	Location
Center pivot system	Soybean[a]	190–233	14%–32% (relative to irrigations based on 100% replenishment of soil water depletion to field capacity)	Reduced irrigations and increased dry grain yields	Texas High Plains
	Cotton[b]	139	26%–32% (relative to irrigations based on 100% replenishment of soil water depletion to field capacity)	Reduced irrigations and increased dry grain yields	Texas High Plains

[a] Peters, R.T. and Evett, S.R., *J. Irrig. Drain. Eng.*, 134, 286–291, 2008.
[b] O'Shaughnessy, S.A. and Evett, S.R., *Agric. Water Manag.*, 97, 1310–1316, 2010.

9.4.8 Plant Modeling

A key subcomponent of decision support are plant growth models, which can be useful in estimating the crop growth stages, the root development, the yield potential, and the soil nutrient and water balances (Hanks and Ritchie 1991). A few examples of generic models include WOFOST (World Food Study; Keulen and Wolf 1986), CropSyst (Cropping Systems Simulator; Stöckle et al. 2003), and AquaCrop, the crop model of the yield response to water of several herbaceous crops (FAO 2010). Other models have been developed for specific species of crop, for example, CERES (Crop Environment Resource Synthesis; Jones and Kiniry 1986), cereal crop growth models for barley, maize, millet, rice, sorghum, and wheat, and SORKAM (Sorghum, Kansas, A&M; Rosenthal et al. 1989) to simulate phenological processes and yield components in sorghum for different environments. Possible SSIM applications for these models include assessing the water limitations, predicting the crop yields at a given geographical location, comparing the attainable yields against the actual yields of a field, identifying the constraints limiting crop production, scheduling deficit and supplemental irrigation, and assessing the actual water productivity.

Strategies for managed DI can also be components of a decision support algorithm. Carefully managed DIs have been widely investigated as a viable and sustainable water conservation technique for many annual and perennial crops in arid climates (Fereres and Soriano 2007; Evans and Sadler 2008; Geerts and Raes 2009; Payero et al. 2009).

9.4.9 Remote Communications for Within-Field Monitoring and Irrigation Control

Increasingly important to SSI is the use of wireless sensors and wireless networks to monitor and control pumps, valves, and irrigation systems and to collect data from remote ground-based and in situ sensors to monitor the crop and soil water status and automatically schedule irrigations. Modern advances in integrated circuits, low-voltage sensors, and battery and wireless radio frequency technologies combined with the accessibility to the Internet offer tremendous opportunities for the development and application of real-time management systems for fixed and moving irrigation systems (Beckwith et al. 2004; Camilli et al. 2007; Liang et al. 2007; Coates and Delwiche 2008, 2009; Kim et al. 2008; Pierce and Elliott 2008; Vellidis et al. 2008). Wireless sensors and wireless sensor networks (WSN) are preferable because they can eliminate the problems and costs associated with deploying and maintaining cables across a field. Although the power requirements for wireless field systems can be a concern, solutions to increase battery longevity are being addressed in terms of either improved density or energy harvesting (Kimball et al. 2009).

A variety of wireless sensors are emerging for agricultural monitoring. Field sensors can include those that measure soil water, air temperature, humidity, precipitation, and surface radiance, including thermometric measurements (infrared). The real-time smart array of soil water sensors developed and deployed by Vellidis et al. (2008) for remotely sensing soil water content and temperature was wireless; Hedley and Yule (2009) used wireless soil moisture sensing networks composed of time-domain transmission sensors, which they strategically located in field-mapped EC zones. Both of these sensor network systems offer the potential for reliable remote monitoring of the spatially variable soil water status in cropped fields and the probable integration with a variable rate irrigation system to provide a closed-loop irrigation system. The Smart Crop system (Mahan et al. 2010) is a recent example of a stationary wireless IRT and data logging system commercialized for canopy temperature monitoring. Various WSNs can also be mounted on sprinkler machines and provide real-time feedback for decision support as the machines move across a field. Center pivot control, automatic irrigation scheduling, and crop canopy monitoring were also accomplished using WSNs composed of a georeferencing sensor and individually powered IRTs mounted on a moving pivot lateral and in the field below (O'Shaughnessy and Evett 2010a; Figure 9.4). Wireless communications allowed the base computer to be located within the control system on the irrigation machine (Peters and Evett 2008; O'Shaughnessy and Evett 2010b) or in a remote location (Kim and Evans 2009).

(a) (b) (c) (d)

FIGURE 9.4 A wireless sensor network system for crop canopy monitoring of a center pivot field: (a) wireless infrared thermometers on masts forward of LEPA drag socks; (b) wireless GPS unit, WAAS corrected; (c) computer embedded at the pivot point; and (d) wireless infrared thermometer location in a stationary position in the field. (Photographs from USDA-ARS, Bushland, TX, summer 2010.)

Decision support frameworks will inevitably rely on local WSNs for real-time, within-field data for crop and soil monitoring and for micro-meteorological measurements to provide continuous feedback. Integrated data sources and networks provide needed information to recalibrate and check various simulation model parameters for on-the-go irrigation scheduling and adjustments (Andrade-Sánchez et al. 2007; Kim et al. 2008, 2009). The integration of these technologies into the irrigation decision-making process can determine when, where, and how much water to apply in real time, which enables the implementation of advanced water conservation measures for economically viable production with limited water supplies, and the opportunity to conserve energy and enhance the environmental benefits.

9.5 LIMITATIONS AND OBSTACLES TO ADOPTION AND COMMERCIALIZATION OF SSIM

The adoption of SSIM technologies is not at the level that it could be (Lamb et al. 2008). However, technology is not the limiting factor; rather there are a number of other barriers to adopting SSIM practices. These include the substantial upfront costs of the equipment (Wichelns and Oster 2006; Thompson et al. 2009), the complexity of the machinery involved (its operation and maintenance), nonintegrated technology for whole system performance at the commercial scale (McBratney et al. 2005), incomplete research and development efforts, and knowledge and expectation gaps between the developers and the users.

Adrian et al. (2005) report that attitudes of confidence toward using precision agriculture technologies, perception of net benefit, farm size and farmer education levels positively influenced the intention to adopt precision agriculture technologies. Yet, profitability was the biggest motivating factor for farmers to purchase PA tools. The profitability of irrigated agriculture can be perceived in a number of ways, but is often compared with the costs of the inputs associated with and the benefits derived from dryland farming, the cost and the availability of water, and its pricing policy and regulatory structure (Molden et al. 2010). The perceptions of both the usefulness and the complexity of information technologies affect an individual's adoption and use of computer technology, a fundamental part of most PA practices. Of all the PA tools, producers ranked the machine guidance systems, yield monitors, and site-specific soil sampling highly, indicating that precision irrigation tools were more likely to be adopted when they demonstrated an expected outcome rather than information alone (Jochinke et al. 2007). Equally important in technology adoption are the avoidance of both product failure and the inappropriate application of a technology. The failure of a new technology at the field level can easily damage a product's reputation if it is not adequately field-tested or appropriately used and therefore does not meet the expectation of the farmer (Lamb et al. 2008).

Despite the barriers surrounding the adoption of precision irrigation technologies, producers do look for ways to minimize labor or improve convenience in daily farming tasks and desire to adopt technology as long as it is reliable and dependable. Lamb et al. (2008) provide additional suggestions for facilitating precision irrigation adoption: (1) engage a broad cross section of potential users, both in attitude and agricultural systems; (2) provide comprehensive operational protocols; (3) understand the changing expectation of the end users as their product moves through an adoption cycle; and (4) establish an evaluation and feedback loop between the end user and the product developer to continuously improve the product performance as it is used in the field.

Other factors impeding effective water management and agricultural development are irrigation supply technology (Birendra et al. 2011), infrastructure, and policies. The full benefits of irrigation scheduling are directly tied to the flexible delivery of available water at both the field-scale and district-scale levels. Farmers need delivery flexibility and water storage as precursors to implementing SSIM. However, in many areas around the world, there are major problems due to undependable water supplies because of drought, competing uses, declining water tables, salt-water intrusion, poor infrastructure, and declining water supplies. If deliveries are restricted in terms of frequency, flow rate, timing, or duration, irrigators are unable to optimize irrigation scheduling. Enabling conditions for farmers and water managers need to be in place to enhance water productivity (Molden et al. 2010).

9.6 SUMMARY AND CONCLUSION

As the availability of agricultural land becomes static or diminishes, methods must be developed to increase yields other than by crop expansion. SSIM represents advancements in irrigation engineering and management. This technology can be conceptualized as a system of tools and a framework for pressurized irrigation systems to help producers optimize the irrigation management of their fields by lowering the inputs and utility costs, by decreasing the time spent on crop management, by improving WUE without significantly impacting profits or yield quality, and by helping to reduce soil and environmental degradation.

Achieving national and global food security will require an intensification of irrigated agricultural technologies for improved land use and WUE while securing safe environmental conditions. Sustainability will require intensification that does not pollute our valuable limited resources of water and land. The foremost need is to blend disparate data sources to estimate the crop water demand over time, to determine when to irrigate and how much to irrigate, and to bring automation and control to pressurized irrigation systems for deliberate and efficient management over time and space. This will require a modification of the irrigation equipment and the synthesis of existing hardware and software technologies by researchers and the irrigation industry and also financial and time commitment from farmers and crop consultants to work with the new technologies. Although the holistic benefits of SSIM technologies have not been scientifically verified (Evans and King 2010), understanding the benefits of the parts and working toward their synthesis is sensible. Integrated approaches will require the combination of various sensor systems (on moving irrigation machines and/or in the field) for control and monitoring, georeferencing sensors, controllers, and computing power. The maximum benefits will be derived from a decision support system when the plant water status in selected areas of a field is monitored by some means to improve the simulation model output and the irrigation scheduling accuracy. Researchers and the irrigation industry must avoid technology overload at the user level by working to increase the usability of SSIM. Increasing the utility of SSIM technologies will increase their cost-effectiveness and commercial potential (King et al. 2009), while research and development will have the role of building confidence in the operability and functionality of these advanced systems.

ABBREVIATIONS

BIOTIC Biologically Identified Optimal Temperature Interactive Console
DI Deficit irrigation
ET Evapotranspiration
ET_c Crop evapotranspiration
GPS Global positioning system
IRTs Infrared thermometers
LAI Leaf area index
LEPA Low energy precision application
PA Precision agriculture
SCADA Supervisory control and data acquisition
SDI Subsurface drip irrigation
SSIM Site-specific irrigation management
TTT Time temperature threshold
UTC Universal coordinated time
WAAS Wide-area augmentation system
WSN Wireless sensor networks
WUE Water-use efficiency

REFERENCES

Adrian, A.M., S.H. Norwood, and P.L. Mask. 2005. Producers' perceptions and attitudes toward precision agriculture technologies. *Comput. Electron. Agric.* 48: 256–271.

Aitkenhead, M.J., I.A. Dalgetty, C.E. Mullins, A.J.S. McDonald, and N.J.C. Strachan. 2003. Weed and crop discrimination using image analysis and artificial intelligence methods. *Comput. Electron. Agric.* 39: 157–171.

Andrade-Sánchez, P., U. Rosa, S.K. Upadhyaya, B.M. Jenkins, J. Aguera, and M. Josiah. 2001. Soil profile force measurements using an instrumented tine. Paper presented at the annual American Society of Agricultural Engineers, Sacramento, CA.

Andrade-Sánchez, P., S.K. Upadhyaya, B.M. Jenkins, and A.G.S. Filho. 2002. Evaluation of the UC Davis compaction profile sensor. Paper presented at the annual American Society of Agricultural Engineers, Chicago, IL.

Andrade-Sánchez, P., F.J. Pierce, and T.V. Elliott. 2007. Performance assessment of wireless sensor networks in agricultural settings. Paper presented at the annual American Society of Agricultural and Biological Engineers, Minneapolis, MN.

Andrade-Sánchez, P., S.K. Upadhyaya, C. Plouffe, and B. Poutre. 2008. Development and field evaluation of a field-ready soil compaction profile sensors for real time applications. *Appl. Eng. Agric.* 24: 743–750.

Aston, A.R. and C.H.M. van Bavel. 1972. Soil surface water depletion and leaf temperature. *Agron. J.* 64: 368–373.

Ayars, J.E., C.J. Phene, R.B. Hutmacher, K.R. Davis, R.A. Schoneman, S.S. Vail, and R.M. Mead. 1999. Subsurface drip irrigation of row crops, a review of 15 years of research at the Water Management Research Laboratory. *Agric. Water Manag.* 42: 1–27.

Bastiaanssen, W.G.M., D.J. Molden, and I.W. Makin. 2000. Remote sensing for irrigated agriculture: Examples from research and possible applications. *Agric. Water Manag.* 46: 137–155.

Bausch, W.C. 1993. Soil background effects on reflectance-based crop coefficients for corn. *Remote Sens. Environ.* 46: 1–10.

Bausch, W.C. 1994. Remote sensing of crop coefficients for improving the irrigation scheduling of corn. *Agric. Water Manag.* 27: 55–68.

Bausch, W.C. and K. Diker. 2001. Innovative remote sensing techniques to increase nitrogen use efficiency of corn. *Commun. Soil Sci. Plant Anal.* 32: 1371–1390.

Beckwith, R., D. Teibel, and P. Bowen. 2004. Report from the field: Results from an agricultural wireless sensor network. Paper presented at the annual IEEE International Conference on Local Computer Networks, Tampa, FL.

Ben-Gal, A., N. Agam, V. Alchanatis, Y. Cohen, U. Yermiyahu, I. Zipori, E. Presnov, M. Sprintsin, and A. Dag. 2009. Evaluating water stress in irrigated olives: Correlation of soil water status, tree water status, and thermal imagery. *Irrig. Sci.* 27: 367–376.

Berni, J.A.J., P.J. Zarco-Tejada, G. Sepulcre-Canto, E. Fereres, and F. Villalobos. 2009. Mapping canopy conductance and CWSI in olive orchards using high resolution thermal remote sensing imagery. *Remote Sens. Environ.* 113: 2380–2388.

Birendra, K.C., B. Schultz, and K. Prasad. 2011. Water management to meet present and future food demand. *J. Irrig. Drain. Eng.* 60: 348–359.

Bordovsky, J.P., W.M. Lyle, and E. Segarra. 2000. Economic evaluation of Texas High Plains cotton irrigated by LEPA and subsurface drip. *Texas J. Agric. Nat. Resour.* 13: 67–73.

Bramley, R. and R. Hamilton. 2004. Understanding variability in wine grape production systems. *Aust. J. Grape Wine Res.* 10: 32–45.

Brase, R. 2006. *Precision Agriculture*. Thomson Delmar Learning, New York, NY.

Camilli, A., C.E. Cugnasca, A.M. Saraiva, A.R. Hirakawa, and P.L.P. Correa. 2007. From wireless sensors to field mapping: Anatomy of an application for precision agriculture. *Comput. Electron. Agric.* 58: 25–36.

Camp, C.R., E.J. Sadler, D.E. Evans, L.J. Usrey, and M. Omary. 1998. Modified center pivot system for precision management of water and nutrients. *Appl. Eng. Agric.* 14: 23–31.

Cardenas-Lailhacar, B. and M.D. Dukes. 2008. Expanding disk rain sensor performance and potential irrigation water savings. *J. Irrig. Drain. Eng.* 134: 67–73.

Chávez, J.L., F.J. Pierce, T.V. Elliott, and R.G. Evans. 2010a. A remote irrigation monitoring and control system (RIMCS) for continuous move systems. Part A: Description and development. *Precis. Agric.* 11: 1–10.

Chávez, J.L., F.J. Pierce, T.V. Elliott, R.G. Evans, Y. Kim, and W.M. Iversen. 2010b. A remote irrigation monitoring and control system (RIMCS) for continuous move systems. Part B: Field testing and results. *Precis. Agric.* 11: 11–26.

Chávez, J.L., F.J. Pierce, and R.G. Evans. 2010c. Compensating inherent linear move water application errors using a variable rate irrigation system. *Irrig. Sci.* 28: 203–210.

Chen, A., W.E. Huffman, and S. Rozelle. 2009. Farm technology and technical efficiency: Evidence from four regions in China. *China Econ. Rev.* 20: 153–161.

Choi, C.Y., I. Song, S. Stine, J. Pimentel, and C.P. Gerba. 2004. Role of irrigation and wastewater reuse: Comparison of subsurface irrigation and furrow irrigation. *Water Sci. Tech.* 50: 61–68.

Chung, S.O., K.A. Sudduth, and J.W. Hummel. 2006. Design and validation of an on-the-go soil strength profile sensor. *Trans. ASABE* 49: 5–14.

Clawson, K.L. and B.L. Blad. 1982. Infrared thermometry for scheduling irrigation of corn. *Agron. J.* 74: 311–316.

Clay, D.E., K. Kim, J. Chang, S.A. Clay, and K. Dalstead. 2006. Characterizing water and nitrogen stress in corn using remote sensing. *Agron. J.* 98: 579–587.

Coates, P.W. and M.J. Delwiche. 2008. Site-specific water and chemical application by wireless valve controller network. Paper presented at the annual American Society of Agricultural and Biological Engineers, Providence, RI.

Colaizzi, P.D., A.D. Schneider, T.A. Howell, and S.R. Evett. 2004. Comparison of SDI, LEPA, and spray irrigation performance for grain sorghum. *Trans. ASAE* 47: 1477–1492.

Corwin, D.L. and S.M. Lesch. 2005a. Characterizing soil spatial variability with apparent soil electrical conductivity I. Survey protocols. *Comput. Electron. Agric.* 46: 103–134.

Corwin, D.L. and S.M. Lesch. 2005b. Apparent soil electrical conductivity measurements in agriculture. *Comput. Electron. Agric.* 46: 11–43.

Cox, M., P. Gerard, M. Warlaw, and M. Abshire. 2003. Variability of selected properties and their relationships with soybean yield. *Soil Sci. Soc. Am. J.* 67: 1296–1302.

Crespo, O., J.E. Bergez, and F. Garcia. 2010. Multiobjective optimization subject to uncertainty: Application to irrigate strategy management. *Comput. Electron. Agric.* 74: 145–154.

Cuenca, R.H. 1989. *Irrigation System Design—An Engineering Approach.* Prentice Hall, Englewood Cliffs, NJ.

DeJonge, K.C., A.L. Kaleita, and K.R. Thorp. 2007. Simulating the effects of spatially variable irrigation on corn yields, costs, and revenue in Iowa. *Agric. Water Manag.* 92: 99–109.

DeTar, W.R. and J.V. Penner. 2007. Airborne remote sensing used to estimate percent canopy cover and to extract canopy temperature from scene temperature in cotton. *Trans. ASABE* 50: 495–506.

Drummond, P.E., C.D. Christy, and E.D. Lund. 2000. Using an automated penetrometer and soil/EC probe to characterize the rooting zone. Paper presented at the 5th International Conference on Precision Agriculture, Madison, WI.

Duke, H.R., D.F. Heermann, and C.W. Fraisse. 1992. Linear move irrigation system for fertilizer management research. Paper presented at the International Exposition and Technical Conference, Falls Church, VA.

Duke, H.R., G.W. Buchleiter, D.F. Heermann, and J.A. Chapman. 1998. Site specific management of water and chemicals using self-propelled sprinkler irrigation systems. Paper presented at the 1st European Conference on Precision Agriculture, Warwick, RI.

Duke, H.R., S.C. Best, and D.G. Westfall. 2000. Spatial distribution of available nitrogen under center pivot sprinklers. Paper presented at the 4th National Irrigation Symposium, Phoenix, AZ.

Dukes, M.D. and C.D. Perry. 2006. Uniformity testing of variable-rate center pivot irrigation control systems. *Precis. Agric.* 7: 205–218.

English, M.J. 2010. A developing crisis in irrigation advisory services. Paper presented at the 5th Decennial Irrigation Conference, Phoenix, AZ.

English, M.J., K.H. Solomon, and G.J. Hoffman. 2002. A paradigm shift in irrigation management. *Irrig. Drain. Eng.* 128: 267–277.

Evans, R.G. and S. Han. 1994. Mapping the nitrogen leaching potential under center pivot irrigation. Paper presented at the annual American Society Agricultural Engineers, Kansas City, MO.

Evans, R.G. and G.B. Harting. 1999. Precision irrigation with center pivot systems on potatoes. Paper presented at the annual meeting of the American Society of Civil Engineers, Seattle, WA.

Evans, R.G. and B.A. King. 2010. Site-specific sprinkler irrigation in a water limited future. Paper presented at the 5th Decennial National Irrigation Symposium, Phoenix, AZ.

Evans, R.G. and E.J. Sadler. 2008. Methods and technologies to improve efficiency of water use. *Water Resour. Res.* 44: 1–25.

Evans, R.G., S. Han, and M.W. Kroeger. 1995. Spatial distribution and uniformity evaluations for chemigation with center pivots. *Trans. ASAE* 38: 85–92.

Evans, R.G., S. Han, S.M. Schneider, and M.W. Kroeger. 1996. Precision center pivot irrigation for efficient use of water and nitrogen. Paper presented at the 3rd International Conference on Precision Agriculture, Madison, WI.

Evans, R.G., G.W. Buchleiter, E.J. Sadler, B.A. King, and G.B. Harting. 2000. Controls for precision irrigation with self-propelled systems. Paper presented at the 4th Decennial National Irrigation Symposium, Phoenix, AZ.

Evans, R.G., W.M. Iversen, and Y. Kim. 2010a. Integrated decision support, sensor networks and controls for wireless site-specific sprinkler irrigation. Paper presented at the 5th Decennial National Irrigation Symposium, Phoenix, AZ.

Evans, R.G., W.M. Iversen, W.B. Stevens, and J.D. Jabro. 2010b. Development of combined site specific MESA and LEPA methods on a linear move sprinkler irrigation system. *Appl. Eng. Agric.* 26: 883–895.

Evett, S.R. 2012. Remote sensing. In *Encyclopedia of Remote Sensing*, ed. E. Njoku. Springer, New York, in press.

Evett, S.R., T.A. Howell, A.D. Schneider, D.R. Upchurch, and D.F. Wanjura. 1996. Canopy temperature based automatic irrigation control. Paper presented at the International Conference of Evapotranspiration and Irrigation Scheduling, San Antonio, TX.

Evett, S.R., T.A. Howell, A.D. Schneider, and D.F. Wanjura. 2002. Automatic drip irrigation control regulates water use efficiency. *Int. Water Irrig. J.* 22: 32–37.

Falkenberg, N.R., G. Piccinni, J.T. Cothren, D.I. Leskovar, and C.M. Rush. 2007. Remote sensing of biotic and abiotic stress for irrigation management of cotton. *Agric. Water Manag.* 87: 23–31.

FAO (Food and Agriculture Organization). 2010. AquaCrop. V 3.1. http://www.fao.org/nr/water/aquacrop.html (accessed September 28, 2010).

Farahani, H.J., G.W. Buchleiter, and M.K. Brodahl. 2005. Characterization of apparent electrical conductivity variability in irrigated sandy and non-saline fields in Colorado. *Trans. ASAE* 48: 155–168.

Farahani, H.J., D.L. Shaner, G.W. Buchleiter, and G.A. Bartlett. 2006. Evaluation of a low volume agro-chemical application system for center pivot irrigation. *Appl. Eng. Agric.* 22: 517–528.

Fereres, E. and M.A. Soriano. 2007. Deficit irrigation for reducing agricultural water use. *J. Exp. Bot.* 58: 147–159.

Ferguson, R., R. Lark, and G. Slater. 2003. Approaches to management zone definition for use of nitrification inhibitors. *Soil Sci. Soc. Am. J.* 67: 937–946.

Fraisse, C.W., D.F. Heermann, and H.R. Duke. 1992. Modified linear move system for experimental water application. Paper presented at the International Conference for Sustainable Irrigation, Leuven, Belgium.

Fraisse, C.W., H.R. Duke, and D.F. Heermann. 1995. Laboratory evaluation of variable water application with pulse irrigation. *Trans. ASAE* 38: 1363–1369.

Fraisse, C., K. Sudduth, and N. Kitchen. 2001. Delineation of site-specific management zones by unsupervised classification of topographic attributes and soil electrical conductivity. *Trans. ASAE* 44: 155–166.

Geerts, S. and D. Raes. 2009. Deficit irrigation as an on-farm strategy to maximize crop water productivity in dry areas. *Agric. Water Manag.* 96: 1275–1284.

Gowda, P.H., J.L. Chávez, P.D. Colaizzi, S.R. Evett, T.A. Howell, and J.A. Tolk. 2008. ET mapping for agricultural water management: Present status and challenges. *Irrig. Sci.* 26: 223–237.

Han, S., R.G. Evans, and S.M. Schneider. 1996. Development of a site specific irrigation scheduling program. Paper presented at the annual American Society of Agricultural Engineers, Tampa, FL.

Han, S., S.M. Schneider, and R.G. Evans. 2003. Evaluating cokriging for improving soil nutrient sampling efficiency. *Trans. ASAE* 46: 845–849.

Han, Y.J., A. Khalili, T.O. Owino, H.J. Farahani, and S. Moore. 2009. Development of Clemson variable rate lateral irrigation system. *Comput. Electron. Agric.* 68: 108–113.

Hanks, R.J. and J.T. Ritchie. 1991. *Modeling Plant and Soil Systems*. ASA, CSSA, SSSA, Madison, WI.

Hedley, C.B. and I.J. Yule. 2009. A method for spatial prediction of daily soil water status for precise irrigation scheduling. *Agric. Water Manag.* 96: 1737–1745.

Hedley, C.B., I.J. Yule, M.P. Tuohy, and I. Vogeler. 2009. Key performance indicators for simulated variable-rate irrigation of variable soils in humid regions. *Trans. ASABE* 52: 1575–1584.

Hornbuckle, J., N. Car, E. Christen, T. Stein, and B. Williamson. 2009. IrriSatSMS: Irrigation water management by satellite and SMS—A utilisation framework. CSIRO Land and Water Science Report No. 04/09. CSIRO Land and Water, Griffith, NSW.

Howell, T.A. 2001. Enhancing water use efficiency in irrigated agriculture. *Agron. J.* 93: 281–289.

Howell, T.A., R.H. Cuenca, and K.H. Solomon. 1990. Crop yield response. In *Management of On-Farm Irrigation Systems*, eds. G.J. Hoffman, T.A. Howell, and K.H. Solomon, pp. 93–122. ASAE, St. Joseph, MI.

Hunsaker, D.J., P.J. Pinter Jr, E.M. Barnes, and B.A. Kimball. 2003. Estimating cotton evapotranspiration crop coefficients with a multispectral vegetation index. *Irrig. Sci.* 22: 95–104.

Hunsaker, D.J., E.M. Barnes, T.R. Clarke, G.J. Fitzgerald, and P.J. Pinter Jr. 2005. Cotton irrigation scheduling using remotely sensed and FAO-56 basal crop coefficients. *Trans. ASAE* 48: 1395–1407.

Idso, S.B., R.D. Jackson, P.J. Pinter, R.J. Reginato, and J.L. Hatfield. 1981. Normalizing the stress degree day for environmental variability. *Agric. Meteorol.* 24: 45–55.

Irmak, A., J.W. Jones, W.D. Batchelor, and J.O. Paz. 2002. Linking multiple layers of information for attribution of causes of spatial yield variability in soybean. *Trans. ASAE* 45: 839–849.

Jabro, J.D., R.G. Evans, Y. Kim, W.B. Stevens, and W.M. Iversen. 2006. Characterization of spatial variability of soil electrical conductivity and cone index using coulter and penetrometer-type sensors. *Soil Sci.* 171: 627–637.

Jackson, R.D., S.B. Idso, R.J. Reginato, and P.J. Pinter. 1981. Crop canopy temperature as a crop water stress indicator. *Water Resour.* 17: 1133–1138.

Jaynes, D., T. Colvin, and T. Kaspar. 2005. Identifying potential soybean management zones from multi-year yield data. *Comput. Electron. Agric.* 46: 309–327.

Jochinke, D.C., B.J. Noonon, N.G. Wachsmann, and R.M. Norton. 2007. The adoption of precision agriculture in an Australian broadacre cropping system—Challenges and opportunities. *Field Crop. Res.* 104: 68–76.

Johnson, R.M. and E.P. Richard Jr. 2005. Sugarcane yield, sugarcane quality, and soil variability in Louisiana. *Agron. J.* 97: 760–771.

Jones, C.A. and J.R. Kiniry. 1986. CERES-Maize: A simulation model of maize growth and development. Texas A&M Univ. Press, College Station, TX.

Keulen, H. Van and J. Wolf. 1986. *Modelling of Agricultural Production: Weather, Soils and Crops*. Simulation Monographs. Center for Agricultural Publishing and Documentation (PUDOC), Wageningen, The Netherlands.

Khan, S., M.A. Khan, M.A. Hanjra, and J. Mu. 2009. Pathways to reduce the environmental footprints of water an energy inputs in food production. *Food Pol.* 34: 141–149.

Kim, Y. and R.G. Evans. 2009. Software design for wireless sensor-based site-specific irrigation. *Comput. Electron. Agric.* 66: 159–165.

Kim, Y., R.G. Evans, and W.M. Iversen. 2008. Remote sensing and control of an irrigation system using a wireless sensor network. *IEEE Trans. Instr. Measure.* 57: 1379–1387.

Kim, Y., R.G. Evans, and W.M. Iversen. 2009. Evaluation of closed-loop site-specific irrigation with wireless sensor network. *J. Irrig. Drain. Eng.* 135: 25–31.

Kimball, J., B.T. Kuhn, and R.S. Balog. 2009. A system design approach for unattended solar energy harvesting supply. *IEEE Trans. Power Electron.* 24: 952–962.

King, B.A. and D.C. Kincaid. 2004. A variable flow rate sprinkler for site-specific irrigation management. *Appl. Eng. Agric.* 20: 765–770.

King, B.A. and R.W. Wall. 2005. Supervisory control and data acquisition system for site specific center pivot irrigation. *Appl. Eng. Agric.* 14: 135–144.

King, B.A., R.A. Brady, I.R. McCann, and J.C. Stark. 1995. Variable rate water application through sprinkler irrigation. In *Site-Specific Management for Agriculture Systems*, eds. P.C. Robert, R.H. Rust, and W.E. Larson, pp. 485–493. ASA-CSSA-SSSA, Madison, WI.

King, B.A., R.W. Wall, D.C. Kincaid, and D.T. Westermann. 1997. Field scale performance of a variable rate sprinkler for variable water and nutrient application. Paper presented at the annual meeting of the American Society of Agricultural Engineers, Minneapolis, MN.

King, B.A., R.W. Wall, and L.R. Wall. 2000. Supervisory control and data acquisition system for closed-loop center pivot irrigation. Paper presented at the annual meeting of the American Society of Agricultural Engineers, Milwaukee, WI.

Klocke, N.L., C. Hunter Jr, and M. Alam. 2003. Application of a linear move sprinkler system for limited irrigation research. Paper presented at the annual meeting of the American Society of Agricultural Engineers, Las Vegas, NV.

Ko, J., G. Piccinni, and E. Steglich. 2009. Using EPIC model to manage irrigated cotton and maize. *Agric. Water Manag.* 96: 1323–1331.

Kock, J., L.P. de Bryun, and J.J. Human. 1990. The relative sensitivity to plant water stress during the reproductive phase of upland cotton (*Gossypium hirsutum* L.). *Irrig. Sci.* 11: 239–244.

Lamb, D.W., P. Frazier, and P. Adams. 2008. Improving pathways to adoption: Putting the P's in precision agriculture. *Comput. Electron. Agric.* 61: 4–9.

Lamm, F.R. and T.P. Trooien. 2003. Subsurface drip irrigation for corn production: A review of 10 years of research in Kansas. *Irrig. Sci.* 22: 195–200.

Liang, Q., D. Yuan, Y. Wang, and H.H. Chen. 2007. A cross-layer transmission scheduling scheme for wireless sensor networks. *Comput. Commun.* 30: 2987–2994.

Lopez-Lozano, R., M.A. Casterad, and J. Herrero. 2010. Site-specific management units in a commercial maize plot delineated using very high resolution remote sensing and soil properties mapping. *Comput. Electron. Agric.* 73: 219–229.

Lyle, W.M. and J.P. Bordovsky. 1983. LEPA irrigation system evaluation. *Trans. ASAE* 24: 1241–1245.

Mahan, J.R., W. Conaty, J. Neilson, P. Payton, and S.B. Cox. 2010. Field performance in agricultural settings of a wireless temperature monitoring system based on a low-cost infrared sensor. *Comput. Electron. Agric.* 71: 176–181.

McBratney, A., B. Whelan, and T. Ancev. 2005. Future directions of precision agriculture. *Precis. Agric.* 6: 7–23.

McCann, I.R., B.A. King, and J.C. Stark. 1997. Variable rate water and chemical application for continuous-move sprinkler irrigation systems. *Appl. Eng. Agric.* 13: 609–615.

McCarthy, A.C., N.H. Hancock, and S.R. Raine. 2010. VARIwise: A general-purpose adaptive control simulation framework for spatially and temporally varied irrigation at sub-field scale. *Comput. Electron. Agric.* 70: 117–128.

Miranda, F.R., R.E. Yoder, J.B. Wilkerson, and L.O. Odhiambo. 2005. An autonomous controller for site-specific management of fixed irrigation systems. *Comput. Electron. Agric.* 48: 183–197.

Mirik, M., G.J. Michels Jr, S. Kassymzhanova-Mirik, and N.C. Elliott. 2007. Reflectance characteristics of Russian wheat aphid (Hemiptera: Aphididae) stress and abundance in winter wheat. *Comput. Electron. Agric.* 57: 123–134.

Molden, D., T. Oweis, P. Steduto, P. Bindraban, M.A. Hanjra, and J. Kijne. 2010. Improving agricultural water productivity: Between optimism and caution. *Agric. Water Manag.* 97: 528–535.

National Agricultural Statistic Service (NASS), USDA, Census of Agriculture. 2009. 2008 Farm and ranch irrigation survey. http://www.agcensus.usda.gov/Publications/2007/Online_Highlights/Farm_and_Ranch_Irrigation_Survey/index.asp (accessed January 12, 2011).

National Research Council. 1996. *A New Era for Irrigation.* National Academy Press, Washington DC.

Neale, C.M.U., W.C. Bausch, and D.F. Heermann. 1989. Development of reflectance-based crop coefficients for corn. *Trans. ASAE* 32: 1891–1899.

Newman, S.C. and J.W. Hummel. 1999. Soil penetration resistance with moisture correction. Paper presented at the annual meeting of the American Society of Agricultural Engineers, Toronto.

Nijbroek, R., G. Hoogenboom, and J.W. Jones. 2003. Optimizing irrigation management for a spatially variable soybean field. *Agric. Syst.* 76: 359–377.

Nyiraneza, J., A. N'Dayegamiye, M. Chantigny, and M. Laverdiere. 2009. Variations in corn yield and nitrogen uptake in relation to soil attributes and nitrogen availability indices. *Soil Sci. Soc. Am. J.* 73: 317–327.

Omary, M., C.R. Camp, and E.J. Sadler. 1997. Center pivot irrigation system modification to provide variable water application depths. *Appl. Eng. Agric.* 13: 235–239.

O'Shaughnessy, S.A. and S.R. Evett. 2010a. Canopy temperature based system effectively schedules and controls center pivot irrigation of cotton. *Agric. Water Manag.* 97: 1310–1316.

O'Shaughnessy, S.A. and S.R. Evett. 2010b. Developing wireless sensor networks for monitoring crop canopy temperature using a moving sprinkler system as a platform. *Appl. Eng. Agric.* 26: 331–341.

Palacin, J., J.A. Salse, X. Slua, J. Arno, R. Blanco, and C. Zanuy. 2005. Center-pivot automation for agrochemical use. *Comput. Electron. Agric.* 49: 491–430.

Payero, J.O., D.D. Tarkalson, S. Irmak, D. Division, and J.L. Petersen. 2009. Effect of timing of a deficit-irrigation allocation on corn evapotranspiration, yield, water use efficiency, and dry mass. *Agric. Water Manag.* 96: 1387–1397.

Pereira, L.S., T. Owens, and A. Zairi. 2002. Irrigation management under water scarcity. *Agric. Water Manag.* 57: 175–206.

Pereira, L.S., P. Paredes, E.D. Cholpankuliv, O.P. Inchenkova, P.R. Teodora, and M.G. Horst. 2009. Irrigation scheduling strategies for cotton to cope with water scarcity in the Fergana Valley, Central Asia. *Agric. Water Manag.* 96: 723–735.

Perry, C., S. Pocknee, and O. Hansen. 2003. A variable rate pivot irrigation control system. Paper presented at the 4th annual European Conference on Precision Agriculture, Berlin.

Perry, C.D., M.D. Dukes, and K.A. Harrison. 2004. Effects of variable-rate sprinkler cycling on irrigation uniformity. Paper presented at the annual meeting of the American Society of Agricultural and Biological Engineers, Ottawa.

Peters, R.T. and S.R. Evett. 2004. Modeling diurnal canopy temperature dynamics using one-time-of-day measurements and a reference temperature curve. *Agron. J.* 96: 1553–1561.

Peters, R.T. and S.R. Evett. 2005. Using low-cost GPS receivers for determining field position of mechanized irrigation systems. *Appl. Eng. Agric.* 21: 841–845.

Peters, R.T. and S.R. Evett. 2007. Spatial and temporal analysis of crop stress using multiple canopy temperature maps created with an array of center-pivot-mounted infrared thermometers. *Trans. ASABE* 50: 919–927.

Peters, T.R. and S.R. Evett. 2008. Automation of a center pivot using the temperature-time threshold method of irrigation scheduling. *J. Irrig. Drain. Eng.* 134: 286–291.

Pierce, F.J. and P. Nowak.1999. Aspects of precision farming. *Adv. Agron.* 67: 1–85.

Pierce, L., R. Nemani, A. Michaelis, and L. Johnson. 2010. GIS and modeling tools—Enhancing water use efficiency. *Water Efficiency* May/June: 42–45.

Pierce, F.J. and T.V. Elliott. 2008. Regional and on-farm wireless sensor networks for agricultural systems in eastern Washington. *Comput. Electron. Agric.* 61: 32–43.

Pierce, F.J., J.L. Chávez, T.V. Elliott, G.R. Matthews, R.G. Evans, and Y. Kim. 2006. A remote real-time continuous move irrigation control and monitoring system. Paper presented at the annual meeting of the American Society of Agricultural and Biological Engineers, Portland. OR.

Ping, J., C. Green, R. Zartman, K. Bronson, and T. Morris. 2008. Spatial variability of soil properties, cotton yield, and quality in a production field. *Commun. Soil Sci. Plant Anal.* 39: 1–16.

Playán, E. and L. Mateos. 2006. Modernization and optimization of irrigation systems to increase water productivity. *Agric. Water Manag.* 80: 100–116.

Price, J.A., F. Workneh, S.R. Evett, D.C. Jones, and C.M. Rush. 2010. Effects of wheat streak mosaic virus on root development and water-use efficiency of hard red winter wheat. *Plant Dis.* 94: 766–770.

Ramos, M.C. 2006. Soil water content and yield variability in vineyards of Mediterranean northeastern Spain affected by mechanization and climate variability. *Hydrol. Process.* 20: 2271–2283.

Rosenthal, W.D., R.L. Vanderlip, B.S. Jackson, and G.F. Arkin. 1989. SORKAM: A grain sorghum crop growth model. Computer Software Documentation Ser. MP 1669. Texas Agric. Exp. Station, College Station, TX.

Sadler, E.J., C.R. Camp, D.E. Evans, and L.J. Usrey. 1996. A site-specific center pivot irrigation system for highly-variable coastal plain soils. Paper presented at the Third International Conference on Precision Agriculture, Minneapolis, MN.

Sadler, E.J., W. Busscher, P. Bauer, and D. Karlen. 1998. Spatial scale requirements for precision farming: A case study in the southeastern USA. *Agron. J.* 90: 191–197.

Sadler, E.J., C.R. Camp, D.E. Evans, and J.A. Millen. 2002. Spatial variation of corn response to irrigation. *Trans. ASAE* 45: 1869–1881.

Sadler, E.J., R.G. Evans, G. Buckleiter, B.A. King, and C.R. Camp. 2000. Design considerations for site specific irrigation. Paper presented at the 4th Decennial National Irrigation Symposium, Phoenix, AZ.

Sadler, E.J., R.G. Evans, K.C. Stone, and C.R. Camp. 2005. Opportunities for conservation with precision irrigation. *J. Soil Water Conserv.* 60: 371–379.

Sadler, E.J., C.R. Camp, and R.G. Evans. 2007. New and future technology. In *Irrigation of Agricultural Crops: Second Edition*, eds. R.J. Lascano and R.E. Sojka, pp. 609–627. ASA, CSSA, SSSA, Madison, VA.

Schneider, A.D. and T.A. Howell. 2000. Surface runoff due to LEPA and spray irrigation of a slowly permeable soil. *Trans. ASAE* 43: 1089–1095.

Schneider, A.D., T.A. Howell, and S.R. Evett. 2001. Comparison of SDI, LEPA, and spray irrigation efficiency. Paper presented at the American Society of Agricultural Engineers, Sacramento, CA.

Smith, R.J., S.R. Raine, A.C. McCarthy, and N.H. Hancock. 2009. Managing spatial and temporal variability in irrigated agriculture though adaptive control. *Aust. J. Multi-Disciplinary Eng.* 7: 70–90.

Snapp, S.S., L.E. Gentry, and R. Harwood. 2010. Management intensity-not biodiversity—The driver of ecosystem services in a long-term row crop experiment. *Agric. EcoSyst. Environ.* 138: 242–248.

Steddom, K., M.W. Bredehoeft, M. Kahn, and C.M. Rush. 2005. Comparison of visual and multispectral radiometric disease evaluations of Cercospora leaf spot of sugar beet. *Plant Dis.* 89: 153–158.

Steddom, K., G. Heidel, D. Jones, and C.M. Rush. 2003. Remote detection of Rhizomia in sugar beets. *Epidemiology* 93: 720–726.

Stöckle, C.O., M. Donatelli, and R. Nelson. 2003. CropSyst, a cropping systems simulation model. *Eur. J. Agron.* 18: 289–307.

Sudduth, K.A. and J.W. Hummel. 1993. Soil organic matter, CEC and moisture sensing with a portable near-infrared spectrophotometer. *Trans. ASAE* 36: 1571–1582.

Sudduth, K.A., N.R. Kitchen, D.F. Hughes, and S.T. Drummond. 1995. Electromagnetic induction sensing as an indicator of productivity on claypan soil. Paper presented at the 2nd International Conference on Site-Specific Management for Agricultural Systems, Minneapolis, MN.

Sudduth, K.A., J.W. Hummel, N.R. Kitchen, and S.T. Drummond. 2000. Evaluation of a soil conductivity sensing penetrometer. Paper presented at the annual meeting of the American Society of Agricultural Engineers, Milwaukee, WI.

Sumner, H.R., P.M. Garvey, D.F. Heermann, and L.D. Chandler. 1997. Center pivot irrigation attached sprayer. *Appl. Eng. Agric.* 13: 323–327.

Tellaeche, A., X.P. BurgosArtizzu, G. Pajares, A. Ribeiro, and C. Fernandez-Quintanilla. 2008. A new vision-based approach to differential spraying in precision agriculture. *Comput. Electron. Agric.* 60: 144–155.

Thomas, A. 2008. Agricultural irrigation demand under present and future climate scenarios in China. *Global Planet. Change* 60: 306–326.

Thompson, T.L., P. Huan-Cheng, and L. Yu-Yi. 2009. The potential contribution of subsurface drip irrigation to water-saving agriculture in the western USA. *Agric. Sci. China* 8: 850–854.

Tilling, A.K., G.J. O'Leary, J.G. Ferwerda, S.D. Jones, G.J. Fitzgerald, D. Rodriguez, and R. Belford. 2007. Remote sensing of nitrogen and water stress in wheat. *Field Crop. Res.* 104: 77–85.

Tolk, J.A. and T.A. Howell. 2008. Field water supply: Yield relationships of grain sorghum grown in three USA southern Great Plains soils. *Agric. Water Manag.* 95: 1303–1313.

Turral, H., M. Svendsen, and J.M. Faures. 2009. Investing in irrigation: Reviewing the past and looking to the future. *Agric. Water Manag.* 97: 551–560.

UNESCO. 2009. Water in a changing world. Report No. 3. World Water Assessment Programme. http://www.unesco.org/water/wwap/wwdr/wwdr3/ (accessed September 30, 2010).

Upadhyaya, S.K., W.J. Chancellor, J.V. Perumpral, R.L. Schafer, W.R. Gill, and G.E. Vandenberg. 1994. *Advances in Soil Dynamics*. ASAE, St. Joseph, MI.

Upchurch, D.R., D.F. Wanjura, J.R. Mahan, J.J. Burke, and J.A. Chapman. 1998. Irrigation system having sensor arrays for field mapping. U.S. Patent Application SN 09/153,652.

Vellidis, G., M. Tucker, C. Perry, C. Kvien, and C. Bednarz. 2008. A real-time wireless smart sensor array for scheduling irrigation. *Comput. Electron. Agric.* 61: 44–50.

Vrindts, E., M. Reyniers, P. Darius, M. Frankinet, B. Hanquet, M. Destain, and J. Baerdemaeker. 2003. Analysis of spatial soil, crop and yield data in a winter wheat field. Paper presented at the annual American Society of Agricultural Engineers, Las Vegas, NV.

Watkins, K.B., Y.C. Lu, and W.Y. Huang. 1998. Economic and environmental feasibility of variable rate nitrogen fertilizer application with carry-over effects. *J. Agric. Resour. Econ. West. Agric. Econ. Assoc.* 23: 401–426.

Watkins, K.B., Y.C. Lu, and W.Y. Huang. 2002. A case study of economic and environmental impact of variable rate nitrogen and water application. *Agric. Eng. J.* 11: 173–185.

West, J.S., C. Bravo, R. Oberti, D. Lemaire, D. Moshou, and H.A. McCartney. 2003. The potential of optical canopy measurement for targeted control of field crop diseases. *Annu. Rev. Phytopathol.* 41: 593–614.

Wichelns, D. and J.D. Oster. 2006. Sustainable irrigation is necessary and achievable, but direct costs and environmental impacts can be substantial. *Agric. Water Manag.* 86: 114–127.

Workneh, F., D.C. Jones, and C.M. Jones. 2009. Quantifying wheat yield across the field as a function of wheat streak mosaic intensity: A state space approach. *Ecol. Epidemiol.* 99: 432–440.

Yang, Z., M.N. Rao, N.C. Elliott, S.D. Kindler, and T.W. Popham. 2009. Differentiating stress induced by greenbugs and Russian wheat aphids in wheat using remote sensing. *Comput. Electron. Agric.* 67: 64–70.

Zhang, H. 2003. Improving water productivity through deficit irrigation: Examples from Syria, the North China Plain and Oregon, USA. In *Water Productivity in Agriculture: Limits and Opportunities for Improvement*, eds. J.W. Kijne, R. Barker, and D. Molden, pp. 301–309. International Water Management Institute, Colombo.

Zhang, N., M. Wang, and N. Wang. 2002. Precision agriculture—A worldwide overview. *Comput. Electron. Agric.* 36: 113–132.

10 Sustainable Soil Water Management Systems

G. Basch, A. Kassam, T. Friedrich, F.L. Santos, P.I. Gubiani, A. Calegari, J.M. Reichert, and D.R. dos Santos

CONTENTS

10.1 INTRODUCTION

During the twentieth century, global water withdrawal increased about sevenfold, from 579 to 3917/km^3 year. In the same period, the share of total water use by agriculture declined from 91% to 66% and it is supposed to decrease to around 61% by 2025 (Ghassemi and White 2007). Over the last decades, there was a considerable decline in the ratio of water consumption (nonrecoverable withdrawn water, i.e., lost by evapotranspiration [ET]) to water withdrawal, from 66% in 1940 to 60% in 2000 (Shiklomanov and Rodda 2003), meaning that water was used more efficiently, especially in the agricultural sector.

Whereas irrigation water use represents almost 70% of total human "blue" water use (water withdrawn from water bodies such as river, lakes, and aquifers) (Rockstrom et al. 2009a), global agricultural blue water consumption, that is, the amount of water that transpires productively through crops or evaporates unproductively from soils, water bodies, and vegetation canopies, is estimated to be even higher (Rost et al. 2008). In addition, nonquantifiable amounts of fossil, nonrenewable groundwater resources, or nonlocal water resources from distant regions are used for irrigation (Vörösmarty et al. 2005). Table 10.1 summarizes the global freshwater pools based on the data compiled by Shiklomanov and Rodda (2003). Although 30% of the global freshwater resources are stored as groundwater, and less than 0.3% in rivers and lakes, the latter are the main sources of fresh water for human use as they represent the dynamic component of the earth's total water resources (Shiklomanov 1993). Whereas global groundwater withdrawals amount to around 20%–25% of the total water withdrawals (Shiklomanov and Rodda 2003; Rosegrant et al. 2002), irrigation in many countries relies heavily on groundwater, with 53% and 46% of irrigation water being pumped from aquifers in countries such as India and the United States, respectively (Shah et al. 2007). Locally, groundwater resources are already overexploited, with the rate of water withdrawal being faster than recharge, causing water tables to fall (Gleick et al. 2002). Globally, about 15%–35% of irrigation withdrawals are estimated to be unsustainable; many of these withdrawals are from groundwater sources (Rosegrant et al. 2009).

TABLE 10.1
Global Distribution of Fresh Water

Global Pools of Freshwater	(10^{12} m^3)	(%)
Fresh groundwater	10530	30.06
Glaciers, permanent snow and permafrost	24364	69.55
Lakes (fresh water)	91	0.26
Marshes and swamps	11.5	0.03
River water	2.12	0.01
Biological water	1.12	0.00
Water in the atmosphere	12.9	0.04
Soil moisture	16.5	0.05
Total	35029.14	100

In addition to the expected increase in agricultural water use to meet the demand for more food, the ongoing urbanization of the world's population will require an increased share of the fresh water available. Based on the estimation of a world population of 9.2 billion people by 2050 and a dietary change driven by higher incomes and urbanization, world food demand is supposed to increase by 70% (or by 100% in the developing world) within less than 50 years (Bruinsma 2009; Thompson 2007). There are only two ways to cope with this increased projected demand, not only for food and feed, but also for fiber and biofuel: by a substantial increase in cropland and/or a large improvement in productivity per unit of land cropped. Based on an optimistic scenario of water availability and management in which a theoretical maximum of 85% of the total ET from cropland and pasture was assumed to be available for plant transpiration and thus biomass production, Rockstrom et al. (2009a) suggested that without improvements in water productivity (WP), a horizontal cropland expansion by about 1000 Mha would be required to produce the food for >10 billion people, which is two-thirds of today's cropland (Ramankutty et al. 2008). In fact, regarding this option, Thompson (2007) considers the possibility of the area of land in farm production being doubled, but only with the massive destruction of forests and the loss of wildlife habitat, biodiversity, and carbon sequestration capacity.

Several authors consider a potential increase in cropland of around 9%–12% as realistic and feasible (Bruinsma 2009; Molden 2007; Thompson 2007), not taking into account that around 10 Mha may be lost every year due to soil erosion, other forms of degradation, and conversion to nonfarm uses (Leach 1995; Pimentel et al. 1995), indicating that cropland expansion could be close to zero (Postel 1998) and that the increasing demand has to be satisfied by higher crop productivities. However, with the exception of sub-Saharan Africa, most of the productivity enhancement potential of the "Green Revolution" technologies has already been exploited (Molden 2007). This means that the gains in agricultural productivity must be achieved through advances in other areas, such as the improvement in soil fertility and water-use efficiency (WUE), and the use of biotechnological innovations to further enhance the efficiency of the already applied inputs and to improve crop performance under stress conditions.

Although water is considered a renewable resource because it depends on rainfall, its availability is finite in terms of the amount available per unit time in any one region (Pimentel et al. 2004). Considering an optimistic scenario, Molden (2007) estimates the agricultural water withdrawal to increase by only 13% by the year 2050. Therefore, and bearing in mind the limited blue water resources and the rapidly increasing competition of other sectors for this resource, the necessary gains in agricultural productivity will depend strongly on improvements in the use of "green" water (precipitation or water that is stored in the root zone) (Rockstrom 2009a). Rost et al. (2008) consider it a misconception to regard agricultural water consumption as dependent primarily on blue water withdrawals. In fact, around 80% of global cropland is rainfed, and 60%–70% of the world's food is produced on rainfed land, that is, by the consumption of precipitation water infiltrated into the soil (Falkenmark and Rockstrom 2004). Further, green water also plays an important role on irrigated land in situations where blue water is only used to supplement the crop water requirement for optimal growth and production. Based on the application of different models, Rost et al. (2008) estimated that a share of over 85% of green water was being consumed on the global cropland. This underpins the statement made by Hoff et al. (2010) that an integrated water resources management (IWRM) relying on blue water only can no longer provide complete sustainable solutions. These authors also refer to the increasing interest in the potential of the "invisible" green water resource for additional crop production.

Managing precipitation as a key resource for production intensification, through integrating green and blue water, has been postulated as the basis of a new paradigm to help close the water gap (Falkenmark and Rockstrom 2004). From a hydrological perspective, in many regions, including those in the semiarid areas, there is enough rainfall to increase crop yields considerably without recourse to the development of large-scale irrigation schemes (Rockstrom et al. 2007). Rockstrom (2000) has recorded estimates of rainfall losses due to unproductive soil evaporation, surface runoff, deep percolation and interception losses of up to 70%–85%. Irrigation plays and will continue to play an important role in feeding the growing world population, and there still exist a need and opportunities for an expansion of the irrigated cropland. However, to avoid at least half of the additional water requirements in agriculture, Rockstrom et al. (2007) have suggested two possibilities: (i) reduction of green water losses, that is, WP improvements in irrigated and rainfed agriculture, and (ii) better use of the local rainfall water.

In this context, soil quality and its management must be considered as key elements for the effective management of water resources, given that the hydrological cycle and land management are intimately linked (Bossio et al. 2007). Bossio et al. (2010) have described soil degradation as the starting point of a negative cycle of soil–water relationships, creating a positive, self-accelerating feedback loop with important negative impacts on water cycling and WP. Therefore, sustainable soil management corresponds to sustainable water management through the improvement of soil water management.

The purpose of this chapter is to review the existing options of soil water management systems and their potential contribution to the improvement of the available soil water in the root zone, WUE, and WP. Ultimately, these water-related aspects seem to be the only solution to producing enough food, feed, fiber, and biofuels for 1.5 times today's world population, without competing excessively with the existing natural ecosystems and their services and the water resources allocated to them and to other human activities.

10.2 PROCESSES AFFECTING SOIL WATER DYNAMICS

Although the earth's land surface (29.2% of total surface) contributes only 14% to total evaporation, it receives around 20% of the precipitation falling on earth (Pimentel et al. 2004). About 115×10^{12} m^3 of precipitation corresponds to about 780 mm of the land surface, on an average (Table 10.2). The transfer of this significant portion of water from oceans to land surface is of vital importance not only to agriculture but also to human life and natural ecosystems. Equally important is the fact that soils are able to store around 20% of the water annually transferred from the land surface to the atmosphere by evapotranspiration.

Globally, it is estimated that out of the total precipitation over the continents, one-third becomes blue water, that is, runoff into rivers and aquifer recharge, and two-thirds infiltrate the soil, forming the so-called "green water" that supports, productively or unproductively, biomass production and returns to the atmosphere as vapor (Hoff 2010). Numerous hydrological models can be found in the literature that attempt to describe water partitioning at different scales, depending

TABLE 10.2
Global Fluxes of Fresh Water (Annual)

Water Flux Component	Units	Amount
Total earth's evapotranspiration	10^{12} m^3	577
Evapotranspiration from oceans (86%)	10^{12} m^3	496.2
Evapotranspiration from land (14%)	10^{12} m^3	80.8
Average rainfall on land surface (20% of global ETP)	mm	780
River runoff (into oceans)	10^{12} m^3	42.8

Source: Shiklomanov, I.A. and Rodda, J.C., *World Water Resources at the Beginning of the Twenty-First Century,* Cambridge University Press, New York, 2003; Shiklomanov, I.A., World freshwater resources. In *Water in Crisis: A Guide to the World's Freshwater Resources*, edited by P. Gleick, pp. 13–24, Oxford University Press, New York, 1993.

on the pretended objectives. Lal (2008b), for example, described the processes of water loss from agricultural watersheds using the equation:

$$P = R + I + D + \Delta\theta + \int Edt + \int Tdt, \quad (10.1)$$

where P is the precipitation, I is the infiltration, R is the runoff, D is the deep drainage, $\Delta\theta$ is the change in soil water storage, T is the transpiration, E is the soil evaporation, and t is the time.

Lal (2008b) considered the sum of the runoff and deep drainage as blue water and the consumptive transpiration water as green water. Ngigi et al. (2006) used the same partitioning components to characterize the available crop water over a season and to describe the aspects of in situ rainwater harvesting and management systems. Still other components are used in water balance equations. For example, to model soil water storage, Makurira et al. (2010) used Equation 10.2, based on a water balance model from Savenije (1997).

$$\frac{dS_s}{dt} + \frac{dS_u}{dt} + \frac{dS_g}{dt} = P - E_T - E_I - E_s - Q_g - Q_s, \quad (10.2)$$

where (all terms in mm/day) P is the precipitation received in the system, E_T is the transpiration, E_I is the evaporation from interception, that is, from the canopy cover and the soil surface, E_s is the evaporation from the soil, Q_s is the net surface runoff, Q_g is the groundwater run off, dS_s/dt is the rate of change of surface water storage, dS_u/dt is the rate of change of water storage in the root zone, and dS_g/dt is the rate of change of groundwater storage.

Compared with Equation 10.1, Equation 10.2, in addition to introducing the component groundwater runoff, considers the evaporation of canopy-intercepted rainwater and the wet soil surface (E_I). According to Savenije (2004), depending on the local conditions, the latter can amount to 40%–50% of the total precipitation. Therefore, this author advocates the abandonment of the term "evapotranspiration" as it only reflects the incapacity to separate the different evaporative processes, that is, evaporation from interception, transpiration, soil evaporation, wet surface, and open-water evaporation, such as water retained at the soil surface or flooded rice fields. Although classified as unproductive flux, evaporation through interception is not regarded as a loss to the water system because it is responsible for the moisture recycling that sustains continental rainfall. Hence, this component cannot be included in the green or blue water fraction.

On cropland, the partitioning of the precipitation into the different components of the water balance (runoff, soil and plant [interception] evaporation, transpiration, deep percolation) may vary tremendously, depending on the agroecological characteristics of the site, mainly the soil type and management, slope, plant cover, and rainfall characteristics. On two different soils in Central India, Laryea et al. (1991) determined over a period of 4 years a blue water share (runoff and percolation) of total rainfall of 37% and 59%. Whereas runoff was similar at both sites (28% and 26%), deep percolation differed considerably, reaching 9% and 33%, respectively. Modeling water partitioning at field scale based on onsite observations at two sites in northern Tanzania, Makurira et al. (2010) also observed big differences in the water partitioning when comparing the traditional and the innovative farming practices, using the runoff diversion to crop plots. They found that the slope of the field and the soil depth contributed decisively to the partitioning of both the rainwater and the diverted runoff and that interception accounted for one-fourth and one-third of the total precipitation for the two sites, respectively. Figure 10.1 gives an overview of the components of the water balance over a landscape or watershed.

Considering the limited land and water resources suitable to produce enough commodities for the growing world population, while sustaining other ecosystem services provided by agriculture, one of the main strategies by which agricultural water management can deal with the large tradeoffs between water uses is improving water management practices on agricultural lands (Gordon

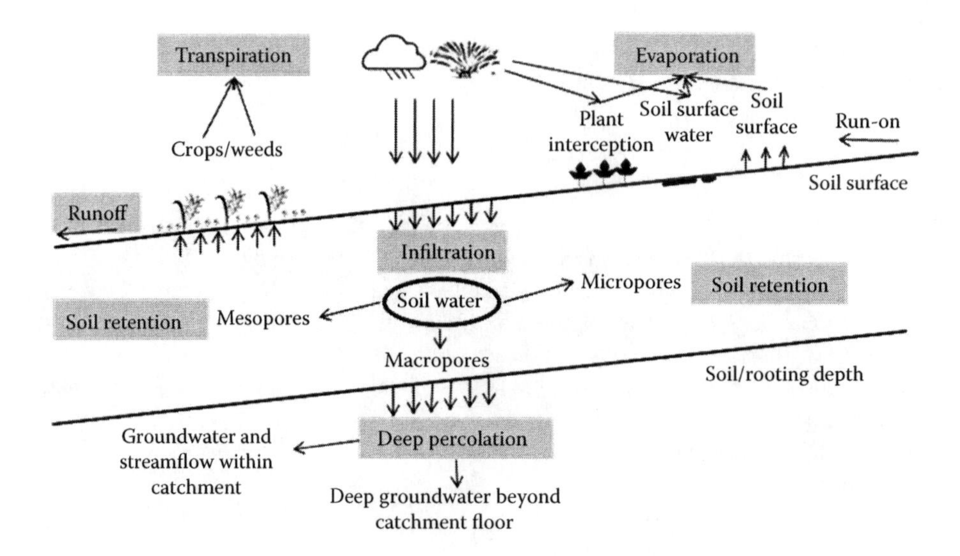

FIGURE 10.1 Water fluxes over a landscape or watershed and the destinations of rain and irrigation water.

et al. 2010). This requires a careful look at the water fluxes described in Figure 10.1 and examining the processes affecting the soil water, which is ultimately the water source for plant growth and biomass production.

Although WP and WUE have been questioned as useful concepts in agricultural water management (Blum 2009; Zoebl 2006), they are the terms commonly used to evaluate the efficiency with which rain and/or irrigation water is transformed into grain yield or biomass production. Without always having the same concept in mind (Ali and Talukder 2008), much has been written in recent years about the ways to improve agricultural WP or WUE (Bossio et al. 2010; Fang et al. 2010b; Liu et al. 2010; Molden et al. 2010; Shaheen et al. 2010; Alvaro-Fuentes et al. 2009; Kang et al. 2009, Katerji and Mastrorilli 2009; Rockstrom et al. 2009b; Evans and Sadler 2008; Khan et al. 2008; Ritchie and Basso 2008; Bluemling et al. 2007; Bouman 2007; Molden et al. 2007; Rockstrom and Barron 2007; Steduto et al. 2007; Adekalu and Okunade 2006; Singh et al. 2006; Zhang et al. 2005).

Regardless of the discussion about terms and definitions, the fundamental question remains: How to produce more with the same or even less amount of water available from rainfall and irrigation? This question is undoubtedly linked to the possibilities of minimizing, at least at field level, unproductive water losses, namely, runoff, evaporation, and deep percolation. Whereas from a water cycle perspective, all these components are also important to replenish the blue water resources, from an agronomist viewpoint, a reduction of these losses must occur, not only to achieve the goal set above of a higher WP, WUE, or, as suggested by Blum (2009), efficient water use (EWU), but also because runoff losses, if uncontrolled, can have other severe and harmful consequences. Albeit the shift from unproductive evaporation losses to an increase in crop transpiration is not expected to alter significantly the return of water vapor to the atmosphere, a considerable blue-to-green water redirection, through either reduced runoff and deep percolation or water withdrawal for irrigation, could involve a corresponding depletion of the stream flow (Falkenmark 2007). These trade-offs should be kept in mind when searching for improved agricultural water resource management systems. They also point, in the first place, at soil and soil water management practices to achieve a sustainable combination of maximizing transpiration and soil water storage and minimizing runoff and evaporation. Technically feasible and cost-effective solutions for this achievement are a high priority (Rockstrom et al. 2002) and form a basis for sustainable soil water management systems. These have to consider the different processes that affect the soil water, as indicated in Figure 10.1, and the parameters that influence these processes (Figure 10.2).

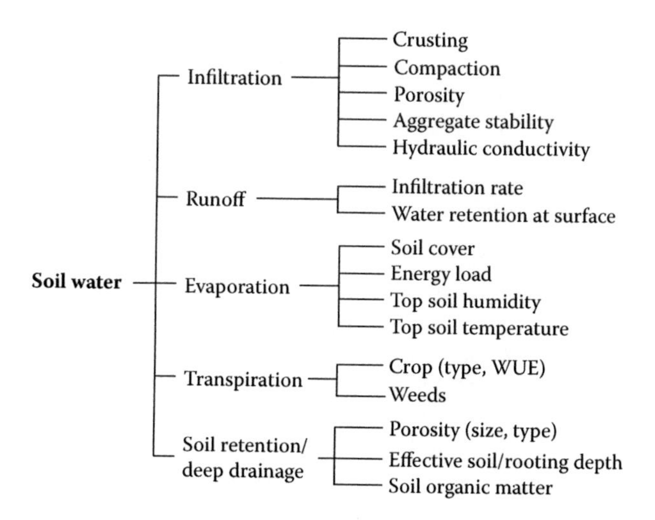

FIGURE 10.2 The processes and parameters affecting soil water.

10.2.1 Infiltration

A first step in converting blue water to green water is to maximize water infiltration into the soil. This process and its variation with time are mainly governed by the initial wetness and suction of the soil, as well as its texture, structure, and the uniformity of its profile. While the soil texture and the initial moisture content can hardly be changed, an enhancement of the amount and the stability of the structural soil aggregates, total porosity and macroporosity, and thus hydraulic conductivity, and a decrease in the surface crusting and compaction are achievable through management practices.

Numerous attempts have been made to describe thoroughly the process of infiltration and many models and equations are used to express infiltrability as a function of time or of the total quantity of water infiltrated into the soil (Hillel 1980). However, beyond the understanding of the process itself, which has been subject to many studies, it is of crucial importance to identify the management practices that are able to enhance the infiltration flux, that is, to act on the factors that can be influenced in a technically feasible and cost-effective way. For that purpose, it is useful to distinguish whether infiltration is supply-, surface-, or profile-controlled. Whereas supply control (amount and intensity) is possible only under irrigated conditions, infiltrability of the soil surface layer (surface controlled) and the subsurface hydraulic conductivity (profile controlled) are manageable conditions, even under a rainfed situation.

Increasing the share of rainfall or irrigation water that infiltrates the soil can be achieved through either an improved infiltrability or an extension of the time period during which water is capable of infiltrating into the soil (surface retention or ponding). Whereas the former is strongly enhanced in the presence of continuous macroporosity (mainly biopores created by macrofauna activity and former root channels), the latter depends on the soil surface roughness and the overall slope of the land, which determine the so-called surface storage capacity. An increase in the surface roughness is often unintentionally attained through any form of soil tillage, or it is intentionally achieved through tillage operations along contour lines or by creating “pockets” or “basins” over the soil surface, especially with row crops.

At the soil surface, the processes defined as particle detachment, sealing, and crusting are strongly influenced by the vulnerability of the soil aggregates to breakdown, which is caused by the kinetic impact of raindrops or the surface irrigation water flow or the mechanical impact through tillage implements or wheeling (Li et al. 2009). Soils prone to crusting and sealing usually show reduced infiltration rates (Ben-Hur and Lado 2008; Lal 2008b; Ramos et al. 2003).

This suggests that either the enhancement of the aggregate stability or the protection of the soil through any form of surface cover could help diminish the extent and impact of these processes on infiltration. Thus, measures that increase the soil organic matter (SOM) content or the application of soil amendments that enhance the aggregate stability contribute to improving the infiltrability of agricultural soils (Lado et al. 2004). In addition to the existing possibilities of reducing the kinetic energy of water applied through overhead irrigation, it is mainly the soil cover, that dissipates the energy load of raindrops before they reach the soil surface. Although a large majority of studies on the effect of mulching report improvements in water infiltration (Hula et al. 2010; Jordan et al. 2010; Roth 1985), some did not find a response of water infiltration to mulching, because of the variability in the soil and subsurface conditions (Blanco-Canqui and Lal 2007a; Singh and Malhi 2006).

In many situations, the subsurface properties, such as the underlying, less conductive, fine-textured layers, may restrict infiltration, meaning that the hydraulic conductivity of the subsurface soil is lower than that of the soil at the surface, which occurs frequently in layered soil profiles or through compacted layers. Subsoiling to break the restrictive soil layers as an emergency repair measure and promoting the development of vertically oriented macropores through earthworm and other macrofauna activities and/or the maintenance of former root channels, both achieved through the absence of soil disturbance as a long-term strategy, are the recommended measures to improve profile-controlled infiltration rates (Tebrugge and During 1999).

10.2.2 Runoff

Surface runoff occurs when the water supply to the soil surface exceeds the rate of infiltration and after the surface storage capacity has reached its upper limit. Two basic types of runoff are normally distinguished: (1) laminar or "sheet" overland flow covering the whole soil surface of areas with little topographic relief and (2) stream flow in channels, also called rills or gullies, which receive the overland flow, generally forming a tree-like pattern down the slope.

Although surface runoff contributes to the recharge of the surface water bodies (blue water), capable of being potentially reused for irrigation, uncontrolled runoff from the land is not desirable as it is strongly associated with soil erosion and the transport of potential water pollutants. Especially in regions where the amount of rainfall is at the margin of being insufficient for crop production, it is of particular importance that as much of the rainfall as possible infiltrates the soil and is held in the root zone.

As mentioned in the previous section, the reduction of the runoff depends essentially on the infiltrability of the soil and its capacity to retain excess water on the surface, thereby reducing the velocity of the runoff and transforming the potential runoff into run-on by increasing the time available for infiltration. Several surface shaping and land configuration techniques, such as tied ridges, contour ridges, flat bed or pit planting (Chiroma et al. 2008; Jadhav et al. 2008), vegetation buffers or hedges (Cullum et al. 2007; Blanco-Canqui et al. 2004), or different types of vegetation cover (Mohammad and Adam 2010) have been found to considerably reduce the surface runoff. On the contrary, Gomez and Nearing (2005), based on rainfall simulation trials, attribute only a temporary effect of the initial surface roughness to the reduction of runoff. Other researchers also report a high variability in the effects of the soil surface characteristics and the conditions on the generation of runoff (Armand et al. 2009; Seeger 2007).

Whereas runoff is generally considered as prejudicial from the viewpoint of soil and water conservation, as it always carries the risk of soil erosion, gully formation, and off-site transport of nutrients and chemicals, in some situations it may be desirable to withdraw excess rainwater from arable land to avoid prolonged waterlogging of the crops (Basch 1988) or to supply adjacent cropping areas with additional water, also designated as runoff diversion (Makurira et al. 2009). The deliberate inducement of runoff through manifold means has been practiced since antiquity in order to increase water availability in designated areas (Hillel 1980).

10.2.3 Soil Water Retention and Deep Drainage

After water infiltrates the soil, its movement in or through the soil is mainly influenced by gravitational forces and suction gradients. Depending on whether these processes start from initially saturated conditions or an unsaturated state, they are called either internal drainages or redistributions (Hillel 1980), and the rate at which they occur are normally described by the saturated and unsaturated hydraulic conductivities of the respective soil. Internal drainage or even unsaturated water flow beyond the effective rooting depth is referred to as deep drainage or deep percolation. The amount of water that is retained in a soil against gravity and does not leave the root zone determines the soil water storage capacity.

Although deep drainage is an important process to allow excess water to drain and avoid waterlogging, and to recharge the groundwater, from an agronomic point of view and especially in regions with a low and erratic rainfall, but even in relatively humid regions, soil water storage is of vital importance for crops to overcome periods without insufficient precipitation. Effective plant-available soil water depends, in the first place, on the porosity (amount and pore size distribution) of a soil. While pore size distribution is strongly affected by the soil properties, such as texture, structure, and SOM, the total pore volume from which plants may extract water depends also on the soil volume that the roots are able to explore. Thus, efforts to increase the amount of plant-available soil water involve the adoption of soil management practices that increase the rooting depth and/or the volume of the pore size range from which plants are able to withdraw water, often referred to as mesopores.

Physical and chemical subsoil constraints may severely affect the rooting depth and restrict water uptake from deeper soil layers (Dang et al. 2010; MacEwan et al. 2010; Passioura and Angus 2010; Rodriguez et al. 2006). Access to and uptake of water from the subsoil can improve crop yields considerably (Kirkegaard et al. 2007), and improving subsoil conditions was found to be justified by higher returns under Mediterranean conditions (Wong and Asseng 2007). Numerous approaches have been suggested to overcome these constraints, which were categorized by Adcock et al. (2007) into three groups: (1) amelioration strategies including any type of deep soil loosening and the application soil amendments, (2) breeding efforts, and (3) avoidance strategies such as raised beds or crop management practices. However, they may be of only temporary effect and limited to the possible depth of intervention (Lopez-Fando et al. 2007).

Changes in the soil porosity to increase the amount of plant-available water will depend on the improvement of the soil structure and aggregation (Horn and Smucker 2005) and the SOM (Abid and Lal 2009; Hudson 1994). However, the relationship of the soil water retention to the organic carbon content is affected by proportions of textural components and is more pronounced at lower matric suctions (Rawls et al. 2003). In particular, coarse-textured soils and soils with low organic carbon contents benefit in their water-holding capacity from increases in organic matter (Rawls et al. 2003).

As previously mentioned, deep drainage must not be considered as a permanent "loss" of water from the root zone once the soil profile is replenished with water reaching the upper limit of the soil's water-holding capacity. The observation that higher infiltration rates where soil conservation measures are used did not result in a better crop performance in South East Queensland is explained by Bell et al. (2005), with the improved macroporosity that may enhance unsaturated flow through the soil profile and contribute to increased deep drainage "losses."

10.2.4 Transpiration

Transpiration by leaves and crop canopies is inseparable from the processes of carbon dioxide assimilation from the atmosphere and biomass production. Its efficiency is generally referred to as biomass transpiration efficiency (TE) or defined as the ratio of total plant biomass (above and below ground) to growing season transpiration (Suyker and Verma 2009), and it has to be distinguished from the terms WUE and WP. WUE is defined as water consumptively used in ET as a proportion of the water applied by either irrigation or rainfall. WP is defined biologically (or economically) in terms of economic yield or total biomass per unit of water consumptively used in ET (Fereres and Kassam, 2003; Sadras and Angus 2006). It is the job of crop management to ensure that WUE and WP are maintained as high as possible for each crop in the cropping system and for the cropping system as a whole within the prevailing agroecological and socioeconomic constraints. Many strategies have been identified to enhance WUE and WP (Bluemling et al. 2007), mainly involving appropriate tillage and soil management (Chiroma et al. 2008; Adekalu and Okunade 2006; Hatfield et al. 2001), irrigation management (amount, timing) (Fang et al. 2010a; Katerji et al. 2010; Liu et al. 2010; Buttar et al. 2007; Adekalu and Okunade 2006), and weed and crop management, including the choice of cultivars and breeding efforts (Fang et al. 2010b; Passioura 2006; Tennakoon and Hulugalle 2006; Zhang et al. 2005; Gregory et al. 2000). Although many authors continue using the concept of WUE to characterize the efficiency with which water is used to produce biomass or yield and to characterize how much biomass or yield is produced per unit of water used in ET, other authors question its usefulness (Zoebl 2006) or prefer alternative approaches such as the "effective use of water" (EUW) as a target for yield improvement in water-limited environments because high WUE, WP, or TE may be achieved at the expense of reduced EUW (Blum 2009). However, it would appear that all the indices have their values, depending on the purpose for which a particular index is being used.

To improve biomass production especially in water-limited environments, it is crucial to increase the total amount of water transpired and/or the TE (Bouman 2007). Whereas water available for transpiration depends on inflows, outflows, and storage capacity, the increase in efficiency with which transpired water is exchanged for CO_2 will require genetic improvements as the ratio of biomass production to transpiration has been shown to be fairly constant for a given species in a given climate (Ehlers and Goss 2003; Steduto et al. 2007). Bennett (2003) has summarized the breeding efforts and bioengineering opportunities to improve TE. Although genetic variations in carbon isotope discrimination as an indicator for differentiated TE have been found between cultivars (Passioura 2006; Richards 2006), most of the breeding efforts are directed toward the avoidance of dehydration, that is, adaptation of the crop cycle to the given hydrological environment (Blum 2009), and the genetic improvement of the harvest index (HI) (Passioura 2006).

The fact that C_4 crops have a more efficient photosynthetic pathway, and thus a higher TE than C_3 plants, has suggested the introduction of their photosynthetic pathway into C_3 plants through genetic engineering (Sheehy 2000), an approach that has been followed over the recent past (Furbank et al. 2009; Hibberd et al. 2008).

10.2.5 Evaporation

Evaporation is called the vaporization of a liquid that occurs at its surface. In a soil–plant–atmosphere system, evaporation occurs from each of the components. Although transpiration is a specific form of evaporation, it is referred to separately or in combination (ET) as it is the productive consumption of water in crop production. All other origins of evaporation, whether from the soil surface (wet soil or free water stored over it), the canopy, or from sprinkler droplets, are considered losses to the soil–plant–atmosphere system. Burt et al. (2005) carried out a comprehensive review of the different evaporative processes and their contribution to the overall evaporation losses. Although, as mentioned previously, the evaporation losses of canopy-intercepted rain or irrigation water can reach considerable percentages of the total water supply (Savenije 2004), they are difficult to be influenced unless through better timing of the overhead irrigation. Therefore, the first option to reduce unproductive evaporation losses is the understanding of the process of soil evaporation in order to minimize it through feasible management practices.

The most important factors that affect evaporation from the soil, apart from the conditions external to the evaporating body (evaporative demand), are its water content, texture, structure, and the degree to which it is covered either by growing vegetation or any form of surface mulch. With regard to the soil water content, three evaporation stages are distinguished: the first stage is when the water supply at/to the surface is sufficient to allow a more or less constant evaporation rate as a function of the evaporative demand; the second stage, the falling-rate stage, depends on the soil properties that are responsible for the delivery of moisture to the evaporation zone; and the third stage depends mainly on the rate of vapor diffusion. Both the soil texture and the structure may affect evaporation due to their influence on the soil hydraulic properties (Ndiaye et al. 2007; Jalota and Arora 2002). Under similar soil structure conditions, finer-textured soils show higher evaporation losses than coarse-textured soils (Jalota and Prihar 1986; Prihar et al. 1996). At low moisture levels when the water is held in coarse-textured soils between soil particles rather than in continuous pores, evaporation occurs mainly through the slower process of diffusion instead of conductance in water-filled pores to reach the zone of evaporation (Ward et al. 2009).

The most notable reduction of evaporation losses and the most easily attained through management practices is through the cover of the soil by vegetation or any form of stubble and mulch, whether of organic (crop residues, waste products, cover crop, etc.) or inorganic (stones, plastic films, etc.) origin. Soil cover interferes with the evaporation process mainly by providing a mechanical barrier or resistance to the removal of moisture over the soil, reducing the energy supply (heat flux) to the zone of evaporation (both lowering the evaporative demand), and decreasing the conductivity or diffusivity of the topsoil layer if superficially incorporated. Numerous studies have been conducted to evaluate the effectiveness of different mulching types and practices (Yuan et al. 2009; Ward et al. 2009; Monzon et al. 2006; Burt 2005; Sauer et al. 1996), and they generally agree that the soil cover is especially effective in reducing the evaporation losses at the first evaporation stage, thereby contributing to a more favorable soil water status. However, under relatively dry soil conditions, a soil cover in the form of standing stubble may sometimes enhance evaporation, which is attributed to the hydraulic redistribution along the senesced roots and stems (Ward et al. 2009; Leffler et al. 2005).

10.3 MANAGEMENT PRACTICES THAT AFFECT SOIL WATER AVAILABILITY

The objectives and arguments for soil tillage are many, including weed control, soil decompaction, crop residue management, and adequate seedbed preparation. However, the results in the literature on the effects of tillage operations on the soil water and its use through crops are highly variable and contradictory. The inconsistency of the effects of different tillage systems on the soil's physical and hydraulic properties is attributed to the transitory nature of the soil structure after tillage, the site history, the initial and the final water contents, the time of sampling, and the extent of soil disturbances (Azooz and Arshad 1996).

Based on the processes that affect soil water (outlined in Figure 10.2), this section reviews the effects of soil tillage practices on these processes and the consequences on crop-available soil water and its use efficiency.

10.3.1 Soil Tillage

10.3.1.1 Effects on Infiltration and Runoff

Tillage practices change the infiltration and runoff components, basically by modifying the soil properties such as the stability of the structural soil aggregates, the total porosity and macroporosity, the hydraulic conductivity, the surface crusting and compaction, and the SOM. Generally, soil aggregation improves with the conversion from conventional soil tillage to no-till (NT) soil management. As a result, pore connectivity takes place, enhancing the soil quality and the water transmission properties.

In the literature, the effect of tillage on infiltration is ambiguous. Reviewing the state-of-science to quantify the agricultural management effects on the soil hydraulic properties, Strudley et al.

(2008) showed that there is a trend of NT systems to promote an increase in the macropore connectivity and the infiltration rate; however, because of the differences in soils, climates, and specific practices of tillage, it is not possible to generalize these results without a detailed knowledge of all the controlling factors. As related to water infiltration, several researches have highlighted the great advantages of the NT systems over conventional tillage (CT) practices. However, the site-specific conditions could indicate the need for a surface, shallow soil disturbance to destroy the surface crust or compacted layers, which can occur as a consequence of little crop residue on the soil surface (Thierfelder and Wall 2009; Singh et al. 2005) or intense tillage and machinery traffic (Reichert et al. 2009b; Mary and Changying 2008; Sasal et al. 2006; Hamza and Anderson 2005). Even considering that this surface disturbance may promote some increase in water evaporation, probably the gain in water infiltration due to the runoff reduction will surpass the aforementioned effect. However, tillage seems to have only a short-lasting effect on the improvement of the infiltration rate (Freese et al. 1993).

Experiments performed in several regions of the world show that NT systems promote soil aggregation (Stone and Schlegel 2010; Rhoton et al. 2002) and water infiltration. One study carried out in southern Brazil showed that rainwater infiltration increased from 20 mm/h under CT to 45 mm/h under NT (which included cover crops and crop rotation) (Calegari et al. 1998). In Kansas, Texas, on a silt loam soil, Stone and Schlegel (2010) observed that the infiltration rate under NT (30.56 mm/h) was 1.99-fold and 2.67-fold greater than in reduced tillage (RT) and in CT, respectively. The infiltration rate was positively correlated with the mean weight diameter (MWD) of the water-stable aggregates (WSA), which, in turn, was also positively correlated with the total soil organic carbon (SOC). The authors attributed this greater steady-state infiltration rate in the NT system to the presence and stability of the surface-connected macropores, the greater concentration of larger, water-stable aggregates in the surface layer, and the reduced surface sealing due to the protection from raindrop impact promoted by the residues. In their review of the effects of tillage systems on the physical properties of the soil and the water content in the Argentine Pampas, Alvarez and Steinbach (2009) clearly document the higher infiltration rates under NT when compared with limited tillage, especially plow tillage (Figure 10.3).

Soil disturbance caused by many tillage practices increases the surface roughness, the macroporosity, and the initial infiltrability, although the infiltration rapidly declines with time as a consequence of aggregate collapse (Guzha 2004). On a silt loam, Wahl et al. (2004) also observed a high infiltration dependence on macroporosity. Their results showed that the infiltration rates measured with a tension infiltrometer were higher in the soil surface layer in CT than in conservation tillage.

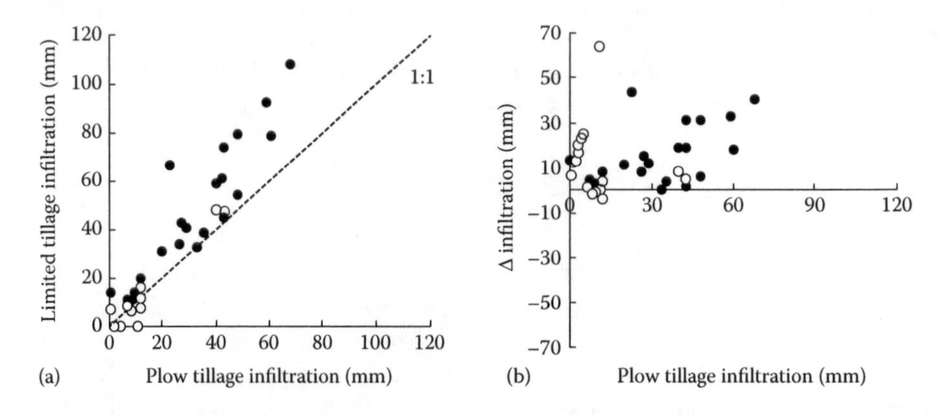

FIGURE 10.3 The relationship between soil infiltration under limited tillage systems and the plow tillage systems (a). The change in soil infiltration (limited tillage–plow tillage) in relation to plow tillage infiltration (b). Full circles—no tillage; empty circles—chisel or disk tillage. (From Alvarez, R. and Steinbach, H.S., *Soil Till. Res.*, 104, 1, 2009.)

At depths greater than 50 cm, the water intake in conservation tillage was higher than in CT and even higher than in the topsoil for both tillage types. The authors attribute these findings to a higher macroporosity under conservation tillage at 20 cm and below and to a much better macropore continuity promoted by the soil fauna activity in the soil profile.

In addition to the importance of macroporosity and its geometry for infiltration and redistribution of the soil water in depth, Sasal et al. (2006) emphasized the importance of the stability of the soil structure and the need for a complete characterization of the soil porosity as essential information to assess the effect of the structural conditions on the soil water dynamics. At two of three study sites, these authors observed a more horizontal orientation of the macropores under zero tillage when compared with the chisel plough, which they ascribe mainly to the pressures generated by repeated traffic and a low volume of crop residues in the soybean-dominated crop rotations. At the same time, they found a very good relationship between the vertically oriented macropores and the infiltration rate, an observation frequently shared by other authors (Imhoff et al. 2010; Buczko et al. 2003; Tebrugge and Abelsova 1999).

An enhanced soil bioporosity through macrofauna, mainly earthworm activity, or created by roots after their decomposition, could compensate for the effect of reduced total porosity or even soil compaction on the water flux within the soil. Although Abreu et al. (2004) could not find an increase in the water infiltration under minimum tillage (MT), they were able to show that saturated hydraulic conductivity measured in the field increased in the soil cultivated with showy crotalaria (*Crotalaria spectabilis* Roth) under MT (Crotalaria), even when the soil mechanical resistance was greater than in other tillage-based systems (Figure 10.4). This result was due to the better pore continuity promoted by the deep root growth of *Crotalaria*, since the total porosity, macroporosity, and bulk density were not different among the tillage systems. Thus, the tillage and cropping systems that enhance vertically oriented bioporosity are likely to increase the amount of water captured and redistributed in the soil. Hartge and Bohne (1983) reported that repacked soil with artificial vertical macropores was more stable than repacked soil with artificial horizontal macropores.

Runoff is the direct consequence of the precipitation intensity being higher than the infiltration rate and the soil surface water storage capacity. Especially in regions where rainfall is low and erratic, runoff is always undesirable as it reduces the amount of soil water available for productive transpiration (Guzha 2004). Soil tillage normally increases the soil surface roughness and the residence time of the water on the soil surface, but the continuous production of earthworm casts on untilled land has also been found to induce a marked surface roughness and to reduce runoff

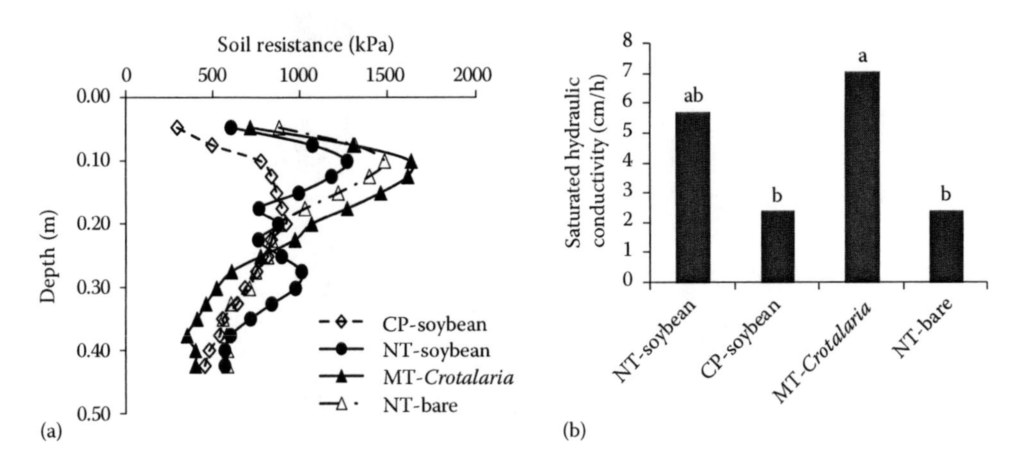

FIGURE 10.4 The soil penetration resistance (a) and the saturated hydraulic conductivity (b) measured at 0.02–0.12 m layer in soils under different management systems. Bars with the same letter are not significantly different (P = 0.5), as determined by a DMS test. CP: chisel plow; NT: no-tillage; MT: minimum tillage. (Adapted from Abreu, S.L., et al., *Rev. Bras. Cienc. Solo*, 28, 519, 2004.)

(Podwojewski et al. 2008). Although soil roughness may play a key role in retaining precipitation in situ and retarding runoff, in warm and dry environments, the soil surface sealing that follows tillage is critical, since the water that ponds on the soil surface for even a short time quickly evaporates, thereby reducing the infiltration and the effective rainfall (Peterson and Westfall 2004).

Rhoton et al. (2002) have comprehensively described the importance of the effects induced by tillage systems on the soil properties and the surface runoff. Based on long-term studies of a silt loam in the southeast of the United States, they correlated the soil properties such as SOM, aggregate stability, total and dispersible clay content, and bulk density to surface runoff. From single, two-, and three-variable regression models performed for depths of 0–1 cm and 1–3 cm, the authors concluded that the most important soil property explaining runoff was bulk density under NT and aggregate stability under CT (chisel plough and disk harrow). Further, they concluded that bulk density is a measure of the surface porosity that is stabilized by a higher SOM content under NT, making other soil properties relatively unimportant. This was not the same for the soil under CT, where the aggregate stability and the water dispersible clay content were identified as the most important properties for controlling runoff.

A reduced aggregate stability and a less favorable soil structure through soil tillage are the most frequent arguments to explain a higher surface runoff when compared with the MT or NT systems. Although tillage temporarily creates more favorable conditions initially for infiltration and less runoff through higher macroporosity and surface retention, soil sealing through the particle detachment of the disturbed soil surfaces contributes to a rapid change in the infiltration conditions (Teixeira et al. 2000, Azooz and Arshad 1996) and to the overall higher runoff and less infiltration of the tilled soil. For example, Castro et al. (1993) observed that NT reduced the runoff water losses by 88% when compared with heavy-disc harrowing, with losses of 13.1, 35.7, and 93 mm/year for NT, the chisel plough, and the disc plough, respectively. Kosgei et al. (2007) evaluated the effects of CT (moldboard plough) and NT on the water losses, infiltration, and rain use efficiency of maize in the 2005/2006 season (Figure 10.5), in a highly diverse catchment in South Africa. NT treatment generated less runoff (22 mm) than CT (37 mm), corresponding to 6% and 8% of the seasonal rainfall (463 mm), respectively. Despite the small difference in the cumulative water intake, the authors ascribe the large differences in the yield to the higher soil water availability under NT throughout the growing season.

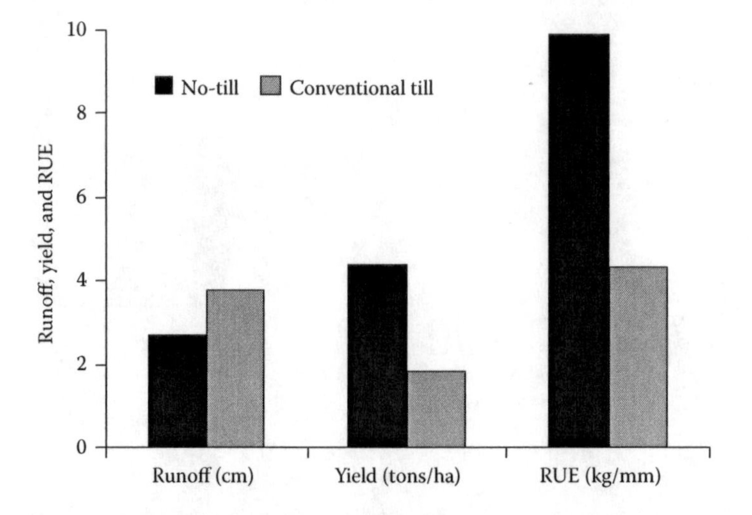

FIGURE 10.5 The seasonal average runoff depths (cm), the maize yields (ton/ha), and the "effective" rain use efficiency (RUE) (kg/mm) in no-tillage and conventional tillage plots in Potshini catchment in the 2005/2006 season. (From Kosgei, J.R., et al., *Phys. Chem. Earth*, 32, 1117, 2007.)

10.3.1.2 Effects on Water Retention Capacity and Deep Drainage

As previously outlined in Section 10.2, plant-available water in a given volume of soil depends on the amount of mesopores that are able to retain water against gravitational forces and to deliver it to the roots on demand. Second, the total soil volume that the roots are able to explore (rooting depth) is equally important as plants can compensate for water stress in the upper, more densely rooted, soil layers by increasing the water uptake from deeper layers (Teuling et al. 2006). Thus, tillage management may directly or indirectly affect the pore size distribution, the pore geometry, and the hydraulic properties of the soil and hence, the plant-available soil water.

Today, it is widely recognized that the absence of soil disturbance improves aggregate stability and promotes SOM accumulation and stabilization. Through the buildup of SOM and the consequent promotion of soil aggregation and structure stabilization, the reduction of tillage intensity contributes to a higher percentage of intermediate pores in relation to the total porosity (Fernandez-Ugalde et al. 2009; da Veiga et al. 2008; Bescansa et al. 2006). Many researchers have found a high correlation between the structural quality of the soil and the SOM content with plant-available water (Imhoff et al. 2010; Abid and Lal 2009; So et al. 2009; Mrabet et al. 2001). Despite a frequently observed reduction of the total porosity in the surface soil layer under NT, the total volume of mesopores is increased in the absence of soil disturbance. After 6 years of differentiated tillage (NT and moldboard plough), Carvalho and Basch (1995) found a lower total and medium-size porosity under NT in the 0–0.1 m soil layer; however, in the 0.1 and 0.3 m soil layers, the total porosity and especially the pore space referring to plant-available water were considerably increased (Table 10.3). There was also a close correspondence to bulk density and the SOM content. Results obtained by other authors corroborate these findings (Fernandez-Ugalde et al. 2009; Bescansa et al. 2006; Bhattacharyya et al. 2006; Rasmussen 1999; Hussain et al. 1998), although, in their review on conservation tillage and the depth stratification of porosity and SOM, Kay and VandenBygaart (2002) identified some cases where tillage-induced changes in mesoporosity did not occur. According to their interpretation, only long-term studies are able to provide consistent information especially with regard to the changes in the SOM and the changes in the pore size fractions.

Chemical and physical subsoil constraints frequently limit the water uptake from the deeper soil layers (Dang et al. 2010; MacEwan et al. 2010; Nuttall and Armstrong 2010). Water that is stored deep

TABLE 10.3
Total Porosity, Pore Size Distribution, Plant-Available Water, and Soil Organic Matter Content in a Vertic Cambisol after 6 Years under No-Till (NT) and Conventional Tillage (CT)

Tillage	Depth (cm)	>50 μm (%)	50–10 μm (%)	10–0.2 μm (%)	<0.2 μm (%)	Total Porosity	Available Water (%)	SOM (g/kg)
NT	10	3.20	2.22	2.7	38.37	46.52	4.92	2.53
	20	0.86	3.91	5.22	36.16	46.15	9.13	2.15
	30	1.86	2.63	11.48	29.44	45.40	14.11	2.25
	0–30	1.97	2.92	6.47	34.66	46.02	9.39	2.31
CT	10	15.08	2.34	4.36	29.95	51.73	6.71	1.58
	20	2.67	1.32	2.31	39.95	42.25	3.63	1.7
	30	1.47	1.56	3.29	35.62	41.94	4.85	1.66
	0–30	6.41	1.74	3.32	35.17	45.31	5.06	1.65

Source: Carvalho, M. and Basch, G., Experience with the applicability of no-tillage crop production in the West-European countries. *Proceedings of the EC-Workshop II*. Wissenschaftlicher Fachverlag, Langgöns, Germany, 1995.

in the soil profile is considered especially valuable to crop yield because it becomes available during grain filling (Passioura and Angus 2010; Kirkegaard et al. 2007). With regard to the physical subsoil constraints, whether of natural or anthropogenic origin (hard setting soils or hard pans), deep soil loosening has become a widely used soil management practice where powerful tractors and subsoiling equipment are available (Hamza and Anderson 2005). Deep ripping and slotting are frequently proposed operations to overcome these constraints and to improve the access of roots to additional soil water and nutrients (MacEwan et al. 2010; Hartmann et al. 2008; Adcock et al. 2007; Sadras et al. 2005; Hamza and Anderson 2003). However, the documented results have been variable and the benefits of deep soil loosening and the consequent crop response may differ from one year to another and from one place to another. Wetter locations and coarser-textured soils are more likely to benefit from subsoiling than fine-textured soils and drier locations (Wong and Asseng 2007). Negative results from deep soil loosening were obtained in dry years or when deep drainage losses occurred (Wong and Asseng 2007). Further, the deep loosening effects are often of short duration, especially if not accompanied by additional measures, such as subsoil conditioners (i.e., gypsum), the installation of primer crops, or reduced or controlled traffic (Lopez-Fando et al. 2007; Hamza and Anderson 2003, 2005; Yunusa and Newton 2003). Adcock et al. (2007) also highlight the need for a careful balance between the expectable increase in economic returns and the costs and duration of any amelioration measure.

In order to establish a relationship between the soil physical constraints, the root growth, and the soil-available water, the use of indicators such as growth-limiting bulk density (GLBD), least-limiting water range (LLWR), and integrated water capacity (IWC) has been proposed. The first one, based on the work of Pierce et al. (1983), provides threshold bulk densities for a given soil texture at which water availability becomes excessively restricted. LLWR serves as an index of the soil structural quality, integrating values of the soil matrix potential, aeration, and the soil strength (Dasilva et al. 1994). The concept of IWC has been proposed as a flexible method to quantify various soil physical limitations when calculating the available water in nonswelling soils and then extended to swelling soils, introducing overburden pressures (Groenevelt et al. 2001). Although considered as useful indicators of soil physical quality for crop production (Leao et al. 2006; Lapen et al. 2004), the applicability of the most widely used LLWR indicator has been questioned by several authors (Asgarzadeh et al. 2010; Kaufmann et al. 2010; Reichert et al. 2009b; Benjamin et al. 2003), disagreeing mainly on the critical values used for some of the soil characteristics, based on bulk soil conditions. Especially under NT, the critical bulk densities do not reveal the pore geometry and continuity and may not necessarily restrict root growth (Reichert et al. 2009b; Bolliger et al. 2006).

Deep drainage, defined as water passing below the potential maximum rooting depth, occurs when the soil volume above is saturated and the wetting front reaches the lower limit of the rooting zone or through preferential pathways even if the soil profile is not saturated. It is an important process to recharge the groundwater and to conduct excess water from the soil profile to the deeper soil layers, thus contributing to reduced surface runoff. In regions with water scarcity, however, deep drainage is often considered to correspond to a loss in the potential crop growth, although for dry environments, deep drainage losses are usually much smaller than the losses through evaporation (Passioura and Angus 2010), except in very sandy soils (Passioura 2006). In some regions, deep drainage is also associated with the problems with secondary salinity (Asseng et al. 2001; Ridley et al. 2001).

According to Strudley (2008), the drainage effects on soils subsequent to tillage have not been studied very extensively, due to their high temporal and spatial variability. In addition to the above-mentioned effects on water retention, tillage may affect deep drainage either through its negative impact on the rooting depth or on the creation and longevity of macropores. It is widely recognized that in the absence of soil disturbance, as is the case in NT systems, biological macropores, earthworm tubes, former root channels, and voids between the soil structural units are preserved, thereby forming preferential pathways for rapid and deep percolation (Verhulst et al. 2010; Cullum 2009; Strudley et al. 2008; Shipitalo et al. 2000; Tebrugge and During 1999). Although Diaz-Ambrona et al. (2005) found that tillage treatments alone made little difference to deep drainage; they highlight the greater soil water storage where residues were maintained, which led to higher drainage losses.

After 18 years of NT and CT on a Vertisol, McGarry et al. (2000) found that in addition to taking a longer time to ponding, final infiltration rate, total infiltration and deep drainage were also enhanced. Through a higher water intake, less evaporation losses in the case of residue maintenance, and a better pore connectivity, deep drainage is more likely to occur under NT than under CT. However, the withdrawal of excess water from the saturated topsoil through deep drainage provides a basis for a positive trade-off between the consequent increase in infiltration and the reduction in runoff.

10.3.1.3 Effects on Soil Evaporation

In arid and semiarid regions, the main unproductive water loss is caused by direct evaporation, especially if there are many low-intensity rainfall events (Lampurlanés and Cantero-Martínez 2006; Passioura 2006). Reducing the evaporation losses is a major challenge for farmers, especially in water-scarce regions where fallow periods are used to accumulate additional soil water. Under rainfed Mediterranean conditions in central Aragon (northeast Spain), based on field measurements and model simulations, Moret et al. (2007) obtained evaporation losses during the fallow period in the range of 55%–91%, whereas deep drainage losses were in the range of 5%–28%. Thus, comparing continuous barley and a barley–fallow crop rotation, Moret et al. (2006) measured only an additional 20 mm of soil water storage through the fallow period.

Evaporation reduction from a bare soil surface during the initial evaporation stage can be attained through a coarse or disturbed layer (or mulch) overlying the wet subsoil. In a conventional system, however, tillage operations with the purpose of reducing evaporation are carried out with the soil at favorable moisture conditions, which is almost at the end of the initial evaporation stage. In addition, soil loosening and exposure boosts water losses from the cultivated soil layer as tillage operations favor heating and the formation of air pockets in which evaporation occurs (Licht and Al-Kaisi 2005). Within 24 h after the primary soil tillage, Moret et al. (2006) measured up to 16 mm of evaporation losses against 2 mm under NT. After the secondary tillage, they still obtained differences of up to 3 mm of water losses between the tilled and the untilled treatments. Therefore, the performance of tillage with the objective of reducing evaporation is the result of a balance between the short-term evaporation losses through enhanced drying of the soil layer disturbed by any form of tillage and the possible long-term gains through the breakup of a faster upward capillary movement in the undisturbed soil.

Tillage may also affect soil evaporation through increased surface roughness, exposing a greater surface to the overlying atmosphere and winds and through a change in the soil surface temperature and albedo. Despite a higher surface albedo on a smooth bare soil when compared with moldboard ploughed soil or CT, Oguntunde et al. (2006) found only small differences in the soil moisture content of the surface layer. On swelling and shrinking soils, evaporation losses may be considerable. According to Ritchie and Adams (1974), evaporation from the cracks near the end of the sorghum growing season was 0.6 mm/day and an additional 15 mm of soil water is lost by evaporation before the soil swells and cracks close from the rains. Mulching or superficial soil tillage could prevent the formation of cracks or at least obliterate them after they have begun to form.

Although the absence of soil disturbance through the practice of NT has been viewed as a method that reduces evaporation losses (Kosgei et al. 2007; Fowler and Rockstrom 2001), information on the tillage-induced effects on water evaporation from bare soils is scarce, as NT treatments are usually associated with the maintenance of some form of residues over the soil surface. The effects of residues on the soil water and especially evaporation, whether in combination with NT practices or not, have been the subject of extensive studies and will be discussed in the following section.

10.3.2 Soil Cover and Residue Management

In natural ecosystems with some minimum rainfall for the growth of vegetation, the soil surface develops some form of organic cover consisting of plants and their residues after senescence. The soil cover controls the flux of energy and water by interacting with components of the atmosphere, the hydrosphere, the biosphere, and the pedosphere (Lal 2009). The conversion of natural environments into agricultural areas leads to a significant change in the partitioning of the water, nutrient, carbon, and energy flow. During and after a rainfall event, rainwater may infiltrate into the soil to replenish the soil water or flow through it to recharge the groundwater, and some may run off as overland flow and evaporate back into the atmosphere (directly from an unprotected soil surface and from plant leaves) (Bot and Benites 2005). The soil cover and residues directly affect the runoff and soil evaporation and indirectly affect deep percolation, all of them representing unproductive water losses. The objectives of sustainable soil and soil water management are to redirect these losses into an increase of soil water storage and availability to plants.

In this section, we address the influence of the soil cover, including the application of organic and inorganic mulching material, cover crops, and crop residues on the soil water, either through their direct impact on infiltration/runoff and evaporation or their indirect effects on the SOM content and macrofauna activity. Additionally, we present evidence on how the soil cover type (including cover crops) and the residue characteristics and their management affect the soil water conservation and the soil productivity.

10.3.2.1 Effects on Infiltration and Runoff

Soil macroaggregate breakdown is seen as the major factor leading to surface pore clogging by primary particles and microaggregates and thus to the formation of surface seals or crusts (Lal and Shukla 2004). The soil cover prevents this breakdown by reducing the kinetic energy with which raindrops reach the soil surface (Ben-Hur and Lado 2008). In addition to the detachment of the soil aggregates through direct raindrop impact and the physicochemical dispersion of the clays, slaking is considered another important process in the disintegration of the aggregates and the consequent seal formation (Lado et al. 2004). The faster the wetting rate of the dry soil, the higher are the slaking forces. As the aggregate breakdown due to slaking is inversely related to the antecedent water content (Haynes 2000), the higher topsoil moisture of the covered soil reduces the slaking forces.

The tendency of a soil to form a surface seal, the resulting decrease in infiltration, and the amount of the resulting runoff and soil loss depend on the aggregate stability (Ben-Hur and Lado 2008). Many reviews have been published on the effects of the soil properties, such as texture, organic matter content, soil mineralogy, and soil salinity and sodicity, on aggregate stability (e.g., Lado and Ben-Hur 2004; Kay and Angers 1999). The amount of crop residues and their management, however, can have a decisive effect on the resilience of the aggregate breakdown. After applying different amounts of wheat straw on an untilled loamy Fluvisol in the southwest of Spain over a period of 3 years, Jordan et al. (2010), using the water-drop test and ultrasonic disruption methods, found a clearly improved aggregate stability with an increase in the amount of straw residues ranging from 0 to 15 Mg/ha (Figure 10.6). However, only the two highest mulching rates provided a significantly

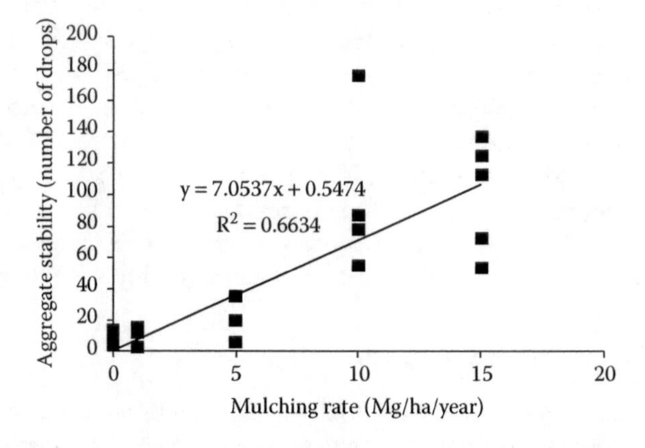

FIGURE 10.6 The relationship between the aggregate stability and the mulching rate. (From Jordan, A., Zavala, L.M., and Gil, J., *Catena*, 81, 77, 2010.)

better aggregate stability after 3 years. A more linear and positive correlation between the amount of wheat straw and the percentage of WSA and the MWD was found by Mulumba and Lal (2008) after 11 years on a stagnic Luvisol in central Ohio. Yet, it seems that crop residues alone are not effective in improving the soil aggregate stability. After 7 years of different residue management, which included NT with residues, residue incorporation through chisel/disk tillage, and residue removal before and replacement after tillage, Wuest (2007) found no differences in the MWD of the aggregates in the 0–5 cm layer between the straw mixed and surface-applied in the tilled treatments; however, under NT, the MWD was more than two times greater. This significant improvement of the aggregate stability under NT, when compared with the mixed treatment, was also expressed in the 5–10 cm soil layer. According to the author, an improved fungal activity might explain the observation of a better aggregate stability in the 5–10 cm layer when the straw was surface-applied after tillage instead of being incorporated into the surface soil. Under NT, this effect would also add to other changes such as an increase in the SOM.

Soil cover with crop residue also promotes topsoil porosity, improving the water entry and transmission into the soil. The continuity of the pores left by decayed roots plays an important role in improving the infiltration rate, particularly in a very fine textured soil.

It is widely accepted that the random roughness of the soil surface created by tillage may contribute to the temporary retardation of the runoff, mainly through an increased depressional storage capacity. However, depending on several soil properties, there is a more or less sharp decline in the depressional storage with the progressive impact of the raindrops (Gomez and Nearing 2005; Guzha 2004). If crop residues are left on the soil surface, or are partially incorporated in the upper soil layer through mesofauna, not only is the impact of the raindrops reduced, but also the stream velocity, as the residues act as a succession of physical barriers (Verhulst et al. 2010; Jin et al. 2009). The residues play a role similar to that of surface roughness, that is, increasing the time for infiltration to take place (Blevins and Frye 1993), with the difference that their effect lasts longer. Therefore, the time lag for runoff generation is also greater when the crop residue is left on the soil surface (Jordan et al. 2010) and the transmission losses (turning small-scale runoff into large-scale runoff) decrease with the increasing vegetation or residue cover (Leys et al. 2010).

When incorporated into the surface soil, the amount of residues also seems to affect the infiltration and runoff. Gimenez and Govers (2008) measured an extra shear stress created by the freshly surface-incorporated residues and a reduced flow velocity, both of which were well correlated with the quantity of residues incorporated. However, at high runoff rates, their effect on reducing the flow hydraulics and erosivity is decreased. Studying the effect of shredded and spring-incorporated corn stalks of different plant populations (0%, 50%, and 100%) on runoff and erosion, Wilson et al. (2008) found a reduction in the average annual soil loss of around 50% for the 50% and 100% plant densities compared with the 0% population with no residues (bare soil), but a very small reduction in the surface runoff of 6.5% and 10.8%, respectively. The 50% and 100% population did not differ in the yield or in the amount of residues left, which was around 7 Mg/ha.

Thus, soil cover and crop residues left at the soil surface seem to be effective in improving infiltration and in reducing the surface runoff and soil loss. It also appears that the amount of residues is closely related to the degree to which the runoff is decreased. After 3 years under different mulching rates of wheat straw, rainfall simulation measurements at intensities of 65 mm/h provided clear differences in the surface runoff between mulching rates (Jordan et al. 2010) (Figure 10.7). In this study, the highest rates of 10 and 15 Mg/ha were necessary to almost completely avoid runoff.

A big difference between high and low standing, surface cut and removed stubbles has been found in regions where the retention of snow is crucial for supplying water to the following crop. Sharratt (2002) reported that taller stubble trapped more snow, reduced the depth of frost penetration, and hastened thawing of the soil profile by at least 25 days, when compared with short stubble or residue removal. Further, the variability in the soil water recharge was closely related to the amount of snow cover.

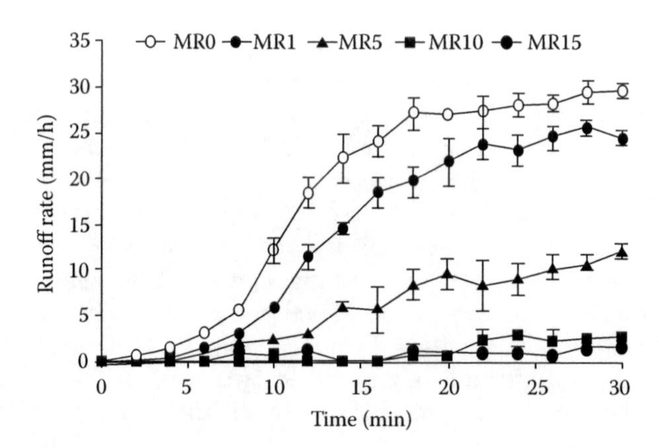

FIGURE 10.7 The variation of the mean runoff rates under different mulching rates. MR0 = control; MR1 = 1 Mg/(ha year); MR5 = 5 Mg/(ha year); MR10 = 10 Mg/(ha year); MR15 = 15 Mg/(ha year). N = 5 for each mulching rate treatment. Vertical bars indicate ± standard deviation. (From Jordan, A., Zavala, L.M., and Gil, J., *Catena*, 81, 77, 2010.)

10.3.2.2 Effects on Evaporation

The transfer of water from the liquid phase to the vapor phase occurs at the expense of the absorbed heat energy and depends on the occurrence of the water vapor deficit in the air above the soil surface and the diffusion resistance along the pathway. The amount of heat energy and the water vapor deficit are increased proportionally due to the absorption of the incident radiant energy from the sun by the soil surface and the resulting increase in the temperature. Furthermore, the evaporation vapor flux from the soil is increased by the wind. Thus, the main approach for reducing water evaporation is by reflecting the incident energy to reduce the energy absorption by the surface, reducing the wind speed at the soil surface, and impeding or reducing the vapor flux from the soil into the atmosphere. The soil cover and residues act on all these processes, but it has been difficult to quantify their effects on the processes separately.

Soils mulched with crop residues or cover crops have a reduced maximum soil temperature and a lower amplitude (Zhang et al. 2009; da Silva et al. 2006; Fabrizzi et al. 2005). The high solar reflectivity and low thermal conductivity of the crop residues prevent an increase in temperature (Shinners et al. 1994; Hillel 1980). On submitting a long-term NT area after a winter cover crop (black oats) to different tillage practices (NT, mouldboard plough, and chisel), da Silva et al. (2006) found that the cover crop residues on the soil surface under NT reduced both the maximum soil temperature and the daily amplitudes. Trevisan et al. (2002) showed a reduction in the soil temperature amplitude down to 20 cm in depth with an oat straw cover throughout the year, compared with soil without a straw cover. Thus, a lower portion of the surface energy balance will be used as latent heat in the system, reducing the evaporation of water from the soil.

Both transpiration and soil evaporation depend on the evaporative demand of the environment. In order to study the interaction between the soil type, the residue cover, and the evaporative demand, Freitas et al. (2006) treated a loamy sand and a heavy clay soil covered with different types and amounts of residues to three different evaporative demands of around 3, 5.2–6, and 7–8 mm/day (Table 10.4).

Whereas the uncovered soil remained in the first evaporation stage only at the evaporative demand of 3 mm/day, both residue-covered treatments maintained this stage over the 21-day trial period for the medium and highest evaporative demands. On average, over both soil and residue types, the highest amounts of residues resulted in total evaporation, which was around 30% of that measured in the treatment without residues. Especially on the loamy sand, evaporation reduction under the highest amount of residues was almost independent of the evaporative demand.

TABLE 10.4
Total Soil Evaporation during 21 Days (after Reaching Field Capacity) for Two Different Soils under Different Types and Amounts of Residues and Different Evaporative Demands

	Residues (kg/ha)		Evaporative Demand (mm/day)					
			Corn			Wheat		
Soil type	Corn	Wheat	8	6	3	7	5.2	3
Loamy sand	0	0	74.2	82	57.2	59.2	68	47.9
Heavy clay	0	0	56.4	74.2	56.4	54.7	59	46.9
Loamy sand	5000	3500	40.2	28.9	19	38	28.4	18.5
Heavy clay	5000	3500	35.7	30.1	22.2	35.2	32	22.8
Loamy sand	10000	7000	20.4	19.8	18.6	20.6	20	16.5
Heavy clay	10000	7000	21.1	18.1	13.6	20.3	17.1	13.1

Source: Adapted from Freitas, P.S.L., et al., *Rev. Bras. Eng. Agr. Ambient.* 10, 104, 2006.

The contribution of residues, whether alone or in combination with NT, to reducing the evaporation component of total ET has been confirmed by numerous studies and under many different conditions. In Punjab, India, Jalota and Arora (2002) observed that straw mulching (6 Mg/ha) substantially reduced the soil water evaporation under medium-textured and coarse-textured soils by 18.5 and 13.1 cm in maize, 23.8 and 16.6 cm in cotton, and 23.6 and 17.6 cm in sugarcane crops, respectively. They concluded that the irrigation requirements of summer crops can be reduced further by mulching with crop residues. Lamm et al. (2009) also suggest that strip tillage and NT, due to the maintenance of the crop residues, should be considered as improved alternatives to CT, particularly when the irrigation capacity is limited. In Texas, Lascano et al. (1994) found in cotton production that the total ET was similar between a conventional and a wheat straw residue–based strip-tillage system. However, they found large differences in the components of ET, with a share of the transpiration of 50% with CT with no residue against 69% under straw mulch, which resulted in a 35% increase in the lint yield. In a recent study using undisturbed mini-lysimeters, Klocke et al. (2009) compared the effect of bare soil with the soil partially or completely covered with wheat stubble and corn stover and with and without the effect of the corn canopy. On average over 3 years, the evaporation in the field study (with canopy and full residue cover) was reduced by almost a half through both types of residues. Even with a surface coverage between 91% and 100%, the higher the amount of residues, the more pronounced was the reduction in evaporation. In the trial without a canopy, evaporation compared with bare soil was reduced by 20% or less by residue treatments with partial cover, but significantly more by the full cover of both wheat and corn residues. Standing wheat stubble surpassed the evaporation reduction effect of the flat corn residue, an observation that the authors attribute to the possible aerodynamic effects of standing straw.

However, the residue management effects on the evaporative water losses may vary with different climatic conditions. For example, in contrast to a possible reduction of the convective component of evaporation through standing wheat stems, advanced by Aiken et al. (2003), Ward et al. (2009), under sandy topsoil conditions, observed an increased evaporation in the presence of standing stubble when compared with cut and removed stems or slightly buried stems. A possible capillary upward movement of water through the senesced roots is provided as a possible explanation. It has also been concluded that residue thickness (volume) is more important than mass per unit area for controlling evaporation (Steiner 1989).

Although high porosity and pore continuity are favorable characteristics for increasing the soil water storage capacity and deep infiltration, they also enhance the upward water movement from the deeper soil layers (Lampurlanes and Cantero-Martínez 2006). Therefore, compared with the retention of a sufficient amount of residues at the soil surface to reduce evaporation effectively, some authors found better results with a very shallow surface incorporation of residues because this is best at breaking the unsaturated hydraulic conductivity, a process that, for Sillon et al. (2003), seems to be more important for evaporation reduction than the differences in albedo and surface roughness. Prihar et al. (1996) found that the benefits of the residue management treatments followed the order of residue-undercut > residue-mulch > residue-incorporated. According to Gill and Jalota (1996), incorporating lower rates of straw mulch into the top few centimeters can be as efficient or more efficient than higher mulch rates at reducing evaporation.

Other types of surface covers have been proposed to reduce unproductive evaporation losses, such as plastic films or sand or gravel mulch (Liu et al. 2009; Yuan et al. 2009; Tao et al. 2006). Despite some positive results with regard to reduced evaporation and improved water storage and productivity, labor and capital investment are clearly the major constraints to the widespread use of these materials, at least with the objective of evaporation reduction.

The process of evaporation and its control remain a complex issue as they strongly depend on the soil and climate conditions and the length of time over which treatments or practices are applied. Nonetheless, it is widely agreed that an increase in the rate of the soil surface mulch can reduce the amount of short-term and probably long-term soil evaporation (Verhulst et al. 2010; Blanco-Canqui and Lal 2009; Singh et al. 2006; Burt et al. 2005). Occasionally, under dry rainfed conditions and on sandy soils, surface mulch may not be effective for evaporation reduction (Ward et al. 2009; Burt et al. 2005). However, under these conditions, the response of the soil evaporation and that of the soil water storage to rainfall are in a phase where all rainfall is evaporated, irrespective of the soil cover (Monzon 2006).

10.3.2.3 Effects on Soil Water through the Increase in Soil Organic Matter and Macrofauna Activity

The removal of crop residues through burning or for fodder and biofuel purposes is considered to be a major threat to soil productivity, environmental quality, and overall sustainable development (Blanco-Canqui and Lal 2009; Hakala et al. 2009; Lal 2009). In addition to the physical protection of the surface soil layer and its impact on infiltration and evaporation, organic residues enhance the buildup of SOM and soil fauna activity, which contribute to improve the soil porosity, soil particle aggregation, soil moisture storage, and deep water infiltration (Lal 2009; Wuest et al. 2005).

The improved pore space is a consequence of the bioturbation activities of earthworms and other macroorganisms and the channels left in the soil by decayed plant roots. Studying the effects of earthworms in Germany, Ernst et al. (2009) found that the soil water was strongly affected by the activity of ecologically different earthworm species. The epigeic *Lumbricus rubellus* tended to enhance the storage of soil moisture in the topsoil, and the endogeic *Aporrectodea caliginosa* strongly improved the water infiltration and hastened the water discharge through the soil. Although the benefits of increased earthworm populations are mainly attributed to the absence of soil disturbance (two to nine times more in NT than under CT [Chan 2001] and relatively less than the amount of residues retained at the soil surface [Eriksen-Hamel et al. 2009]), Blanco-Canqui and Lal (2007b) found a strong effect of corn stover removal on the reduction in the number of earthworms. On all three soils studied, stover removal at rates above 25% drastically reduced the number of earthworms and, on the occasionally anaerobic clayey soil, stover removal above 50% eliminated the earthworms. At a different site after 10 years of applying 0, 8, and 16 Mg/ha year of wheat straw without crop and cultural operations, Blanco-Canqui and Lal (2007a) found 158 $\pm$ 52 earthworms per square meter in the medium and 267 $\pm$ 58 earthworms per square meter in the highest mulching treatment, whereas no earthworms were present in the zero mulch level.

Whereas the authors associate the higher water infiltration rates obtained with less or no stover removal in the study under three different soils (Blanco-Canqui and Lal 2007b) to the greater number of surface-connected earthworm burrows and other biopores, they found no difference in the infiltration rates between the residue levels at the other site (Blanco-Canqui and Lal 2007a). In both studies, however, they measured significantly higher soil water retention under the higher amount of residues, although this was confined to the surface soil layer. While some reports indicate that an abundance of earthworms has a strong influence on the soil porosity and the consequent water infiltration in mulched NT soils, Bottinelli et al. (2010), Kladivko et al. (1997), and Blanco-Canqui and Lal (2007a) concluded that increases in the earthworm population by mulching does not always increase the water infiltration rate in all soils, depending on the dominating type of earthworms.

Increases in the surface mulch or the residues incorporated into the soil tend to increase the SOM (Wuest et al. 2005; Sharma and Acharya 2000). Even under NT, the amount, type, and management of the residues play an important role in the evolution of the SOM. Basch et al. (2010) compared the residues of chickpea and the different amounts and management of wheat straw with regard to the changes in the SOM under Mediterranean conditions. After 3 years, they had already found significant differences in the SOM in the following order (letters indicate differences $P \leq 0.05$): chickpea residues (c) >stubble only (bc) >in-field grazing of straw and stubble (b) >>straw retained (a) >2 × straw retained (a).

SOM promotes soil biological activities and processes, resulting in more bacterial waste products, organic gels, fungal hyphae (polysaccharides), and worm secretions and casts (Wuest et al. 2005), which improve the aggregate stability and porosity. Directly or indirectly, these organic compounds are related to the water-holding capacity, although it is the total SOC or organic matter that is usually considered as an important aggregate indicator in a discussion on water retention pedofunctions (Rawls et al. 2003). Evaluating the efficiency of the pedotransfer functions to estimate water retention in 725 soil samples from the state of Rio Grande do Sul, Brazil, covering all types of soil textures, Reichert et al. (2009a) concluded that organic matter must be included as an independent variable, because it had an individual positive effect on the field capacity and the plant-available water. Sharma and Acharya (2000) found that the application of fresh lantana (*Lantana camara* L.) at a rate of 8 Mg DM/ha either as a surface mulch or incorporated over 4 years significantly increased the SOM content in the layer 0–15 cm. At the different sowing dates, the mulched treatments compared with the unmulched treatment showed a higher amount of stored soil water (between 15.1 and 22 mm) in the 0–45 cm layer. From the third year onward, both mulch treatments yielded significantly higher than the unmulched treatment, and in the fourth year the yield in the surface-applied mulch surpassed those in the incorporated mulch.

Crop residue incorporation is not the best residue management practice because it implies soil disruption and eliminates the beneficial effects of the residues retained on the soil surface. Even so, in a long-term experiment, Singh et al. (2005) found that rice straw incorporation was less detrimental to the soil physical and hydraulic properties than the burned or removed rice straw. Whereas straw removal compared with the other residue management systems performed worst with regard to SOM and soil aggregation, straw burning led to reduced water retention due to an increased water repellency of the soil surfaces. Comparing the effect of rice straw incorporation plus 60% of the mineral fertilizer–only treatment over 8 years in a sweet potato–rapeseed rotation, Zhu et al. (2010) found a 13% increase in the SOM with the rice straw application and a significant increase in the water-holding capacity. The correlation between these two parameters was highly significant.

10.3.2.4 Influence of Type of Soil Cover and Residues and Their Management on Soil Water and Crop Productivity

As shown in the previous sections, the soil cover has a decisive effect on the soil water dynamics and contributes to enhancing the green water component and promoting WP. The possibilities and the choice of the soil cover and its impact depend, in addition to the main objective behind it, on the climate, the soil properties, the management and cropping system (Wilhelm et al. 2004), and the alternative uses of biomass (Lal 2009), among others.

Whereas the use of cover crops is mainly restricted to humid or subhumid regions, in semiarid environments, soil cover through crop residues is the most commonly used option to improve the use efficiency of the main limiting factor to crop productivity. On a very limited scale, other materials such as plastic films, gravel or sand, or organic waste products are used for mulching to protect the soil and enhance the green water component. The use and the effectiveness of these materials have been reported mainly from Asian countries and are considered an option for reducing the soil evaporation, thereby increasing the infiltration of rainwater and soil water retention (Liu et al. 2010; Ghosh et al. 2006; Ramakrishna et al. 2006). In studies using soil cover with plastic film, Wang et al. (2009) found that transpiration was the main component of total ET. From a 2-year study, Liu et al. (2010) reported an increase of 19%–24% in maize yield and 23%–25% in WUE in soil covered with plastic film compared with rainfed bare soil. Total and partial covers of plain soil or ridges and furrows with different materials (straw, plastic film, gravel-sand) and their combinations have been proposed and studied with regard to WUE and crop performance (Liu et al. 2010; Wang et al. 2009; Yuan et al. 2009; Zhou et al. 2009). Although these techniques have been found to be more or less effective in reducing evaporation and runoff, improving infiltration and soil temperature, halting wind and water erosion, and enhancing biological activity and soil fertility (Li 2003), their use on extensive agricultural land can be seriously questioned for several reasons.

Therefore, in large-scale agriculture under semiarid conditions, crop residues, including those from cover crops, seem to be the only technically feasible and economically viable option to cover and protect the soil, while improving the soil water and WUE. Conservation of the soil moisture is one of the major advantages of the mulch farming systems (Mulumba and Lal 2008; Baumhardt and Jones 2002). In semiarid environments with rainfall above the minimum threshold for a benefit in terms of water storage, straw mulching generally increases yields by enhancing the soil water storage (Bescansa et al. 2006; Monzon et al. 2006), but in poorly drained soils or in temperate climates with suboptimal springtime temperatures, residue retention may sometimes reduce yield below optimal levels due to the decreases in the soil temperature (Lal 2008a; Fabrizzi et al. 2005; Anken et al. 2004) and the soil nitrogen (Gao and Li 2005).

As already discussed, the impact of the residues on soil water conservation may depend on their composition, management, and amount (Leys et al. 2010; Ward et al. 2009; Blanco-Canqui and Lal 2007b; Sauer et al. 1996; Steiner 1989). Although the possible impacts of the crop residues on the hydrophysical characteristics of the soil are well studied and it is widely recognized that the management systems that retain crop residues at the surface deliver the highest benefit in terms of soil water availability (Coppens et al. 2006; Burt et al. 2005), studies that relate long-term residue cover to soil water availability and crop productivity under field conditions are scarce and are sometimes inconclusive or contradictory (Blanco-Canqui et al. 2006) as the benefits of the residue cover in terms of soil fertility and water availability might be offset mainly by lower soil temperatures during the initial crop stages and weed and pest problems (Liu et al. 2004; Mann et al. 2002). The increasing demand for residues for biofuel production (Graham et al. 2007; Wilhelm et al. 2004) is raising concerns regarding excessive residue removal (Lal 2009) and that the benefits of long-term NT management may be lost by removing the crop residues (Dabney et al. 2004). Studying the different percentages of corn stover removal over 2 years on three long-term NT sites in Ohio, Blanco-Canqui and Lal (2007b) found a decrease in the plant-available water with an increase in the percentage of stover removal. However, this was reflected in higher crop yields only at one site, well-drained but erosion-prone. They concluded that soils with different characteristics might reveal yield effects if stover removal above a certain threshold level was continued over a longer time period and that site-specific determination of these threshold levels was urgently needed. However, these thresholds should also be assessed with regard to other ecosystem services provided by retaining crop residues, such as offsetting CO_2 emissions and maintaining the overall soil quality (Lal 2005).

Cover crops are grown for multiple reasons and their use may present advantages and disadvantages, as comprehensively reviewed by Dabney et al. (2001). With regard to soil moisture conditions for the main crop, the benefits may derive from higher water infiltration, less evaporation losses through an increased residue cover, an increase in SOC, improved soil physical properties (Lu et al. 2000), or removal of the excess water from a wet soil to allow timely establishment of the next crop (Unger and Vigil 1998). However, the reduction in soil moisture is the main reason why cover crops are more suited to subhumid and humid regions, unless irrigation is available to compensate for the extra water consumption by the cover crop. The use and the choice of cover crop species are highly site-specific and depend on the main objective to be achieved. Short-cycle and early maturing species or a premature interruption of the cover crop cycle have been proposed to reduce competition with the main crop (Whish et al. 2009; Salako and Tian 2003; Zhu et al. 1991). In semiarid regions with summer or winter rainfall, normally a single cash or food crop is produced during the growing season, often followed by fallow. In some regions, more than one-third of the agricultural land may be under fallow. With the NT system of soil and crop management, it has been shown that introducing cover crops (for forage or grain) in rotation can reduce the fallow land and simultaneously improve the soil cover, rainwater infiltration, soil water storage, biological nitrogen fixation (in case of legumes), and SOM and fertility (Goddard et al. 2008; Crabtree 2010), while reducing the soil evaporation as already indicated from crop residues (Jalota and Arora 2002). This has been shown to work in semiarid regions in many parts of the world, including North Africa (Mrabet 2008), Canada (Baig and Gamache 2009; Lindwall and Sonntag 2010), the United States (Ransom et al. 2007), Australia (Flower et al. 2008), and Eurasia (Gan et al. 2008). Similarly, with irrigated systems, off-season cover crops provide similar advantages.

10.4 PRODUCTION SYSTEM MANAGEMENT

10.4.1 Crop Management

This section focuses on the different ways that crops can be managed within production systems to improve the soil water availability, WUE, and WP, apart for cultural practices related to soil tillage, residue management, and soil cover, which were dealt with in the earlier sections. These constitute a large range and include crop and cultivar choice, crop establishment and yield response to water, crop genetic improvement, pest management, fertilization and nutrient management, crop phenotypic expression, and crop rotation and intensification.

However, it must be stressed that individual practices that form a constituent part of good crop management and good production system management for optimizing the use of rainfall or irrigation water are often interrelated in terms of their effects on the final outcome. The interactions among practices can work synergistically to produce outcomes in terms of soil moisture availability, WUE, and WP, in which the "whole is larger than the sum of the parts." For example, for a given amount of rainfall, the soil moisture availability to plants depends on how the soil surface, the SOM, and the plant root systems are being managed. Also, high water productivities under a good soil moisture supply are only possible when plant nutrition is adequate. Similarly, no amount of fertilizer application and use of modern varieties will improve the WUE and the WP if the soil has a 20–30 cm hard plough pan 15–20 cm below the surface; and worse, if the soil has no organic matter and life in it to build and maintain a good soil structure and porosity for maximum moisture storage and root growth. Equally, without the maintenance of a good water infiltration status of the soil and without the soil cover to minimize soil evaporation, it is not possible to fully optimize and maximize water use and WP.

Thus, all else being equal, soils that are maintained in good health and quality will offer the possibility of making the maximum amount of soil moisture available for crop production and the possibility of optimal water-use efficiencies and water productivities through good agronomic manipulation or good crop management. However, good crop management is not an independent variable but a function of how sustainably the production system as a whole is managed in order to sustain or intensify production while harnessing the desired ecosystem services. It is with this concept in mind that the following sections discuss some of the key elements of crop and production system management in relation to soil water availability, WUE, and WP.

10.4.1.1 Crop and Cultivar Choice

The choice of adapted crops and cultivars in irrigated or rainfed production systems, from a moisture viewpoint, is dictated primarily by the nature of the water supply (amount, frequency, and variability) and the type of production system deployed (tillage system or NT system; also generally known as the conservation agriculture [CA] system) (Friedrich et al. 2009; Kassam et al. 2009). Production systems define the possible biological space–time relationships with the prevailing environment and resource use and have an overriding influence on crop agronomy or crop management and cultivar choice, whereas the economic and environmental objectives of the producer will dictate which adapted crops and their cultivars can best fit into the cropping system in space and time. For example, relatively early sowing is possible with NT production system, with improved WUE and WP, compared with tillage systems. The NT system can also offer the opportunity to introduce crop cultivars of longer maturity and higher yield potential or to include a shorter maturity relay crop variety for food or as cover crop.

The water relationships of crops depend on many attributes of the crop and the soil, but they depend, in the case of rainfed crops, even more on the seasonal climate and the weather conditions of the place where it is grown—which determine how much water the crop will receive and when, and how fast the water will be used, and how much of water can be stored in the root zone. It is therefore important that the environmental physiology of the crops and the crop cultivars fit appropriately into the *time available* for crop growth and phenological development and that the crops and their cultivars participating in the cropping system are able to adjust their life cycles to match the unpredictable year-to-year variations in the length of the growing period and in the soil moisture balance. The ability to withstand diurnal water deficits and to survive dry periods in a state of physiological dormancy seems likely to be important during this stage under both rainfed and irrigated conditions (Blum 2009; Soriano et al. 2004; Bunting and Kassam 1988).

Within any irrigated or rainfed production system, only a portion of the soil-available water (between field capacity and wilting point) is *readily available* to crops, which is equal to the level of the maximum depletion of the soil water that a crop can tolerate without a decrease in the plant growth rate. This varies with the type of crop as well as the cultivar. The value of readily available water for production depends in part on the crop cultivar, the quality of the soil, and the evaporative demand of the atmosphere. All these factors, including the crop and the cultivar environmental adaptability requirements, and in combination with economic factors alongside the length of time that the water supply from irrigation or rainfall will be available and its reliability, will influence the choice of crops and cultivars that might be considered for the cropping system (Kassam et al. 2007; Gregory et al. 2000; Bunting and Kassam 1988; FAO 1978–1981; Doorenbos and Kassam 1979). Some crops, such as potato, onion, and strawberry, require the soil to be continuously moist if they are to produce good yields; others, such as cotton, wheat, sorghum, safflower, and olive, will tolerate drier soil conditions. However, the level of depletion that a crop will tolerate varies greatly with their stage of development; most grain crops are vulnerable at the time of germination or planting, particularly under rainfed conditions, and, once established, prefer a smaller depletion during changes from the vegetative to reproductive growth, or in the case of cereals, during the period of panicle initiation, heading, and flowering to fruit and seed setting.

Further, crops vary in the extent to which the leaf water potential can fall without interrupting transpiration or doing damage to the leaves or other parts of the plant. For a given soil type or quality and level of evaporative demand, differences in the root system properties, the leaf and tissue water relations, and the crop development characteristics are all important in determining the differences between crops and among cultivars in the magnitude and time course of the readily available soil water. Doorenbos and Pruitt (1977) have reviewed the general information for different crops on the rooting depth and on the readily available water for different soil types and evaporative demand. Such information together with the information on the yield response to water provides a basis for designing cropping systems that can optimize the available water and offer best water productivities, including under deficit irrigation (FAO 1992; Doorenbos and Kassam 1979).

The WUE and WP of rainfed crops can be improved through crop and cultivar choice by ensuring a good fit between the crop growth cycle and the length of the prevailing rainfed growing period across the different climatic zones and also ensuring that the chosen crop cultivars have access to adequate nutrients and pest control (including weeds) to offer best WUE and WP. For example, in the warm tropical climatic zones with rainfall between 400 and 600 mm, annual grain crops of similar maturity are selected to fit the moisture regime, but there are specific component crops included in the crop association that allow for fuller use of the end of wet season moisture. In areas of higher rainfall up to 1000 mm, crop mixtures of grain crops with some root and tuber crops, especially those involving different maturities, are common. In areas with above 1000 mm of rainfall, crops and their cultivars are selected to fit into multiple cropping systems that are based on both the simultaneous (intercropping) and sequential (relay cropping) principles to maximize the use of the available soil water (Bunting and Kassam 1988; Kowal and Kassam 1978; Andrews and Kassam 1976).

In warmer regions with a long wet season as in the humid tropics, or a shorter wet season as in the seasonally dry tropics, with irrigation facilities, crops can be grown year round. Once crop cultivars of a certain duration have been selected to match the prevailing moisture regime, and barring other constraints such as poor soil health, soil compaction, and limited soil rooting volume, WP improvement is a function of good crop nutrition and protection, and ensuring minimum soil evaporation losses and the maximum proportion of available water consumed as transpiration (Passioura and Angus 2010), aspects that are discussed later in this section. Under drought-prone environments, WP can be improved or maintained by selecting cultivars that have an effective dehydration avoidance ability (Blum 2009) so that they can endure or withstand a dry period. Usually, this is based on the cultivars' ability to extract more stored water from the soil profile, by developing a bigger working range in the water potential in leaves and other plant parts through osmotic adjustment and by storing water in their tissues so that wilting is delayed (Chimenti et al. 2006; Blum 2009; Sellin 2001; Ali et al. 1999; Ludlow and Muchow 1988; Kassam et al. 1979).

10.4.1.2 Crop Establishment and Yield Response to Water

Good and timely crop establishment is essential for achieving high WUE and WP. However, crop establishment can be a precarious or a vulnerable stage in a crop's life, particularly if the crop must be established with soil moisture derived from rainfall. This is because not only must the soil moisture supply be adequate for the seed to germinate, but it must also continue to supply the seedling roots with water and nutrients for growth. Under rainfed conditions, in a seasonally dry climate, whether in the tropics, subtropics, or a temperate climate, every year the farmer and the crop must cope with the variability of the soil moisture supply around the onset of the rains, and therefore at the start of the growing period. Each year, the start of the growing period can be different. However, for the seasonally dry tropics, it has been shown that an adequate soil moisture supply for crop establishment is reached when the rainfall is around 0.5 ET, increasing subsequently to meet the actual crop water requirement of the growing crop as its leaf area increases (FAO 1978–1981; Kowal and Kassam 1978). The actual crop water requirement is dictated by the evaporative demand of the atmosphere and the crop growth stage, in particular the crop leaf area. Dry spells soon after germination can be harmful if the soil moisture supply drops below 0.5 ET.

It is possible to make practical estimates of actual evapotranspiration (ETa), and hence the crop water requirement, from computed ET using empirically derived crop coefficients (kc), such as Eta = kc ET. Values of kc for different crops at different growth stages are given in Doorenbos and Kassam (1979). As indicated, for many dryland crops, kc at the time of crop emergence and establishment is 0.4–0.6, increasing to a maximum of 1.0–1.3 when the crop canopy covers most or all of the ground and is able to intercept most or all of the incoming radiation. This occurs in many crops and environments when the leaf area index (LAI) is around 3 (Stewart 1991; Bunting and Kassam 1988; Kowal and Kassam 1978). The relationship between relative ET (ETa/ET) for several field crops shows that at a given LAI, crops of markedly different canopy structures (e.g., sorghum, cotton, groundnut, pearl millet, and maize) use water at very similar rates (Kowal and Kassam 1978). Thus, factors that control the leaf area, particularly the nutrient fertility and the plant population, will dictate the time course of ETa, or WUE, and yield or WP.

In general, the relationship between yield (Y) and ET is linear and that each cultivar has its own ration of yield decline to ET deficit, provided water is the only limiting factor (Stewart 1991) and the required inputs of nutrients were used and weeds were controlled, etc. However, a water deficit of a given magnitude may occur either continuously over the total growing period of the crop or it may occur during any one of the individual growth periods,that is, establishment, vegetative, flowering, yield formation, or ripening. The effects of a water shortage on yield at the different growth stages of a number of crops are reviewed in Doorenbos and Kassam (1979), where the response of the yield to the water supply was quantified through the yield response factor (ky), which relates the relative yield decrease to the relative ET deficit. In the case of deficits occurring continuously over the total growing period, the effects of increasing water deficits on yield were less (ky < 1) for alfalfa, groundnut, safflower, and sugar beet than for banana, maize, and sugarcane (ky > 1). In the case of deficits occurring during the individual growth periods, the effect on yield is relatively small for the vegetative and ripening periods and relatively large for the flowering and yield formation periods.

This means that when water and crop management are not limiting, an analysis of the crop water production functions when performed for a range of crops can serve to identify those crops and cultivars that are best suited ecologically to the prevailing or expected water regime from rainfall or irrigation. They also help identify what crops and cultivars should be selected for the different seasonal moisture expectations from rainfall or irrigation. When the effects of the management decisions (such as plant population and fertilizer application levels) are simulated in the analysis, optimal management practices for different types of rainfall and irrigated moisture regimes can be identified. They can thus provide the basis for an economic evaluation for better estimates of production capabilities (Stewart 1991).

For irrigated conditions, crop management for the optimal use of water (i.e., to achieve best WUE and WP) can be simulated against particular objective functions, and actual crop management can follow the planned simulations. In the case of rainfed conditions, the rainfall probability analysis and the associated soil water balance analysis are required to quantify the probabilities of different rainfall amounts in selected time periods. This also quantifies the dates when the rainy period may begin and end and reveals the probability of dry (or wet) spells in specific time periods. This provides a basis for broad-based planning, including an analysis of the risk, allowing reference crop and cultivar mix and cropping systems to be identified. Linking such an analysis to an additional analysis of rainfall predictions, as is done in the case of response farming, allows crop management decisions regarding crop and variety types, planting dates, plant densities, and fertilizer levels and application to be made in response to the upcoming season (Stewart 1991). Crop water management for improved WUE and WP based on response farming relies on the notion that just prior to the start of each season, it is possible to exclude a significant portion of the probabilities (from the total range of probabilities) and have new probabilities assigned to the remainder. The key principle of response farming, as elaborated by Stewart (1991), is the reduction of the effective variability through an improved rainfall prediction, which does not mean pinpointing what is to occur, but, rather, identifying a portion of the range of recorded happenings that should not need to be considered as possibilities in the current season. This concept is based on the findings in different locations that there is a relationship between the time the rainfall season begins (date at which a particular soil moisture supply may be reached) and the rainfall amount and duration thereafter (Stewart 1991; Stewart and Kashasha 1984; Kowal and Kassam 1978).

Thus, from the above, it is clear that the crop management strategy for improved or optimal WUE and WP requires attention to a whole suite of elements. There are additional factors that have a significant impact on the overall water-related performance. For example, with tillage-based sowing and crop establishment, time is required to prepare the seedbed at the start of the rains. Also, actual rainfall is needed for germination, since after preparation, the seedbed dries out and often loses its capillary contact to the deeper soil water. As a result, moisture and time are spent that can delay sowing and crop establishment, as well as expend energy that may be saved or spent on something more productive. Also, the effective rainfall is reduced, thereby decreasing the potential WP, as well as WUE or effective water use (Blum 2009; Soriano et al. 2004).

The key to the effective use of soil water under rainfed conditions is to be able to plant the crop as early as possible. Any delay in crop establishment usually leads to a loss in yield in the case of rainfed crops. Where the average length of the growing period is short, as in the case of the semiarid regions, early sowing reduces the chances of late season water deficit. Given the rainfall variability at the start of the rainy season, often it is not possible to take full advantage of early sowing with tillage-based approaches in which the soil moisture that is available at the beginning of the season is used unproductively in land preparation through tillage for subsequent sowing.

An alternate approach to sowing in tilled soil is the possibility of sowing early into dry soil or just at the time of the onset of rain, if the soil has some moisture. This is only feasible under CA, which involves direct seeding into a soil with an organic mulch cover that allows, as seen earlier, maximum infiltration and therefore maximum effective rainfall. Where the rainfall climate is semiarid savannah with less than 90–120 days or it is a dryland type with no humid period during the rainy season, an adaptation such as dry sowing in mulch-covered microbasins or pits (called *likoti*, *tassa*, and *zai*) help achieve maximum infiltration and early sowing and crop establishment (Marongwe et al. 2011; Owenya et al. 2011; Silici 2010). In undisturbed dry soils, germination can occur on the basis of the available humidity in the soil pores from subsoil moisture, even when the bulk soil is below the permanent witling point. Similarly, in rainfall climates that are humid, a mulch-covered NT permanent bed system provides a good basis for crop establishment and for achieving higher WUE and WP (Govaerts et al. 2007). This is because the soil moisture in undisturbed mulch-covered soils is still at much higher levels and closer to the seeding soil horizons than in fully tilled soils.

Where the rainfall season is longer, it is possible to increase the WUE and the WP through an increase in the cropping intensity as well. Early sowing and crop establishment of the first crop allows a second crop to be fitted into the cropping system more optimally, and in certain cases, can even create time for a third short-season crop to be fitted into the cropping system. This has happened in Brazil in the Cerrados with the maize–soybean cropping system (Landers 2007) and in the Indo-Gangetic Plains with the wheat–rice cropping system (Hobbs et al. 2008; Hobbs 2007).

10.4.1.3 Crop Genetic Improvement

Physiologically, an improvement of the genetic yield potential, and therefore WP, with modern cultivars has been achieved through improving the HI by improving the sink capacity. At the same time, to achieve higher WUE and WP, the root system must be able to exploit the largest possible soil rooting volume for available water. Also, key phenological and physiological processes that determine sink size and yield formation, for example, panicle initiation, flowering and seed setting in cereals or tuber initiation in tuber crops, and yield components such as the number of head-bearing tillers, seeds per spikelet, or seeds per cob, etc., are protected against drought or extreme temperatures as much as possible. Thus, the selection of improved WUE and WP has tended to lead to a larger root system and a higher HI, but also to physiological resilience to drought and temperature stress. The HI, WUE, and WP are indices and represent the outcomes of a series of crop ecophysiological processes operating in the right way under normal circumstances as well as under situations of stress causing water deficits and under situations of heat stress. Outcome-related indices are not helpful as explicit targets of breeding and genetic improvement programs. Instead, for water-limited and drought-prone environments to target plant adaptive characteristics including dehydration avoidance traits that can (1) enable the crop to establish as early as possible and reach and exploit the maximum amount of soil-available water for transpiration and the maximum photosynthesis and desired biomass partitioning and (2) help minimize the impact of dehydration under water deficit conditions (Blum 2009).

This is supported by the fact that in water-limited environments, the timing of flowering is perhaps the most important trait for breeders to select in order to achieve a good balance between the water used during canopy development and the water used during seed setting and grain filling (Passioura and Angus 2010; Fischer 1979). This is because the yield is correlated with the soil moisture available from the soil storage and is supplemented by the rainfall or irrigation during the yield formation period for all crops.

However, because of the yearly variation in the length of the rainfed growing period, the crop in which yield formation begins early may do better in one season whereas the crop with a later set of yield formation may do better in another season. Under water deficit or drought situations, many short-duration modern cultivars are less able to cope due to the lack of elasticity and, combined with the high-density close spacing approach, often fail completely to produce a yield. Local cultivars, on the other hand, often have a better ability to respond to drought with a reduction in yield rather than complete failure.

The ability to withstand, tolerate, and recover from drought depends on the extent to which the crop can adjust its solute potential to maintain turgor in the roots and in the shoots and leaves (Passioura and Angus 201; Blum 2009; Ali et al. 1999; Ludlow and Muchow 1988; Kassam et al. 1979). Thus, it should be possible to produce cultivars that have a full complement of drought-tolerant and drought-resistant genes introgressed through marker-assisted breeding as well as through gene transformation including trait-specific genes from novel sources. Such drought tolerance would also impart salinity tolerance, making possible the more effective use of saline water.

However, it must be emphasized that the best drought proofing cannot be achieved through genetic improvement alone. In the final analysis, adaptability to drought is a production system responsibility in which agronomic manipulation and the management of all the different components of the soil–plant–nutrient–water system have an influence on the final outcome in terms of WUE and WP. Often, the agronomic manipulation and the soil management to improve the root formation and the rooting depth, as in the case of CA or the system of rice intensification (SRI) in uncompacted and well-structured soils with deep reaching biopores, is the best foundation layer of resilience against drought that can be deployed.

10.4.1.4 Pest (Weeds, Insects, and Pathogens) Management

Unhealthy and weak plants in degraded agroecosystems tend to succumb to infestation by pests of all kinds, thereby reducing both the WUE, or EUW, and WP. The reductions in WUE and WP can occur mainly through a reduction in the photosynthesis and the growth of the crop plants, including the root system, due to competition from weeds or an attack by insect pests or pathogens.

In the case of weeds, the decrease in WUE and WP occurs because water that would otherwise be available for crop growth is transpired by weeds. The loss of water through weeds can occur at any stage in the cropping cycle, but this has to be balanced with evaporation loss from the bare soil surfaces. Weeds also compete with crops for nutrients and light, thereby reducing their growth. In the case of semiparasitic weeds such as *Striga*, the host plant becomes stunted, thereby decreasing both WUE and WP. Where cropping relies on stored water in the soil in water-limited environments, WUE and WP can be increased by keeping the land weed free through the entire cropping season, as was shown by Anderson and Greb (1987) for proso millet grown in dryland agriculture in the Great Plains of the United States. Similarly, in the case of summer fallow periods to accumulate and conserve soil moisture for subsequent cropping, weed growth during the fallow period is reduced or avoided by using herbicides.

Herbicide technology eliminates the need for tillage in many cropping systems (Shear 1985). However, tillage is still common in many regions of the world, and where it is practiced, WUE and WP are lower as a result of the loss of soil moisture in land preparation and also due to the delay in sowing. Many weed seeds are relatively small and can only thrive because of tillage, which creates improved seedling establishment conditions. Where crop residues are used to develop a mulch soil cover, many weeds are disadvantaged. Thus, integrated weed management involving the use of NT and mulch cover offers an important opportunity for weed suppression (Liebman and Mohler 2001), thereby increasing WUE and WP.

In situations where there is no alternative use of soil water, growing spontaneous vegetation can have a positive impact on the overall WUE and WP because the biomass generated can be used to develop mulch cover as well as protect the soil from erosion. Such vegetation can also include plants normally regarded as weeds, provided their further propagation is avoided by adequate measures.

Insect pest and diseases usually reduce the crop capacity to protect itself against unproductive water loss. This can occur because of a loss in the leaf surface area from attack by leaf-eating insects and by pathogens that cause leaf spots, leaf streaks, and crinkling, thereby reducing photosynthesis. The damage caused to the root systems by soil diseases and nematodes leads to a reduced ability to fully explore and utilise soil water, thereby reducing WUE and WP. Such damage can be greater under conditions of cereal monocropping. Losses can be reduced by using nonhost crops in rotation with cereals, as is occurring in southern Australia, Canada, and Eurasia (Baig and Gamache 2009; Flower et al. 2008; Gan et al. 2008; Goddard et al. 2008; Blackshaw et al. 2008).

10.4.1.5 Fertilization and Nutrient Management

Plant nutrients play an important role in determining the growth of roots and the yield (Rockstrom and Barron 2007) because the source of the substrate for root growth is photosynthesis, which depends on the unit leaf rate as well as on the leaf area, both of which are nutrient-dependent, as well as age-dependent. The leaf area directly affects the transpirational losses, and there is a linear relationship between the transpiration and the biomass that a crop produces, but the slope of the line depends on the nutrient availability. However, the portion of the biomass that is harvested as yield (HI) is a feature of the crop type or variety and of the moisture regime and the sensitivity of the crop growth stage to water deficit and nutrient stress (Doorenbos and Kassam 1979).

An adequate and balanced nutrient supply from a healthy soil is a prerequisite for good growth of the roots and the aboveground plant parts, for yields, and therefore for WUE and WP (Ali and Talukder 2008; Hatfield et al. 2001; Ryan 2000; Liu et al. 1998). For example, when the roots are not impaired by pathogens, the higher N status of the crop leads to a larger root system and to more soil water extraction (Deng et al. 2003; Angus and van Herwaarden 2001; Liu et al. 1998). However, as indicated earlier, under the variable rainfed conditions of the semiarid tropics and subtropics, both summer and winter rainfall, effective nutrient management for improved WUE and WP can be achieved through the practice of response farming in which risk can be minimized by delaying the decision to apply fertilizer, and how much, until later in the season when it becomes possible to predict what kind of moisture season it is most likely to be (Stewart 1991). In southern Australia, this tactic has also been shown to work and the advantage of delaying the decision to top dress is in the saving on the cost of fertilizer and avoiding yield loss by not applying fertilizer if the season is dry (Passioura and Angus 2010; Angus 2001; Angus and Fischer 1991). In practice, effective nutrient management must be seen in terms of the nutrient needs of the crops within the cropping systems in space and time so that the overall production system deployed is also conducive to efficient nutrient productivity alongside the aim of maintaining desirable levels of WUE and WP. Thus, nutrient management under the CA system for improved WUE and WP is a fundamentally different nutrient management strategy compared with that under a tillage-based system (Kassam and Friedrich 2009). Under the CA systems, the WP and the nutrient productivities are higher, and often less mineral fertilizer is needed because of greater biological nitrogen fixation and improved nutrient conservation within the cropping system (Baig and Gamache 2009; Friedrich et al. 2009; Goddard et al. 2008).

Vegetative growth has a direct relation to water use, as well as to the yield and WP for a given supply of soil water. In the case of cereals, this is because the vegetative biomass at the time of anthesis is related to the number of grains per unit area. Similarly for legume crops, biomass at the onset of flowering and subsequent biomass growth during further flowering determine the numbers of flowers, pods, and seeds produced per unit area. In the case of root and tuber crops, biomass at the time of tuber initiation and the subsequent growth of the crop determine the number of tubers per unit area, the number that actually bulk, and the extent of bulking. Assuming healthy crop roots, vegetative growth and the formation of reproductive- or yield-forming parts depend on the nutrient status of the soil and of the plants. Too little vegetative growth, and therefore suboptimal WUE and WP, can be caused by insufficient nutrients, late sowing, and suboptimal plant density. On the other hand, early sowing, excessive nitrogen, and high plant density cause excessive vegetative growth. In areas that suffer from end-of-season drought, excessive growth can lead to the exhaustion of the soil water, leaving insufficient soil water for transpiration and grain filling (Passioura and Angus 2010). There is also evidence that excessive nitrogen can lead to greater structural carbohydrate rather than stored carbohydrate that can be translocated to the grain together with nitrogen during grain filling, thus reducing WP. Further, excessive nitrogen can lead to foliar diseases and insect attack (Kitchen et al. 2008; Chaboussou 2004) and crop lodging, all of which can lower WUE and WP.

The above effects from an excessive nitrogen supply have been recorded when using mineral sources of nutrients under production systems involving tillage over many years so that the soil health is often in a suboptimal condition from compaction, poor infiltration, and low SOM. Results can also be in the opposite direction when organic sources of nutrients are used or when inorganic and organic sources are used in combination. For example, with maize, an increase in the WUE and WP was recorded when the ratio of the organic to the inorganic nitrogen fertilizer was 1:2 (Xiaobin et al. 2001). Larger root systems are produced when there is an organic source of nutrients and where the SOM content is higher and the soil microorganisms are more active and diverse (Uphoff et al. 2006). In this regard, the behavior of the rice grown under mostly aerobic soil conditions, as is the case under the SRI methods, is of particular interest. Under the SRI approach, some 20%–30% less fertilizer is required compared with irrigated flooded rice grown under the best management practice, and 40%–50% less water is required to produce a full crop. Because of the greater yields with SRI and the reduced water requirement, both WUE and WP are higher, and nutrient productivity is superior (see the SRI case description for more details).

Examples of soil nutrient deficiencies affecting WUE and WP also relate to the zinc deficiency in wheat in Turkey (Cakmak et al. 1996) and the sulfur deficiency in groundnuts in India (Patel et al. 2008). The role of calcium and magnesium in improving the pH, the soil structure, and the water-holding capacity and, consequently, WUE and WP is well known. Similarly, several researchers (e.g., Cakmak 2005) have recorded the role of the potassium nutritional status in alleviating the detrimental effects of abiotic stresses through osmotic adjustment.

Evidence shows that mineral fertilization requirements, particularly of N and P, decrease in soils that have been under the CA system for extended periods of time (Landers 2007), and the problem of low availability or immobilized P in soil is ameliorated, even when soil analyses do not show high quantities of soluble P (FAO 2008; Turner et al. 2006). Thus, combined water and nutrient productivity improved over time in CA systems, whereas with tillage-based production systems, nutrient and total productivity including WP remained at a suboptimal level.

10.4.1.6 Agronomic Manipulation for Best Phenotypic Expression

Much of our scientific thinking about agronomic practices and crop production has been based on the assumptions that a crop can be best produced with soils that must be tilled year after year and with increasing tillage intensity in many cases; that soil microorganisms and the SOM are not essential to soil fertility or to the maintenance of soil health and ecosystem health; that plant root systems and their interactions with the soil microorganisms can be ignored in studies aimed at understanding the ecophysiological basis of nutrient- and water-use efficiencies and productivity; that soil mulch cover and crop rotation can be considered as optional in the maintenance of soil, crop, and ecosystem health and in the optimization of the use of resources such as water and nutrients; that there is only one standard way of agronomically manipulating the crop–soil–nutrient–water parameters; that the so-called undefined and unbridled quest for genetic improvement must continue to override improvements that are possible through alternative crop production practices.

For example, the CA and SRI approaches to crop production show us a different way forward. CA and SRI are both works in progress and their concepts and methods are being extended to more crops and more agroecologies, for small-scale and large-scale production. These systems are harnessing an agronomic performance that cannot be predicted by current models or the scientific knowledge generated through the reductionist scientific research approaches that have characterized much of the agricultural research during the last century and still continue to do so. It would appear that there has been a "closure of the mind" in the last three to four decades, particularly within the global public research system, with regard to the additional opportunities that exist in improving WUE and WP through agronomic manipulation of soil–plant–water–nutrient relationships as well as the manner in which the soil health and root systems are managed. Systems such as the CA, SRI, and CA–SRI have not been receiving the kind of attention they should from the scientific community. Given that such systems and agronomic manipulation can help small farmers to improve their overall and factor productivity and livelihood, this lack of attention is a serious gap in the current knowledge system.

While early planting with CA and SRI permits better WUE and WP because of improved soil moisture, upon which the nutrient productivity depends, optimal spacing appears to depend on the soil fertility conditions. Although, generally, a high seed rate and closer spacing have been the dominant approach with modern cultivars that are selected within such conditions, this may not always be optimal, as has been recently shown by the SRI approach for rice, as well as with other crops such as sugarcane, wheat, and finger millet (Uphoff and Kassam 2009). The high-density seed rate appears to have been favored over the past three to four decades, but the SRI approach shows that it is possible to improve the genetics × environment (G × E) interactions and achieve higher WUE and WP through the integration and manipulation of a crop establishment strategy with crop nutrition and weed management. In fact, CA and SRI have revealed a whole new set of opportunities to improve WUE and WP based on alternative approaches to crop and water management (as elaborated in the CA and SRI case details elsewhere).

10.4.1.7 Crop Rotation and Intensification

Many advantages and benefits are associated with crop rotations, including the possibility of higher WUE and WP for the individual crops participating in the rotation and for the cropping system as a whole, when compared with monocropping. In environments of variable rainfall, crop rotations with crops of different maturity allow the reduction of climatic risk because in poor years, not all crops are affected equally and there are positive effects between crops in the rotation, involving cereal and legume crops, from the yield viewpoint, and therefore improved WUE and WP (Tanaka et al. 2005). Equally, rotations also reduce the risk of attack by insect pests and diseases (Chabossou 2004; Krupinsky et al. 2002), thus maintaining WP. Rotations involving high biomass legume crops also allow the in situ production of functional biomass in terms of crop residues and green manure crops and can help add organic matter to deeper layers in the soil as well as increase the soil biopores (Friedrich et al. 2009; Shaxson et al. 2008) Mixed sequences of crops, plus the presence of a permanent soil cover, tend to inhibit the buildup of specific weed species that would thrive under less varied or monocrop conditions and reduce WUE and WP.

The rotation of crops involves the rotation in sequence of several species of crops, including legumes as symbiotic (plant × rhizobia) sources of plant-fixed atmospheric N, and other usable green manure cover crops, for maintaining the soil cover at all times, as well as the provision of labile organic residues both at and below the surface. It is important that the nutrient balances in the soil are maintained from one rotation cycle to the next. C accumulation only seems to occur when there is a legume in the system that fixes more N than is removed in the crop products or is otherwise lost from the system (Boddey et al. 2006; Uphoff et al. 2006).

Crop intensification involves making fuller use of the time available within the annual cropping cycle by introducing additional crops within and between seasons, thereby making fuller use of the soil water while keeping the ground covered for longer periods. According to Gan et al. (2008), long-term studies in Kazakhstan have shown that reducing and gradually eliminating summer fallow are feasible (Suleimenov and Akshalov 2007), thus improving WUE and WP. Similarly, studies in the Canadian prairie have indicated that conventional summer fallow can be replaced using annual grain legumes or green manure crops (Gan et al. 2008), and similarly in North Dakota in the United States (Ransom et al. 2006). Such replacement in the rotational system has been shown to improve the overall farm productivity as well as profitability and improving WUE and WP by 30% (Gan and Goddard 2008; Peterson and Westfall 2004).

The greater the range of plants grown, in mixtures or in sequence, the more varied will be the biodiversity of associations of organisms above ground and inhabiting the rooting depth, and the greater the competition that can suppress those that may be detrimental to the root function and thus considered weeds or pests. A crop rotation will further help in interrupting the infection chain of diseases and might have other insect pest–repellent and insect pest–suppressing characteristics. For the alterations in cropping systems to be worthwhile to farmers, there need to be local uses and/or markets for additional outputs generated by improved crop sequences and mixtures.

10.4.2 Irrigation Management

Irrigation plays and will continue to play an important role in global food security, and the need and the opportunities for expansion of irrigated cropland still exist (Oweis and Hachum 2003; Seckler et al. 2003). However, and in agreement with Rockstrom and Barron (2007), to minimize further blue water withdrawals and increase WP, a reduction in green water (water that is stored in the root zone) losses is critical. In irrigated agriculture, the fundamental question still lingers: How to produce more with the same or even less amount of water? The answer to this question is still, undoubtedly, linked to the possibilities of minimizing unproductive water losses, namely, runoff (tail water), evaporation, and deep percolation (Rockstrom et al. 2002). Sustainable management practices and technically feasible and cost-effective solutions to maximize crop transpiration and soil water storage and minimize runoff and evaporation, as well as irrigation systems to carry out such efforts, are examined here. They have to consider the different processes that affect the soil water (Figure 10.1) and the parameters that influence the processes (Figure 10.2).

10.4.2.1 Irrigation Performance

Worldwide, irrigation schemes are often designed and managed to maximize irrigation efficiencies and minimize labor and capital requirements. For this multiobjective goal, one major challenge that confronts every designer and irrigator is that the soil that conveys the water over the field has properties that are highly variable both spatially and temporally, creating an engineering problem in which at least two of the primary variables, discharge and time of application, must be estimated not only at the field layout stage, but must also be judged by the irrigator prior to the start of every irrigation event (Trout et al. 1992; Keller and Bliesner 2001; Walker and Skogerboe 1987). Recent developments in surface irrigation technology, with its array of automating devices, have largely caught up with the irrigation efficiency advantages of the sprinkler and microirrigation systems (Duke et al. 1992; Heerman et al. 1992). Thus, while it is possible for the new generation of surface irrigation systems to be attractive alternatives to the sprinkler and drip systems, their associated design and management practices are much more difficult to define and implement (de Sousa et al. 1999; Clemmens et al. 1998; Clemmens and Dedrick 1982; Heerman et al. 1992).

Among the factors that are used to judge the performance of an irrigation system or its management, the most common are efficiency and uniformity (Clemmens and Molden 2007; Hamdy 2007; Santos 1996a, 1998; Heerman et al. 1992; FAO 1989). These parameters have been subdivided and defined in a multitude of ways and have been named in various manners (Hamdy 2007; Bos and Nugteren 1990; ASCE 1978; ICID 1978). However, there are other factors influencing irrigation efficiency, building a chain of efficiency steps (Hsiao et al. 2007), and irrigation efficiency at a field or farm level may not be the same as at water basin level (Jensen 2007). In agriculture, farmers,

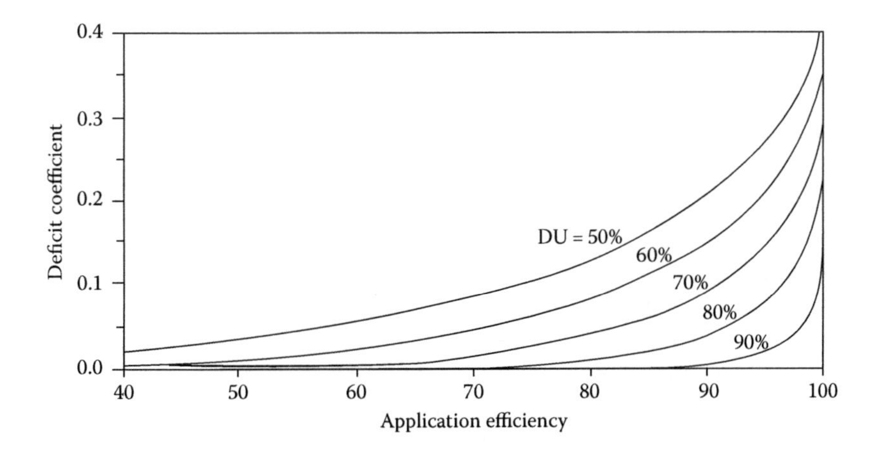

FIGURE 10.8 The relationship between the deficit coefficient, the application efficiency, and the distribution uniformity (DU), assuming normal distribution of the infiltrated applied water. The deficit coefficient is the fraction of the root zone that has not undergone irrigation. (From Playan, E. and Mateos, L., *Agr. Water Manage.*, 80, 100, 2006. With permission.)

irrigation project managers, and river basin authorities may define WUE quite differently, consisting of various components and taking into account losses during storage, conveyance, and application to irrigation plots (Hamdy 2007; ICID 1978). More consensually, uniformity (distribution uniformity) is used to express the variation in the depths of application or supplied volumes (ICID 1978; Christiansen 1942). Conceptually, the adequacy of on-farm irrigation (field level) depends on how much water is stored within the crop root zone, the losses percolating below the root zone, the losses occurring as surface runoff or tail water, the uniformity of the applied water, and the remaining deficit or underirrigation within the soil profile following irrigation (Fereres and Soriano 2007; Hamdy 2007; Heerman et al. 1992; Bos and Nugteren 1990; Losada et al. 1990). Assuming that the statistical distribution of the infiltrated water follows a normal distribution (Santos 1996a,b, 1998; Losada et al. 1990; Till and Bos 1985), Figure 10.8 illustrates the relationships between uniformity, the water deficit, and the percolation (Playan and Mateos 2006). For a given target deficit coefficient, the lower the distribution uniformity, the lower is the application efficiency.

With proper and careful design and operation, high on-farm irrigation efficiency and uniformity can be achieved directly with systems such as sprinkler and microirrigation systems (Keller and Bliesner 2001; Solomon and Keller 1978; Hart and Heerman 1976) that do not depend on the soil surface for water distribution. The issue is more challenging for surface irrigation systems that depend on the soil to convey water and where the depth of water infiltrated (defining the distribution uniformity) is a function of the opportunity time, the length of time for which water is present on the soil surface to infiltrate (Heerman et al. 1992; FAO 1989). It is worthwhile remembering that the practice of surface irrigation is thousands of years old and, collectively, it still represents perhaps as much as 95% of the common irrigation activity of today (Oweis and Hachum 2003; FAO 1989; Walker and Skogerboe 1987).

With the two sources of surface irrigation system inefficiency in mind, deep percolation and surface runoff or tail water, a very large number of causes of poor on-farm irrigation performance have been outlined in the technical literature (Hamdy 2007; Heerman et al. 1992; Trout et al. 1992). They range from inadequate design and management at the farm level to inadequate operation of the upstream water supply facilities (Walker and Skogerboe 1987). Nonetheless, since the depth of the water infiltrated at several locations in the field is commonly taken as a function of the opportunity time for water to infiltrate, in terms of a root cause, it is a most often accepted fact that the soil physical conditions and characteristics, primarily the soil infiltration capacity, constrain the sustainable performance of irrigation and the economical production of irrigated crops (Tarboton and Wallender 1989). Management practices that can eliminate or at least mitigate these constraints are reviewed in Heerman et al. (1992), Trout et al. (1992), and FAO (1989), among others. Outlined management options include both cultural practices that alter the undesirable soil condition and irrigation practices that minimize or avoid the constraints. The underpinning conclusion is that soil must absorb adequate water during irrigation to meet the crop water requirements between irrigations, with water absorption depending on the soil infiltration characteristics, the irrigation system, and the system's management.

10.4.2.2 Infiltration

Infiltration changes a great deal from irrigation to irrigation (temporal variability), from soil to soil (excessive, inadequate, and inherent spatial variability) and is neither predictable nor effectively manageable. Thus, the infiltration rate is an unknown variable in irrigation practice (Tarboton and Wallender 1989; Walker and Skogerboe 1987). Soil infiltration varies both locally and with time, the former resulting from a nonuniform soil texture and structure, topography, soil cover rate, tillage, and wheel traffic, and the latter from soil structure changes caused by SOM accumulation or depletion, frost action, tillage, consolidation from wetting and drying, surface sealing due to drop impact and overland flow, soil animal and microorganism activity, and changes in the ionic soil composition (Trout et al. 1992; Tarboton and Wallender 1989; Undersander and Regier 1988). Soils that absorb water rapidly (excessive infiltration) or slowly (inadequate infiltration) or store only limited quantities in the soil profile (inadequate water-holding capacity) often increase the costs and/or decrease the efficiency of irrigation.

Soils that slowly absorb water constrain the irrigation process by requiring low application rates to avoid water wastage (redistribution and runoff) and long application times or short irrigation intervals to maintain adequate soil moisture in the root zone. Management strategies that increase infiltration require determining the location and the nature of the restricting layer and the process that created it (Trout et al. 1992). The agronomic remedial actions that are required to improve the existing conditions are (1) NT to promote the formation and maintenance of biopores created by the macrofauna activity and the former root channels, in combination with surface residue retention (Jordan et al. 2010; Tebrugge and During 1999; Miller et al. 1987); (2) reduced or controlled traffic to decrease the formation of dense tillage pans and compaction of the tillage layer (Fornstrom et al. 1985; Eisenhauer et al. 1982); (3) deep, vertical, noninversion subsoiler to break tillage hard pans; (4) increased organic matter content or a decrease in the proportion of sodium in the soil, to enhance soil aggregate stability (Ben-Hur and Lado 2008); (5) use of chemical soil stabilizers such as polyacrylamide (PAM) to maintain soil stability (Orts et al. 2000; Lentz and Sojka 1994); (6) reduction of clay dispersion by calcium addition (Trout et al. 1992); and (7) use of plants and residues for the protection of the soil surface aggregates from water drop impact (pressurized systems) and the shear force of the overland flow (Silva 2010; Cary 1986). As already outlined in previous sections, there are ways to approach agricultural production systems, whether rainfed or irrigated, to conditions that are close to those of natural ecosystems in terms of hydrophysical conditions, which per se show the most favorable, site-specific behavior in terms of water infiltration. However, in tillage-based production systems where soil infiltrability is below the necessary rate, the irrigation system and the system's management must be adapted to the low rate, to improve the existing conditions, with (1) use of long and frequent irrigations and systems (level-basin, surge flow, cablegation) that allow for the rapid advance of surface flows and uniform infiltration opportunity times (de Sousa et al. 1999; Clemmens 1998; Shahidian et al. 1998; Clough and Clemmens 1994; Kemper et al. 1987); (2) use of sprinkler spray heads on drop tubes (Thompson and James 1985); (3) use of spray booms or long throw nozzles to reduce sprinkler application rates (Solomon et al. 1985); (4) conversion of center pivots to lateral move or stationary systems that allow for the application of smaller application depths with an increased irrigation frequency (Trout et al. 1992; Solomon et al. 1985); (5) use of microbasin or reservoir (Garvin et al. 1986) to pond and hold water until it can infiltrate; and (6) conversion to or use of microirrigation (drip irrigation), which allows low application rates to match the low soil infiltrability (Keller and Bliesner 2001; Solomon and Keller 1978). Cablegation systems are automated surface irrigation gated-pipe systems (de Sousa et al. 1999; Shahidian et al. 1998; Kemper et al. 1987) that inherently provide for cutbacks in the furrow stream and the subsequent reduction in runoff, potentially increasing the irrigation efficiency in low infiltrating soils. Surge flow (surge irrigation) (de Sousa et al. 1999; Miller et al. 1987) is a process by which an irrigation is accomplished through a series of individual pulses of water onto the field such that the flow interruption is long enough to infiltrate all surface water.

Soils with excessive infiltration are usually coarse-textured soils, freshly tilled soils that develop large voids between aggregates following tillage, and shrinking/swelling clays. Reducing their high infiltration rate is difficult, with irrigation usually increasing the cost and/or decreasing the WUE (Trout et al. 1992). A mix of agronomic and irrigation management practices is needed to cope with the conditions of excessive infiltration: (1) compaction (Khalid and Smith 1978) and compact furrows with equipment and/or pacing wheels (Fornstrom et al. 1985; Musick et al. 1985); (2) NT or a reduction of the depth and disturbance of the tillage to improve the soil aggregation and to reduce the creation of interaggregates; (3) surge irrigation (de Sousa et al. 1999; Miller et al. 1987); (4) high surface irrigation applications, level basin systems (Clemmens 1998; Santos 1996b; Clough and Clemmens 1994), and a reduced field length to decrease the time required to spread the water across the field and thus improve the water distribution uniformity; and (5) conversion from surface to sprinkler or drip systems that do not depend on the soil surface for water distribution, therefore circumventing the problem.

All soils exhibit some degree of soil infiltration variability, locally (spatial) and with time (temporal). Spatial infiltration variability results from a nonuniform soil texture and structure (inherent variability), topography, tillage, and wheel traffic (Miller et al. 1987; Fornstrom et al. 1985; Trout and Kemper 1983), while temporal infiltration variation results from structural changes caused by distinct causes. Identifiable, large-scale spatial variability is best dealt with through (1) differential application of the residue and other organic matter that counteracts it and (2) subdivision of large fields into management subunits based on infiltration. Inherent soil variability is difficult to ameliorate. However, spatial variability resulting from tillage and wheel traffic can be ameliorated through (1) management of tillage and equipment traffic to reduce uneven soil compaction; (2) even traffic across all or alternate furrows; and (3) wheel compaction (better used with surge irrigation) to reduce the subsurface texture or structure of nonuniformity soils (Purkey and Wallender 1989). As far as delivering irrigation water is concerned, the fix is more challenging and complex because the means to deal with both the spatial and temporal variabilities in infiltration is to monitor the irrigation, to adjust the application rates, and to set times to obtain acceptable performances (Trout et al. 1992; Walker and Skogerboe 1987). Manual adjustments are critical, but costly. Feedback control systems that automatically adjust the irrigation application rates and times based on automatically sensed advance rates and tail water runoff have been effectively used (de Sousa et al. 1999; Purkey and Wallender 1989). In such cases, according to Purkey and Wallender (1989), surge irrigation decreased the effect of the infiltration variability by as much as 50%. Precision irrigation has also been advocated (Sadler et al. 2005). Sprinkler and drip irrigation systems that do not depend on the soil surface infiltration rate for water distribution are the next best option to deal with the infiltration variability. As long as their water application rates do not exceed the infiltration rates, water will be absorbed into the soil, counteracting the infiltration variability problems (Silva 2010; Keller and Bliesner 2001; Solomon et al. 1985; Solomon and Keller 1978).

Control systems, water supply management, and precision irrigation certainly present real opportunities to handle the uncertainty associated with variable soil infiltration and to apply water to croplands uniformly and efficiently (Heerman et al. 1992; FAO 1989). The literature suggests that in all cases where high levels of uniformity and efficiency were achieved, irrigators utilized one or more of the following practices: (1) precise and careful field preparation; (2) timely irrigation scheduling; (3) regulated inflow discharges; and (4) tail water runoff restrictions, reduction, or reuse. Opportunities for water conservation with such precision irrigation and cutting-edge soil management practices are discussed in Sadler et al. (2005).

10.4.2.3 Soil Water Storage

The water storage capacity varies primarily with the texture, the SOM (Rawls et al. 2003), the inherent restrictive layers, and the compacted soils layers formed by tillage and equipment traffic (Voorhees et al. 1986), limiting the maximum amount of water that can be efficiently applied and the allowable interval between irrigations. The worse situation is when irrigation management must adapt to spatial soil variations in infiltration alongside variations in the soil water storage capacity. Either or both of those characteristics being lower in one location than in the bulk of the field can cause runoff from that location, despite the irrigation system being optimally designed for the bulk of the field (Sadler et al. 2005). Runoff water collecting within the irrigated area or leaving the field damages crops, wastes water, and moves sediments, nutrients, and biocides.

The frequent, light irrigation applications required on soils with a low-water-holding capacity increase the labor costs (except for mechanized irrigation systems) and decrease the water distribution uniformity of the surface systems (Trout et al. 1992; Walker and Skogerboe 1987). Management practices, such as restricted traffic; lightweight tillage or no-tillage, harvesting, and transport equipment; and the avoidance of traffic in moist soil, have been successfully used to slow the creation of compacted layers (Musick et al. 1985; Kaddah 1976). Since short or frequent irrigation intervals require the systems to apply small amounts of irrigation water efficiently, conversion to automated, mechanized, or microirrigation systems is advocated when possible (Sadler et al. 2005; Buchleiter et al. 2000; Camp et al. 1998; Batchelor et al. 1996; Duke et al. 1992).

10.4.2.4 Soil Crusts

Soil crusts occur over a wide range of soils due to the action of rainfall and irrigation water and are more prevalent in soils with a low organic matter and a high silt content (Ben-Hur and Lado 2008; Lado et al. 2004; Bjorneberg et al. 2003; Ramos et al. 2003; Santos and Serralheiro 2000; Miller and Gifford 1974). Created when the water-drop impact and the overland flow break down the surface structure and rearrange particles into a denser, amorphous, and hard mass, the crusts impede seedling emergence and impact the exchange of water, air, and heat between the soil and the atmosphere, thereby substantially lowering the infiltration (Trout et al. 1992; Miller and Gifford 1974). Irrigation management practices comparable to ones used to deal with low infiltration soils (described above) are advocated (Ben-Hur and Lado 2008; Lado et al. 2004). Reduced and especially NT systems that leave enough crop residues and promote the accumulation of organic matter at the surface provide the effect of shielding the soil surface from those destructive forces and are the first option to be considered in preventing soil sealing and crusting (Lado et al. 2004; Rawls et al. 2003; Miller and Gifford 1974). Comparing different irrigation methods, sprinkler systems are the main culprit in causing surface crusts. Minimum sprinkler application—in amount, intensity, and kinetic energy breakdown—with reduced sprinkler height and droplet sizes can decrease the soil collapse and crust formation (Silva 2010; Bjorneberg et al. 2003). Soil conditioners, such as PAM, also tend to stabilize the soil aggregates from the destructive impact energy of the sprinkler irrigation systems' water droplets (Bjorneberg et al. 2003; Sojka and Bjorneberg 2002) and the surface irrigation shear forces of the overland flow (Sojka and Bjorneberg 2002; Santos and Serralheiro 2000).

10.4.2.5 Irrigation-Induced Soil Erosion

An overview of water erosion from irrigation by Koluvek et al. (1993) indicates that measured annual sediment yields from furrow-irrigated fields often exceed 20 t/ha with some fields exceeding 100 t/ha. Under the center pivot, sediment yields as high as 33 t/ha were measured, with annual sediment yields as high as 4.5 t/ha also reported from irrigation tracts. Typically, overland flow applies shear forces to the soil surface, which causes particle detachment and movement (Sojka and Bjorneberg 2002; Koluvek et al. 1993). As flow velocities increase, shear forces increase and eventually exceed the shear stress required to overcome the cohesive forces between the soil particles. Under surface irrigation, as the water infiltrates the soil, the sediments deposit at the furrow surface to form a thin seal or depositional layer (Orts et al. 2000; Trout and Neibling 1993). The process is potentially halted if the depositional seal formation is slowed down and high infiltration is maintained through the use of known erosion control practices coupled with minimum soil disturbance and the selection of the appropriate cropping sequences (Lado et al. 2004; Koluvek et al. 1993; Trout et al. 1992). Furrow erosion has been reduced using various approaches, including straw placed in furrows (Brown 1985) and sodded furrows (Cary 1986). With a large percentage of the total seasonal furrow erosion occurring during the first irrigation following tillage (Lentz et al. 1992), PAM with an 18% negative charge density injected in the irrigation furrow advance water has also been used to reduce furrow erosion (Orts et al. 2000; Lentz and Sojka 1994; Lentz et al. 1992). Santos and Serralheiro (2000) showed that PAM increased the cumulative infiltration by 15%–20% on furrow-irrigated Mediterranean soils. Permanent ridges for furrow irrigation systems and crop establishment under NT have been successfully applied (Cahoon et al. 1999) and could substantially reduce furrow erosion.

Silva (2010) has reviewed the factors affecting runoff and control practices under sprinkler irrigation that cause erosion only if the application rate exceeds the soil infiltration rate, resulting in water ponding and subsequent surface flow (Lyle and Bordovsky 1983). The soil and topographic variations, along with the water supply and economic constraints, often compromise system designs, and repeatedly, the application rates exceed soil infiltration rates, primarily under the outer spans of the center pivots irrigation system and with the use of low-pressure nozzles that have smaller wetted diameters (Silva 2010; Bjorneberg et al. 2003; Sojka and Bjorneberg 2002; Trout et al. 1992). With an improper average operating pressure as the most common cause for poor sprinkler system performance (Heerman et al. 1992), reducing the sprinkler application rate or increasing the soil infiltration capacity (described above) reduces or eliminates runoff. Tillage practices, such as basin or reservoir tillage, increase the surface storage to prevent overland flow (Garvin et al. 1986) and erosion. Sprinklers applying high molecular weight, water-soluble, anionic PAM were shown to improve the soil infiltration rate and reduce soil erosion (Santos et al. 2003; Aase et al. 1998). PAM applied to the soils through the irrigation water acted as a binding and settling agent to increase the soils aggregate stability and infiltration and reduce runoff and sediment losses (Santos et al. 2003; Bjorneberg and Aase 2000; Aase et al. 1998).

10.4.2.6 Deficit Irrigation

The inherent and management-induced nonuniformity of the irrigation systems implies that some water deficit and/or percolation must occur even with the best irrigation schedule. With volumes of irrigation less than the volume of water needed for ET (crop water requirement), under deficit irrigation, all of the applied water remains in the root zone and may be used in ET (Fereres and Soriano 2007). Evidently, the whole field will have some soil water deficit after irrigation and there will be areas with a level of deficit that may be detrimental for production (Fereres and Soriano 2007; Zwart and Bastiaanssen 2004), emphasizing the need for irrigation systems that can deliver high uniformity water applications (Figure 10.8). In the process, the WP (either taken as yield or net income per unit of water used in ET) of the applied irrigation water under deficit irrigation, that is, the application of water below the full crop water requirements or ET, is higher than under full irrigation (water satisfying the full crop water requirement) (Fereres and Soriano 2007).

Broadly, two types of deficit irrigation, sustained (SDI) and regulated (RDI), are assumed (Ramos and Santos 2009, 2010; Fereres and Soriano 2007; Santos et al. 2007; Shatanawi 2007). In SDI, the irrigation is reduced during the whole season while RDI starts with normal irrigation and then the irrigation is gradually reduced. In RDI, the deficit irrigation strategy is based on limiting the nonbeneficial water losses by reducing the amount of water for the crop during the noncritical phenological stages. The deficit irrigation is controlled during times when the adverse effects on productivity are minimized. As summarized in Aboukeira (2010), Geerts and Raes (2009), and Fereres and Soriano (2007), field results from both these practices in annual crops and fruit trees and vines show that deficit irrigation can reduce irrigation water use and raise crop WP in a number of crops. Globally, the potential benefits of deficit irrigation derive from three factors: reduced costs to production, greater irrigation WUE, and the opportunity costs of water (Aboukeira 2010).

Accomplishments in the irrigation of fruit trees and vines with an innovative technique of imposing deficit irrigation by alternating drip irrigation on either side of the fruit tree and vine row (partial root zone drying; PRD) are summarized in Fereres and Soriano (2007), dos Santos et al. (2003), and Goldhamer et al. (2002). In Perry et al. (2009), Ali and Talukder (2008), and Bouman (2007), the factors affecting WP in crop production and techniques to increase WP are analyzed.

10.4.2.7 Evaporation

Reducing evaporation while increasing productive transpiration can enhance WP if there is adequate plant nutrition. Evaporation varies with agricultural practices (Burt et al. 2005) and ranges from 4% to 15%–25% in sprinkler irrigation systems (Burt et al. 2001) where wind is the major concern (Playan and Mateos 2006). The adverse effects of an incremental wind drift increase the evaporation losses and sharply reduce irrigation uniformity. The amount of evaporation depends on the climate, the soils, and the extent of the mulch cover and of the crop canopy that shades the soil, with evaporation claiming a very high share of ET with low plant densities. As for the rainfed systems, evaporation losses under irrigation can be drastically reduced by both the tillage system and stubble or mulch. In a furrow-irrigated cotton crop with 325 mm of rain plus irrigation water, Lascano et al. (1994) measured 100 mm of evaporation under NT with standing stubble against 160 mm under CT without residues. The importance of the surface mulch rates was reported by Hares and Novak (1992), who found 1-day evaporation losses of 1.9, 1.7, 0.6, and 0.3 mm with 0, 907, 9070, and 18140 kg/ha spread straw, respectively. Burt et al. (2005) report that drip and sprinkler irrigation systems do not necessarily result in less evaporation than good surface irrigation systems. Nonetheless, the decreased area of surface wetting obtained with microirrigation is a distinct advantage to minimize the evaporation from the soil surface (Pereira 2007; Batchelor et al. 1996). Burt et al. (2005) also highlight that frequent microspray irrigation and rapid cycling of the center pivots can result in a high percentage of soil/plant surface evaporation.

10.4.3 Case Studies on Improved Production System Management

The following two case studies have been chosen to illustrate how changes in the production system concepts and the associated management practices can improve land productivity through better use of water and improvements in the soil quality.

10.4.3.1 Soil Tillage Systems in the Central Great Plains

A field study was set up in 1989 by the Kansas State University near Tribune, a region with a semiarid continental climate (mean annual precipitation = 425 mm, mean annual air temperature = 11.2°C), on a deep and well-drained loess-derived silt loam, very characteristic of the west-central Great Plains. Three different tillage systems were established on a virgin, native grass prairie area with a 3-year rotation of wheat–sorghum–fallow (WSF) under rainfed conditions. The CT system was based on a sweep plow, also used for the necessary weed control during the fallow period (three to four operations). RT used a combination of tillage and herbicides to control weeds during fallow, whereas in the NT system, weed control was entirely based on herbicides both during fallow and between crops. In all three systems, in-crop weed control was done by herbicides as needed. Fertilization was identical for the three systems, as well as the maintenance of the crop residues in the field. The only difference between the tillage treatments was that the row spacing and the drill system used in the wheat crop was 30.5 cm (hoe drill) for CT and RT and 19.1 cm (single-disk opener drill) for NT (Stone and Schlegel 2010).

This study continues today and the effects of the tillage systems on the soil physical properties are described in several publications. Based on soil samples taken in 2000, Stone and Schlegel (2010) measured the bulk density, the total N and C, the water content at –1.5 MPa matric potential, and the aggregate stability, the latter through the determination of the concentration of the WSA and the MWD. The ponded, steady-state infiltration rate was also measured in 2000 using double-ring

TABLE 10.5
Some Soil Parameters after 10 Years under Different Tillage Systems

Parameter	Units	Tillage System			
		NT	RT	CT	NP
Total carbon	g/kg	19.3	18.1	17.5	20.1
Water content at −1.5 Mpa	kg/kg	0.131	0.124	0.122	0.145
Mean weight diameter	Mm	1.55	0.66	0.57	3.78
Ponded steady-state infiltration	mm/h	30.6	15.3	11.4	24.3

Source: Adapted from Stone, L.R. and Schlegel, A.J., *Agron. J.*, 102, 483, 2010.
Note: NT: no-till; RT: reduced tillage; CT: conventional tillage; NP: native prairie.

infiltrometers. The results of these measurements are summarized in Table 10.5, and the respective yield data can be found in the Report of Progress 997 of Kansas State University Southwest Research-Extension Center (2008), available online through the Kansas State University library. More recent measurements regarding this experimental site and other long-term tillage studies in the central Great Plains have been published by Blanco-Canqui et al. (2009a,b). Those studies focus on the aggregate properties with regard to soil erodibility and SOC, maximum bulk density (BD_{max}), and critical water content (CWC).

The data presented by Stone and Schlegel (2010) on some soil parameters after 10 years of differentiated tillage treatments indicate that of the three tillage treatments, NT maintains soil conditions closest to those determined under native prairie. Although the water content at −1.5 MPa matric potential is not an indicator for plant-available water, the authors interpret its good correlation to SOC as a strong reason for the water-holding capacity of the soil. Together with the much higher infiltration rate under NT, which surpassed even the infiltration capacity of the native soil, there is strong evidence that the decrease in tillage intensity improves plant-available soil water. The results of the parameters MWD and the concentration of WSA from samples taken in 2000 are a clear indicator for better aggregate stability under NT when compared with the two tilled treatments. These results were confirmed 19 years after the start of the study by Blanco-Canqui et al. (2009a,b), who found that 4.75–8 mm aggregates from the NT treatment required a significantly higher kinetic energy to be disintegrated than the aggregates from RT and CT. This behavior was corroborated by the water-drop penetration test and measurements of BD_{max} and CWC. Although soil erosion by water is certainly not a major issue on the plain and the permeable soil of this study, other areas with even gentle slopes may lose part of the scarce precipitation through runoff.

The grain yields obtained in this study show a clear benefit of the conservation tillage systems over CT, which was more pronounced in grain sorghum. On average, over 17 years, CT produced 75% and 50% and RT 87% and 79% of NT yields, for wheat and sorghum, respectively. Both graphs of Figure 10.9 also indicate a trend for the differences increasing with time.

Although no in-field soil moisture data are available from this study, Stone and Schlegel (2010) conclude from their results that the better conditions of aggregation and water infiltrability under NT management are indicators for a better water intake and therefore enhanced precipitation use efficiency. In fact, the considerable differences in yield between the tillage systems, especially in the summer rotation crop, corroborate the interpretation of water availability being the main responsible factor for the differences in crop performance.

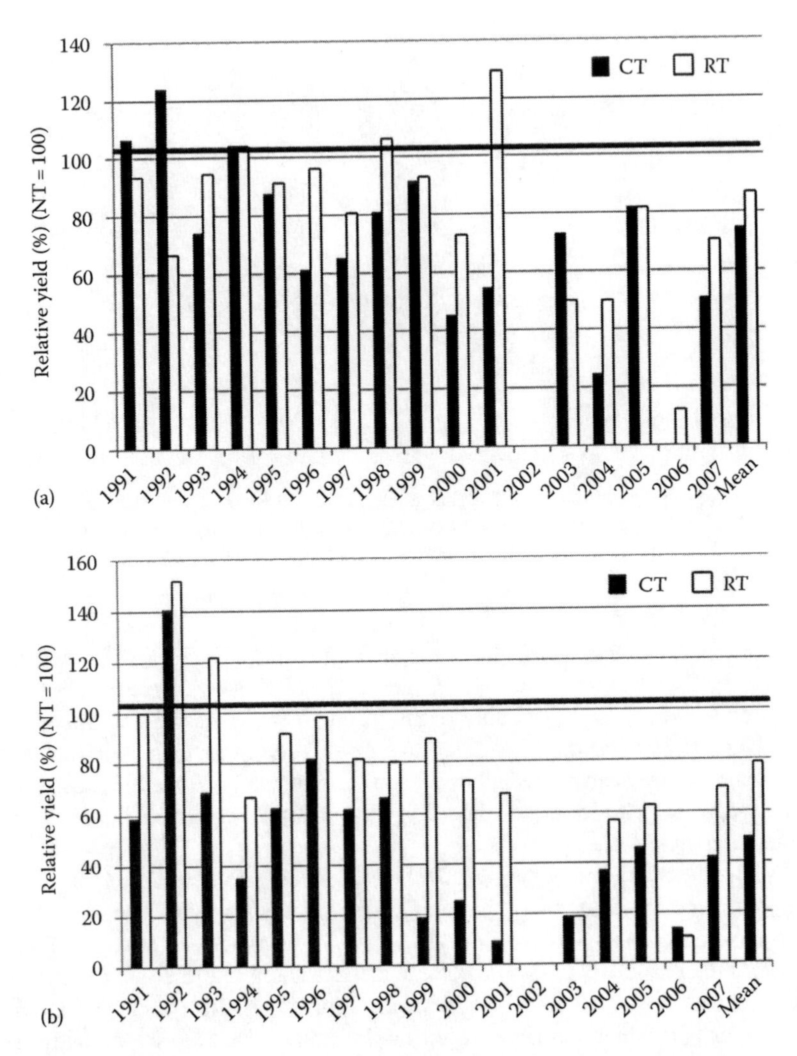

FIGURE 10.9 Relative grain yields of (a) wheat and (b) sorghum of conventional (CT) and reduced tillage (RT) as compared with no-till (NT). (From Kansas State University Southwest Research-Extension Center, 2008.)

10.4.3.2 The System of Rice Intensification

The SRI—a rice production system based on alternative ideas about crop and water management practices—has taken root on an international scale, moving far beyond its origins in Madagascar. At the same time, the diversity of reports shows that SRI is "not yet finished," it is still evolving and changing. The productivity gains, including WP and a decrease in the water requirement, from SRI changes in the management of crops, soil, water, and nutrients have now been demonstrated in more than 40 countries by farmers and a diverse group of stakeholders who support resource-limited, small-scale rice farmers in raising their output and incomes by using locally available resources as productively as possible.

Over recent years, the merits of the SRI system as compared with the recommended anaerobic (flooded) rice production systems have become better understood, based on both scientific and empirical data. The SRI production concept has been defined on the basis of six agronomic practices: (1) the use of very young—about 10 days old—seedlings for transplanting; (2) single transplant/hill; (3) wide spacing of transplants, from 20 × 20 cm to 50 × 50 cm depending on the variety and the soil fertility; (4) mainly moist (not saturated and flooded) soil water regimes kept through intermittent irrigation; (5) regular weeding through a rotary hoe to also facilitate soil aeration; and (6) liberal use of organic fertilizers. These practices were first described in detail some 30 years ago by Henri de Laulanié, a Jesuit priest, who recognized that small rice farmers in Madagascar simply lacked the resources to invest in intensifying their rice cultivation practices through the recommended "modern" technological package based on costly (and unavailable) external inputs and inadequate or nonexistent extension support.

de Laulanié paid little attention to the issue of genetically improved and input responsive modern varieties (the backbone of "modern" rice production and indeed of industrialized agriculture in general). Yet, by manipulating the other crop management factors, including their interactions, he recorded a large decrease in the water requirement and spectacular yield increases, for the local varieties. This corresponded to large water savings as well as greatly increased WP. In essence, SRI crop management and water management at the level of practice represents an "integrated" agronomy. Through integrated management of its various crop–soil–soil biota–water–nutrient–space–time components, SRI seeks to capitalize on a number of basic agronomic principles aimed at optimizing the aboveground as well as the belowground plant growth and development and the performance of the crop as a whole.

Of particular interest here is the SRI recommendation of keeping the soil just moist but not continuously flooded, either by making minimum daily applications of water or by intermittent irrigation. SRI practices of single seedling per hill and wider spacing together with aerobic soil conditions are reported to increase the yields of irrigated rice by 25%–75% or even more with an even greater increase in WP and a reduction in the water requirement by 40%–50%, in seeds by 80%–90%, in the cost of production by 20%, and in the use of fertilizer by some 50%.

Thus, SRI offers an opportunity to reduce water demand while enhancing yields and WP. As has been shown in several studies, the most evident phenotypic difference with SRI is in the plant root growth. Direct measurements confirm that the SRI methods induce both greater and deeper root growth, which contributes to increased WUE and nutrient uptake throughout the crop cycle, compared with the shallower rooting and shorter duration of root functioning under continuous flooding. Rice plants grown with the SRI methods take up more macronutrients than the roots of conventionally managed plants.

Evidence is accumulating that making the changes in the rice-growing practices that constitute SRI can result in win-win outcomes—for farmers, consumers, and the environment—in terms of WP as well as water savings. These gains are possible across a wide range of agroecosystems and are not limited to smallholders. Although the greatest benefits come from using the full set of practices, and using them as recommended, there are demonstrable advantages from "partial SRI." Based on the results of large-scale factorial trials in Asia and Africa, one can predict that in most of the cases reported, there are opportunities to achieve still-greater benefits from SRI methods.

SRI methods, with appropriate adaptations, are effective in a wide variety of environments: tropical humid ecology (Panama), a semiarid region on the edge of the desert (Mali), midaltitude subhumid tropical environment (Madagascar), sandy–marshy regions (southern Iraq), various dry and humid environments in Asia (India, Pakistan, and Indonesia), and even mountainous areas with a short growing season (northern Afghanistan). In each of these environments, farmers have found it possible through their modifications of standard rice-growing practices, according to the SRI principles, to create microenvironments that are favorable to a more beneficial expression of rice genetic potential. A crop management and water management strategy such as SRI is *not an alternative* to getting and using genotypes best suited to a given production situation; rather, it is a way to make the most of any given variety's production capability.

The success of SRI is not dependent on using more modern rice cultivars, although most of the highest SRI yields have come from combining its practices with high-yielding varieties or hybrids. Plant breeding has been, and will continue to be, successful in improving yield and other crop potentials. It is true, however, that SRI methods can also raise the yields of most indigenous varieties, and where these command a higher market price because of consumer preferences, farmers may find these "unimproved" varieties more profitable. This can help with the conservation of rice biodiversity.

Another important consideration is that SRI phenotypes are widely reported by farmers and observers to be less susceptible to pest and disease damage. In 2005–2006, a systematic evaluation was carried out in eight provinces in Vietnam, comparing SRI plots with neighboring farmer-practice plots. This found the prevalence of major rice diseases and pests (sheath blight, leaf blight, small leaf-folder, and brown planthopper) to be 55% less on SRI plants in the spring season and 70% less in summer (National IPM Program, 2007). Farmers frequently say that with SRI management, their rice plants are resistant enough to crop damage that agrochemical protection is unnecessary or it gives them no net economic benefit. The SRI approach is an example of a paradigm shift, to more biologically driven, agroecological strategies for crop production, in contrast to chemically dependent ones.

With any agricultural strategy, we should be concerned about the genetic potentials, as these are the starting points for all life. However, the SRI experience is showing that better optimizing management of the environment for growth can achieve a fuller expression of these potentials while using overall water use and maximizing crop WP.

10.5 CONCLUSION

At the end of the first decade of the twenty-first century, there is more awareness than ever before regarding the future need and the importance of producing more food, feed, fiber, and biofuels that must be attained through a 70% increase in total global output based on increased productivity per unit of land and production inputs rather than by extending agricultural production to so far untouched terrestrial ecosystems. The production inputs used to push forward the "Green Revolution" in the 1960s and 1970s, mainly based on high yielding "modern" varieties, more and better use of fertilizers, plant protection products, and tillage, contributed decisively to production increases over the past few decades to keep pace with population growth. Today, there are many voices highlighting the fact that the potential productivity gains through increases in the HI and in the use of water, agrochemicals, and tillage have been met and that a new kind of Green Revolution is needed to match the increasing demand for agricultural commodities while conserving and enhancing natural as well as altered ecosystems and environmental quality. Additionally, such a Green Revolution must address the challenges of increasing food, energy and input costs, pervasive poverty, water scarcity, land degradation, loss of biodiversity, and climate change.

In this context, two aspects are of fundamental importance for agricultural intensification: soil quality and the EUW. Within the ecosystem, both soil resources and water resources are inextricably linked, and so is their management for agricultural and nonagricultural uses. However, the expansion of irrigated land and the withdrawal of blue water for irrigation purposes are reaching their normal exploitable limits, thus making further improvements in WUE and WP a necessity in both irrigated and rainfed agriculture. This entails increasing the productive use of rainfall that infiltrates the soil and is accessible to plants, for use in transpiration in support of biomass growth and harvestable yields. The latter also applies to irrigated lands, because a larger green water share in the soil water balance effectively reduces the amount of supplementary irrigation.

The key message of this chapter, based on substantial empirical and scientific evidence, is that it is perfectly possible to design and put into practice sustainable production systems, both rainfed and irrigated, that are simultaneously productive, profitable, resource conserving, and environmentally protective. In such production systems, the management of the soil water balance in favor of sustainable intensification and therefore the optimization of rainfall infiltration, soil water storage, WUE and WP, as well as all the ecosystem services required by society, can be achieved, provided the three principles of CA are applied simultaneously: minimum soil disturbance, permanent organic soil cover, and diversified cropping system. Similarly, SRI agronomy and water management show that there is a large scope for improving WUE and WP in conventional irrigated or flooded rice systems.

However, CA and SRI (or CA–SRI) systems are "works in progress" and their development has been led largely by farmers. These systems deserve much greater attention from the scientific research community and policy makers. Improved modern varieties and irrigation systems can be important in enhancing WUE and WP, but in themselves, they can only do so much. CA and SRI provide excellent examples of how to obtain "more output for less input" from most adapted cultivars, traditional or modern. They show that when production systems pay attention to ecosystem services, it is possible to achieve sustainable intensification. While CA and SRI are not organic, they can be; they are probiotic and promote biodiversity in all parts of the production systems. They can maintain high overall farm productivity as well as individual factor productivities by promoting soil life and biodiversity, large root systems, organic matter as a substrate for soil micro-organisms and soil organic cover, and species diversification in the cropping systems. These attributes strongly suggest that the principles of the CA and CA–SRI systems need to be better understood and spread over ever-larger areas to meet the future global food security and ecosystem service needs. They embody the notion of sustainable soil water management.

REFERENCES

Aase, J. K., D. L. Bjorneberg, and R. E. Sojka. 1998. Sprinkler irrigation runoff and erosion control with polyacrylamide—Laboratory tests. *Soil Science Society of America Journal* 62 (6):1681–1687.

Abid, M. and R. Lal. 2009. Tillage and drainage impact on soil quality: II. Tensile strength of aggregates, moisture retention and water infiltration. *Soil & Tillage Research* 103 (2):364–372.

Aboukeira, A. A. A. 2010. Deficit irrigation as an agricultural water management system for corn: A review. In *Proceedings CIGR XVIIth World Congress*. Québec: CSBE100254.

Abreu, S. L., J. M. Reichert, and D. J. Reinert. 2004. Mechanical and biological chiseling to reduce compaction of a sandy loam alfisol under no-tillage. *Revista Brasileira De Ciencia Do Solo* 28 (3):519–531.

Adcock, D., A. M. McNeill, G. K. McDonald, and R. D. Armstrong. 2007. Subsoil constraints to crop production on neutral and alkaline soils in south-eastern Australia: A review of current knowledge and management strategies. *Australian Journal of Experimental Agriculture* 47 (11):1245–1261.

Adekalu, K. O. and D. A. Okunade. 2006. Effect of irrigation amount and tillage system on yield and water use efficiency of cowpea. *Communications in Soil Science and Plant Analysis* 37 (1–2):225–237.

Aiken, R. M., D. C. Nielsen, and L. R. Ahuja. 2003. Scaling effects of standing crop residues on the wind profile. *Agronomy Journal* 95 (4):1041–1046.

Ali, M., C. R. Jensen, V. O. Mogensen, M. N. Andersen, and I. E. Henson. 1999. Root signaling and osmotic adjustment during intermittent soil drying sustain grain yield of field grown wheat. *Field Crops Research* 62 (1):35–52.

Ali, M. H. and M. S. U. Talukder. 2008. Increasing water productivity in crop production—A synthesis. *Agricultural Water Management* 95 (11):1201–1213.

Alvarez, R. and H. S. Steinbach. 2009. A review of the effects of tillage systems on some soil physical properties, water content, nitrate availability and crops yield in the Argentine Pampas. *Soil & Tillage Research* 104 (1):1–15.

Alvaro-Fuentes, J., J. Lampurlanes, and C. Cantero-Martinez. 2009. Alternative crop rotations under Mediterranean no-tillage conditions: Biomass, grain yield, and water-use efficiency. *Agronomy Journal* 101 (5):1227–1233.

Anderson, R. L. and B. W. Greb. 1987. Residual herbicides for weed-control in proso millet (*Panicum miliaceum* L.). *Crop Protection* 6 (1):61–63.

Andrews, D. J. and A. H. Kassam. 1976. The importance of multiple cropping in increasing world food supplies. In *Multiple Cropping*, edited by R. I. Papendick, A. Sanchez, and G. B. Triplett, pp. 1–10. Madison, WI: American Society of Agronomy.

Angus, J. F. 2001. Nitrogen supply and demand in Australian agriculture. *Australian Journal of Experimental Agriculture* 41 (3):277–288.

Angus, J. F. and R. A. Fischer. 1991. Grain and protein responses to nitrogen applied to wheat growing on a red earth. *Australian Journal of Agricultural Research* 42 (5):735–746.

Angus, J. F. and A. F. van Herwaarden. 2001. Increasing water use and water use efficiency in dryland wheat. *Agronomy Journal* 93 (2):290–298.

Anken, T., P. Weisskopf, U. Zihlmann, H. Forrer, J. Jansa, and K. Perhacova. 2004. Long-term tillage system effects under moist cool conditions in Switzerland. *Soil & Tillage Research* 78 (2):171–183.

Armand, R., C. Bockstaller, A. V. Auzet, and P. Van Dijk. 2009. Runoff generation related to intra-field soil surface characteristics variability application to conservation tillage context. *Soil & Tillage Research* 102 (1):27–37.

ASCE. 1978. Describing irrigation efficiency and uniformity. *ASCE Journal of Irrigation and Drainage Division* 104 (IR1):35–41.

Asgarzadeh, H., M. R. Mosaddeghi, A. A. Mahboubi, A. Nosrati, and A. R. Dexter. 2010. Soil water availability for plants as quantified by conventional available water, least limiting water range and integral water capacity. *Plant and Soil* 335 (1–2):229–244.

Asseng, S., F. X. Dunin, I. R. P. Fillery, D. Tennant, and B. A. Keating. 2001. Potential deep drainage under wheat crops in a Mediterranean climate. II. Management opportunities to control drainage. *Australian Journal of Agricultural Research* 52 (1):57–66.

Azooz, R. H. and M. A. Arshad. 1996. Soil infiltration and hydraulic conductivity under long-term no-tillage and conventional tillage systems. *Canadian Journal of Soil Science* 76 (2):143–152.

Baig, M. N. and P. M. Gamache. 2009. The economic, agronomic and environmental impact of no-till on the Canadian prairies. *Alberta Reduced Tillage Linkages* (http://www.reducedtillage.ca/docs/Impactnotillrtlaug2009.pdf (accessed November 2, 2010)).

Basch, G. 1988. Alternativen zum traditionellen Landnutzungssystem im Alentejo/Portugal, unter besonderer Berücksichtigung der Bodenbearbeitung, Göttinger Beiträge zur Land- und Forstwirtschaft in den Tropen und Subtropen, 31, 1–188, Göttingen.

Basch, G., M. Carvalho, J. F. C. Barros, and J. M. G. Calado. 2010. The importance of crop residue management for carbon sequestration under no-till. In *Proceedings of the European Congress on Conservation Agriculture—Towards Agro-environmental, Climate and Energetic Sustainability*, edited by Ministerio de Medio Ambiente e Medio Rural e Marino, pp. 241–248, Madrid.

Batchelor, C., C. Lovell, and M. Murata. 1996. Simple microirrigation techniques for improving irrigation efficiency on vegetable gardens. *Agricultural Water Management* 32 (1):37–48.

Baumhardt, R. L. and O. R. Jones. 2002. Residue management and paratillage effects on some soil properties and rain infiltration. *Soil & Tillage Research* 65 (1):19–27.

Bell, M. J., B. J. Bridge, G. R. Harch, and D. N. Orange. 2005. Rapid internal drainage rates in Ferrosols. *Australian Journal of Soil Research* 43 (4):443–455.

Ben-Hur, M. and M. Lado. 2008. Effect of soil wetting conditions on seal formation, runoff, and soil loss in arid and semiarid soils—A review. *Australian Journal of Soil Research* 46 (3):191–202.

Benjamin, J. G., D. C. Nielsen, and M. F. Vigil. 2003. Quantifying effects of soil conditions on plant growth and crop production. *Geoderma* 116 (1–2):137–148.

Bennett, J. 2003. Opportunities for increasing water productivity of CGIAR crops through plant breeding and molecular biology. *Water Productivity in Agriculture: Limits and Opportunities for Improvement* 1:103–126.

Bescansa, P., M. J. Imaz, I. Virto, A. Enrique, and W. B. Hoogmoed. 2006. Soil water retention as affected by tillage and residue management in semiarid Spain. *Soil & Tillage Research* 87 (1):19–27.

Bhattacharyya, R., V. Prakash, S. Kundu, and H. S. Gupta. 2006. Effect of tillage and crop rotations on pore size distribution and soil hydraulic conductivity in sandy clay loam soil of the Indian Himalayas. *Soil & Tillage Research* 86 (2):129–140.

Bjorneberg, D. L. and J. K. Aase. 2000. Multiple polyacrylamide applications for controlling sprinkler irrigation runoff and erosion. *Applied Engineering in Agriculture* 16 (5):501–504.

Bjorneberg, D. L., F. L. Santos, N. S. Castanheira, O. C. Martins, J. L. Reis, J. K. Aase, and R. E. Sojka. 2003. Using polyacrylamide with sprinkler irrigation to improve infiltration. *Journal of Soil and Water Conservation* 58 (5):283–289.

Blackshaw, R. E., K. N. Harker, J. T. O'Donovan, H. J. Beckie, and E. G. Smith. 2008. Ongoing development of integrated weed management systems on the Canadian prairies. *Weed Science* 56 (1):146–150.

Blanco-Canqui, H., C. J. Gantzer, S. H. Anderson, E. E. Alberts, and A. L. Thompson. 2004. Grass barrier and vegetative filter strip effectiveness in reducing runoff, sediment, nitrogen, and phosphorus loss. *Soil Science Society of America Journal* 68 (5):1670–1678.

Blanco-Canqui, H. and R. Lal. 2007a. Impacts of long-term wheat straw management on soil hydraulic properties under no-tillage. *Soil Science Society of America Journal* 71 (4):1166–1173.

Blanco-Canqui, H. and R. Lal. 2007b. Soil and crop response to harvesting corn residues for biofuel production. *Geoderma* 141 (3–4):355–362.

Blanco-Canqui, H. and R. Lal. 2009. Crop residue removal impacts on soil productivity and environmental quality. *Critical Reviews in Plant Sciences* 28 (3):139–163.

Blanco-Canqui, H., R. Lal, W. M. Post, and L. B. Owens. 2006. Changes in long-term no-till corn growth and yield under different rates of stover mulch. *Agronomy Journal* 98 (4):1128–1136.

Blanco-Canqui, H., M. M. Mikha, J. G. Benjamin, L. R. Stone, A. J. Schlegel, D. J. Lyon, M. F. Vigil, and P. W. Stahlman. 2009a. Regional study of no-till impacts on near-surface aggregate properties that influence soil erodibility. *Soil Science Society of America Journal* 73 (4):1361–1368.

Blanco-Canqui, H., L. R. Stone, A. J. Schlegel, D. J. Lyon, M. F. Vigil, M. M. Mikha, P. W. Stahlman, and C. W. Rice. 2009b. No-till induced increase in organic carbon reduces maximum bulk density of soils. *Soil Science Society of America Journal* 73 (6):1871–1879.

Blevins, R. L. and W. W. Frye. 1993. Conservation tillage—An ecological approach to soil-management. *Advances in Agronomy* 51:33–78.

Bluemling, B., H. Yang, and C. Pahl-Wostl. 2007. Making water productivity operational—A concept of agricultural water productivity exemplified at a wheat-maize cropping pattern in the North China plain. *Agricultural Water Management* 91 (1–3):11–23.

Blum, A. 2009. Effective use of water (EUW) and not water-use efficiency (WUE) is the target of crop yield improvement under drought stress. *Field Crops Research* 112 (2–3):119–123.

Boddey, R. M., J. R. A. Bruno, and S. Irquiaga. 2006. Leguminous biological nitrogen fixation in sustainable tropical agroecosystems. In *Biological Approaches to Sustainable Soil Systems*, edited by N. Uphoff, A. S. Ball, E. Fernandes, et al. Boca Raton, FL: CRC Press, Taylor & Francis Group.

Bolliger, A., J. Magid, T. J. C. Amado, F. S. Neto, M. D. D. Ribeiro, A. Calegari, R. Ralisch, and A. de Neergaard. 2006. Taking stock of the Brazilian "zero-till revolution": A review of landmark research and farmers' practice. *Advances in Agronomy* 91:47–110.

Bos, M. G. and J. Nugteren. 1990. *On Irrigation Efficiencies*, 4th edition, ILRI Publication No. 19. Wageningen: ILRI.

Bossio, D., W. Critchley, K. Geheb, G. van Lynden, and B. Mati. 2007. Conserving land-protecting water. In *Water for Food, Water for Life: A Comprehensive Assessment of Water Management in Agriculture*, edited by D. Molden, pp. 551–583. London, Colombo: Earthscan, International Water Management Institute.

Bossio, D., K. Geheb, and W. Critchley. 2010. Managing water by managing land: Addressing land degradation to improve water productivity and rural livelihoods. *Agricultural Water Management* 97 (4):536–542.

Bot, A. and J. Benites. 2005. Importance of soil organic matter. Key to drought-resistant soil and sustained food production. In *Soils Bulletin No. 80*, edited by FAO. Rome: FAO.

Bottinelli, N., T. Henry-des-Tureaux, V. Hallaire, J. Mathieu, Y. Benard, T. D. Tran, and P. Jouquet. 2010. Earthworms accelerate soil porosity turnover under watering conditions. *Geoderma* 156 (1–2):43–47.

Bouman, B. A. M. 2007. A conceptual framework for the improvement of crop water productivity at different spatial scales. *Agricultural Systems* 93 (1–3):43–60.

Brown, M. J. 1985. Effect of grain straw and furrow irrigation stream size on soil-erosion and infiltration. *Journal of Soil and Water Conservation* 40 (4):389–391.

Bruinsma, J. 2009. The resource outlook to 2050: By how much do land, water use and crop yields need to increase by 2050? Paper read at the Expert Meeting on "How to Feed the World in 2050," June 24–26. Rome: FAO.

Buchleiter, G. W., C. R. Camp, R. G. Evans, and B. A. King. 2000. Technologies for variable water application with sprinklers. In *Proceedings 4th Decennial National Irrigation Symposium,* edited by R. G. Evans, B. L. Benham, and T. P. Trooien, pp. 316–321. St. Joseph, MI: American Society of Agricultural Engineers.

Buczko, U., O. Bens, E. Hangen, J. Brunotte, and R. F. Huttl. 2003. Infiltration and macroporosity of a silt loam soil under two contrasting tillage systems. *Landbauforschung Volkenrode* 53 (2–3):181–190.

Bunting, A. H. and A. H. Kassam. 1988. Principles of crop water use, dry matter production and dry matter partitioning that govern choices of crops and systems. In *Drought Research Priorities for the Dryland Tropics*, edited by F. R. Bidinger and C. Johansen, pp. 43–61. Hyderabad: ICRISAT.

Burt, C. M., D. J. Howes, and A. Mutziger. 2001. Evaporation estimates for irrigated agriculture in California. In *Conference Proceedings of the Annual Irrigation Association Meeting*, San Antonio, TX, pp. 103–110. Falls Church, VA: The Irrigation Association.

Burt, C. M., A. J. Mutziger, R. G. Allen, and T. A. Howell. 2005. Evaporation research: Review and interpretation. *Journal of Irrigation and Drainage Engineering-ASCE* 131 (1):37–58.

Buttar, G. S., M. S. Aujla, H. S. Thind, C. J. Singh, and K. S. Saini. 2007. Effect of timing of first and last irrigation on the yield and water use efficiency in cotton. *Agricultural Water Management* 89 (3):236–242.

Cahoon, J. E., D. E. Eisenhauer, R. W. Elmore, F. W. Roeth, B. Doupnik, R. A. Selley, K. Frank, R. B. Ferguson, M. Lorenz, and L. J. Young. 1999. Corn yield response to tillage with furrow irrigation. *Journal of Production Agriculture* 12 (2):269–275.

Cakmak, I. 2005. The role of potassium in alleviating detrimental effects of abiotic stresses in plants. *Journal of Plant Nutrition and Soil Science-Zeitschrift Fur Pflanzenernahrung Und Bodenkunde* 168 (4):521–530.

Cakmak, I., A. Yilmaz, M. Kalayci, H. Ekiz, B. Torun, B. Erenoglu, and H. J. Braun. 1996. Zinc deficiency as a critical problem in wheat production in Central Anatolia. *Plant and Soil* 180 (2):165–172.

Calegari, A., M. R. Darolt, and M. Ferro. 1998. Towards sustainable agriculture with a no-tillage system. *Advances in GeoEcology* 31:1205–1209.

Camp, C. R., E. J. Sadler, D. E. Evans, L. J. Usrey, and M. Omary. 1998. Modified center pivot system for precision management of water and nutrients. *Applied Engineering in Agriculture* 14 (1):23–31.

Carvalho, M. and G. Basch. 1995. Effects of traditional and no-tillage on physical and chemical properties of a Vertisol. In *Experience with the Applicability of No-tillage Crop Production in the West-European Countries. Proceedings of the EC-Workshop II*, edited by F. Tebrugge and A. Bohrnsen, pp. 17–23. Langgöns, Germany: Wissenschaftlicher Fachverlag.

Cary, J. W. 1986. Irrigating row crops from sod furrows to reduce erosion. *Soil Science Society of America Journal* 50 (5):1299–1302.

Castro, O. M., A. C. R. Severo, and E. J. B. N. Cardoso. 1993. Avaliação da atividade de microorganismos do solo em diferentes sistemas de manejo da soja. *Scientia Agricola* 50 (2):212–219.

Chaboussou, F. 2004 (1985). *Healthy Crops: A New Agricultural Revolution; Orig. Santé des Cultures: Une Révolution Agronomique*. Charlbury, UK/Paris: Jon Carpenter Publications/Flammarion La Maison Rustique.

Chan, K. Y. 2001. An overview of some tillage impacts on earthworm population abundance and diversity—Implications for functioning in soils. *Soil & Tillage Research* 57 (4):179–191.

Chimenti, C. A., M. Marcantonio, and A. J. Hall. 2006. Divergent selection for osmotic adjustment results in improved drought tolerance in maize (*Zea mays* L.) in both early growth and flowering phases. *Field Crops Research* 95 (2–3):305–315.

Chiroma, A. M., A. B. Alhassan, and B. Khan. 2008. Yield and water use efficiency of millet as affected by land configuration treatments. *Journal of Sustainable Agriculture* 32 (2):321–333.

Christiansen, J. E. 1942. *Irrigation by Sprinkling*. Berkeley, CA: University of California.

Clemmens, A. J. 1998. Level-basin design based on cutoff criteria. *Irrigation and Drainage Systems* 12 (2):85–113.

Clemmens, A. J. and A. R. Dedrick. 1982. Limits for practical level-basin design. *Journal of the Irrigation and Drainage Division-ASCE* 108 (2):127–141.

Clemmens, A. J. and D. J. Molden. 2007. Water uses and productivity of irrigation systems. *Irrigation Science* 25 (3):247–261.

Clemmens, A. J. 1998. Level-basin design based on cutoff criteria. *Irrigation and Drainage Systems* 12 (2):85–113.

Clough, M. R. and A. J. Clemmens. 1994. Field evaluation of level-basin irrigation. In *ASAE Annual Meeting Paper No. 94-2136*, edited by ASAE. St. Joseph, MI: ASAE.

Coppens, F., P. Garnier, S. De Gryze, R. Merckx, and S. Recous. 2006. Soil moisture, carbon and nitrogen dynamics following incorporation and surface application of labelled crop residues in soil columns. *European Journal of Soil Science* 57 (6):894–905.

Crabtree, B. 2010. *Search for Sustainability with No-Till Bill in Dryland Agriculture*. Beckenham, WA: Crabtree Agricultural Consulting.

Cullum, R. F. 2009. Macropore flow estimations under no-till and till systems. *Catena* 78 (1):87–91.

Cullum, R. F., G. V. Wilson, K. C. McGregor, and J. R. Johnson. 2007. Runoff and soil loss from ultra-narrow row cotton plots with and without stiff-grass hedges. *Soil & Tillage Research* 93 (1):56–63.

da Silva, V. R., J. M. Reichert, and D. J. Reinert. 2006. Soil temperature variation in three different systems of soil management in blackbeans crop. *Revista Brasileira De Ciencia Do Solo* 30 (3):391–399.

da Veiga, M., D. J. Reinert, J. M. Reichert, and D. R. Kaiser. 2008. Short and long-term effects of tillage systems and nutrient sources on soil physical properties of a southern Brazilian Hapludox. *Revista Brasileira De Ciencia Do Solo* 32 (4):1437–1446.

Dabney, S. M., J. A. Delgado, and D. W. Reeves. 2001. Using winter cover crops to improve soil and water quality. *Communications in Soil Science and Plant Analysis* 32 (7–8):1221–1250.

Dabney, S. M., G. V. Wilson, K. C. McGregor, and G. R. Foster. 2004. History, residue, and tillage effects on erosion of loessial soil. *Transactions of the ASAE* 47 (3):767–775.

Dang, Y. P., R. C. Dalal, S. R. Buck, et al. 2010. Diagnosis, extent, impacts, and management of subsoil constraints in the northern grains cropping region of Australia. *Australian Journal of Soil Research* 48 (2):105–119.

Dasilva, A. P., B. D. Kay, and E. Perfect. 1994. Characterization of the least limiting water range of soils. *Soil Science Society of America Journal* 58 (6):1775–1781.

de Sousa, P. L., L. L. Silva, and R. P. Serralheiro. 1999. Comparative analysis of main on-farm irrigation systems in Portugal. *Agricultural Water Management* 40 (2–3):341–351.

Deng, X. P., L. Shan, S. Z. Kang, S. Inanaga, and M. E. K. Ali. 2003. Improvement of wheat water use efficiency in semiarid area of China. *Agricultural Science in China* 2 (1):35–44.

Diaz-Ambrona, C. G. H., G. J. O'Leary, V. O. Sadras, M. G. O'Connell, and D. J. Connor. 2005. Environmental risk analysis of farming systems in a semi-arid environment: Effect of rotations and management practices on deep drainage. *Field Crops Research* 94 (2–3):257–271.

Doorenbos, J. and A. H. Kassam. 1979. Yield response to water. In *FAO Irrigation and Drainage Paper 33*, edited by FAO. Rome: FAO.

Doorenbos, J. and W. Pruitt. 1977. Guidelines for predicting crop water requirements. In *Irrigation and Drainage Paper 24 (Rev.)*, edited by FAO. Rome: FAO.

dos Santos, T. P., C. M. Lopes, M. L. Rodrigues, C. R. de Souza, J. P. Maroco, J. S. Pereira, J. M. Ricardo-da-Silva, and M. M. Chaves. 2003. Partial rootzone drying: Effects on growth and fruit quality of field-grown grapevines (*Vitis vinifera*). *Functional Plant Biology* 30:663–671.

Duke, H. R., L. E. Stetson, and N. C. Ciancaglini. 1992. Irrigation systems control. In *Management of Farm Irrigation Systems, ASAE Monograph No. 9*, edited by G. J. Hoffman, T. A. Howell, and K. H. Solomon. St. Joseph, MI: ASAE.

Ehlers, W. and M. Goss. 2003. *Water Dynamics in Plant Production*. Wallingford, UK: CABI Publishing.

Eisenhauer, D. E., E. C. Dickey, P. E. Fischbach, and K. D. Frank. 1982. Influence of reduced tillage on furrow irrigation infiltration. In *ASAE Annual Paper No. 82-2587*, edited by ASAE. St. Joseph, MI: ASAE.

Eriksen-Hamel, N. S., A. B. Speratti, J. K. Whalen, A. Legere, and C. A. Madramootoo. 2009. Earthworm populations and growth rates related to long-term crop residue and tillage management. *Soil & Tillage Research* 104 (2):311–316.

Ernst, G., D. Felten, M. Vohland, and C. Emmerling. 2009. Impact of ecologically different earthworm species on soil water characteristics. *European Journal of Soil Biology* 45 (3):207–213.

Evans, R. G. and E. J. Sadler. 2008. Methods and technologies to improve efficiency of water use. *Water Resources Research* 44:15.

Fabrizzi, K. P., F. O. Garcia, J. L. Costa, and L. I. Picone. 2005. Soil water dynamics, physical properties and corn and wheat responses to minimum and no-tillage systems in the southern Pampas of Argentina. *Soil & Tillage Research* 81 (1):57–69.

Falkenmark, M. 2007. Shift in thinking to address the 21st century hunger gap—Moving focus from blue to green water management. *Water Resources Management* 21 (1):3–18.

Falkenmark, M. and J. Rockstrom. 2004. *Balancing Water for Humans and Nature*. London: Earthscan.

Fang, Q., L. Ma, Q. Yu, L. R. Ahuja, R. W. Malone, and G. Hoogenboom. 2010a. Irrigation strategies to improve the water use efficiency of wheat-maize double cropping systems in North China Plain. *Agricultural Water Management* 97 (8):1165–1174.

Fang, Q. X., L. Ma, T. R. Green, Q. Yu, T. D. Wang, and L. R. Ahuja. 2010b. Water resources and water use efficiency in the North China Plain: Current status and agronomic management options. *Agricultural Water Management* 97 (8):1102–1116.

FAO. 1978–1981. *Agroecological Zones Project Report. Methodology and Results for Africa (Vol.1), West Asia (Vol.2), South and Central America (Vol.3), Southeast Asia (Vol.4)*. In *World Soil Resources Report 48*. Rome: FAO.

FAO. 1989. Guidelines for designing and evaluating surface irrigation systems. In *Irrigation and Drainage Paper 45*. Rome: FAO.

FAO. 1992. CROPWAT—A computer program for irrigation planning and management. In *Irrigation and Drainage Paper No. 26*. Rome: FAO.

FAO. 2008. Investing in sustainable crop intensification: The case for soil health. In *Report of the International Technical Workshop, FAO, Rome, July. Integrated Crop Management Vol. 7*. Rome: FAO.

Fereres, E. and A. Kassam. 2003. Water and the CGIAR: A strategic framework. *Water International* 28 (1):122–129.

Fereres, E. and M. A. Soriano. 2007. Deficit irrigation for reducing agricultural water use. *Journal of Experimental Botany* 58 (2):147–159.

Fernandez-Ugalde, O., I. Virto, P. Bescansa, M. J. Imaz, A. Enrique, and D. L. Karlen. 2009. No-tillage improvement of soil physical quality in calcareous, degradation-prone, semiarid soils. *Soil & Tillage Research* 106 (1):29–35.

Fischer, A. 1979. Growth and water limitations to dryland wheat yield in Australia: A physiological framework. *Journal of the Australian Institute of Agricultural Science* 45:83–94.

Flower, K., B. Crabtree, and G. Butler. 2008. No-till cropping systems in Australia. In *No-Till Farming Systems, Special Publication No. 3*, edited by T. Goddard, M. Zoebisch, Y. Gan, W. Ellis, A. Watson, and S. Sombatpanit, pp. 457–467. Bangkok: World Association of Soil and Water Conservation (WASWAC).

Fornstrom, K. J., J. A. Michel, J. Borrelli, and G. D. Jackson. 1985. Furrow firming for control of irrigation advance rates. *Transactions of the ASAE* 28 (2):519–531.

Fowler, R. and J. Rockstrom. 2001. Conservation tillage for sustainable agriculture—An agrarian revolution gathers momentum in Africa. *Soil & Tillage Research* 61 (1–2):93–107.

Freese, R. C., D. K. Cassel, and H. P. Denton. 1993. Infiltration in a piedmont soil under 3 tillage systems. *Journal of Soil and Water Conservation* 48 (3):214–218.

Freitas, P. S. L., E. C. Mantovani, G. C. Sediyama, and L. C. Costa. 2006. Influência da cobertura de resíduos de culturas nas fases da evaporação direta da água do solo. *Revista Brasileira de Engenharia Agrícola e Ambiental* 10 (1):104–111.

Friedrich, T., A. H. Kassam, and F. Shaxson. 2009. Conservation agriculture. In *Agriculture for Developing Countries. Science and Technology Options Assessment (STOA) Project*, edited by R. Meyer. Karlsruhe: European Parliament, European Technology Assessment Group.

Furbank, R. T., S. von Caemmerer, J. Sheehy, and G. Edwards. 2009. C-4 rice: A challenge for plant phenomics. *Functional Plant Biology* 36 (10–11):845–856.

Gan, Y. and T. Goddard. 2008. Roles of annual legumes in no-till farming systems. In *No-Till Farming Systems, Special Publication No. 3*, edited by T. Goddard, M. Zoebisch, Y. Gan, W. Ellis, A. Watson, and S. Sombatpanit, pp. 279–288. Bangkok: World Association of Soil and Water Conservation (WASWAC).

Gan, Y., K. N. Harker, B. McConkey, and M. Suleimanov. 2008. Moving towards no-till practices in northern Eurasia. In *No-Till Farming Systems, Special Publication No. 3*, edited by T. Goddard, M. Zoebisch, Y. Gan, W. Ellis, A. Watson, and S. Sombatpanit, pp. 179–195. Bangkok: World Association of Soil and Water Conservation (WASWAC).

Gao, Y. J. and S. X. Li. 2005. Cause and mechanism of crop yield reduction under straw mulch in dry land. *Transactions of the Chinese Society of Agricultural Engineers* 21:15–19 (in Chinese with English abstract).

Garvin, P. C., J. R. Busch, and D. C. Kincaid. 1986. Reservoir tillage for reducing runoff and increasing production under sprinkler irrigation. In *ASAE Annual Paper No. 86-2093*, edited by ASAE. St. Joseph, MI: ASAE.

Geerts, S. and D. Raes. 2009. Deficit irrigation as an on-farm strategy to maximize crop water productivity in dry areas. *Agricultural Water Management* 96 (9):1275–1284.

Ghassemi, F. and I. White. 2007. *Inter-Basin Water Transfer: Case Studies from Australia, United States, Canada, China and India.* Cambridge: Cambridge University Press.

Ghosh, P. K., D. Dayal, K. K. Bandyopadhyay, and K. Mohanty. 2006. Evaluation of straw and polythene mulch for enhancing productivity of irrigated summer groundnut. *Field Crops Research* 99 (2–3):76–86.

Gill, B. S. and S. K. Jalota. 1996. Evaporation from soil in relation to residue rate, mixing depth, soil texture and evaporativity. *Soil Technology* 8 (4):293–301.

Gimenez, R. and G. Govers. 2008. Effects of freshly incorporated straw residue on rill erosion and hydraulics. *Catena* 72 (2):214–223.

Gleick, P. H., E. L. Wolff, and R. R. Chalecki. 2002. *The New Economy of Water: The Risks and Benefits of Globalization and Privatization of Freshwater.* Oakland, CA: Pacific Institute for Studies in Development, Environment, and Security.

Goddard, T., M. Zoebisch, Y. Gan, W. Ellis, A. Watson, and S. Sombatpanit. 2008. *No-Till Farming Systems, Special Publication No. 3.* Bangkok: World Association of Soil and Water Conservation (WASWAC).

Goldhamer, D. A., M. Salinas, C. Crisosto, K. R. Day, M. Soler, and A. Moriana. 2002. Effects of regulated deficit irrigation and partial root zone drying on late harvest peach tree performance. *Acta Horticulturae* 592:343–350.

Gomez, J. A. and M. A. Nearing. 2005. Runoff and sediment losses from rough and smooth soil surfaces in a laboratory experiment. *Catena* 59 (3):253–266.

Gordon, L. J., C. M. Finlayson, and M. Falkenmark. 2010. Managing water in agriculture for food production and other ecosystem services. *Agricultural Water Management* 97 (4):512–519.

Govaerts, B., K. D. Sayre, K. Lichter, L. Dendooven, and J. Deckers. 2007. Influence of permanent raised bed planting and residue management on physical and chemical soil quality in rain fed maize/wheat systems. *Plant and Soil* 291 (1–2):39–54.

Graham, R. L., R. Nelson, J. Sheehan, R. D. Perlack, and L. L. Wright. 2007. Current and potential U.S. corn stover supplies. *Agronomy Journal* 99 (1):1–11.
Gregory, P. J., L. P. Simmonds, and C. J. Pilbeam. 2000. Soil type, climatic regime, and the response of water use efficiency to crop management. *Agronomy Journal* 92 (5):814–820.
Groenevelt, P. H., C. D. Grant, and S. Semetsa. 2001. A new procedure to determine soil water availability. *Australian Journal of Soil Research* 39 (3):577–598.
Guzha, A. C. 2004. Effects of tillage on soil microrelief, surface depression storage and soil water storage. *Soil & Tillage Research* 76 (2):105–114.
Hakala, K., M. Kontturi, and K. Pahkala. 2009. Field biomass as global energy source. *Agricultural and Food Science* 18 (3–4):347–365.
Hamdy, A. 2007. Water use efficiency in irrigated agriculture: An analytical review. *Options Méditerranéennes Series B* (57):9–19.
Hamza, M. A. and W. K. Anderson. 2003. Responses of soil properties and grain yields to deep ripping and gypsum application in a compacted loamy sand soil contrasted with a sandy clay loam soil in Western Australia. *Australian Journal of Agricultural Research* 54 (3):273–282.
Hamza, M. A. and W. K. Anderson. 2005. Soil compaction in cropping systems—A review of the nature, causes and possible solutions. *Soil & Tillage Research* 82 (2):121–145.
Hares, M. A. and M. D. Novak. 1992. Simulation of surface-energy balance and soil-temperature under strip tillage. 2. Field-test. *Soil Science Society of America Journal* 56 (1):29–36.
Hart, W. E. and D. F. Heerman. 1976. Evaluating water distribution of sprinkler irrigation systems. In *Tech. Bulletin No. 128*, edited by C. S. University. Fort Collins, CO: CSU Press.
Hartge, K. H. and H. Bohne. 1983. Effect of pore geometry on compressibility of soil and development of rye seedlings. *Zeitschrift Fur Kulturtechnik Und Flurbereinigung* 24 (1):5–10.
Hartmann, C., R. Poss, A. D. Noble, A. Jongskul, E. Bourdon, D. Brunet, and G. Lesturgez. 2008. Subsoil improvement in a tropical coarse textured soil: Effect of deep-ripping and slotting. *Soil & Tillage Research* 99 (2):245–253.
Hatfield, J. L., T. J. Sauer, and J. H. Prueger. 2001. Managing soils to achieve greater water use efficiency: A review. *Agronomy Journal* 93 (2):271–280.
Haynes, R. J. 2000. Interactions between soil organic matter status, cropping history, method of quantification and sample pretreatment and their effects on measured aggregate stability. *Biology and Fertility of Soils* 30 (4):270–275.
Heerman, D. F., W. W. Wallender, and M. G. Bos. 1992. Irrigation efficiency and uniformity. In *Management of Farm Irrigation Systems,* ASAE Monograph No. 9, edited by G. J. Hoffman, T. A. Howell, and K. H. Solomon. St. Joseph, MI: ASAE.
Hibberd, J. M., J. E. Sheehy, and J. A. Langdale. 2008. Using C-4 photosynthesis to increase the yield of rice—rationale and feasibility. *Current Opinion in Plant Biology* 11 (2):228–231.
Hillel, D. 1980. *Applications of Soil Physics*. New York: Academic Press.
Hobbs, P. R. 2007. Conservation agriculture: What is it and why is it important for future sustainable food production? *Journal of Agricultural Science* 145:127–137.
Hobbs, P. R., K. Sayre, and R. Gupta. 2008. The role of conservation agriculture in sustainable agriculture. *Philosophical Transactions of the Royal Society B-Biological Sciences* 363 (1491):543–555.
Hoff, H. 2010. The global water challenge—Modeling green and blue water preface. *Journal of Hydrology* 384 (3–4):175–176.
Hoff, H., M. Falkenmark, D. Gerten, L. Gordon, L. Karlberg, and J. Rockstrom. 2010. Greening the global water system. *Journal of Hydrology* 384 (3–4):177–186.
Horn, R. and A. Smucker. 2005. Structure formation and its consequences for gas and water transport in unsaturated arable and forest soils. *Soil & Tillage Research* 82 (1):5–14.
Hsiao, T. C., P. Steduto, and E. Fereres. 2007. A systematic and quantitative approach to improve water use efficiency in agriculture. *Irrigation Science* 25 (3):209–231.
Hudson, B. D. 1994. Soil organic-matter and available water capacity. *Journal of Soil and Water Conservation* 49 (2):189–194.
Hula, J., P. Kovaricek, and M. Kroulik. 2010. Water infiltration into the soil and surface water run-off in wide-row crops. *Listy Cukrovarnicke a Reparske* 126 (1):22–26.
Hussain, I., K. R. Olson, and J. C. Siemens. 1998. Long-term tillage effects on physical properties of eroded soil. *Soil Science* 163 (12):970–981.
ICID. 1978. Standards for the calculation of irrigation efficiencies. *ICID Bulletin* 27:91–101.
Imhoff, S., P. J. Ghiberto, A. Grioni, and J. P. Gay. 2010. Porosity characterization of Argiudolls under different management systems in the Argentine Flat Pampa. *Geoderma* 158 (3–4):268–274.

Jadhav, M. G., V. G. Maniyar, D. N. Gokhale, and S. R. Nagargoje. 2008. Impact of land use pattern and land configurations on surface run-off and soil erosion during south-west monsoon at Parbhani. *Journal of Agrometeorology* 10:38–41.

Jalota, S. K. and V. K. Arora. 2002. Model-based assessment of water balance components under different cropping systems in north-west India. *Agricultural Water Management* 57 (1):75–87.

Jalota, S. K. and S. S. Prihar. 1986. Effects of atmospheric evaporativity, soil type and redistribution time on evaporation from bare soil. *Australian Journal of Soil Research* 24 (3):357–366.

Jensen, M. E. 2007. Beyond irrigation efficiency. *Irrigation Science* 25 (3):233–245.

Jin, K., W. M. Cornelis, D. Gabriels, et al. 2009. Residue cover and rainfall intensity effects on runoff soil organic carbon losses. *Catena* 78 (1):81–86.

Jordan, A., L. M. Zavala, and J. Gil. 2010. Effects of mulching on soil physical properties and runoff under semi-arid conditions in southern Spain. *Catena* 81 (1):77–85.

Kaddah, M. T. 1976. Subsoil chiseling and slip plowing effects on soil properties and wheat grown on a stratified fine sandy soil. *Agronomy Journal* 68 (1):36–39.

Kang, Y. H., S. Khan, and X. Y. Ma. 2009. Climate change impacts on crop yield, crop water productivity and food security—A review. *Progress in Natural Science* 19 (12):1665–1674.

Kansas State University Southwest Research-Extension Center. 2008. Reducing tillage intensity in a wheat-sorghum-fallow rotation. Report of Progress 997, pp. 13–15 (http://www.ksre.ksu.edu/library/crpsl2/SRP997.pdf (accessed November 02, 2010)).

Kassam, A., T. Friedrich, F. Shaxson, and J. Pretty. 2009. The spread of conservation agriculture: Justification, sustainability and uptake. *International Journal of Agricultural Sustainability* 7 (4):292–320.

Kassam, A. H., H. Doggett, and D. J. Andrews. 1979. Cereal physiology in relation to genetic improvement at ICRISAT with some reference to drought endurance. In *Proceedings of Physiology Programme Formulation Workshop*, edited by K. J. Treharne, pp. 45–51. Ibadan: IITA.

Kassam, A. H. and T. Friedrich. 2009. Perspectives on nutrient management in conservation agriculture. In *Lead Papers of the IV World Congress on Conservation Agriculture*, pp. 85–92. New Delhi.

Kassam, A. H., D. Molden, E. Fereres, and J. Doorenbos. 2007. Water productivity: Science and practice—Introduction. *Irrigation Science* 25 (3):185–188.

Katerji, N. and M. Mastrorilli. 2009. The effect of soil texture on the water use efficiency of irrigated crops: Results of a multi-year experiment carried out in the Mediterranean region. *European Journal of Agronomy* 30 (2):95–100.

Katerji, N., M. Mastrorilli, and H. E. Cherni. 2010. Effects of corn deficit irrigation and soil properties on water use efficiency. A 25-year analysis of a Mediterranean environment using the STICS model. *European Journal of Agronomy* 32 (2):177–185.

Kaufmann, M., S. Tobias, and R. Schulin. 2010. Comparison of critical limits for crop plant growth based on different indicators for the state of soil compaction. *Journal of Plant Nutrition and Soil Science* 173 (4):573–583.

Kay, B. D. and D. A. Angers. 1999. Soil structure. In *Handbook of Soil Science*, edited by M. E. Sumner, pp. 229–269. Boca Raton: CRC Press.

Kay, B. D. and A. J. VandenBygaart. 2002. Conservation tillage and depth stratification of porosity and soil organic matter. *Soil & Tillage Research* 66 (2):107–118.

Kay, B. D. and D. A. Angers. 1999. Soil structure. In *Handbook of Soil Science*, edited by M. E. Sumner, pp. 229–269. Boca Raton: CRC Press.

Keller, J. and R. Bliesner. 2001. *Sprinkler and Trickle Irrigation*. London: The Blackburn Press.

Kemper, W. D., T. J. Trout, and D. C. Kincaid. 1987. Cablegation: Automate supply for surface irrigation. In *Advances in Irrigation*, edited by D. Hillel. Orlando, FL: Academic Press.

Khalid, M. and J. L. Smith. 1978. Control of furrow infiltration by compaction. *Transactions of the ASAE* 21 (4):655–657.

Khan, S., M. M. Hafeez, T. Rana, and S. Mushtaq. 2008. Enhancing water productivity at the irrigation system level: A geospatial hydrology application in the Yellow River Basin. *Journal of Arid Environments* 72 (6):1046–1063.

Kirkegaard, J. A., J. M. Lilley, G. N. Howe, and J. M. Graham. 2007. Impact of subsoil water use on wheat yield. *Australian Journal of Agricultural Research* 58 (4):303–315.

Kitchen, N. R., K. W. T. Goulding, and J. F. Shanahan. 2008. Proven practices and innovative technologies for on-farm crop nitrogen management. In *Nitrogen Environment: Sources, Problems, and Management*, edited by J. L. Hatfield and T. R. F. Follet. Boston, MA: Academic Press.

Kladivko, E. J., N. M. Akhouri, and G. Weesies. 1997. Earthworm populations and species distributions under no-till and conventional tillage in Indiana and Illinois. *Soil Biology & Biochemistry* 29 (3–4):613–615.

Klocke, N. L., R. S. Currie, and R. M. Aiken. 2009. Soil water evaporation and crop residues. *Transactions of the ASABE* 52 (1):103–110.

Koluvek, P. K., K. K. Tanji, and T. J. Trout. 1993. Overview of soil-erosion from irrigation. *Journal of Irrigation and Drainage Engineering-ASCE* 119 (6):929–946.

Kosgei, J. R., G. P. W. Jewitt, V. M. Kongo, and S. A. Lorentz. 2007. The influence of tillage on field scale water fluxes and maize yields in semi-arid environments: A case study of Potshini catchment, South Africa. *Physics and Chemistry of the Earth* 32 (15–18):1117–1126.

Kowal, J. M. and A. H. Kassam. 1978. *Agricultural Ecology of Savanna: A Study of West Africa*. Oxford: Clarendon Press.

Krupinsky, J. M., K. L. Bailey, M. P. McMullen, B. D. Gossen, and T. K. Turkington. 2002. Managing plant disease risk in diversified cropping systems. *Agronomy Journal* 94 (4):198–209.

Lado, A. and A. Ben-Hur. 2004. Soil mineralogy effects on seal formation, runoff and soil loss. *Applied Clay Science* 24 (3–4):209–224.

Lado, M., A. Paz, and M. Ben-Hur. 2004. Organic matter and aggregate size interactions in infiltration, seal formation, and soil loss. *Soil Science Society of America Journal* 68 (3):935–942.

Lal, R. 2005. World crop residues production and implications of its use as a biofuel. *Environment International* 31 (4):575–584.

Lal, R. 2008a. Crop residues as soil amendments and feedstock for bioethanol production. *Waste Management* 28 (4):747–758.

Lal, R. 2008b. Managing soil water to improve rainfed agriculture in India. *Journal of Sustainable Agriculture* 32 (1):51–75.

Lal, R. 2009. Soil quality impacts of residue removal for bioethanol production. *Soil & Tillage Research* 102 (2):233–241.

Lal, R. and M. K. Shukla. 2004. *Principles of Soil Physics*. New York: Marcel Dekker.

Lamm, F. R., R. M. Aiken, and A. A. Abou Kheira. 2009. Corn yield and water use characteristics as affected by tillage, plant density, and irrigation. *Transactions of the ASABE* 52 (1):133–143.

Lampurlanes, J. and C. Cantero-Martinez. 2006. Hydraulic conductivity, residue cover and soil surface roughness under different tillage systems in semiarid conditions. *Soil & Tillage Research* 85 (1–2):13–26.

Landers, J. 2007. Tropical crop-livestock systems in conservation agriculture: The Brazilian experience. In *Integrated Crop Management,* edited by FAO, Vol. 5. Rome: FAO.

Lapen, D. R., G. C. Topp, E. G. Gregorich, and W. E. Curnoe. 2004. Least limiting water range indicators of soil quality and corn production, eastern Ontario, Canada. *Soil & Tillage Research* 78 (2):151–170.

Laryea, K. B., P. Pathak, and M. C. Klaij. 1991. Tillage systems and soils in the semiarid tropics. *Soil & Tillage Research* 20 (2–4):201–218.

Lascano, R. J., R. L. Baumhardt, S. K. Hicks, and J. L. Heilman. 1994. Soil and plant water evaporation from strip-tilled cotton—Measurement and simulation. *Agronomy Journal* 86 (6):987–994.

Leach, G. 1995. *Global Land and Food in the 21st Century: Trends & Issues for Sustainability*. Stockholm: Stockholm Environment Institute.

Leao, T. P., A. P. da Silva, M. C. M. Macedo, S. Imhoff, and V. P. B. Euclides. 2006. Least limiting water range: A potential indicator of changes in near-surface soil physical quality after the conversion of Brazilian savanna into pasture. *Soil & Tillage Research* 88 (1–2):279–285.

Leffler, A. J., M. S. Peek, R. J. Ryel, C. Y. Ivans, and M. M. Caldwell. 2005. Hydraulic redistribution through the root systems of senesced plants. *Ecology* 86 (3):633–642.

Lentz, R. D., I. Shainberg, R. E. Sojka, and D. L. Carter. 1992. Preventing irrigation furrow erosion with small applications of polymers. *Soil Science Society of America Journal* 56 (6):1926–1932.

Lentz, R. D. and R. E. Sojka. 1994. Field results using polyacrylamide to manage furrow erosion and infiltration. *Soil Science* 158 (4):274–282.

Leys, A., G. Govers, K. Gillijns, E. Berckmoes, and I. Takken. 2010. Scale effects on runoff and erosion losses from arable land under conservation and conventional tillage: The role of residue cover. *Journal of Hydrology* 390 (3–4):143–154.

Li, X. Y. 2003. Gravel-sand mulch for soil and water conservation in the semiarid loess region of northwest China. *Catena* 52 (2):105–127.

Li, Y. X., J. N. Tullberg, D. M. Freebairn, and H. W. Li. 2009. Functional relationships between soil water infiltration and wheeling and rainfall energy. *Soil & Tillage Research* 104 (1):156–163.

Licht, M. A. and M. Al-Kaisi. 2005. Strip-tillage effect on seedbed soil temperature and other soil physical properties. *Soil & Tillage Research* 80 (1–2):233–249.

Liebman, M. and C. L. Mohler. 2001. Weeds and the soil environment. In *Ecological Management of Agricultural Weeds*, edited by M. Liebman, C. L. Mohler, and C. P. Staver. Cambridge: Cambridge University Press.

Lindwall, C. W. and B. Sonntag (eds). 2010. *Landscape Transformed: The History of Conservation Tillage and Direct Seeding. Knowledge Impact in Society*. Saskatoon, SK: University of Saskatchewan.

Liu, C. A., S. L. Jin, L. M. Zhou, Y. Jia, F. M. Li, Y. C. Xiong, and X. G. Li. 2009. Effects of plastic film mulch and tillage on maize productivity and soil parameters. *European Journal of Agronomy* 31 (4):241–249.

Liu, W. D., M. Tollenaar, G. Stewart, and W. Deen. 2004. Response of corn grain yield to spatial and temporal variability in emergence. *Crop Science* 44 (3):847–854.

Liu, Y., S. Q. Li, F. Chen, S. J. Yang, and X. P. Chen. 2010. Soil water dynamics and water use efficiency in spring maize (*Zea mays* L.) fields subjected to different water management practices on the Loess Plateau, China. *Agricultural Water Management* 97 (5):769–775.

Liu, Z. M., L. Shan, X. P. Deng, S. Inanaga, W. Sunohara, and J. Harada. 1998. Effects of fertilizer and plant density on the yields of root system and water use of spring wheat. *Research of Soil and Water Conservation* 5 (1):70–75.

Lopez-Fando, C., J. Dorado, and M. T. Pardo. 2007. Effects of zone-tillage in rotation with no-tillage on soil properties and crop yields in a semi-arid soil from central Spain. *Soil & Tillage Research* 95 (1–2):266–276.

Losada, A., L. Juana, and J. Roldan. 1990. Operation diagrams for irrigation management. *Agricultural Water Management* 18 (4):289–300.

Lu, Y. C., K. B. Watkins, J. R. Teasdale, and A. A. Abdul-Baki. 2000. Cover crops in sustainable food production. *Food Reviews International* 16 (2):121–157.

Ludlow, M. M. and R. C. Muchow. 1988. Critical evaluation of the possibilities for modifying crops for high production per unit of precipitation. In *Proceedings of Physiology Programme Formulation Workshop*, edited by K. J. Treharne, pp. 45–51. Ibadan: IITA.

Lyle, W. M. and J. P. Bordovsky. 1983. Lepa irrigation system evaluation. *Transactions of the ASAE* 26 (3):776–781.

MacEwan, R. J., D. M. Crawford, P. J. Newton, and T. S. Clune. 2010. High clay contents, dense soils, and spatial variability are the principal subsoil constraints to cropping the higher rainfall land in south-eastern Australia. *Australian Journal of Soil Research* 48 (2):150–166.

Makurira, H., H. H. G. Savenije, and S. Uhlenbrook. 2010. Modelling field scale water partitioning using on-site observations in sub-Saharan rainfed agriculture. *Hydrology and Earth System Sciences* 14 (4):627–638.

Makurira, H., H. H. G. Savenije, S. Uhlenbrook, J. Rockstrom, and A. Senzanje. 2009. Investigating the water balance of on-farm techniques for improved crop productivity in rainfed systems: A case study of Makanya catchment, Tanzania. *Physics and Chemistry of the Earth* 34 (1–2):93–98.

Mann, L., V. Tolbert, and J. Cushman. 2002. Potential environmental effects of corn (*Zea mays* L.) stover removal with emphasis on soil organic matter and erosion. *Agriculture, Ecosystems & Environment* 89 (3):149–166.

Marongwe, L. S., K. Kwazira, M. Jenrich, C. Thierfelder, A. Kassam, and T. Friedrich. 2011. An African success: The case of conservation agriculture in Zimbabwe. *International Journal of Agricultural Sustainability* 9 (1):153–161.

Mary, G. R. and J. Changying. 2008. Influence of agricultural machinery traffic on soil compaction patterns, root development, and plant growth, overview. *American-Eurasian Journal of Agricultural and Environmental Science* 3 (1):49–62.

McGarry, D., B. J. Bridge, and B. J. Radford. 2000. Contrasting soil physical properties after zero and traditional tillage of an alluvial soil in the semi-arid subtropics. *Soil & Tillage Research* 53 (2):105–115.

Miller, D. E., J. S. Aarstad, and R. G. Evans. 1987. Control of furrow erosion with crop residues and surge flow irrigation. *Soil Science Society of America Journal* 51 (2):421–425.

Miller, D. E. and R. O Gifford. 1974. Modification of soil crusts for plant growth. In *Soil Crusts*, edited by D. D. E. Cary. J.W. Tucson, AZ: University of Arizona Technical Bulletin No. 214.

Mohammad, A. G. and M. A. Adam. 2010. The impact of vegetative cover type on runoff and soil erosion under different land uses. *Catena* 81 (2):97–103.

Molden, D. (ed.) 2007. *Water for Food, Water for Life: A Comprehensive Assessment of Water Management in Agriculture*. London, Colombo: Earthscan, International Water Management Institute.

Molden, D., D. Bin, R. Loeve, R. Barker, and T. P. Tuong. 2007. Agricultural water productivity and savings: Policy lessons from two diverse sites in China. *Water Policy* 9:29–44.

Molden, D., T. Oweis, P. Steduto, P. Bindraban, M. A. Hanjra, and J. Kijne. 2010. Improving agricultural water productivity: Between optimism and caution. *Agricultural Water Management* 97 (4):528–535.

Monzon, J. P., V. O. Sadras, and F. H. Andrade. 2006. Fallow soil evaporation and water storage as affected by stubble in sub-humid (Argentina) and semi-arid (Australia) environments. *Field Crops Research* 98 (2–3):83–90.

Moret, D., J. L. Arrue, M. V. Lopez, and R. Gracia. 2006. Influence of fallowing practices on soil water and precipitation storage efficiency in semiarid Aragon (NE Spain). *Agricultural Water Management* 82 (1–2):161–176.

Moret, D., I. Braud, and J. L. Arrue. 2007. Water balance simulation of a dryland soil during fallow under conventional and conservation tillage in semiarid Aragon, northeast Spain. *Soil & Tillage Research* 92 (1–2):251–263.

Mrabet, R. 2008. No-till practices in Morocco. In *No-Till Farming Systems, Special Publication No. 3*, edited by T. Goddard, M. Zoebisch, Y. Gan, W. Ellis, A. Watson, and S. Sombatpanit, pp. 393–412. Bangkok: World Association of Soil and Water Conservation (WASWAC).

Mrabet, R., N. Saber, A. El-Brahli, S. Lahlou, and F. Bessam. 2001. Total, particulate organic matter and structural stability of a Calcixeroll soil under different wheat rotations and tillage systems in a semiarid area of Morocco. *Soil & Tillage Research* 57 (4):225–235.

Mulumba, L. N. and R. Lal. 2008. Mulching effects on selected soil physical properties. *Soil & Tillage Research* 98 (1):106–111.

Musick, J. T., F. B. Pringle, and P. N. Johnson. 1985. Furrow compaction for controlling excessive irrigation water-intake. *Transactions of the ASAE* 28 (2):502–506.

National IPM Program. 2007. *SRI Application in Rice Production in Northern Ecological Areas of Vietnam.* Report to the Council of Science and Technology, Ministry of Agriculture and Rural Development, Hanoi, Vietnam.

Ndiaye, B., J. Molenat, V. Hallaire, C. Gascuel, and Y. Hamon. 2007. Effects of agricultural practices on hydraulic properties and water movement in soils in Brittany (France). *Soil & Tillage Research* 93 (2):251–263.

Ngigi, S. N., J. Rockstrom, and H. H. G. Savenije. 2006. Assessment of rainwater retention in agricultural land and crop yield increase due to conservation tillage in Ewaso Ng'iro river basin, Kenya. *Physics and Chemistry of the Earth* 31 (15–16):910–918.

Nuttall, J. G. and R. D. Armstrong. 2010. Impact of subsoil physicochemical constraints on crops grown in the Wimmera and Mallee is reduced during dry seasonal conditions. *Australian Journal of Soil Research* 48 (2):125–139.

Oguntunde, P. G., A. E. Ajayi, and N. van de Giesen. 2006. Tillage and surface moisture effects on bare-soil albedo of a tropical loamy sand. *Soil & Tillage Research* 85 (1–2):107–114.

Orts, W. J., R. E. Sojka, and G. M. Glenn. 2000. Biopolymer additives to reduce erosion-induced soil losses during irrigation. *Industrial Crops and Products* 11 (1):19–29.

Oweis, T. Y. and A. Y. Hachum. 2003. Improving water areas of west Asia productivity in the dry and North Africa. *Water Productivity in Agriculture: Limits and Opportunities for Improvement* 1:179–198.

Owenya, M. Z., W. L. Mariki, J. Kienzle, T. Friedrich, and A. Kassam. 2011. Conservation Agriculture (CA) in Tanzania: The case of the Mwangaza CA farmer field school (FFS). *International Journal of Agricultural Sustainability* 9 (1): 145–152.

Passioura, J. 2006. Increasing crop productivity when water is scarce—From breeding to field management. *Agricultural Water Management* 80 (1–3):176–196.

Passioura, J. B. and J. F. Angus. 2010. Improving productivity of crops in water-limited environments. *Advances in Agronomy* 106:37–75.

Patel, G. N., P. T. Patel, and P. H. Patel. 2008. Yield, water use efficiency and moisture extraction pattern of summer groundnut as influenced by irrigation schedules, sulfur levels and sources. *Journal of SAT Agricultural Research* 6:1–4.

Pereira, L. S. 2007. Relating water productivity and crop evapotranspiration. *Options Méditerranéennes Series B* (57):31–49.

Perry, C., P. Steduto, R. G. Allen, and C. M. Burt. 2009. Increasing productivity in irrigated agriculture: Agronomic constraints and hydrological realities. *Agricultural Water Management* 96 (11):1517–1524.

Peterson, G. A. and D. G. Westfall. 2004. Managing precipitation use in sustainable dryland agroecosystems. *Annals of Applied Biology* 144 (2):127–138.

Pierce, F. J., W. E. Larson, R. H. Dowdy, and W. A. P. Graham. 1983. Productivity of soils—Assessing long-term changes due to erosion. *Journal of Soil and Water Conservation* 38 (1):39–44.

Pimentel, D., B. Berger, D. Filiberto, M. Newton, B. Wolfe, E. Karabinakis, S. Clark, E. Poon, E. Abbett, and S. Nandagopal. 2004. Water resources: Agricultural and environmental issues. *Bioscience* 54 (10):909–918.

Pimentel, D., C. Harvey, P. Resosudarmo, et al. 1995. Environmental and economic costs of soil erosion and conservation benefits. *Science* 267 (5201):1117–1123.

Playan, E. and L. Mateos. 2006. Modernization and optimization of irrigation systems to increase water productivity. *Agricultural Water Management* 80 (1–3):100–116.

Podwojewski, P., D. Orange, P. Jouquet, C. Valentin, V. T. Nguyen, J. L. Janeau, and D. T. Tran. 2008. Land-use impacts on surface runoff and soil detachment within agricultural sloping lands in northern Vietnam. *Catena* 74 (2):109–118.

Postel, S. L. 1998. Water for food production: Will there be enough in 2025? *Bioscience* 48 (8):629–637.

Prihar, S. S., S. K. Jalota, and J. L. Steiner. 1996. Residue management for reducing evaporation in relation to soil type and evaporativity. *Soil Use and Management* 12 (3):150–157.

Purkey, D. R. and W. W. Wallender. 1989. Surge flow infiltration variability. *Transactions of the ASAE* 32 (3):894–900.

Ramakrishna, A., H. M. Tam, S. P. Wani, and T. D. Long. 2006. Effect of mulch on soil temperature, moisture, weed infestation and yield of groundnut in northern Vietnam. *Field Crops Research* 95 (2–3):115–125.

Ramankutty, N., A. T. Evan, C. Monfreda, and J. A. Foley. 2008. Farming the planet: 1. Geographic distribution of global agricultural lands in the year 2000. *Global Biogeochemical Cycles* 22 (1):19.

Ramos, A. F. and F. L. Santos. 2009. Water use, transpiration, and crop coefficients for olives (cv. Cordovil), grown in orchards in southern Portugal. *Biosystems Engineering* 102 (3):321–333.

Ramos, A. F. and F. L. Santos. 2010. Yield and olive oil characteristics of a low-density orchard (cv. Cordovil) subjected to different irrigation regimes. *Agricultural Water Management* 97 (2):363–373.

Ramos, M. C., S. Nacci, and I. Pla. 2003. Effect of raindrop impact and its relationship with aggregate stability to different disaggregation forces. *Catena* 53 (4):365–376.

Ransom, J. K., G. J. Endres, and B. G. Schatz. 2007. Sustainable improvement of wheat yield potential: The role of crop management. *Journal of Agricultural Science* 145:55–61.

Rasmussen, K. J. 1999. Impact of ploughless soil tillage on yield and soil quality: A Scandinavian review. *Soil & Tillage Research* 53 (1):3–14.

Rawls, W. J., Y. A. Pachepsky, J. C. Ritchie, T. M. Sobecki, and H. Bloodworth. 2003. Effect of soil organic carbon on soil water retention. *Geoderma* 116 (1–2):61–76.

Reichert, J. M., J. A. Albuquerque, D. R. Kaiser, D. J. Reinert, F. L. Urach, and R. Carlesso. 2009a. Estimation of water retention and availability in soils of rio grande do sul. *Revista Brasileira De Ciencia Do Solo* 33 (6):1547–1560.

Reichert, J. M., Leas Suzuki, D. J. Reinert, R. Horn, and I. Hakansson. 2009b. Reference bulk density and critical degree-of-compactness for no-till crop production in subtropical highly weathered soils. *Soil & Tillage Research* 102 (2):242–254.

Rhoton, F. E., M. J. Shipitalo, and D. L. Lindbo. 2002. Runoff and soil loss from midwestern and southeastern U.S. silt loam soils as affected by tillage practice and soil organic matter content. *Soil & Tillage Research* 66 (1):1–11.

Richards, R. A. 2006. Physiological traits used in the breeding of new cultivars for water-scarce environments. *Agricultural Water Management* 80 (1–3):197–211.

Ridley, A. M., B. Christy, F. X. Dunin, P. J. Haines, K. F. Wilson, and A. Ellington. 2001. Lucerne in crop rotations on the Riverine Plains 1. The soil water balance. *Australian Journal of Agricultural Research* 52 (2):263–277.

Ritchie, J. T. and J. E. Adams. 1974. Field measurement of evaporation from soil shrinkage cracks. *Soil Science Society of America Journal* 38 (1):131–134.

Ritchie, J. T. and B. Basso. 2008. Water use efficiency is not constant when crop water supply is adequate or fixed: The role of agronomic management. *European Journal of Agronomy* 28 (3):273–281.

Rockstrom, J. 2000. Water resources management in smallholder farms in eastern and southern Africa: An overview. *Physics and Chemistry of the Earth Part B-Hydrology Oceans and Atmosphere* 25 (3):275–283.

Rockstrom, J. and J. Barron. 2007. Water productivity in rainfed systems: Overview of challenges and analysis of opportunities in water scarcity prone savannahs. *Irrigation Science* 25 (3):299–311.

Rockstrom, J., J. Barron, and P. Fox. 2002. Rainwater management for increased productivity among smallholder farmers in drought prone environments. *Physics and Chemistry of the Earth* 27 (11–22):949–959.

Rockstrom, J., M. Falkenmark, L. Karlberg, H. Hoff, S. Rost, and D. Gerten. 2009a. Future water availability for global food production: The potential of green water for increasing resilience to global change. *Water Resources Research* 45:16.

Rockstrom, J., P. Kaurnbutho, J. Mwalley, A. W. Nzabi, M. Temesgen, L. Mawenya, J. Barron, J. Mutua, and S. Damgaard-Larsen. 2009b. Conservation farming strategies in east and southern Africa: Yields and rainwater productivity from on-farm action research. *Soil & Tillage Research* 103 (1):23–32.

Rockstrom, J., M. Lannerstad, and M. Falkenmark. 2007. Assessing the water challenge of a new green revolution in developing countries. *Proceedings of the National Academy of Sciences of the United States of America* 104 (15):6253–6260.

Rodriguez, D., J. Nuttall, V. O. Sadras, H. van Rees, and R. Armstrong. 2006. Impact of subsoil constraints on wheat yield and gross margin on fine-textured soils of the southern Victorian Mallee. *Australian Journal of Agricultural Research* 57 (3):355–365.

Rosegrant, M. W., X. Cai, and S. A. Cline. 2002. *World Water and Food to 2025: Dealing with Scarcity*. Washington, DC, London: IFPRI, International Water Management Institute.

Rosegrant, M. W., C. Ringler, and T. J. Zhu. 2009. Water for agriculture: Maintaining food security under growing scarcity. *Annual Review of Environment and Resources* 34:205–222.

Rost, S., D. Gerten, A. Bondeau, W. Lucht, J. Rohwer, and S. Schaphoff. 2008. Agricultural green and blue water consumption and its influence on the global water system. *Water Resources Research* 44 (9):17.

Roth, C. H. 1985. Infiltrabilität von Latossolo-Roxo-Böden in Nordparaná, Brasilien, in Feldversuchen zur Erosionskontrolle mit verschiedenen Bodenbearbeitungssystemen und Rotationen. Göttinger Bodenkundliche Berichte 83, pp. 1–104, Göttingen.

Ryan, J. 2000. Plant nutrient management under pressured irrigation systems in the Mediterranean region. In *Proceedings of the IMPHOS International Workshop Organized by the World Phosphate Institute*. Aleppo: ICARDA.

Sadler, E. J., R. G. Evans, K. C. Stone, and C. R. Camp. 2005. Opportunities for conservation with precision irrigation. *Journal of Soil and Water Conservation* 60 (6):371–379.

Sadras, V. O. and J. F. Angus. 2006. Benchmarking water-use efficiency of rainfed wheat in dry environments. *Australian Journal of Agricultural Research* 57 (8):847–856.

Sadras, V. O., G. J. O'Leary, and D. K. Roget. 2005. Crop responses to compacted soil: Capture and efficiency in the use of water and radiation. *Field Crops Research* 91 (2–3):131–148.

Salako, F. K. and G. Tian. 2003. Soil water depletion under various leguminous cover crops in the derived savanna of West Africa. *Agriculture, Ecosystems & Environment* 100 (2–3):173–180.

Santos, F. L. 1996a. Evaluation and adoption of irrigation technologies. 1. Management-design curves for furrow and level basin systems. *Agricultural Systems* 52 (2–3):317–329.

Santos, F. L. 1996b. Quality and maximum profit of industrial tomato as affected by distribution uniformity of drip irrigation system. *Irrigation and Drainage Systems* 10:281–294.

Santos, F. L. 1998. Evaluation of alternative irrigation technologies based upon applied water and simulated yields. *Journal of Agricultural Engineering Research* 69 (1):73–83.

Santos, F. L., J. L. Reis, O. C. Martins, N. L. Castanheira, and R. P. Serralheiro. 2003. Comparative assessment of infiltration, runoff and erosion of sprinkler irrigated soils. *Biosystems Engineering* 86 (3):355–364.

Santos, F. L. and R. P. Serralheiro. 2000. Improving infiltration of irrigated Mediterranean soils with polyacrylamide. *Journal of Agricultural Engineering Research* 76 (1):83–90.

Santos, F. L., P. C. Valverde, A. F. Ramos, J. L. Reis, and N. L. Castanheira. 2007. Water use and response of a dry-farmed olive orchard recently converted to irrigation. *Biosystems Engineering* 98:102–114.

Sasal, M. C., A. E. Andriulo, and M. A. Taboada. 2006. Soil porosity characteristics and water movement under zero tillage in silty soils in Argentinian Pampas. *Soil & Tillage Research* 87 (1):9–18.

Sauer, T. J., J. L. Hatfield, and J. H. Prueger. 1996. Corn residue age and placement effects on evaporation and soil thermal regime. *Soil Science Society of America Journal* 60 (5):1558–1564.

Savenije, H. H. G. 1997. Determination of evaporation from a catchment water balance at a monthly time scale. *Hydrology and Earth System Sciences* 1:93–100.

Savenije, H. H. G. 2004. The importance of interception and why we should delete the term evapotranspiration from our vocabulary. *Hydrological Processes* 18 (8):1507–1511.

Seckler, D., D. Molden, and R. Sakthivadivel. 2003. The concept of efficiency in water-resources management and policy. *Water Productivity in Agriculture: Limits and Opportunities for Improvement* 1:37–51.

Seeger, M. 2007. Uncertainty of factors determining runoff and erosion processes as quantified by rainfall simulations. *Catena* 71 (1):56–67.

Sellin, A. 2001. Hydraulic and stomatal adjustment of Norway spruce trees to environmental stress. *Tree Physiology* 21 (12–13):879–888.

Shah, T., J. Burke, and K. Villholth. 2007. Groundwater: A global assessment of scale and significance. In *Water for Food, Water for Life: A Comprehensive Assessment of Water Management in Agriculture*, edited by D. Molden, pp. 396–432. London, Colombo: Earthscan, International Water Management Institute.

Shaheen, A., M. A. Naeem, G. Jilani, and M. Shafiq. 2010. Integrated soil management in eroded land augments the crop yield and water-use efficiency. *Acta Agriculturae Scandinavica Section B-Soil and Plant Science* 60 (3):274–282.

Shahidian, S., R. P. Serralheiro, and L. L. Silva. 1998. Real time management of furrow irrigation with a cablegation system. In *Water and the Environment: Innovation Issues in Irrigation and Drainage*, edited by L. S. Pereira and J. W. Gowing. pp. 149–155. London: ICID-CIID, E & Spon.

Sharma, P. K. and C. L. Acharya. 2000. Carry-over of residual soil moisture with mulching and conservation tillage practices for sowing of rainfed wheat (*Triticum aestivum* L.) in north-west India. *Soil & Tillage Research* 57 (1–2):43–52.

Sharratt, B. S. 2002. Corn stubble height and residue placement in the northern U.S. Corn Belt Part I. Soil physical environment during winter. *Soil & Tillage Research* 64 (3–4):243–252.

Shatanawi, M. R. 2007. Future options and research needs of water uses for sustainable agriculture. *Options Méditerranéennes Series B* (57):21–29.

Shaxson, F., A. H. Kassam, T. Friedrich, B. Boddey, and A. Adekunle. 2008. Underpinning the benefits of conservation agriculture: Sustaining the fundamental of soil health and function. In *Workshop on Investing in Sustainable Crop Intensification: The Case of Soil Health*. Rome: FAO.

Shear, G. M. 1985. Introduction and history of limited tillage. In *Weed Control in Limited-Tillage Systems*, edited by A. F. Wiese. Champaign, IL: Weed Science Society of America.

Sheehy, J. E. 2000. Limits to yield for C-3 and C-4 rice: An agronomist's view. *Redesigning Rice Photosynthesis to Increase Yield* 7:39–52.

Shiklomanov, I. A. 1993. World freshwater resources. In *Water in Crisis: A Guide to the World's Freshwater Resources*, edited by P. Gleick, pp. 13–24. New York: Oxford University Press.

Shiklomanov, I. A. and J. C. Rodda. 2003. *World Water Resources at the Beginning of the Twenty-First Century*. New York: Cambridge Univiversity Press.

Shinners, K. J., W. S. Nelson, and R. Wang. 1994. Effects of residue-free band-width on soil-temperature and water-content. *Transactions of the ASAE* 37 (1):39–49.

Shipitalo, M. J., W. A. Dick, and W. M. Edwards. 2000. Conservation tillage and macropore factors that affect water movement and the fate of chemicals. *Soil & Tillage Research* 53 (3–4):167–183.

Silici, L. 2010. Conservation agriculture and sustainable crop intensiification in lesotho. In *Integrated Crop Management Vol. 10*, edited by Plant Production and Protection Division. Rome: FAO.

Sillon, J. F., G. Richard, and I. Cousin. 2003. Tillage and traffic effects on soil hydraulic properties and evaporation. *Geoderma* 116 (1–2):29–46.

Silva, L. L. 2010. *Runoff under Sprinkler Irrigation: Affecting Factors and Control Practices*. New York: Nova Science Publishers.

Singh, B. and S. S. Malhi. 2006. Response of soil physical properties to tillage and residue management on two soils in a cool temperate environment. *Soil & Tillage Research* 85 (1–2):143–153.

Singh, G., S. K. Jalota, and B. S. Sidhu. 2005. Soil physical and hydraulic properties in a rice-wheat cropping system in India: Effects of rice-straw management. *Soil Use and Management* 21 (1):17–21.

Singh, R., J. C. van Dam, and R. A. Feddes. 2006. Water productivity analysis of irrigated crops in Sirsa district, India. *Agricultural Water Management* 82 (3):253–278.

Six, J., C. Feller, K. Denef, S. M. Ogle, J. C. D. Sa, and A. Albrecht. 2002. Soil organic matter, biota and aggregation in temperate and tropical soils—Effects of no-tillage. *Agronomie* 22 (7–8):755–775.

So, H. B., A. Grabski, and P. Desborough. 2009. The impact of 14 years of conventional and no-till cultivation on the physical properties and crop yields of a loam soil at Grafton NSW, Australia. *Soil & Tillage Research* 104 (1):180–184.

Sojka, R. E. and D. L. Bjorneberg. 2002. Erosion, controlling irrigation-induced. In *Encyclopedia of Soil Science*, edited by R. Lal. New York: Marcel Dekker, Inc.

Solomon, K. and J. Keller. 1978. Trickle irrigation uniformity and efficiency. *ASCE Journal of the Irrigation and Drainage Division* 104 (IR3):293–306.

Solomon, K. H., D. C. Kincaid, and J. C. Bezdek. 1985. Drop size distributions for irrigation spray nozzles. *Transactions of the ASAE* 28 (6):1966–1974.

Soriano, M. A., F. Orgaz, F. J. Villalobos, and E. Fereres. 2004. Efficiency of water use of early plantings of sunflower. *European Journal of Agronomy* 21 (4):465–476.

Steduto, P., T. C. Hsiao, and E. Fereres. 2007. On the conservative behavior of biomass water productivity. *Irrigation Science* 25 (3):189–207.

Steiner, J. L. 1989. Tillage and surface residue effects on evaporation from soils. *Soil Science Society of America Journal* 53 (3):911–916.

Stewart, J. L. 1991. *Principles and Performance of Response Farming*. Wallingford, CT: CAB International.

Stewart, J. L. and D. A. R. Kashasha. 1984. Rainfall criteria to enable response farming through crop-based climate analysis. *East African Agricultural and Forestry Journal* 44:58–79.

Stone, L. F. and P. M. da Silveira. 1999. Effects of soil tillage on soil compaction, available soil water, and development of common bean. *Pesquisa Agropecuaria Brasileira* 34 (1):83–91.

Stone, L. R. and A. J. Schlegel. 2010. Tillage and crop rotation phase effects on soil physical properties in the West-Central Great Plains. *Agronomy Journal* 102 (2):483–491.

Strudley, M. W., T. R. Green, and J. C. Ascough. 2008. Tillage effects on soil hydraulic properties in space and time: State of the science. *Soil & Tillage Research* 99 (1):4–48.

Suleimenov, M. and K. Akshalov. 2007. Eliminating summer fallow in black soils of northern Kazakhstan. In *Climate Change and Terrestrial Carbon Sequestration in Central Asia*, edited by R. Lal, M. Suleimenov, B. A. Stewart, D. O. Hansen, and P. Doraiswamy. New York: Taylor and Francis Group.

Suyker, A. E. and S. B. Verma. 2009. Evapotranspiration of irrigated and rainfed maize-soybean cropping systems. *Agricultural and Forest Meteorology* 149 (3–4):443–452.

Tanaka, D. L., R. L. Anderson, and S. C. Rao. 2005. Crop sequencing to improve use of precipitation and synergize crop growth. *Agronomy Journal* 97 (2):385–390.

Tao, H. B., H. Brueck, K. Dittert, C. Kreye, S. Lin, and B. Sattelmacher. 2006. Growth and yield formation of rice (*Oryza sativa* L.) in the water-saving ground cover rice production system (GCRPS). *Field Crops Research* 95 (1):1–12.

Tarboton, K. C. and W. W. Wallender. 1989. Field-wide furrow infiltration variability. *Transactions of the ASAE* 32 (3):913–918.

Tebrugge, F. and J. Abelsova. 1999. Biopores increase seepage—The influence of soil tillage on biogenic pores and on unsaturated infiltration capacity of soils. *Landtechnik* 54 (1):13–15.

Tebrugge, F. and R. A. During. 1999. Reducing tillage intensity—A review of results from a long-term study in Germany. *Soil & Tillage Research* 53 (1):15–28.

Teixeira, F., G. Basch, and M. J. Carvalho. 2000. Tillage effects on splash detachment, overland flow and interril erosion. In *Proceedings of the 4th International Conference on Soil Dynamics*, edited by Agricultural Machinery Research and Design Centre, pp. 307–314. University of South Australia: Adelaide, SA.

Tennakoon, S. B. and N. R. Hulugalle. 2006. Impact of crop rotation and minimum tillage on water use efficiency of irrigated cotton in a Vertisol. *Irrigation Science* 25 (1):45–52.

Teuling, A. J., R. Uijlenhoet, F. Hupet, and P. A. Troch. 2006. Impact of plant water uptake strategy on soil moisture and evapotranspiration dynamics during drydown. *Geophysical Research Letters* 33 (3): L03401, 5 pp.

Thierfelder, C. and P. C. Wall. 2009. Effects of conservation agriculture techniques on infiltration and soil water content in Zambia and Zimbabwe. *Soil & Tillage Research* 105 (2):217–227.

Thompson, A. L. and L. G. James. 1985. Water droplet impact and its effect on infiltration. *Transactions of the ASAE* 28 (5):1506–1510.

Thompson, R. L. 2007. Sustainability in the food & agricultural sector: The role of the private sector and government. In *Proceedings from the 40th IPC Seminar*, Stratford-upon-Avon, UK (http://www.agritrade.org/events/documents/2007SustainabilitySeminar_002.pdf (accessed November 2, 2010)).

Till, M. R. and M. G. Bos. 1985. The influence of uniformity and leaching on the field application efficiency. *ICID Bulletin* 34 (1):32–36.

Trevisan, R., F. G. Herter, and I. S. Pereira. 2002. Variação da amplitude térmica do solo em pomar de pessegueiro cultivado com aveia preta (*Avena* sp.) e em sistema convencional. *Revista Brasileira de Agrociencia* 8:155–157.

Trout, T. J. and W. D. Kemper. 1983. Factors which affect furrow intake rates. In *Proceedings ASAE on Advances in Infiltration*, edited by American Society of Agricultural Engineers, pp. 302–312. St. Joseph, MI: ASCE.

Trout, T. J. and W. H. Neibling. 1993. Erosion and sedimentation processes on irrigated fields. *Journal of Irrigation and Drainage Engineering-ASCE* 119 (6):947–963.

Trout, T. J., R. E. Sojka, and L. I. Okafor. 1992. Soil management. In *Management of Farm Irrigation Systems, ASAE Monograph No. 9*, edited by G. J. Hoffman, T. A. Howell, and K. H. Solomon. St. Joseph, MI: ASAE.

Turner, B. L., E. Frossard, and A. Oberson. 2006. Enhancing phosphorus availability in low-fertility soils. In *Biological Approaches to Sustainable Soil Systems*, edited by N. Uphoff, A. S. Ball, E. Fernandes, et al. Boca Raton, FL: CRC Press.

Undersander, D. J. and C. Regier. 1988. Effect of tillage and furrow irrigation timing on efficiency of preplant irrigation. *Irrigation Science* 9 (1):57–67.

Unger, P. W. and M. F. Vigil. 1998. Cover crop effects on soil water relationships. *Journal of Soil and Water Conservation* 53 (3):200–207.

Uphoff, N., A. S. Ball, E. Fernandes, et al. (eds). 2006. *Biological Approaches to Sustainable Soil Systems*. Boca Raton, FL: CRC Press.

Uphoff, N. and A. H. Kassam. 2009. System of rice intensification (SRI). In *Agriculture for Developing Countries. Science and Technology Options Assessment (STOA) Project*, edited by R. Meyer. Karlsruhe: European Parliament, European Technology Assessment Group.

Verhulst, N., B. Govaerts, E. Verachtert, A. Castellanos-Navarrete, M. Mezzalama, P. Wall, J. Deckers, and K. D. Sayre. 2010. Conservation agriculture, improving soil quality for sustainable production systems? In *Advances in Soil Science: Food Security and Soil Quality*, edited by R. Lal and B. A. Stewart, pp. 137–208. Boca Raton, FL: CRC Press.

Voorhees, W. B., W. W. Nelson, and G. W. Randall. 1986. Extent and persistence of subsoil compaction caused by heavy axle loads. *Soil Science Society of America Journal* 50 (2):428–433.

Vörösmarty, C. J., C. Lévêque, and C. Revenga. 2005. Fresh water, in ecosystems and human well-being: Current states and trends. In *Millenium Ecosystem Assessment Report*, edited by R. C. C. Bos, J. Chilton, E. M. Douglas, M. Meybeck, and D. Prager, pp. 165–207. Washington, DC: Island Press.

Wahl, N. A., O. Bens, U. Buczko, E. Hangen, and R. F. Huttl. 2004. Effects of conventional and conservation tillage on soil hydraulic properties of a silty-loamy soil. *Physics and Chemistry of the Earth* 29 (11–12):821–829.

Walker, W. R. and G. V. Skogerboe. 1987. *Surface Irrigation: Theory and Practice*. New York: Prentice-Hall Press.

Wang, Y. J., Z. K. Xie, S. S. Malhi, C. L. Vera, Y. B. Zhang, and J. N. Wang. 2009. Effects of rainfall harvesting and mulching technologies on water use efficiency and crop yield in the semi-arid Loess Plateau, China. *Agricultural Water Management* 96 (3):374–382.

Ward, P. R., K. Whisson, S. F. Micin, D. Zeelenberg, and S. P. Milroy. 2009. The impact of wheat stubble on evaporation from a sandy soil. *Crop & Pasture Science* 60 (8):730–737.

Whish, J. P. M., L. Price, and P. A. Castor. 2009. Do spring cover crops rob water and so reduce wheat yields in the northern grain zone of eastern Australia? *Crop & Pasture Science* 60 (6):517–525.

Wilhelm, W. W., J. M. F. Johnson, J. L. Hatfield, W. B. Voorhees, and D. R. Linden. 2004. Crop and soil productivity response to corn residue removal: A literature review. *Agronomy Journal* 96 (1):1–17.

Wilson, G. V., K. C. McGregor, and D. Boykin. 2008. Residue impacts on runoff and soil erosion for different corn plant populations. *Soil & Tillage Research* 99 (2):300–307.

Wong, M. T. F. and S. Asseng. 2007. Yield and environmental benefits of ameliorating subsoil constraints under variable rainfall in a Mediterranean environment. *Plant and Soil* 297 (1–2):29–42.

Wuest, S. B. 2007. Surface versus incorporated residue effects on water-stable aggregates. *Soil & Tillage Research* 96:124–130.

Wuest, S. B., T. C. Caesar-TonThat, S. F. Wright, and J. D. Williams. 2005. Organic matter addition, N, and residue burning effects on infiltration, biological, and physical properties of an intensively tilled silt-loam soil. *Soil & Tillage Research* 84 (2):154–167.

Xiaobin, W., C. Dianxiong, and Z. Jingqing. 2001. Land application of organic and inorganic fertilizer for corn in dryland farming region of North China. In *Sustaining the Global Farm*, edited by D. E. Scott, R. H. Mohtar, and G. C. Steinhardt, Selected Papers from the 10th International Soil Conservation Organization Committee Meeting Held May 24–29, 1999 at Purdue University and the USDA-ARS National Soil Erosion Research Laboratory. West Lafayette, IN: International Soil Conservation Organization.

Yuan, C. P., T. W. Lei, L. L. Mao, H. Liu, and Y. Wu. 2009. Soil surface evaporation processes under mulches of different sized gravel. *Catena* 78 (2):117–121.

Yunusa, I. A. M. and P. J. Newton. 2003. Plants for amelioration of subsoil constraints and hydrological control: The primer-plant concept. *Plant and Soil* 257 (2):261–281.

Zhang, S. L., L. Lovdahl, H. Grip, Y. N. Tong, X. Y. Yang, and Q. J. Wang. 2009. Effects of mulching and catch cropping on soil temperature, soil moisture and wheat yield on the Loess Plateau of China. *Soil & Tillage Research* 102 (1):78–86.

Zhang, X. Y., S. Y. Chen, M. Y. Liu, D. Pei, and H. Y. Sun. 2005. Improved water use efficiency associated with cultivars and agronomic management in the North China Plain. *Agronomy Journal* 97 (3):783–790.

Zhou, L. M., F. M. Li, S. L. Jin, and Y. J. Song. 2009. How two ridges and the furrow mulched with plastic film affect soil water, soil temperature and yield of maize on the semiarid Loess Plateau of China. *Field Crops Research* 113 (1):41–47.

Zhu, H. H., J. S. Wu, D. Y. Huang, Q. H. Zhu, S. L. Liu, Y. R. Su, W. X. Wei, J. K. Syers, and Y. Li. 2010. Improving fertility and productivity of a highly-weathered upland soil in subtropical China by incorporating rice straw. *Plant and Soil* 331 (1–2):427–437.

Zhu, J. C., C. J. Gantzer, S. H. Anderson, P. R. Beuselinck, and E. E. Alberts. 1991. Water-use evaluation of winter cover crops for no-till soybeans. *Journal of Soil and Water Conservation* 46 (6):446–449.

Zoebl, D. 2006. Is water productivity a useful concept in agricultural water management? *Agricultural Water Management* 84 (3):265–273.

Zwart, S. J. and W. G. M. Bastiaanssen. 2004. Review of measured crop water productivity values for irrigated wheat, rice, cotton and maize. *Agricultural Water Management* 69 (2):115–133.

11 Sustainable Management of Brackish Water Agriculture

Paramjit Singh Minhas

CONTENTS

11.1 INTRODUCTION

Land irrigation is playing a major role in enhancing food and livelihood security the world over, especially three-fourths of the area that is present in developing countries. About two-fifths of the world's total food and fiber output is contributed by irrigated agriculture, although its area is only 17%. The FAO (2003) estimates that ~70% of the water withdrawn from rivers, lakes, and aquifers (~$820 \times 10^7 m^3$/day) is used for irrigation. In fact, the productivity of irrigated areas in arid and semiarid regions largely depends upon the ability to enlarge this resource base by better rainwater management and/or development of groundwater. Globally, the aquifer withdrawal has increased manifold during the second half of the last century. For example, in the United States the share of groundwater used for irrigation has increased from 23% in 1950 to 42% in 2000. In the Indian subcontinent, groundwater use soared from 10–20 km^3 in 1950 to 240–260 km^3 during 2000. Nevertheless, a typical scenario in the groundwater-irrigated regions has emerged: the areas characterized by water scarcity also usually have underlying aquifers of poor quality. These areas often have the greatest need for economic development, public welfare, and more food to supply the growing populations and regional conflicts over water and environmental degradation. But, driven by the pressure to produce more, even the brackish groundwater is being increasingly diverted to irrigate agricultural lands. The use of such saline or alkali water to produce many conventional grain, forage, and feed crops as well as salt-tolerant plants and trees is prevalent particularly in Bangladesh, China, Egypt, India, Iran, Pakistan, Syria, and the United States (Tanwar 2003). The overexploitation of good-quality water in many developing countries and the alarming rate of decline in groundwater levels are also putting aquifers at risk of contamination from adjoining poor-quality aquifers. Moreover, irrigation efficiency in most of the world's irrigated areas is on the order of 50%, suggesting substantial secondary salinization from seeped water. About 20% of the globally irrigated area is affected by varying levels of secondary salinity and sodicity (Ghassemi et al. 1995). The most technical method to combat irrigation-induced salinity being installation of expensive drainage systems, large amounts of drainage effluents of poor quality are produced in areas covered with subsurface/surface drainage systems. In addition, recent trends in climate change and salt-water intrusion suggest the influence of even greater volumes of these waters in agricultural production in coastal areas in the coming years.

Indiscriminate use of brackish waters in the absence of proper soil–water–crop management strategies poses grave risks to soil health and environment (Bouwer 2000; Minhas and Bajwa 2001; Minhas and Samra 2003). Development of salinity, sodicity, and toxicity problems in soils not only reduces crop productivity and quality, but also limits the choice of crops. Its management signifies those methods, systems, and techniques of water conservation, remediation, development, application, use, and removal that provide for a socially and environmentally favorable level of water regime to agricultural production systems at the least economic cost (Hillel 2000). Possibilities have now emerged to safely use waters otherwise designated unfit if the characteristics of water, soil, and intended usages are known (Minhas and Gupta 1992c; Qadir et al. 2003). This has led to replacement of too conservative water quality standards with site-specific guidelines, where factors like soil texture, rainfall, and crop tolerance have been given due consideration. The increased scientific use of these "degraded" waters such as brackish groundwater, saline drainage water, and treated wastewaters therefore offers opportunities to address the current and future shortage (O'Connor et al. 2008). The opportunities include (i) substituting for the applications of those that do not require high-quality water, (ii) augmenting water supplies and providing alternate sources of supply to assist in meeting present and future needs, (iii) protecting ecosystems, (iv) reducing the need for additional water control structures, and (v) complying with environmental responsibilities and social needs in terms of food and livelihood security for rapidly growing populations in developing countries. This chapter briefly outlines several remedial management actions at the crop, root zone, and farm and irrigation system level strategies available for alleviating the hazards of brackish waters. Although recent research focus has shifted from salinity to other environmental problems, such as concerns related to As, B, F, Cd, NO_3, Pb, Se, and so on (Minhas and Samra 2003; Qadir et al. 2007b; Corwin et al. 2008), for the sake of brevity, only the recent advances on the management of typical saline and alkali groundwaters are included in the following sections.

11.2 SALINITY/ALKALINITY HAZARDS

The total salt concentration and the proportion of sodium (Na) have long been recognized as key parameters in characterizing brackish waters. The quantity of salts dissolved in water is usually expressed in terms of electrical conductivity (EC), mg/L (ppm.), or meq/L, the former being most popular because of ease and precision in its measurement. The cations Na^+, Ca^{2+}, and Mg^{2+} and the anions Cl^-, SO_4^{2-}, HCO_3^-, and CO_3^{2-} are the major constituents of saline water. Plant growth is affected adversely with saline irrigation, primarily through the impacts of excessive salts on osmotic pressures of the soil solution, though the excessive concentration and absorption of individual ions, for example, Na, Cl, and B, may prove toxic to plants and/or retard the absorption of other essential plant nutrients. The reduced water availability at high salinity thus causes water deficits for plants, and plant growth becomes inhibited when the soil solution concentration reaches a critical concentration, referred to as the threshold salinity. Under field situations, the first reaction of plants to the use of saline water is reduction in the germination, but the most conspicuous effect is the growth retardation of crops. A general conclusion can be that the detrimental effects of salinity include reduced initial growth resulting in smaller plants. These smaller plants with less leaf area in turn are able to produce fewer assimilates for conversion to seeds. In other terms, a complementary development of vegetative and reproductive phases is necessary for higher yields, as translocation of assimilates once developed may remain unaffected by salinity provided the environmental factors remain favorable during flowering.

Experimental evidence indicates that an interplay of factors like nature and content of soluble salts, soil type, rainfall, water table conditions, nature of crops grown, and the water management practices followed governs the resultant salinity buildup vis-à-vis crop performance. Under field conditions, the distribution of salts is neither uniform with soil depth nor constant with time. The nonuniformity of salinity distribution is usually affected by the irrigation and leaching practices followed to control salt gradients in the root zone. In the monsoon climate, sowing time salinity for winter crops is higher in lower soil depths due to the displacement of salts with rains in well-drained soils, whereas inverted salinity profiles develop with the movement of salts toward the surface during postrainy season in high water table areas. Again, the rate of salinization during irrigation to winter crops and final salinity buildup may also vary depending upon the salt loads of irrigation waters, conjunctive use modes of fresh and brackish waters, irrigation needs, moisture extraction patterns of crops, and so on. On the contrary, plants are also known to exercise control over root growth and adjust to meet water requirements consistent with water availability vis-à-vis salinity distribution in different zones. Analysis of experiments in pots, lysimeters, and fields by Meiri and Plaut (1985) showed that effective salinity is the temporal and spatial mean of the salinity of the root zone. But most of these experiments were related to steady-state conditions where differential salinities were created either by varying the salt inputs or growing the crops in nonsaline conditions until their establishment, and then rapidly exposing them to specified salinity that was kept uniform with depth by maintaining 50% leaching fraction (LF) at each irrigation event. Because of frequent irrigation, fluctuations in osmotic and matric potentials were minimized. For the situations representing nonsteady-state conditions, Minhas and Gupta (1992a) reported the results of an experiment where wheat responses to initially variable salinity profiles superimposed by various patterns of salinization were evaluated. Although the total salt with which the wheat roots interacted during the growth period was kept constant, threefold variations in its yield were observed (Figure 11.1). Among the various indices of salinity, yields were best related with weighted average root zone salinity, calculated by giving weight according to relative root density and then averaging it over time. Independent estimates of response to salinity that existed down to rooting depth at different stages of wheat showed $EC_{e_{50}}$ (EC_e for 50% yield reduction) to increase from 9.1 until crown rooting to 13.2 dS/m at dough stage. It is thus implied that for nonsteady-state conditions, as exist in the monsoon climate, the salt tolerances at critical stages of crop plants change in response to salinity with modes of salinization, and initial distribution of salinity needs to be considered for effective description of crop responses to salinity.

Some brackish waters, when used for irrigation of crops, have a tendency to produce alkalinity/sodicity hazards, depending upon the absolute and relative concentrations of specific cations and anions. The parameters determining the potential of irrigation waters to create these hazards are sodium adsorption ratio [$SAR = (Na)/\sqrt{(Ca + Mg)/2}$]; residual sodium carbonate [$RSC = (CO_3^{2-} + HCO_3^-) - (Ca^{2+} + Mg^{2+})$], concentrations expressed in me/L and adjusted.SAR [$adj.SAR = Na/\sqrt{[(Cax + Mg)/2}$, where Cax represents the calcium (Ca) in applied water modified due to salinity (ionic strength) and HCO_3^-/Ca^{2+} ratio]. Irrigation with sodic water contaminated with Na^+ relative to Ca^{2+} and Mg^{2+} and high carbonate (CO_3^{2-} and HCO_3^-) leads to an increase

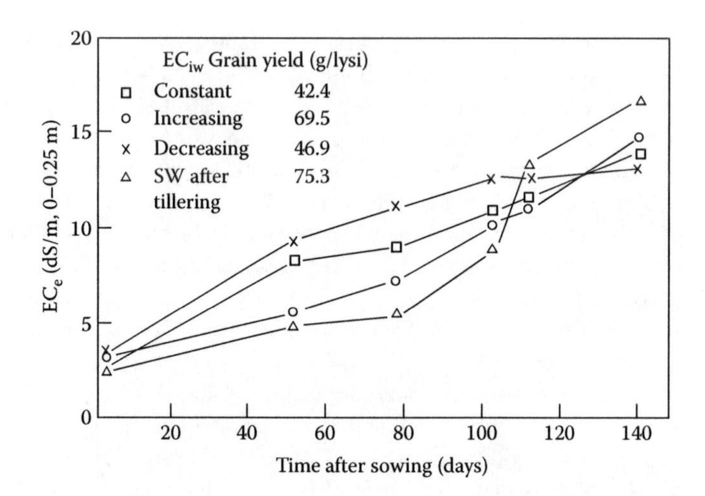

FIGURE 11.1 Salinity buildup and wheat yields with the application of water of constant (SW/NSW 3:7 throughout), increasing (SW/NSW 1:9, 2:8, 3:7, 4:6, 5:5), and decreasing EC_{iw}, and when SW was introduced at tillering (SW/NSW 0:1, 0:1, 2:5, 6.25:3.75, 6.25:3.75), but with similar total salt input. SW = 400 meq/L (EC_w 34.2 dS/m), NSW = 0.4 dS/m. (From Minhas, P.S. and Gupta, R.K., *Agric. Water Manag.* 23, 125–137, 1992.)

in alkalinity and Na saturation in soils. In the early stages of sodic irrigation, large amounts of divalent cations are released into the soil solution from exchange sites. Several reports on the sodication of soils due to irrigation with waters having residual alkalinity have come up, especially from the north-west parts of India (Bajwa et al. 1983a,b, 1986, 1993; Bajwa and Josan 1989a,b; Minhas et al. 2007a,b). The buildup of sodicity (ESP), especially in upper soil layers, was sharper under the paddy–wheat cropping system, obviously due to the larger number of irrigation and thus higher quantities of applied water when compared with the upland crops like cotton, maize, and pearl millet in rotation with wheat (Figure 11.2). With sodicity-induced reduction in water infiltration (relative infiltration rate, RIR = 0.3 at an ESP > 20), the opportunity for alkali waters to penetrate deeper is reduced. Therefore, the alkali solutions further induce sodicity in the upper layers when concentrated through loss of water due to evapotranspiration. Such conditions do not allow for the achievement of steady-state conditions that have been the basis for the development of various earlier indices of sodicity (Bower et. al. 1968; Rhoades 1968). For these reasons, the field results are contradictory to those predicted with the above indices that sodicity buildup should decline with leaching fractions (LF). Thus, rather than $1/\sqrt{LF}$ that has been most commonly used to define the concentration factors, the general experience is that although steady-state conditions are never reached in a monsoonal climate, a quasistable salt balance is reached within 4–5 years of sustained sodic irrigation, when the further rise in pH and ESP becomes low (Minhas and Gupta 1992c). On the basis of a large number of longer-term experiments (>5 years; n = 100), sodicity buildup was analyzed to be directly related to the annual quantities of alkali waters applied (D_{iw}), the rainfall (D_{rw}) at the site, and the evapotranspiration demands of the crops grown in sequence (ET) (Minhas and Sharma 2006). The sodicity (ESP) buildup could be adequately predicted ($R^2 = 0.69$) as $ESP = (D_{iw}/D_{rw})\ (\sqrt{}\ (1 + D_{rw}/ET)$ (adj. R_{Na}). Thus, based upon the ion chemistry of water (R_{Na}), parameters like D_{iw}, D_{rw}, and ET of crops and their sodicity tolerance, cropping patterns can be appropriately adjusted.

The consequence of an increase in exchangeable sodium percentage (ESP) is that it adversely affects soil physical properties as manifested through increased surface crusting, which impacts seedling emergence, reduced infiltration affecting water-holding capacity of soil profile, increased soil strength impacting root penetration, and reduced aeration resulting in anoxic conditions for roots. Due to these effects, the tillage and sowing operation becomes more difficult (Oster and Jaywardane 1998). Several instances have been documented in the literature since the 1950s that the tendency for swelling, aggregate failures, and dispersion increases with increase in ESP and

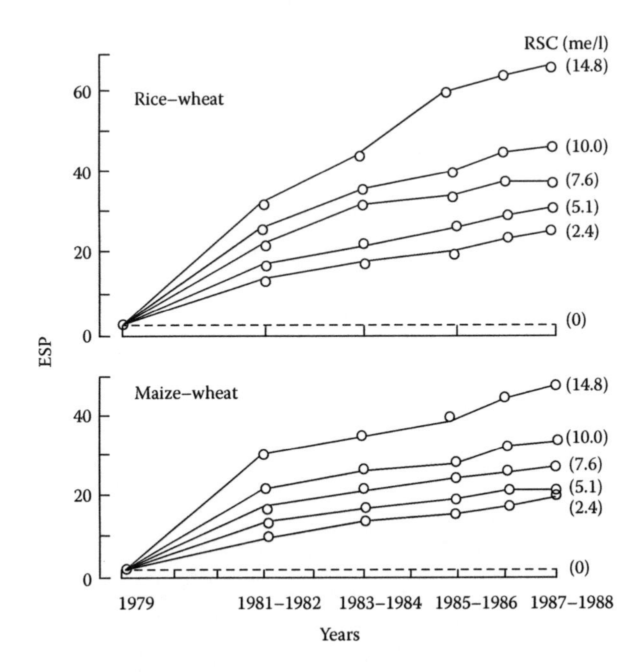

FIGURE 11.2 Successive buildup in ESP following irrigation with alkali waters in rice–wheat and maize–wheat rotations. (From Bajwa, M.S. and Josan, A.S., *Agric. Water Manag.*, 16, 227–228, 1989; Bajwa, M.S. and Josan, A.S., *Agric. Water Manag.* 16, 53–61, 1989; Bajwa, M.S. and Josan, A.S., *Exp. Agric.* 25, 199–205, 1989.)

decline in salinity and even the nonsodic soils with ESP < 3 may behave like sodic soils at very low electrolyte concentrations (Shainberg and Letey 1984; Minhas and Sharma 1986; Oster and Schroer 1979; Oster et al. 1999; Sumner et al. 1998). There is a salinity–sodicity continuum and the higher sodicity soils require high salinities to stabilize them. Soil's water intake in terms of infiltration/permeability values should decline in most of the soils when electrolytic concentration of permeating water is insufficient to compensate for deteriorating effects of Na^+. However, under real field situations, the rainfall and irrigation water infiltration alternate, especially during the monsoon season. It is a general observation that the upland crops suffer the most due to water stagnation problems during the rainy season and even crusts may be formed due to rain on normal soils. After simulating such monsoonal conditions, data showed that drastic reductions in rainwater infiltration occur even at SAR/ESP around 5 even in the absence of alkaline carbonates, that is, in soils with neutral salts (Minhas and Sharma 1986), and the changes are irreversible. Several other workers (Shainberg and Letey 1984; Oster et al. 1999; Sumner et al. 1998) have reported that slaking upon wetting and thereafter dispersion and movement of clay particles are the main causes of limiting infiltration of rainfall water. Minhas et al. (1994), while monitoring the hydraulic properties of a sandy loam soil irrigated with various EC and SAR waters for the last 8 years, established that in the monsoon climate, dispersion and movement of clays with the traction of infiltrating water during the rainy season leads to development of a zone with ingressed clay (i.e., below the plow layer where remixing of move-in clay does not occur), which ultimately starts controlling steady-state infiltration rates. This was further confirmed with a laboratory experiment (Minhas et al. 1999) where the "washed-in" subsoil became restrictive and controlled K-values even with saline water under alterations of saline and simulated rainwater (Figure 11.3). Thus, the dynamic equilibrium between the ESP of the soil, inherent infiltration characteristics (as determined by texture and mineralogy), and salt release in relation to rainfall should determine the amount and depth to which dispersed clay can migrate and consequently cause permeability problems. Therefore, a rethinking is required on water quality guidelines such that structural changes in soils could be accurately described on the

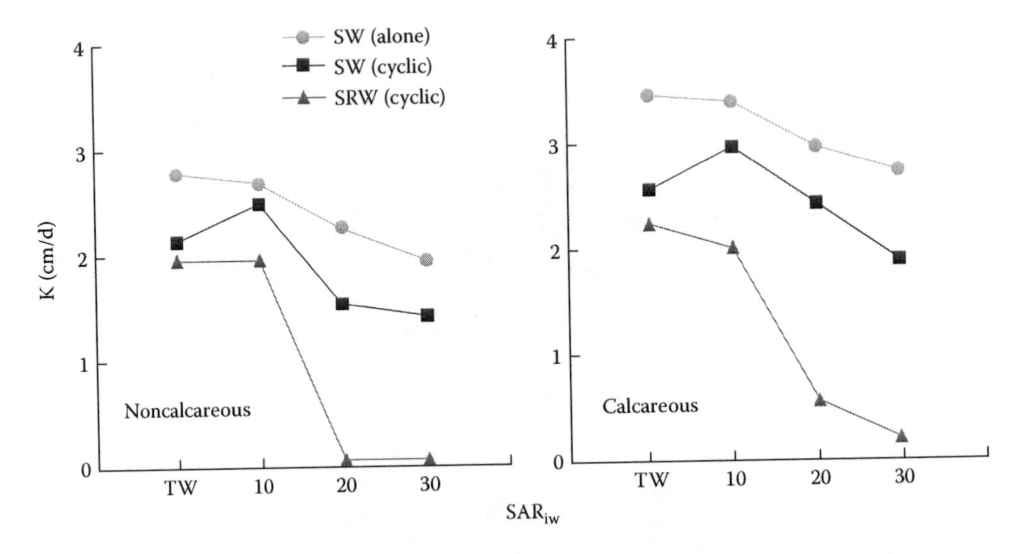

FIGURE 11.3 Saturated conductivity in calcareous (9.7% $CaCO_3$) and noncalcareous soils (0.8%) after six cycles of irrigation with saline water of varying SAR (SW) and simulated rainwater (SRW). (From Minhas, P.S., Singh, Y.P., Chhabba, D.S., and Sharma, V.K. *Irrig. Sci.*, 18, 199–203, 1999.)

basis of quality parameters of waters as well as the soil, climate, and other management parameters. Nevertheless, Minhas et al. (1999) have proposed that for evaluating infiltration hazards upon irrigation with saline–sodic waters, measurements of K-values after consecutive cycles of saline and simulated rainwater can serve as a better diagnostic criterion. Therefore, to achieve optimum plant growth, adequate physical properties of soils must be maintained by using various combinations of crop, soil, and water amendments. The primary concerns are water movement into and through soils and the ability to prepare seedbeds with a tilth that fosters seed germination and emergence, the first crucial step in plant growth. Adverse effects on crop growth are further supplemented through the surface buildup of salts (Minhas et al. 2003).

11.3 MANAGEMENT OF SALINE AND ALKALI WATERS

The management practices for optimal crop production with saline/alkali water irrigation must aim at preventing the buildup of salinity, sodicity, and toxic ions in the root zone to levels that limit the productivity of soils, control the salt balances in the soil–water system, as well as minimize the damaging effects of salinity on crop growth. The latter has not been that successful because of the complexity of crop salt tolerance (Läuchli and Grattan 2007). The development of management options requires the analysis of sensitivity parameters between salinity and crop yield (Zeng et al. 2001). With scientific advances, the principles to produce crops in saline and alkali environment are now well understood and advocate the adoption of special management practices. But most of the past research efforts have treated saline–alkali water use in the context of root zone salinity/sodicity management involving the application or withholding of irrigation to maintain an environment favorable to crop production (Minhas and Gupta 1992c; Qadir et al. 2006). This has led to the development of management practices at the field level without considering their implications and practicality at the farm/irrigation system/river basin levels. However, for sustainable agricultural production, a salinity balance has to be maintained at the irrigation and basin levels. Conjunctive use, water table management, rainwater conservation, and chemical amelioration of alkali waters are some of the important practices to achieve these objectives (Minhas et al. 2003; Sharma and Minhas 2005). In calcareous soils under alkali water irrigation, amelioration can be achieved by mobilizing native $CaCO_3$ by the application of organic materials (Sekhon and Bajwa 1993; Minhas et al. 1995; Choudhary et al. 2004) or through phytoremediation (Qadir et al. 2007a). It has been established that success with saline irrigation can only be achieved if factors such as rainfall, climate, and water table and water quality characteristics on soils and crops are integrated with appropriate crop and irrigation management practices. The available management options mainly include irrigation, crop, chemical, and other cultural practices, but there seems to be no single management measure to control salinity and sodicity of irrigated soils. However, several practices interact with each other and should be considered in an integrated manner. For a better understanding of the subject, each management option has been described separately. Nevertheless, the crop production on saline water– and alkali water–irrigated soils is generally more costly per unit area of land, whereas crop yields are usually low. Hence, the profit margins are also less, whereas risk of crop failures may always continue to be there.

11.3.1 Crop Management

11.3.1.1 Selection of Crops

Crops differ considerably in their ability to tolerate salinity. These intergenic differences can be exploited for selecting those crops that produce satisfactory yield under the given levels of root zone salinity (Minhas and Gupta 1992c; Maas and Grattan 1990; Katerji et al. 2000; Koyama et. al. 2001). The values of salinity for obtaining specific crop yields are mostly computed as piecewise response equation as (Equation 11.1)

$$RY = 100 - S\left(EC_e - EC_t\right), \tag{11.1}$$

where EC_t is threshold salinity. Crops requiring less water, such as oilseeds, can tolerate high levels of irrigation water salinity. Most of the pulses and vegetables are sensitive to salts. Cotton is a salt-tolerant crop, but it is sensitive at the germination stage. The general recommendations are that for successful use of saline waters, crops that are semitolerant to tolerant (mustard, wheat, cotton) as well as with low requirement of water should be grown, whereas crops like rice, sugarcane, and forages requiring liberal use of water should be avoided (Minhas and Bajwa 2001). In low rainfall areas (<400 mm), mono-cropping is recommended for maintaining salt balances.

Similarly, plants also vary in their tolerance to soil sodicity/alkalinity. Gupta and Abrol (1990) compiled earlier data on sodicity tolerance of crops as emerging from alkali soils under reclamation with gypsum and these may form the basis of selection of crops for irrigation with waters having residual alkalinity. However, the comparisons under the two situations (i.e., alkali soils undergoing reclamation process versus the soils being sodicated with the use of alkali waters for irrigation) (Table 11.1) show the lower tolerance (ESP_t) of crops under the latter. The sodicity tolerances of crops are still lower in sodic vertisols, which are usually heavy-textured, and where depth to root penetration is much shallower than the alluvial soils (Sharma et al. 1998). This is unexplainable on the basis of ESP profiles because in soils under a reclamation process, sodicity increases with depth (Abrol and Bhumbla 1979; Gupta and Abrol 1990), whereas with the use of alkali waters, it is maximum at the surface and decreases with depth (Bajwa et al. 2003; Bajwa and Josan 1989a,b). The differential availability of Ca and toxicity of HCO_3 ions to crop root seems to play an important role. During the reclamation of alkali soils, Ca is furnished through gypsum and other amendments to the soil solution for reducing its SAR vis-à-vis ESP, whereas in soils being sodicated, Ca is knocked out of soil solution as a consequence of calcite precipitation. Besides increasing sodicity and pH of surface soil, use of alkali waters reduces the infiltration rate with the result that salts added through irrigation also start concentrating in the surface soil layers. The simultaneous buildup of salts consequent to the use of alkali waters further enhances the stress and thus influences the sodicity tolerance (Sharma and Mondal 1982). However, it may be pointed out that cultivation of high water-requiring crops like rice and sugarcane should be avoided with alkali waters, as these aggravate the sodicity problems. For example, the average sodicity (ESP) buildup in the most important surface

TABLE 11.1
ESP Tolerance of Crops in Alluvial Alkali, Sodic Vertisols under Reclamation, and Alluvial Soils Being Sodicated with Alkali Waters

Crop	Y_{max}	ESP_t	Slope	ESP_{75}	ESP_{50}
Alluvial Sodic Soils under Reclamation					
Rice	6.9 (7.0)[a]	24.4	0.9	52.1	80.0
Wheat	3.9 (4.5)[a]	16.1	2.1	28.0	40.2
Pearl millet	1.9 (3.0)[a]	13.6	2.6	23.2	32.8
Sodic Vertisols under Reclamation					
Rice	6.0 (6.0)	3.1	1.2	23.9	44.8
Wheat	4.0 (5.0)	1.6	2.0	14.0	26.6
Cotton	1.5 (1.9)	4.0	2.6	13.6	23.2
Alluvial Soils under Alkali Water Irrigation					
Rice	6.5	20.1	1.6	35.7	51.3
Wheat	5.6	16.2	1.9	29.3	42.5
Cotton	1.9	14.9	1.3	34.1	53.4
Pearl millet	2.8	6.1	1.3	25.3	44.5

Source: Minhas, P.S., *J. Indian Soc. Soil Sci.* 58(1), 12–24, 2010.
Note: ESP_{75} and ESP_{50} denote ESP for 75% and 50% of the maximum yield (Y_{max} fitted).
[a] Figures in parentheses are yields in normal soils.

layer (0–30 cm) almost equals adj.SAR of the alkali water used in millet/maize–wheat rotation, but the values for rice–wheat are almost 2.6 times higher, indicating greater soil deterioration in the latter (Minhas and Bajwa 2001).

In fact, the accumulation of salts/ESP vis-à-vis tolerance limits to the use of saline/alkali waters gets modified with soil texture, annual rainfall, and ionic constituents of salinity. In addition, the changes in tolerance of crops to osmotic stress can also occur due to several factors, for example, ageing, crop cultivars, presence of other toxic constituents along with salinity, and so on (Minhas 1996; Katerji et al. 2000). The relative effects of various management and other factors on the values of tolerance parameters of some crops were compiled by Minhas (1996) and have been summarized in Table 11.2. These have been differentiated into the possible four types of modifications in piecewise linear response curve as pointed out by Meiri and Plaut (1985). These include (Case 1) simultaneous change in threshold salinity (EC_t) and slope (S) while maintenance of zero yield salinity (EC_0); (Case 2) simultaneous change in EC_t, S, and EC_0; (Case 3) change in S and EC_0 only; and (Case 4) change in EC_t and EC_0 only. In some experiments, there were several interacting variables, and colinearity existed in salinity data as the salinity of the succeeding stage was also dependent upon the salinity of the previous stage. Therefore, to remove the multicolinearity of salinity and other variables over the years, independent estimates of responses to salinity were derived from multiple regressions with dummy variables. Details of the usefulness of such an analysis in making management decisions are described in the following sections.

11.3.1.2 Growth Stages

All crops do not tolerate salinity equally at different stages of their growth. During the initial stages, the interacting zone of roots is limited to the few centimeters at the surface, where most salts concentrate from the evaporating soils. Hence, in most crops, germination and early seedling establishment are the most critical stages. Therefore, to increase the plant stands, strategies for minimizing the salinity of the seeding zone should be followed. The other critical periods for crops are phase

TABLE 11.2
Modifications of Crop Responses to Salinity by Management and Other Factors

Factor Modified	Crop	Salinity Considered	Response Function	EC_0 (dS/m)	EC_{50} (dS/m)	Case	Reference
			(a) Growth Stages				
Combined use of saline and nonsaline water	Wheat	Time averaged	$RY = 100-4.1 (EC_e-3.8)$	28.4	16.0	2	Naresh et al. (1993)
		Sowing time	$RY = 109.9-6.2\ EC_e$	17.3	9.7		
		Midseason	$RY = 115.7-5.5\ EC_e$	21.0	11.9		
		Harvest time	$RY = 106.7-3.4\ EC_e$	31.1	16.7		
	Mustard	Time averaged	$RY = 100-8.5 (EC_e-3.8)$	15.6	9.7	2	Naresh et al. (1993)
		Sowing time	$RY = 115.6-8.2\ EC_e$	14.1	8.0		
		Midseason	$RY = 168.0-12.6\ EC_e$	13.3	9.4		
		Harvest time	$RY = 106.6-3.3\ EC_e$	32.3	17.1		
Growth stages	Corn	Constant salinity	$RY = 100-10.0 (EC_e-5.5)$	15.5	10.5	2	Maas et al. (1983)
		Increased salinity after tesseling	$RY = 100-6.0 (EC_e-6.4)$	21.6	14.7		
			(b) Irrigation				
Irrigation method	Potato	Sprinkler	$RY = 100-11.7 (EC_e-1.1)$	10.1	5.4	2	Meiri and Plaut (1985)
		Drip	$RY = 100-6.3 (EC_e-2.6)$	17.1	10.5		
			(c) Agro-Ecological Conditions				
Temperature and evaporative demand	Wheat (Agra)	Colder climate	$RY = 100-4.1 (EC_e-3.8)$	28.4	16.0	1	AICRP-Saline Water
	Dharwad	Warmer climate	$RY = 100-4.9 (EC_e-0.9)$	21.2	11.0		
Soil texture	Wheat	Loamy sand	$RY = 100-4.4 (EC_w-6.0)$		17.5	1	
		Sandy loam	$RY = 100-3.9 (EC_w-4.0)$		16.8		
		Silty clay loam	$RY = 100-4.2 (EC_w-1.0)$		12.9		
	Mustard	Loamy sand	$RY = 100-2.2 (EC_w-2.0)$		24.9	1	
		Sandy loam	$RY = 100-6.9 (EC_w-5.1)$		12.3		
		Silty clay loam	$RY = 100-3.7 (EC_w-1.1)$		14.7		

(continued)

TABLE 11.2 (Continued)
Modifications of Crop Responses to Salinity by Management and Other Factors

Factor Modified	Crop	Salinity Considered	Response Function	EC_0 (dS/m)	EC_{50} (dS/m)	Case	Reference
			(d) Ionic Constituents/Applied Nutrients				
RSC water	Wheat		$Y = 10.66-0.081(pH)^2-0.018EC_e \times SAR_e+0.88 \times 10^{-3}(SAR_e)^2$				
Levels of RSC			EC_e for Y = 3 Mg/ha				
Neutralization with		$pH_{8.5}SAR_{e10}$	$RY = 4.90-0.18\ EC_e$	10.5		4	Sharma et al. (1993)
gypsum		$pH_{8.5}SAR_{e30}$	$RY = 5.60-0.54\ EC_e$	4.8			
		$pH_{9.0}SAR_{e10}$	$RY = 4.19-0.18\ EC_e$	6.6			
		$pH_{9.0}SAR_{e30}$	$RY = 4.89-0.54\ EC_e$	3.5			
EC_w and SAR_w	Wheat	$RY = 98.14-0.54EC_w-SAR_w(0.10EC_w-0.45)-0.01(SAR_w)^2$					
			EC_{iw} for RY = 90%				
		$SAR_w = 5$	$RY = 100.14-1.04\ EC_w$	7.8		3	Singh et al. (1992)
		=10	$RY = 101.64-1.54\ EC_w$	7.6			
		=20	$RY = 103.14-2.54\ EC_w$	5.2			
		=30	$RY = 103.64-3.54\ EC_w$	3.6			
		=40	$RY = 106.14-4.54\ EC_w$	2.2			
Cl/SO_4 ratio (R)	Wheat	$Y = 3.144-0.047EC_w-1.115 \times 10^{-2}P-0.229 \times 10^{-4}P^2-0.036R^2-0.035EC_w \times R-6.167 \times 10^{-4}EC \times P-0.41 \times 10^{-3}R \times P$					
Applied P (kg/ha)			EC_{iw} for Y = 3.5 Mg/ha				
		$P_{26}R_{0.3}$	$Y = 4.501-0.130\ EC_w$	7.7		3+4	Chauhan et al. (1991)
		$P_{26}R_{3.0}$	$Y = 4.328-0.121\ EC_w$	6.8			
		$P_{26}R_{5.0}$	$Y = 3.858-0.114\ EC_w$	2.1			
		$P_{39}R_{0.3}$	$Y = 5.040-0.173\ EC_w$	8.9			
		$P_{39}R_{3.0}$	$Y = 4.943-0.164\ EC_w$	8.8			
		$P_{39}R_{5.0}$	$Y = 4.530-0.157\ EC_w$	6.7			

*Functions not written as per Equation 11.1 were derived from multiple regression with/without dummy variables.

changes from vegetative to reproductive, that is, heading and flowering to seed setting. An increase in zero yield salinity and slope of salinity response function with the plant development in crops like wheat, mustard, and mung bean has been reported by Naresh et al. (1992a,b) and Minhas et al. (1990a,b). These differences in salt sensitivity of crops at various growth stages (ontogeny) should help in planning appropriate irrigation management measures, especially where both saline and nonsaline waters are to be utilized. The use of saline waters can be avoided at some of the sensitive stages to minimize the salinity damage.

11.3.1.3 Crop Cultivars

In addition to intergenic variations of different crops to tolerate salinity/sodicity, there is also a wide variation in the inherent salt tolerance of the crop cultivars. Most of the research endeavors until now have been aimed at identifying the genotypes and breeding new varieties of crops for normal soil conditions, although limited efforts have also been made in this respect for saline environments. Usually there is a negative correlation (Case 2) between tolerance of varieties and their potential yields. Hence, there are not many varieties which are both tolerant to salinity and produce economic yield, which is a major consideration for most farmers. To cite examples, cultivars like "Damodar" in rice and "Kharchia" in wheat are well documented for their salinity tolerance, but have low yield potentials. Farmers prefer vigorously growing and high-yielding varieties like "Jaya" of rice and "HD-2304" of wheat even though they have low tolerance for salinity, simply because these may still out-yield their tolerant counterparts. Cultivars like "HD-2560" of wheat, "CS-52" of mustard, and "MESR-16" of cotton suggest that it is possible to breed cultivars both with high yield potential as well as higher salt tolerances. However, the selection of crop cultivars showing stability under salinity should be preferred. Studies on Na tolerance under alkali water–irrigated conditions (Choudhary et al. 1994, 1996) have revealed that tolerant wheat and barley varieties should possess deep and penetrative root systems, which enable them to produces higher amount of spikes per unit area with bolder grains and result in high yield even under high level of sodication. Crop varieties having higher tolerance are also able to maintain low Na/K ratio in shoots through restricting Na uptake (Gill and Qadir 1998). The traditional breeding approach of developing crop varieties best suited to brackish water irrigation has been largely an indirect one. The major development which has recently been prominently featured is the role of biotechnology that has opened up the era of new gene technologies of transgenic crops. These technologies are paving the way for further breakthroughs in increasing production and productivity with minimal cost. The "stacking" (use of more than one trait in a single crop) of these traits in specific varieties would help in tackling multiple constraints in crop production, such as tolerances to drought and salinity coupled with existing traits. Assuming genetic engineering for production of salt tolerance, transgenic crops will be successful in the near future; it will provide us with crop plants that show superior productivity on salt-affected soils in comparison with their existing varieties.

11.3.1.4 Environmental Factors

One of the typical situations with continental monsoon climate is the concentration of rains in a short span of 2–3 months (July–September). Thus, if the water penetrating into soils during this period exceeds the evapotranspiration demands of crops, it induces leaching of salts added through saline irrigation to winter crops or in low rainfall regions. Thus, farmers resort to fallowing during monsoon rain for achieving salt balances in semiarid areas. In arid parts of Rajasthan, cultivation of lands every alternate year or after 2–3 years to allow the accumulated salts to leach out by rain before the next crop is sown is a common practice. The amount and frequency of rains not only govern the crops grown in the area but also the associated salt dynamics vis-à-vis limits of salinity in waters that could be used for raising crops; for example, in areas with annual rainfall less than 250 mm, saline waters of EC about 4 dS/m will cause salt toxicity in most of the crops, whereas in areas where annual rainfall exceeds 500 mm, waters up to an EC of 16 dS/m could be gainfully utilized for crops like wheat and barley when grown in coarse-textured soils (Manchanda et al. 1989).

Other environmental factors like temperature and atmospheric evaporativity also markedly influence the selection and salt tolerance of crops. Increased EC_t and EC_0 and decreased "S" (Case 1) when wheat was grown under comparatively cooler climate, that is, low ET demands prevailing during growth period in northern India when compared with southern parts (Table 11.2). The studies by Sinha and Singh (1976a,b) showed that salt content of the soil closely adhering to the roots was much higher than the bulk soil (1.3–2.0 times) and was linearly related to the total amount of water transpired by maize and wheat plants, as well as the water transpired per unit root length. Based upon these studies, it was pointed out that the salt stress to which the plants will be subjected is determined by transpiration rates vis-à-vis evaporative demand for water.

11.3.1.5 Soil Texture

The dynamics of salt in soil are also affected by texture. Though the amount and frequency of rain basically govern the salt leaching occurring during the monsoon season, soil texture has also been shown to influence leaching. Predictions (Minhas and Gupta 1992c) show that removal of 80% of the salts accumulated during the period preceding monsoons would require 1.85, 0.95, and 0.76 cm of rainwater per centimeter of soil depth in fine-, medium-, and coarse-textured soils, respectively. Because of low infiltration rates of fine-textured soils (having high clay content), rainwater either tends to runoff or evaporate from stagnated water at the surface and this reduces water availability for displacing the salts downward. Moreover, the water requirements for displacing salts are also affected by their water retention capacity. Thus, as a "rule of thumb," accumulation of salts (EC_e) in saline water–irrigated soils is nearly one-half the salinity of irrigation water in coarse-textured soils (loamy sand and sand). It is equal to that of irrigation water in medium-textured sandy loam to loam soils and more than two times that in fine-textured soils (clay and clay loam). In other terms, irrigation with water having salinity of 8 dS/m would result in soil salinity of about 4, 8, and 16 dS/m in loamy sand, sandy loam, and clay loam soils, respectively. Thus, waters of salt concentration as high as an EC of 12 dS/m can be used for growing tolerant and semitolerant crops in coarse-textured soils (Table 11.2), provided the annual rainfall is not less than 400 mm. But in fine-textured soils, waters with EC more than 2 dS/m would often create salinity problems. Miyamoto and Chacon (2006) have also reported that the concentration factors increase exponentially with texture in golf courses, parks, and sports fields of Texas and New Mexico under sprinkler irrigation.

11.3.1.6 Ionic Constituents of Salinity

In addition to total electrolyte contents, the plant responses are also governed by the concentrations of different ions in soil solution. The associated cations and anions of salinity influence the tolerance of crops by (1) governing effective salinity of soil solution with which the plant roots interact through their control over precipitation and dissolution reactions, leaching and dispersive behavior of soils, and so on; and (2) direct toxicity due to excessive accumulation of ions in the plant tissues, thus causing nutritional imbalances. Some corrective measures to antagonize the latter effects through chemical fertilizers/amendments are discussed in the following section on "chemical management approach." Examples of interaction between Na and Ca under high-SAR and alkali water–irrigated conditions (Singh et al. 1992; Sharma et al. 1998), along with chloride and sulfate dominance in salinity (Chauhan et al. 1991), are included in Table 11.5. The tolerance of wheat to salinity decreased with increase in SAR, pH, or Cl contents (Cases 3 and 4). Manchanda et al. (1991) have also shown that pulse crops like chickpea, faba bean, and peas performed better under sulfate than chloride-dominated saline conditions at comparable EC_e levels. Rhoades et al. (1992) viewed that if soil is saline or if the Ca concentration exceeds about 2 mmol/L, even a high level of SAR will have little adverse effect on most crops, as distinguishable from salinity, and can be ignored. Thus, the major problem with respect to Na toxicity or Ca nutrition issues should occur under relatively less saline but sodic and alkaline pH conditions when Na concentration is high and Ca concentration is low and/or where the Mg/Ca ratio exceeds 3. Otherwise, the prognosis of reduced salt tolerance therefore lies with structure deterioration, leading to poor physical

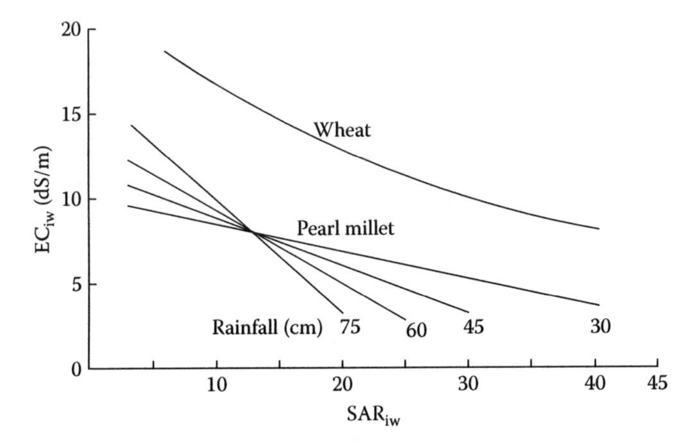

FIGURE 11.4 Predicted EC_w for 75% of potential yields of wheat and pearl millet as affected by SAR_w and rainfall. (From Singh, R.B., Minhas, P.S., Chauhan, C.P.S., and Gupta, R.K., *Agric. Water Manag.* 21, 93–105, 1992.)

conditions. However, in a long-term experiment, sodicity-induced accumulation of salts with the use of high SAR waters was observed to be the main cause of yield reductions (Chauhan et al. 1991). Further analysis revealed rainfall dependence as a cause for reduced yields of pearl millet grown during monsoon (Figure 11.4). Expectedly, rainwater must have reduced the effects of salinity due to dilution, but increased reductions in pearl millet yields with increased SAR_{iw} were ascribed to water stagnation problems.

11.3.2 Water Management

11.3.2.1 Leaching Requirement for Salt Balance

The traditional salinity management approach (USSL 1954) assumed the steady-state conditions to exist in the long run, which implied that the economic way for controlling soil salinity is to ensure net downward flow of water through the root zone. Therefore, leaching requirement was defined as the minimal fraction of the total water applied that must pass through the root zone to prevent reductions in crop yield below the acceptable level. The leaching requirements for any acceptable yield (Y_a) can be calculated using the equation as

$$RY = Y_a/Y_{max} = 1 - S\ (EC_a - EC_t)$$

$$EC_a = EC_t + \left(\left(1 - Y_a/Y_{max}\right)/S\right)$$

$$LR = EC_{iw}/2\ EC_a.$$

The concept of leaching requirement is mainly of practical importance for the situations of no or very low rains, where the steady state can nearly be achieved. However, in continental monsoon climates, the concentration of rains in a short span of 2–3 months is the most uncontrolled factor causing nonsteady-state salinity. It leaches down the salts when infiltrating down the soil, and gets stored in the soil profile to be carried over until it either gets mixed with the irrigation water applied or is consumed by winter crops. In addition to monsoons taking care of a part of leaching, surface irrigation systems in India are quite inefficient (with farm irrigation efficiency of only 60%–70%) and, thus, these also inadvertently provide for the leaching requirements. Nevertheless, a large number of experiments (Minhas and Gupta 1992c) have shown that the practices of providing leaching requirements to crops are not very beneficial. Further studies by Minhas and Gupta (1992b) have pointed out that a practice that applies water in amounts excessive to the need of the crops displaces high-quality precipitation water from the root zone that otherwise would have been used by the growing crop. Therefore, few, if any, additional advantages are expected from applying extra saline water to meet leaching requirements. Rather, a better strategy would be to apply saline water (if $EC_e > EC_{iw}$) to boost the antecedent moisture contents and reduce the salinity levels (equaling EC_{iw}) before the onset of monsoon rains. The refill of the topsoil with water just before the onset of monsoon will enhance salt leaching during *kharif* rains. Or, if the monsoon rains are subnormal and not sufficient to leach the salts of the seeding zone, a heavy preplant irrigation with better-quality water should help (Rhoades 1999). Goyal et al. (1999a,b) also reported that the feasibility of crop irrigation with saline water, however, needs to be evaluated on a long-term basis for each crop species with allowance of leaching of soil between cropping seasons to control soil salinity. Similarly, under alkali water irrigation, the development of sodicity has been observed to depend more on the quantities of water applied and the annual rainfall at the site rather that the leaching fractions attained, for example, the buildup of sodicity (ESP) in rice–wheat system (LF ~0.7) is almost 2.4 times higher than in millet–wheat (LF 0.4) because of increased input of alkali water. Experimental evidence further pointed out to 30%–50% higher salinity buildup even in light-textured soil, when 50% extra saline water (EC_{iw} 3.2 dS/m and RSC 4 meq/L) was applied to meet the leaching requirement in rice–wheat and maize–wheat systems (Bajwa et al. 1986). Thus, the concept of LR is also invalid for irrigation with waters having residual alkalinity.

11.3.2.2 Farm Irrigation Management

On farm irrigation management with saline and alkali waters should involve those irrigation schedules that minimize irrigations, eliminate salinity buildup, and also assure optimal crop production. This has led the researchers to probe the methods and frequency of irrigation, the total amount of irrigation water to be applied, and ways to make judicious use of multiquality waters.

Irrigation interval: Under saline conditions, irrigation should meet both the water requirements of crops and the leaching requirements to maintain a favorable salt balance in the root zone. During the intervening periods between the irrigation cycles, evapotranspiration by crops reduces the soil water contents, which in turn decreases the matric (due to attraction of soil matrix for water and water molecules for each other) as well as solute (due to increase of soil solution concentration) potentials. The rate of ET and the soil water characteristic curve [$\theta = f(\Psi)$] determine the rate of fall of the two components of total soil water potentials, but as a consequence the water uptake by crops and hence the yields are expected to suffer. Therefore, irrigation in saline soils should be more frequent because it reduces the cumulative water deficits (both matric and osmotic) between the irrigation cycles. But such an opinion is still controversial as small irrigation intervals subsequently induce water uptake from shallow soil layers, increase unproductive evaporative losses from soil surface and with saline irrigations, and increase the salt load of soils. Moreover, the nonsaline soil water carried over from the monsoon rains may also be displaced beyond the reach of plant roots by the added saline solutions (Minhas and Gupta 1992c). Sinha and Singh (1976a,b) have shown that soil solution concentrations adjacent to growing roots in saline soils are 1.5–2-fold higher than the bulk soil. The wetter the soil and the higher the transpiration rate, the larger are the differences, indicating that keeping the soil wet by decreased irrigation intervals may enhance the adverse effects of salinity. Extended irrigation intervals, on the other hand, usually result in deeper roots and larger proportions of water extractions from deeper zones. Reductions in water uptake and thus evapotranspiration losses occur under saline conditions. This in turn means higher-salinity soils will retain more water than the low-salinity ones in between the irrigations, and such a situation should moderate the total water stress and thus reduce the inhibitory effects of increase in solution concentration on growth. The net results of the above counteracting processes still await further experimentation, but based upon model predictions, Minhas and Gupta (1992b) have shown that depth of applied water should be simultaneously reduced if higher benefits from small intervals of irrigations are to be accrued. Because the infiltration rate controls the application depths, it is difficult to apply <25 mm water with surface methods and too frequent irrigations may in fact lead to aeration problems. This illustrates the usefulness of microirrigation systems where scheduling is typically at very high-frequency applications at shallow depths. This aspect is discussed in a later section on "methods of irrigation." Similarly, under alkali water–irrigated conditions in wheat and maize (fodder) crops grown during winter and monsoon seasons, respectively, all irrigation intervals produced similar yields, whereas in maize (fodder) grown during hot-dry summer season, shorter irrigation interval lowered the soil temperature and hence improved dry matter yield (Bajwa and Josan 1989c). The buildup of salts and ESP under three irrigation frequencies was, however, similar. The frequency of irrigation did not appreciably alter the effectiveness of applied gypsum in wheat and millet but in maize (fodder), the gypsum treatment was more effective under more frequent irrigation schedules.

Preirrigation: Primary objectives of presowing irrigation are the creation of optimal soil moisture conditions to facilitate tillage and seedbed preparation and to recharge the projected root zone with water for germination and later ET needs of crops. In saline soils, these should further include the leaching of soluble salts below the seeding zone, as the germination and seedling establishment are most critical, and the failures at this stage cannot be rectified later on. Plants are also known to tolerate salinity better with aging. Crops like mung bean, sorghum, and mustard could tolerate higher salinity once the nonsaline water was substituted for presowing irrigation to leach out the salts of the seeding zone (Minhas et al. 1989, 1990a,b). This substitution enhanced germination, crop growth, and yields markedly, and also resulted in better utilization of soil water, even from the lower soil layers (Table 11.3).

TABLE 11.3
Yields and Water Extraction Patterns Following the Use of Different Salinity Waters

	Seed Yield	Water Extracted (cm) from Layer (cm)				
EC_{iw} (dS/m)	(Mg/ha)	0–30	30–60	60–90	90–150	Total
Mung Bean						
0.3 (throughout)	2.52	27.8	9.7	4.0	3.3	44.8
4.7 (throughout)	0.27	16.6	5.8	0.2	–	22.6
4.7 (PInsw)	1.56	23.4	9.7	4.2	0.7	38.1
Sorghum[a]						
0.3 (throughout)	9.70	18.4	7.7	2.6	2.3	31.0[b]
4.7 (throughout)	6.50	17.0	5.1	2.0	0.5	24.7
4.7 (PInsw)	8.50	19.1	6.9	3.7	2.0	31.7
Indian Mustard						
0.3 (throughout)	2.32	19.5	9.0	6.2	2.2	36.9
12.3 (throughout)	1.05	10.7	5.1	1.8	0.5	18.1
12.3 (PInsw)	1.80	13.7	7.7	4.8	1.7	27.9

Source: Minhas, P.S., Sharma, D.R., and Khosla, B.K., *J. Indian Soc. Soil Sci.*, 37, 140–146, 1989; Minhas, P.S., Sharma, D.R., and Khosla, B.K., *Indian J. Agric. Sci.*, 57, 343–346, 1990; Minhas, P.S., Sharma, D.R., and Khosla, B.K., *Irrig. Sci.*, 11, 57–62, 1990.

Note: PInsw = Presowing irrigation with nonsaline water.

[a] dry forage yield.

[b] up to the last irrigation only.

Naresh et al. (1992a,b) and Sharma et al. (1994) have made similar observations with wheat and mustard. The presowing irrigation assumes a still more critical role for the success of summer crops.

Multiquality irrigation practices: Under most saline situations, canal water supplies are either unsure or inadequate such that farmers are forced to pump saline/alkali waters to meet the crop water requirements. These waters from the two sources can be applied either separately or mixed together. Mixing of waters to acceptable quality for crops also results in improving stream size and thus enhances the uniformity in irrigation, especially for the surface method practiced on sandy soils. Allocation of the two waters separately, if available on demand, can be done either to different fields, seasons, or crop growth stages so that higher salinity water is not applied to sensitive crops/growth stages (Minhas et al. 2007a). As pointed out earlier, the germination and seedling establishment has been identified as the most sensitive stage in most crops. Therefore, better-quality water should be utilized for presowing irrigation and early stages of crop growth. Then one can switch over to poor-quality water later, when the crops can tolerate higher salinity. Rhoades et al. (1992) have also advocated the seasonal cyclic use, called "Dual Rotation," strategy where nonsaline water is used for salt-sensitive crops/initial stages of tolerant crops to leach out the accumulated salts from irrigations with salty waters to previously grown tolerant crops. Such a management strategy may work better for arid climates with very low rainfall, but it is of natural occurrence in the monsoon climate. Thus, the options of utilizing the multiquality waters have to be either mixing or cyclic use, mainly during the growth of *rabi* crops. If it is presumed that the prerequisite facilities for blending exist and different qualities of waters are simultaneously available on demand, then the question arises as to which option should be followed. Analysis of a large number of experiments (Minhas and Gupta 1992c) showed that at the same level of EC_{iw} (weighted average salinity), the yields for different cyclic use modes were higher than the estimated yields for mixing. The advantage from various cyclic irrigation modes followed the order: (2S:1S) > (1C:1S) > (1C:2S); canal/saline water irrigations. Differences between the observed and estimated yields were greater at low relative yields, indicating increased benefits from cyclic use at higher EC_{iw}. This provides useful evidence that multisalinity waters should be used cyclically where canal water is applied at early stages and the use of saline waters should be delayed to later stages. In addition to better performance of crops, the cyclic uses have operational advantages over mixing which demand for the creation of infrastructure for mixing the two supplies in desired proportions. Further, experiments (Naresh et al. 1992a,b; AICRP-Saline Water 2000) where combined use of saline (EC_{iw} 8–12 dS/m) and canal waters was made for cotton–wheat, pearl millet–mustard, and mustard–sunflower rotations (Table 11.4), and others (Sharma et al. 1994) where drainage (EC_{iw} 12.5–14.5 dS/m) and canal waters were used in pearl millet–wheat rotation also support the creditability of the above cyclic use strategy. Surveys of farmers using brackish waters (Bouwmans et al. 1988) indicated that farmers alternating canal and saline waters were getting higher production of cotton and millets than those using mixed water, whereas mixing proved quite beneficial for wheat and mustard. In later studies by Malash et al. (2005) and Ragab et al. (2005) with a shallow rooted tomato crop using saline drainage water of comparatively lower salinity (4.2–4.5 dS/m), a mixed water management practice produced higher growth and yields than alternate irrigation either using drip or furrow method of irrigation.

However, in the case of alkali waters, the strategy that would either minimize the precipitation of Ca or maximize the dissolution of precipitated Ca can be expected to be better. Usually both canal water and groundwater are in equilibrium with calcite, the former at the pCO_2 of the atmosphere and the latter at a much higher pCO_2. The relation between concentration of Ca^{2+} and pCO_2 is not linear and is governed by the following relation:

$$MCa^{2+} = \frac{Kh\,PCO_2\,K_1\,KCal}{4K_2\,\lambda Ca^{2+}\,\lambda HCO_3}$$

TABLE 11.4
Crops Yields (Mg/ha) under Varying Modes of Combined Use of Canal and Saline Irrigation Waters

Treatments	Cotton	Wheat	Pearl Millet	Mustard	Mustard	Sunflower
Canal water (CW)	1.63	4.88	3.15	2.07	2.42	1.34
Saline water (SW)[a]	0.46	3.59	2.91	1.18	2.52	0.29
			Cyclic Mode			
1CW/RSS	0.98	4.05	2.99	1.88	2.25	0.71
1SW/RSW	—	—	—	—	2.39	0.99
1SW/1CW/RSW	0.72	4.08	2.80	1.67	—	—
1CW/1SW	1.23	4.72	2.96	1.96	2.54	0.99
2CW/2SW	1.28	4.62	—	—	—	—
2CW/1SW	—	—	—	—	2.47	0.98
1SW/1CW	0.76	4.02	—	—	2.31	0.81
2SW/1CW	—	—	2.91	1.41	—	—
			Mixing Mode			
1CW/1SW	1.04	4.37	2.80	1.81	—	—
1CW/2SW	—	—	—	—	2.60	0.72
2CW/1SW	—	—	—	—	2.50	0.89
LSD (p = 0.05)	0.03	0.35	NS	0.36	NS	0.15

Source: Compiled by Minhas, P.S., Sharma, D.R., and Chauhan, C.P.S., *Advances in Sodic Land Reclamation*, UPCAR, Lucknow, 2003.

[a] EC_{sw} 9, 12, and 8 dS/m for cotton–wheat, pearl millet–mustard, and mustard–sunflower, respectively. RSW denotes rest with saline water.

In the above equation, MCa^{2+} refers to the concentration of Ca^{2+}(g/L). K_1 and K_2 represent the first and second dissociation constants of carbonic acid and Kh is Henry's gas constant. λCa and λHCO_3 are the activity coefficients of ions, while pCO_2 is partial pressure of CO_2. Therefore, it seems that mixing of surface waters with groundwaters of higher alkalinity and low Ca would result in undersaturation with respect to calcite. Consequently, the blended water will have the tendency to pick up Ca through the dissolution of native Ca. Benefits that can be accrued from such a preposition are, however, yet to be quantified. Bajwa and Josan (1989c) reported that irrigation of sandy loam soil (18%–26.8% clay) with alkali water (EC_w 1.35 dS/m, RSC 10.1 meq/L, SAR 13.5 adj.SAR 26.7) increased the pH and ESP of the surface layers and reduced its infiltration rate to 14%. The yields of rice and wheat decreased progressively with time and were 62% and 57%, respectively, of the potential yield, that is, that obtained under canal irrigation during 6 years. However, when the alkali water was used in cyclic mode with canal water, yields of both the crops were maintained on par with canal water, except in the CW-2AW mode. Cyclic use of two waters decreased sodication of soils. Interestingly, after accounting for rainfall and canal water in estimating the adj SAR, ESP of the surface soil was 1.2–1.5 times compared with a factor of 1.8 observed with alkali waters alone. In another experiment (Minhas et al. 2007b) where alkali water (EC 2.3 dS/m, RSC 11.3 meq/L) and good-quality tubewell water (EC 0.5 dS/m, RSC nil) were used for 6 years, cyclic modes (2TW:1AW, 1TW:1AW, 1TW:2AW) with a decline in yield in the range of 16%–20% and 6%–12% in the case of paddy and wheat, respectively, performed slightly better than their counter-mixing modes where the decline ranged between 19%–23% and 9%–14%, respectively (Table 11.5). Dilution with monsoons helped to induce greater use of alkali water in paddy. Similar results were reported by Choudhary et al. (2007, 2008) for cotton and wheat and by Chauhan et al. (2007) for potato–sunflower–sesbania crop rotations (Table 11.6). Thus, alternating alkali and canal waters

TABLE 11.5
Crop Yields (Mg/ha)[a] under Mixing and Cyclic Modes of Irrigation with Alkali and Good-Quality Water

Treatment	Paddy[a]	Wheat[a]	Cotton	Wheat	Potato	Sunflower	Sesbania
Good water (GW)	0.80	0.59	1.32	5.20	35.0	1.54	22.3
Alkali water (AW)[a]	0.52	0.48	0.95	4.43	11.9	0.49	11.9
			Blending				
2GW/1AW	0.64	0.55			28.9	1.24	20.2
1GW/1AW	0.63	0.53					
1GW/2AW	0.61	0.50			23.0	1.09	19.2
			Cyclic Use (Irrigationwise)				
2GW/1AW	0.67	0.57	1.26	5.10			
1GW/1AW	0.65	0.55	1.21	4.95	29.8	1.44	21.2
1GW/2AW	0.63	0.51	1.15	4.70			
2CW/2AW[b]			1.22	4.82	28.4	1.28	20.3
2AW/2CW[b]			1.08	4.70	22.7	1.01	18.2
4AW/2CW[b]			1.02	4.75	14.0	0.62	14.8
			(Seasonwise)				
AWp/GWw	0.55	0.52			28.0	1.00	19.1
GWp/AWw	0.66	0.52					
LSD (p = 0.05)	0.02	0.01	0.18	0.21	2.4	0.19	0.9

Source: Minhas, P.S., Dubey, S.K., and Sharma, D.R., *Agric. Water Manag.*, 87, 83–90, 2007; Choudhary, O.P., Ghuman, B.S., Josan. A.S., and Bajwa, M.S., *J. Sust. Agric.* 32, 269–286, 2008; Chauhan, S.K., Chauhan, C.P.S., and Minhas, P.S., *Irrig. Sci.*, 26, 81–89, 2007.

[a] Yield in kg/lysimeter. RSC 11.3, 10.1, and 15 me/L for paddy–wheat, cotton–wheat, and potato–sunflower–sesbania.

[b] CW/2AW, AW/2CW, and 2AW/CW for cotton–wheat.

can be considered a practical way to alleviate sodicity problems caused by the use of alkali water. Field observations in Kaithal area further point out that those farmers who are usually getting some canal water supplies are able to sustain yields of rice–wheat crops, whereas yields of these crops decline on farmers' fields that do not receive canal water (Minhas et al. 1995).

Methods of irrigation: The distribution of water and salts in soils varies with the method of irrigation. Therefore, the methods followed should create and maintain favorable salt and water regimes in the root zone such that water is made readily available to plants for their growth and without any damage to the yield. The specific advantages and disadvantages of some of the most important irrigation methods for application of water, that is, flooding (checks, border strips, and furrows), sprinkling, and the drip system, are summarized here.

The surface irrigation methods, including border strips, check basins, and furrows, are the oldest and most commonly practiced in most parts of India. These irrigation methods, even after following the best design criteria, generally result in excessive irrigation and nonuniformity in water application. Consequently, the on-farm irrigation efficiency is low (60%–70%). However, properly designed and operated surface irrigation methods can maintain the salt balance and minimize salinity hazards. To meet these twin objectives, land needs to be properly leveled to ensure even distribution of water. Parameters such as the length of the water run, stream size, slope of the soil, and cutoff ratio, which influence the uniformity and the depth of water application for a given soil type, should be as per the desired specifications.

TABLE 11.6
Yield and Water-Use Efficiency of Crops under Different Irrigation Methods

Crop	Average Yield (Mg/ha) for Irrigation Method			
	Surface Method		Sprinkler Method	
	CW	SW	CW	SW
Wheat (1976–1979)[a]	4.00 (97)	3.62 (83)	3.69 (107)	3.54 (97)
Barley (1980–1982)	3.51 (147)	2.32 (98)	3.48 (159)	2.59 (117)
Cotton (1980–1982)	2.30	1.71	2.28	1.34
Pearl millet (1976–1978)	2.38	2.07	2.54	1.50
	Drip Method			
	Surface	Subsurface	Furrow/Surface	
Radish (EC_w 6.5 dS/m)[b]	15.7 (17.5)	23.6 (26.2)	9.9 (8.7)	
Potato (4 dS/m)	30.5 (93.5)	20.8 (78.5)	19.2 (53.6)	
Tomato (10 dS/m)	59.4	43.9		
Tomato (4 dS/m)[c]	42.6		36.9	
(8 dS/m)	28.0		24.5	
Okra (3.0)[d]	4.4		2.7	
(6.0)	3.0		1.8	

[a]Aggarwal, M.C. and Khanna, S.S., Bulletin of HAU, Hisar, p. 118, 1983; [b]Singh, S.D., Gupta, J.P., and Singh, P., *Agron. J.*, 70, 948–951, 1978; [c]AICRP-Saline Water. Annual Progress Reports. CSSRI, Karnal, 1972–2002; [d]Phogat, V., Sharma, S.K., Kumar, S., Stayvan and Gupta, S.K., Bulletin, CCS HAU, Hisar, 72 p., 2010.
*Figures in parentheses denote water-use efficiency (kg/ha cm).

High-energy pressurized irrigation methods such as sprinkler and drip are typically more efficient as the quantity of water to be applied can be adequately controlled, but the initial investment and maintenance costs of such systems are high. Application of highly saline (EC_{iw} = 12 dS/m) water through sprinkler to pearl millet and cotton is detrimental, whereas it can be safely used for wheat and barley (Aggarwal and Khanna 1983). Water-use efficiency, although decreased with salinity of water (Table 11.6), was higher when the water was applied by using sprinkler than by surface method to winter crops (wheat and barley). For saline water use, sprinklers should be better operated in the evening/nighttime when evaporation rates are low. Sprinklers also ensure uniform distribution of water even on undulating and sandy terrains and can even help in better leaching of the salts. The lower pore water velocity and the water content at which water moves in soil under sprinkler methods reduce the preferential flow and increase the efficiency of salt leaching. Saline water use through sprinklers, however, may cause leaf burning and toxicity when used in some sensitive crops (Figure 11.5).

The application of irrigation waters through drip systems has revolutionized the production of some high-value crops and orchards in countries like Israel and elsewhere, especially when using saline waters. Though the drip irrigation method has still to pick up in India, the system has a great potential in the arid and semiarid regions, particularly for light-textured soils. As regular and frequent water supply is possible with the drip system of irrigating crops, it has been observed to enhance the threshold limits of their salt tolerance (Table 11.5, Case 2, as described later) by modifying the patterns of salt distribution and maintenance of constantly higher matric potentials (Meiri and Plaut 1985). Due to enhanced leaching and accumulation of the salts at the wetting front and the soil between the drip laterals, the salt accumulation below the drippers remains very low, whereas the water contents are maintained at higher levels at the latter sites. As crop roots are known to follow the path of least resistance, most roots are found below the surface drippers. Hence the drip

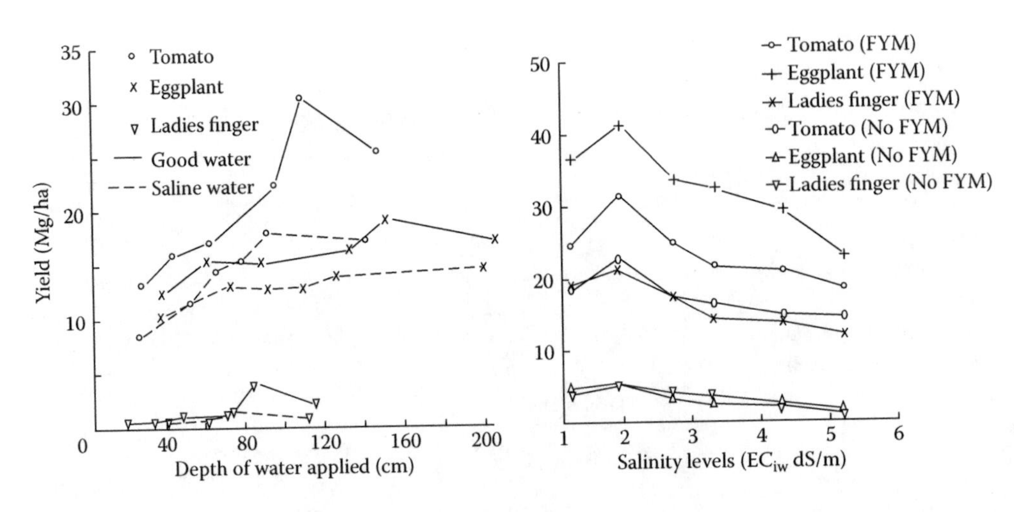

FIGURE 11.5 Crop water production function with normal and saline water. (From AICRP-Saline Water. (1972–2006). Annual Progress Reports. All India Co-ordinated Research Project on Management of Salt-affected Soils and Use of Saline Water in Agriculture, CSSRI, Karnal.)

system seems to be the best method of saline water application as it avoids leaf injury to plants, as with sprinklers, and maintains optimum conditions for water uptake by plant roots. Even with the use of saline water, Singh et al. (1978), Aggarwal and Khanna (1983), and Rajak et al. (2006) have reported superiority in yield and water-use efficiency as well as size and quality of vegetables (Table 11.6). Nevertheless, not much advantage of drip irrigation was observed for crops grown during high evaporative demands with excessive loss of water from the wetted soil surface. The major drawback of irrigation with drippers is the high salt concentration that develops at the wetting front. Accumulated salts cause difficulties in the planting of subsequent crops because effective leaching of salts would require the use of flood or sprinkler irrigation.

11.3.3 Chemical Management

11.3.3.1 Fertilizer Use

The accumulated salts in saline soils can affect the nutrient availability for plants in the following ways: by changing the forms in which the nutrients are present in soils; by increasing the losses through leaching when the saline soils are leached heavily (or as in nitrogen (N) through denitrification) or by precipitation in soils; through interactive effects of cations and anions; and through the effects of complementary (nonnutrient) ions on nutrient uptake. By and large, most soils in India are deficient in N, which needs to be supplemented through fertilizer sources. Urea is by far the most widely used N source for crops. Urea is first hydrolyzed to ammonia and carbon dioxide by the enzyme urease and the process has the most commonly expressed disadvantage of loss of N via NH_3 volatilization. Following the application of N through inorganic fertilizer sources, there is a sudden burst in microbial activity and a large pool of NH_4^+ is generated. Thus, ammonia volatilization is extensive in salt-affected soils, which leads to low N use efficiency by crops. Proper splitting of fertilizer N doses so as to meet crop demands, deep incorporation, slow-release N fertilizers, application of urease inhibitors, and use of organic N sources have all been reported to increase N use efficiency by reducing N losses.

Interactions between fertilizers and salinity have been studied at large. However, the evaluation of the concept of alleviating salinity stress through enhanced fertility reveals that such a strategy of additional application of fertilizer N to reduce/overcome the adverse effect of salts may not pay off well. In general, when salinity is not a yield-limiting factor, the applied nitrogenous fertilizers will increase the yields of crops proportionately more than when the salinity becomes a limiting factor

(Dhir et al. 1977; Dayal et al. 1994). A better strategy for improving N use efficiency therefore seems to be to substitute a part of inorganic fertilizer requirements through organic materials. Experiments on the use of organic materials have been conducted in the network trials on different crops and the results have been summarized in Table 11.7. At a given salinity level, increasing application rates of organic materials improved yields of all the tested crops. However, when salinity of the irrigation water was higher, the percent response was reduced when referenced to yields where no organics were applied. It seems that addition of organic materials temporarily immobilize the NH_4-N and subsequently release the organically bound N to crops during the growing season. Increased responses to N fertilizers in the presence of organic materials suggest its role in reducing the volatilization losses and enhancing the N use efficiency under saline environment. A combination of organic and inorganic sources reduced N losses by 50% in *rabi* and by 25% in *kharif.*

On the other hand, increasing the level of phosphorus over the recommended dose seemed to mitigate the adverse effects of salinity. Type of salinity has also been observed to influence the response of crops to phosphorus application. When wheat and barley crops were irrigated with chloride-dominated waters, the yield response to phosphate application was higher compared to sulfate-dominated waters (Manchanda et al. 1982; Chauhan et al. 1991). Results presented in Table 11.5 show that the application of phosphatic fertilizers most likely will improve the threshold limits of crops to the use of chloride-dominated saline waters. The generalization of results with fertilizer use under saline conditions seems difficult, but it can be stated that in most cases, moderate levels of salinity can perhaps be compensated by increased fertilizer doses so long as salinity levels are not excessively high and the crops under consideration are salt sensitive.

11.3.3.2 Organic/Green Manures

The beneficial effects of organic/green manure as a source of nutrients and on improvement of soil structure and permeability are well known. Thus, in addition to better leaching of salts during the

TABLE 11.7
Effect of Nitrogen Levels and Organic Materials on Yield of Crops (Mg/ha)

	Rabi		Agra			Gangawati		
Kharif (Inorg. N)	Inorg. N (% RDN)	Org. Mat	Mustard	Sorghum	OC (%)[b]	Wheat	Maize	OC (%)[b]
Nil	Nil	Nil	0.66	17.4	0.25	0.96	1.16	0.40
50	50	Nil	1.45	23.9	0.33	1.96	2.21	0.48
100	100	Nil	1.93	28.4	0.34	2.39	3.27	0.50
125	125	Nil	2.17	30.6	0.34	2.52	3.52	0.48
75	Nil	GM_1 (10t/ha)	1.39	26.8	0.42	1.56	3.15	0.54
75	Nil	GM_2 (10t/ha)	1.30	27.2	0.43	1.47	3.10	0.56
75	Nil	OM_1 (15t/ha)	1.44	29.6	0.54	1.47	3.25	0.56
75	Nil	OM_2 (5t/ha)	0.89	24.3	0.39	1.22	3.07	0.57
50[a]	50	GM_1	1.93	28.1	0.43	2.35	3.37	0.51
50[a]	50	GM_2	1.76	28.7	0.42	2.21	3.19	0.51
50[a]	50	OM_1	2.04	31.5	0.54	2.31	3.23	0.54
75	50	OM_2	1.46	25.7	0.42	1.99	2.98	0.54
	LSD (p = 0.05)		0.26	3.7		0.41	0.52	

Source: Minhas, P.S., Sharma, D.R., and Chauhan, C.P.S., *Advances in Sodic Land Reclamation*, UPCAR, Lucknow, 2003.

Note: GM_1 Dhaincha, GM_2 Subabul for Agra and Glyricidia at Gangawati; OM_1 FYM, OM_2 Paddy straw.

[a] 75% at Gangawati.

[b] organic carbon determined after 5–6 years.

monsoon season, the incorporation of organic manures may have advantages in saline and alkali soil environments. As stated earlier, the losses through NH_3 volatilization are aggravated in salt-affected soils. Thus, it can serve as a temporary binding agent for the ammoniacal pool of N and reduce its losses. Experiments show an increased response to N fertilizers in the presence of organic materials, suggesting their role in reducing the volatilization losses and enhancing N use efficiency. Because of small and less active microflora in saline soils, the mineralization of organic nutrient fractions is comparatively lower. So the retention of nutrients in organic forms for longer periods will guard against their leaching and other losses from the mostly sandy nature of soils irrigated with saline water. Finally, farmyard manure (FYM) has a beneficial acidifying effect on the soil's sodicity both through the action of organic acids formed during its breakdown and because the Ca + Mg that FYM contains replaces the Na from the exchange complex. It is generally accepted that additions of organic materials improve sodic soils through mobilization of inherent Ca^{2+} from $CaCO_3$ and other minerals by organic acids (formed during its breakdown) and increased pCO_2 in soils (Qadir et al. 2007a). The solubilized Ca^{2+} in soil replaces Na^+ from the exchange complex. However, there is some disagreement in the literature concerning short-term effects of organic matter on the dispersion of sodic soil particles. Poonia and Pal (1979) studied the Na–(Ca + Mg) exchange equilibrium on sandy loam soil treated with or without FYM, and reported that variations in the proportions of Ca/Mg in the equilibrium solutions only slightly improved the Na^+ selectivity of the soils over the soils treated with FYM. In another study, Poonia et al. (1980) observed that the applied organic matter apparently had a grater preference for divalent cations than that present in natural forms in the soils. However, Gupta et al. (1984) cautioned against the use of organic manure on the soils undergoing sodication process through irrigation with alkali waters. Organic matter was shown to enhance dispersion of soils due to greater interparticle interactive forces at high pH. Sharma and Manchanda (1989) studied the effect of irrigation with alkali water (EC_{iw} 4 dS/m, SAR 26, and RSC 15 meq/L) on the growth of pearl millet and sorghum crops with and without gypsum and FYM on a noncalcareous sandy clay loam soil. The soil was previously deteriorated due to irrigation with alkali water. Six-year results with fallow–wheat rotation showed that the use of FYM alone further decreased the crop yields and the permeability of the soils.

In a long-term experiment on a soil that received alkali waters (RSC 2.4–16 meq/L) without additions of FYM, the infiltration rate, pH, and wheat yield were 5.2 mm/h, 10.34, and 2.7 Mg/ha, respectively. These values improved to 8.1 mm/h, 9.7, and 3.14 Mg/ha, respectively, for soils receiving FYM (Dhanker et al. 1990). The response to FYM, however, decreased with increase in RSC of irrigation water. Thus it may be opined that the addition of organic materials for use of alkali waters should be preceded by gypsum application when upland *kharif* crops are taken. Nevertheless, short-term reduction in permeability may be rather beneficial for paddy that requires submerged conditions for its growth. As the additions of FYM decreased soil pH and sodicity and improved soil fertility, the yields of rice and wheat improved by 8%–10% on a soil that received irrigation with an alkali water (EC_{iw} 3.2 dS/m, RSC 5.6, meq/L, SAR 11.3) (Minhas et al. 1995). Recently, Choudhary et al. (2004) have reported the synergetic effects of adding FYM and gypsum in improving sugar yield when applied to alkali water–irrigated soil (8.6–12.3 t/ha) compared to soil irrigated with saline–sodic water (7.4–10.7 Mg/ha). In the case of saline–sodic irrigation, sugar yield under FYM treatment (10.8 t/ha) was significantly higher than that under gypsum (9.1 Mg/ha) and was on par with gypsum plus FYM treatment. Sekhon and Bajwa (1993) reported the salt balances in soil under rice–wheat–maize system irrigated with alkali waters (RSC 6.0 and 10.6 meq/L) from a greenhouse experiment. Incorporation of organic materials decreased the precipitation of Ca^{2+} and carbonates, increased removal of Na in drainage waters, decreased soil pH and ESP, and improved crop yields. The effectiveness follows the order: paddy straw > green manure > FYM. It can therefore be concluded that with the mobilization of Ca^{2+} during decomposition of organic materials, the quantity of gypsum required for controlling the harmful effects of alkali water irrigation can be considerably decreased. Thus, occasional application of organic materials should help in sustaining yields of rice–wheat system receiving alkali waters. Other reports (Yaduvanshi and Swarup 2005; Murtaza et al. 2009; Phogat et al. 2010) further support the above results where synergetic effects of combined use of organic and inorganic amendments in improving crop yields were reported.

11.3.3.3 Use of Amendments

The presence of excess Na in relation to Ca content in soils increases the pH and ESP, which in turn decreases the soil permeability to water and can also cause nutritional imbalance within plants. The adverse effects of high Na on physical and chemical properties of soils can be mitigated by the use of amendments which contain Ca (e.g., gypsum). Acids or acid-forming substances such as sulfuric acid or pyrites, which on reaction with soil $CaCO_3$ release Ca^+ in solution, can also be used. Whether or not to use amendments for saline–sodic conditions should be judged from their effectiveness in improving soil properties and crop growth in relation to the cost involved. It is usually opined that Ca contents in highly saline soils will always be more that the critical (>2 mmol/L) contents required for plants, and desodication occurs simultaneous to desalinization when such soils are leached. But there are instances where leaching of saline–sodic soils leads to arise in their pH, dispersion, and disaggregation (Sharma and Khosla 1984; Minhas and Sharma 1989). Moreover, the high-SAR saline soils are prone to infiltration and water stagnation, problems mainly during monsoon rains (Minhas and Sharma 1986), and the changes are irreversible when long-term consequences of using high-SAR saline waters are considered (Minhas et al. 1994, 1999). Such soils require small additions of amendments like gypsum to maintain electrolyte concentrations for the stability of aggregates and hence help in avoiding or alleviating problems of such reduced infiltrability. In experiments on pearl millet–wheat irrigated with saline (EC_{iw} 8 dS/m) waters of varying SAR (10–40 mmol/L), gypsum application at 25% GR improved the average yields (1999–2002) of pearl millet by 5%–23% and 6%–18% under conditions when stagnating water was allowed as such or removed after heavy rainfall events, respectively (AICRP-Saline Water 1998; Table 11.8). Response to gypsum was observed only during the year when heavy rainfall and consequent water stagnation problem occurred during its initial stages, and the overall effects of applied gypsum were higher at SAR_{iw} of 30 and 40 mmol/L. The yields further improved with surface draining of stagnated water during the monsoon (2%–11%). However, the long-term consequences of such a practice of removing rain-stagnated water that is expected to reduce the water available for salt leaching.

Since the application of amendments is a recurring need under alkali water–irrigated conditions, the effects of various amendments, their doses, modes, and frequency of application have been studied at large. No response to gypsum has been reported on light-textured (loamy sand–sandy

TABLE 11.8
Effect of Applied Gypsum on Grain Yield (Mg/ha) of Pearl Millet Grown on Soils Irrigated with Saline Waters of Varying SAR

	With Surface Drainage[a]			Without Drainage		
SAR_{iw}	GR_0	GR_{25}	Mean	GR_0	GR_{25}	Mean
10	2.58	2.78	2.68	2.51	2.71	2.61
20	2.27	2.43	2.35	1.96	2.28	2.12
30	1.36	2.03	1.70	1.11	1.64	1.38
40	1.10	1.79	1.45	0.89	1.31	1.10
Mean	1.83	2.26	2.05	1.62	1.99	1.81

Source: AICRP-Saline Water. (1972–2006). Annual Progress Reports. All India Co-ordinated Research Project on Management of Salt-affected Soils and Use of Saline Water in Agriculture, CSSRI, Karnal.

[a] Surface stagnating water removed after heavy rainfall events: GR indicates gypsum requirement of soil and 0 and 25 are nil and 25% of GR.

loam) soils when irrigated with waters having RSC up to 10 meq/L under wheat–fallow rotation (AICRP-Saline Water 1985). In a soil already deteriorated (SAR_e 48.5) due to irrigation with alkali water (EC_{iw} 2.6 dS/m, SAR_{iw} 20.5, and RSC 9.5 meq/L), application of gypsum did not affect the rice yields, but the yield of the succeeding wheat crop increased significantly (Sharma and Mondal 1982). Application of gypsum in fallow–wheat system also improved the yield of wheat. Even a small dose of gypsum (25% GR) improved the wheat yield from almost nil (0.06) to 2.67 Mg/ha in a highly deteriorated sandy loam soil (pH 10, ESP 92, and infiltration rate <2 mm/h) with the use of an alkali water (TSS 1000 ppm, RSC 10 meq/L). When the gypsum dose was increased to 100% GR, the yield increased to 6.33 Mg/ha (Manchanda et al. 1985). Later, Sharma and Manchanda (1989) concluded that guar/pearl millet–wheat crops can be successfully grown in rotation with alkali waters provided the ESP of the soils is maintained below 15 and 20 with addition of gypsum at 100% GR of the soil. Joshi and Dhir (1991) studied the response of crops to the application of gypsum on an abandoned land in the arid climate of Rajasthan as a result of irrigation with high RSC waters (7.2–8.9 meq/L). Application of gypsum (equaling 100% GR) plus that required to neutralize RSC in applied irrigation water during 2 years resulted in a moderate production of wheat (2.61 Mg/ha) and mustard (2.0 Mg/ha) in the second year. Yadav and Kumar (1994) reported that addition of gypsum at 50% GR to a loamy sand soil (pH 9.6–9.7) irrigated with alkali water (EC 1.93 dS/m, RSC 12 meq/L) was appropriate for growing *kharif* crops like pearl millet, *urd* bean, *mung* bean, cowpea, and pigeonpea, whereas cluster bean responded to gypsum up to 100% GR of soil. Amongst *rabi* crops, response of mustard to gypsum was more than wheat and barley. Gypsum to supply 2.5 and 5.0 meq/L to alkali irrigation water for wheat and rice, respectively, was sufficient for the maintenance of higher yields (Bajwa and Josan 1989a). The response to the use of gypsum to mitigate the adverse effects of high RSC waters has been variable depending on the salt status of the deteriorated soil (Minhas et al. 2005). For experiments where RSC water was applied on saline–sodic soils, it was concluded that (i) under the rice–wheat system, pH and sodicity determine the wheat yields and it responded to the application of gypsum in almost all the experiments, except when RSC < 5 meq/L; (ii) wheat yields under fallow–wheat system were mainly governed by soil salinity and the response to gypsum was erratic; and (iii) the response of wheat to gypsum in sorghum–wheat rotation is masked by the interactive effects of EC_e, SAR_e, and pH of soils, indicating that at a given pH as SAR_e increases, EC_e should be lower. Thus, in saline–sodic soils developed with the use of alkali waters under high water table conditions, high levels of EC_e, SAR_e, and pH together affect plant growth and need to be considered simultaneously for evaluating the salinity and sodicity tolerance in plants. In experiments where RSC water was applied on nonsaline sodic soils, the response of gypsum, however, remained low (Sharma et al. 2001). The need to add gypsum for sustained crop production, especially of rice–wheat, when irrigated with waters having high RSC was clearly evident from the above studies.

Mode and time of gypsum application: Bajwa et al. (2003) observed that gypsum applied at each irrigation was more effective for increasing maize yields in maize–wheat sequence irrigated with RSC water (8 meq/L) as compared to its single dose applied annually. Later, Bajwa and Josan (1989a) reported that gypsum improved the soil properties and significantly increased the yields of rice and wheat crops irrigated with water of RSC 6.8 meq/L, EC_{iw} 0.85 dS/m. Response to gypsum, either applied annually as one dose or at each irrigation, remained the same. With higher RSC (10.3 meq/L) water, the improvement in wheat yields was similar for the two modes of gypsum application, although rice responded better to gypsum application when applied before irrigation. This was because more water was applied to rice which led to an appreciable increase in soil sodicity during the season, affecting rice yields. The depth of irrigation water applied for wheat being less, the increase in Na saturation was not sufficient to adversely affect the wheat yields. Minhas et al. (2003) reported that gypsum/pyrite, when applied every year, gave higher yield of paddy and wheat compared to its application after every 3 years. While comparing the time of application of gypsum, Yadav and Kumar (1994) observed that its application before the onset of monsoons was better than

TABLE 11.9

Average Yields (Mg/ha) under Paddy–Wheat and Mustard–Sorghum (1993–2003) and Soil Properties[a] as Affected by Equivalent Doses of Gypsum Applied Either to Soil or Passing Alkali Water through Gypsum Beds

Treatment	Paddy	Wheat	pH	ESP	Mustard	Sorghum	PH	ESP
Control (T_1)	3.08	2.68	9.6	66	2.27	1.18	9.5	61
			Gypsum Through Beds					
3.3 me/l (T_2)	3.97	3.73	8.0	19	3.06	1.98	8.0	25
5.2 me/l (T_4)	4.24	3.93	8.0	18	3.18	21.3	8.0	24
			Equivalent Soil Application					
As in T_2 (T_4)	3.91	3.71	8.2	20	2.86	1.92	8.0	26
As in T_3 (T_5)	4.11	3.89	8.1	20	3.00	2.05	8.1	24
LSD (p = 0.05)	0.43	0.46			0.38	0.24		

Source: Minhas, P.S., Sharma, D.R., and Chauhan, C.P.S., *Advances in Sodic Land Reclamation*, UPCAR, Lucknow, 2003.

[a] At the harvest of rabi (2002–2003) crops.

its application before presowing irrigation of the *rabi* crops and at each irrigation. Pyrites have also been used for amending the deleterious effects of high RSC waters. Pyrite application once before the sowing of wheat has proved better than its split application at each irrigation or mixing it with irrigation water (Chauhan et al. 1986). Pyrites application was also found better for paddy nursery growth compared to gypsum, press mud, and FYM (Sharma and Yaduvanshi 2002).

Gypsum beds: Results of the above studies indicate that application of gypsum at each irrigation either proved to be better or at least was equal to soil application in alleviating the deleterious effects of alkali waters in rice–wheat system. Translation of these results in practical terms requires some mechanism for dissolution of gypsum in the irrigation water itself. Such a practice will also eliminate the costs involved in powdering, bagging, and storage before its actual use. In view of the costs involved, the dissolution of gypsum directly in water through gypsum beds or its application to the irrigation channels appears to be an economical proposition. Passing irrigation water (EC_{iw} 1.83 dS/m, RSC 15 meq/L) through a bed containing gypsum in assorted (2–50 mm) clods decreased its alkalinity (Pal and Poonia 1979).

It should be noted, however, that the gypsum bed water quality improvement technique may not dissolve >8 meq/L of Ca^{2+}. The response of paddy and wheat to the application of equivalent amounts of gypsum, either by passing the water (RSC 9 meq/L) through gypsum beds where the thickness of the bed was maintained at 7 and 15 cm, or the soil application of gypsum, is presented in Table 11.9 (AICRP-Saline Water 2002). Though crops under both the rotations (paddy–wheat, sorghum–mustard) responded to the application of gypsum through either of the methods, overall response of crops was more in the case of alkali water, which was ameliorated (3–5 meq/L) after passing through gypsum beds. Thus, it seems that the gypsum bed technique can help with the efficient utilization of gypsum.

11.3.4 Cultural Practices

11.3.4.1 Planting Procedures and Tillage Practices

Failure to achieve satisfactory germination and thus the required plant population is the major factor limiting crop production with saline water. Crops like wheat, barley, and safflower can tolerate fairly high levels of salinity, but they are very sensitive at germination and at early seedling stages. Thus, once the plants are established, the salinity effects are substantially reduced. One of the alternatives to

TABLE 11.10
Effect of Seeding Method on Crop Stand and Yield of Indian Mustard Irrigated with Saline Water

	Plant Stand (No./m)			Seed Yield (kg/ha)		
EC_{iw} (dS/m)	SM_1	SM_2	SM_3	SM_1	SM_2	SM_3
3	10.4	10.6	10.0	1670	1560	1510
7	9.8	10.7	9.7	1740	1620	1570
11	2.4	9.3	7.4	590	1540	1380
16	1.4	6.3	2.5	190	1290	750

Source: Minhas, P.S. and Gupta, R.K., *Quality of Irrigation Water—Assessment and Management*, 123, ICAR, New Delhi, 1992c.
Note: SM_1 = Seeding after conventional presowing irrigation; SM_2 = Dry seeding followed by postsowing irrigation; SM_3 = 1/2 presowing and 1/2 postsowing irrigation.

overcome this problem is to use higher seed rates. However, this does not always work out as desired. Other alternatives are the planting practices, which would ensure suitable environment in the seeding zone during the germination and seedling emergence periods. Conventional seeding of most crops is done when optimum moisture conditions for tillage and seedbed preparation are attained following a presowing irrigation. After the application of presowing irrigation, the movement of salts toward the surface via evaporative drying, both up to seeding and during the periods of germination and emergence, exposes the seeds to soil water of higher salinity, especially when saline irrigation is practiced ($EC_{sw} > EC_{iw}$). This makes the seed germination and emergence even more critical, especially for summer crops seeded under high evaporative conditions. Therefore, the objectives of presowing irrigation should include leaching out the salts of the seeding zone by a heavy application of nonsaline water wherever possible. The other technique, which seems safe to establish crops, is to apply a postsowing irrigation to push the salts deeper and to maintain better moisture conditions (Minhas et al. 1988). But the timing of this irrigation should be such so as to avoid the subsequent crusting problem. In a field experiment, Indian mustard was seeded with saline waters used in presowing and postsowing irrigation modes (Table 11.10). Compared with the potential (BAW), the seed yield in postsowing irrigation with saline water following dry seeding was sustained up to 11 dS/m. Yadav and Kumar (1994) reported beneficial effects of furrow planting in mustard and sorghum over the flooding of saline waters. Furrow irrigation and bed planting (FIRB) system has been compared with conventional planting for cotton/pearl millet–wheat rotations for 3 years (AICRP-Saline Waters; Table 11.11) and showed

TABLE 11.11
Yields (Mg/ha; Mean of 3 Years) of Crops under Furrow Irrigation and Ridge Bed (FIRB) and Conventional Planting Systems

EC_{iw}	Cotton		Wheat		EC_{iw}	Pearl Millet		Wheat	
(dS/m)	Conv.	FIRB	Conv.	FIRB	(dS/m)	Conv.	FIRB	Conv.	FIRB
BAW	1.29	1.77	3.18	3.63	BAW	2.71	3.11	3.96	4.36
4	1.10	1.67	3.24	3.61	6	2.40	2.99	3.39	4.01
8	0.06	0.55	2.59	3.10	12	1.83	2.30	3.02	3.60
12	Nil	Nil	0.23	2.67					

Source: AICRP-Saline Water. (1972–2006). Annual Progress Reports. All India Co-ordinated Research Project on Management of Salt-affected Soils and Use of Saline Water in Agriculture, CSSRI, Karnal.

an overall improvement in yields under the FIRB system. In addition to few waterlogging effects during monsoon, the advantage of such a system was low irrigation water requirements during *rabi* season. Nevertheless, during the deficit rainfall years, more salt accumulated toward the center of the beds, thus affecting the growth of the central row.

With the development of sodicity in the surface soil, the clay particles in alkali water–irrigated soil become prone to dispersion and displacement, and thus the possibility of formation of dense subsoil layers (plow sole) increases. Moreover, such soils become very hard and dense (hard setting soils) on drying. Both these factors retard root proliferation and poor crop yields are mainly ascribed to this. Therefore, deep plowing/chiseling can be considered as a short-term measure to overcome physical hindrances in such soils. Wheat crop responds to deep tillage, and the average yield increase was on the order of 2–4 Mg/ha (Minhas and Bajwa 2001).

11.3.4.2 Row Spacing/Plant Density

As described in the earlier sections, stunted growth and poor tillering of crops are the major causes of yield reduction in saline environment. Hence, the crop yield, which is the product of stand density (number of plants or tillers per unit area) and yield per plant or tiller, in saline soils should increase if density of stunted plants is increased. This can be achieved by narrowing the interrow and/or intrarow spacing of row crops. Studies with wheat at Agra (AICRP-Saline Waters 1993) have shown 10%–15% improvements in grain yield when 25% extra seeds were planted and plants later thinned to a uniform population.

11.3.4.3 Rainwater Conservation

Since monsoon rains play a crucial role in salt leaching and thus maintaining salt/sodicity balances, the emphasis should be to maximize the infiltration of rainwater into soil and minimize its losses due to runoff and evaporation during the periods in between. To achieve this, the fields should be properly leveled and bunded, and the surface soil kept open and protected against the beating action of raindrops. This can be achieved through plowing in between the rains and by adopting other water conservation practices. Besides increasing the intake of rainwater, plowing also helps in controlling the unproductive losses of water through weeds and evaporation. This practice will also reduce the upward movement of salts between rainfall events and increase salt removal by rains. Creation of soil mulch during the redistribution periods was observed to enhance the leaching of surface applied salts by 10%–13% (Minhas et al. 1986; Minhas and Khosla 1987). Use of straw mulches can also enhance leaching of salts by rainfall, but shortage of straw in saline areas is a serious impediment in adopting this practice. Singh et al. (1994) reported marked improvements in the yield of saline (EC_{iw} 12 dS/m) water–irrigated mustard (82% and 54%) with mulch and fallow than in sorghum grown during the monsoon season during a deficit rainfall year (1989–1990), whereas no response was observed during above-normal rainfall year (1990–1991). Performance of mustard when seeded with conserved moisture so as to avoid saline irrigation at critical germination and establishment stages was considerably better than with the normal practice of seeding after a presowing irrigation with saline water (Chauhan and Singh 1993). In a similar experiment (Dayal et al. 1994), response to applied N ($R^2 = 0.76$) could be explained by the relation $Y = 531 + W (0.27 N - 0.61 S)$, where Y is the yield, W the total extractable water in soil to a depth of 1.2 m plus irrigation and rainfall (cm), N the applied nitrogen, and S the time-averaged salinity in the soil to a depth of 0.3 m. This indicated an increase in marginal productivity of mustard with increase in water supply and decrease in salinity.

In addition to the amount and frequency of rainfall and the soil texture, the anionic constituents of saline irrigation waters also affect leaching of salts during monsoons. In a sandy loam soil irrigated with saline water (EC_{iw} 16 dS/m), higher salt leaching with monsoon rains was observed when irrigation waters had dominance of chlorides compared with sulfate ions (Chauhan et al. 1991; Singh et al. 1994). The amounts of rainwater to leach out 80% of salts were 0.60, 0.89, and 0.92 cm/cm depth of soils irrigated with waters having Cl/SO_4 ratios of 3:1, 1:1, and 1:3, respectively. While the soil profile was almost free of Cl^- (as highly soluble salts are leached easily), some of

SO_4^{2-} was held back because precipitated salts of sulfate in soils (e.g., relatively insoluble gypsum) continued to dissolve with passage of each parcel of rainwater. Associated cations of SO_4^{2-} were both $Ca^{2+}+Mg^{2+}$ and Na^+. The results indicated that solubilized Ca^{2+} from gypsum was replacing Na from the soil's exchange complex and increasing the concentration of the latter in the solution. This resulted in maintenance of higher SAR_e in solution of lower layers of soils irrigated with high SO_4 waters. Sharma and Manchanda (1996) have also reported that desodication upon leaching the soil columns of sandy loam soil with 40–60 cm of deionized water showed that there was a predominance of SO_4^{2-} rather than Cl, while the reverse was the case with desalinization. Studies on the leaching behavior of high-SAR saline/sodic water–irrigated soils (Sharma and Khosla 1984; Singh et al. 1992) have shown that during leaching, pH increases and clay particles become vulnerable to dispersion and movement. Thus, salts are held back and such soils require almost double the quantity of water than that required for leaching of waterlogged saline soils. Under such a situation, the addition of gypsum to prevent surface sealing and to enhance infiltrability of rainwater is advocated.

11.4 ALTERNATE LAND USES

In some cases, it is neither feasible nor economical to use highly saline waters for crop production, especially on lands that are already degraded. Best land use under such situations is to retire such areas to permanent vegetation. To establish plantations and improve biomass production from such lands, a system of planting "SPFIM" (subsurface planting and furrow irrigation method) has been devised (Tomar et al. 1994; Minhas et al. 1997a,b). It not only saves irrigation time and labor, but also leads to addition of lesser salts in the soil profile since irrigation is applied only to furrows covering one-fifth to one-tenth of the total area. Quantities equaling 10% of the open pan evaporation sufficed for the optimal growth of several tree species of arid and semiarid areas. In addition to the creation of favorable water regimes in the rooting zone during irrigation to furrow-planted tree saplings, this method showed the advantage of pushing the salts toward interrow areas with monsoon rains. Preferred choices for tree species include *Tamarix articulata*, *Prosopis juliflora*, *Acacia nilotica*, *Acacia tortilis*, *Feronia limonia*, *Acacia farnesiana*, and *Melia azadirach* (Tomar et al. 2002). Halophytic species like *Salvadora* and *Sueda* have been identified for bio-saline agriculture. In California, the sequential reuse of drainage water involving the use of trees, shrubs, and grasses has only been partly successful (Tanji and Kajreh 1993; Oster et al. 1999). Here *Eucalyptus camaldulensis* was grown with subsurface drainage water collected from nearby cropland (EC 10 dS/m, SAR 11), while the effluent from eucalyptus and the perimeter interceptor drain (EC 32 dS/m; SAR 69) were used to irrigate Atriplex species.

Moreover, the degraded lands in arid and semiarid regions are traditionally left for pastures, but their forage productivity is low, unstable, and unremunerative. Usually there are acute shortages of fodder during the postmonsoon period. When the limited ($D_{iw}/CPE = 0.4$) saline groundwater resources were utilized to supplement rainwater supplies, Tomar et al. (2003) observed that forage grasses like *Panicum laevifolium* (3.43–4.23 Mg/ha/year) followed by *P. maximum* (both local wild and cultivated) outperformed the other grasses. Saline irrigation not only improved their productivity threefold to fourfold, but fodder (about 30%) could also be made available during the scarce months of April–June when most nomads are forced to move toward the adjoining irrigated areas in search of fodder. Similarly, Oster et al. (1999) have reported that Bermuda grass can be grown with saline–sodic waters having EC up to 17 dS/m and SAR > 17.

11.5 GUIDELINES FOR USING SALINE AND ALKALI WATERS

It is evident from the above discussion that, apart from its composition, determination of suitability of specific water requires that specifications of conditions of its use (soil, climate, crops, etc.), irrigation, and other management practices be followed. Because of inherent problems in integrating the effects of the above factors, it is difficult to develop rigid standards for universal use. Therefore, broad

TABLE 11.12
Guidelines for Using Saline Irrigation Waters

Soil Texture (% Clay)	Crop Tolerance	EC_w (dS/m) Limit for Rainfall Region (mm) <350	350–550	>550
	(a) Saline Waters (RSC < 2.5 me/L)			
Fine (>30)	Sensitive	1.0	1.0	1.5
	Semitolerant	1.5	2.0	3.0
	Tolerant	2.0	3.0	4.5
Moderately fine (20–30)	Sensitive	1.5	2.0	2.5
	Semitolerant	2.0	3.0	4.5
	Tolerant	4.0	6.0	8.0
Moderately coarse (10–20)	Sensitive	2.0	2.5	3.0
	Semitolerant	4.0	6.0	8.0
	Tolerant	6.0	8.0	10.0
Coarse (<10)	Sensitive	—	3.0	3.0
	Semitolerant	6.0	7.5	9.0
	Tolerant	8.0	10.0	12.5

(b) Alkali Waters (RSC > 2.5, EC_w < 4.0 dS/m)

Soil Texture (% Clay)	Upper Limit of SAR_w	Upper Limit of RSC	Remarks
Fine (>30)	10	2.5–3.5	1. Limits pertain to *kharif* fallow–*rabi* crop rotation when annual rainfall is 350–550 mm
Moderately fine (20–30)	10	3.5–5.0	2. When the waters have Na < 75%, Ca+Mg < 25%, or rainfall is >550 mm, upper limit of RSC range is safe
Moderately coarse (10–20)	15	5.0–7.5	3. For double cropping, RSC neutralization with gypsum is essential based on the quantity of water used during *rabi* season; grow less-water-requiring crops during *kharif*; avoid growing rice
Coarse	20	7.5–10.0	

Source: Minhas, P.S. and Gupta, R.K., *Quality of Irrigation Water—Assessment and Management*, p. 123, ICAR, New Delhi, 1992c.

Note: Textural criteria should be applicable for all soil layers down to at least 1.5 m depth.
In areas where groundwater table reaches within 1.5 m at any time of the year or a hard subsoil layer is present in the root zone, the limits of the next finer textural class should be used.
Fluorine is at times a problem and limits should be worked out.

guidelines for assessing suitability of irrigation waters have been suggested from time to time for average use conditions (USSL 1954; Ayers and Westcot 1985). However, it is widely acknowledged that the earlier guidelines were very conservative since these were based mainly on the response of soils rather than the crop responses under diverse field conditions. Based upon the field experiences and the results of long-term experimentation, guidelines were recommended (Table 11.12) for their wider applicability in different agro-ecological zones of India (Minhas and Gupta 1992c). Though developed for the monsoonal climate, these can also be applied to areas with seasonal rainfall. For meeting site-specific water quality objectives, factors like water quality parameters, soil texture, crop tolerances, and rainfall have been given due consideration. Some of the addendums added to these guidelines include use of gypsum for saline water having SAR > 20 and/or Mg/Ca > 3 and rich in silica; fallowing during rainy season when SAR > 20 and higher salinity waters are used in low rainfall areas; additional phosphorous application, especially when Cl/SO_4 > 2.0, using canal water preferably at early growth stages, including presowing irrigation for conjunctive use with saline

TABLE 11.13
Permissible Limits of adj.R_{Na} in Irrigation Water for Sustaining Yields under Different Cropping Sequences

Cropping Sequence	Permissible adj.R_{Na} for Sustainable Yields for Rainfall Zone (cm)		
	<40	40–60	>60
Fallow–wheat	16	21	27
Maize/millet–wheat	14	17	23
Paddy–wheat	6	9	14
Cotton–wheat	14	20	26

Source: Minhas, P.S. and Sharma, D.R., *J. Indian Soc. Soil Sci.*, 54, 331–338, 2006.

waters; putting 20% extra seed rate and a quick postsowing irrigation (within 2–3 days) to help better germination when $EC_w < EC_e$ (0–45 cm soil at harvest of *rabi* crops); saline water irrigation just before the onset of monsoons to lower soil salinity and raise the antecedent soil moisture for greater salt removal by rains; use of organic materials in saline environment to improve crop yields (for soils having (i) shallow water table (within 1.5 m in *kharif*), and (ii) hard subsoil layers, the next lower EC_{iw}/alternate modes of irrigation (canal/saline) is applicable). Similarly, the guidelines for alkali waters were based upon RSC/SAR and texture of soils for varying cropping intensity and amendments. These were later modified by Minhas and Sharma (2006), who proposed permissible limits in terms of adj.RNa for the sustainable use of alkali waters for different rainfall regions and the cropping sequences (Table 11.13). As expected, waters with a relatively high adj.R_{Na} could be utilized for sustained irrigation in fallow/cotton–wheat followed by millet–wheat rotations, whereas paddy–wheat seems to be the most unsustainable system, since the permissible values of adj.R_{Na} were just about half of the former. Similarly, the role of rainfall in enhancing the use of alkali water is also evident; that is, on average, it should be possible to use water with 1.8 *adj.R_{Na} under conditions where the annual rainfall is >600 mm compared to drier regions (<400 mm). Suitability is further expected to vary with soil type and the associated anions, but present predictions can serve as a useful tool to attain the desired level of production under various cropping sequences and climatic situations.

REFERENCES

Abrol, I.P. and Bhumbla, D.R. (1979). Crop responses to differential gypsum application in a highly sodic soil and the tolerance of several crops to exchangeable sodium under field conditions. *Soil Sci.* 127: 79–85.

Aggarwal, M.C. and Khanna, S.S. (1983). Efficient soil and water management in Haryana. Bulletin of HAU, Hisar, p. 118.

AICRP-Saline Water. (1972–2006). Annual Progress Reports. All India Co-ordinated Research Project on Management of Salt-affected Soils and Use of Saline Water in Agriculture, CSSRI, Karnal.

Ayers, R.S. and Westcot, D.W. (1985). Water quality for agriculture. Irrigation and Drainage Paper No. 29, Rev. 1, FAO, Rome, Italy, p. 174.

Bajwa, M.S. and Josan, A.S. (1989a). Prediction of sustained sodic irrigation effects on soil sodium saturation and crop yields. *Agric. Water Manag.* 16: 227–228.

Bajwa, M.S. and Josan, A.S. (1989b). Effect of gypsum and sodic water irrigation on soil and crop yields in a rice-wheat rotation. *Agric. Water Manag.* 16: 53–61.

Bajwa, M.S. and Josan, A.S. (1989c). Effect of alternating sodic and non-sodic irrigation on build up of sodium in soil and crop yield in northern India. *Exp. Agric.* 25: 199–205.

Bajwa, M.S., Hira, G.S., and Singh, N.T. (2003). Effect of sodium and bicarbonate irrigation waters on sodium accumulation and maize and wheat yields in northern India. *Irrig. Sci.* 4: 191–199.

Bajwa, M.S., Hira, G.S., and Singh, N.T. (1986). Effect of sustained saline irrigation on soil salinity and crop yields. *Irrig. Sci.* 7: 27–34.

Bajwa, M.S., Choudhary, O.P., and Josan, A.S. (1992). Effect of continuous irrigation with sodic and saline sodic water on soil properties and crop yields under cotton-wheat rotation in northern India. *Agric. Water Manag.* 22: 345–350.

Bajwa, M.S., Josan, A.S., and Chaudhary, O.P. (1993). Effect of frequency of sodic and saline-sodic irrigation and gypsum on the build up of sodium in soil and crop yields. *Irrig. Sci.* 13: 21–26.

Bouwer, H. (2000). Groundwater problems caused by irrigation with sewage effluent. *J. Environ. Health* 63: 17–20.

Bouwmans, J.H., van Hoorn, J.W., Cruiseman, G.P., and Tanwar, B.S. (1988). Water table control, reuse and disposal of drainage water in Haryana. *Agric. Water Manag.* 14: 537–545.

Bower, C.A., Ogata, G., and Tucker, J.M. (1968). Sodium hazards of irrigation waters as influenced by leaching fraction and by precipitating or selection of calcium carbonates. *Soil Sci.* 106: 29–34.

Chauhan, C.P.S. and Singh, R.B. (1993). Mustard performs well even with saline irrigation. *Indian Farm.* 42(12): 17–20.

Chauhan, R.P.S., Chauhan, C.P.S., and Singh, V.P. (1986). Use of pyrites in minimising the adverse effects of sodic waters. *Indian J. Agric. Sci.* 56: 717–721.

Chauhan, C.P.S., Singh, R.B., Minhas, P.S., Agnihotri, A.K., and Gupta, R.K. (1991). Response of wheat to irrigation with saline waters varying in anionic constituents and phosphorous application. *Agric. Water Manag.* 20: 223–231.

Choudhary, O.P., Bajwa, M.S., and Josan, A.S. (1994). Characteristics of different wheat cultivars for tolerance to soil sodium saturation. *Trans. World Soil Sci. Cong.* 3(b): 368–369.

Choudhary, O.P., Josan, A.S., and Bajwa, M.S. (1996). Rooting and yield relationships in different barley cultivars grown under increasing sodicity stress conditions. *Crop Improve.* 23: 1–11.

Choudhary, O.P., Josan, A.S., and Bajwa, M.S. (2001). Yield and fibre quality of cotton cultivars as affected by the build-up of sodium in soils with sustained sodic irrigations under semi-arid conditions. *Agric. Water Manag.* 49: 1–9.

Choudhary, O.P, Josan, A.S., Bajwa, M.S., and Kapur, M.L. (2004). Effect of sustained sodic and saline-sodic irrigations and application of gypsum and farmyard manure on yield and quality of sugarcane under semi-arid conditions. *Field Crops Res.* 87: 103–116.

Choudhary, O.P., Ghuman, B.S., Josan, A.S., and Bajwa, M.S. (2006). Effect of alternating irrigation with sodic and non-alkali waters on soil properties and sunflower yield. *Agric. Water Manag.* 85: 151–156.

Chauhan, S.K., Chauhan, C.P.S., and Minhas, P.S. (2007). Effect of cyclic use and blending of alkali and good quality waters on soil properties, yield and quality of potato, sunflower and Sesbania. *Irrig. Sci.* 26: 81–89.

Choudhary, O.P., Ghuman, B.S., Josan, A.S., and Bajwa, M.S. (2008). Cyclic use of sodic and non-sodic canal waters for irrigation in cotton-wheat cropping system in a semi-arid region. *J. Sust. Agri.* 32: 269–286.

Corwin, D.L., Lesch, S.M., Oster, J.D., and Kaffaka, K. (2008). Short term sustainability of drainage water reuse: Spatio-temporal impacts on soil chemical properties. *J. Environ. Qual.* 37: S8–S24.

Dayal, B., Minhas, P.S., Chauhan, C.P.S., and Gupta, R.K. (1994). Effect of supplementary saline irrigation and applied nitrogen on performance of dryland mustard (*Brassica juncea* L). *Exp. Agric.* 31: 423–428.

Dhanker, O.P., Yadav, H.D., and Yadav, O.P. (1990). Long term effect of sodic water on soil deterioration and crop yields in loamy sand soil of semiarid regions. In *National Symposiumon Water Resource Conservation Recycling and Reuse*, Nagpur, February 3–5, pp. 57–60.

Dhir, R.P., Bhola, S.N., and Kolarkar, A.S. (1977). Performance of 'Kharchia 65' and 'Kalyan Sona' wheat varieties at different levels of water salinity and nitrogenous fertilizers. *Indian J. Agric. Sci.* 47: 244–248.

FAO. (2003). World Agriculture: Towards 2015/2030. Available at http://www.fao.org/docrep/005/ Y4252e00. htm, FAO, Rome.

Ghassemi, F., Jakeman, A.J., and Nix, H.A. (1995). *Salinization of Land and Water Resources: Human, Causes, Extent, Management and Case Studies.* CABI Publishing, Wallingford.

Gill, K.S. and Qadir, A. (1998). Physiological aspects of salt tolerance. In Tyagi, N.K. and Minhas, P.S. (eds), *Agricultural Salinity Management in India.* CSSRI, Karnal, pp. 243–260.

Goyal, S.S., Sharma, S.K., Rains, D.W., and Läuchli, A. (1999a). Long-term reuse of drainage waters of varying salinities for crop production in a cotton-safflower rotation system in the San Joaquin valley of California—A nine year study. I: Cotton (*Gossypium hirustum* L.). *J. Crop Prod.* 2: 181–213.

Goyal, S.S., Sharma, S.K., Rains, D.W., and Läuchli, A. (1999b). Long-term reuse of drainage waters of varying salinities for crop production in a cotton-safflower rotation system in the San Joaquin valley of California—A nine year study. II: Safflower (*carthamus tinctorus* L.). *J. Crop Prod.* 2: 215–227.

Gupta, R.K. and Abrol, I.P. (1990). Salt affected soils—Their reclamation and management for crop production. *Adv. Soil Sci.* 12: 223–275.

Gupta, R.K. and Abrol, I.P. (2000). Salinity build-up and changes in the rice-wheat system of the Indo-Gangetic plains. *Exp. Agric.* 36: 273–284.

Gupta, R.K., Bhumbla, D.R., and Abrol, I.P. (1984). Effect of soil pH, organic matter and calcium carbonate on the dispersion behaviour of alkali soils. *Soil Sci.* 137: 245–251.

Hillel, D. (2000). *Salinity Management for Sustainable Agriculture.* The World Bank, Washington, DC.

Joshi, D.C. and Dhir, R.P. (1991). Rehabilitation of degraded sodic soil in an arid environment by using residual Na-carbonate water for irrigation. *Arid Soil Res. Rehab.* 5: 175–185.

Katerji, N., Van Hoorn, J.W., Hamdy, A., and Mastrorilli, M. (2000). Salt tolerance classification of crops according to soil salinity and to water stress day index. *Agric. Water Manag.* 43: 99–109.

Koyama, M.L., Levesley, A., Koebner, R.M.A., Flowers, T.J., and Yeo, A.R., (2001). Quantitative trait loci for component physiological traits determining salt tolerance in rice. *Plant Physiol.* 125: 406–422.

Läuchli, A. and Grattan, S.R. (2007). Plant growth and development under salinity stress. In Jenks, M.A., Hasegawa, P.A., and Jain, S.M. (eds), *Advances in Molecular-Breeding: Towards Salinity and Drought Tolerance.* Springer-Verlag, pp. 1–31.

Maas, E.V. and Grattan, S.R. (1990). Crop yield as affected by salinity. In Skaggs, R.W. and van Schilfgaarde, J. (eds), *Agricultural Draiange.* ASA-CSSA-SSSA, Madison, WI, pp. 55–108.

Malash, N., Flowers, T.J., and Ragab, R. (2005). Effect of irrigation systems and water management practices using saline and non-saline water on tomato production. *Agric. Water Manag.* 78: 25–38.

Manchanda, H.R., Verma, S.L., and Khanna, S.S. (1982). Identification of some factors for use of sodic waters with high residual sodium carbonate. *J. Indian Soc. Soil Sci.* 30:353–360.

Manchanda, H.R., Garg, R.N., Sharma, S.K., and Singh, J.P. (1985). Effect of continuous use of sodium and bicarbonate rich irrigation water with gypsum and farm-yard manure on soil properties and yield of wheat in a fine loamy soil. *J. Indian Soc. Soil Sci.* 33: 876–883.

Manchanda, H.R., Gupta, I.C., and Jain, B.L. (1989). Use of poor quality waters. In *Review of Research on Sandy Soils in India*, International Symposium on Managing Sandy Soils. CAZRI, Jodhpur, February 6–10, 1989, pp. 362–383.

Manchanda, H.R., Sharma, S.K., and Mor, R.P. (1991). Relative tolerance of pulses for chloride and sulphate salinity. *Indian J. Agric. Sci.* 61: 20–26.

Meiri, A. and Plaut, Z. (1985). Crop production and management under saline conditions. *Plant Soil.* 89: 253–271.

Minhas, P.S. (1996). Saline water management in India. *Agric. Water Manag.* 30: 1–24.

Minhas, P.S. and Bajwa, M.S. (2001). Use and management of poor quality waters in rice-wheat production system. *J. Crop Prod.* 4: 273–306.

Minhas, P.S. and Samra, J.S. (2003). *Quality Assessment of Water Resources in the Indo-Gangetic Basin Part in India.* Technical Bulletin No. 2/2003, 68p. Central Soil Salinity Research Institute, Karnal.

Minhas, P.S. and Gupta, R.K. (1992a). Conjunctive use of saline and non-saline waters. I: Response of wheat to initially variable salinity profiles and modes of salinisation. *Agric. Water Manag.* 23: 125–137.

Minhas, P.S. and Gupta, R.K. (1992b). Conjunctive use of saline and non-saline waters. III: Validation and applications of a transient state model for wheat. *Agric. Water Manag.* 23: 149–160.

Minhas, P.S. and Gupta, R.K. (1992c). *Quality of Irrigation Water—Assessment and Management.* ICAR, New Delhi, p. 123.

Minhas, P.S. and Khosla, B.K. (1987). Leaching of salts as affected by the method of water application and atmospheric evaporativity under shallow and saline water-table conditions. *J. Agric. Sci. (Camb)* 109: 415–419.

Minhas, P.S. and Sharma, D.R. (1986). Hydraulic conductivity and clay dispersion as affected by the application sequence of saline and simulated rainwater. *Irrig. Sci.* 7: 159–167.

Minhas, P.S. and Sharma, D.R. (1989). Salt displacement in a saline-sodic and amended soil using a low electrolyte water. *J. Indian Soc. Soil Sci.* 37: 435–440.

Minhas, P.S. and Sharma, D.R. (2006). Predictability of existing indices and an alternative coefficient for estimating sodicity build-up using adj.R_{Na}, and permissible limits for crops grown on soils irrigated with waters having residual alkalinity. *J. Indian Soc. Soil Sci.* 54: 331–338.

Minhas, P.S., Dubey, S.K., and Sharma, D.R. (2007a). Effects of soil and paddy-wheat crops irrigated with waters containing residual alkalinity. *Soil Use Manag.* 23: 254–261.

Minhas, P.S., Dubey, S.K., and Sharma, D.R. (2007b). Comparative effects of blending, intra/inter-seasonal cyclic uses of alkali and good quality waters on soil properties and yields of paddy and wheat. *Agric. Water Manag.* 87: 83–90.

Minhas, P.S., Khosla, B.K., and Prihar, S.S. (1986). Evaporation and redistribution of salts as affected by tillage induced soil mulch. *Soil Tillage Res.* 7: 301–13.

Minhas, P.S., Sharma, D.R., and Khosla, B.K. (1988). Effect of postsowing irrigation and planting techniques on germination of sorghum irrigated with saline water. *J. Indian Soc. Soil Sci.* 36: 584–587.

Minhas, P.S., Sharma, D.R., and Khosla, B.K. (1989). Response of sorghum to the use of saline waters. *J. Indian Soc. Soil Sci.* 37: 140–146.

Minhas, P.S., Sharma, D.R., and Khosla, B.K. (1990a). Effect of alleviating salinity stress at different growth stages of Indian mustard (*Brassica juncea*). *Indian J. Agric. Sci.* 57: 343–346.

Minhas, P.S., Sharma, D.R., and Khosla, B.K. (1990b). Mungbean response to irrigation with waters of different salinity. *Irrig. Sci.* 11: 57–62.

Minhas, P.S., Naresh, R.K., Chauhan, C.P.S., and Gupta, R.K. (1994). Field determined hydraulic properties of a sandy loam soil irrigated with various salinity and SAR waters. *Agric. Water Manag.* 24: 93–104.

Minhas, P.S., Sharma, D.R., and Singh, Y.P. (1995). Response of paddy and wheat to applied gypsum and FYM on an alkali water irrigated soil. *J. Indian Soc. Soil Sci.* 43: 452–455.

Minhas, P.S., Sharma, D.R., Sharma, D.K. (1996). Perspective of alkali water management for paddy-wheat cropping system, *J. Indian Water Res. Soc.* 2: 57–61.

Minhas, P.S., Singh, Y.P., Tomar, O.S., Gupta, R.K., and Gupta, R.K. (1997a). Saline-water irrigation for the establishment of furrow planted trees in north-western India. *Agrofor. Syst.* 35: 177–186.

Minhas, P.S., Singh, Y.P., Tomar, O.S., Gupta, R.K., and Gupta, R.K. (1997b). Effect of saline irrigation and its schedules on survival, growth, biomass production and water use by *Acacia* and *Dalbergia* on a highly calcareous soil. *J. Arid Environ.* 36: 181–192.

Minhas, P.S., Singh, Y.P., Chhabba, D.S., and Sharma, V.K. (1999). Changes in hydraulic conductivity of a highly calcareous and non-calcareous soil under cycles of irrigation with saline and simulated rainwater. *Irrig. Sci.* 18: 199–203.

Minhas, P.S. (2010). A relook on diagnostic criteria for salt affected soils in India. *J. Indian Soc. Soil Sci.* 58(1): 12–24.

Miyamoto, S. and Chacon, A. (2006). Soil salinity of urban turf areas irrigated with saline water. II: Soil factors. *Landsc Urban Plan.* 77: 28–38.

Murtaza, G., Ghafoor, A., Owens, G., Qadir, M., and Kahlon, U.Z. (2009). Environmental and economic benefit of saline-sodic reclamation using low quality water and soil amendments in conjunction with rice-wheat cropping system, *J. Agron. Crop Sci.* doi. 10.111/j.1439-037X.2008.00350.x.

Naresh, R.K., Minhas, P.S., Goyal, A.K., Chauhan, C.P.S., and Gupta, R.K. (1992a). Conjunctive use of saline and non-saline waters. II: Field comparisons of mixing and cyclic use modes for winter wheat. *Agric. Water Manag.* 23: 139–148.

Naresh, R.K., Minhas, P.S., Goyal, A.K., Chauhan, C.P.S., and Gupta, R.K. (1992b). Production potential of cyclic irrigation and mixing of canal and saline waters in Indian mustard–pearl millet rotation. *Arid Soil Res. Rehab.* 7: 103–111.

O'Connor, G.A., Elliott, H.A., and Bastian, R.K. (2008). Degraded water reuse: An overview. *J. Environ. Qual.* 37: 157–168.

Oster, J.D. and Jaywardane, N.S. (1998). Agricultural management of sodic soils. In Sumner, M.E. and Naidu, R. (eds), *Sodic Soils: Distribution, Properties, Management and Environmental Consequences.* Oxford University Press, New York, pp. 124–147.

Oster, J.D. and Schroer, F.W. (1979). Infiltration as influenced by irrigation water quality. *J. Soil Sci. Soc. Am.* 43: 444–447.

Oster, J.D., Shainberg, I., and Abrol, I.P. 1999. Reclamation of salt affected soils. In Skaggs, R.W. and van Schilfgaarde, J. (eds), *Agricultural Drainage.* ASA-CSSA-SSSA, Madison, WI, pp. 659–691.

Pal, R. and Poonia, S.R. (1979). Dimension of gypsum bed in relation to residual sodium carbonate of irrigation water, size of gypsum fragments and flow velocity. *J. Indian Soc. Soil Sci.* 27: 5–10.

Phogat, V., Sharma, S.K., Kumar, S., Stayvan and Gupta, S.K. (2010). Vegetable cultivation with poor quality water. Bulletin, CCS HAU, Hisar, 72p.

Pitman, M.G. and Läuchli, A. (2002). Global impacts of salinity and agricultural ecosystems. In Läuchli, A. and Lüttage, U. (eds), *Salinity: Environments–Plants–Molecules.* Kluwer, Dordrecht, Netherlands, pp. 3–20.

Poonia, S.R. and Pal, R. (1979). Effect of organic manuring and water quality on water transmission parameters and sodication of a sandy loam soil. *Agric. Water Manag.* 2: 163–175.

Poonia, S.R., Mehta, S.C., and Pal, R. (1980). Calcium-sodium, magnesium-sodium exchange equilibria in relation to organic matter in soils. In *Proceedings of International Symposium on Salt-Affected Soils*, CSSRI, Karnal, pp. 135–142.

Qadir, M., Boers, Th.M., Schubert, S., Ghafoor, A., and Murtaza, G. (2003). Agricultural water management in water starved countries. *Agric. Water Manag*. 62: 165–185.

Qadir, M., Noble, A.D., Schubert, S., Thomas, R.J., and Arslan, A. (2006). Sodicity-induced land degradation and its sustainable managements: Problems and prospects. *Land Degrad. Dev*. 17: 661–676.

Qadir, M., Oster, J.D., Schubert, S. Noble, A.D., and Sahrawat, K.L. (2007a). Phytoremediation of sodic and saline-sodic soils. *Adv. Agron*. 96: 197–247.

Qadir, M., Wichelns, D., Raschid-Sally, L., Minhas, P.S., Drechsel, P., and Bahri, A., et al. (2007b). Agricultural use of marginal quality water: Opportunities and challenges. In: Molden, D. (ed.), *Water for Food, Water for Life: A Comprehensive Assessment on Water Management in Agriculture*. Earthscan, London, pp. 425–457.

Ragab, R., Malash, N., Abdel-Gawad, G., Arslan, A., and Ghaibeh, A. (2005). A holistic generic integrated approach for irrigation, crop and filed management 2. The SALTMED model validation using filed data of five growing seasons from Egypt and Syria. *Agric. Water Manag*. 78: 89–107.

Rajak, D., Manjunatha, M.V., Rajkumar, G.R., Hebbara, M., and Minhas, P.S. (2006). Comparative effects of drip and furrow irrigation on the yield and water productivity of cotton (*Gossypium hirustum* L.) in a saline and water logged vertisol. *Agric. Water Manag*. 83: 30–36.

Rhoades, J.D. (1968). Mineral weathering correction for estimating sodium hazard of irrigation waters. *Soil Sci. Soc. Amer. Proc*. 32: 648–652.

Rhoades, J.D. (1999). Use of saline waters in irrigation. In Skaggs, R.W. and van Schilfgaarde, J. (eds), *Agricultural Drainage*. ASA-CSSA-SSSA, Madison, WI, pp. 615–657.

Rhoades, J.D., Kandiah, A., and Mishali, A.M. (1992). The use of saline waters for crop production. FAO's Irrigation and Drainage Paper No. 48, FAO, Rome, Italy, p. 133.

Sekhon, B.S. and Bajwa, M.S. (1993). Effect of incorporation of organic materials and gypsum in controlling sodic irrigation effects on soil properties under rice-wheat-maize system. *Agric. Water Manag*. 24: 5–25.

Shainberg, I. and Letey, J. (1984). Response of soils to sodic and saline conditions. *Hilgardia* 52: 1–57.

Sharma, B.R. and Minhas, P.S. (2005). Strategies for managing saline/alkali waters for sustainable agricultural production in South Asia. *Agric. Water Manag*. 78: 136–151.

Sharma, D.R. and Khosla, B.K. (1984). Leaching a sodic water deteriorated salty soil. *J. Indian Soc. Soil Sci.* 32: 344–348.

Sharma, D.R. and Minhas, P.S. (1997). Effect of irrigation with sodic waters of varying EC, RSC and SAR/adj. SAR on soil properties and yield of cotton –wheat. *J. Indian Soc. Soil Sci.* 46: 116–119.

Sharma, D.R. and Mondal, R.C. (1982). Effect of irrigation of sodic water and gypsum application on soil properties and crop yields. In *12th International Congress of Soil Science*, Abst. No. 604. February 8–12, 1982, New Delhi, p.170.

Sharma, D.R. and Yaduvanshi, N.P.S. (2002). Effect of Amendments on the growth and yield of paddy nursery raised with alkali water. In *National Seminar on Developments in Soil Science 2002*, 67th Annual Convention of ISSS, November 11–15, Udaipur, India, pp. 215–216.

Sharma, D.P., Singh, K.N., Rao, K.V.G.K., Kumbhare, P.S., and Oosterbaan, R.J. (1994). Conjunctive use of saline and non-saline irrigation waters in semi-arid regions. *Irrig. Sci.* 15: 25–33.

Sharma, D.R., Minhas, P.S., and Sharma, D.K. (2001). Response of rice–wheat to sodic water irrigation and gypsum application. *J. Indian Soc. Soil Sci.* 49(2): 324–327.

Sharma, O.P., Sharma, D.N., and Minhas, P.S. (1998). Reclamation and management of alkali soils. In *25 Years of Research on Management of Salt-Affected Soils and Use of Saline Water in Agriculture*, CSSRI, Karnal, pp. 64–85.

Sharma, S.K. and Manchanda, H.R. (1989). Using sodic water with gypsum for some crops in relation to soil ESP. *J. Indian Soc. Soil Sci.* 37: 135–139.

Sharma, S.K. and Manchanda, H.R. (1996). Influence of leaching with different amounts of water on desalinisation and permeability behaviour of chloride and sulphate-dominated saline soils. *Agric. Water Manag* 31: 225–235.

Singh, M., Poonia, S.R., and Pal, R. (1986). Improvement of irrigation water by gypsum beds. *Agric. Water Manag* 11: 293–301.

Singh, R.B., Minhas, P.S., Chauhan, C.P.S., and Gupta, R.K. (1992). Effect of high salinity and SAR waters on salinisation, sodication and yields of pearl millet-wheat. *Agric. Water Manag*. 21: 93–105.

Singh, R.B., Minhas, P.S., Chauhan, C.P.S., and Gupta, R.K. (1994). Salt leaching with monsoons and yield of mustard as affected by saline irrigation waters of varying Cl/SO4 ratios. *J. Indian Soc. Soil Sci.* 42: 436–441.

Singh, S.D., Gupta, J.P., and Singh, P. (1978). Water economy and saline water use by drip irrigation. *Agron. J*. 70: 948–951.

Sinha, B.K. and Singh, N.T. (1976a). Magnitude of chloride accumulation near corn roots under different transpiration, soil moisture and salinity regimes. *Agron. J.* 60: 346–348.

Sinha, B.K. and Singh, N.T. (1976b). Salt distribution around roots of wheat under different transpiration rates. *Plant Soil.* 44: 141–147.

Sumner, M.E., Rengasamy, P., and Naidu, R. (1998). Sodic soils: A reappraisal. In Sumner, M.E. and Naidu, R. (eds), *Sodic Soil: Distribution, Management and Environmental Consequences.* Oxford University Press, New York, pp. 3–17.

Tanji, K.K. and Kajreh, F.F. (1993). Saline water reuse in agroforesrty systems. *J. Irrig. Drain. Eng. ASCE,* 119: 170–180.

Tanwar, B.S. (2003). Saline water management for irrigation. International Commission on Irrigation and Drainage (ICID), New Delhi, India, p. 123.

Tomar, O.S., Minhas, P.S., and Gupta, R.K. (1994). Potentialities of afforestation of water-logged soils. In Singh, P., Pathak, P.S., and Roy, M.M. (eds), *Agroforestry Systems for Degraded Lands,* Vol.1. Oxford and IBH Publ. Co. Pvt. Ltd., New Delhi, pp. 111–20.

Tomar, O.S., Minhas, P.S., Sharma, V.K., Singh, Y.P., and Gupta, R.K. (2002). Performance of 32 tree species and soil conditions in a plantation established with saline irrigation. *Forest Ecol. Manag.* 177(1–3): 333–346.

Tomar, O.S., Minhas, P.S., Sharma, V.K., and Gupta, Raj K. (2003). Response of 9 forage grasses to saline irrigation and its schedules in a semi-arid climate of north-west India. *J. Arid Environ.* 55(3): 533–544.

U.S. Salinity Laboratory Staff. (1954). *Diagnosis and Improvement of Saline and Alkali Soils,* USDA Handbk No. 60, p. 160.

Wichelns, D. and Oster, J.D. (2006). Sustainable irrigation is necessary and achievable, but direct costs and environmental impacts can be substantial. *Agric. Water Manag.* 86: 114–127.

Yadav, H.D. and Kumar, V. (1994). Management of sodic water in light textured soils. In *Proceedings of Seminar on Reclamation and Management of Waterlogged Saline Soils* CSSRI, Karnal, April 5–8, pp. 226–241.

Yaduvanshi, N.P.S. and Swarup, A., (2005). Effect of continuous use of sodic irrigation water with and without gypsum, farmyard manure, pressmud and fertilizer on soil properties and yields of rice and wheat in a long term experiment. *Nutr. Cyclic. Agroecosyst.* 73: 111–118.

Zeng, L., Shannon, M.C., and Lesch, S.M., (2001). Timing of salinity stresses affects rice growth and yield components. *Agric. Water Manag.* 48: 191–206.

12 China's Food Security and Soil Water Management

A Green Water and Blue Water Approach

Li Baoguo and Huang Feng

CONTENTS

12.1 INTRODUCTION

12.1.1 Soil Water: A Central Node Linking FWB and SPAC

Soil water plays a crucial role in regulating farm field water balance (FWB) and soil–plant–atmosphere continuum (SPAC), in which precipitation, surface water, groundwater, irrigation, soil evaporation, and plant transpiration are received and converted by soils of farm fields and exert interactive influences on one another (Heilig et al. 2000; Li et al. 2000). The ultimate goal of soil water management is to increase the efficiency of output for per unit of water, that is, economic output produced by depleting unit volume of soil water. All the water that is consumed by crop growth and development occurs through the uptake of soil water by plant rooting systems, which in turn exert considerable influences on the changes and conversion of soil water. Soil water is thus affected by precipitation, irrigation, percolation, and groundwater. Conversely, soil moisture determines both the intensity and the difficulty of crops' uptake, thereby affecting crop growth and development and eventually the economic output of crops. Hence, soil water is a key to converting all water components involved in both FWB and SPAC.

Investigating soil water in the context of FWB and SPAC is a scale-sensitive issue, with different focuses of study at different spatial scales, that is, pedon, farm field, and region (Table 12.1).

During the past two decades (1990s and 2000s), research and investment communities have been undergoing a paradigm shift in reforming and renovating many long-standing and widely held concepts and ideas in water resource development and management. Principal among these are green and blue water (GBW) (Falkenmark 1997), real water savings (Seckler 1996), water accounting (Molden 1997), net or effective irrigation efficiency (Hsiao et al. 2007), crop water productivity (CWP) (Droogers and Kite 1999), and virtual water flow and trade (Allan 1998; Hoekstra and Hung 2002). The concept of GBW is at the center to differentiate the roles played by natural precipitation and human-withdrawn irrigation water. From the perspective of GBW, soil water is both a receiver and a converter of GBW (Figure 12.1).

This chapter attempts to integrate and synthesize the two paradigms with the objective of tackling soil water and associated food security issues at and across multiple spatial scales, ranging from SPAC to the field, region, and, eventually, at the national level (Figure 12.1). The primary goal of soil water management with respect to crop production and food security is to retain more available soil moisture while minimizing the nonproductive parts of green water flow (i.e., soil evaporation) and maximizing the productive part of it (i.e., plant transpiration).

TABLE 12.1
Research Focuses of Farm Field Water Balance (FWB) and Soil Water Management at Three Spatial Scales

Scale	Area (m^2)	Spatial Distribution	Climatic Factors	Land Use Factors	Geologic Factors
Pedon	10^0–10^1	N.A.	Temporal variation	Irrigation and cropping system	Texture and stratification of soil profiles. On-site groundwater table
Field	10^2–10^6	Spatial variation of soil	Temporal variation	Irrigation and cropping system	Configuration of soil profile and soil types. Multisite groundwater dynamics
Region	>10^6	Spatial variation	Spatial–temporal variation when area >100 km^2	Irrigation/drainage and cropping system	Changes in soil and landforms; regional groundwater

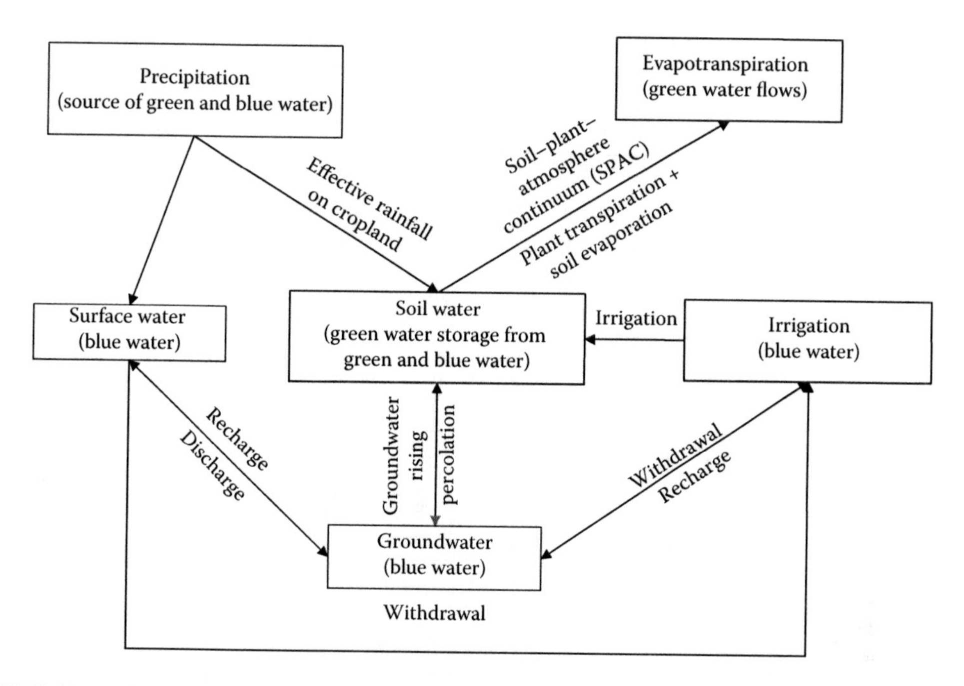

FIGURE 12.1 Illustration of farm field water balance (FWB) and soil–plant–atmosphere continuum (SPAC) by using a green and blue water (GBW) approach. Soil water lies at the center of both FWB and SPAC, with the former being a macro water cycle providing a platform for a micro water cycle of the latter. Both FWB and SPAC start from precipitation, as the ultimate source of GBW, falling on natural water bodies, that is, streams, lakes, and ponds, to form surface blue water. The remaining precipitation falls on various types of terrestrial land covers including cropland, the effective part of which finds its way into soil, together with irrigated blue water withdrawn from either surface or ground sources, to replete the green water storage for a crop's consumption in the form of transpired green water flows, the center of SPAC. Here, soil water is a converging point to receive and convert precipitation, surface water, groundwater, and irrigation water. (Adapted from Li, B.G., Gong, Y.S., and Zuo, Q., *Dynamic Modeling on Farmland Soil Water and Its Applications*. Science Press, Beijing, 2000.)

12.1.2 China's Food and Water Problems

China's food security has long been a controversial topic and at the forefront of the research and consultancy agendas of both global organizations and domestic research and governmental agencies since the sensational proposition raised by Lester Brown (*Who Will Feed China*) in 1994 (Brown 1995). Even though Brown's thesis was based on an incomplete picture and sporadic and scattered information regarding China's crop production, he nonetheless gave a wake-up call to alert China's food security under rising industrialization, urbanization, ecosystem preservation, and, more importantly, growing resource constraint in food production, especially the rapidly shrinking cropland area and growing water scarcity. Water may be among the first and foremost resource constraints limiting China's crop output, which has been researched and recognized over decades (Heilig 2000; Wallace 2000; Varis and Vakkilainenr 2001; Rosegrant et al. 2003; Huang and Li 2010a; Li et al. 2000). Hence, *Who Will Feed China* can be translated into *Who Will Water China*. More frequent droughts in breadbasket regions across the country, especially in typically water-abundant regions and provinces (e.g., the widespread drought in southwestern China in early 2009 and the most recent drought in the five breadbasket provinces [BPs] in the middle and lower reaches of the Yangtze River), are serious concerns. There are numerous alleged reasons accounting for ever-severe drought, and the climate change may be the biggest factor. Whatever the reasons, the trend of growing water scarcity and the resultant threat to China's economic development and food security is apparent and must be addressed.

Thus, the objective of this chapter is to review and report on China's water for food in the recent decade from 1998 to 2008. The specific goal is to discuss the water-for-food problem under the ever growing blue water scarcity by using a conceptual analytical framework established on a soil-water-centered GBW approach and outline a research method coupling a hydrological model with crop and water statistics.

12.1.3 Framework and Concepts in Approaching Water for Food

The framework begins with GBW. The concept of a broadly defined agricultural water resource (BAWR) is proposed as an alternative to the traditional agricultural water resource that only focuses on the "blue" portion of water potentially available for crop use (Figure 12.2). At the

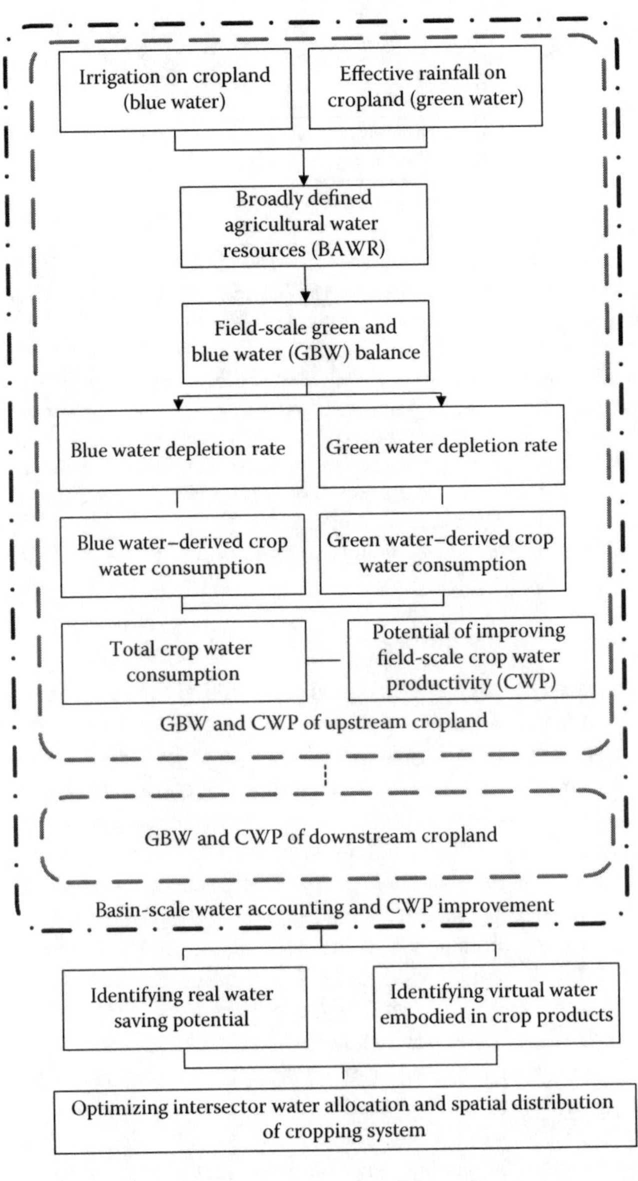

FIGURE 12.2 An integrated framework addressing agricultural water use at multiple spatial scales. (From Huang, F. and Li, B., *Agricultural Water Management*, 97, 1077–1092, 2010a.)

field scale, an insight into FWB is of necessity in allocating and converting more gross input water to crop production. The conversion entails the enhancement of the productive use rates of both irrigated "blue" water and rainfall "green" water. The strategy is to increase the depleted fraction of both blue water and green water as the centerpiece to improve the field-scale CWP. Moving beyond the field scale, the water "lost" from one specific tract of farm field located upstream may be captured and reused at another tract of field downstream. Thus, a system-wide approach to enhance CWP and increase the productive rates of irrigated and rainfall water is essential to quantify the "real water saving" potential at the watershed scale. After determining the watershed-scale CWP and real water saving potential, further investigations into the intersector water allocation and virtual water flows across regions can be realized. Finally, at the national level, more reasonable and optimized water resource allocation schemes and water-limiting cropping systems could thus be formulated (Huang and Li 2010a). This chapter uses this framework as a reference system to conceptualize the issue, analyze problems, and present results regarding China's food and water issues.

This chapter outlines and defines a series of key concepts in approaching agricultural water management issues in China based on GBW and CWP in the framework (Li and Huang 2010).

12.1.3.1 Broadly Defined Agricultural Water Resources

BAWRs encompass the total water that is potentially available for use by crop production (i.e., total soil water supply). It primarily consists of irrigated "blue" water and effective rainfall or soil-held "green" water. It is calculated by using

$$Q_{bawr} = Q_{bw} + Q_{gw}, \tag{12.1}$$

where Q_{bawr} is the total amount of BAWR (km^3/year), Q_{bw} is the irrigated blue water diverted for cropland use (km^3/year), and Q_{gw} is the effective rainfall green water falling on cropland (km^3/year).

Q_{bw} is estimated by using

$$Q_{bw} = Q_{ag} \times p_{ir}, \tag{12.2}$$

where Q_{ag} is the total water withdrawn by agriculture (km^3/year) and p_{ir} is the irrigated water diverted for cropland use as percentage of total agricultural water withdrawal. Total agricultural water withdrawal is obtained from the China Water Resource Bulletin. Of all the agricultural withdrawal, the diverted water for cropland irrigation accounts for a large fraction, ranging from 90% to 95%, varying over space (provinces and basins) and time (years). The remaining 5%–10% is used by livestock, aquaculture, forestry, and domestic activities in rural areas.

By contrast, the estimation of green water is more difficult because of the complexity involved in quantifying the effective rainfall received by a cropland covering a vast area with diverse conditions, that is, location, climate, topography, cropping system, and most importantly, soil chemical, physical, and biological properties. Considering the complexities, a simplified method is proposed to estimate green water that effectively falls on croplands. The effective rainfall that finds its way into the soil body as "green water storage" can be calculated by using

$$Q_{gw} = P_{cr} - R_{cr} - D_{cr}, \tag{12.3}$$

where P_{cr} is the precipitation falling on the cropland (km^3/year), R_{cr} is the runoff on cropland (km^3/year), and D_{cr} is the water drainage percolating out from the soil of the cropland (km^3/year).

P_{cr} can be estimated by using

$$P_{cr} = P_t \times \frac{A_{cr}}{A_{ld}}, \quad (12.4)$$

where P_t is the total precipitation (km^3), A_{cr} is the area of cropland (10^3 ha), A_{ld} is the total land area (10^3 ha), and A_{cr}/A_{ld} is the cropland area as the percentage of total land area. It is assumed that precipitation falls and is distributed evenly on land cover/use types, and hence the amount of precipitation that falls on one specific type of land cover/use is proportional to its share of the total land area.

The cropland runoff R_{cr} in Equation 12.3 is estimated by using

$$R_{cr} = P_{cr} \times \frac{R_t}{P_t}, \quad (12.5)$$

where R_t is the total surface runoff in depth (mm/year) and P_t is the total precipitation in depth (mm/year); thus, R_t/P_t is the coefficient of runoff. It is assumed that the coefficient of runoff is applicable for croplands. The cropland drainage D_{cr} in Equation 12.3 is calculated by using a hydrological modeling approach (Huang and Li 2010a).

12.1.3.2 GBW Share in a Crop's Output

Two fractions constitute the total crop water consumption, that is, blue water–derived actual evapotranspiration (ET_a) and green water–derived ET_a. The relative contribution of both components is thus calculated by using

$$p_{bw} = \frac{ET_{bw}}{ET_a} \times 100\%, \quad (12.6)$$

where p_{bw} is the blue water share in total crop water consumption (%), ET_a is the total actual evapotranspiration consumed to produce the crop's economic yield (mm/year), and ET_{bw} is the blue water–derived actual evapotranspiration (mm/year).

$$p_{gw} = \frac{ET_{gw}}{ET_a} \times 100\%, \quad (12.7)$$

where p_{gw} is the green water share in total water consumption. Similar to Equation 12.6, ET_a is the total evapotranspiration consumed to produce the crop's economic yield, while ET_{gw} is the green water–derived evapotranspiration (mm/year). The calculations of ET_{gw}, ET_{bw}, and ET_a in Equations 12.6 and 12.7 are based on a hydrological modeling approach, and the detailed procedure and results are presented by Huang and Li (2010a).

12.1.3.3 Match of Cropland Acreage and Blue Water and Green Water

The match of cropland and water resources refers to the water resources shared by unit area of cropland, indicating the water endowments possessed by a cropland. It is defined by

$$MAT_{bwr} = \frac{Q_{bwr}}{A_{cr}}, \quad (12.8)$$

where MAT_{bwr} is the match of cropland and water resources (m^3 ha/year) and Q_{bwr} is the total internally renewable water resource (IRWR) (km^3/year). Similarly, the match of cropland and irrigated water is calculated by using

$$MAT_{bw} = \frac{Q_{bw}}{A_{cr}}, \tag{12.9}$$

where MAT_{bw} is the match of cropland and irrigated blue water (m^3 ha).

Both indicators cannot, however, reflect the complete picture of match of cropland and the water resources that are potentially used by a cropland. Hence, a novel indicator, the match of cropland and BAWR and MAT_{bawr} (m^3/hm^2), is proposed in

$$MAT_{bawr} = \frac{Q_{bawr}}{A_{cr}}, \tag{12.10}$$

where MAT_{bawr} encompasses both the green and blue water components that are potentially used by a cropland, since its blue water parts come from the actual volume of irrigation diverted for a cropland and its green water parts originate from the direct calculation of effective rainfall received by a cropland. The three indicators are crucial in revealing the relationship between water and cropland.

12.1.3.4 Crop Water Productivity

CWP in this study is defined according to

$$CWP_{bs} = \frac{Y_c}{ET_a}, \tag{12.11}$$

where CWP_{bs} is the crop water productivity at basin scale (kg/m^3), Y_c (kg) is the crop yield under study in the basin, and ET_a is the actual evapotranspiration consumed by crop production in the basin (m^3). ET_a consists of two components: one is the actual evapotranspiration arising from irrigated water or blue water–derived ET_a and the other is the actual evapotranspiration from effective rainfall or green water–derived ET_a.

12.2 METHOD, STUDY AREA, AND DATA

12.2.1 Method

A hydro-model-coupled-statistics approach was developed to analyze grain production and associated GBW cycles (Huang and Li 2010a). The method is based on a hydrological modeling soil and water assessment tool (SWAT) (Arnold et al. 1998) and crop production and water-use statistics. SWAT was employed as the modeling tool to investigate the water cycles that are associated with crop production, that is, rainfall, runoff, evapotranspiration, available soil water, etc. Statistics on crop production were collected and compiled. A geographic information system (GIS)-based spatial aggregation and disaggregation method was developed to group the provincial statistics to basin level. Through dividing basin-level crop statistics by basin-scale crop's actual evapotranspiration, basin-scale CWP was computed. The method itself has been subjected to extensive calibration and validation. The results show that SWAT can simulate basin-level water cycles, that is, precipitation, runoff, and soil water, with acceptable accuracy after calibration and validation using monthly river discharge and soil moisture monitoring across the country. The calculated CWPs were also validated by comparing them with the results of similar studies. The results of such a comparison showed that the calculated CWPs were within reasonable ranges. In summary, the method developed herein is appropriate for estimating basin-scale CWP and associated GBW use and depletion (Huang and Li 2010a).

12.2.2 Study Area

The study area covers all administrative provinces of China, with special attention given to 13 BPs (Figure 12.3) that play pivotal roles in China's grain production and food security (Table 12.2). As for the top four grains of China (i.e., rice, wheat, maize, and soybean), 13 BPs produce the majority of outputs on a large fraction of the area sown. For the top four crops combined, the BPs produce 74.2% of the total outputs on 72.2% of the total sown areas. Hence, the 13 BPs have and will continue to exert enormous influence on China's food production, and their water situations will directly affect China's food security.

12.2.3 Data

The research requires an enormous amount of data, covering biophysical, socioeconomic, and officially released statistics data. A more detailed description of the collection and compiling is presented in the research work done by Huang and Li (2010a). This chapter primarily focuses on the description of the water-use bulletin data. The China Ministry of Water Resource (MOWR) has officially released Water Resource Bulletin of China (WRB) on a yearly basis since 1998. The administrative agencies of nine highest-order river basins and provincial water resources management agencies also release yearly WRBs. China irrigation monitoring dataset contains over 300 monitoring stations scattered across the country, in which the irrigation volume for major food and fiber crops (i.e., rice, wheat, maize, and cotton) is documented.

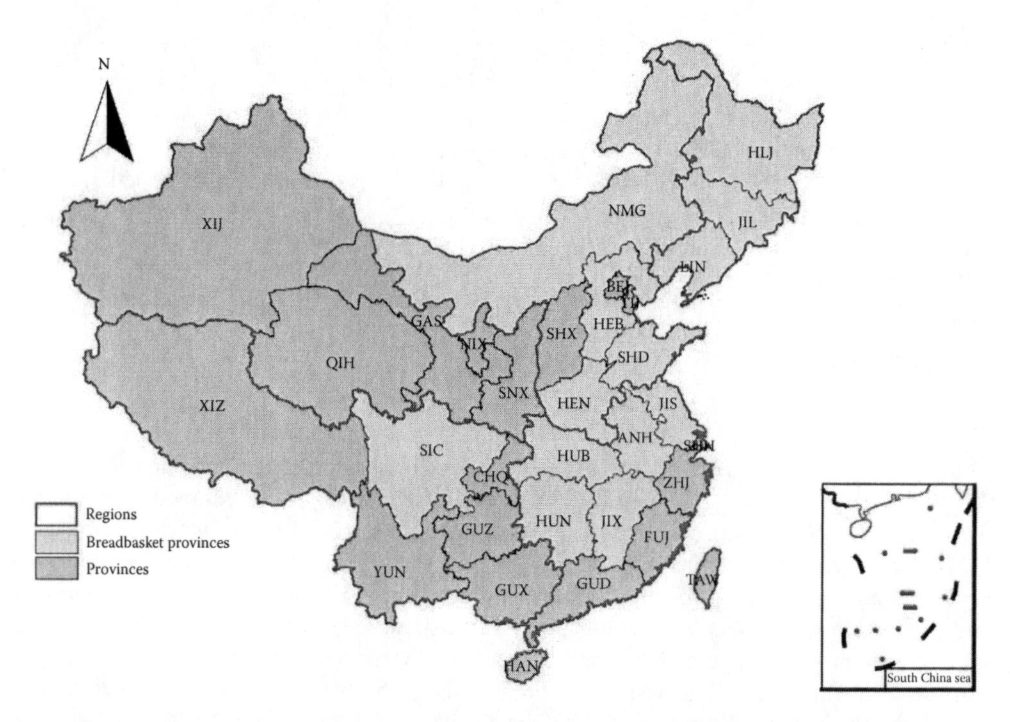

FIGURE 12.3 China's regional delineation and breadbasket provinces. Northeast China covers *Heilongjiang (HLJ)*, *Jilin (JIL)*, and *Liaoning (LIN)*. North China includes Beijing (BEJ), Tianjin (TIJ), *Hebei (HEB)*, Shanxi (SHX), *NeiMengGu (NMG*, or *Inner Mongolia)*, *Henan (HEN)*, and *Shandong (SHD)*. Southeast China embraces Shanghai (SHH), *Jiangsu (JIS)*, *Anhui (ANH)*, Zhejiang (ZHJ), *Jiangxi (JIX)*, *Hubei (HUB)*, *Hunan (HUN)*, Fujian (FUJ), Guangdong (GUD), and Hainan (HAN). Southwest China encompasses Guangxi (GUX), Yunan (YUN), *Sichuan (SIC)*, Chongqing (CHQ), Guizhou (GUZ), and Xizang (XIZ, Tibet). Northwest China comprises Shannxi (SHX), Gansu (GAS), Ningxia (NIX), Qinghai (QIH), and Xinjiang (XIJ). The *italics* symbolize the 13 breadbasket provinces (BPs).

TABLE 12.2
Role Played by Breadbasket Provinces (BPs) and Non-Breadbasket Provinces (NBPs) in China's Grain Crop Production

	A_r (%)	O_r (%)	A_w (%)	O_w (%)	A_m (%)	O_m (%)	A_s (%)	O_s (%)	A_g (%)	O_g (%)
BPs	65.0	68.0	75.0	80.7	74.1	76.7	82.3	84.8	72.2	74.2
NBPs	35.0	32.0	25.0	19.3	25.9	23.3	17.7	15.2	27.8	25.8

Note: A denotes area; O: output; r: rice; w: wheat; m: maize; s: soybean; g: grain. Grain refers to the four crops combined.

12.3 RESULTS AND DISCUSSION

12.3.1 China's Water Resources between 1998 and 2008

Precipitation over the last decade declined slightly. Consequently, the IRWR also declined nonlinearly with precipitation, but at a more rapid pace (Figure 12.4). Precipitation is the ultimate source of all GBW. If the declining trend of precipitation and IRWR continues into the next decade, it will definitely affect the agricultural production that relies heavily on precipitation and an IRWR from which the irrigated blue water is withdrawn. Moreover, the relationship between precipitation and the IRWR is nonlinear, with the falling rate of the IRWR being faster than that of precipitation due to the complexity of the climatic and land use/cover change underpinning the basic mechanisms of renewable water resources.

The higher interannual variability of both variables may further exacerbate the problems of their declining trend. For the coefficient of variation (c.v.) of precipitation, six BPs fall at or well below the median c.v., 0.13, with the remaining seven BPs falling above the median value (Figure 12.5). But most of the BPs are either at and below or above the median (0.14, 0.15), with only NMG, HEN, and SHD being far above the median, with values of 0.20, 0.21, and 0.22, respectively. BPs with above-median c.v.s are also those with annual precipitation less than 600 mm, with the exception

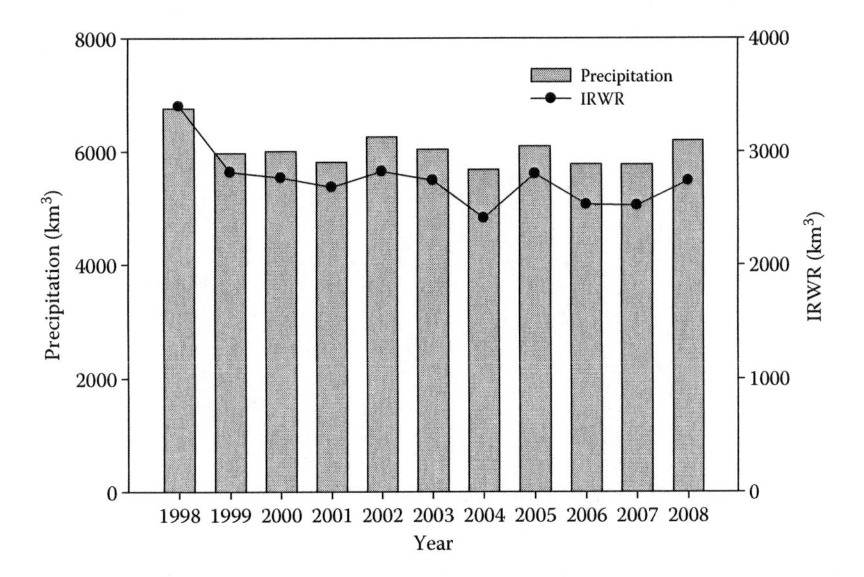

FIGURE 12.4 Precipitation and internally renewable water resources (IRWRs) of China during 1998–2008.

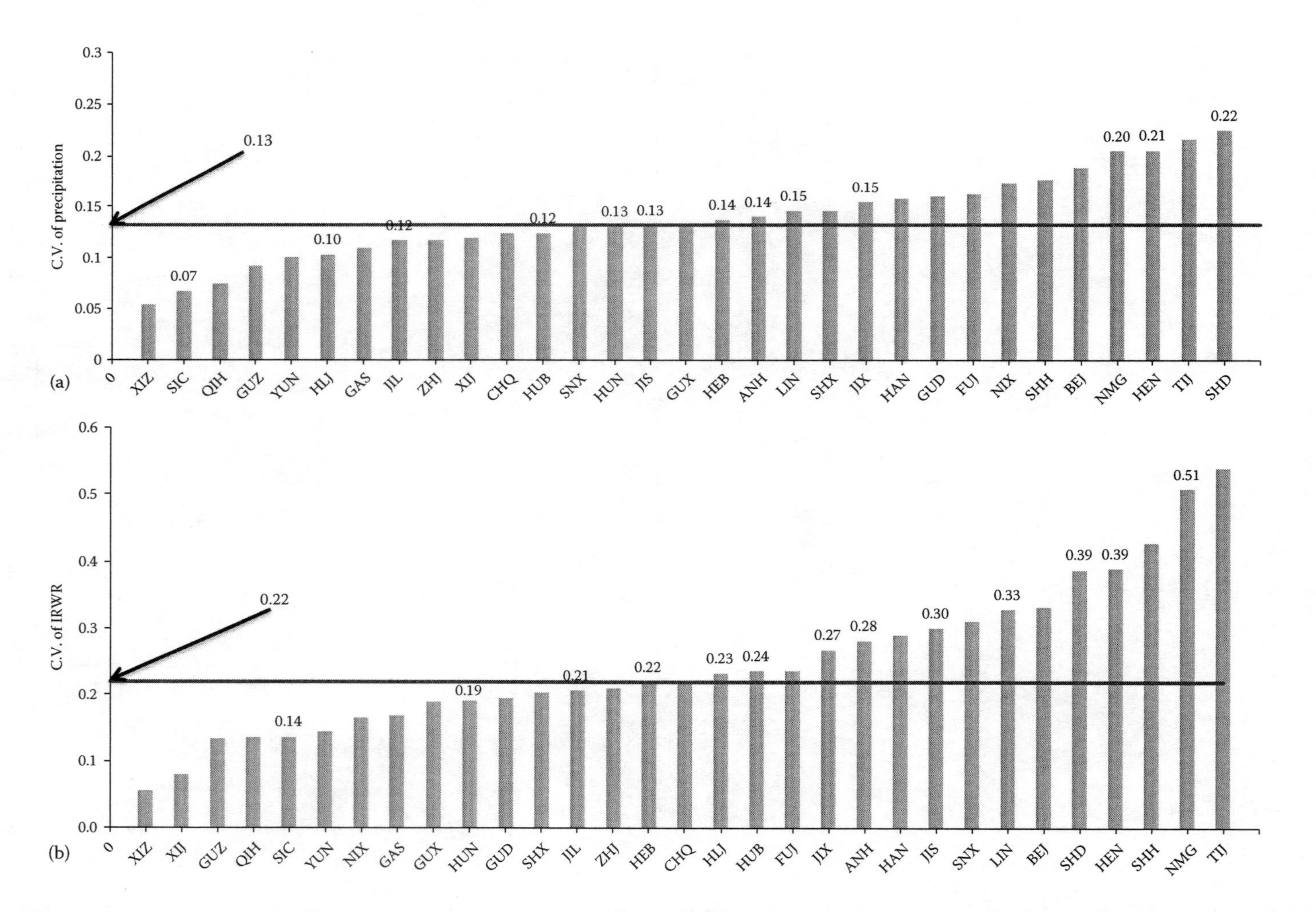

FIGURE 12.5 Coefficient of variation (c.v.) for provincial-level precipitation and IRWR. The bars topped by numerical labels indicate the c.v. of breadbasket provinces. Horizontal lines are the median of precipitation c.v. and IRWR, respectively. The black arrows indicate the median values of c.v. of precipitation and IRWR, respectively, in panels (a) and (b).

of ANH and JIX, which are all located in the water-rich Huai and Yangtze Basins. But ANH and JIX need no attention because of their already high level of rainfalls, all exceeding 1000 mm/year.

As for the c.v. of the IRWR, most of the BPs are well and far above the median level, with only four BPs being below the median c.v. that is, SIC, 0.14; HUN, 0.19; JIL, 0.21; and HEB, 0.22. HEB is a crucial BP in North China. Its precipitation is around 450–600 mm, and it has a high interannual variability. Its IRWR is relatively stable because it is at a fairly low level. As presented later in Section 12.3.1.3, the water supply in HEB relies primarily on the source. When combining the c.v. of precipitation and the IRWR, the NMG, HEN, and SHD are major issues because they all are highly variable. In sum, China's ultimate source of GBW is already showing a declining trend and high variability.

Agricultural water withdrawal is the blue portion of the water that is potentially used by the agricultural sector. Over the last decade, total water withdrawal in China increased continuously and will continue to increase in the coming decades with the rising demands from industrialization, urbanization, ecosystem preservation, and environmental protection. The share of industrial and domestic withdrawal increased continuously between 1998 and 2008, whereas the share of agricultural withdrawal declined at a more rapid pace (Figure 12.6). In 2004, the China Water Resource Bulletin first reported water withdrawal for ecosystem and environmental use. Since then, water for ecosystem preservation has increased dramatically. Though the volume of agricultural water withdrawal fluctuates around 380 km^3, its share in total water withdrawal definitely declined from 69.3% in 1998 to 62.0% in 2008, with an average decline of 10.5%. During the same period, both industrial and domestic withdrawal experienced increased growth rates of 16.3% and 22.1%, respectively.

12.3.1.1 Groundwater Use

At the country level, the relative share of surface water and groundwater in the total water supply for the decade of 1998–2008 remained fairly constant, being approximately 81% and 19%, respectively. Groundwater has been the most important source of water supply in grain production in China, especially BPs located in north and northeast China (Figure 12.7).

HEB had an extremely high level of groundwater supply, and consequently its grain outputs relied heavily on the over tapped groundwater. The widespread expansion of drawdown cones posed great threats to HEB's sustainable crop production, and hence it will affect food security not only for itself but also for many other provinces relying on its food exports. The No. 2 groundwater supplier in BPs was HEN, followed by NMG and SHD, all severely water-scarce provinces (the three provinces also showed the highest variability in precipitation and IRWR). Most BPs in the Yangtze Basin relied principally on surface water sources due to their benign surface water conditions and unfavorable conditions to form aquifers and tap groundwater.

12.3.2 BAWR and Crop Production

12.3.2.1 BAWR and Its Blue/Green Water Components

As defined in Section 12.1.2, BAWR broadened the horizons of water that is potentially used for crop production. The share of GBW in BAWR represents the relative contributions of the two components in agricultural production. The year 2008 is taken as a case study to illustrate the calculation of BAWR and its green/blue water share. In 2008, China as a whole possessed 786.5 km^3 of BAWR, of which the effective precipitated green water falling on cropland was 456.1 km^3, or 58.0% of BAWR, and irrigated blue water diverted to cropland was 330.4 km^3, or 42.0% of the total BAWR (Table 12.3).

Regional results also revealed that the share of green water in most regions exceeded that of blue water, with the only exception being northwest China, where the extremely arid climate produced much less natural rainfalls on cropland (Table 12.3). Irrigated water withdrawn primarily from

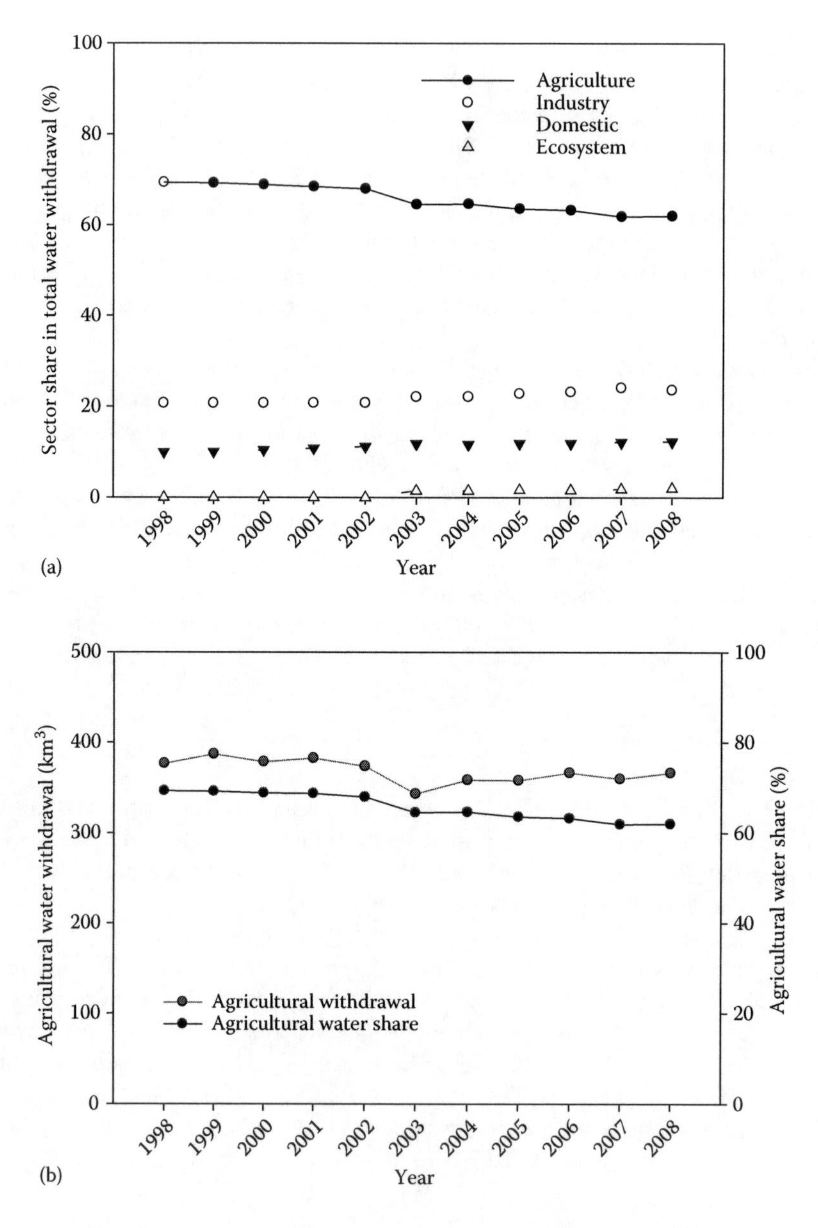

FIGURE 12.6 (a) Sector-based share in total water withdrawal and (b) agricultural withdrawal volume and share.

either geo-historic aquifers or snowmelts of snow-capped mountains is the first and foremost water source for crop growth in northwest China.

Match of the cropland and water resources reflects to some extent the need for water requirements in croplands. Two levels of match between water and cropland (i.e., MAT_{bwr} and MAT_{bw}) are widely used. However, they reflect only the match between water and cropland (i.e., blue water portion) while neglecting the green water portion and its role in meeting cropland water requirements. The national match of water and cropland reveals that there exist major gaps among three indictors. If measured by MAT_{bwr}, the national value is 20,752 m^3/ha; if measured by MAT_{bw}, the value is 2670 m^3/ha; and if measured by MAT_{bgw}, the value is 6213 m^3/ha.

Matching croplands and waters resources in the form of either blue water or blue plus green water can be interpreted from another perspective, that is, the gap between cropland share and

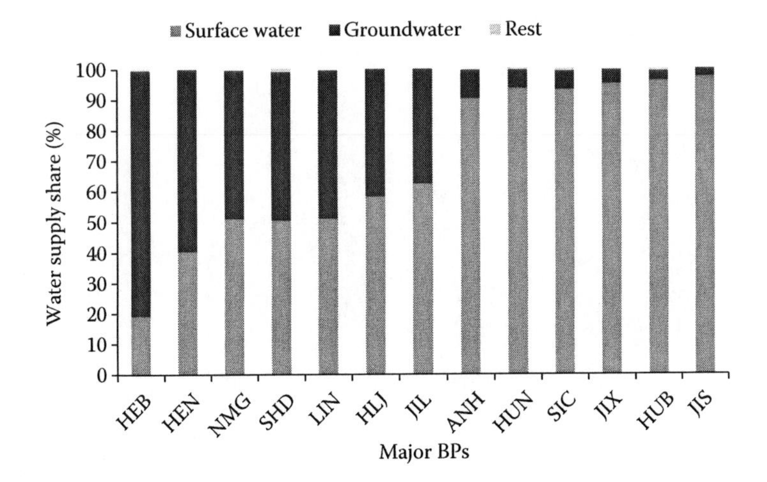

FIGURE 12.7 Relative shares of surface, ground, and rest sources of water supply for 13 BPs. The rest sources of water supply primarily refer to treated reused water and desalinized water. Abbreviations of BPs (breadbasket provinces) refer to Hebei (HEB), Henan (HEN), NeiMengGu (NMG, or inner Mongolia), Shandong (SHD), Liaoning (LIN), Heilongjiang (HLJ), Jilin (JIL), Anhui (ANH), Hunan (HUN), Sichuan (SIC), Jiangxi (JIX), Hubei (HUB), and Jiangsu (JIS).

water share for one specific region (Figure 12.8). If measured by cropland and IRWR, north China sustained 28% of the country's total cropland by accessing only 5% of the total IRWR of the country, which is a huge gap between the cropland and its supporting water endowments. If, however, measured by MAT_{gbw}, north China cultivated the same amount of cropland by using 22% of the country's total BAWR, and the gap is dramatically closed. The same is also true of northeast China, where 18% of the cropland is sustained by 13% of the BAWR rather than 4% of the IRWR. Hence, the core of BAWR is that it takes both green and blue portions of the water that is potentially accessed and used for crop production into consideration, thus reflecting a more complete picture of crop water use.

12.3.2.2 Crop Water Consumption and Its Blue/Green Water Components

As defined previously, green water share in a crop's output refers to the percentage of green water–derived (available soil moisture) evapotranspiration (ET) relative to crop-output-associated ET, while blue water share refers to the percentage of irrigated blue water–derived ET in the total crop

TABLE 12.3
Country- and Region-Level BAWRs and Their Constituents in 2008

Region	Precipitation (mm)	Cropland Area (10^3 ha)	Cropland Precipitation (km³)	Cropland Irrigation (km³)	BAWR (km³)	Green Water Share (%)	Blue Water Share (%)
Country	654.8	121,715.9	456.1	330.4	786.5	58.0	42.0
North	405.1	33,634.8	113.1	58.4	171.4	66.0	34.1
Northeast	528.9	21,450.0	64.8	37.3	102.1	63.5	36.5
Southeast	1481.1	28,827.7	148.0	123.2	271.3	54.6	45.4
Southwest	960.2	23,319.8	103.8	49.7	153.5	67.6	32.4
Northwest	238.0	14,483.5	26.5	61.8	88.3	30.0	70.0

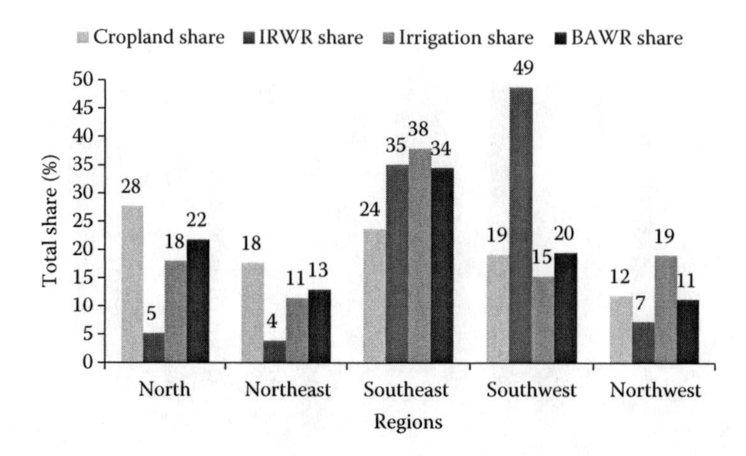

FIGURE 12.8 Match of cropland area and water resources as measured by the gap between cropland share and water share in various forms.

ET. The result showed that green water represents most (59%) of the total ET consumed in forming the crop's output (Figure 12.9), corresponding to the relative share of GBW in BAWR.

12.3.2.3 Grain-Associated Water Consumption and Water Productivity

China's grain production in the decade 1998–2008 experienced dramatic changes: after touching a record high in 1998, grain output has slumped since then down to the lowest value of the decade in 2003, after which crop production regained growth, and the growing trend continued until 2008, when the total output surpassed that of 1998. Correspondingly, the total water consumption accompanying that output also touched a record high in 2008, while the lowest value was also observed in 2003 (Figure 12.10). Overall, grain-derived water consumption had a good match with grain production, which had significant implications for the crop–water relation in that the additional grain output does necessarily require additional water consumption.

CWP defined herein is the unit weight of grains produced by unit volume of actual evapotranspiration consumed by crops. For the decade 1998–2008, China's CWP remained almost constant, with a slightly rising trend. There were numerous control factors determining CWP, with yield being one of the most important controls (Huang and Li 2010b). CWP and yield were strongly correlated for the period 1998–2008 (Figure 12.11). The gains in CWP accompany considerable water savings achieved by improved efficiency of water use, that is, producing more grains by using per unit of depleted or consumed water in the form of ET.

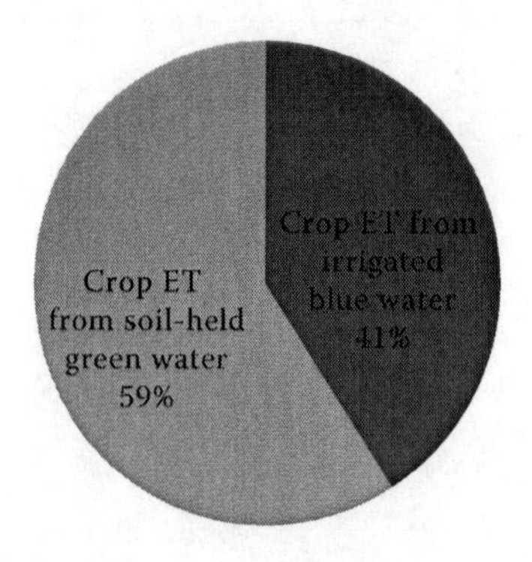

FIGURE 12.9 The share of green and blue water in total crop water consumption.

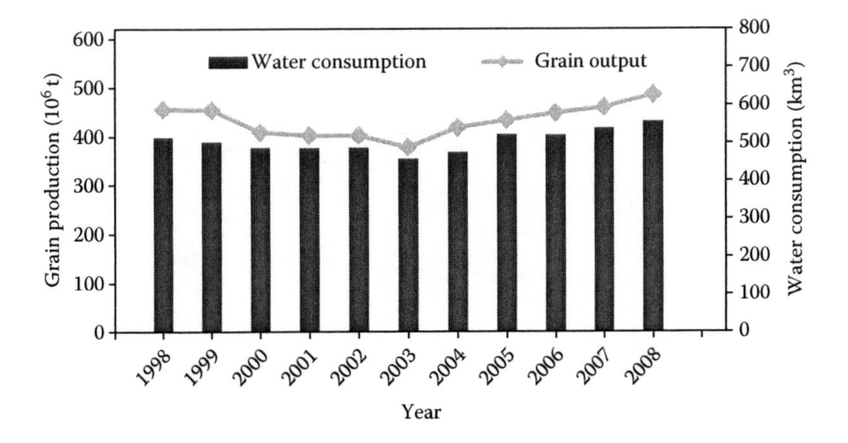

FIGURE 12.10 Grain output and associated total water consumption during 1998–2008.

12.3.3 Blue and Green Water Management in China

12.3.3.1 Water-Saving Practices for Irrigated Blue Water

China's irrigated cropland area expanded substantially over 1992–2008, but its share in the total cropland area declined modestly over the same period, indicating a growing scarcity of blue water and consequentially a shrinking share and slowing growth rate of irrigated cropland area.

Over the last decade, engineering-based water-saving practices were widely adopted and the use of such practices in irrigated cropland increased substantially (Figure 12.12). Widely used technologies are sprinkler irrigation, drip irrigation, micro-irrigation, and canal lining.

At the country level, engineering-based water-saving practices accounted for 35% of the total cropland area under irrigation (Figure 12.13). It is understandable that northwest China, the most arid region, had the highest adoption rate of engineering-based water-saving technologies, followed by north China, which is the second driest region of the country. Ironically, BPs had an overall lower adoption rate of water-saving practices than those of non-breadbasket provinces (NBPs), possibly and primarily because most Yangtze Basin–located BPs had quite a low adoption rates of water-saving practices.

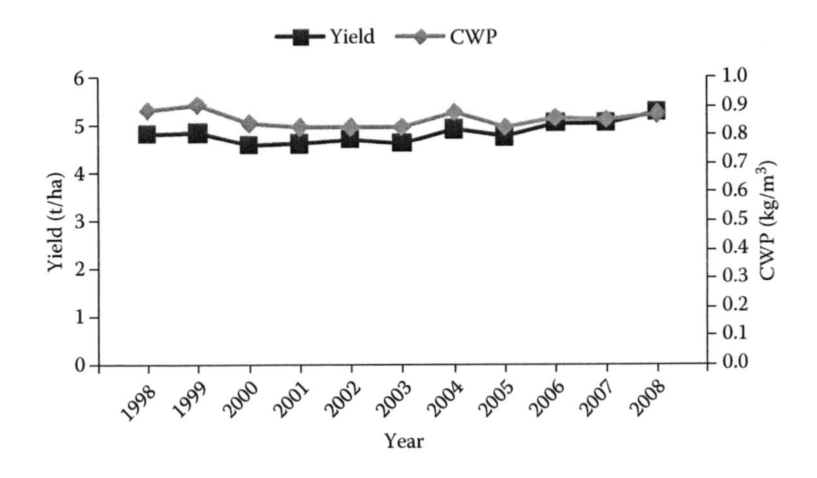

FIGURE 12.11 Crop water productivity changes during 1998–2008.

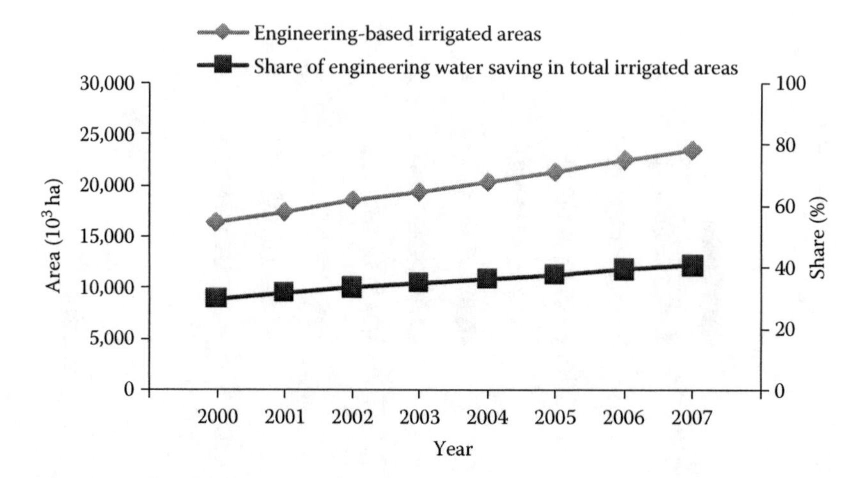

FIGURE 12.12 Adoption area and its share in the total irrigated acreage of engineering-based water-saving practices. The engineering-based water-saving practices included (1) sprinkler irrigation, (2) micro-irrigation, (3) drip irrigation, (4) canal lining, and (5) other type of engineering-based technologies.

12.3.3.2 Water-Saving Practices for Soil-Held Green Water

In contrast to engineering-based systems, soil-crop-management-based (or management-based) water-saving practices are commonly referred to as low-cost and handy-to-operate on-farm technologies or techniques requiring no large-scale infrastructure investment. Another distinct feature of management-based practices, as the term implies, is that they place great importance on adapting and adjusting crop and soil management practices to suit the crop's water requirements and to retain more available soil moisture for crop use by taking the crop's biological, physiological, and genetic traits into consideration. The representative management-based water-saving techniques collected and summarized herein may result in more efficient use of water through (1) improving the drought-resistant ability of crop varieties (i.e., drought-resistant varieties), (2) altering and enhancing the soil texture to hold more soil-available moisture by altering tilling practices (i.e., mechanical deepening and loosening tillage, ditch and ridge tillage), (3) covering the soil surface to

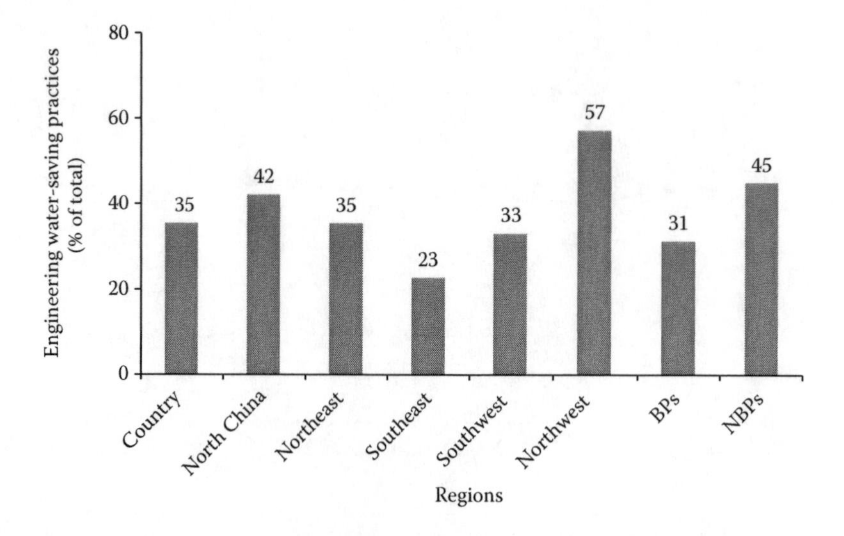

FIGURE 12.13 The regional adoption rate of engineering-based water-saving practices.

prevent the nonproductive evaporative loss of soil moisture (i.e., straw and stalks coverage, plastic sheet coverage), (4) rescheduling the time and volume of irrigation to match exactly with the crop's water requirements (i.e., rescheduling irrigation, supplemental irrigation, sowing with seed-specific watering), and (5) restructuring the cropping systems to match with natural rainfall (i.e., restructuring the cropping pattern).

Of all the management-based water-saving techniques, drought-resistant varieties received the most widespread application, followed by tillage- and coverage-type water-saving practices (Figure 12.14). Irrigation- and watering-related techniques came at the fourth place in technology as measured by the adoption rate. Regionally, northeast China had the highest application rate, while southeast China had the lowest, with north China (21.2%), southwest China (13.9%), and northwest China (11.1%) sequentially lying in-between. A comparatively low adoption level of management-based techniques in northwest China relative to its high application rate of engineering-based techniques (57%) may be due to (1) its poorer soil and water conditions that are vulnerable to perturbations by tillage practices and (2) its low biomass not enough to cover soil surface.

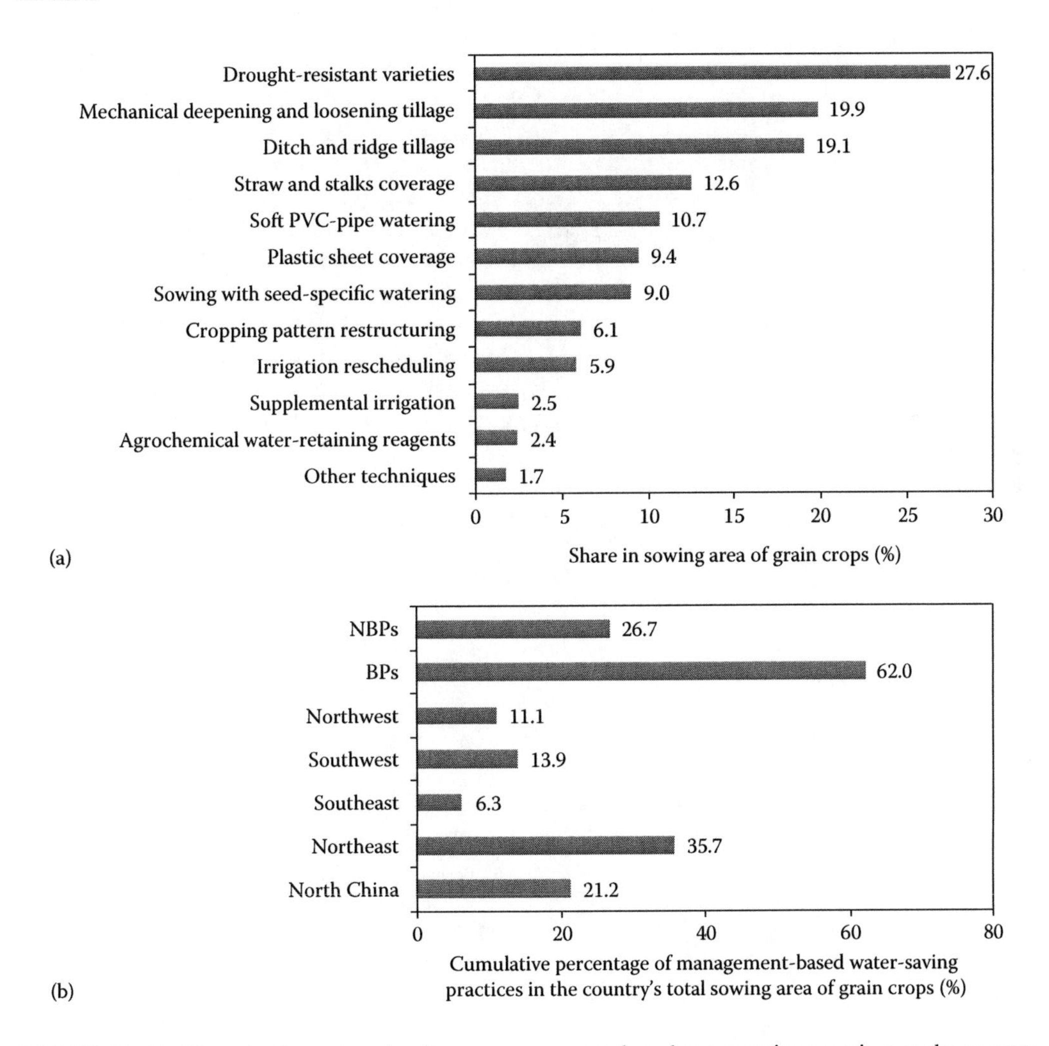

FIGURE 12.14 The adoption rates of soil-crop-management-based water-saving practices at the country, region, BP, and NBP levels. Panel (a) indicates the technology-based distributions of these practices and panel (b) indicates the region-based adoption rates as measured by the cumulative percentage of management-based water-saving practices in the total sowing area of grain crops.

12.3.4 Discussion

The year 2030 may be marked as a watershed year for China's development because of the projected peak in its population at 1.6 billion. Thus, the first three decades of the twenty-first century are crucial to China's economic development, societal progress, and improved living standards for its large population. The same is true of China's food security, achieving it on ever-diminishing natural resource endowments (i.e., cropland, water, nutrient, and energy), of which water may be the largest constraint limiting China's crop production. During the decade of 1998–2008, precipitation declined slightly, and correspondingly the IRWR also decreased albeit at a somewhat faster rate. Both the absolute volume of agricultural withdrawal and its share of total withdrawal continuously declined, demonstrating that China's crop production is already facing increasing competition from other sectors, especially from industrialization, urbanization, and ecosystem restoration and environmental protection. It can be expected that the share of agricultural withdrawal would steadily decrease with the volume of withdrawal remaining constant or increasing slightly in the years to come. Groundwater has been extensively exploited and withdrawn in China, especially for irrigated crop production in water-scarce northern China (i.e., north China, northeast China, and northwest China). As mentioned by Postel (2008), cereal output relying on geo-historic groundwater was analogous to the sub-loan crises, the bubbles of which would eventually burst due to the unsustainable tapping and the consequential depletion of deep aquifers. Hence, in terms of blue and green water, the blue water resource will decline for the agricultural sector in general and crop production in particular. Thus, there is a strong need for an even view of the strength of sustainable management of agricultural water resources (i.e., soil water management aside from irrigation management).

By differentiating GBW, a BAWR is proposed to account for all potential available water that might be used for crop production. Of all BAWRs, green water accounted for 58% and blue irrigated water for the remainder 42%, indicating the major role played by green water in crop production. The analysis of data on consumptive water use and its blue/green water share supports the conclusion arrived at by BAWR partitioning that the soil-held green water–derived ET represents 59% of the total ET and the irrigated blue water–derived ET accounts for the remaining 41%. Both BAWR and water consumption analysis demonstrated the crucial roles of soil-held green water in crop production.

The most striking finding from the BAWR-based cropland–water match is the fact that the gaps or deficits between cropland and water were not as severe as previously estimated and conceived in most water-stressed regions of China when the addition of green water is taken into consideration. For instance, north China has always been considered to support 28% of the total cropland by using only 5% of the IRWR and 18% of the total irrigation. But in fact, north China sustained that same share of cropland by using 22% of the total BAWR, indicating that water resources in forms of both blue water and green water provided a firm foundation for its grain production. Though the gap between land and water still exists, it is not as large as previously estimated. Hence, BAWR is an appropriate indicator to assess water endowments for crop production in a specific region.

Grain production requires an enormous amount of water consumption. The period between 1998 and 2008 was typical of China's grain output and associated water consumption, which matched well with the fluctuation of grain outputs and illustrated that the additional grain production does imply extra water consumption. These data have strong implications for China's food security and water savings. To feed its 1.3 billion people at present and 1.6 billion in 2030, China must continue to increase its crop yields and total grain production. Given the ever-shrinking resource endowments allocated to crop production, more efforts must be directed toward the best management of soil, water, nutrients, and energy. Low-cost and handy-to-operate management-based water-saving practices must be strengthened in the future. Management-based water savings have been widely practiced on most of the grain crops throughout the BPs. In the meantime, the management of irrigated blue water is equally important due to the mismatch between natural rainfalls and water demands at different stages of crop growth. Irrigated blue water will continue to play a pivotal role in stabilizing and enhancing crop yield and total production. Over the last two decades, the adoption rate of engineering-based water-saving practices has steadily increased in China's cropland, with the canal lining being the most promising technology. But from the point of view of a larger spatial scale (i.e., watershed, river basin, and region), water saved by canal lining is a myth rather than "real" water savings because the water "saved" by upstream users may be reused by downstream users. Therefore, scientists, engineers, and decision makers in irrigation water management should direct more efforts toward water-saving technologies to ensure real water savings.

The major pathway of realizing real water saving is to increase the output per unit of input or water productivity. Between 1998 and 2008, even with the growing scarce precipitation and IRWR, China has been able to feed its large population. Since 1998, CWP experienced a modest increase, with a slightly improved crop yield. CWP of major grain crops was 0.854 kg/m^3. North China has the highest CWP of 1.22 kg/m^3, followed by that of southeast China at 0.896 kg/m^3. Northwest China has the lowest CWP of 0.394 kg/m^3. The years 2005 and 2006 may illustrate typically the real water savings achieved by CWP gains. The yield increased from 4.77 t/ha in 2005 to 5.07 t/ha in 2006, with a 0.30 t/ha of improvement, while the CWP improved from 0.832 to 0.861 kg/m^3, with a 0.03 kg/m^3 improvement. The total grain output in 2005 was 434 Mt, with a corresponding water consumption of 522 km^3, and the total output in 2006 was 448 Mt, with an associated 521 km^3 of water consumption. If the CWP of 2006 had not increased to 0.861 kg/m^3, the water consumption accompanying the grain output would be 539.7 km^3. Hence, the real water saving from raising the CWP is 18.81 km^3, which is realized through reducing 539.70–520.89 km^3.

12.4 CONCLUSION

Despite numerous constraints (i.e., climate change, industrialization, urban expansion, and ecosystem and environmental protection), China's water use for food may definitely fall in the years to come, especially the irrigated blue water in BAWR. Hence, the focus of water for food should be broadened to embrace not only blue water but also green water.

BAWR constitutes both green and blue water components, and green water represents a large part of all the water that is potentially available for crop use. And a match of BAWR and cropland can reflect comprehensively land and water endowments for crop production, confirming the strategy that "land use decision is a water use decision." The protection of cropland areas has severe implications to crop water use in general, and green water in particular, since croplands are the most important receivers and converters of naturally precipitated and soil-held green water.

On the basis of the unit weight of the crop produced, green water represents the largest component. Hence, strengthening green water management must be given a high priority in the years to come. Additional grain production implies additional water consumption. An efficiency-oriented real water-saving pathway (i.e., gains in CWP) is a viable option in achieving food security while slowing down the pace of growth in water consumption. In summary, a win-win solution for China's food and water security may lie in the best management of soil water, incorporating both blue and green water components.

ABBREVIATIONS

BAWR	Broadly defined agricultural water resources
BPs	Breadbasket provinces
BW	Blue water
CWP	Crop water productivity
FWB	Farm field water balance
GBW	Green and blue water
GW	Green water

IRWR	Internally renewable water resources
MOWR	China Ministry of Water Resources
NBPs	Non-breadbasket provinces
SPAC	Soil–plant–atmosphere continuum
SWAT	Soil and water assessment tool
WRB	Water Resource Bulletin of China

REFERENCES

Allan, J.A. 1998. Virtual water: A strategic resource global solutions to regional deficits. *Groundwater* 36: 545–546.

Arnold, J.G., Srinivasan, R., and Muttiah, R.S. 1998. Large-area hydrologic modeling and assessment: Part I. Model development. *Journal of American Water Resource Association* 34(1): 73–89.

Brown, L. 1995. *Who Will Feed China?— A Wake-Up Call for a Small Planet.* World Watch Environmental Alert Series. New York: W.W. Norton & Company.

Droogers, P. and Kite, G. 1999. Water productivity from integrated basin modeling. *Irrigation and Drainage Systems* 13(3): 275–290.

Falkenmark, M. 1997. Meeting water requirements of an expanding world population. *Philosophical Transaction of Royal Society of London Series B* 352: 929–936.

Heilig, G.K. et al. 2000. Can China feed itself?: An analysis of China's food prospects with special reference to water resources. *Int. J. Sustain. Dev. World Ecol.* 7: 153–172.

Hoekstra, A.Y. and Hung, P.Q. 2002. Virtual water trade: A quantification of virtual water flows between nations in relation to international crop trade. Value of Water Research Report Series No. 11. The Netherlands: UNESCO-IHE, Delft.

Hsiao, T.C., Steduto, P., and Fereres, E. 2007. A systematic and quantitative approach to improve water use efficiency in agriculture. *Irrigation Science* 25: 209–231.

Huang, F. and Li, B. 2010a. Assessing grain crop water productivity of China using a hydro-model-coupled-statistics approach. Part I: Method development and validation. *Agricultural Water Management* 97: 1077–1092.

Huang, F. and Li, B. 2010b. Assessing grain crop water productivity of China using a hydro-model-coupled-crop-statistics approach. Part II: Application in breadbasket basins of China. *Agricultural Water Management* 97: 1259–1268.

Li, B.G. and Huang, F. 2010. Trends in China's agricultural water use during recent decade using the green and blue water approach. *Advances in Water Science* 21: 574–583 (in Chinese).

Li, B.G., Gong, Y.S., and Zuo, Q. 2000. *Dynamic Modeling on Farmland Soil Water and Its Applications.* Science Press, Beijing (in Chinese).

Molden, D.J. 1997. Accounting for water use and productivity, SWIM paper 1, system-wide initiative for water management. Colombo, Sri Lanka: International Water Management Institute.

Postel, S. 2008. Avoiding a water crisis. *CSA News.* V53N12, 2–5.

Rosegrant, M.W., Cai, X.M., and Cline, S.A. 2003. *World Water and Food to 2025: Dealing with Scarcity.* Washington, DC: International Food Policy Research Institute.

Seckler, D. 1996. The new era of water resources management: From 'dry' to 'wet' water savings. Research Report 1. Colombo: International Irrigation Management Institute.

Varis, O. and Vakkilainenr, P. 2001. China's 8 challenges to water resources management in the first quarter of the 21st century. *Geomorphology* 41: 93–104.

Wallace, J.S. 2000. Increasing agricultural water use efficiency to meet future food production. *Agriculture, Ecosystems and Environment* 82: 105–119.

Section IV

Agronomic Management of Soil and Crop

13 Sustainable Management of Scarce Water Resources in Tropical Rainfed Agriculture

Suhas P. Wani, Kaushal K. Garg, Anil Kumar Singh, and Johan Rockström

CONTENTS

13.1 ACHIEVING GLOBAL FOOD SECURITY IS A CHALLENGE

Ensuring global food security for the ever-growing population that will cross 9 billion by 2050 and reducing poverty are challenging tasks. Growing per capita income in the emerging giant economies such as Brazil, Russia, India, and China (BRIC) implies increased additional pressure on global food production due to changing food habits. The increased food production has to come from the available and limited water and land resources, which are finite. The quantity of neither available water nor land has increased since 1950, but the availability of water and land per capita has declined significantly due to increase in global human population. For example, in India, per capita water availability has decreased from 5177 m^3 in 1951 to 1820 m^3 in 2001 due to increase in population from 361 million in 1951 to 1.02 billion in 2001, which is expected to rise to 1.39 billion by 2025 and 1.64 billion by 2050 with associated decrease in per capita water availability of 1341 m^3 by 2025 and 1140 m^3 by 2050, respectively. Distribution of water and land varies differently in different countries and regions in the world as also the current population and anticipated growth, which is likely to be more in developing countries. In 2009, more than 1 billion people went undernourished; it is not because of shortage of food (availability), but because people are too poor to buy (accessibility). Although the percentage of hungry people in the developing world had been dropping for decades (Figure 13.1), the absolute number of hungry people worldwide has barely dipped. The recent food price crises in 2008 reversed the decades of gains (*Nature* 2010). In this chapter, we analyze the current status of agricultural water use in the tropical rainfed areas, assess the potential, and propose a new paradigm to manage agricultural water efficiently through a holistic watershed management approach and operationalize the integrated water resource management (IWRM) strategy for harnessing the untapped potential of rainfed agriculture in the tropics to increase food production and improve the livelihoods of people with finite and scarce water resource.

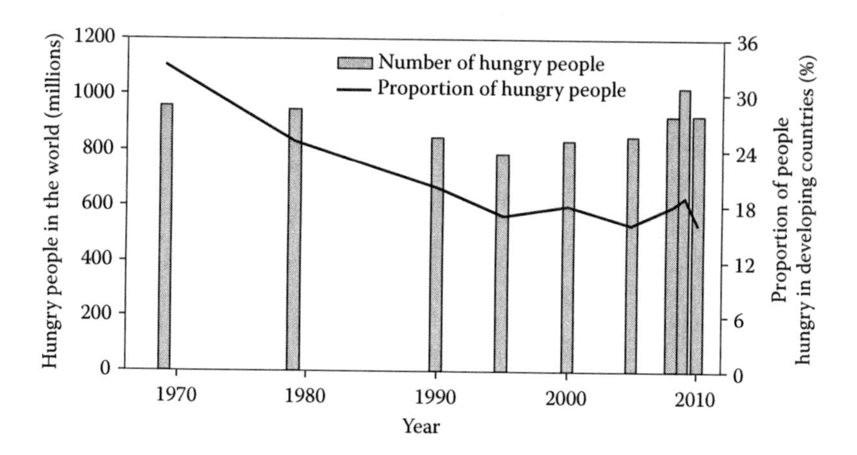

FIGURE 13.1 Total number of hungry people in the world and the proportion in developing countries. (Data from FAOSTAT. http://hungerreport.org/2011/data/hunger.)

13.2 FINITE AND SCARCE FRESHWATER RESOURCES

Water, a natural resource, is a finite one and keeps circulating through the hydrological cycle of evaporation, transpiration, and precipitation mainly driven by various climatic and land management factors (Falkenmark 1997). The total water on earth is 1385.5 million km^3 (Shiklomanov 1993), out of which 97.3% is salt water in oceans. Fresh water constitutes only 2.7% of total global water resource and is the lifeline of the biosphere where forests, woodlands, wetlands, grasslands, and croplands are the major biomes (Postel et al. 1996; Rockström et al. 1999). Rockström et al. (1999) reported that about 35% of annual precipitation (110,305 km^3) received on the earth's surface returns to the oceans as surface runoff (38,230 km^3) while the remaining 65% is converted into water vapor flow. Moreover, major terrestrial biomes, that is, forests, woodlands, wetlands, grasslands, and croplands, together consume almost 98% of the global green water flow (Figure 13.2) and generate essential ecosystem services (Rockström et al. 1999; Rockström and Gordon 2001). Freshwater availability for producing a balanced food diet (i.e., 3000 Kcal/person/day) under the present conditions concomitant with increasing population pressure is an important concern. Figure 13.2 shows that on an average, 6,700 and 15,100 km^3/year of consumptive fresh water is used by croplands and grasslands, which generate food and animal proteins for feeding humanity, respectively (Rockström and Gordon 2001). This quantity is 30% of the total green water flux on the earth.

13.2.1 Green and Blue Water

Water resources are classified into green water and blue water resources (Falkenmark 1995); rainfall is partitioned into blue and green water resources through an important hydrological process (Figure 13.3). Green water is the large fraction of precipitation, which is held in the soil and available for plants' consumption on-site and it returns to the atmosphere through the process of evapotranspiration (ET). A fraction of green water that is consumed by plants is referred to as transpiration and the amount that returns to the atmosphere directly from water bodies and soil surface is labeled as evaporation. Blue water is the portion of precipitation that enters into streams and lakes and also recharges groundwater reserves. Human beings can directly consume blue water for their domestic and industrial uses and also for food production off-site (away from the area where it originates).

Freshwater consumption for major biomes assessed by Rost et al. (2008), however, is comparable with the estimates by Rockström et al. (1999), but this value for grasslands is dissimilar (8258 km^3/year by Rost et al. 2008 compared to 15,100 km^3/year by Rockström et al. 1999) probably due to difference in the methodologies adopted.

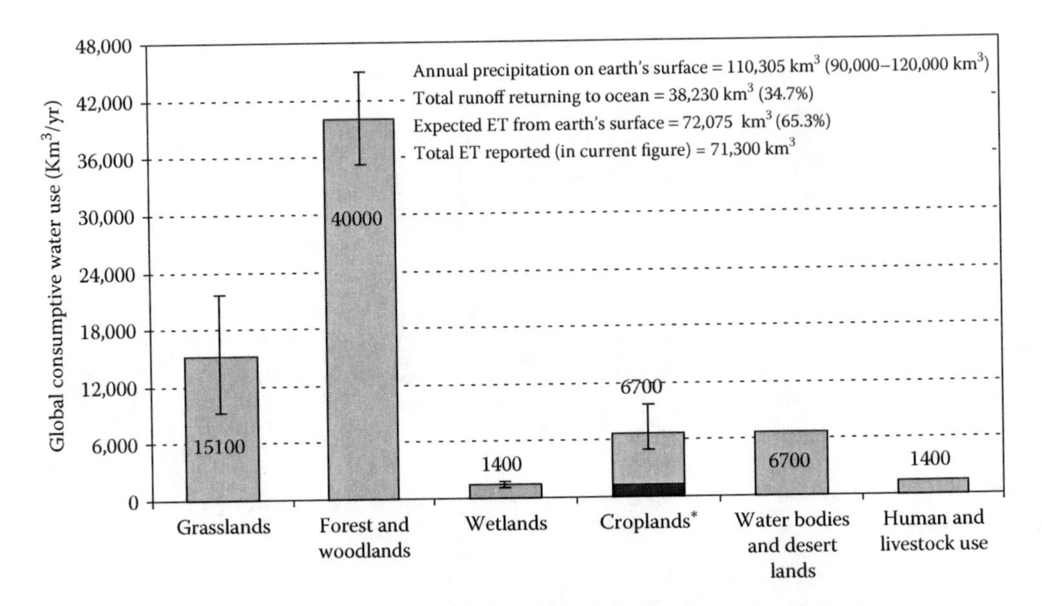

FIGURE 13.2 Global annual consumptive water use of major terrestrial biomes. (Data from Rockström, J., Gordon, L., Folke, C., et al., *Conservation Ecology*, 3(2), 5, 1999; Rockström, J. and Gordon, L., *Physics and Chemistry of the Earth*, 8(26)(11–12), 843–851, 2001.) *Consumptive water used by croplands is partitioned as (1) ET for productive use (upper portion) and (2) ET in noneconomic vegetation including weeds and vegetation in open drainage ditches, green enclosures, and wind breaks (lower portion).

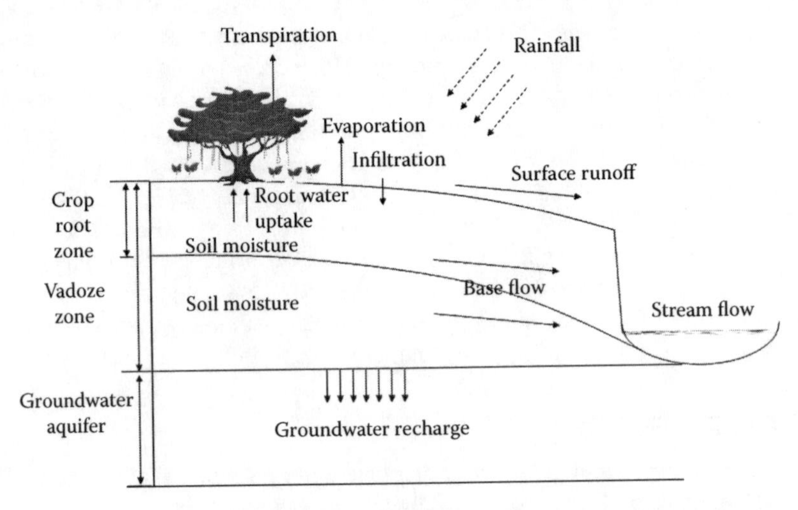

FIGURE 13.3 Conceptual representation of the hydrological cycle and different hydrological components.

Figure 13.4 shows consumptive use of blue and green water from croplands and grasslands (Rost et al. 2008), and the share of green water (adding part one and two) is about 85% of total consumptive freshwater use in cropland and 98% in grassland in the entire globe. Although the contribution of green water in generating global food production is significantly high (Rockström et al. 1999; Rost et al. 2008; Hoff et al. 2010), traditionally, emphasis has been given on augmenting blue water resources (Molden et al. 2007; Falkenmark and Molden 2008; Sulser et al. 2010), and green water potential has not been harnessed properly (Falkenmark et al. 2009; Wani et al. 2009a, 2011a). Large dams/reservoirs were constructed on every important river basin for harvesting river water (Falkenmark and Molden 2008). Figure 13.5 shows the global blue water withdrawal, its

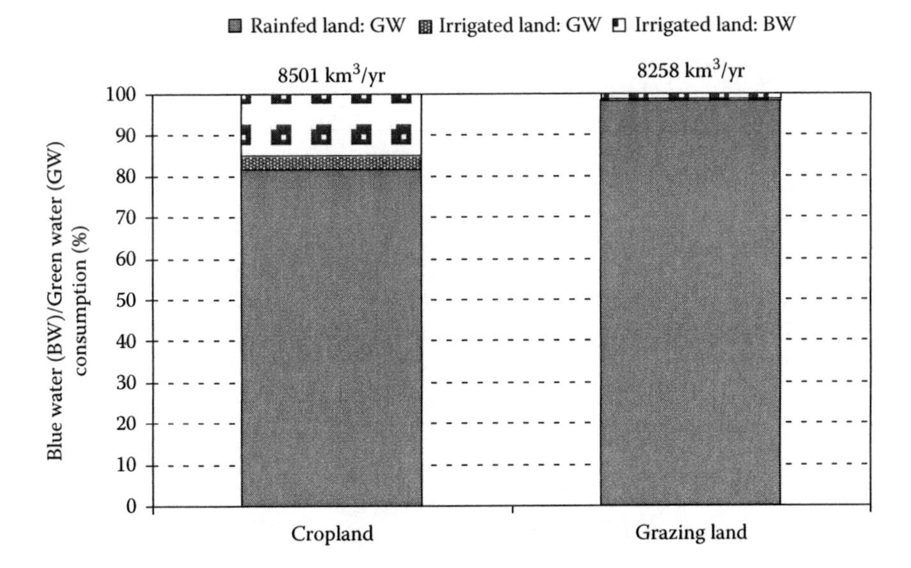

FIGURE 13.4 Blue and green water contribution of consumptive use in cropland and grazing land. (Data from Rost, S., Gerten, D., Bondeau, A., Luncht, W., Rohwer, J., Schaphoff, S., et al., *Water Resources Research,* 44, W09405, doi:10.1029/2007WR006331, 2008.)

consumptive use for domestic and irrigation purpose, and the expansion of cropland and pasture land since 1900. It is clear from the figure that total blue water withdrawal at present has increased by 350% (3800 km^3/year) compared with that in the 1940s, and there is not much scope left to harvest blue water further (Scanlon et al. 2007). With increasing food demand, huge land areas were converted from forest/woodlands to croplands and grasslands, which resulted in reduction in ET by 4% (equivalent to 3000 km^3/year) globally compared with its original native stage. On the other

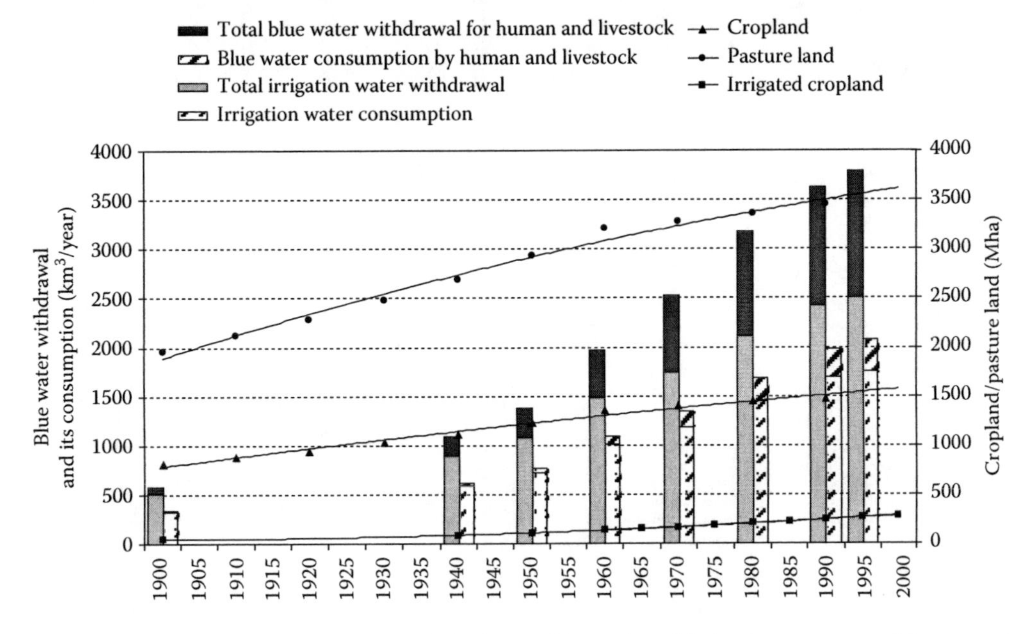

FIGURE 13.5 Total blue water withdrawal for human/livestock and irrigation purpose and its consumptive use since 1900 onward; Expansion of total cropland, pasture land and irrigated land globally since 1900 onward. (Data from Scanlon, B.R., Jolly, I., Sophocleous, M., et al., *Water Resource Research,* 43, W03437, doi:10.1029/2006WR005486, 2007.)

hand, developed water resource projects have enhanced vapor flow by 2600 km³/year in subsequent years (Gordon et al. 2005). However, the net change in global vapor flows is negligible, but differential spatial distribution of deforestation and irrigation has led to change in ecosystems and rainfall pattern at the local, regional, and global scales (Gordon et al. 2005).

13.2.2 Zooming in on Freshwater Resources in India

Out of the annual average precipitation of 4000 km³ over the country, 1120 km³ is partitioned as blue water (690 and 430 km³ surface and groundwater resources, respectively) and the remaining 2880 km³ is available as green water. Land use in India in 2001–2002 shows that 49% of total geographical area is cultivable, 22% area is under forest, 20% area is under wasteland and fallow category, and 9% land is for other uses and not available for cultivation. At present, a total of 142 mha (43% of total geographical area) is the net cultivated area under agricultural use; within that, 40% is irrigated and 60% used for rainfed farming.

From 1950 to 2000, the gross cultivated area (rainfed and irrigated) has increased from 130 mha to 190 mha (Figure 13.6a), whereas the net sown area has remained virtually constant for the last

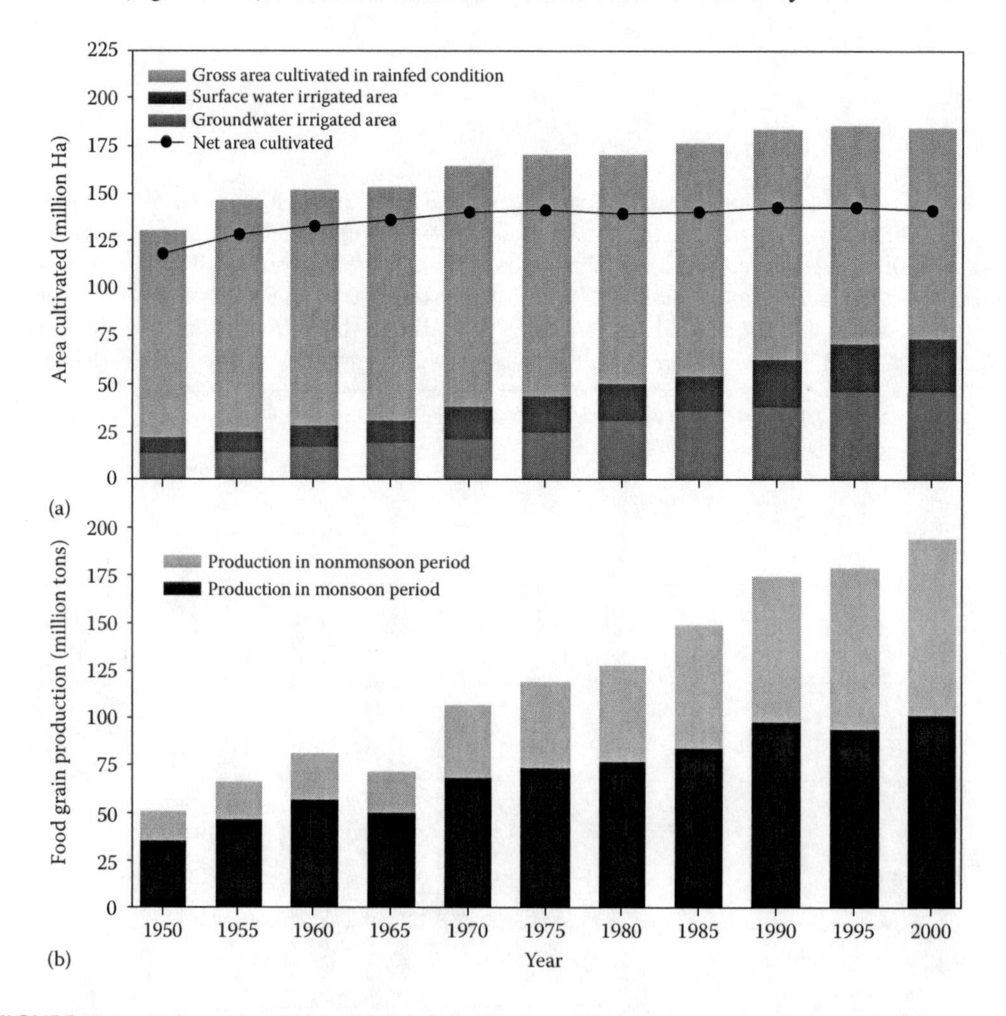

FIGURE 13.6 (a) Gross area cultivated in rainfed and irrigated (groundwater and surface water-irrigated area) croplands and net cultivated area in India; and (b) total food production (during monsoon and postmonsoon period) in India. (Data from Centre Water Commission, *Hand Book of Water Resources Statistics*, 2005. http://www.cwc.nic.in/main/webpages/publications.html.)

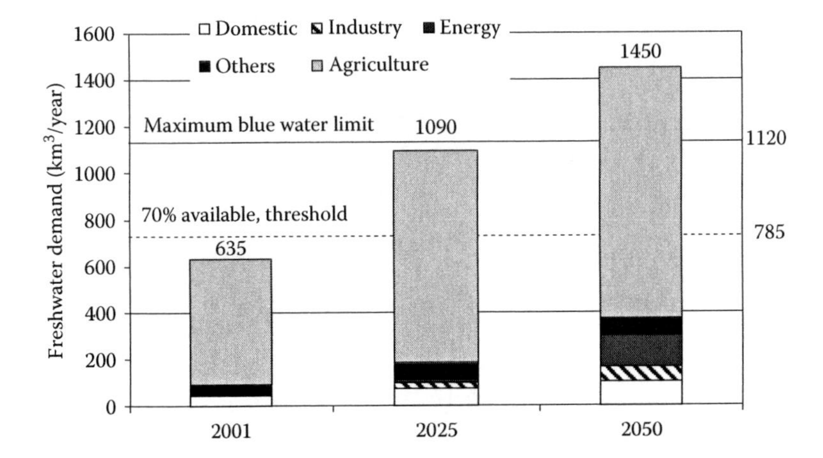

FIGURE 13.7 Present and anticipated future freshwater demand for food production and other uses in India; solid and dashed horizontal lines show the maximum and sustainable (70% of maximum) blue water thresholds. This scenario assumes that water productivity from rainfed and irrigated agriculture will remain same in the future as the current production system. *Note*: Consumptive green water use from croplands has not been reported in this figure. (Data from Centre Water Commission, *Hand Book of Water Resources Statistics*, 2005. http://www.cwc.nic.in/main/webpages/publications.html.)

four decades. The cropping intensity of the current production system is 135%. Irrigated area has increased from 17% to 40% (0.8% expansion per year) in a span of 50 years. Within irrigated agriculture, the area irrigated by groundwater is 65% and surface water is 35%.

Food grain production in India during monsoonal and nonmonsoonal periods is shown in Figure 13.6b. The green revolution in the 1970s significantly increased crop productivity and total grain production, which resulted in food self-sufficiency. Moreover, development of canal command areas (major and minor irrigation projects), village electrification, development of irrigation technology, and infrastructure all together converted substantial fraction of rainfed land into irrigated agriculture (Figure 13.6a). Available fresh water, however, is sufficient enough to meet the current food demand in the country but it will fall severely short with the increasing population pressure in the future. Figure 13.7 shows anticipated freshwater demand (in different sectors: domestic, agriculture, industry, energy, and others) in 2025 and 2050 and also explains maximum and sustainable blue water thresholds. This analysis assumes that water productivity (WP) of rainfed and irrigated agriculture in the future will remain the same as of the current production system. Under this scenario, all blue water will have to be harvested (Table 13.1) and diverted for human consumption by 2025, which may jeopardize social fabric in the society, environment, and ecosystems. Moreover, freshwater demand in 2050 will be much higher than maximum available blue water resources, clearly suggesting that blue water resource alone will not be sufficient to satisfy future water needs in India. The vast untapped potential of rainfed agriculture will have to be harnessed to meet future food and water demands of the country (Wani et al. 2003a, 2008, 2009a, 2011a; Rockström et al. 2007, 2010; Sharma et al. 2010).

13.2.3 Competing Demands for Limited Available Water from Different Sectors

Water scarcity is particularly acute in many developing countries where there is an urgent need to eradicate poverty and improve quality of life for people to exist. River flows are declining with increasing water resources development, which has led to serious transboundary issues and conflicts among different stakeholders in addition to a growing concern over the social and environmental impacts (Landell-Mills and Porras 2002; Bunn and Arthington 2002). Moreover, great uncertainty is arising on future water availability due to upcoming climate changes (IPCC 2007). Extreme events such as flash floods or longer dry spells, more number of dry or wet years, change in crop

TABLE 13.1
Surface and Groundwater Potential: Current and Future Utilization in India

Surface Water Resources	Fresh Water (km^3)
Utilizable average surface water (per year)	690
Reservoir storage capacity	213
Projects under construction	76
Projects for further consideration	108
Groundwater Resources	
Replenishable groundwater	430
Available for agricultural use	360
Net draft at present	115

Source: Data from Centre Water Commission, *Hand Book of Water Resources Statistics*, 2005. http://www.cwc.nic.in/main/webpages/publications.html.

water demand, temperature change, and pest/disease infestation are the various characteristics driven by the climate change phenomenon.

As stated earlier, water availability for croplands and grasslands is becoming less with increasing population pressure and changing food habits (Rockström et al. 1999, 2009). Figure 13.8 shows the present and anticipated future food demands (Figure 13.8a) in developing and developed countries and corresponding total freshwater requirements (Figure 13.8b for developing countries and Figure 13.8c for the entire globe) if the current trend of WP continues in the future as well (Rockström et al. 2007). It is anticipated that total food demand in 2050 will be approximately 11,200 million tons, out of which 9300 million tons of food will be required for developing countries (de Fraiture et al. 2007; Rockström et al. 2007; Khan and Hanjra 2009; Hanjra and Qureshi 2010).

Blue water in most of the river basins (except sub-Saharan Africa [SSA]) has already been diverted for domestic/industrial use and also in irrigated agriculture for food production (Figure 13.5), with little scope left for further harvest. There are two alternatives for meeting increasing food demand: (i) improvement in WP with existing croplands (both rainfed and irrigated) and grasslands and (ii) expansion in agriculture areas by clearing some fraction of forest/woodlands and wetlands into croplands; or a combination of these two. Several examples/studies show that change in land use from forestlands to crop/grasslands, however, increased food production but developed imbalance in the traditional terrestrial ecosystem and feedback mechanism, with the loss of ecosystem resilience and also various other ecosystem services. This also led to climate change from local to regional/global level and reduction in overall water availability (Gordon et al. 2005; Hoff et al. 2010). For example, the mass clearing of *Eucalyptus mallee* forest to croplands and pasture lands in Australia in the late 1800s and early 1900s initially increased the groundwater table, which subsequently created waterlogging and soil salinization problems over the landscape (Scanlon et al. 2007). Similarly, conversion of natural savannas into millet-growing rainfed land in Niger, Africa, enhanced surface runoff, resulting in soil loss and primary gully formations (Leduc et al. 2001; Massuel et al. 2006; Scanlon et al. 2007).

13.3 UNDERSTANDING WATER SCARCITY

Assessment of the amount of renewable surface and groundwater per capita (i.e., the so-called blue water) suggests that water stress is increasing in a number of countries, as we understand

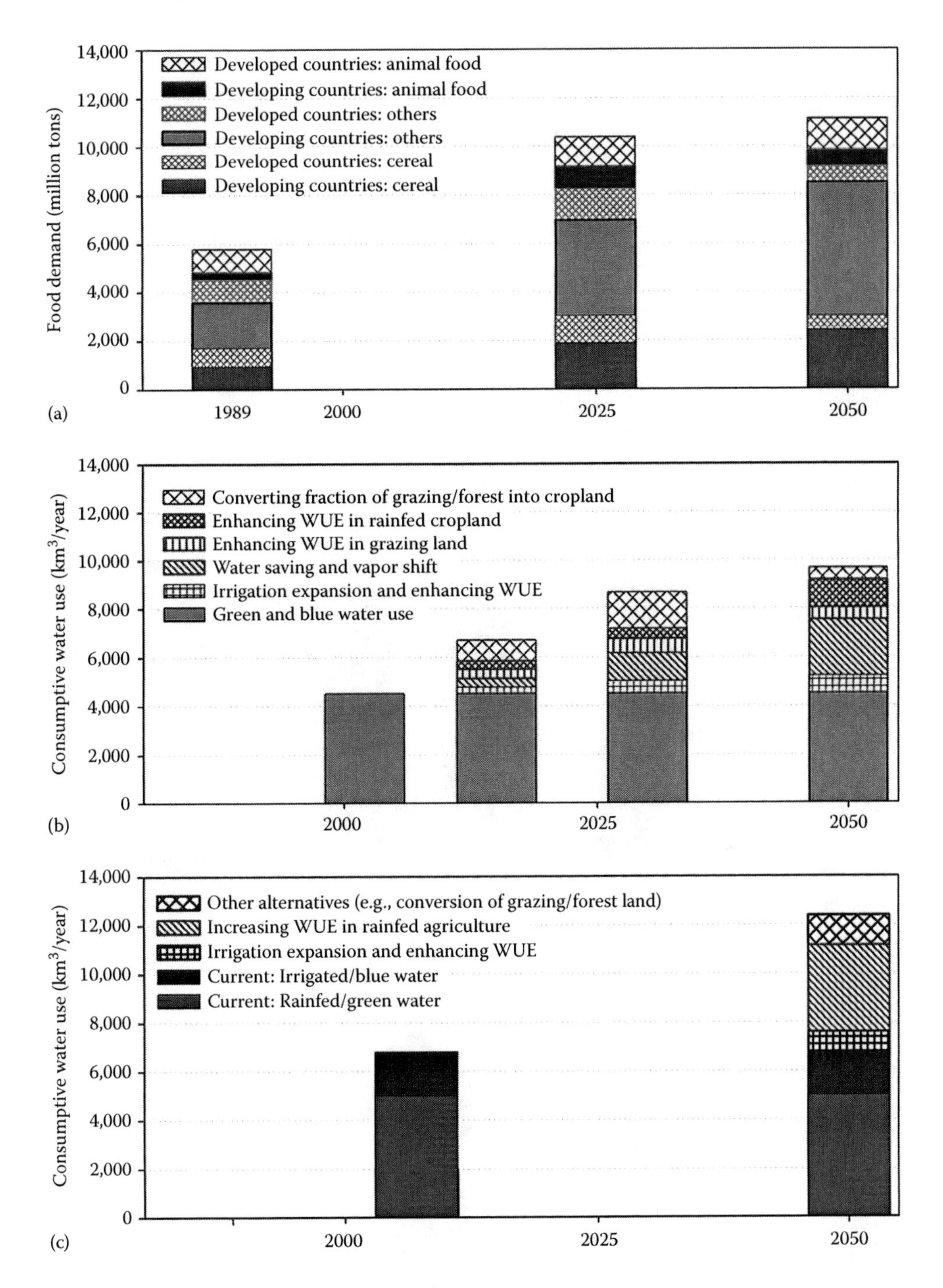

FIGURE 13.8 (a) Present and anticipated future global food demand; present and future fresh water required for food production and possible source to fill up demand gap (b) in developing countries; and (c) both in developing and developed countries. (Data from Rockström, J., Hatibu, N., Oweis, T., et al., In *Water for Food, Water for Life: A Comprehensive Assessment of Water Management in Agriculture*, pp. 315–348, Earthscan, London and International Water Management Institute (IWMI), Colombo, Sri Lanka, 2007; de Fraiture, C., Wichelns, D., Rockström, J., et al., In *Comprehensive Assessment of Water Management in Agriculture, Water for Food, Water for Life: A Comprehensive Assessment of Water Management in Agriculture*, pp. 91–145, International Water Management Institute, Colombo, Sri Lanka and Earthscan, London, UK, 2007; Khan, S. and Hanjra, M.A., *Food Policy*, 34(2), 130–140, 2009; Hanjra, M.A. and Qureshi, M.E., *Food Policy* 35, 365–377, 2010.)

conventionally. However, water scarcity is a relative concept and water is not equally scarce in all parts of the world. As Figure 13.9a illustrates, South Asia (SA), East Asia (EA), and the Middle East North Africa (MENA) regions are the worst affected in terms of blue water scarcity. However, this picture may be misleading because these water quantities only include blue water and full resource, notably rainwater "green water," that is, soil moisture used in rainfed cropping and natural vegetation is not included. Further, the average amount of water per capita in each pixel could obscure large differences in actual access to a reliable water source. In a recent assessment that included both green and blue water resources, the level of water scarcity changed significantly for many countries (Figure 13.9b) and suggested that large opportunities are still possible in the management of rainfed areas, that is, the green water resources in the landscape (Rockström et al. 2009; Wani et al. 2009a, 2011b). The current global population that has blue water stress is estimated to be 3.17 billion and is expected to reach 6.5 billion in 2050. If both green and blue water are considered, the number currently experiencing absolute water stress is a fraction of this (0.27 billion) and will only marginally exceed today's blue water stress in 2050.

Absolute water stress is found most notably in arid and semiarid regions with high population densities such as parts of India, China, and the MENA region. The MENA region is increasingly

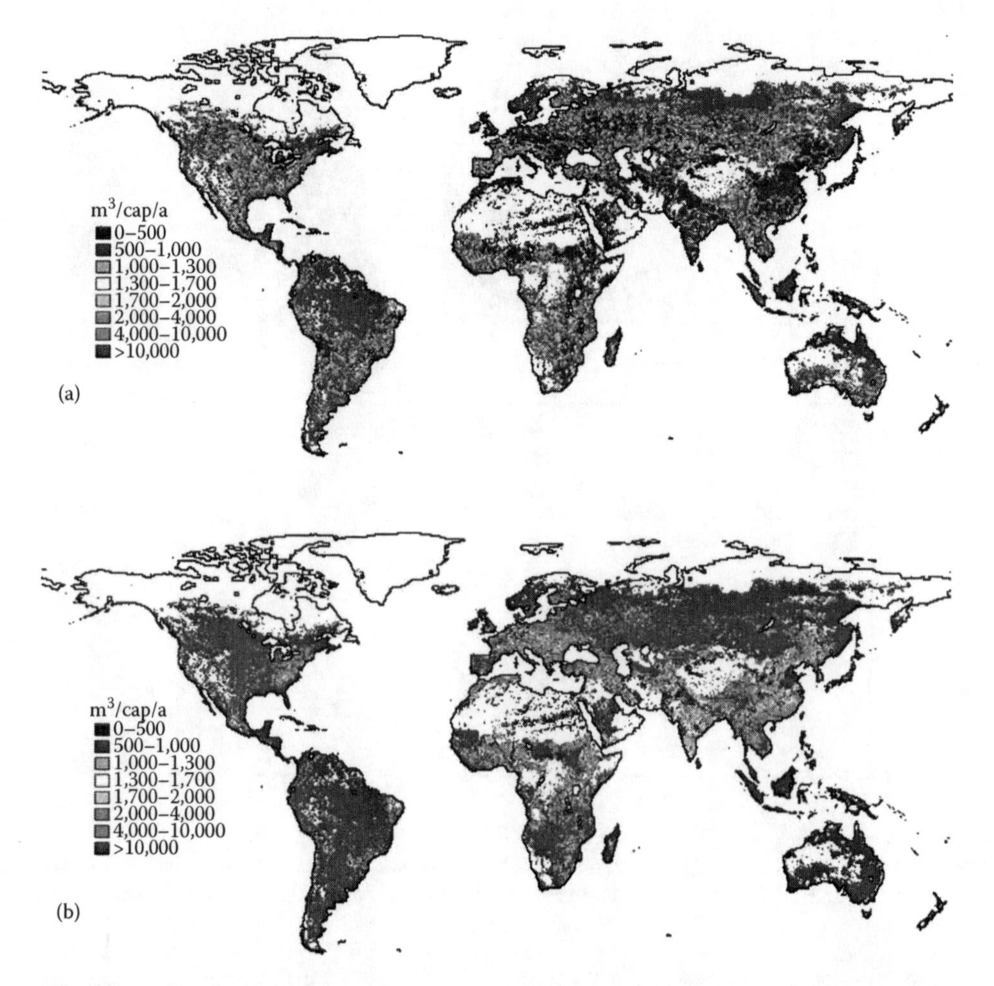

FIGURE 13.9 (a) Renewable liquid freshwater (blue) stress per capita (m^3/cap/a) using LPJ dynamic modeling year 2000. (b) Renewable rainfall (green and blue) water stress per capita (m^3/cap/a) using LPJ dynamic modeling year 2000. (From Rockström, J., Falkenmark, M., Karlberg, L., et al., *Water Resources Research*, 45, W00A12, doi:10.1029/2007WR006767, 2009.)

unable to produce the food required locally due to increasing water stress from a combination of population increase, economic development, and climate change and will have to rely more and more on food (and virtual water) imports.

Among the regions that are conventionally (blue) water-scarce but still have sufficient green and blue water to meet the water demand for food production are large parts of SSA, India, and China. If green water (on current agricultural land) for food production is included, per capita water availability in countries such as Uganda, Ethiopia, Eritrea, Morocco, and Algeria more than doubles or triples. Moreover, low ratios of transpiration to evapotranspiration (T/ET) in countries such as Bangladesh, Pakistan, India, and China indicate high potential for increasing WP through vapor shift (Rockström et al. 2009).

Considering the vast rainfed areas (1.25 billion hectares) covering 80% of cultivated land and 85% of consumptive use of fresh water in agricultural land, agricultural water management is larger than irrigation (blue water). There is an urgent need to make all the stakeholders understand the need to consider large quantities of available green water globally and the potential to enhance water-use efficiency (WUE) for food production. Not only is water availability for food production restricted to blue water but green water also needs to be brought into the ambit for management and harnessing the potential.

Given the increasing pressures on water resources and the increasing demands for food, fiber, and biofuel crops for energy, the world must succeed in producing more food with less water. Hence, it is essential to increase WP in humid, semiarid, and arid regions. Some describe the goal as increasing the "crop per drop" (more crops per drop) or the "dollars per drop" (more income per drop) produced in agriculture. Regardless of the metric, it is essential to increase the productivity of water and other inputs in agriculture. Success on this front will generate greater agricultural output, while enhancing water availability in other sectors and contributing to environmental quality. There are several field and simulation studies showing huge untapped potential of rainfed and irrigated areas (Wani et al. 2003a, 2008, 2009a, 2011b; Rockström et al. 2007; Fisher et al. 2009; Kijne et al. 2009; Sahrawat et al. 2010a). The main reasons for poor WUE in rainfed areas are land degradation, water scarcity, lack of knowledge among farmers, low and inappropriate input use, and climatic variability (Barron et al. 2003; Kijne et al. 2003; Molden et al. 2007; Wani et al. 2003a, 2007, 2009a; Sharma et al. 2010). Water availability in irrigated areas, especially in canal command areas, is good but poor water management, lack of institutional arrangements, and faulty government policies (e.g., subsidy on canal water use and free electricity for groundwater pumping) are the main reasons for poor WUE (Molden et al. 2007). Overdrafting and more water inputs are the common practices in irrigated areas (e.g., in India), which leads to waterlogging and soil salinity problem (Khare et al. 2007; Shah et al. 2007) and declining productive status of the landscape subsequently (Manjunatha et al. 2004; Rajak et al. 2006).

13.3.1 Water Scarcity and Poverty in the Tropical Regions

There is a correlation between poverty, hunger, and water stress (Falkenmark 1986). A recent study by Rockström and Karlberg (2009) mapped hot spots of poverty in SSA, SA, and EA for bridging the yield gaps in rainfed areas where agriculture is the principal source of economy and livelihood of millions of people in developing countries. Poor investment/capacity, poor financial structures, and poor extension support are the major reasons keeping rainfed farming at subsistence level. Furthermore, landholdings are becoming smaller, and consequently land share and livelihood opportunities are reducing (Wani et al. 2011b). The UN Millennium Development Project has identified the "hot spot" countries in the world suffering from the largest dominance of malnourishment. These countries coincide closely with those located in the semiarid and dry subhumid hydroclimates in the world, that is, savannahs and steppe ecosystems, where rainfed agriculture is the dominating source of food and where water constitutes a key limiting factor to crop growth (SEI 2005). Following this, we strongly make an evidence-based case for harnessing the full potential of vast rainfed areas through operationalizing the IWRM framework for enhancing crop yields through increasing WP.

13.4 CHALLENGES AND OPPORTUNITIES FOR WATER MANAGEMENT IN RAINFED AGRICULTURE

For obvious reasons, water is the primary limiting factor in dryland agriculture (Falkenmark and Rockström 2008). Rainfall in dry land areas is characterized by erratic and nonuniform distribution, which results in frequent dry spells at different time periods during the monsoon. Barron et al. (2003) studied dry spell occurrence in semiarid locations in Kenya and Tanzania and found that meteorological dry spells of >10 days occurred in 70% of seasons during the flowering stage of the crop (maize), which is very sensitive to water stress. Regions with similar seasonal rainfall can experience different dry spell occurrence. In the semiarid Nandavaram watershed, Andhra Pradesh, India, with approximately 650 mm of rainfall, there is a high risk of dry spell occurrence (>40% risk) during the vegetative and flowering stages of the crop, compared with semiarid Xiaoxingcun, southern China, receiving similar rainfall, but with only a 20% risk of early season dry spells (Rao et al. 2007).

For achieving better crop growth and yield, a certain amount of water is essentially required to meet plant metabolic and evaporative demands (Stewart et al. 1975). There exists a direct relationship between consumptive water use (ET) and crop growth/yield. Rockström et al. (2007) described that if all the green water captured in the root zone is utilized fully by crop, a yield of 3 t/ha in rainfed agriculture is achievable. If water that is lost as deep percolation and surface runoff is also made available to crop, then production level would reach 5 t/ha and further up to 7.5 t/ha. All the above such conditions assume that nutrient availability for plant is nonlimiting. In reality, only a small fraction of rainfall is used by the plant (through transpiration) while the rest is channelized through nonproductive use and lost from crop production system. A water stress situation, especially during critical growth stages, reduces crop yield and may even seriously damage the entire crop. Numerous data on productivity enhancement studies from Africa and Asia demonstrate huge potential to enhance green WUE as well as increasing availability of green water (Wani et al. 2002, 2003a, 2008, 2009b, 2011c; Rockström et al. 2007, 2010; Barron and Keys 2011).

13.4.1 Importance of Green Water Management in Rainfed Agriculture

Most of the 1338 million poor people in the world live in the developing countries of Asia and Africa, more so in drylands/rainfed areas (Rockström et al. 2007; Wani et al. 2009a, 2011b). Approximately 50% of total global land area is located under dry and arid regions (Karlberg et al. 2009). The importance of rainfed agriculture varies regionally, but it produces most food for poor communities in developing countries (Rockström et al. 2007; Wani et al. 2011a). In SSA more than 95% of the farmed land is rainfed, while the corresponding figure for Latin America is almost 90%, for South Asia about 60%, for EA 65%, and for the Near East and North Africa 75% (FAOSTAT 2010) (Table 13.2). A large fraction of the global expansion in the total cropland since 1900 is in rainfed regions (Figure 13.6). Native vegetation such as forests and woodlands were converted into croplands (mostly into rainfed agriculture) and grasslands, which produced more staple food and animal proteins but also, in the event of severe land degradation, depletion of soil nutrients and loss of biodiversity, which resulted in poor productive status as well as loss in system resilience and ecosystem services (Gordon et al. 2005). Most countries in the world depend primarily on rainfed agriculture for their grain food and a great number of poor families in many developing countries such as Africa and Asia still face poverty, hunger, food insecurity, and malnutrition, where rainfed agriculture is the main agricultural activity. These problems are exacerbated by adverse biophysical growing conditions and the poor socioeconomic infrastructure in many areas in the arid, semiarid tropics (SAT), and the subhumid regions (Wani et al. 2011a). In other words, where water limits crop production, poverty is strongly linked to variations in rainfall and to the farmers' ability to bridge intraseasonal dry spells (Karlberg et al. 2009).

TABLE 13.2
Global and Continentwise Rainfed Area and Percentage of Total Arable Land

Continent Regions	Total Arable Land (million hectares)	Rainfed Area (million hectares)	Percentage of Rainfed Area
World	**1551.0**	**1250.0**	**80.6**
Africa	*247.0*	*234.0*	*94.5*
Northern Africa	28.0	21.5	77.1
Sub-Saharan Africa	218.0	211.0	96.7
Americas	*391.0*	*342.0*	*87.5*
Northern America	253.5	218.0	86
Central America and Caribbean	15.0	13.5	87.7
Southern America	126.0	114.0	90.8
Asia	*574.0*	*362.0*	*63.1*
Middle East	64.0	41.0	63.4
Central Asia	40.0	25.5	63.5
Southern and Eastern Asia	502.0	328.0	65.4
Europe	*295.0*	*272.0*	*92.3*
Western and Central Europe	125.0	107.5	85.8
Eastern Europe	169.0	164.0	97.1
Oceania	*46.5*	*42.5*	*91.4*
Australia and New Zealand	*46.0*	*42.0*	*91.3*
Other Pacific Islands	*0.57*	*0.56*	*99.3*

Source: FAO. AQUASTAT database. 2010. http://www.fao.org/nr/aquastat; FAO. FAOSTAT database. 2010. http://www.faostat.fao.org/.

13.4.2 Vast Potential to Enhance Water Productivity in the Tropics

A linear relationship is generally assumed between biomass growth and vapor flow (ET), which describes WP in the range between 1000 and 3000 m^3/t for grain production (Rockström 2003) (Figure 13.10). Increasingly, it is recognized that this linear relationship does not hold true for yields up to 3 t/ha, which exactly coincide with yield levels of small and marginal farmers in dryland/rainfed areas. The reason is that improvements in agricultural productivity, resulting in yield increase and denser foliage, will involve a vapor shift from nonproductive evaporation (E) in favor of productive transpiration (T) and a higher T/ET as transpiration increases (essentially linearly) with higher yield (Stewart et al. 1975; Rockström et al. 2007). Therefore, this is a huge scope for improving WP through green water management especially at lower yield level (Figure 13.10), and agricultural water interventions can help in reducing the water stress situation by enhancing green water availability. Evidence from water balance analyses on farmers' fields around the world shows that only a small fraction, less than 30% of rainfall, is used as productive green water flow (plant transpiration) supporting plant growth (Rockström 2003). In arid areas typically as little as 10% of the rainfall is consumed as productive green water flow (transpiration), while 90% of the flows constitute nonproductive evaporation flow, that is, no or very limited blue water generation (Oweis and Hachum 2001). In temperate arid regions, such as West Africa and North Africa, a large portion of the rainfall is generally consumed in the farmers' fields as productive green water flow (45%–55%), which results in higher yield levels (3–4 t/ha as compared with 1–2 t/ha) and 25%–35% of the rainfall flows as nonproductive green water flow while the remaining 15%–20% generates blue water flow. Agricultural water interventions in the watershed in Indian SAT reduced runoff amount by 30%–50%, depending on the rainfall distribution and converted more of it into green water (Figure 13.11; Garg et al. 2011a).

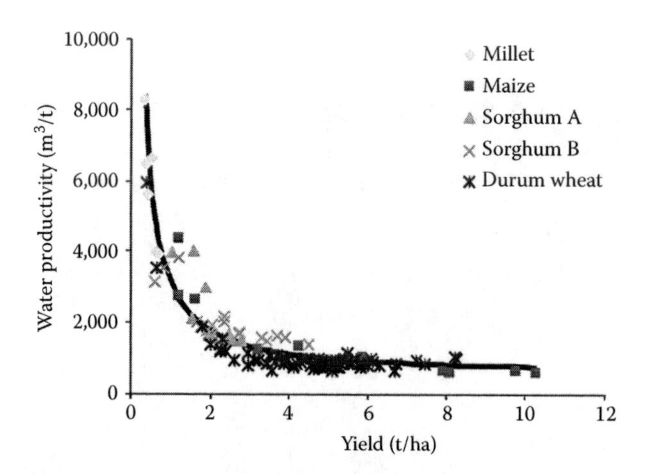

FIGURE 13.10 Dynamic relationship between green water productivity and yield for cereal crops in different climatic conditions and management. (Data from Rockström et al. (1998) (Millet); Stewart (1988) (Maize); Dancette (1983) (Sorghum A); Pandey et al. (2000) (Sorghum B); and Zhang and Oweis (1999) (Durum Wheat). Regression line after Rockström (2003). (From Karlberg, L., Rockström, J., Falkenmark, M., et al., In *Rainfed Agriculture: Unlocking the Potential*, pp. 1–310, The Comprehensive Assessment of Water Management in Agriculture Series, Volume 7, CABI, Wallingford, UK, 2009.)

There is a vast untapped potential in rainfed areas with appropriate soil and water interventions (Rockström and Falkenmark 2000; Wani et al. 2003a, 2009a, 2011a,b,c; Rockström et al. 2007, 2010; Figures 13.12 and 13.13).

Even in tropical regions, particularly in the subhumid and humid zones, agricultural yields in commercial rainfed agriculture exceed 5–6 t/ha (Rockström and Falkenmark 2000; Wani et al. 2003a,b; Figure 13.13). At the same time, the dry subhumid and semiarid regions have experienced the lowest yields and the weakest yield improvements per unit land. Here, yields oscillate between 0.5 and 2 t/ha, with an average of 1 t/ha in SSA and 1–1.5 t/ha in South Asia, Central Asia, West Asia, and North Africa for rainfed agriculture (Rockström and Falkenmark 2000; Wani et al. 2003a,b). Data of a long-term experiment at the International Crops Research Institute for the Semi-Arid Tropics (ICRISAT's) Heritage watershed site (Figure 13.13) has conclusively established

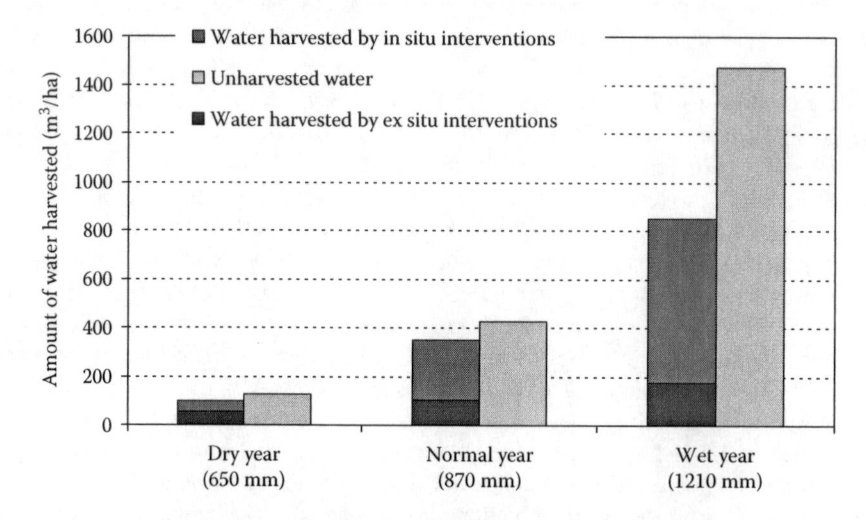

FIGURE 13.11 Fraction of runoff harvested in blue and green water from unharvested amount by implementing various agricultural water interventions compared to nonintervention stage.

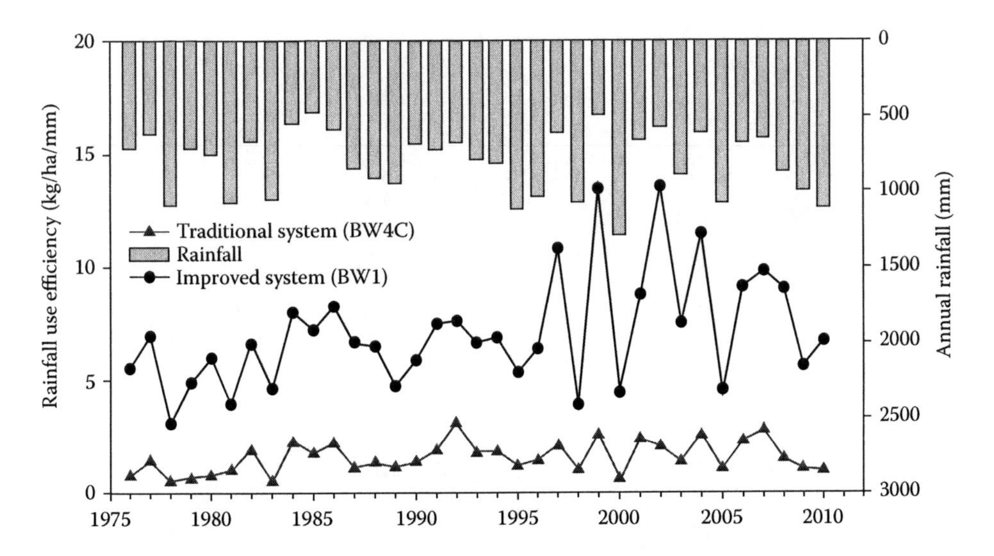

FIGURE 13.12 Increased rainwater-use efficiency in low rainfall years in a long-term experiment at Heritage watershed site, ICRISAT, Patancheru, India.

that integrated IWRM interventions' average crop yield is fivefold higher compared with traditional practices (Wani et al. 2003a, 2011a,b). Similar results were also recorded at Kothapally watershed where implementing IWRM interventions enhanced crop yields almost two to three times as compared with that in 1998 prior to such interventions (Wani et al. 2003a; Sreedevi et al. 2004).

Yield gap analyses carried out for comprehensive assessment, for major rainfed crops in semiarid regions in Asia and Africa and rainfed wheat in West Africa and North Africa, revealed large yield gaps with farmers' yields being a factor of 2–4 times lower than achievable yields for major rainfed crops (Figures 13.14 and 13.15 and Table 13.3). Detailed yield gap analyses of major rainfed crops in different parts of the world have been discussed by Fisher et al. (2009) and Singh et al. (2009). In eastern and southern African countries, the yield gap is very large (Figure 13.15). Similarly, in many countries in

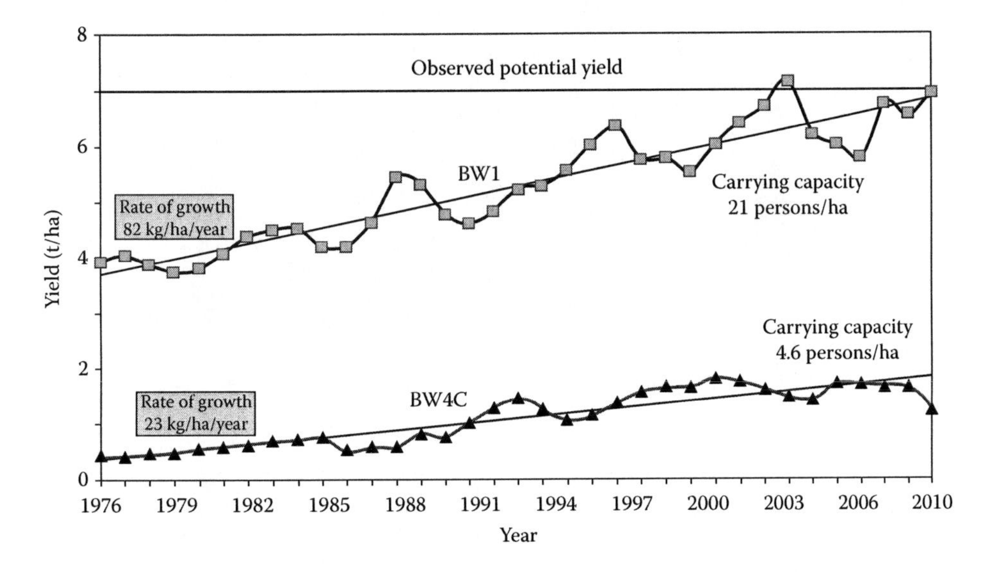

FIGURE 13.13 A comparison of harvested grain yield by implementing IWRM techniques in BW1 Vertisol watershed at ICRISAT with traditional farmers' practices at BW4C; results are shown since 1976 onward.

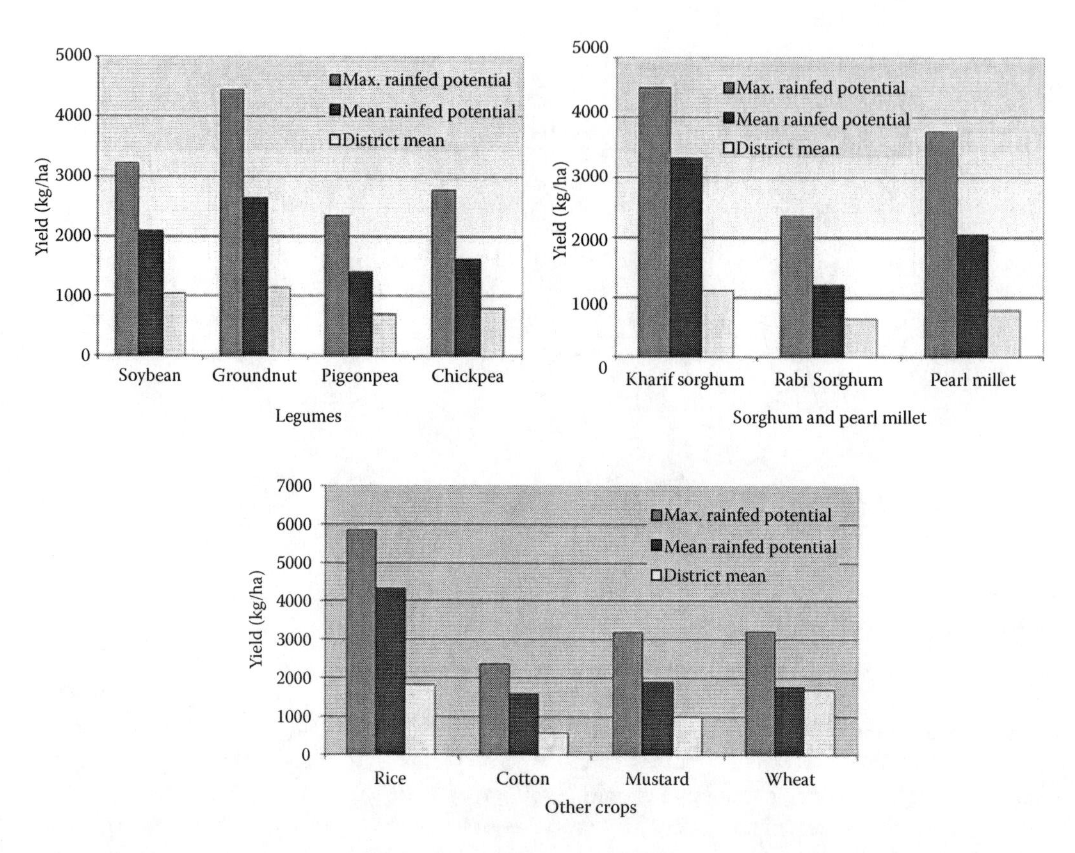

FIGURE 13.14 Rainfed potential yields and yield gaps of crops in India. (From Singh, P., Aggarwal, P.K., Bhatia, V.S., et al., *Yield Gap Analysis: Modeling of Achievable Yields at Farm Level in Rain-Fed Agriculture: Unlocking the Potential*, pp. 81–123, CAB International, Wallingford, UK, 2009.)

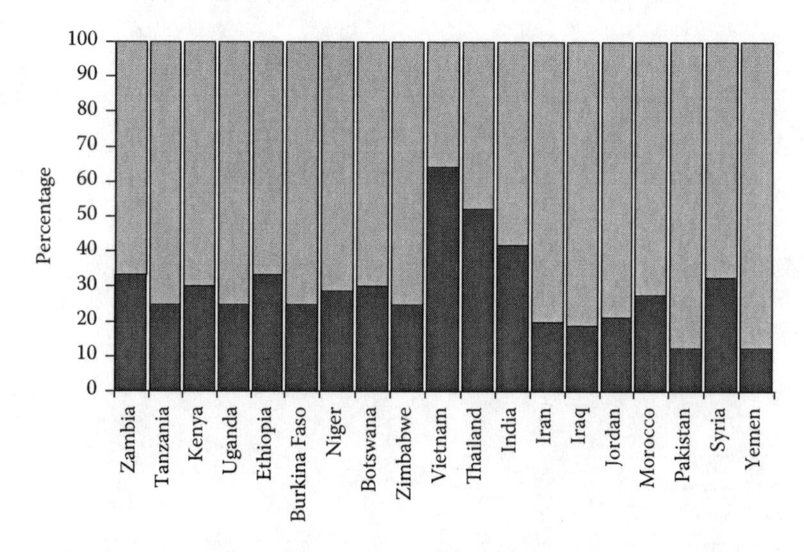

FIGURE 13.15 Examples of observed yield gap (for major grains) between farmers' yields and achievable yields (100% denotes achievable yield level, and columns actual observed yield levels). (From Rockström, J., Hatibu, N., Oweis, T., et al., In *Water for Food, Water for Life: A Comprehensive Assessment of Water Management in Agriculture*, Earthscan, London and International Water Management Institute (IWMI), Colombo, Sri Lanka, 2007.)

TABLE 13.3
Yield Gap Analysis of Soybean Crop in Selected Benchmark Location in India

Seasonal Rainfall (mm)	Crop Analyzed	Number of Benchmark Location Analyzed	Observed Crop Yield (kg/ha)		Rainfed Yield (Simulated) Potential (kg/ha)		References
			Mean	Maximum	Mean	Maximum	
600–700	Soybean	2	730	910	1200	3190	Singh et al. (2001)
700–800		7	840	1000	1930	3070	
800–900		2	860	840	1750	3110	
900–1000		10	790	930	1950	3330	
1000–1100		5	820	860	2200	3350	
1100–1200		2	770	770	1960	3200	
300–400	Groundnut	2	1045	1390	1020	3495	Bhatia et al. (2009)
400–500		2	615	730	2050	4710	
500–600		3	1417	1790	2860	4897	
600–700		5	900	1120	2642	5030	
700–800		4	1150	1550	3425	4978	
800–900		2	820	860	3935	5655	
300–400	Pigeonpea	1	310	310	920	1810	Bhatia et al. (2006)
400–500		3	350	470	1083	2130	
500–600		2	310	430	1490	2305	
600–700		6	647	910	1260	2198	
700–800		7	478	1040	1681	1963	
800–900		8	513	1140	1790	2405	
900–1000		3	623	930	1453	2140	
>1000		5	306	640	1856	2110	
Postmonsoon crop	Chickpea	26	715 (330–1050)	1050	1130 (490–2030)	2470 (1090–4300)	Bhatia et al. (2006)

Source: Singh, P., Vijaya, D., Srinivas, K., et al., Potential productivity, yield gap, and water balance of soybean-chickpea sequential system at selected benchmark sites in India. Global Theme 3: Water, Soil, and Agrobiodiversity Management for Ecosystem Heath. Report no.1. Patancheru 502 324, International Crops Research Institute for the Semi-Arid Tropics, Andhra Pradesh, India, 2001.

West Asia, farmers' yields are less than 30% of achievable yields, while in some Asian countries the figure is closer to 50%. Historic trends present a growing yield gap between farmers' practices and farming systems that benefit from management advances (Wani et al. 2003b, 2009a, 2011a).

13.5 NEW PARADIGM TO OPERATIONALIZE IWRM IN RAINFED AREAS

Business as usual to manage rainfed agriculture as subsistence agriculture with low resource use efficiency cannot sustain economic growth and is needed for ensuring food security to the growing population with increasing incomes (Wani et al. 2002, 2009a, 2011a; Molden et al. 2007; Rockström et al. 2007). There is an urgent need to develop a new paradigm for operationalizing the IWRM framework to harness the untapped potential of rainfed agriculture. The conventional sectoral approach to water management produced low WUE, resulting in increased demand for water to produce food while also causing degradation of natural resources. We need to have a holistic approach based on the convergence of all the necessary aspects of natural resource conservation, their efficient use, production functions, and income enhancement avenues through the value chain and enabling policies and much-needed investments in rainfed areas.

The policy on water resource management for agriculture conventionally remains focused on irrigation, and the framework for IWRM at catchment and basin scales is primarily concentrated on allocation and management of blue water (irrigation water) in rivers, groundwater, and lakes. The evidence from the comprehensive assessment indicated that water for agriculture is more than for irrigation, and there is an urgent need for a widening of the policy scope to include explicit strategies for water (green and blue) management in rainfed agriculture including grazing and forest systems. Effective integration is necessary to focus on the investment options on water management across the continuum (range) from rainfed to irrigated agriculture. This is the time to abandon the obsolete sectoral divide between irrigated and rainfed agriculture, which would place water resource management and planning more centrally in the policy domain of agriculture at large, and not as today, as a part of water resource policy (Molden et al. 2007).

Furthermore, the current focus on water resource planning at the river basin scale is not appropriate for water management in rainfed agriculture, which overwhelmingly occurs on farms of <5 ha at the scale of small catchments, below the river basin scale. Therefore, focus should be on managing water at the catchment scale (or small tributary scale of a river basin) and initiating the much-needed investments in water resource management also in rainfed agriculture (Wani et al. 2002, 2009a, 2011a; Rockström et al. 2007, 2010; Kijne et al. 2009; Wilson 2011).

The world's available land and water resources can satisfy future demands by taking the following steps (Molden et al. 2007):

- Upgrading rainfed agriculture by investing more in rainfed agriculture to enhance agricultural productivity (rainfed scenario)
- Discarding the artificial divide between rainfed and irrigated agriculture and adopting the IWRM approach for enhancing resource efficiency and agricultural productivity
- Investing in irrigation for expanding irrigation where scope exists and improving efficiency of the existing irrigation systems (irrigation scenario)
- Recycling wastewater (gray water) for fodder and food production after suitable treatment
- Conducting agricultural trade within and between countries (trade scenario)
- Reducing gross food demand by influencing diets and reducing postharvest losses, including industrial and household waste

To upgrade rainfed agriculture in the developing countries, community participatory and integrated watershed management approach is recommended and success has been proved as evidenced from a number of islands of Asia and Africa (Wani et al. 2002, 2003a, 2009a, 2011a; Rockström et al. 2007; Wilson 2011). In the rainfed areas of the tropics, water scarcity and growing land degradation cannot be tackled through farm-level interventions alone and community-based management of natural resources for enhancing productivity and improving rural livelihoods is urgently needed (Wani et al. 2002, 2009a; Rockström et al. 2007, 2010). A major research and development challenge to upgrade rainfed agriculture is to bring in convergence among different stakeholders and scientific disciplines by coming out of disciplinary silos and to translate available blueprints into operational plans and implement them (Wani et al. 2003a, 2006, 2009a, 2011a; Rockström et al. 2007, 2010). We know what to do but the challenge is how to do it (Wani et al. 2008, 2011a).

The community-based management of natural resources calls for new approaches (technical, institutional, and social) that are knowledge-intensive and need strong capacity development (more than training of human resources) for all the stakeholders including policy makers, researchers, development agents, and farmers. The small and marginal farmers are deprived of the new knowledge and materials produced by the researchers. There are several disconnects between the farmers and the researchers as the extension systems in most developing countries are not functioning to the desired level. There is an urgent need to bring in the changes in the ways we are addressing the issues of rainfed agriculture to achieve food security and alleviate poverty to meet the Millennium Development Goals (MDGs) (Rockström et al. 2007; Wani et al. 2008, 2009, 2011a,b; Wilson 2011).

13.5.1 Need for Holistic Integrated Approach to Harness the Full Potential

Farmers who are solely dependent on agriculture, especially in dry lands, face a high level of uncertainty and risk of failure due to various extreme climatic events, pest and disease attack, and market shocks. Therefore, integration of agriculture (on-farm) and nonagriculture (off-farm) activities is required for generating consistent source of income and support for livelihood. For example, agriculture, livestock production, and dairy farming system together can be more resilient and sustainable compared with adopting agriculture practice alone. The product or by-product of one system could be utilized for the other and vice versa.

This approach suggests the integration of technologies within the natural boundaries for optimum development of land, water, and plant resources to meet the basic needs of people and animals in a sustainable manner. The holistic approach focuses on (i) conservation, upgradation, and utilization of natural endowments such as land, water, plant, animal, and human resources in a harmonious and integrated manner with low-cost, simple, effective, and replicable technology; and (ii) reduction of inequalities between irrigated and rainfed areas and poverty alleviation. Thus, this approach aims to improve the standard of living of common people by increasing their earning capacity by making available all facilities required for optimum production and disposal of marketable surplus (Wani et al. 2006b). This approach suggests adopting land and water conservation practices, water harvesting in ponds, and recharging of groundwater for increasing the potential of water resources, and emphasizes on crop diversification, use of improved variety of seeds, integrated nutrient management (INM), and integrated pest management (IPM) practices.

13.5.2 Integrated Watershed Management for Sustainable Intensification of Rainfed Agriculture

It is well documented (Wani et al. 2007, 2008; Joshi et al. 2008) that the watershed management program is one of the most suitable options for increasing WUE and also as an adaptive strategy to cope with climate change impact in rainfed areas (Wani et al. 2002, 2009a, 2011a; Mujumdar 2008; Batisani and Yarnal 2010; Feng et al. 2010; Hanjra and Qureshi 2010; Barron and Keys 2011; Wilson 2011). The watershed development program recorded increased soil and water conservation with concomitant retention of more rainwater through several in situ (green water) and ex situ interventions of blue water at the farm (micro) and watershed/catchment (meso) scale and augmented its use within the boundary of the landscape (Samra and Eswaran 2000; Wani et al. 2008, 2011a,b; Barron and Keys 2011; Wilson 2011). Wani et al. (2009a) described the watershed scale as the "entry point" for effective management of smallholder agroecosystems for improving livelihoods. Wilson (2011) described in detail the integrated watershed management for improving livelihoods and integrated rural development in developing countries, particularly in Asia and possibly in Africa. Further, Barron and Keys (2011) interpreted successes in watershed case studies in terms of overall agroecosystem stability, described watershed management through resilience, and suggested that "entry point" refer to a specific point of entry for managers or farmers to actively intervene in the dynamic smallholder rainfed agroecosystems.

Implementing watershed activities at smaller landscape levels probably may not realize actual benefits, as was clearly visible at the mesoscale level, as Joshi et al. (2005) observed that watersheds >1000 ha were more effective in economic, equity, and sustainability parameters. It is quite likely that farm pond/check dams built at one location may benefit groundwater recharge beyond the boundary of the implementation. Similarly generated groundwater recharge/water table may increase base flow at a further downstream location (Sreedevi et al. 2004; Wani et al. 2011a). The national program of watershed management in India has realized the scale issue as recommended (Wani et al. 2008) and has adopted 1000–5000 ha of watershed area for implementing the program with new common watershed guidelines (GoI 2008).

13.5.3 Learnings from Meta-Analyses of Watershed Case Studies from India

A descriptive summary of multiple benefits derived from 636 watersheds revealed that watershed programs are silently bringing about a revolution in rainfed areas with a mean benefit–cost (B/C) ratio of 2.0 with the benefits ranging from 0.82 to 7.30 (Table 13.4) and >99% of projects were economically remunerative. About 18% watersheds generated a B/C ratio above 3, which is fairly modest (Figure 13.16a). However, it also indicated a large scope to enhance the impact of 68% of watersheds that performed below an average B/C of 2.0. Merely 0.6% of the watersheds failed to commensurate with the cost of the project (Joshi et al. 2008).

The mean internal rate of return of 27.43% was significantly high and comparable with any successful government program (Table 13.4). The internal rates of return in 41% of watersheds were in the range of 20%–30%, whereas about 27% of watersheds yielded IRR of 30%–50% (Figure 13.16b). The watersheds with IRR below 10% were only 1.9%. Watershed programs generated significant and substantial employment opportunities in the watershed areas (Table 13.4), which means raising their purchasing power, resulting in alleviating rural poverty and income disparities. This has an important implication in the sense that the watershed investment may be considered as a poverty alleviation program in the fragile ecosystem areas (Joshi et al. 2008).

The estimates show that watershed programs were quite effective in addressing the problems of land degradation due to soil erosion and loss of water due to excessive runoff. Soil loss of about 1.12 t/ha/year was prevented due to interventions in the watershed framework. Conserving soil means raising farm productivity, increasing WUE, and preserving the good soils for the next generation. It was noted that on average, about 38 ha-m (10^4 cubic meters) additional water storage capacity was created in a watershed of 500 ha as a result of the watershed program. Augmenting water storage capacity contributed to (i) reducing rate of runoff by 46% and (ii) increasing groundwater recharge by 3.6 m on an average in the watershed areas. These had a direct impact on expanding the irrigated area, increasing cropping intensity, and diversifying systems with high-value crops. On an average, the irrigated area increased by about 52%, while the cropping intensity increased by 35.5%. In some cases the irrigated area increased up to 204% while the cropping intensity increased by 283%. Such an impressive increase in the cropping intensity was not realized in many surface-irrigated areas in the country. These benefits confirm that the watershed programs perform as a viable strategy to overcome several externalities arising due to soil and water degradation (Joshi et al. 2008).

The above evidence suggests that watershed programs, which have been specifically launched in rainfed areas with the sole objective of improving the livelihood of poor rural households in a sustainable manner, have paid rich dividends and were successful in raising income levels, generating employment opportunities, and augmenting natural resources in the rainfed areas. These benefits have far-reaching implications for rural masses in the rainfed environment, and watershed management is recommended as a growth engine for the rural development of rainfed areas (Wani et al. 2008).

The results of meta-analysis regression further showed that the benefits vary depending upon the location, size, type, rainfall pattern, implementing agency, and people's participation. It is also important to state that the focus of the watershed program, status of the target population, and people's participation are some of the critical factors that play a deterministic role in the performance and efficiency of watersheds (Joshi et al. 2008). The drivers of success of watershed programs through increased efficiency (Wani et al. 2008) are discussed below:

- Macro watersheds (>1200 ha) achieved better impact than micros of 500 ha. Development activities need to be undertaken in clusters of at least four to six micro watersheds (2000–3000 ha).
- Available technologies are effective between 700 mm and 1100 mm of rainfall zone and the principle of "one size fits all" does not work. There is an urgent need to evaluate technologies for <500 and >1100 mm annual rainfall zones.

TABLE 13.4
Summary of Benefits from the Sample Watersheds Using Meta-Analysis

	Particulars	Unit	Number of Studies	Mean	Mode	Median	Minimum	Maximum	t-Value
Efficiency	B/C	Ratio	311	2.01	1.70	1.70	0.82	7.30	35.09
	IRR	percent	162	27.43	25.90	25.00	2.03	102.70	21.75
Equity	Employment	person days/ha/year	99	154.53	286.67	56.50	0.05	900.00	8.13
Sustainability	Increase in irrigated area	percent	93	51.55	34.00	63.43	1.28	204.00	10.94
	Increase in Cropping intensity	percent	339	35.51	5.00	21.00	3.00	283.00	14.96
	Runoff reduced	percent	83	45.72	43.30	42.53	0.38	96.00	9.36
	Soil loss saved	t/ha/year	72	1.12	0.91	0.99	0.11	2.05	47.21

Source: Joshi, P.K., Jha, A.K., Wani, S.P., et al., Impact of watershed program and conditions for success: A meta-analysis approach. In *Global Theme on Agroecosystems*, Report no. 46. Patancheru 502 324, International Crops Research Institute for the Semi-Arid Tropics, Andhra Pradesh, India, 2008.

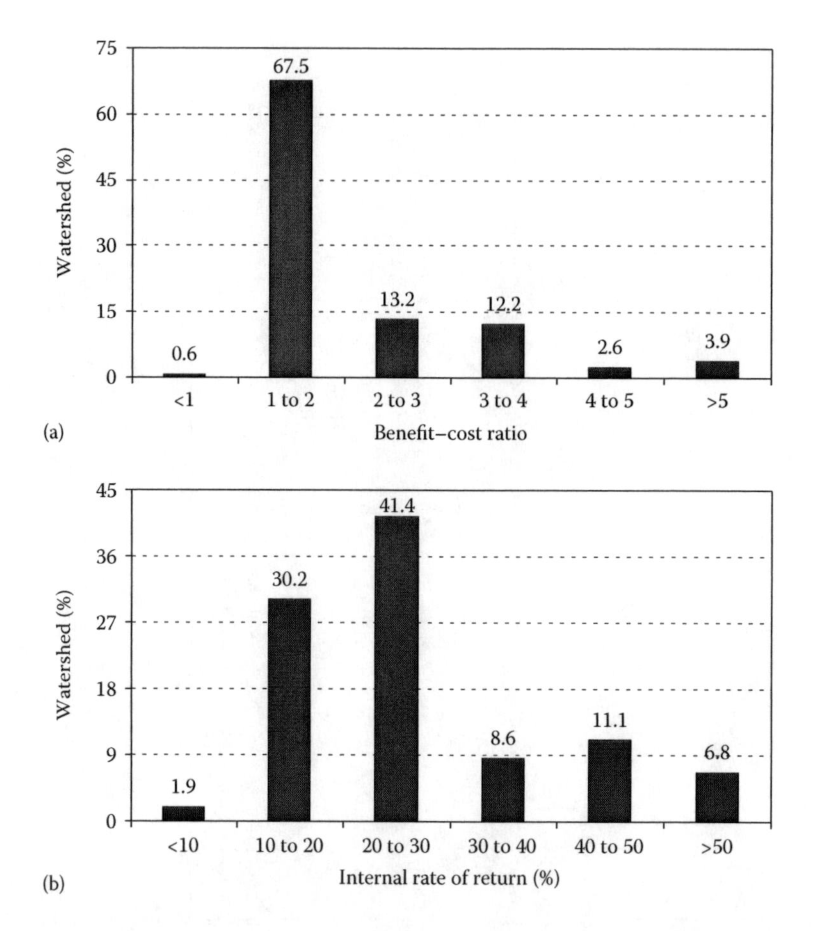

FIGURE 13.16 (a) Distribution (%) of watersheds according to benefit–cost ratio (BCR). (b) Distribution (%) of watersheds according to internal rate of return (IRR). (From Joshi, P.K., Jha, A.K., Wani, S.P., et al., Impact of watershed program and conditions for success: A metaanalysis approach. In *Global Theme on Agroecosystems,* Report no. 46. Patancheru 502 324, International Crops Research Institute for the Semi-Arid Tropics, Andhra Pradesh, India, 2008.)

- Use of new scientific tools such as crop simulation and water balance models, GIS, remote sensing and information and communication technology (ICT), participatory research and development (PR&D), and collective action for planning, implementation, monitoring and evaluation (M&E) are needed to manage natural resources more efficiently and sustainably in the watersheds.
- The drivers of success are tangible economic benefits to a large number of people; empowerment through knowledge sharing; equal partnership, trust, and shared vision; good local leadership; transparency and social vigilance in financial dealings; equity through low-cost structures; predisposition to work collectively; activities targeted at the poor and women; increased drinking water availability; and income-generating activities for women.
- The current allocations are insufficient to "treat" a complete watershed or to adopt the livelihood approach. Higher investments are a must to make watersheds engines of growth. The Government of India (GoI) has increased investments in new integrated watershed management programs (IWMP) from (Indian rupees) 6,000 (USD 133) to 12,000 (USD 266) per ha in plains and 15,000 (USD 333) in hilly areas (GoI 2008) and has adopted a livelihood approach to ensure tangible economic benefits to people in a watershed.

- Reduction of costs through convergence of action to avoid duplication, costs of environment deterioration, and enhancing efficiency of interventions.
- Interventions to benefit women and vulnerable groups have developed social capital and increased sustainability.
- Impact on production, poverty, the environment, and community involvement was achieved through capacity building. In order to effectively implement programs, the implementing agencies need to expand and broaden their capacities, skills and reach; and communities need to strengthen their institutions and skills. This will require a longer implementation period of 7–8 years with more time spent in preparation and in postintervention support. There is a need for additional funds and more flexibility in using budgets, as well as the engagement of specialist service providers. New common guidelines (GoI 2008) have addressed these recommendations and the project duration is increased up to 7 years with 5% of the total budget earmarked for capacity building using the services of quality service providers.
- New technologies and technical backstopping improved the performance of watershed programs. Forming consortia and employing agencies to provide specialist technical backstopping through a National Support Group (NSG) are needed.
- Improved and concurrent M&E and constant feedback improved performance. Detailed monitoring of one or two representative watersheds in each district for a broad range of technical and socioeconomic parameters measured provided a scientific benchmark and a better economic valuation of impact through scaling-up using bioeconometric and social models.

13.5.4 Business Model

Watersheds should be seen and developed as a business model. This calls for a shift in approach from subsidized activities to knowledge-based entry points and from subsistence to marketable surplus, ensuring tangible economic benefits for the population of the watershed at large. This is being done with productivity enhancement, diversification to high-value enterprises, income-generating activities, market links, public–private partnerships, microentrepreneurship, and broad-based community involvement. Strengths of rainfed areas using available water resources efficiently through involvement of private entrepreneurs and value addition can be harnessed by linking small and marginal farmers to markets through a public–private partnership business model for watershed management (Wani et al. 2008).

13.5.5 Rainwater Conservation and Harvesting: An Entry Point for Sustainable Intensification

In situ interventions and land management such as field and contour bunding, conservation agriculture (CA), and minimum tillage practices can enhance infiltration capability and convert more rainfall into green water (Wani et al. 2003a, 2008, 2009; Rockström et al. 2007; Garg et al. 2011a). In addition, soil organic matter augmentation, improved crop agronomy options, balanced plant nutrition, improved crops and crop varieties, crop protection, crop intensification through double cropping, contingency cropping, and reduction of rainy season fallows and rice fallows play an important role in enhancing green WUE by plants (Wani et al. 2009a, 2011b; Singh et al. 2011).

Agricultural water interventions, especially ex situ interventions, are helpful in enhancing blue water resources in watersheds as well as downstream areas (Wani et al. 2003a; Pathak et al. 2009, 2011; Glendenning and Vervoort 2011). Rainwater harvesting (RWH) has great potential of contributing to poverty reduction efforts by improving agricultural productivity and profitability in rainfed areas in Africa and Asia (Wani et al. 2002, 2009a, 2011a,b; Rockström et al. 2007; Pathak et al. 2009, 2011; Oweis and Hachum 2009; Sharma et al. 2010; Mati et al. 2011). Low-cost water-harvesting structures such as check dams and farm ponds could be constructed using available local expertise and materials (Wani et al. 2003b, 2011a; Pathak et al. 2007). This water could directly be used for supplemental irrigation or to enhance groundwater recharge, and with increased water availability, farmers can shift from low-value crops to cultivate high-value vegetables, fruit trees, and other cash crops (Wani et al. 2009a, 2011a). Moreover, it reduces flash flood, enhances nonerosive base flow, and also helps in reducing soil and nutrient loss.

Unlike the green revolution in Asia, the African agricultural sector is predominantly rainfed, even in ecological zones, which by necessity should be fully or partially irrigated. Currently, 4% of water resources have only been developed for agriculture, water supply, and hydropower use in Africa compared with 70%–90% in Asia and developed countries (Mati 2010). Moreover, RWH techniques build reliance against extreme events such as long dry spells upstream and flood-type situations downstream (Reij et al. 1996; Mati 2005; Mati et al. 2011; Wani et al. 2006b, 2011a).

13.5.6 Strategies for Enhancing Water Productivity in Rainfed Areas

There are several climatic and land management factors responsible for crop growth, crop yield, and crop WP in dryland agriculture. For example, soil water availability, nutrient/fertility status, selection of right crop/variety, supplemental irrigation, and pest and disease infestation are among a few. Selection of crop/variety should be based on the length of growth period such that it has high probability to attain production successfully. Various agricultural water interventions increase soil moisture availability and are particularly helpful during long dry spells.

13.5.7 Field-Scale Interventions to Shift Water Vapor Losses through Evapotranspiration

13.5.7.1 Crop Intensification through Land Surface Management

Intercropping or mixed cropping systems are more resilient compared with monocropping system in rainfed areas due to efficient and better utilization of resources such as green water, soil nutrients, and light. These systems are also stable under adverse weather and pest/disease situations. Land smoothening and forming of field drains are basic components of land and water management for conservation and safe removal of excess water in a guided manner. Broad bed and furrow (BBF) system is an improved in situ soil and moisture conservation and drainage technology for clayey soils with low infiltration rate as soil profile gets saturated and waterlogged with the progression of the rainy season (El-Swaify et al. 1985).

Data from long-term research trials at ICRISAT show that management of Vertisols with improved management options and interventions improved soil physical, chemical, and biological properties of micro watersheds. Field-scale intervention of improved management comprises sowing of crops on graded BBF of 45 cm as practice for in situ soil and water conservation and safe disposal of excess runoff during heavy downpour. The rainy season crops (sole and intercrops) along with pigeonpea/maize/sorghum/soybean/green gram were sown in the dry bed prior to the onset of monsoon rains, and two crops were grown annually in rotation. Fertilizer management involved the application of 80 kg N and 40 kg P_2O_5 per hectare. Under traditional practice, the seedbed was kept flat, and one crop, either sorghum or chickpea, was grown during the postrainy season utilizing the stored soil moisture in the profile. No mineral fertilizers were added, and farmyard manure (FYM) was added at 10 t/ha every 2 years. Results show that improved management significantly increased soil porosity, infiltration rate, and carbon content compared with traditionally managed fields (Table 13.5). Such changes in the biophysical properties also led to changes in the hydrological cycle as runoff was reduced in BBF fields and stored more rainfall into green water form. A significant amount of total rainfall is used in productive transpiration; therefore, crop yields in BBF fields were found consistently higher than 4.5 t/ha, irrespective of several deficit and surplus water years (Wani et al. 2003a, 2011b; Pathak et al. 2005). On the other hand, average crop yield in traditionally managed fields was found to be 0.9 t/ha. Average crop WP of BBF fields was found to be 0.65 kg/m^3 compared with 0.15 kg/m^3 in traditionally managed fields (Table 13.5).

TABLE 13.5
Effects of Long-term Landform Treatment on Physical, Chemical, and Biological Soil Properties of Micro Watershed and Its Impact on Hydrology, Crop Yield, and Water Productivity at the ICRISAT, Heritage Watershed Site in Patancheru, India (1976 and 1998)

Parameter	Improved System	Traditional System
Land management practices	Broad bed and furrow	Flat land
Cropping system and its rotation	*First year:* maize followed by chickpea *Second year:* sorghum intercropped with pigeonpea	Sorghum or chickpea
Fertilizer application per hectare	80 kg N and 40 kg P_2O_5	FYM every 2 years (10 t/ha)
Biophysical properties of soil		
Bulk density of surface soil (g/cm^3)	1.2	1.5
Air-filled porosity (%)	41	33
Penetration resistance (M Pa)	1.1	9.8
Sorptivity (mm/30 min)	121	100
Cumulative infiltration in 1 h (mm)	347	205
Chemical properties of soil		
Organic C in 0–60 cm soil (t/ha)	27.4	21.4
Total N (kg/ha)	2684	2276
Organic carbon content in 0–120 cm soil (t/ha)	46.8	39.5
Biological properties in 0–60 cm soil		
Soil respiration (kg C/ha per 10 days)	723	260
Microbial biomass C (kg C/ha)	2676	2137
Microbial biomass N (kg N/ha)	86.4	39.2
Hydrology and soil loss		
Average annual rainfall in 1974–1982 (mm)	823	823
Surface runoff (mm)	112 (13.6%)	207 (25.1 %)
Soil loss (t/ha)	1.5	6.5
Crop yield and water productivity		
Grain yield between 1976 and 2006 (t/ha)	4.5	0.9
Increasing average yield rate (kg/ha/year)	82	23
Carrying capacity (person/year)	21	4.6
Crop water productivity (kg/m^3)	0.65	0.15

On-farm trials on land management of Vertisols of central India revealed that the BBF system resulted in a 35% yield increase in soybean during the rainy season and yield advantage of 21% in chickpea during postrainy season compared with the farmers' practice. A similar yield advantage was recorded in maize and wheat rotation under the BBF system (Table 13.6a). Yield advantage of 15%–20% was recorded in maize, soybean, and groundnut with conservation furrows on Alfisols over farmers' practices at Haveri, Dharwad, and Tumkur watersheds in Karnataka (Table 13.6a).

TABLE 13.6a
Impact of IWRM-Based Intervention on Crop Yields at Different Benchmark Locations and Farm Fields in India and Elsewhere

Study Location/Benchmark Site	Interventions Made	Parameter Identified/ Estimated	Before/without Interventions	After/with Interventions	Impact Achieved	Data Source
Sujala Watershed, Karnataka, India	Contour cultivation along with conservation furrows	Crop yields (t/ha)	1.7 (1.2–3.4)	2.0 (1.4–3.9)	20% increased	Sujala-ICRISAT watershed project, Terminal Report (2008)
Vidisha, Sagar, Guna, Sehore and Raisen (MP, India) (170 farmers)	Land form treatment (bbf) + micro nutrient application	Soybean yield (t/ha)	1.9 (1.5–2.5)	2.3 (1.7–2.9)	20% increased	water-use efficiency project, Completion Report (2009)
Ginchi, Akaki in Ethiopia	Land form treatment (bbf)	Wheat yield (t/ha)	0.8–0.9	1.2–1.5	60% increased	Srivastava et al. (1993)
Sahel (1998–2000)	Supplemental irrigation and fertilizer application	Sorghum yield (t/ha)	0.45 (0.25–0.65)	1.4 (0.9–1.8)	210% increased	Fox and Rockstrom (2003)
Jhansi, Bengaluru and Indore, India	One supplemental irrigation of 40 mm in monsoon	maize, millet, soybean yield (t/ha)	2.2 (average)	2.8 (average)	30% increased	Vijayalakshmi (1987)
Andhra Pradesh, India (2002–2004)	Micro-nutrient s, b, zn + n p application	Maize yield (t/ha)	2.6	4.3	65% increased	Rego et al. (2005)
	Micro-nutrient s, b, zn + n p application	Groundnut yield (t/ha)	0.75	1.1	55% increased	
Vietnam (2000)	Mulching in groundnut	Groundnut yield (t/ha)	5.3	6.3	19% increased	Ramakrishna et al. (2006)
Vietnam (Spring 2001)	Nutrient management	Groundnut yield (t/ha)	5.5	6.6	20% increased	Ramakrishna et al. (2006)
		Biomass yield (t/ha)	9.5	11.3	19% increased	
Haveri, Karnataka, India	Contour cultivation (year 2006–2008)	Maize yield (t/ha)	3.35	3.89	16% increased	ICRISAT (2008) and Pathak et al. (2011)
Dharwad, India	Contour cultivation (year 2006–2008)	Soybean yield (t/ha)	1.47	1.8	23% increased	ICRISAT (2008) and Pathak et al. (2011)
Kolar, India	Contour cultivation (year 2006–2008)	Groundnut yield (t/ha)	1.23	1.43	16% increased	ICRISAT (2008) and Pathak et al. (2011)
Tumkur, India	Contour cultivation (year 2006–2008)	Finger millet yield (t/ha)	1.28	1.59	24% increased	ICRISAT (2008) and Pathak et al. (2011)

Guna	BBF + improved crop	Soybean yield (t/ha)	1.46	1.70	Increased (%)	ICRISAT (2008)
Raisen	varieties + application of		1.56	2.28	16%	
Videsha	balanced fertilizer (total 140		1.72	2.23	45%	
Indore	farmers fields (covering 17		2.51	2.90	30%	
Sehore	village in Madhya Pradesh,		2.09	2.50	15%	
(during year 2007–2009)	India))				19%	
Ginchi, Ethiopia	Raised BBF	Wheat yield (t/ha)	0.83 (±0.08)	1.2 (±0.05)	46% increased	Srivastava et al. (1993)
Akaki, Ethiopia	Raised BBF	Wheat yield (t/ha)	0.96 (±0.06)	1.5 (±0.07)	54% increased	
Bellary, Karnataka, India 1988–1996	Vegetative barrier on resource conservation (land slope 1.5%)	Sorghum yield (t/ha)	0.47	0.78	35% increased	Rao et al. (2003)
Sahel 1998–2000	Supplemental irrigation	Sorghum yield (t/ha)	0.45 (±0.23)	0.71 (±0.32)	60% increased	Fox and Rockstrom (2003)
	Fertilizer application		0.45 (±0.23)	0.98 (±0.40)	120% increased	
	Supplemental irrigation + fertilizer application		0.45 (±0.23)	1.40 (±0.36)	210% increased	
Short duration rainy season	Supplemental irrigation (cm)	Yield (t/ha)			Increased (%)	Vijayalakshmi et al. (1987)
Hyderabad, India	1.6	Sorghum	0.38	2.51	560	
Jhansi, India	1.0	Maize	2.31	2.66	15	
Jhansi, India	2.0	Maize	3.16	4.43	40	
Bengaluru, India	5.0	Finger millet	1.56	2.23	43	
Indore, India	8.0	Soybean	1.80	2.05	14	
Long duration rainy season	Supplemental irrigation (cm)	Yield (t/ha)			Increased (%)	Vijayalakshmi et al. (1987)
Hyderabad, India	5.0	Castor	1.01	1.32	31	
Jhansi, India	3.0	Pigeonpea	0.05	0.17	240	
Jhansi, India	5.0	Pigeonpea	0.05	0.33	560	
Dantiwada, India	4.0	Tobacco	0.82	1.30	58	
Postrainy season	Supplemental irrigation (cm)	Yield (t/ha)			Increased (%)	Vijayalakshmi et al. (1987)
Dehradun, India	2.0	Wheat	1.17	1.58	35	
Dehradun, India	4.0	Wheat	1.17	2.06	78	
Dehradun, India	6.0	Wheat	1.17	2.60	123	
Ranchi, India	1.0	Rape seed	0.25	0.35	40	
Ranchi, India	3.0	Rape seed	0.25	0.46	84	
Ranchi, India	5.0	Rape seed	0.25	0.54	116	

(continued)

TABLE 13.6a (Continued)
Impact of IWRM-Based Intervention on Crop Yields at Different Benchmark Locations and Farm Fields in India and Elsewhere

Study Location/Benchmark Site	Interventions Made	Parameter Identified/ Estimated	Before/without Interventions	After/with Interventions	Impact Achieved	Data Source
ICRISAT, Patancheru, India	Supplemental irrigation (cm)	Chickpea yield (t/ha)			Increased (%)	Pathak et al. (2009)
	6.3		0.69	0.92	32	
	4.6		0.69	0.91	32	
ICRISAT, Patancheru, India	Supplemental irrigation	Yield (t/ha)			Increased (%)	Pathak et al. (2009)
Vertisol watershed		Maize chickpea	1.04	1.54	47	
		Mung-chilli	1.00	1.33	32	
		Maize-safflower	1.07	1.24	15	
Semiarid tropics, Andhra Pradesh, India	Balanced nutrient management	Yield (t/ha)			Increased (%)	Rego et al. (2007)
(results based on total 286 farmers field during year 2002–2004)		Maize	2.4–2.7	4.2–4.8	72	
		Castor	0.5–0.9	0.8–1.3	52	
		Mung bean	0.7–0.9	1.1–1.5	58	
		Groundnut	0.8–1.3	1.4–1.8	47	
		Pigeonpea	0.5–1.0	0.8–1.5	72	
Semiarid tropics, Karnataka, India (results based on total 992 farmers field during year 2005–2009)	Balanced nutrient management	Yield (t/ha)			Increased (%)	ICRISAT 2008
		Maize	4.0–5.6	5.4–8.7	44	
		Finger millet	1.6–2.1	2.1–3.2	49	
		Groundnut	0.9–1.8	1.4–2.1	35	
		Soybean	1.3–2.1	1.6–3.4	60	
Madhya Pradesh, India	Balanced nutrient management	Yield (t/ha)			Increased (%)	ICRISAT 2008
(results based on total 286 farmers		Soybean	1.49	1.84	23	
field during year 2008–2009)		Chickpea	1.25	1.44	15	
Rajasthan, India	Balanced nutrient management	Yield (t/ha)			Increased (%)	ICRISAT2008
(results based on total 33 farmers		Maize	2.7	2.9	20	
field during year 2008)		Pearl millet	2.3	2.5	20	

Yield advantage and rainfall use efficiency (RUE) were also reflected in cropping systems involving soybean–chickpea, maize–chickpea, and soybean/maize–chickpea under improved land management systems. The RUE ranged from 10.9 to 11.6 kg/ha/mm under BBF systems across various cropping systems compared with 8.2–8.9 kg/ha/mm with flat-on-grade system of cultivation on Vertisols.

13.5.7.2 Rainy Season Fallow Management

Vertisols and associated soils, which occupy large areas globally (approximately 257 mha; Dudal 1965), are traditionally cultivated during postrainy season on stored soil moisture due to waterlogging-associated risks during the rainy season caused by poor infiltration rates. The practice of fallowing Vertisols and associated soils in Madhya Pradesh, India, was perceived to be decreased after the introduction of soybean; however, 2.02 mha of cultivable land is still kept fallow in central India, during the kharif season (Wani et al. 2002; Dwivedi et al. 2003). However, the survey also indicated that rainy season fallows of soybean-replaced sorghum remained fallow because rainy season crop delays the sowing of postrainy (rabi) crop, forcing the farmers to keep the cultivable lands fallow, thus reducing WUE and enhancing soil erosion. Through watershed on-farm participatory research, ICRISAT demonstrated the avoidance of waterlogging during initial crop growth periods on Vertisols by preparing the fields as BBF along with grassed waterways. Simulation studies using the SOYGRO model showed that early sowing of soybean in 7 out of 10 years was possible by which soybean yields can be increased threefold along with appropriate nutrient management. Hence, evolving timely sowing with short-duration soybean genotypes could pave the way to successful postrainy season crop where the moisture-carrying capacity is sufficiently high to support it. On-farm soybean trials conducted by ICRISAT involving improved land configuration (BBF) and short-duration soybean varieties along with fertilizer application (including micronutrients) showed a yield increase of 1300–2070 kg/ha compared with 790–1150 kg/ha in Guna, Vidisha, and Indore districts of Madhya Pradesh. Increased crop yields (40%–200%) and incomes (up to 100%) were realized with landform treatment, new varieties, and other best-bet management options (Wani et al. 2008).

13.5.7.3 Rice Fallow Management for Crop Intensification

A considerable amount of green water is available after the monsoon, especially in rice–fallow systems, which could easily be utilized by introducing a short-duration legume crop with simple seed priming and micronutrient amendments (Subbarao et al. 2001; Kumar Rao et al. 2008; Wani et al. 2009a; Singh et al. 2010). About 14.29 mha (30% of rice-growing area) rice–fallows are available in the Indo-Gangetic Plains (IGP) spread over Bangladesh, Nepal, Pakistan, and India, out of which 11.4 mha (82%) are in the states of Bihar, Madhya Pradesh, Chhattisgarh, Jharkhand, West Bengal, Orissa, and Assam in India (Subbarao et al. 2001). Taking advantage of sufficient available soil moisture in the soil after harvesting rice crop during the cool season in eastern India, growing of early maturing chickpea in rice–fallow areas with best-bet management practices (minimum tillage for chickpea, seed priming of chickpea, 4–6 h with the addition of sodium molybdate to the priming water at 0.5 g/L/kg seed and *Rhizobium* inoculation at 5 g/L/kg seed, micronutrient amendments, and use of short-duration rice cultivars during rainy season) resulted in chickpea yields of 800–850 kg/ha (Harris et al. 1999; Kumar Rao et al. 2008). An economic analysis has shown that growing legumes in rice fallows is profitable for the farmers with a B/C ratio exceeding 3.0 for many legumes. Also, utilizing rice–fallows for growing legumes could result in the generation of 584 million person-days employment for South Asia.

In a number of villages in the states of Chhattisgarh, Jharkhand, and Madhya Pradesh in India, on-farm farmers' participatory action research trials sponsored by the Ministry of Water Resources, GoI, showed significantly enhanced RUE through cultivation of rice–fallows with a total production of 5600–8500 kg/ha for two crops (rice + chickpea), benefiting the farmers with increased average net income of Indian rupees 51,000–84,000 (USD 1130–1870/ha) (Singh et al. 2010).

13.5.7.4 Soil Organic Matter Management

In addition to its importance for sustainable crop production, low soil organic matter in tropical soils is a major factor contributing to their poor productivity (Lee and Wani 1989; Bationo and Mokwunye 1991; Syers et al. 1996; Edmeades 2003; Katyal and Rattan 2003; Bationo et al. 2008; Ghosh et al. 2009; Materechera 2010). Management practices that augment soil organic matter and maintain it at a threshold level are needed. Sequestration of carbon in soil has attracted the attention of researchers and policy makers alike as an important mitigation strategy for minimizing the impacts of climate change (Velayutham et al. 2000; Lal 2004; ICRISAT 2005; Bhattacharya et al. 2009; Srinivasa Rao et al. 2009), which also serves the purpose of enhancing soil moisture storage. Agricultural soils are among the earth's largest terrestrial reservoirs of carbon and hold potential for expanded C sequestration (Lal 2004). Improved agricultural management practices in the tropics such as intercropping with legumes, horticultural crop systems, application of balanced plant nutrients, suitable land and water management, and use of stress-tolerant high-yielding cultivars improved soil organic C content and also increased crop productivity (Lee and Wani 1989; Wani et al. 1995, 2003a, 2005, 2007; ICRISAT 2005; Srinivasa Rao et al. 2009) and enhanced soil moisture storage capacity (Lee and Wani 1989; Wani et al. 1994; Pathak et al. 2005, 2009, 2011). Farm bunds and degraded common lands in the villages could be productively used for growing nitrogen (N)-fixing shrubs and trees to generate N-rich loppings. For example, growing *Gliricidia sepium* at close spacing of 75 cm on farm bunds could provide 28–30 kg N per hectare in addition to valuable organic matter (Wani et al. 2009a, 2011c). Also, through vermicomposting as a microenterprise by women self-help groups (SHGs), large quantities of farm residues and other organic wastes are converted into valuable sources of plant nutrients and organic matter, enhancing agricultural productivity (Nagavallama et al. 2005; Sreedevi et al. 2007; Wani et al. 2008; Sreedevi and Wani 2009).

13.5.7.5 Minimum Tillage or Conservation Agriculture

As mentioned earlier, there is a direct relationship between consumptive water use (ET) and crop yield. ET comprises two major processes: nonproductive evaporation and productive transpiration. Evaporation, however, cannot be avoided completely, but it can be minimized through various field-scale management practices. The three basic elements of CA are (i) no or minimal tillage without significant soil inversion, (ii) retention of crop residues on the soil surface, and (iii) growing crops in rotation appropriate to the soil–climate environment and socioeconomic conditions of the region (crop diversification). Mulching by crop residue (CA), minimal or no tillage, mixed cropping system, and practicing agroforestry are some of the examples that cover the soil surface partially and reduce evaporation. Consequently, the same amount of water could be utilized by plant transpiration, leading to more biomass and crop yield.

Conservation tillage, an essential component of CA, constitutes land cultivation techniques that try to reduce labor, promote soil fertility, and enhance soil moisture conservation. CA is now recognized as the missing link between sustainable soil management and reduced cost of labor, especially during land preparation, and holds the potential to increase crop production and reduce soil erosion. On Alfisols at ICRISAT, Yule et al. (1990) while comparing the effects of tillage (i.e., no-till, 10 cm deep till, 20 cm deep till), amendments (i.e., bare soil, rice straw mulch applied at 5 t/ha, FYM applied at 15 t/ha), and the use of perennial species (e.g., perennial pigeonpea, *Cenchrus ciliaris,* and *Stylosanthes hamata* alone or in combination) on runoff and infiltration found that straw mulch consistently reduced runoff compared with bare plots. Tillage produced variable responses in their study. Runoff was reduced for about 20 days after tillage, but the tilled plots had more runoff than no-tilled treatments during the remainder of the cropping season, suggesting some structural breakdown of the soil aggregates in the tilled plots. On an average, straw mulch and tillage increased annual infiltration by 127 and 26 mm, respectively. These results of Yule et al. (1990) indicate that mulching or keeping the soil covered (as in the case of *Stylosanthes*) should be an important component in the cropping systems of the SAT.

Studies conducted in the semiarid regions of Africa also indicate that some of the conservation tillage systems, particularly no-till techniques, give lower yield than conventional tillage methods. For example, Huxley's (1979) no-till experiments at Morogoro in Tanzania showed that no-tilled maize yielded two-thirds to three-quarters the amount of that in cultivated soil. Furthermore, Nicou and Chopart (1979) conclude in their studies in Senegal, West Africa, that in order to be effective, straw mulch in conservation tillage systems needs to be applied in sufficient quantity to cover the surface of the soil completely so that it can fully protect the soil against evaporation and runoff. It has been gaining acceptance in countries such as Tanzania, Madagascar, Zambia, and Zimbabwe in Africa (Biamah et al. 2000; Nyagumbo 2000).

Kajiru and Nkuba (2010) reported that by adopting CA techniques in Tanzania's Bukoba and Missenyi districts of Kagera region, average maize yield increased from 2.50 t/ha to 3.40 t/ha by smallholder farmers. Tanzania has been fostering the adoption of CA because of its potential to address three areas of crucial importance to smallholder farmers: demand on household labor, food security through increased and sustainable crop yields, and household income (Mariki 2004; Lofstrand 2005). Some form of CA is practiced on 40% of the rainfed farm lands in the United States and is also becoming popular in several Latin American countries (Landers et al. 2001; Derpsch 2005). Examples from SSA show that converting from plough to CA resulted in yield improvements ranging between 20% and 120%, with WP enhancement ranging from 10% to 40% (Rockström et al. 2009). On the Loess Plateau, CA increased wheat productivity and WUE by up to 35% compared with conventional tillage, especially in the low rainfall years, suggesting benefits of CA in dry farming areas of northern China (Li HongWen et al. 2007; Wang et al. 2007). For the best results, CA practices such as mulching must be accompanied by requisite agronomic practices such as use of fertilizers, manures, pesticides, and high-quality seed, as well as proper water application and management. The potential disadvantages of CA are higher costs of pests and weed control, the cost of acquiring new management skills, and investments in new planting equipment. CA can be practiced on all soils, especially light soils. It increases the productivity, sustainability, and efficient use of natural resources (Rockström et al. 2009). Straw tends to be used for animal feed in most parts of the SAT, particularly in India, Senegal, and Mali. Therefore, while mulches appear to be useful theoretically, from a practical point of view it is difficult to see how they can be used in the present conditions of SAT agriculture. It is even debatable if production of more biomass through breeding will induce farmers in the region to apply residues to their soils or induce them to sell their extra residues in view of the attractive prices offered for fodder during the dry season.

13.5.8 Runoff Harvesting, Groundwater Recharge, and Supplemental Irrigation for Enhancing Rainwater Productivity

Rainfall in dry lands is highly erratic and nonuniform, which often leads to dry spells of longer duration. Various land and water interventions alleviate water stress to a certain extent, but supplemental irrigation can sometimes be extremely essential to save a crop. Crop intensification with the help of supplemental irrigation is also an important option for better use of available water resources and enhancement of income in rainfed regions.

Sharma et al. (2010) recently showed that the rainfed districts in India receiving rainfall in the range of 400–1600 mm covering 39 mha generate on an average 115 km^3/year surface runoff in a normal year. Twenty percent of harvested runoff can provide 100 mm of supplemental irrigation for 25 mha rainfed lands and the remaining 80% could contribute to meet river/environmental flow and other requirements for downstream locations. Figure 13.17 showed an average increase of 50% in total production through increased WP with one supplemental irrigation and improved management compared with the traditional practice. Several studies showed that water harvesting and supplemental irrigation are economically viable at the national level (Joshi et al. 2005, 2008; Wani et al. 2008, 2011a,b; Pathak et al. 2009, 2011; Sharma et al. 2010).

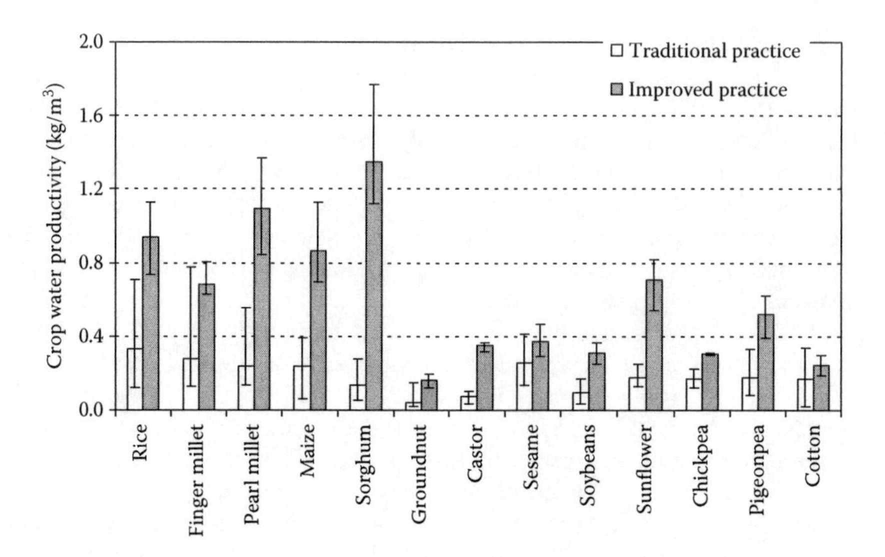

FIGURE 13.17 Crop water productivity of rainfed agriculture under traditional practices and improved technology situation in India; column in figure shows average crop yields and bars show their maximum and minimum range. (Data from Sharma, B.R., Rao, K.V., Vittal, K.P.R., et al., *Agricultural Water Management,* 97, 23–30, 2010.)

13.5.8.1 Water Harvesting and Groundwater Augmentation

RWH in watersheds is a basic activity and clear impacts of runoff harvesting through various types of structures in terms of increased groundwater availability, increased irrigated area, and increased cropping intensity are well documented in a meta-analysis result of 636 case studies reported by Joshi et al. (2008). Similar results have been also reported from a number of watersheds in India, Thailand, Vietnam, and China (Wani et al. 2003a, 2008, 2009).

13.5.9 Ex Situ Soil and Water Conservation

13.5.9.1 Runoff Harvesting and Supplemental Irrigation

The mean annual rainfall in most rainfed regions is sufficient for raising one, or in some cases, two good crops in a year. However, the onset of rainfall and its distribution are erratic, and prolonged droughts are frequent. A large part of rain occurs as high-intensity storms, resulting in sizable runoff volumes. In most rainfed regions, harvesting of excess runoff and storage into appropriate structures as well as recharging groundwater are very much feasible and a successful option for increasing and sustaining the productivity of rainfed agriculture through timely and efficient use of supplemental irrigation. In the areas with annual rainfall >500 mm, this approach could be widely adopted to enhance the cropping intensity, diversify the system into high-value crops, increase productivity and income from rainfed agriculture, and at the same time, create assets in the villages (Pathak et al. 2009, 2011; Sharma et al. 2010). Different types of runoff harvesting and groundwater-recharging structures are currently used in various regions. Some of the most commonly used structures are earthen check dams, masonry check dams, farm ponds, tanks, sunken pits, recharge pits, loose boulders, gully checks, drop structures, and percolation ponds (Figure 13.18).

Designing runoff harvesting and groundwater-recharging structures requires estimates of runoff volume, peak runoff rate, and other hydrological parameters, which are generally not available in most of the rainfed regions. Due to nonavailability of the data, many times these structures are

FIGURE 13.18 Commonly used water harvesting and groundwater recharging structures. (From Pathak, P., Sahrawat, K.L., Wani, S.P., et al., In *Rainfed Agriculture: Unlocking the Potential*, pp. 197–221, Comprehensive Assessment of Water Management in Agriculture Series. CAB International, Wallingford, UK, 2009.)

constructed without being properly designed, resulting in higher costs and often failure of the structures. Studies conducted by ICRISAT scientists have shown that the cost of water harvesting and groundwater-recharging structures varies considerably with the types of structures (Figure 13.19a) and the selection of appropriate location. Selection of appropriate location for structures can also play a very important role in reducing the cost of the structures (Figure 13.19b).

Pathak et al. (2009) reported that considerable information on various aspects of runoff water harvesting and supplemental irrigation could be obtained by using various models (Pathak et al. 1989; Ajay Kumar 1991), namely, runoff model, water harvesting model (Sireesha 2003), and model for optimizing the tank size (Sharma and Helweg 1982; Arnold and Stockle 1991). These models can assess the prospects of runoff water harvesting and possible benefits from irrigation. They can also be used to estimate the optimum tank size, which is very important for the success of the water-harvesting system. The information generated can also help in developing strategies for scheduling supplemental irrigation, particularly in cases where drought occurs more than once during the cropping season.

Rainfed agriculture has traditionally been managed at the field scale. Supplemental irrigation systems, with storage capacities generally in the range of 20–100 mm of irrigation water, even though small in comparison to irrigation storage, require planning and management at the catchment scale, as capturing local runoff may impact other water users and ecosystems. Legal frameworks and water rights pertaining to the collection of local surface runoff are required, as are human capacities for planning, constructing, and maintaining storage systems for supplemental irrigation, and moreover, farmers must be able to take responsibility for the operation and management of the systems. Supplemental irrigation systems also can be used in small vegetable gardens

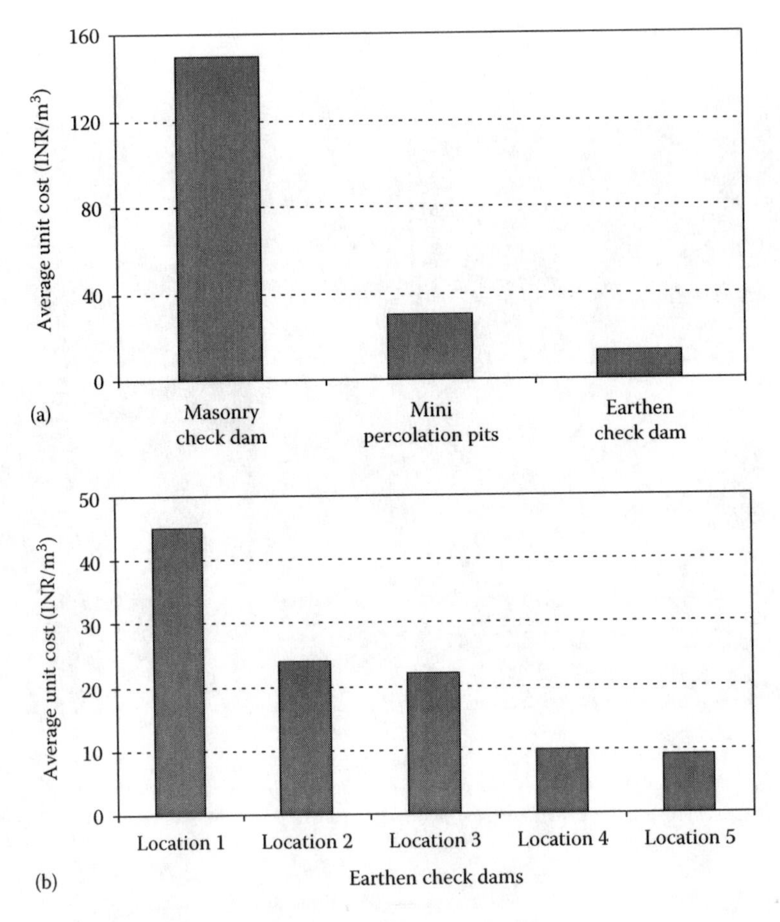

FIGURE 13.19 (a) Cost of harvesting water in different structures at Kothapally watershed, Andhra Pradesh, India. (b) Cost of water harvesting at different locations in Lalatora watershed, Madhya Pradesh, India. (From Pathak, P., Sahrawat, K.L., Wani, S.P., et al., In *Rainfed Agriculture: Unlocking the Potential*, pp. 197–221, Comprehensive Assessment of Water Management in Agriculture Series. CAB International, Wallingford, UK, 2009.)

during the dry seasons to produce fully irrigated cash crops. It is a key strategy, still underused, for unlocking the rainfed productivity potential and WP.

13.5.10 Increasing Water Use and Water-Use Efficiency

13.5.10.1 Efficient Supplemental Irrigation

In the semiarid and subhumid agroecosystems, dry spells occur in almost every season. These dry spells need to be mitigated to save the crop from drought and minimize the climate risks to crop production in rainfed systems. Supplemental irrigation is also used to secure harvests or to provide irrigation to the second crop during the postrainy season. Supplemental irrigation systems are ex situ water-harvesting systems comprising surface ponds or recharged groundwater. Efficient use of water involves both the timing of irrigation to the crop and efficient water application methods. Broadly, the methods used for application of irrigation water can be divided into two types: surface irrigation systems (border, basin, and furrow) and pressurized irrigation systems (sprinkler and drip). In the surface irrigation system, the application of irrigation water can be divided into two parts: (1) conveyance of water from its source to the field and (2) application of water in the field.

13.5.10.2 Conveyance of Water to the Field

In most SAT areas, water is carried to cultivated fields through open channels, which are usually unlined, and therefore, a large amount of water is lost through seepage. On the SAT vertisols, generally there is no need of lining the open field channels as the seepage losses in these soils are low mainly due to very low saturated hydraulic conductivity in the range of 0.3–1.2 mm/h (El-Swaify et al. 1985). On alfisols and other sandy soils having more than 75% sand, the lining of open field channel or use of irrigation pipes is necessary to reduce the high seepage water losses. The use of closed conduits (plastic, rubber, metallic, and cement pipes) are becoming popular, especially with farmers growing high-value crops, namely vegetables and horticultural crops (Pathak et al. 2009).

13.5.10.3 Methods of Application of Supplemental Water on SAT Vertisols

Formation of deep and wide cracks during soil drying is a common feature of the SAT Vertisols. The abundance of cracks is responsible for high initial infiltration rates (as high as 100 mm/h) in dry Vertisols (El-Swaify et al. 1985). This specific feature of Vertisols makes efficient application of limited supplemental water to the entire field a difficult task. As compared with narrow ridge and furrow, the BBF system saved 45% of the water without affecting crop yields on Vertisols. Compared with narrow ridge and furrow and flat systems, the BBF system had higher water application efficiency (WAE), water distribution uniformity, and better soil wetting pattern (Pathak et al. 2009). Studies conducted to evaluate the effect of shallow cultivation in furrow on the efficiency of water application showed that the rate of water advance was substantially higher in cultivated furrows as compared with that in uncultivated furrows. Shallow cultivation in moderately cracked furrows before the application of irrigation water reduced the water required by about 27% with no significant difference in chickpea yields.

13.5.10.4 Scheduling of Irrigation and Deficit Irrigation

Srivastava et al. (1985) studied the response of postrainy season crops to supplemental irrigation of maize or mung bean grown on a vertisol. The highest WAE was recorded for chickpea (5.6 kg/mm/ha), followed by chili (4.1 kg/mm/ha) and safflower (2.1 kg/mm/ha) (Table 13.6a). It was concluded that a single presowing irrigation to the sequential crops of chickpea and chili was profitable on Vertisols. Average additional gross returns due to supplemental irrigation were about USD 36/ha for safflower, USD 175/ha for chickpea, and USD 324/ha for chili.

Impressive benefits were reported from supplemental irrigation of rainy and postrainy season crops on Alfisols at ICRISAT, Patancheru, India (El-Swaify et al. 1985; Pathak and Laryea 1991). The average WAE for sorghum (14.9 kg/mm/ha) was more than that for pearl millet (8.8–10.2 kg/mm/ha) (Table 13.6b). An intercropped pigeonpea responded less to irrigation, and the average WAE ranged from 5.3 to 6.7 kg/mm/ha for both sorghum–pigeonpea and pearl millet–pigeonpea intercrop systems. Tomato responded very well to water application with an average WAE of 186.3 kg/mm/ha (Table 13.6b).

For the sorghum–pigeonpea intercrop, two irrigations of 40 mm each gave an additional gross return of USD 217/ha. The highest additional gross return of USD 1296/ha from supplemental irrigation was obtained with tomato.

The best responses to supplemental irrigation were obtained when irrigation water was applied at critical stages. To get the maximum benefit from the available water, growing high-value crops (namely, vegetables and horticultural crops) is becoming popular even with poor farmers (Pathak et al. 2009). According to Oweis (1997), supplemental irrigation of 50–200 mm can bridge critical dry spells and stabilize yields in arid to dry subhumid regions. The potential yield increase in supplemental irrigation varies with rainfall. An example from Syria illustrates that improvements in yields can be more than 400% in arid regions (Oweis 1997). Several studies indicate that supplemental irrigation systems are affordable by small-scale farmers (Fan et al. 2000; Fox et al.

TABLE 13.6b
Grain Yield Response of Cropping Systems to Supplemental Irrigation on an Alfisol Watershed at ICRISAT, Patancheru, Andhra Pradesh, India, 1981–1982

Yield with Irrigation (kg/ha)	Yield Increase (kg/ha)	WAE (kg/ha mm)	Yield with Irrigations (kg/ha)	Yield Increase (kg/ha)	WAE (kg/ha mm)	Combined WAE (kg/ha mm)
Intercropping System						
	Pearl millet			pigeonpea		
2353	403	10.0	1,197	423	5.3	6.8
	Sorghum			pigeonpea		
3155	595	14.9	1,220	535	6.7	9.4
Sequential Cropping System						
	Pearl millet			cowpea		
2577	407	10.2	735	425	5.3	6.9
	Pearl millet			tomato		
2215	350	8.8	26,250	14,900	186.3	127.1

Source: Pathak, P. and Laryea, K.B., Prospects of water harvesting and its utilization for agriculture in the semi-arid tropics. In *Proceedings of the Symposium of the SADCC Land and Water Management Research Program Scientific Conference,* October 8–10, 1990, pp. 253–268. Gaborone, Botswana, 1991.

Note: Irrigation of 40 mm each was applied.

$$\text{Water Application Efficiency (WAE)} = \frac{\text{Increase in yield due to irrigation}}{\text{Amount of irrigation applied}}$$

2005). However, policy framework, institutional structure, and human capacity similar to those for full irrigation infrastructure are required to successfully apply supplemental irrigation in rainfed agriculture.

13.5.11 Water Alone Cannot Do It

Water indeed is the primary element for crop growth, but water alone cannot bring production to its potential level; balanced nutrients (macro and micro), genetically improved stress-tolerant and high-yielding cultivars, and a pest- and disease-free environment are equally important.

13.5.11.1 Balanced Plant Nutrition

Along with water scarcity, soil fertility management in particular needs to be paid due attention alongside water stress management in view of the fragile nature of the soil resource base (Wani et al. 2009a; Sahrawat et al. 2010a,b). Moreover, it is commonly believed that at relatively low yields of crops in the rainfed systems, the deficiencies of major nutrients, especially N and P, are important for the SAT soils (El-Swaify et al. 1985; Rego et al. 2003; Sharma et al. 2009), and little attention was given to diagnose the extent of deficiencies of the secondary nutrients such as S and micronutrients in various crop production systems (Rego et al. 2005; Sahrawat et al. 2007, 2010a, 2011) on millions of small and marginal farmers' fields. Since 1999, ICRISAT and its partners have been conducting systematic and detailed studies on the diagnosis and management of nutrient deficiencies in the semiarid regions of Asia with emphasis on the semiarid regions of India under the IWMP (Wani et al. 2009a). These studies revealed widespread deficiencies of multiple nutrients including micronutrients such as boron, zinc, and the secondary nutrient sulfur in 80%–100% of farmers' fields (Rego et al. 2005; Sahrawat et al. 2007, 2010b, 2011). On-farm trials conducted in several states of India (Andhra Pradesh, Madhya Pradesh, Rajasthan, Karnataka, Maharashtra, Uttar Pradesh, Jharkhand, Tamil Nadu, Chhattisgarh, and Gujarat) showed significantly increased yields by 30%–120% in different crops with amendment of soils with the deficient micronutrients and secondary nutrients over the farmers' practice, resulting in overall increase in WUE and nutrient use efficiency (Table 13.6a) (Wani et al. 2006b, 2009a, 2011c; Rego et al. 2007). For example, Singh et al. (2009, 2011) reported that the application of S, B, and Zn over the FI treatment in on-farm trials in the SAT regions of India (states of Andhra Pradesh and Madhya Pradesh) increased the productivity of rainfed crops, resulting in increased RUE. The RUE of maize for grain production under FI was 5.2 kg/mm ha water compared with 9.2 kg/mm ha water with the combined application of S, B, and Zn over the FI treatment (Table 13.6c). The best results in terms of RUE for maize and several other crops, however, were obtained under the BN treatment when N and P were added along with S, B, and Zn. These results are in agreement with those reported by Rego et al. (2007), who found that farmers were applying sub-optimum quantity of major nutrients, especially N and P, and thus the applications of NP along with SBZn (NP + SBZn) gave the best results in terms of crop yield, biomass production, and nutrient uptake.

In an on-farm study conducted for three seasons (2005–2007) in the SAT region of Karnataka, Rajashekhara Rao et al. (2010) reported that balanced nutrient application not only increased grain and stover yield of rainfed maize (see results in Table 13.6a) but also increased partial factor productivity (grain yield in fertilized plot = [grain yield in absolute control + yield increase due to treatment] × amount of nutrient applied), agronomic efficiency (the incremental efficiency of applied nutrients over the control), B/C ratio ([grain yield of fertilized plot × price of grain] : [amount of nutrient applied × price of the applied nutrient inputs]), and RUE (grain yield/rainfall received during the growing season) for maize production (Table 13.6a).

Thus, soil quality or health is a major driver of enhanced RUE and productivity in rainfed systems and needs an implementing strategy in which balanced nutrients are integrated with soil and water conservation and management (Wani et al. 2009b).

13.5.11.2 Genetically Improved Crop Cultivars

The adoption of improved varieties always generates significant field-level impact on crop yield and stability. The yield advantage through the adoption of improved varieties has been recognized undoubtedly in farmer participatory trials across India under rainfed systems. Recent trials during the rainy season conducted across the Kolar and Tumkur districts of Karnataka, India, revealed that a mean yield advantage of 52% in finger millet was achieved with the use of high-yielding varieties such as GPU 28, MR 1, HR 911, and L 5 under farmer nutrient inputs and traditional management compared with use of local variety and farmer management. These results showed that the efficient use of available resources by the improved varieties reflected in the grain yields under given situations. However, a yield advantage of 103% was reported in finger millet due to improved varieties under best-bet management practices (balanced nutrition including the application of Zn, B, and S and crop protection). Similarly, the use of improved groundnut variety ICGV 91114 resulted in pod yield of 2.32 t/ha under farmer management compared with the local variety under similar inputs. The yields of improved varieties further improved by 83% over the local variety with improved management that included balanced nutrient application (Sreedevi and Wani 2009).

13.5.11.3 Integrated Pest Management

Introduction of IPM in cotton and pigeonpea substantially reduced the number of chemical insecticidal sprays in Kothapally, India, during the season and thus reduced the pollution of water bodies with harmful chemicals. Introduction of IPM and improved cropping systems decreased the use of pesticides worth USD 44–66/ha (Ranga Rao et al. 2007). The IPM practices, which brought into use

TABLE 13.6c
Impact of IWRM-Based Intervention on Surface Runoff, Soil Loss, Cropping Intensity and Change in Land Use at Different Benchmark Locations and Farm Fields in India and Elsewhere

Study Location/ Benchmark Site	Interventions Made	Parameter Identified/ Estimated	Before/without Interventions	After/with Interventions	Impact Achieved	Data Source
ICRISAT, Patancheru, India	Land form treatment in Alfisol	Soil loss (t/ha)	5.6	3.3	40% decreased	Pathak and Laryea (1995)
	Land form + Surface mulching	Soil loss (t/ha)	5.6	1.4	4 folds decreased	
ICRISAT Patancheru, India	Land form treatment in Vertisol	Runoff (% of rainfall received)	27	10	63% decreased	Pathak et al. (1985)
	Land form treatment in Vertisol	Soil loss (t/ha)	6.7	0.6	90% decreased	
Bellary, Karnataka, India	In situ moisture conservation practices	Water productivity (Kg/m³)	0.73	0.84	15% increased	Patil (2003)
Bellary, Karnataka, India 1988–1996	Vegetative barrier (land slope 1.5%)	Average runoff (mm) during rainfall of 100 mm intensity	59	44	36% decreased	Rama Mohan Rao et al. (2000) and Pathak et al. (2011)
		Soil loss (t/ha)	1.6	0.9	41% decreased	
Shekta watershed, Maharashtra (MH), India	IWRM-based interventions	Waste land rehabilitation	8% of total area	Nil	Improved landscape	Sreedevi et al. (2008)
	IWRM-based interventions	Cropping intensity (%)	95	123	30% increased	Wani et al. (2011)
Watershed in Ghod catchment, MH, India (area: 1333 ha)	IWRM-based interventions	Landuse change	Wasteland: 62% Ag Land: 38% Double crop: 11%	Wasteland: 48% Ag Land: 52% Double crop: 18%	Resilient and productive land use	Wani et al. (2005)
Tad Fa, NE, Thailand	IWRM-based interventions	Runoff mm	364 (28% of rainfall)	169 (13% of rainfall)	Reduced by 54%	Wani et al. (2011)
		Soil loss (t/ha)	31.2	4.2	Decreased by 87%	
Andhra Pradesh, India	Effect of micronutrient application	Rain use efficiency (kg/ha/mm)			Increased (%)	Singh et al. (2009)
		Maize	5.2	9.2	77	
		Groundnut	1.6	2.8	75	
		Mung bean	1.7	2.9	71	
		Sorghum	1.7	3.7	118	
Madhya Pradesh, India		Soybean	1.4	2.7	93	

local knowledge of using insect traps of molasses, light traps, and tobacco waste, led to extensive vegetable production in Xiaoxingcun (China) and Wang Chai (Thailand) watersheds (Wani et al. 2006b).

13.5.12 Water–Energy Nexus

Efficient use of water for irrigation, particularly groundwater, is closely related to assured supply of power for pumping out water from wells (open and bore wells). In India, above 44% of 142 mha arable land (62 mha) is irrigated, out of which 65% (37 mha) is irrigated with groundwater from 22 million wells powered largely with electrical pump sets. Most state governments in India have subsidized or provided free electricity for running pump sets in agricultural use. However, as the large demand for power cannot be met, as rural areas face severe power cuts and receive low-quality/low-voltage power for a limited time. As a result of free/heavily subsidized and insecure supply of low-quality power, farmers adopt the practice of leaving their pumps on continually for irrigating their fields whenever power is available. This results in low WUE, as irrespective of the plants' need, fields are irrigated. Generally, farmers irrigate the soil and not the plants.

Assured power supply is very closely related with efficient use of power as well as water in the agricultural sector. In Gujarat, the government has provided separate feeders and transformers to supply good-quality, assured power supply through a scheme called "Jyoti Gram," which has shown very good results in terms of efficient use of power as well as water. Alternatively, decentralized bioenergy produced in rural areas can also power the rural pump sets to irrigate the fields as and when needed (D'Silva et al. 2004). As in many countries including India, biofuels are considered an option for addressing the energy security concerns (Achten et al. 2010a), while also responding to the challenges of climate change mitigation (Phalan 2009). Programs for stimulating complementary use of biodiesel to displace petroleum-based diesel primarily focused on biodiesel production based on nonedible oil seeds produced on marginal or degraded lands (Wani et al. 2007, 2008).

Other than agricultural land, wasteland in the watersheds has the potential to grow trees and bioenergy crops such as *Jatropha* and *Pongamia* (Sreedevi and Wani 2009; Wani et al. 2009b), which can enhance RUE and also protect the environment. A substantial wasteland area consists of degraded lands that are deteriorating due to lack of appropriate soil and water management, or due to natural causes, which can be brought into more productive use. In India, roughly 40% of the wasteland area has been estimated as available for forestation (Sathaye et al. 2001) and about 14 mha is considered suitable for cultivating biofuel feedstocks, such as *Jatropha* (Wani et al. 2009b). Establishment of biofuel plantations is considered an option for rehabilitating wastelands, enhancing energy security, and providing employment opportunities and better livelihoods in rural areas (Wani and Sreedevi 2005; Wani et al. 2006b, 2009b; Phalan 2009; Sreedevi et al. 2009b; Achten et al. 2010b). In Powerguda hamlet in Adilabad district of Andhra Pradesh, which is inhabited by indigenous people, women SHGs have achieved through collective action a feat of extracting nonedible oil from *Pongamia pinnata* seeds collected from the existing trees in the forest using their right to harvest nontimber produce from the forest. The farmers from Kistapur have used a common bore well for pumping water using *Pongamia* oil in a diesel pump set and shared the bore well water among 12 small farmers. This initiative implemented by ICRISAT was funded by the United States Agency for International Development (USAID) for enhancing WUE through assured supply of power and sharing a common bore well along with other crop productivity enhancement options (Wani et al. 2009b).

However, to assess the impact of developing degraded lands in a watershed with biodiesel plantations, Garg et al. (2011b) investigated the opportunities and trade-offs of *Jatropha* cultivation on wastelands from a livelihood and environmental perspective, with soil and water as the critical resources. The water balance for fallow wasteland and *Jatropha*-cultivated land from a site located in Andhra Pradesh, southern India, showed reduced runoff from 43% to 31% following cultivation of *Jatropha* in fallow wasteland. Correspondingly, green water consumption increased from 52% to

TABLE 13.7
Annual Water Budget of Wasteland under Two Different Land Uses during 2009 (Velchal Village, Andhra Pradesh, India)

Water Balance Component	Fallow Land	*Jatropha* Land with Land Management Practices
Rainfall (mm)	896	896
Outflow (mm)	393 (43%) Erosive runoff	274 (31%) Less erosive
E or ET (mm)	460 (52%) (nonproductive)	200 (E) + 380 (T) = 580 (64%) (productive use)
GW recharge (mm)	43 (5%)	42 (5%)

Watershed water balance: rainfall = outflow (surface runoff) + evapotranspiration (ET) + groundwater recharge

64% due to a shift from soil evaporation to crop ET without affecting the groundwater recharge in both the scenarios (Garg et al. 2011a; Yeh et al. 2011; Table 13.7).

In fallow wasteland, a large fraction of rainfall absorbed by the soil (in the form of soil moisture) was lost through soil evaporation in monsoon and nonmonsoon periods. Diversion of water from runoff and evaporation to ET led to increased plant growth. This benefited the landscape by increasing soil moisture content and reducing soil erosion and nutrient losses. Measured agronomical data show that *Jatropha* produced approximately 1–1.5 t/ha of seed biomass annually, and biomass containing 1 tC/ha per annum was added to soil during dormancy (leaf fall and pruned plant parts). Thus, *Jatropha* could be a suitable candidate for sequestering carbon and rehabilitating wasteland into productive lands with increased water-holding capacity of soil over a long time period (Wani et al. 2009b; Yeh et al. 2011). At the subbasin scale, reductions in runoff as a result of converting wastelands to biofuel plantations may pose problems for downstream ecosystems and water users if implemented on a large area; however, base flow actually improved with biofuel cropping while storm flows and sedimentation loads were lower. On the other hand, the risk from flooding and soil loss was reduced with less runoff from the upstream land. The net impact of these changes depended on the characteristics of downstream water users and ecosystems (Garg et al. 2011a).

13.5.13 Water Augmentation and Demand Management Must Go Hand in Hand

Water scarcity symbolizes a situation (gap) when water is not sufficient to meet the entire demand. Water scarcity is the issue not only in dry land areas but sometimes also in higher rainfall regions (rainfall > 1500–2000 mm). Water scarcity could be physical, economical, and institutional (Rijsberman 2006); therefore, water augmentation and demand management must go together to bridge this gap.

In agriculture, timely and exact quantity of water application can enhance WP and simultaneously reduce water losses. Improved methods of irrigation application can further reduce water demand. WUE in most of the command areas are below 30% (Ray et al. 2002; Khare et al. 2007; Garg et al. 2011c). Water is lost through poor conveyance methods right from canal release to water application in the field. Excess water, however, returns to downstream or groundwater recharges, but a significant amount is also lost as unproductive evaporation losses. Infrastructure development, institutional arrangement, and appropriate water policy can help in demand management. Demand management in the domestic and industrial sectors is also important. Roof water harvesting can enhance safe and good-quality drinking water availability to the downstream user and cut the domestic water demand.

13.5.14 Gray Water Recycling for Demand Management

Wastewater and gray water recycling and its reuse are emerging as an integral part of demand management (Al-Jayyousi and Odeh 2003; Al-Hamaiedeh and Bino 2010). Gray water is defined as wastewater generated from domestic activities such as dish washing, laundry, and bathing, whereas black water consists of toilet water. Gray water is a large potential source of water and could be diverted for toilet flushing, irrigation in parks, school yards, golf areas, car washing, and fire protection, which can reduce freshwater demand up to 30% in cities (Christova-Boal et al. 1996; Dixon et al. 1999; Eriksson et al. 2002; Lu and Leung 2003; Al-Hamaiedeh and Bino 2010).

With rapid expansion of cities and domestic water supply, the quantity of wastewater is also increasing in the same proportion. Almost 90% of total water supplied for domestic use gets generated as wastewater, which is used for irrigation in agricultural areas located near the city and where freshwater availability is limited. Wastewater availability remains consistent throughout the years, which drives farmers to make use of wastewater. It could be utilized as irrigation source for rice, vegetable, and fodder production (Buechler and Scott 2006). Other than agriculture, the activities directly dependent on wastewater are practiced by different social groups on a small, medium, or large scale and include, for example, livestock rearing, aquaculture, and floriculture (Buechler 2004; Buechler and Scott 2006).

There are several benefits and challenges on gray water and wastewater use. Judicious use of gray water reuse in Australia has reduced freshwater demand, strain on wastewater treatment plants, and energy consumption. Aquifer recharge has improved due to increased infiltration flows from gray water use (Raschid 2004; Madungwe and Sakuringwa 2007). In Lebanon, gray water is a valuable resource for encouraging plant growth because of its higher nutrient content (Madungwe and Sakuringwa 2007). Gray water reuse in agriculture contributes significantly to the supply of fresh fruits and vegetables to urban markets in Latin America and in the Caribbean. The problem of blue green algae in sewage ponds and water reservoirs is significantly reduced by household reuse of gray water in Mexico (Madungwe and Sakuringwa 2007). Approximately 16,000 ha of land in and downstream of Hyderabad (India) is irrigated with wastewater or with a combination of wastewater and groundwater (Buechler and Devi 2005). Along the 10 km stretch of the Musi River (southern India) where wastewater from Hyderabad is disposed of, year-round employment is generated on wastewater-irrigated fields for female and male agricultural laborers to cultivate fodder grass or vegetables for sale in nearby markets or for use by their livestock (Buechler and Scott 2006). However, there are also higher risks associated with human health and the environment on use of wastewater, especially in developing countries, where rarely the wastewater is treated and large volumes of untreated wastewater are being reused in agriculture (Buechler and Scott 2006).

Wastewater is more saline due to dissolved solids originating in urban areas and concentrated further through high evaporation in arid, tropical climates. Heavy use of wastewater in agriculture may cause a salinity problem and can decrease the land productivity. Several types of grass fodder can be grown with saline wastewater; therefore this water is more likely to be used for fodder production, particularly where demand for dairy products is high (Buechler and Scott 2006). With the use of wastewater-generated products and exposure to animals, the health of the livestock can be at risk and the quality of their milk may decline, which can transfer the health risks to humans who consume the milk (Buechler and Scott 2006). Health problems can pose a serious hazard for agricultural workers due to pathogenic bacteria, viruses, and parasites present in the wastewater as well as for consumers of wastewater-irrigated produce, particularly if the produce is not cooked before it is consumed. Hookworm infections are more common in agricultural workers who go barefoot in wastewater-irrigated fields (Hoek et al. 2002; Buechler and Scott 2006). Gray water and wastewater, however, are potential sources of water but they have to be used very cautiously in different sectors.

13.5.15 Linking Scales through Watershed Management

Rainfed areas predominate in generating global food production and providing several ecosystem services essential for humanity. A watershed is a spatial unit containing diverse natural resources

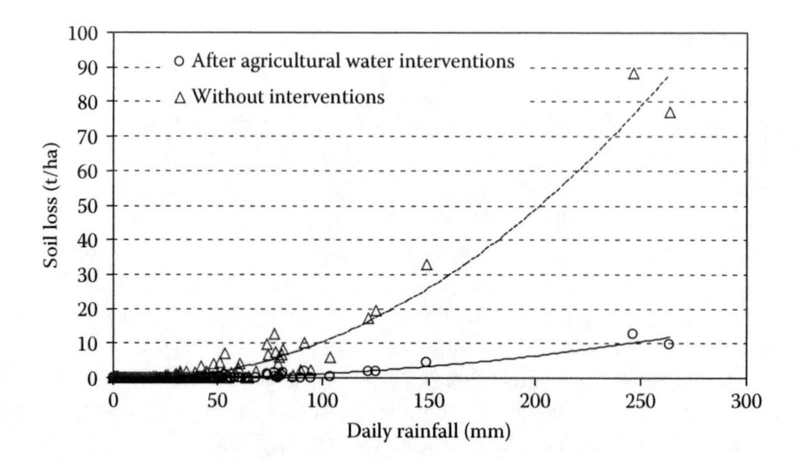

FIGURE 13.20 Impact of agricultural water interventions on soil loss in the Kothapally watershed (17°22°N latitude, 78°07°E longitude), Andhra Pradesh, India. Average annual rainfall of the study area is 850 mm.

that are unevenly distributed within a given geographical area and are ecologically complex; they are geologically and socially shared by temporal and spatial interdependence among resources and resource users (Wani et al. 2011d). The water flow (surface and subsurface) interconnects upstream and downstream areas and provides life support to people holding unequal use rights (Wani et al. 2006c). Watersheds are also inhabited by socially and economically heterogeneous groups of people located at different points along the terrain, creating potential conflicts among users of the same resources. A multitude of resources and processes that are supplied by the natural ecosystem can be strengthened by implementing IWRM. IWRM is not only helpful in enhancing the crop production and income of smallholder farmers but also in improving the water quality of groundwater wells and downstream water bodies, as well as better soil quality through C sequestration, protecting biodiversity, and minimizing soil loss. Figure 13.20 shows that soil loss was drastically reduced by implementing various water interventions in the Kothapally watershed compared with the degraded stage.

13.5.15.1 On-Site and Off-Site Impacts and Trade-Offs of Watershed Management

The principal users of the water flows are the agriculture, both rainfed and irrigated, and the ecosystem services that rely on the water quantity and quality for their functions. The IWRM approach of water management is therefore considered as an effective method in alleviating the water stress situation (Rockström et al. 2007, 2010; Rockström and Barron 2007; Wani et al. 2008, 2011b; Barron and Keys 2011). Green and blue water management at various scales not only increases food production, but has a number of social, economic, and environmental cobenefits such as protection of the environment, increase in biodiversity, and improvement in the livelihood status of local communities (Wani et al. 2003a, 2008; Rockström et al. 2007, 2010). In the IWRM approach, agricultural water interventions and in situ and ex situ practices allow more rainwater to infiltrate and enhance soil moisture (green water) and groundwater (blue water) availability. Adopting suitable cropping systems such as mixed cropping pattern (e.g., maize–pigeonpea intercropping) can enhance WP by utilizing more green water within the monsoon and postmonsoon periods.

The Kothapally watershed in Andhra Pradesh, southern India, is a classic example showing the success of IWRM where the community has moved from subsistence farming to a market-driven agriculture stage after implementation in 1999. Sreedevi et al. (2004), Wani et al. (2006b), and Garg et al. (2011a) reported that water availability and crop yield have substantially improved after the

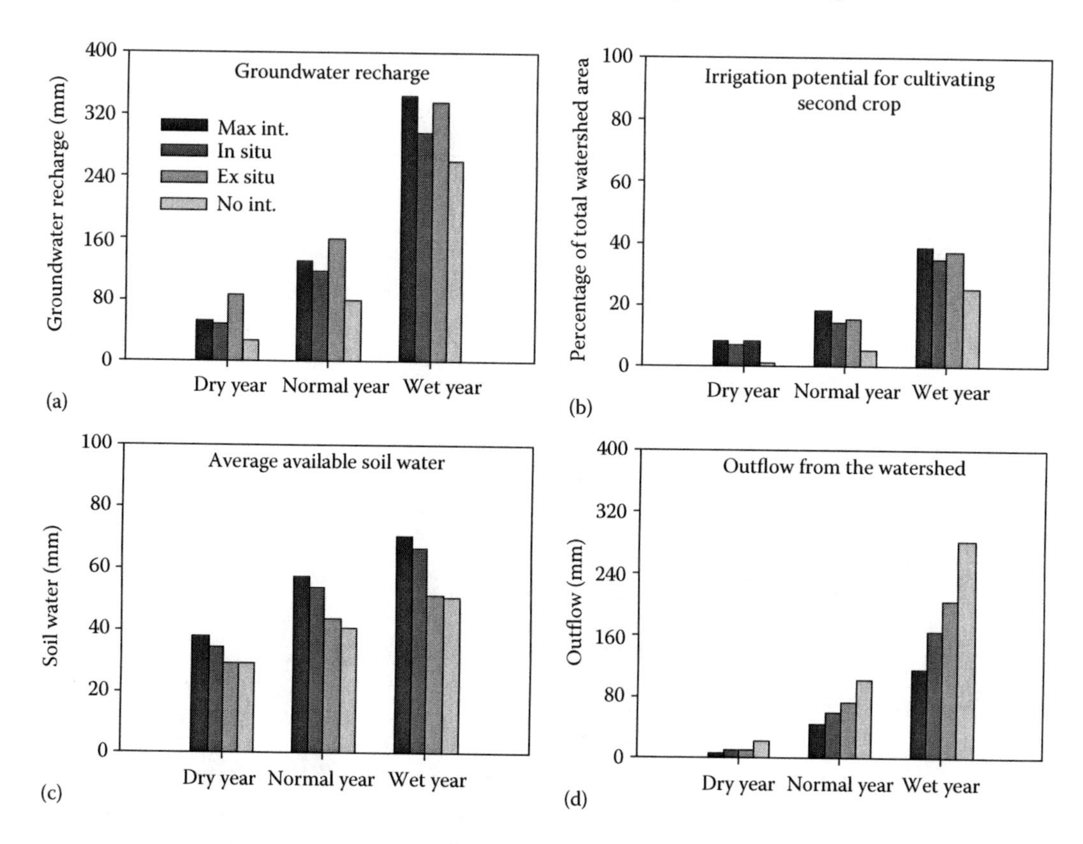

FIGURE 13.21 Comparison of (a) groundwater recharge, (b) developed irrigation potential, (c) average available soil water during crop period, and (d) outflow amount in different land management scenarios during dry, normal, and wet years.

IWRM supportive interventions. Since 1999, several shallow wells that had low groundwater levels have reverted into active wells for irrigation. The cropping pattern has changed in recent years as a consequence of improved soil moisture availability and irrigation access. Farmers who were cultivating cotton of traditional varieties, sorghum, maize, paddy, onion, and chilies before the onset of the watershed development program have switched to cultivating higher-yielding cash crops such as *Bt* cotton and vegetables. Along with in situ and ex situ agricultural water management interventions, farmers have also adopted better nutrient and pest management as well as better timely operations (Sreedevi et al. 2004), which further improves agricultural productivity.

Different agricultural water interventions (shown by four scenarios) in the Kothapally watershed impact as groundwater recharge, its availability for cultivating second crop, average available soil moisture, and amount of surface runoff from watershed boundary during dry, normal, and wet years (Figure 13.21) (Garg et al. 2011a). During dry years, water management interventions became particularly important for groundwater recharge, which was more than twice as high for both ex situ and in situ interventions compared with the degraded state. Groundwater availability impacts the potential to grow a second, fully irrigated crop during the dry season (Figure 13.21b). The irrigation potential is found to have more than doubled with water management interventions during dry and normal years. In situ water management resulted in higher soil moisture availability (Figure 13.21c). Outflow varies significantly between years and with water management interventions (Figure 13.21d). Outflow was more than ten times higher during wet years compared with dry years. With maximum water interventions, outflow from the watershed was more than halved compared with the degraded state.

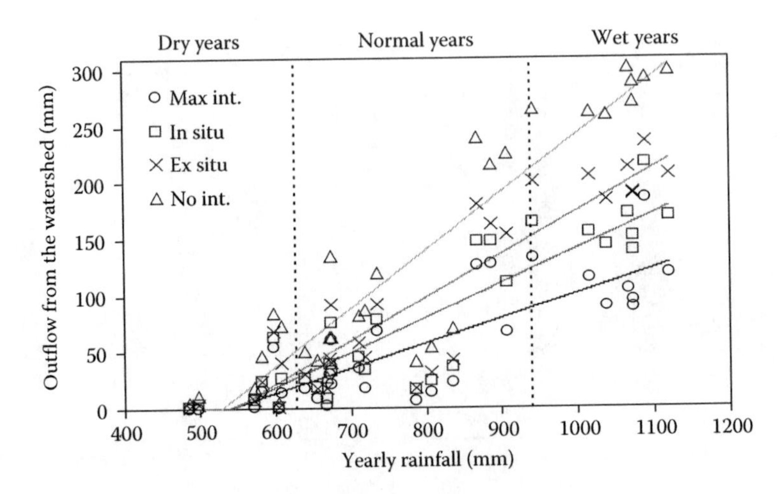

FIGURE 13.22 Rainfall–runoff relationship for the four different water management scenarios in a micro-watershed at Kothapally located in SAT, southern India. Results are based on 31 years of simulation run (SWAT, a hydrological model) from 1978 to 2008. Max int.: in situ + check-dams; In situ: in situ + no check-dams; Ex situ: no in situ + check-dams; No Int.: no in situ + no check-dams.

Figure 13.22 shows a linear relationship between the rainfall amount and the outflow of water from the watershed on a yearly time scale, but varied with water management interventions on the field scale. The lowest outflow was generated with both check dams and in situ water management in place (Max int.), while the no-interventions scenario (No int.) generated the highest outflow per rainfall event. Moreover, the results show that runoff losses were smaller for in situ management (In situ) compared with ex situ interventions (Ex situ), indicating that practicing in situ management caused larger outflow reductions from the fields than check dams in this case. This harvested amount was available in green and blue form, which helps in reducing crop water stress.

Long-term trade-off analysis on various aspects is helpful in understanding the overall benefits or losses if the IWMP program is implemented on a larger scale. It is generally assumed that IWRM does enhance water resources availability at farm and community scales at the upstream location but leads to a negative impact at downstream water bodies. It is important to analyze various ecosystem trade-offs at upstream and downstream locations before any decision making, for example, (i) increase in water resource availability, crop production, and total income developed at upstream could be compared with downstream water availability and its benefits/loss; (ii) water resource availability at upstream and downstream locations needs to be analyzed for dry, normal, and wet years; (iii) impact of soil and nutrient loss on crop production in upstream and deposition/accumulation of soil/pollutant on river beds and at downstream water bodies has to be analyzed; (iv) water quality at upstream and downstream; and (v) comparison of ecosystem services at upstream/downstream location are a matter of important concern.

Bouma et al. (2011) indicated that the capital invested under various water interventions for the Upper Musi subbasin is not remunerative and recommended the development of various infrastructures (road, school, hospital, etc.). Watershed benefits are far larger than the economic benefits, as evidence has convincingly shown that watershed development addresses the issues of minimizing land degradation, enhancing green WUE, and increasing equity for landless and women's groups and, more so, building the social capital in the rural community (Wani et al. 2003a, 2011; Bouma et al. 2011; Wilson 2011), but considering only economic returns has overlooked the issue of green WUE as well as equity concerns for the upland areas (Rockström et al. 2007, 2010; Wani et al. 2008; Kijne et al. 2009; Barron and Keys 2011).

Our analysis for the same area showed positive economic trade-offs by implementing the watershed development program in the Osman Sagar catchment area and subsequent increase in income

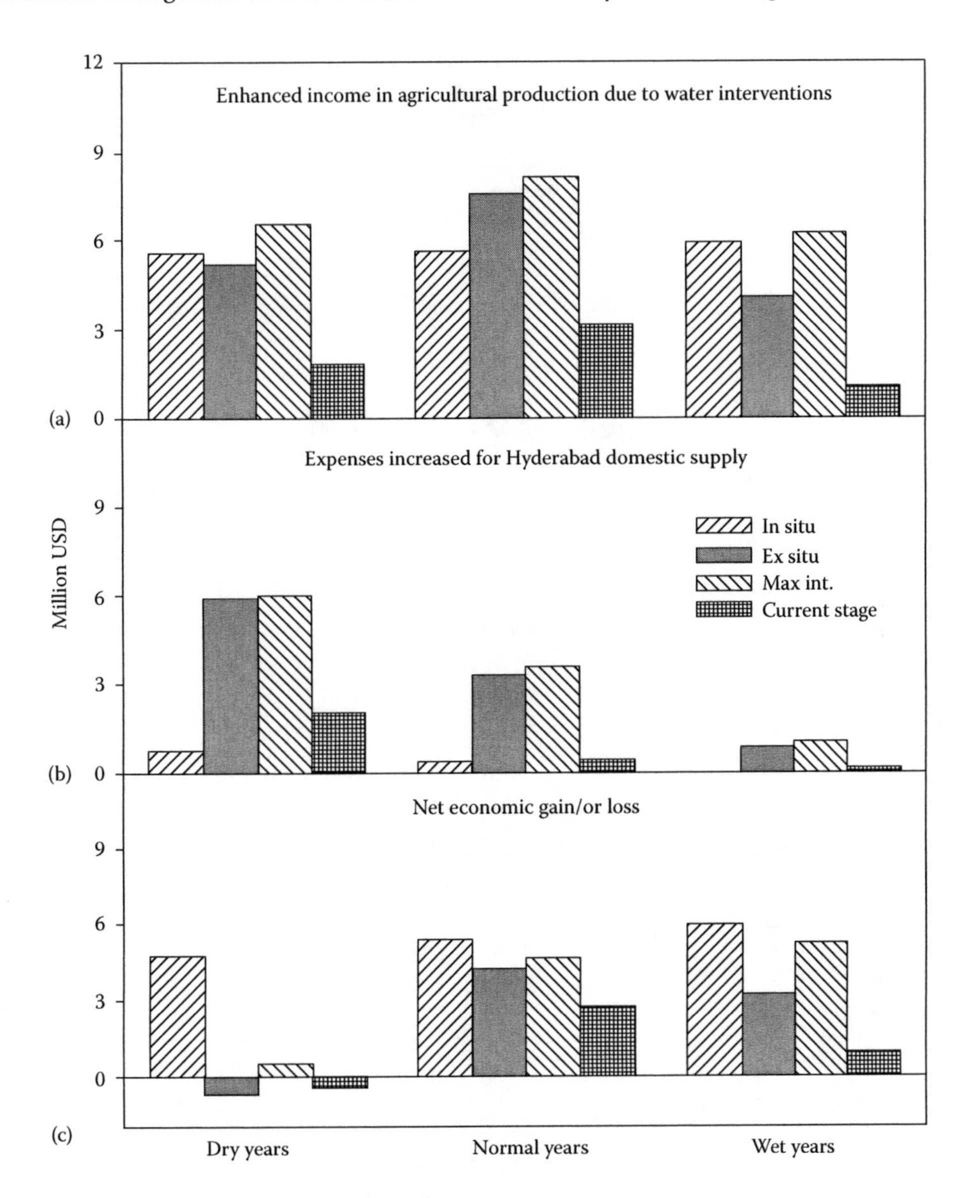

FIGURE 13.23 Trade-off analysis of (a) enhanced agricultural incomes, (b) increased costs for domestic water supply for domestic use in Hyderabad city, and (c) net economic returns/losses for three water interventions and base line scenarios compared to no interventions, under-dry, normal, and wet years.

compared with downstream water supply for Hyderabad city. It is clear from Figure 13.23 that merely by accounting the yield benefit in economic terms against the costs needed to meet water demand under varying climatic conditions, we showed a net benefit. We ascribe the differences in result to the use of an improved modeling approach more effectively representing both water and sediment flows, as well as crop yields, under varying climatic conditions (dry, normal, and wet years; Garg et al. in press). If our analyses were to include various social and environmental gains/benefits as described in the previous meta-analyses of watershed programs in India (Joshi et al. 2008; Wani et al. 2011c), the outcome of this analysis would be many more benefits in addition to economic benefits. However, as Joshi et al. (2008) concluded, there are a range of social and environmental benefits that also need to be addressed and valued for obtaining a strong case in water allocation between different users and uses in catchments and basins under watershed interventions.

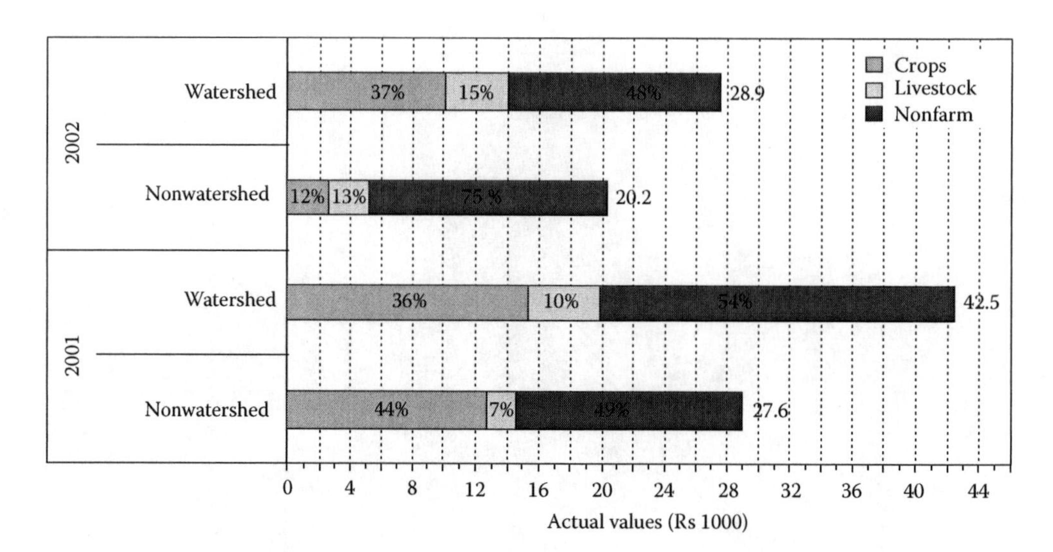

FIGURE 13.24 Income stability and resilience effects during drought year (2002) in the Adarsha watershed, Kothapally (Andhra Pradesh, India) compared to closely located nonintervention village; economic values in given figures are in Indian currency rupees (approximately 1 USD = 45 Indian rupees).

13.5.16 Building Resilience of Communities and Natural Resources against Climate Change Impacts

IWRM-based interventions are helpful in building resilience of natural resources and of communities against future changes including climatic variability and shocks by reducing uncertainty of crop failure and by providing better income stability (Wani et al. 2006c, 2008, 2009a; Barron and Keys 2011). Total income generated and sources of income were compared between the Adarsha watershed, Kothapally, which was transformed by IWRM-based interventions since 1999 and a nearby located nonintervention village during dry (2002) and normal rainfall years (2001). Figure 13.24 shows that average annual income of the Kothapally farmers is 45% and 55% higher than in the nonintervention village in dry and normal years, respectively, and income from crop husbandry was similar (36%–37%) to the total income in the case of the Adarsha watershed, Kothapally, during drought and normal rainfall years (showing the resilience effect of the interventions made in the watershed) (2001). However, the income from farming had drastically reduced to 12% in the nonwatershed village during the drought year (2002). During the same period (drought year), the share of income from nonfarm activities was more in nonwatershed village total income compared with that in the watershed village and people had to migrate out of nonwatershed villages in search of livelihood (Shiferaw et al. 2009).

13.5.17 Opportunities for Enhancing Ecosystem Services through Watershed Management

Ecosystem services were classified into four broad categories: (i) provisioning (ecosystem goods such as fuel, food, and timber); (ii) regulating (e.g., climatic regulation, pest control, and pollination); (iii) cultural (providing humans with recreational, spiritual, and aesthetic values); and (iv) supporting services (basic ecological properties/processes such as soil formation) (Millennium Ecosystem Assessment 2005; Gordon et al. 2010). Conversion of forest and woodlands into agricultural lands, however, increased the total food production to meet global food demand (provisioning ecosystem services) but at the same time led to the development of serious complications at local, regional, and global scales such as climate change, land and environmental degradation, and loss of other ecosystem services. Moreover, the feedback mechanism indicated a negative impact on agricultural productivity compared with its original stage. For example, clearing the forestland declined biodiversity, which resulted in reducing the overall pollination process (regulating ecosystem services) that is important for agriculture itself. Similarly, mass clearing of forestland increased blue water availability, which created (groundwater table) waterlogging and soil salinity, which in turn reduced the productivity of landscape and increased the risk of crop failure in Australia (Gordon et al. 2003, 2008).

A landscape that is already degraded and almost about to cross the tipping points is generally located in dryland regions. However, to rehabilitate a degraded landscape to its original state is an expensive affair, but IWRM-based interventions provide an opportunity to build, rehabilitate, and protect the ecosystem from further degradation. In this context, water management especially in agriculture plays an important role in solving some of the most pressing trade-offs between an increase in agricultural production that can contribute to food security and economic growth on the one hand, and dealing with the losses of important ecosystem benefits that also sustain human well-being and livelihoods on the other (Gordon et al. 2010). Tables 13.6a, 13.6b, and 13.6c show the impact of IWRM-based interventions on various biophysical and economic variables identified by different case studies and research findings at benchmark locations and farmers' fields in India and elsewhere. IWRM-based interventions reduced surface runoff and decreased soil loss; increased ground and surface water availability and enhanced crop yield; and increased cropping intensity and rehabilitated wastelands in a more sustainable and productive manner.

Use of fertilizers has brought major benefits to agriculture, but has also led to widespread contamination and eutrophication of surface water and groundwater (Verhoeven et al. 2006). For example, the flux of reactive N to the oceans has increased by nearly 80% between 1860 and 1990 (Millennium Ecosystem Assessment 2005). Eutrophication is usually followed by a loss of ecosystem services, such as loss of recreational values and fish production through the development of algal blooms, anoxia, and the decline of aquatic macrophytes and fisheries (Verhoeven et al. 2006; Gordon et al. 2010; Barron and Keys 2011). There are many instances where poor water management practices in agriculture have contributed to a decline in the human well-being and health (Finlayson and D'Cruz 2005; Gordon et al. 2010). Figure 13.20 shows that various agricultural water interventions reduced soil loss and runoff, which were directly associated with nitrate losses from agricultural lands (Wani et al. 2009a) by many times compared with the nonintervention stage. Similarly, it is expected that implementation of various agricultural water interventions would also help in reducing nutrient loss from agricultural fields, creating a win-win situation for both the agricultural farm at the upstream location and the water quality at the downstream level. Many water-related diseases could be successfully controlled through water management either specifically or by thoughtful approaches (e.g., watershed development) in agriculture (Coravalan et al. 2005).

Commercial agriculture has tended to favor conversion of ecosystems into monocropping (or low diversity of crops) with management focusing on a single or a few provisioning ecosystem services, such as food, timber, or fish (Gordon et al. 2010), whereas the watershed development approach promotes crop intensification along with crop diversification such as intercropping, agroforestry, floriculture by adopting various soil and water conversion measures, as well as IPM and INM practices. Such improved technologies not only maintain the productive status of the landscape but also improve the physical, chemical, and biological properties of the landscape, enhance carbon (Tables 13.5, 13.6a, 13.6b, and 13.6c) subsequently, and build system resilience against external stocks.

13.5.18 Link ESS with Economic Drivers for the Community to Enhance Sustainability

Tangible economic benefits to all stakeholders (community in case of watersheds) are a must for community participation, which is the primary pillar of sustainability (Wani et al. 2002, 2006c, 2009a). Sustainability of watershed interventions is an important issue (Pangare 1998; Kerr et al.

2002; Wani et al. 2002, 2006, 2008,), and several assessment studies have highlighted an urgent need for improving sustainability through economic benefits. The economic drivers play an important role as they involve people in generating revenue to sustain their livelihood. The urban migration to seek jobs, especially when income and livelihood opportunities in agriculture are not sufficient to fulfill their family needs, is a typical example of this. Implementing watershed development programs enhances provisioning ecosystem services (e.g., crop, fodder, wood production) and could be linked with higher economic gain. In addition to the on-site services, there are a number of off-site ecosystem services such as reducing flooding, siltation of downstream water bodies, as well as reduced eutrophication of water bodies, improved groundwater availability, water quality, and carbon sequestration. The environmental trade-off could be directly or indirectly beneficial at both upstream and downstream locations.

Participatory biodiversity conservation enables the poor to manage their natural resources better. In Govardhanpura and Gokulpura villages in Bundi district of eastern Rajasthan, India, a participatory community initiative regenerated half of the degraded common pool resources or grazing area by adopting appropriate social and biophysical interventions. This ensured the availability of fodder for all households and an income of USD 1670 annually for the SHGs through the sale of surplus grass to surrounding villages, as for villagers in the watershed, grass on a cut-and-carry system was available freely. In Thanh Ha watershed, Vietnam, the introduction of legumes saw a jump in crop diversity factor from 0.25 in 1998 to 0.6 in 2002. In Kothapally watershed in Andhra Pradesh, India, farmers now grow 22 crops in a season with a shift in cropping pattern from cotton to a maize/pigeonpea intercrop system. More legumes are now grown in Vietnam and Thailand, reducing the need for fertilizer N (Wani et al. 2006c). Similarly, converting wasteland into *Jatropha*-cultivating land in Velchal village provided an additional source of income for marginal and landless laborers (Wani et al. 2006a,b) and also showed positive soil and hydrological trade-offs (Garg et al. 2011b).

There is a need for developing a mechanism for monitoring ESS to ensure that ESS providers from upland areas are rewarded suitably to enhance sustainability of watersheds. Benefit of the ecosystem services for individual farmer/community is relatively dependent on their socioeconomic factors and access to that service. Valuation of ecosystem services is a cumbersome task due to complex and nonlinear relationships among various interventions and ecosystem responses. However, identification and valuation of ecosystem services are important and is a challenging but essential task. There is a need to develop assessment methods, valuation, institutional mechanisms, and financial instruments.

13.5.19 Empowering Women as Water Resource Managers

Women constitute more than 50% of the world's population and 550 million women live below the poverty line as reported by the World Food Program. Two-thirds of the illiterates in the world are women, without any property rights, and have no economic independence (70% of the world's poor are women) (UNEP 1997). According to the "Draft National Policy for Women in Agriculture (2008)" in India, women constitute 40% of the agricultural workforce and this share is increasing. Currently, 53% of all male workers are in agriculture, while 75% of all female workers and 85% of all rural female workers are in agriculture. Women as economic income providers, caregivers, and household managers are responsible for ensuring that their families have basic resources for daily living. They are often the managers of community natural resources and have learned to protect these resources in order to preserve them for future generations (managers of sustainability) (ecosystem service providers). Although women play a pivotal role in agriculture development, more than 55% of female agricultural workers are considered laborers rather than being the owners themselves, even when their family owns land. Participation of women and resource-poor individuals is of paramount importance for the effective implementation of IWRM programs, so that they become effective vehicles for the integrated development of communities and sustainable impacts and continuing ecosystem services. In drought-prone rainfed areas, watersheds are recognized as growth engines for agricultural as well as overall development to achieve food security (Wani et al. 2008). Community participation is an important aspect of watershed development programs, and it is necessary to include equity and gender parity into the program design itself. Inclusion of women and those who are resource-poor is of paramount importance for watershed development to become truly participatory in both implementation and impact (Sreedevi et al. 2009a). Creating awareness about IWRM and water management is important among women groups and could be achieved with capacity building and women empowerment. New watershed common guidelines include microenterprises for generating income for women (GoI 2008) as the economic security/independence is already associated with the decision-making power in the house and community (Sreedevi et al. 2009a; Wani et al. 2009a).

13.5.20 Public–Private People-Centric Partnership for Water Management

With the basic objective of improving rural livelihoods through sustainable management of natural resources in the watersheds, it is imperative that watersheds produce marketable surplus to come out of the subsistence agriculture. To achieve the tangible economic benefits through increased productivity and diversification with high-value crops, there is a need to adopt and operationalize a holistic approach through convergence and collective action (Wani et al. 2003a, 2009a). To achieve the goal of sustainable intensification, as indicated earlier, water alone cannot do the job and it needs backward and forward linkages in terms of providing necessary inputs (seeds, fertilizers, pesticides, machineries, credit and insurance, etc.) and for value addition and linking farmers to the markets to enhance income and agricultural productivity in rural areas. There is ample space and scope to bring in public–private partnership in the consortium (Wani et al. 2006b, 2008, 2011b). For example, widespread deficiencies of micronutrients in the soils of farmers' fields were recorded and once the benefits of soil test–based applications were demonstrated, farmers were in need of the fertilizer formulations containing micronutrients as well as seeds of improved cultivars in their region.

The issue of availability of boron and other micronutrients in remote villages was resolved for thousands of farmers in different villages by building partnership with Borax Morarji Limited, a producer of B fertilizer in India as the consortium partner to link with SHGs and farmers' cooperatives. Adarsha watershed in Kothapally serves as an example of livestock-based microenterprises. Once the milk production in the village increased through animal breed improvement activity and improved/increased fodder availability, Reliance Company came forward to establish a milk procurement center in the village to buy the marketable surplus quantity. It also provided technical support and inputs for animal feed and health for ensuring increased milk production. The public–private partnership in the area of IWMP is also envisaged in the new common watershed guidelines (GoI 2008). A pilot program of public–private partnership for IWMP has been initiated in Madhya Pradesh; earlier, in Rajasthan, Indian Tobacco Company (ITC) had joined hands with the Government of Rajasthan to develop watersheds through public–private partnership in the consortium. ICRISAT has worked with Confederation of Indian Industry (CII) with other corporates such as Coca-Cola, SAB Miller in Rajasthan, Andhra Pradesh, and Karnataka to address the issues of water conservation and enhancing agricultural productivity through sustainable intensification and diversification with high-value crops. There are a number of isolated examples of public–private partnership in various areas of IWMP. For example, GIZ (GTZ), Southern Online Bio-Technology (SBT), and ICRISAT had a public–private partnership project under which SBT operated a 40 kl/day biodiesel plant in Nalgonda district, Andhra Pradesh, with German technology provided by Lurgi and ICRISAT provided technical support to the farmers for cultivating biodiesel plantations and facilitating buyback arrangements between the farmers and SBT (Kashyap 2007). A public–private partnership has to be a win-win proposition for the industries/corporate houses as well as the implementing agencies of the IWMP and the farmers, which is possible through a business model of watershed development. In addition, a number of corporates such as Sir Dorabji Tata Trust

(SDTT, Mumbai), Sir Ratan Tata Trust (SRTT, Mumbai), TVS Foundation (Chennai), Coca-Cola Foundation (United States and India), SAB Miller, India Cements Limited, and their formal associations such as Confederation of Indian Industries (CII) and Federation of Indian Chambers and Commerce Industries (FICCI) are collaborating with the ICRISAT-led consortium for fulfilling their corporate social responsibility mandate. However, to make public–private partnership a norm rather than an exception, there is a need to promote public–private partnership for harnessing the full potential of rainfed agriculture for improving rural livelihoods through sustainable management of natural resources and through enabling policies and institutional arrangements.

13.5.21 Awareness Building among All Stakeholders

Creating awareness among all the stakeholders starting from policy makers, researchers, extension officers to the end users–farmers is important. The stakeholders may collectively derive some synergetic benefits from being able to integrate their efforts. Effective participation and collective action in resource management, however, depend on the degree of awareness of important technical considerations. This awareness is possible with capacity-building programs and knowledge dissemination at every level. IWRM requires multiple interventions that jointly enhance the resource base and livelihoods of rural people. Capacity building is a process to strengthen the abilities of people to make effective and efficient use of resources in order to achieve their own goals on a sustainable basis (Wani et al. 2008). Awareness-building programs ranging from seminars, workshops, training programs to one-to-one interaction on a regular basis are required at different stakeholder group levels. Demonstration of advanced irrigation techniques such as drip and sprinklers have to be conducted on farmers' fields. Awareness-building programs on water-related issues such as groundwater augmentation and its proper utilization are helpful for long-term sustainability of water resources at the community/village scale.

13.5.22 Enabling Policies and Institutions for IWRM in Rainfed Areas

Rising demand for food and environmental water supplies presents a challenge to prevailing institutional arrangements governing freshwater access and use. With increasing water scarcity and food demand challenging the development paradigm, right policies and institutional arrangements are essential at various levels to ensure efficient management of resources. For example, in India, the focus primarily was on augmenting blue water resources. However, the GoI realized the importance of rainfed areas and therefore made significant investments (USD 6 billion until 2006) on watershed development programs at the national level since the 1970s (Wani et al. 2008). Initially, the watershed development program in India was mainly oriented toward constructing water-harvesting structures for soil and water conservation and some productivity enhancement. Subsequently, policies and institutional arrangements were modified as per the needs of the changing development scenario through lessons learned from past experiences, and observations are currently aimed at the holistic development of the rural community where watershed management is considered an entry point for improving livelihoods of people through natural resource management. However, a huge scope still exists for water management in rainfed and fallow wastelands, which requires further policy support.

Energy subsidy for tubewell irrigation in India enhanced groundwater use many times (approximately 240–260 km^3 per year in 2000) compared to the 1950s level (10–12 km^3) (Shah et al. 2005). Rural India at present uses subsidized energy worth an equivalent of USD 4.5–5.0 billion per year to pump almost 150 km^3 of water for agricultural use. Groundwater management is also a major concern for achieving sustainable development. The Ministry of Water Resources, GoI, debated the groundwater bill (control and regulation) in 1970 and revalidated it in 1992 to regulate and control the overexploitation of groundwater. The bill was circulated to all state governments to prepare similar bills to keep a check on groundwater overexploitation because water is a state subject (Singh

1995). However, as of today, only a few states have regularized groundwater bills, as the remaining states have not implemented it due to various economic and political reasons. Electric supply and pricing policy, however, offer a powerful tool for groundwater management indirectly, but most state governments are unable to implement it due to stiff resistance from farmers' groups. As the free power supply or subsidized power supply drives the groundwater exploitation movement in the country, there are a few examples of policy changes that exemplify a break from the traditional vote bank-oriented policies to ensure sustainable management of resources by taking users into confidence. The "Jyotigram scheme" is one such scheme of the Government of Gujarat, India, which is an example of comanagement of electric power and sustainable groundwater use by implementing the right policy targeting rural people. Rather than supplying free but poor-quality electric power, the Government of Gujarat, introduced (i) 24 h three-phase power supply for domestic and village industries, all subjected to metered tariff; and (ii) 8 h good-quality (uninterrupted, full voltage) assured power supply for running tube wells for agriculture use in 2003. This reduced unwanted groundwater pumping and improved the life quality of village people and their economic status (Shah and Verma 2008).

13.6 WAY FORWARD FOR IMPROVING SUSTAINABLE MANAGEMENT OF WATER RESOURCES IN THE TROPICAL REGIONS

As discussed in an earlier section, rainfed agriculture holds a huge potential to meet the future food demand. In order to achieve these targets, IWRM is the promising framework for managing water and natural resources effectively. To meet the challenges of the twenty-first century for producing more food from limited finite water and land resources, there is an urgent need to bring in a shift in managing agricultural water in the world, particularly so in the developing world. Traditionally, water management dealt with irrigated agriculture; however, as shown by the comprehensive assessment of water for food and water for life (Molden et al. 2007), agricultural water management has a wider meaning than just irrigation and the vast untapped potential of 1.2 billion hectares of rainfed agriculture needs to be harvested (Rockström et al. 2007; Wani et al. 2009a). For harnessing the potential of rainfed agriculture, the large portion of green water that is underutilized at present needs to be improved substantially. The shift in water vapor from croplands from nonproductive evaporation loss to productive ET needs to be improved, and a large scope exists for enhancing green WUE from 30%–35% to 65%–95% in rainfed areas. Appropriate policy and institutional support to decentralized water management in rainfed areas are the urgent need. Increased investments and credit support for the small and marginal farmer are required, to shift them from growing low-value crops to high-value crops through inclusive market-oriented development (IMOD).

In the initial stage of IMOD, small and marginal farmers will need incentives and enabling policies and institutions to slowly innovate to produce marketable surplus and invest further in sustainable intensification so that they can grow and prosper by intensifying rainfed agriculture. The weak link between the research and development organizations—the farmers—needs to be strengthened for efficient knowledge/technology transfer.

Use of new ICT tools not only as a means of knowledge exchange but also as a source of livelihood for the educated youth in the rural areas has to be worked out. There exists a large space for public–private partnership in the area of agricultural water management in developing countries. However, it has to be a win-win proposition for all the stakeholders/partners. Small and marginal farmers' interests must be at the center while devising public–private partnership policies. However, the operationalization of an integrated holistic strategy through a consortium approach calls for a change in the mindset of the various actors such as researchers, policy makers and development workers, farmers, and private industries.

Currently, most of the players feel comfortable while working in their own compartments/silos and there is a reluctance to work together for achieving the common goal of improving livelihoods of small and marginal farmers. However, successful examples such as the consortium approach for watershed management in Asia, developed and adopted by ICRISAT, National Agricultural Innovation Project (NAIP) of Indian Council of Agricultural Research (ICAR), GoI, as well as the Bhoochetana initiative of Government of Karnataka with the ICRISAT-led consortium, have shown very good results, and the various actors are realizing the benefits of working together in a holistic manner for a win-win proposition, which could become a powerful trigger to an operationalized holistic IWRM framework to harness the vast untapped potential of rainfed agriculture in developing countries.

ABBREVIATIONS

B/C ratio	Benefit–cost ratio
BBF	Broad bed and furrow
CA	Conservation agriculture
EA	East Asia
ET	Evapotranspiration
FAO	Food and Agriculture Program of the United Nations
FYM	Farmyard manure
GoI	Government of India
ICRISAT	International Crops Research Institute for the Semi-Arid Tropics
ICT	Information and Communication Technology
IGP	Indo-Gangetic Plains
IMOD	Inclusive market-oriented development
INM	Integrated nutrient management
IPM	Integrated pest management
IRR	Internal rate of return
IWMP	Integrated Watershed Management Programs
IWRM	Integrated Water Resource Management
M&E	Monitoring and evaluation
MA	Millennium Ecosystem Assessment
MDG	Millennium Development Goal
MENA	Middle East North Africa
NSG	National Support Group
PR&D	Participatory Research and Development
RUE	Rainfall use efficiency
RWH	Rainwater harvesting
S, B, and Zn	Sulfur, boron, and zinc
SA	South Asia
SAT	Semiarid tropics
SHG	Self-help group
SSA	Sub-Saharan Africa
T/ET	Transpiration to evapotranspiration ratio
UNDP	United Nations Development Programme
UNEP	United Nations Environment Programme
USAID	U.S. Agency for International Development
WAE	Water application efficiency
WP	Water productivity
WUE	Water-use efficiency

REFERENCES

Achten, W.M.J., J. Almeida., V. Fobelets, et al. 2010a. Life cycle assessment of *Jatropha* biodiesel as transportation fuel in rural India. *Applied Energy* 87: 3652–3660.

Achten, W.M.J., W.H. Maes., R. Aerts, et al. 2010b. *Jatropha*: From global hype to local opportunity. *Journal of Arid Environments* 74: 164–165.

Ajay Kumar. 1991. Modeling runoff storage for small watersheds. MTech Thesis, Asian Institute of Technology, Bangkok, Thailand.

Al-Hamaiedeh, H., and M. Bino. 2010. Effect of treated grey water reuse in irrigation on soil and plants. *Desalination* 256: 115–119.

Al-Jayyousi., and R. Odeh. 2003. Greywater reuse: Towards sustainable water management. *Desalination* 156(1–3): 181–192.

Arnold, J.G., and C. Stockle. 1991. Simulation of supplemental irrigation from on-farm ponds. *Journal of Irrigation and Drainage Engineering* 117(3): 408–424.

Barron, J. 2004. Dry spell mitigation to upgrade semi-arid rainfed agriculture: Water harvesting and soil nutrient management. PhD thesis, Natural Resources Management, Department of Systems Ecology, Stockholm University, Stockholm, Sweden.

Barron, J. and P. Keys. 2011. Watershed management through a resilience lens. In *Integrated Watershed Management*, eds. S.P. Wani, J. Rockström, and K.L. Sahrawat, Chapter 12, pp. 391–420. CRC Press, The Netherlands.

Barron, J. and J. Rockström. 2003. Water harvesting to upgrade smallholder farming: Experiences from on-farm research in Kenya and Burkina Faso. RELMA, Nairobi, Kenya.

Barron, J., J. Rockström., F. Gichuki, et al. 2003. Dry spell analysis and maize yields for two semi-arid locations in East Africa. *Agriculture and Forest Meteorology* 117(1–2): 23–37.

Bationo, A. and A.U. Mokwunye. 1991. Role of manures and crop residue in alleviating soil fertility constraints to crop production: With special reference to the Sahelian and Sudanian zones of West Africa. *Fertilizer Research* 29: 117–125.

Bationo, A., J. Kihara, B. Vanlauwe, J. Kimetu, B.S. Waswa, and K.L. Sahrawat. 2008. Integrated nutrient management: Concepts and experience from sub-Saharan Africa. In *Integrated Nutrient Management for Sustainable Crop Production*, eds. M.S. Aulakh and C.A. Grant, pp. 467–521. The Haworth Press, Taylor and Francis Group, New York.

Batisani, N. and B. Yarnal. 2010. Rainfall variability and trends in semi-arid Botswana: Implications for climate change adaptation policy. *Applied Geography* 30(4): 483–489.

Bhatia, V.S., P. Singh., A.V.R.K. Rao, et al. 2009. Analysis of water non-limiting and water limiting yields and yield gaps of groundnut in India using CROPGRO-Peanut Model. *Journal of Agronomy & Crop Science* 195: 455–463.

Bhatia V.S., P. Singh, S.P., Wani, A.V.R.K. Rao, and K. Srinivas. 2006. Yield gap analysis of soybean, groundnut, pigeonpea and chickpea in India using simulation modeling. *Global Theme on Agroecosystems,* Report No. 31. Patancheru 502 324, Andhra Pradesh, India: International Crops Research Institute for the Semi-Arid Tropics.

Bhattacharya, T., S.K. Ray., D.K. Pal, et al. 2009. Soil carbon stocks in India—Issues and priorities. 2009. *Journal of the Indian Society of Soil Science* 57(4): 461–468.

Biamah, E.K., J. Rockström, and J. Okwach. 2000. Conservation tillage for dryland farming. Technological options and experiences in eastern and southern Africa. RELMA, Nairobi, Kenya.

Bouma, J.A., T.W. Biggs., L.M. Bouwer, et al. 2011. The downstream externalities of harvesting rainwater in semi-arid watersheds: An Indian case study. Paper submitted to Agricultural Water Management.

Buechler, S. 2004. A sustainable livelihoods approach for action research on wastewater use in agriculture. In *Wastewater Use in Irrigated Agriculture: Confronting the Livelihood and Environmental Realities*, eds. C. Scott, N. Faruqui, and L. Rachid-Sally, pp. 25–40. CAB International in Association with International Water Management Institute and International Development Research Center, Wallingford, UK.

Buechler, S., and G. Devi. 2005. Local responses to water resource degradation in India: Groundwater farmer innovations and the reversal of knowledge flows. *The Journal of Environment and Development* 14(4): 410–438.

Buechler, S. and C. Scott. 2006. Wastewater as a controversial, contaminated yet coveted resource in South Asia. Human Development Report Office Occasional paper. UNDP.

Bunn, S.E. and A.H. Arthington. 2002. Basic principles and ecological consequences of altered flow regimes for aquatic biodiversity. *Environmental Management* 30(4):492–507.

Centre Water Commission, 2005. Ministry of water resources, Government of India. *Hand Book of Water Resources Statistics*, 2005. http://www.cwc.nic.in/main/webpages/publications.html.

Christova-Boal, D., R.E. Evans., S. McFarlane, et al. 1996. An investigation into greywater reuse for urban residential properties. *Desalination* 106: 391–397.

Coravalan, C., S. Hales., A. McMichael, et al. 2005. Ecosystems and human well-being: Health synthesis. World Health Organization, Geneva, 53 p.

D'Silva, E., S.P. Wani., B. Nagnath, et al. 2004. The making of new Powerguda: Community empowerment and new technologies transform a problem village in Andhra Pradesh. Global Theme on Agroecosystems Report No. 11. Patancheru 502 324, International Crops Research Institute for the Semi-Arid Tropics, Andhra Pradesh, India.

Dancette, C. 1983. Estimation des besoins en eau des principales cultures pluviales en zone Soudano-Sahélienne. *L'Agronomie Tropicale* 38(4): 281–294.

de Fraiture, C., D. Wichelns., J. Rockström, et al. 2007. Looking ahead to 2050: Scenarios of alternative investment approaches. In *Comprehensive Assessment of Water Management in Agriculture, Water for Food, Water for Life: A Comprehensive Assessment of Water Management in Agriculture*, ed. Molden, D., Chapter 3, pp. 91–145. International Water Management Institute, Colombo, Sri Lanka and Earthscan, London, UK.

Derpsch, R. 2005. The extent of conservation agriculture adoption worldwide: Implications and impact. Keynote paper at the 3rd World Congress on Conservation Agriculture, October 3–7, 2005, RELMA/ICRAF, Nairobi, Kenya.

Dixon, A., D. Butler., A. Fewkes et al. 1999. Water saving potential of domestic water reuse systems using greywater and rainwater in combination. *Water Science and Technology* 39(5): 25–32.

Dudal, R. 1965. Dark clay soils of tropical and subtropical regions. *FAO Agricultural Development Paper* 83: 161.

Dwivedi, R.S., K.V. Ramana., S.P. Wani., and P. Pathak. 2003. Use of satellite data for watershed management and impact assessment. In *Integrated Watershed Management for Land and Water Conservation and Sustainable Agricultural Production in Asia,* eds. S.P. Wani, A.R. Maglinao, A. Ramakrishna, and T.J. Rego, pp. 149–157. Proceedings of the ADB-ICRISAT-IWMI Project Review and Planning Meeting, Hanoi, Vietnam, December 10–14, 2001. Patancheru, 502 325, International Crops Research Institute for the Semi-Arid Tropics, Andhra Pradesh, India.

Edmeades, D.C. 2003. The long-term effects of manures and fertilizers on soil productivity and quality: A review. *Nutrient Cycling in Agroecosystems* 66: 165–180.

El-Swaify, S.A., P. Pathak, T.J. Rego, and S. Singh. 1985. Soil management for optimized productivity under rainfed conditions in the semi-arid tropics. *Advances in Soil Science* 1: 1–64.

Eriksson, E., K. Auffarth., M. Henze, et al. 2002. Characteristics of grey wastewater. *Urban Water* 4: 85–104.

Falkenmark, M. 1986. Fresh water—Time for a modified approach. *Ambio* 15(4): 192–200.

Falkenmark, M. 1989. The massive water scarcity now threatening Africa—Why isn't it being addressed?, *Ambio* 18: 112–118.

Falkenmark, M. 1995. FAO Land and Water Bulletin Number 1. *Land and Water Integration and River Basin Management*, pp. 15–16. FAO, Rome, Italy.

Falkenmark, M. 1997. Meeting water requirements of an expanding world population. *Philosophical Transactions of the Royal Society of London* 352: 929–936.

Falkenmark, M. and D. Molden. 2008. Wake up to realities of river basin closure. *Water Resources Development* 24 (2): 201–215.

Falkenmark, M. and J. Rockström. 2008. Building resilience to drought in desertification-prone savannas in sub-Saharan Africa: *The water perspective, Natural Resources Forum* 32: 93–102.

Falkenmark, M., J. Rockström, L., Karlberg, et al. 2009. Present and future water requirements for feeding humanity. *Food Security* 1: 59–69.

Fan, S., P. Hazell, and P. Haque. 2000. Targeting public investments by agro-ecological zone to achieve growth and poverty alleviation goals in rural India. *Food Policy* 25(4): 411–428.

FAOSTAT. 2005. http://faostat.fao.org/

FAOSTAT. 2010. http://faostat.fao.org/site/567/default.asp.ancor. Accessed date April 2010.

Feng, S., A.B. Krueger., M. Oppenheimer, et al. 2010. Linkages among climate change, crop yields and Mexico–U.S. cross-border migration. *Proceedings of National Academy of Science USA* 107: 14.257–14.262.

Finlayson, C.M. and R. D'Cruz. 2005. Inland water systems. In *Millennium Ecosystem Assessment, Conditions and Trends.* Island Press, Washington, DC.

Fisher, G., V.V. Harrij, H. Eva, et al. 2009. Potentially obtainable yields in the semiarid tropics. *Global Theme on Agroecosystems* Report no. 54. Patancheru 502 324, International Crops Research Institute for the Semi-Arid Tropics, Andhra Pradesh, India.

Fox, P. and J. Rockström. 2003. Supplemental irrigation for dry-spell mitigation of rainfed agriculture in the Sahel. *Agricultural Water Management* 1817: 1–22.

Fox, P., J. Rockström., and J. Barron. 2005. Risk analysis and economic viability of water harvesting for supplemental irrigation in semi-arid Burkino Faso and Kenya. *Agricultural Systems* 83(3): 231–250.

Garg, K.K., S.P. Wani., J. Barron, et al. (submitted). Up scaling potential impacts on water flows from agricultural water interventions: Opportunities and trade-offs in the Osman Sagar catchment, Musi sub basin, India Water Resource Management.

Garg, K.K., L. Karlberg., J. Barron, et al. 2011a. Assessing impacts of agricultural water interventions in the Kothapally watershed, Southern India. *Hydrological Processes*. DOI: 10.1002/hyp.8138.

Garg, K.K., L. Karlberg., J. Barron, et al. 2011b. Biofuel production on wastelands in India: Opportunities and trade-offs for soil and water management at the watershed scale. *Biofuels, Bioproducts & Biorefining* 5(4): 410–430.

Garg, K.K., L. Bharati., A. Gaur, et al. 2011c. Spatial mapping of agricultural water productivity using SWAT model in Upper Bhima Catchment, India. *Irrigation and Drainage*. DOI: 10.1002/ird.618.

Ghosh, K., D.C. Nayak., and N. Ahmed. 2009. Soil organic matter. *Journal of the Indian Society of Soil Science* 57: 494–501.

Glendenning, C.J. and R.W. Vervoort. 2011. Hydrological impacts of rainwater harvesting (RWH) in a case study catchment: The Arvari River, Rajasthan, India. Part 2. Catchment—scale impacts. *Agricultural Water Management* 98: 715–730.

Gordon, L., M. Dunlop., B. Foran, et al. 2003. Land cover change and water vapour flows: Learning from Australia. *Philosophical Transactions of the Royal Society B* 358(1440): 1973–1984.

Gordon, L.J., W. Steffen., B.F. Jonsson, et al. 2005. Human modification of global water vapor flows from the land surface. *Proceedings of the National Academy of Sciences of United States of America* 102: 7612–7617.

Gordon, L.J., G.D. Peterson., E. Bennett, et al. 2008. Agricultural modifications of hydrological flows create ecological surprises. *Trends in Ecology and Evolution* 23(4): 11–219.

Gordon, L.J., C.M. Finlayson., M. Falkenmark, et al. 2010. Managing water in agriculture for food production and other ecosystem services. *Agricultural Water Management* 97: 512–519.

Government of India (GoI). 2008. Common guidelines for watershed Development Projects. Department of Land Resources, Ministry of Rural Development, Government of India, New Delhi.

Hanjra, M.A. and M.E. Qureshi. 2010. Global water crisis and future food security in an era of climate change. *Food Policy* 35: 365–377.

Harris, D., A. Joshi., P.A. Khan., P. Gothkar., and P.S. Sodhi. 1999. On-farm seed priming in semi-arid agriculture: Development and evaluation in maize, rice and chickpea in India using participatory methods. *Experimental Agriculture* 35: 15–29.

Hoek, V., M.U. Hassan., J.H.J. Ensink, et al. 2002. Urban wastewater: A valuable resource for agriculture—A case study from Haroonabad, Pakistan. IWMI-Research Report 63. IWMI, Colombo, Sri Lanka.

Hoff, H., M. Falkenmark., D. Gerten, et al. 2010. Greening the global water system. *Journal of Hydrology* 384: 177–186.

Holger, H. 2009. Global water resources and their management. *Current Opinion in Environmental Sustainability* 1: 141–147.

Huxley, P.A. 1979. Zero-tillage at Morogoro, Tanzania. In *Soil Tillage and Crop Production*, ed. R. Lal, pp. 259–265. Proceedings No.2. International Institute of Tropical Agriculture, Ibadan, Nigeria.

ICRISAT (International Crops Research Institute for the Semi-Arid Tropics). 2005. Identifying systems for carbon sequestration an increased productivity in semiarid tropical environments. A Project Completion Report funded by National Agricultural Technology Project, Indian Council of Agricultural Research.

IPCC. 2007. Climate change-impacts: Adaptation and vulnerability. In Technical summary of Working Group II to Fourth Assessment Report Intergovernmental Panel on Climate Change, eds. M.L. Parry, O.F. Canziani, J.P. Paultikof, P.J. van der Linden, and C.E. Hanon, pp. 23–78. Cambridge University Press, Cambridge.

Joshi, P.K., A.K. Jha., S.P. Wani., L. Joshi., and R.L. Shiyani. 2005. Meta-analysis to assess impact of watershed program and people's participation. In *Watershed Management Challenges: Improved Productivity, Resources and Livelihoods*, eds. Bharat R. Sharma, J.S. Samra, C.A. Scott, and S.P. Wani. Comprehensive assessment Research Report 8, Comprehensive Assessment Secretariat, Colombo, Sri Lanka. Colombo, Sri Lanka: IWMI.

Joshi, P.K., A.K. Jha, S.P. Wani, et al. 2008. Impact of watershed program and conditions for success: A meta-analysis approach. In *Global Theme on Agroecosystems,* Report no. 46. Patancheru 502 324, International Crops Research Institute for the Semi-Arid Tropics, Andhra Pradesh, India.

Kajiru, G.J. and Nkuba, J.M. 2010. Assessment of runoff harvesting with 'majaluba' system for improved productivity of smallholder rice in Shinyanga, Tanzania. Agricultural water management interventions delivers returns on investment in Africa. A compendium of 18 case studies from six countries in eastern and southern Africa, ed. B.M. Mati, pp. 65–75. VDM Verlag.

Karlberg, L., J. Rockström., M. Falkenmark, et al. 2009. Water resource implications of upgrading rainfed agriculture—Focus on green and blue water trade-offs. In *Rainfed Agriculture: Unlocking the Potential*, eds. S.P. Wani, J. Rockström, and T. Oweis, pp. 1–310. The Comprehensive Assessment of Water Management in Agriculture Series, Volume 7. CABI, Wallingford, UK.

Kashyap, D. 2007. PPP on farmers' participation in the value chain of biodiesel production: German Development Cooperation's experience. Presented at GTZ-ICRISAT Meeting on Public Private Partnership (PPP), September 14, 2007, ICRISAT. ICRISAT, Patancheru, India.

Katyal, J.C. and R.K. Rattan. 2003. Secondary and micronutrients: Research gaps and future needs. *Fertility News* 48(4): 9–14, 17–20.

Kerr, J., Pangare, G., and V. Lokur. 2002. Watershed development projects in India. Research Report 127. IFPRI, Washington, DC.

Khan, S. and M.A. Hanjra. 2009. Footprints of water and energy inputs in food production-global perspectives. *Food Policy* 34(2): 130–140.

Khare, D., M.K. Jat., J.D. Sunder, et al. 2007. Assessment of water resources allocation options: Conjunctive use planning in a link canal command. *Resources, Conservation and Recycling* 51(2): 487–506.

Kijne, J.W., R. Barker., D. Molden, et al. 2003. *Water Productivity in Agriculture: Limits and Opportunities for Improvement.* CAB International, Wallingford, UK.

Kijne, J., J. Barron., H. Holger, et al. 2009. Opportunities to increase water productivity in agriculture with special reference to Africa and South Asia. In Stockholm Environment Institute Project Report 2009. Sweden.

Kumar Rao, J.V.D.K., D. Harris., M. Kankal., and B. Gupta. 2008. Extending *rabi* cropping in rice fallows of eastern India. In *Improving Agricultural Productivity in Rice-Based Systems of the High Barind Tract of Bangladesh*, eds. C.R. Riches, D. Harris, D.E. Johnson, and B. Hardy, pp. 193–200. International Rice Research Institute, Los Banos, Philippines.

Lal, R. 2004. Soil carbon sequestration to mitigate climate change. *Geoderma* 123: 1–22.

Landell-Mills, N. and T.I. Porras. 2002. *Silver Bullet or Fools' Gold? A Global Review of Markets for Forest Environmental Services and their Impact on the Poor.* Instruments for Sustainable Private Sector Forestry Series. London, UK: International Institute for Environment and Development. (http://www.iied.org/pubs/pdf/full/9066IIED.pdf).

Landers, J.N., H. Mattana Saturnio, P.L. de Freitas, and R. Trecenti. 2001. Experiences with farmer clubs in dissemination of zero tillage in tropical Brazil. In *Conservation Agriculture, a Worldwide Challenge*, eds. L. García-Torres, J. Benites, and A. Martínez-Vilela, pp. 71–76. Food and Agriculture Organization of the United Nations (FAO), Rome, Italy.

Leduc, C., G. Favreau., P. Schroeter, et al. 2001. Long-term rise in a Sahelian water-table: The continental terminal in south-west Niger. *Journal of Hydrology* 243: 43–54.

Lee, K.K. and S.P. Wani 1989. Significance of biological nitrogen fixation and organic manures in soil fertility management. In *Soil Fertility and Fertility Management in Semiarid Tropical India*, ed. C.B. Christianson, pp. 89–108. IFDC, Muscle Shoals, AL.

Li, H.W., H. Gao, H. Wu, L.W. Ying, H.Y. Wang, and J. He. 2007. Effects of 15 years of conservation tillage on soil structure and productivity of wheat cultivation in northern China. *Australian Journal of Soil Research* 45(5): 344–350.

Lofstrand, F. 2005. Conservation agriculture in Babati district, Tanzania. Impacts of conservation agriculture for small-scale farmers and methods for increasing soil fertility. Master of Science Thesis, Swedish University of Agricultural Science, Department of Soil Science, Uppsala, Sweden.

Lu, W. and A.Y.T. Leung. 2003. A preliminary study on potential of developing shower/laundry wastewater reclamation and reuse system. *Chemosphere* 52: 1451–1459.

MA. 2005. Millennium ecosystem assessment, ecosystems and human well-being: Current status and trends. In *Fresh water*, Vol. 1, eds. R. Hassan, R. Scholes, and N. Ash. Isaland Press, Washington, USA. http://www.millenniumassessment.org/en/Products.Global.Condition.aspx.

Madungwe, E. and S. Sakuringwa. 2007. Greywater reuse: A strategy for water demand management in Harare? *Physics and Chemistry of the Earth* 32: 231–1236.

Manjunatha, M.V., R.J. Oosterbaan, S.K. Gupta, et al. 2004. Performance of subsurface drains for reclaiming waterlogged saline lands under rolling topography in Tungabhadra irrigation project in India. *Agricultural Water Management* 69(1): 69–82.

Mariki, W.L. 2004. The impact of conservation tillage and cover crops on soil fertility and crop production in Karatu and Hanang districts in northern Tanzania. TFSC/GTZ Technical Report 1999–2003. TFSC/GTZ, Arusha, Tanzania.

Massuel, S., G. Favreau., M. Descloitres., Y.L. Troquer., Y. Albouy., and B. Cappelaere. 2006. Deep infiltration through a sandy alluvial fan in semiarid Niger inferred from electrical conductivity survey, vadose zone chemistry and hydrological modeling. *Catena* 67: 105–118.

Materechera, S.A. 2010. Utilization and management practices of animal manure for replenishing soil fertility among smallscale crop farmers in semi-arid farming districts of the North West Province, South Africa. *Nutrient Cycling in Agroecosystems* 87: 415–428.

Mati, B.M. 2005. Overview of water and soil nutrient management under smallholder rain-fed agriculture in East Africa. Working Paper 105. International Water Management Institute, Colombo, Sri Lanka, www.iwmi.cgiar.org/pubs/working/WOR105.pdf.

Mati, B.M. 2010. Agricultural water management delivers returns on investment in eastern and southern Africa: A regional synthesis. In *Agricultural Water Management Interventions Delivers Returns on Investment in Africa. A Compendium of 18 Case Studies from Six Countries in Eastern and Southern Africa*, ed. B.M. Mati, pp. 1–29. VDM Verlag.

Mati, B.M., N. Hatibu, I.M.G. Phiri, and J.N. Nyanoti. 2007. Policies and institutional frameworks impacting on agricultural water management in Eastern and Southern Africa (ESA). Synthesis report of a rapid appraisal covering nine countries in the ESA. IMAWESA, Nairobi, Kenya.

Mati, B.M., W.M. Mulinge, E.T. Adgo, G.J. Kajiru, J.M. Nkuba, and T.F. Akalu. 2011. Rainwater harvesting improves returns on investment in smallholder agriculture in sub-Saharan Africa. In *Integrated Watershed Management*, eds. S.P. Wani, J. Rockström, and K.L. Sahrawat, Chapter 8, pp. 249–280. CRC Press, The Netherlands.

Molden, D., K. Frenken, R. Barker, et al. 2007. Trends in water and agricultural development. In *Water for Food, Water for Life—A Comprehensive Assessment of Water Management in Agriculture,* ed. D. Moden, pp. 57–89. International Water Management Institute (IWMI), Colombo, Sri Lanka.

Mujumdar, P.P. 2008. Implications of climate change for sustainable water resources management in India. *Physics and Chemistry of the Earth* 33(5): 354–358.

Nagavallemma, K.P., S.P. Wani, S. Lacroix, et al. 2005. Vermicomposting: Recycling wastes into valuable organic fertilizer. *Journal of Agriculture and Environment for International Development* 99: 188–204.

Nature. 2010. 466**,** 546–547. doi:10.1038/466546a.

Nicou, R. and J.L. Chopart. 1979. Water management methods in sandy soil of Senegal. In *Soil Tillage and Crop Production*, ed. R. Lal, pp. 248–257. Proceedings No. 2. International Institute of Tropical Agriculture, Ibadan, Nigeria.

Nyagumbo, I. 2000. Conservation technologies for smallholder farmers in Zimbabwe. In *Conservation Tillage for Dryland Farming. Technological Options and Experiences in Eastern and Southern Africa*, eds. E.K. Biamah, J. Rockström, and J. Okwach, pp. 70–86. RELMA, Nairobi, Kenya.

Oweis, T. 1997. Supplemental irrigation: A highly efficient water-use practice. International Center for Agricultural Research in the Dry Areas, Aleppo, Syria.

Oweis, T. and A. Hachum. 2001. Reducing peak supplemental irrigation demand by extending sowing dates. *Agricultural Water Management* 50, 109–123.

Oweis, T. and A. Hachum. 2009. Supplemental irrigation for improved rainfed agriculture in WANA region. In *Rainfed Agriculture: Unlocking the Potential*, eds. S.P. Wani, J. Rockström, and T. Oweis, pp. 1–310. The Comprehensive Assessment of Water Management in Agriculture Series, Volume 7. CABI Publishing, Wallingford, UK.

Pandey, S. 2001. Adoption of soil conservation practices in developing countries: Policy and institutional factors. In *Response to Land Degradation*, eds. E. Bridges, I. Hannam, I. Oldeman, Penning de Vries, F. Scherr, S. Sombatpanit, pp. 66–73. Science Publishers Inc, Enfield, New Hampshire.

Pandey, R.K., J.W. Maranville, A. Admou, et al. 2000. Deficit irrigation and nitrogen effects on maize in a Sahelian environment: I. Grain yield and yield components. *Agricultural Water Management* 46(1): 1–13.

Pangare, V.L. 1998. Gender issues in watershed development and management in India. In: *Agricultural Research & Extension Network Paper* No. 88a: 1–9.

Pathak, P. and K.B. Laryea. 1991. Prospects of water harvesting and its utilization for agriculture in the semi-arid tropics. In *Proceedings of the Symposium of the SADCC Land and Water Management Research Program Scientific Conference*, October 8–10, 1990, pp. 253–268. Gaborone, Botswana.

Pathak, P. and K.B. Laryea. 1995. Soil and water conservation in the Indian SAT: Principles and improved practices. In *Sustainable Development of Dryland Agriculture in India*, ed. R.P. Singh, pp. 83–94. Scientific Publishers, Jodhpur, India.

Pathak, P., S.M. Miranda, and S.A. El-Swaify. 1985. Improved rainfed farming for the semi-arid tropics: Implications for soil and water conservation. In *Soil Erosion and Conservation*, eds. S.A. El-Swaify, W.C. Moldenhauer, and L. Andrew, pp. 338–354. Soil Conservation Society of America, Ankeny, Iowa.

Pathak, P., K.B. Laryea, and R. Sudi. 1989. A runoff model for small watersheds in the semi-arid tropics. *Transactions of American Society of Agricultural Engineers* 32: 1619–1624.

Pathak, P., K.L. Sahrawat, T.J. Rego, et al. 2005. Measurable biophysical indicators for impact assessment: Changes in soil quality. In *Natural Resource Management in Agriculture: Methods for Assessing Economic and Environmental Impacts*, eds. B. Shiferaw, H.A. Freeman, and S.M. Swinton, pp. 53–74. CAB International, Wallingford, UK.

Pathak, P., S.P. Wani., and R. Sudi. 2007. Rural prosperity through integrated watershed management: A case study of Gokulpura-Goverdhanpura in eastern Rajasthan. Global Theme on Agroecosystems Report No. 36. Patancheru 502 324, International Crops Research Institute for the Semi-Arid Tropics, Andhra Pradesh, India.

Pathak, P., K.L. Sahrawat, S.P. Wani,et al. 2009. Opportunities for water harvesting and supplemental irrigation for improving rain-fed agriculture in semi-arid areas. In *Rainfed Agriculture: Unlocking the Potential.*eds. S.P. Wani, J. Rockström, and T. Oweis, pp. 197–221. Comprehensive Assessment of Water Management in Agriculture Series. CAB International, Wallingford, UK.

Pathak, P., P.K. Mishra, S.P. Wani., and R. Sudi. 2011. Soil and water conservation for optimizing productivity and improving livelihoods in rainfed areas. In *Integrated Watershed Management*, eds. S.P. Wani, J. Rockström, and K.L. Sahrawat, Chapter 7, pp. 205–248. CRC Press, The Netherlands.

Patil, S.L. 2003. Effect of moisture conservation practices and nitrogen application on growth and yield of winter sorghum in vertisols of semi-arid tropics of South India. *Special International Symposium on Transactions in Agriculture for Enhancing Water Productivity*, September, 23–25, 2003, pp. 70–71. TNAU, Tamil Nadu, India.

Phalan, B. 2009. The social and environmental impacts of biofuels in Asia: An overview. *Applied Energy* 86: S21–S29.

Postel, S.L., G.C. Daily., P.R. Ehlich, et al. 1996. Human appropriation of renewable fresh water. *Science* 271: 785–788.

Rajak, D., M.V. Manjunatha., G.R. Rajkumar, et al. 2006. Comparative effects of drip and furrow irrigation on the yield and water productivity of cotton (*Gossypium hirsutum* L.) in a saline and waterlogged vertisol. *Agricultural Water Management* 83(1–2): 30–36.

Rajashekhara Rao, B.K., K.L. Sahrawat, S.P. Wani, and G. Parthasaradhi. 2010. Integrated nutrient management to enhance on-farm productivity of rainfed maize in India. *International Journal of Soil Science* 5: 216–225.

Ramakrishna, A., T. Hoang Minh., Wani, S.P., and T.D. Long. 2006. Effect of mulch on soil temperature, moisture, weed infestation and yield of groundnut in northern Vietnam. *Journal of Field Crops Research. Elsevier Publishers* 95: 115–125.

Rao, K.V., B. Venkateswarlu, K.L. Sahrawath, S.P. Wani, P.K. Mishra, S. Dixit, K.S. Reddy, M. Kumar, and U.S. Saikia, 2010. Proceedings of National Workshop-cum-Brainstorming on Rainwater Harvesting and Reuse through Farm Ponds: Experiences, Issues and Strategies. April 21–22, 2009, CRIDA, CRIDA, Hyderabad.

Rao, A.V.R.K., S.P. Wani, P. Singh, I. Ahmed and K. Srinivas. 2007. Agroclimatic characterization of APRLP–ICRISAT nucleus watersheds in Nalgonda, Mahabubnagar and Kurnool districts. *Journal of SAT Agricultural Research* 3(1): 1– 55.

Raschid, L. 2004. IWMI assessment of wastewater irrigation practices in selected cities of less developed regions, Sri Lanka.

Ray, S.S., V.K. Dadhwal., R.R. Navalgund, et al. 2002. Performance evaluation of an irrigation command area using remote sensing: A case study of Mahi command, Gujarat, India. *Agricultural Water Management* 56(2): 81–91.

Rego, T.J., V.N. Rao, B. Seeling, G. Pardhasaradhi, and J.V.D.K. Kumar Rao. 2003. Nutrient balances—A guide to improving sorghum and groundnut-based dryland cropping systems in semi-arid tropical India. *Field Crops Research* 81: 53–68.

Rego, T.J., Wani, S.P., Sahrawat, K.L., et al. 2005. Macro-benefits from boron, zinc, and sulphur application in Indian SAT: A step for grey to green revolution in agriculture. Global Theme on Agroecosystems Report no. 16. ICRISAT, Patancheru, Andhra Pradesh, India.

Rego, T.J., K.L. Sahrawat, S.P. Wani, et al. 2007. Widespread deficiencies of sulfur, boron and zinc in Indian semi-arid tropical soils: On-farm crop responses. *Journal of Plant Nutrition* 30: 1569–1583.

Reij, C., I. Scoones., and C. Toulmin (eds) 1996. *Sustaining the Soil. Indigenous Soil and Water Conservation in Africa.* Earthscan, London, UK.

Rijsberman, F.R. 2006. Water scarcity: Fact or fiction? *Agricultural Water Management* 80(1–3): 5–22.

Rockström, J. 2003. Water for food and nature in the tropics: Vapour shift in rain-fed agriculture. Invited paper to the Special issue 2003 of *Royal Society Transactions B Biology, Theme Water Cycle as Life Support Provider* 358(1440): 1997–2009.

Rockström, J. and Falkenmark, M. 2000. Semiarid crop production from a hydrological perspective: Gap between potential and actual yields. *Critical Reviews in Plant Science* 19(4): 319–346.

Rockström, J. and J. Barron. 2007. Water productivity in rainfed systems: Overview of challenges and analysis of opportunities in water scarcity prone savannahs. *Irrigation Science* 25(3): 299–311.

Rockström, J. and Gordon, L. 2001. Assessment of green water flows to sustain major biomes of the world: Implications for future ecohydrological landscape management. *Physics and Chemistry of the Earth* 8(26)(11–12): 843–851.

Rockström, J. and L. Karlberg. 2009. Zooming in on the global hotspots of rainfed agriculture in water-constrained environments. In *Rainfed Agriculture: Unlocking the Potential*, eds. S.P. Wani, J. Rockström, and T. Oweis, pp. 1–310. The Comprehensive Assessment of Water Management in Agriculture Series, Volume 7. CABI, Wallingford, UK.

Rockström, J., P.E. Jansson., J. Barron, et al. 1998. Estimates of on-farm rainfall partitioning in pearl millet field with run-on and run-off flow based on field measurements and modelling. *Journal of Hydrology* 210: 68–92.

Rockström, J., L. Gordon., C. Folke, et al. 1999. Linkages among water vapor flows, food production, and terrestrial ecosystem services. *Conservation Ecology* 3(2): 5.

Rockström, J., N. Hatibu, T. Oweis, et al. 2007. Managing water in rain-fed agriculture. In *Water for Food, Water for Life: A Comprehensive Assessment of Water Management in Agriculture*, ed. D. Molden, pp. 315–348. Earthscan, London and International Water Management Institute (IWMI), Colombo, Sri Lanka.

Rockström, J., M. Falkenmark, L. Karlberg, et al. 2009. Future water availability for global food production: The potential of green water for increasing resilience to global change. *Water Resources Research* 45: W00A12. doi:10.1029/2007WR006767.

Rockström, J., L. Karlberg, S.P. Wani, et al. 2010. Managing water in rainfed agriculture—The need for a paradigm shift. *Agricultural Water Management* 97: 543–550.

Rost, S., Gerten, D., Bondeau, A., Luncht, W., Rohwer, J., Schaphoff, S., et al. 2008. Agricultural green and blue water consumption and its influence on the global water system. *Water Resources Research* 44: W09405. doi:10.1029/2007WR006331.

Sahrawat, K.L., Wani, S.P., Rego, T.J., et al. 2007. Widespread deficiencies of sulphur, boron and zinc in dryland soils of the Indian semi-arid tropics. *Current Science* 93(10): 1–6.

Sahrawat, K.L., S.P. Wani., P. Pathak, et al. 2010a. Managing natural resources of watersheds in the semi-arid tropics for improved soil and water quality: A review. *Agricultural Water Management* 97: 375–381.

Sahrawat, K.L., S.P. Wani., G. Parthasaradhi, et al. 2010b. Diagnosis of secondary and micronutrient deficiencies and their management in rainfed agroecosystems: Case study from Indian semi-arid tropics. *Soil Science and Plant Analysis* 41: 1–15.

Sahrawat, K.L., S.P. Wani, A. Subba Rao, and G. Pardhasaradhi. 2011. Management of emerging multinutrient deficiencies: A prerequisite for sustainable enhancement of rainfed agricultural productivity. In *Integrated Watershed Management*, eds. S.P. Wani, J. Rockström, and K.L. Sahrawat, Chapter 9, pp. 281–314. CRC Press, The Netherlands.

Samra, J.S. and H. Eswaran. 2000. Challenges in ecosystem management in a watershed context in Asia. In *Integrated Watershed Management in the Global Ecosystem*, ed. Ratan Lal, pp. 19–33. CRC Press, Boca Raton.

Sathaye, J., W. Makundi., K. Andrasko, et al. 2001. Management of forests for mitigation of greenhouse gas emissions. *Mitigation and Adaptation Strategies for Global Change* 6(3–4): 83–184.

Scanlon, B.R., I. Jolly, M. Sophocleous, et al. 2007. Global impacts of conversions from natural to agricultural ecosystems on water resources: Quantity versus quality. *Water Resource Research* 43: W03437. doi:10.1029/2006WR005486.

Shah, T., and S. Verma. 2008. Co-management of electricity and groundwater: An assessment of Gujarat's Jyotigram scheme. *Economic and Political Weeks* 59–66.

Shah, T., J. Burke and K. Villholth. 2007. Groundwater: A global assessment of scale and significance. In *Water for Food, Water for Life*, ed. D. Molden, pp. 395–423. Earthscan, London, UK and IWMI, Colombo, Sri Lanka.

Shah, T., I. Makin., R. Sakthivadivel, et al. 2005. Limits to leapfrogging: Issues in transposing successful river basin management institutes in the developing world. In *Irrigation and River Basin Management: Options for Governess and Institutes*, ed. M. Svendsen, pp. 31–49. CAB International, Wallingfors, UK.

Shah, A., S.P. Wani, T.K. Sreedevi, et al. (eds). 2009. Impact of watershed management on women and vulnerable groups. Proceedings of the Workshop on Comprehensive Assessment of Watershed Programs in India, July 25, 2007. ICRISAT, Patancheru 502 324, International Crops Research Institute for the Semi-Arid Tropics, Andhra Pradesh, India.

Sharma, P.N. and O.J. Helweg. 1982. Optimal design of a small reservoir system. *ASCE Journal of Irrigation and Drainage* IR4: 250–264.

Sharma, K.L., J.K. Grace, K. Srinivas, et al. 2009. Influence of tillage and nutrient sources on yield sustainability and soil quality under sorghum-mung bean system in rainfed semi-arid tropics. *Communications in Soil Science and Plant Analysis* 40: 2579–2580.

Sharma, B.R., K.V. Rao., K.P.R. Vittal, et al. 2010. Estimating the potential of rainfed agriculture in India: Prospects for water productivity improvements. *Agricultural Water Management* 97: 23–30.

Shiferaw, B., J. Okello., and V.R. Reddy. 2009. Challenges of adoption and adaption of land and water management options in smallholder agriculture: Synthesis of lessons and experience. In *Rainfed Agriculture: Unlocking the Potential*, eds. S.P.Wani, J. Rockström, and T. Oweis, pp. 1–310. The Comprehensive Assessment of Water Management in Agriculture Series, Volume 7. CABI Publishing, Wallingford, UK.

Shiklomanov, I. 1993. World fresh water resources. In *Water in Crisis: A Guide to the World's Fresh Water Resources*, ed. P. H. Gleick, pp. 13–24. Oxford University Press, New York.

Singh, K. 1995. Co-operative property rights a instrument for managing groundwater, In *Groundwater Law: The Growing Debate. Monograph*, ed. M. Moench. VIKSAT—Natural Heritage Institute, Ahmedabad, India.

Singh, P., D. Vijaya, K. Srinivas, et al. 2001. Potential productivity, yield gap, and water balance of soybean-chickpea sequential system at selected benchmark sites in India. Global Theme 3: Water, Soil, and Agrobiodiversity Management for Ecosystem Health. Report no.1. Patancheru 502 324, International Crops Research Institute for the Semi-Arid Tropics, Andhra Pradesh, India.

Singh, P., P.K. Aggarwal, V.S. Bhatia, et al. 2009. *Yield Gap Analysis: Modeling of Achievable Yields at Farm Level in Rain-Fed Agriculture: Unlocking the Potential*, eds. S.P. Wani, J. Rockstorm, and T. Oweis, pp. 81–123. CAB International, Wallingford, UK.

Singh, P., P. Pathak, S.P. Wani, and K.L. Sahrawat. 2010. Integrated watershed management for increasing productivity and water use efficiency in semi-arid tropical India. In *Water and Agricultural Sustainability Strategies*, ed. M. S. Kang, 181–205. CRC Press, The Netherlands.

Singh, P., S.P. Wani., P. Pathak, K.L. Sahrawat., and A.K. Singh. 2011. Increasing crop productivity and water use efficiency in rainfed agriculture. In *Integrated Watershed Management*, eds. S.P.Wani, J. Rockström, and K.L. Sahrawat, Chapter 10, pp. 315–348. CRC Press, The Netherlands.

Sireesha, P. 2003. Prospects of water harvesting in three districts of Andhra Pradesh. MTech Thesis, Center for Water Resources, Jawaharlal Nehru Technological University (JNTU), Hyderabad, India.

Sreedevi, T.K. and S.P. Wani, 2009. Integrated farm management practices and up-scaling the impact for increased productivity of rain-fed systems. In *Rain-Fed Agriculture: Unlocking the Potential*, eds. S.P. Wani, J. Rockström, and T. Oweis, pp. 222–257. Comprehensive Assessment of Water Management in Agriculture Series. CAB International, Wallingford, UK.

Sreedevi, T.K., B. Shiferaw, S.P. Wani, et al. 2004. Adarsha watershed in Kothapally, Understanding the drivers of higher impact. *Global Theme on Agroecosystems* Report No. 10. International Crops Research Institute for the Semi-Arid Tropics, Patancheru, Andhra Pradesh, India.

Sreedevi, T.K., Wani, S.P., Sudi, R., et al. 2006. On-site and off-site impact of watershed development: A case study of Rajasamadhiyala, Gujarat, India. Global Theme on Agroecosystems Report No. 20. ICRISAT, Patancheru, Andhra Pradesh, India.

Sreedevi, T.K., S.P. Wani, and P. Pathak. 2007. Harnessing gender power and collective action through integrated watershed management for minimizing land degradation and sustainable development. *Journal of Financing Agriculture* 36: 23–32.

Sreedevi, T.K., S.P. Wani., V. Nageswara Rao, et al. 2009a. Empowerment of women for equitable participation in watershed management for improved livelihoods and sustainable development: An analytical study. In *Impact of Watershed Management on Women and Vulnerable Groups*, eds. A. Shah, S.P. Wani, T.K. Sreedevi, Proceedings of the Workshop on Comprehensive Assessment of Watershed Programs in India, July 25, 2007, ICRISAT, Patancheru 502 324, ICRISAT, Andhra Pradesh, India.

Sreedevi, T.K., S.P. Wani., Ch. Srinivasa Rao et al. 2009b. *Jatropha* and *Pongamia* rainfed plantations on wastelands in India for improved livelihoods and protecting environment. Sixth International Biofuels Conference, March 4–5, 2009.

Srinivasa Rao, Ch., K.P.R. Vittal, B. Venkateswarlu, et al. 2009. Carbon stocks in different soil types under diverse rainfed production systems in tropical India. *Communications in Soil Science and Plant Analysis* 40: 2338–2356.

Srivastava, P., 2005. Poverty targeting in India. In *Poverty Targeting in Asia*, ed. J. Weiss, pp. 34–78. Edward Eigar, Cheltenham.

Srivastava, K.L., P. Pathak, J.S. Kanwar, and R.P. Singh. 1985. Watershed-based soil and rainwater management with special reference to vertisols and alfisols. Presented at the National Seminar on *Soil Conservation and Watershed Management,* September 5–7, 1985, Soil Conservation and Watershed Management, New Delhi, India.

Srivastava, K.L., A. Astatke, T. Mam, et al. 1993. Land, soil and water management. In *Improved Management of Vertisols for Sustainable Crop–Livestock Production in the Ethiopian Highlands: Synthesis Report,* eds. T. Mamo, A. Astatke, K.L. Srivastava, and A. Dibabe, pp. 75–84. Technical Committee of the Joint Vertisol Project, Addis Ababa, Ethiopia.

Stewart, J.I. 1988. *Response Farming in Rainfed Agriculture*. The WHARF Foundation Press, Davis, CA.

Stewart, J.I., R.D. Misra, W.O. Pruitt, et al. 1975. Irrigating corn and grain sorghum with a deficient water supply. *Transactions of American Society of Agricultural Engineers* 18: 270–280.

Stockholm Environment Institute (SEI). 2005. Sustainable pathways to attain the millennium development goals—Assessing the role of water, energy and sanitation. Document prepared for the UN World Summit, Sept 14, 2005, SEI, New York, NY, Stockholm. http://www.sei.se/mdg.htm.

Subbarao, G.V., J.V.D.K. Kumar Rao, J. Kumar, et al. 2001. Spatial distribution and quantification of rice-fallows in South Asia—Potential for legumes. International Crops Research Institute for the Semi-Arid Tropics, Patancheru, Andhra Pradesh, India.

Sujala-ICRISAT watershed Project, Terminal Report, 2008. Establishing participatory research-cum-demonstrations for enhancing productivity with sustainable use of natural resources in Sujala watersheds of Karnataka, 2005–2008.

Sulser, T.B., C. Ringler., T. Zhu, et al. 2010. Green and blue water accounting in the Ganges and Nile basins: Implications for food and agricultural policy. *Journal of Hydrology* 384(3–4): 276–291.

Syers, J.K, J. Lingard, J. Pieri, et al. 1996. Sustainable land management for the semiarid and sub-humid tropics. *Ambio* 25: 484–491.

United Nations Environment Programme (UNEP). 1997. *World Atlas of Desertification*, 2nd edn. UNEP, Nairobi, Kenya.

Velayutham, M., D.K. Pal, and T. Bhattacharyya, 2000. Organic carbon stock in soils of India. In *Advances in Soil Science. Global Climate Change and Tropical Ecosystems,* eds. R. Lal, J.M. Kible, and B.A. Stewart, pp. 71–95. CRC Press, Boca Raton.

Verhoeven, J.T.A., B. Arheimer., C. Yin, et al. 2006. Regional and global concerns over wetlands and water quality. *Trends in Ecology and Evolution*, 21(2): 96–103.

Wang, X.B., D.X. Cai, W.B. Hoogmoed, O. Oenema, and U.D. Perdok. 2007. Developments in conservation tillage in rainfed regions of North China. *Soil and Tillage Research* 93(2): 239–250.

Wani, S.P. and J. Rockström. 2011. Watershed development as a growth engine for sustainable development of dryland areas. In *Integrated Watershed Management in Rainfed Agriculture*, eds. S.P. Wani, J. Rockström, and K.L. Sahrawat, pp. 35–52. CRC Press, The Netherlands.

Wani, S.P. and T.K. Sreedevi. 2005. *Pongamia*'s journey from forest to micro-enterprise for improving livelihood. *International Crop Research Institute for the Semi-Arid Tropics, Research Report, Global Theme of Agroecosystems.* ICRISAT, Patancheru, Andhra Pradesh, India.

Wani, S.P., W.B. McGill, J.A. Robertson, K.L. Haugen Kozyra, and J.J. Thurston. 1994. Improved soil quality, increased barley yields with fababeans and returned manure in a crop rotation on a gray luvisol. *Canadian Journal of Soil Science* 74: 75–84.

Wani, S.P., T.J. Rego, S. Rajeswari, et al. 1995. Effect of legume-based cropping systems on nitrogen mineralization potential of Vertisol. *Plant and Soil* 175: 265–274.

Wani, S.P., P. Pathak, H.M. Tam, et al. 2002. Integrated watershed management for minimizing land degradation and sustaining productivity in Asia. In *Integrated Land Management in Dry Areas*, ed. Zafar Adeel, pp. 207–230. Proceedings of a joint UNU-CAS international workshop, September 8–13, 2001, Beijing, China.

Wani, S.P., Pathak, P., Jangawad, L.S, et al. 2003a. Improved management of vertisols in the semi-arid tropics for increased productivity and soil carbon sequestration. *Soil Use and Management* 19: 217–222.

Wani, S.P., P. Pathak, T.K. Sreedevi, et al. 2003b. Efficient management of rainwater for increased crop productivity and groundwater recharge in Asia. In *Water Productivity in Agriculture: Limits and Opportunities for Improvement*, eds. J.W. Kijne, R. Barker, and D. Molden, pp. 199–215. CAB International, Wallingford, UK and IWMI, Colombo, Sri Lanka.

Wani, S.P., P. Singh, R.S. Dwivedi, et al. 2005. Biophysical indicators of agro-ecosystem services and methods for monitoring the impacts of NRM technologies at different scale. In *Natural Resource Management in Agriculture: Methods for Assessing Economic and Environmental Impacts*, eds. B. Shiferaw, H.A. Freemen, and S.M. Swinton, pp. 97–123. CAB International, Wallingford, UK.

Wani, S.P., M. Osman., E. D'Silva, et al. 2006a. Improved livelihoods and environmental protection through biodiesel plantations in Asia, *Asian Biotechnology and Development Review* 8(2): 11–29.

Wani, S.P., Y.S. Ramakrishna, T.K. Sreedevi, et al. 2006b. Issues, concepts, approaches and practices in integrated watershed management: Experience and lessons from Asia. In: *Integrated Management of Watersheds for Agricultural Diversification and Sustainable Livelihoods in Eastern and Central Africa,* 17–36. Proceedings of the international workshop held at ICRISAT, December 6–7, 2004, Nairobi, Kenya. ICRISAT, Patancheru, Andhra Pradesh, India.

Wani, S.P., Y.S. RamaKrishna, T.K. Sreedevi, et al. 2006c. Greening drylands and improving livelihoods. International Crops Research Institute for the Semi Arid Tropics, Patancheru 502 324, ICRISAT, Andhra Pradesh, India.

Wani, S.P., K.L. Sahrawat., T.K. Sreedevi, et al. 2007. Efficient rainwater management for enhanced productivity in Arid and Semi-arid drylands. *Journal of Water Management* 15(2): 126–140.

Wani, S.P., P.K. Joshi, K.V. Raju, et al. 2008. Community watershed as a growth engine for development of dryland areas. A comprehensive assessment of watershed programs in India. Global Theme on Agroecosystems Report No. 47. International Crops Research Institute for the Semi-Arid Tropics, Andhra Pradesh, India.

Wani, S.P., T.K. Sreedevi., J. Rockström, et al. 2009a. Rain-fed agriculture—Past trend and future prospects. In *Rain-Fed Agriculture: Unlocking the Potential*, eds. S.P. Wani, J. Rockström, and T. Oweis, pp. 1–35. Comprehensive Assessment of Water Management in Agriculture Series. CAB International, Wallingford, UK.

Wani, S.P., T.K. Sreedevi., S. Marimuthu et al. 2009b. Harnessing the potential of *Jatropha* and *Pongamia* plantations for improving livelihoods and rehabilitating degraded lands. 6th International Biofuels Conference, March 4–5, 2009.

Wani, S.P., K.H. Anantha, T.K. Sreedevi, et al. 2011a. Assessing the environmental benefits of watershed development: Evidence from the Indian semi-arid tropics. *Journal of Sustainable Watershed Science & Management* 1(1): 10–20.

Wani, S.P., J. Rockström, B. Venkateswarlu, et al. 2011b. New paradigm to unlock the potential of rainfed agriculture in the semiarid tropics. In *World Soil Resources and Food Security*, eds. R. Lal and B.A. Steward, pp. 419–470. CRC Press, UK.

Wani, S.P., B. Venkateswarlu, and V.N. Sharda. 2011c. Watershed development for rainfed areas: Concept, principles, and approaches. In *Integrated Watershed Management in Rainfed Agriculture*, eds. S.P. Wani, J. Rockström, and K.L. Sahrawat, Chapter 3, pp. 53–86. CRC Press, The Netherlands.

Water Use Efficiency Project Report, 2009. Kharif fallow management for improving rainfall use efficiency in Madhya Pradesh. Project Completion Report 2007–2009, submitted to the Ministry of Water Resources, Government of India. Global Theme on Agro eco systems, ICRISAT, Patancheru, Andhra Pradesh, India.

Wilson, M.J. 2011. Improving livelihoods in rainfed areas through integrated watershed management: A development perspective. In *Integrated Watershed Management in Rainfed Agriculture*, eds. S.P. Wani, J. Rockström, and K.L. Sahrawat, Chapter 1, pp. 1–34. CRC Press, The Netherlands.

Yeh, S., G. Berndes, G.S. Mishra, et al. 2011. Evaluation of water use for bioenergy at different scales. *Biofuels, Bioproducts & Biorefining* 5(4): 361–374.

Yule, D.F., G.D. Smith, and P.J. George. 1990. Soil management options to increase infiltration in alfisols. In *Proceedings of the International Symposium on Water Erosion, Sedimentation and Resource Conservation*, pp. 180–187. Central Soil and Water Conservation Research and Training Institute, Dehradun, Uttar Pradesh, India.

Zhang, H. and T. Oweis. 1999. Water–yield relations and optimal irrigation scheduling of wheat in the Mediterranean region. *Agricultural Water Management* 38: 195–211.

14 Manipulating Crop Geometries to Increase Yields in Dryland Areas

B.A. Stewart and Rattan Lal

CONTENTS

14.1 INTRODUCTION

Drylands comprise regions characterized by different moisture regimes on the basis of the rainfall received. A region is termed hyperarid if the annual rainfall is <200 mm, arid if <200 mm during winter and <400 mm during summer, semiarid if 200–500 mm during winter and 400–600 mm during summer, and dry subhumid if 500–700 mm during winter and 600–800 mm during summer (FAO 1993). On the basis of the length of the growing season, considering favorable water balance and temperature regime, a region is categorized arid if the growing season is <75 days, semiarid if 75–120 days, and dry subhumid if 120–150 days (FAO 1993). These regions are also classified on the basis of the aridity index (AI), defined as the ratio of precipitation (P) to potential evapotranspiration (PET). The region is termed hyperarid if AI is <0.05, arid if 0.05–0.2, semiarid if 0.2–0.5, and dry subhumid if 0.5–0.65. Globally, these regions occupy 1.96 billion hectares (Bha) in Africa, 1.95 Bha in Asia, 0.66 Bha in Australasia, 0.3 Bha in Europe, 0.74 Bha in North America, and 0.54 Bha in South America (Table 14.1). Thus, drylands occupy a total of 6.15 Bha or 47.1% of the earth's land area. Principal soils consist of Aridisols (2.1 Bha), Entisols (2.3 Bha), and Alfisols (0.38 Bha) (Table 14.2). Most soils (except Vertisols and Mollisols) are coarse-textured and low in soil organic matter content and inherent soil fertility. Drought stress, low nutrient reserves, and susceptibility to erosion (by water and wind) and secondary salinization are principal soil-related constraints to achieving high biomass production and agronomic yields.

During the past 50 years, world cereal production increased about 2.7 times compared to 2.3 times for world population. This increased production was a remarkable achievement and the result of many factors. However, increased irrigated areas and increased use of chemical fertilizers are clearly two of the most important reasons. Irrigated areas more than doubled and fertilizer consumption increased several fold. Irrigated lands and favorable rainfed areas benefitted greatly

TABLE 14.1
Continental Distribution of the World's Drylands

Region	Area (Bha)					% of Earth's Land Area
	Hyperarid	Arid	Semiarid	Dry Subarid	Total	
Africa	0.67	0.5	0.51	0.27	1.96	15.0
Asia	0.28	0.63	0.69	0.35	1.95	14.9
Australasia	0	0.3	0.31	0.05	0.66	5.1
Europe	0	0.01	0.11	0.18	0.30	2.3
North America	0.003	0.08	0.42	0.23	0.74	5.6
South America	0.03	0.05	0.27	0.21	0.54	4.2
Total	0.98	1.57	2.31	0.30	6.15	
% of earth's area	7.5	12.0	17.7	9.9		47.1

Source: Adapted from Middleton, N.J. and Thomas, D.S.G., *World Atlas of Desertification*, Edward Arnold/UNEP, Seven Oaks, TN, 1992; Noin, D. and Clark, J.I., In: *Population and Environment in Arid Regions*, eds. J. Clark and D. Noin, MAB/UNESCO, Vol. 19, The Parthenon Publishing Group, New York, 1997; Food and Agriculture Organization, *Land Resource Potential and Constraints at Regional and Country Levels*, World Soil Resources Report 90, FAO, Rome, Italy, 2000.

Note: Bha = 10^9 ha = billion hectares.
Earth's total land area = 13.06 Bha.

from chemical fertilizers and also from improved plant cultivars that resulted from plant breeding and molecular biology. Rainfed areas are generally considered areas that are not irrigated but that receive enough precipitation to allow the cultivation of crops. The climates for these areas, however, are highly variable ranging from semiarid to humid. A lack of water limits the yield in nearly all climates unless irrigation water is added, but the extent of the limitation varies greatly. It can be anticipated in humid regions that there will be sufficient precipitation to avoid drought that

TABLE 14.2
Continental Distribution of Major Dryland Soils

Soil Order	Area (Bha)							% of Earth's Land Area
	Africa	Asia	Australasia	North America	South America	Europe	Total	
Alfisols	0.235	—	0.0464	0.0294	0.0706	?	0.3815	2.9
Aridisols	0.549	0.799	0.2917	0.3312	0.1520	?	2.1229	16.3
Entisols	1.136	0.663	0.2453	0.0589	0.2281	?	2.3313	17.9
Mollisols	0.0196	0.3898	—	0.3018	0.0923	?	0.8035	6.2
Vertisols	0.0169	0.0975	0.0796	0.0147	—	?	0.2087	1.6
Total	1.959	1.949	0.663	0.736	0.543	0.30	6.15	
% of earth's land area	15.0	14.9	5.1	5.6	4.2	2.30		47.1

Source: Recalculated from Dregne, H.E., *Soils of the Arid Regions*, Elsevier, Amsterdam, 1976; Noin, D. and Clark, J.I., In: *Population and Environment in Arid Regions*, eds. J. Clark and D. Noin, MAB/UNESCO, Vol. 19, The Parthenon Publishing Group, New York, 1997.

threatens the survival of a crop, but complete crop failure often occurs in semiarid regions. There are different approaches used for classifying climates, and Stewart et al. (2006) present and discuss some of these. Although rainfed areas include all of the cultivated areas that are not irrigated, many workers tend to separate rainfed areas into more favorable and less favorable areas. The less favored areas are commonly called dryland areas. Hargreaves (1957) defined dryland farming as agriculture without irrigation in regions of scanty precipitation. Stewart and Burnett (1987) stated that dryland farming emphasizes water conservation in every practice throughout the year. Bowden (1979) further divided semiarid lands into the semiarid tropics and the mid-latitude steppes. However, he listed four keys that are unique and that apply to all semiarid lands.

- *Key 1*: No growing season is or will be nearly the same in precipitation amount, kind, or range, or in temperature average, range, or extremes, as the previous growing season. Although this key is critical in any rainfed system, it requires absolute attention in dryland farming. Crop cultivation requires an adjustment every year, which leads to the second key.
- *Key 2*: Crops cannot be planned or managed to be the same from season to season. Most of the world's agricultural practices in either humid or arid areas have some predictability on an annual basis. In semiarid climates, however, even highly mechanized, technically advanced, commercial farms such as those in the High Plains of North America or the outback of Western Australia do not have sufficiently stable production for the individual or government to count on a given production figure for the following season.
- *Key 3*: The soil and water resource does not remain the same for any long period of time once agriculture is introduced into a semiarid region. A generalization necessary to support this key is that soils of most semiarid lands developed under grass on relatively flat topography. The competition for water and nutrients to produce crops requires removal of the protective grass cover. Because the crops are annual and dependent on precipitation, severe drought often leaves the soil highly vulnerable to wind erosion.
- *Key 4*: There is abundant sunshine due to many cloud-free days. This key has potential benefit and is shared with most arid climates. Abundant sunshine means higher temperatures that induce rapid growth, but it also creates a situation that demands careful management of soil water. Warm seasons, abundant sunshine, and cloud-free conditions stimulate growth but also increase evaporation and transpiration. It is possible for a grain crop to mature rapidly due to several weeks of sun-drenched, rainless conditions and desiccate just days before ripening. It is equally possible for a few millimeters of precipitation to occur at almost the last moment and produce a good crop.

14.2 EXAMPLE SEMIARID REGION

Dryland areas occur in many parts of the world. By the Food and Agriculture Organization (FAO) classification system that is based on the number of growing days (FAO 2000), 45% of the world's total land area is considered drylands. In this treatise, only the dryland area of the Central and Southern Great Plains of the United States will be discussed in detail, but the principles and technologies presented will be applicable to other areas. The similarities and differences of the climate in this area compared to other dryland regions of the world have been presented and discussed by others (Peterson et al. 2012, Chapter 16; Stewart 1988; Unger et al. 2006). Akron, CO, and Big Spring, TX, are the locations where USDA Agricultural Research Service has conducted experiments since the early 1900s. The average monthly PET values are much higher than the average monthly P values (Figure 14.1). In fact, there is not a single month in the year where the average P even reaches one-half of the PET value. The approximate location of these sites within the Great Plains is shown in Figure 14.2. While the entire Great Plains is considered semiarid, the degree of aridity increases moving from north to south because the temperature gradient increases in that direction, while the precipitation increases mostly from west to east. In general terms, the precipitation is about 400 mm

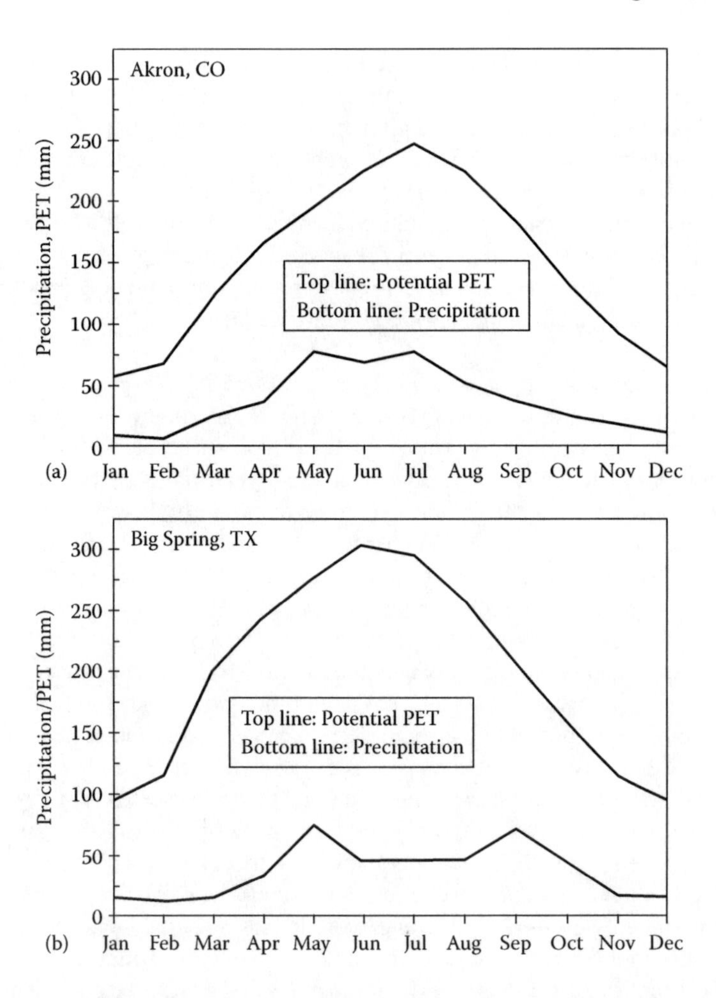

FIGURE 14.1 (a, b) Monthly average potential evapotranspiration (Penman–Monteith equation) and precipitation for two locations in the U.S. central and southern Great Plains (see Figure 14.2).

on the west side of the Great Plains and about 650 mm on the east side. The low precipitation amounts on the west side of the Great Plains and the high temperatures in the southern portions make these areas very close to arid. Therefore, dryland crop production generally requires fallowing the land for several months between crops to store sufficient water in the soil profile to supplement the growing season precipitation. Wheat (*Triticumaestivum* L.), grain sorghum (*Sorghum bicolor* (L.) Moench), and cotton (*Gossypiumhirsutum*) are the most widely grown dryland crops; maize (*Zea mays*) and sunflower (*Helianthus annuus*) are sometimes included in a cropping system.

The proportion of precipitation received during the fallow period stored in the soil profile for use by a subsequent crop was historically only 15%–20%. However, the use of limited and no-tillage systems has increased this significantly. Peterson et al. (2010, Chapter 16) have summarized some recent studies and have shown that 40% or more of the precipitation can be stored, and this has greatly increased the dryland crop production.

14.3 MANIPULATING CROP GEOMETRIES

Although a tremendous amount of research has been conducted during the past three or four decades on using crop residues to increase soil water storage, there has been little or none on manipulating crops to increase the efficiency of using this additional water. Loomis (1983) states that crop manipulations involve variations in the choice of species and cultivar, timing of events,

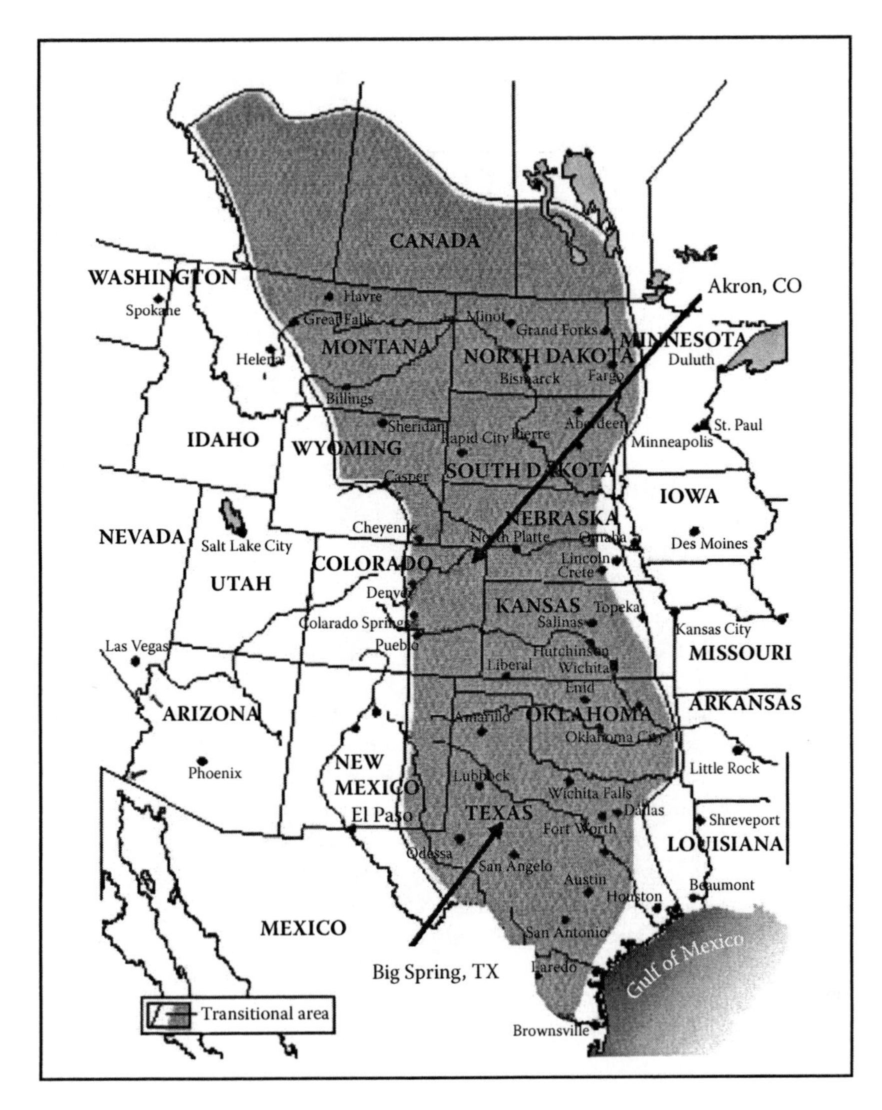

FIGURE 14.2 The Great Plains of the United States and Canada (approximate locations of Big Spring, TX, and Akron, CO, are shown by the arrows).

plant density, fertility status, provision for irrigation, and other factors. He emphasizes that where water is limiting, several basic strategies need to be followed to bring crops to maturity within the available supply: (1) ensure that a large proportion of the available water goes to transpiration, (2) achieve a high level of production per unit of transpiration, and (3) achieve a balance between seasonal use of water and seasonal supply. In principle, plant spacing can be varied in a way that influences the time when stored moisture is used. Roots can reach all of the soil mass earlier in the season with a regular spacing than with a clumped pattern. Where plants are closely spaced within rows (or hills), but widely spaced between rows, roots may reach the interrow soil mass much later in the season (Loomis 1983). Crop yields may or may not be increased by different arrangements depending on the stored soil water supply, time and amount of growing season precipitation, depth and lateral extent of rooting, and other factors. Loomis stated that a useful generalization is that where water is limiting, nonuniform treatment of the land or the crop can be an advantage, but where water is not limiting, uniform cropping will provide the greatest efficiency in light interception and photosynthesis.

In dryland areas such as described earlier, water is always limited. However, since dryland crops are commonly grown on land that has been fallowed for several months, there is generally considerable plant available water in the soil profile at the time of seeding. Careful management and well-designed strategies are necessary to prevent most or all of this water from being utilized for vegetative growth because this will almost certainly lead to a severe lack of water during the later growth stages that determine the quantity and quality of the yield. It is critical, particularly for grain crops, that plants have available water during the reproductive and grain-filling periods. Craufurd et al. (1993) reported that water stress during booting and flowering stages resulted in a grain yield reduction of up to 85%.

14.3.1 Vapor Pressure Deficit Effect on Transpiration

Plant growth is directly related to water availability (Sinclair 2009a). Sinclair states that the difference in vapor pressure inside and outside a leaf (VPD) is what controls the rate of water loss through the stomata. The VPD in arid regions is large because the vapor pressure in the atmosphere is low. For a given environment, the VPD cannot be controlled—it is what it is. Sinclair (2009b) stated that "Despite claims that crop yields will be substantially increased by the application of biotechnology, the physical linkage between growth and transpiration imposes a barrier that is not amenable to genetic alteration."

Sinclair and Weiss (2010) state that a C_4 crop growing in a 2 kPa transpiration environment will have a transpiration rate of approximately 220 g water/day for each gram of plant growth. By contrast, a C_3 crop will require 330 g water/day for each gram of plant growth. The water requirement, however, is greatly influenced by the climatic variables. The relationship between transpiration and biomass production for three different climates is shown in Figure 14.3. The amount of biomass produced from 400 mm water can be almost two times more for plants growing in a humid environment compared to an arid environment. The microclimate of plants, specifically the vapor pressure deficit, is determined by temperature, radiation, humidity, and wind, or in simple terms: (1) how hot it is, (2) how sunny it is, (3) how dry the air is, and (4) how windy it is. These characteristics are usually measured in the environment close to where plants are growing, but the vapor pressure deficit of the plant leaves is generally not measured. Therefore, an important link is generally missed. It is the vapor pressure deficit of the plant leaves that determines the transpiration rate—not the surrounding environment. Even though these may be somewhat similar, the plants often grow in a microclimate

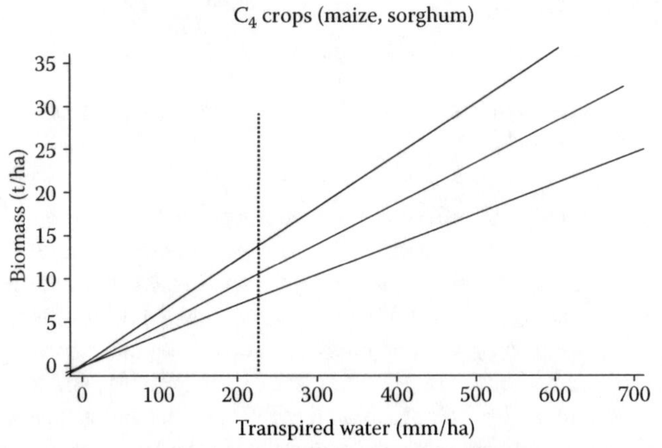

FIGURE 14.3 The effect of different climates on the amount of water transpired for each gram of above ground biomass produced. (Figure based on information from Sinclair, T.R. and Weiss, A., *Principles of Ecology in Plant Production*, 2nd edn, CAB International, Cambridge, MA, 2010.)

FIGURE 14.4 Maize plants growing in very different microclimates but in the same general environment.

that is significantly different from the surrounding environment. Figure 14.4 shows two fields of maize very close to each other, but the microclimate is very different for the plants in the different fields. Even if the plants are genetically the same, the transpiration efficiency will differ because the temperature, radiation, humidity, and wind speed will be different at the leaf surface.

Historically, plants were often grown in clumps, or hills, where there would be several plants close together and then open space between the clumps. William Brown, who later became the President of Pioneer Hi-Bred International, Inc., investigated during the mid-1950s the origin, evolution, and culture of some of the important crops that furnish the major sources of food for certain native Americans of the desert area of southwestern United States (Brown 1985). In parts of what are now Arizona and New Mexico, a number of Indian tribes and their ancestors learned and practiced the art of dryland agriculture with a high degree of success. The way maize was grown would not be recognized in most areas today, but it was based on methods that had sustained their needs for centuries. Some is still being grown similarly today in areas using seed that has been naturally selected over many years. The corn is planted in hills about 2 m apart, and the hills are in rows, also equally widely spaced (Figure 14.5). A hill contains 10–12 or even more plants. The Indians learned that the clumping of plants within a relatively small space reduces the desiccation of the foliage, anthers, and silk that allows normal fertilization to occur in an extremely arid environment. Singularly spaced plants growing in the same environment seldom produce an ear.

Recent studies (Bandaru et al. 2006; Kapanigowda et al. 2010; Krishnareddy et al. 2010) have shown the cultivation of grain sorghum and maize plants in clumps compared to equally spaced geometries. For each crop, the number of plants m^2 was equal. Bandaru et al. (2006) grew grain sorghum under semiarid conditions at Bushland, TX, and Tribune, KS, in clumps of either three or four plants and compared vegetative growth, tillering, harvest index, and grain yield to plants uniformly spaced. A schematic showing the treatments in this study is presented in Figure 14.6, and the results from the experiments conducted in 2004 on the no-tilled study area at Bushland are presented in Table 14.3. Bandaru et al. (2006) concluded that growing plants in clumps compared to uniformly spaced plants reduces the number of tillers and vegetative growth. This preserves the soil water until reproductive and grain-filling growth stages, which increases the grain yield. There are marked differences in the plant architecture of uniformly spaced plants compared to clumped plants. Uniformly spaced plants produce more tillers, and the leaves on both the main stalk and

FIGURE 14.5 Dryland maize in 2005 growing on the Hopi Indian Reservation in Arizona in an arid environment.

the tillers grow outward, exposing essentially all of the leaf area to sunlight and wind. By contrast, clumped plants grow upward with the leaves partially shading on another and reducing the effect of wind, thereby reducing water use. The benefit of clumps decreased as grain yields increased, and there was even a slight decrease when yields exceeded 6000 kg/ha. However, grain yields were increased by clump planting by as much as 100% when yields were in the 1000 kg/ha range and 25%–50% in the 2000–3000 kg/ha range.

Kapanigowda et al. (2010) grew maize under dryland conditions at Bushland, TX, and also when small amounts (50–125 mm) of seasonal irrigation water were added. Two planting geometries, clumps of three plants 1 m apart and equidistant-spaced plants in 75 cm rows, were compared at plant populations of 39,000 plants/ha. Clump planting produced significantly greater grain yields (321 vs. 225 g/m^2 and 454 vs. 292 g/m^2 during 2006 and 2007, respectively) and harvest indexes (0.54 vs. 0.49 and 0.52 vs. 0.39 during 2006 and 2007, respectively) compared with equidistant plants under dryland conditions. Water-use efficiency measurements in 2007 indicated that clumps had a lower evapotranspiration (ET) threshold for initiating grain production. The authors also concluded that mutual shading may have played a role in reducing transpiration and plants growing close to each other could have possibly reduced the effect of wind and lowered transpiration rates. Tillers, which will be discussed later, also played a part. Although maize does not produce nearly as many tillers as grain sorghum, tillers are often formed on maize plants when population density is low as is often the case in dryland conditions. Kapanigowda et al. (2010) found in 2007 that tillers accounted for 10% of the stover produced by equidistant-spaced maize plants, but less than 3% of the grain.

Recent unpublished data (MS theses in 2010 by Bharath Reddy and Brijesh Angira, West Texas A&M University, Canyon, TX) also support the hypothesis that plants growing in close proximity to one another change the microclimate enough to significantly affect the transpiration rate of maize and grain sorghum. Reddy grew grain sorghum and Angira grew maize, but the studies were conducted simultaneously in the same greenhouse, so the environmental conditions were the same. Reddy grew six grain sorghum plants in 1 m long boxes containing 93 kg soil (dry weight) maintained at two different water levels (soil water maintained between 75% and 100% plant available water and between 25% and 75% plant available water). The six plants in each box were arranged

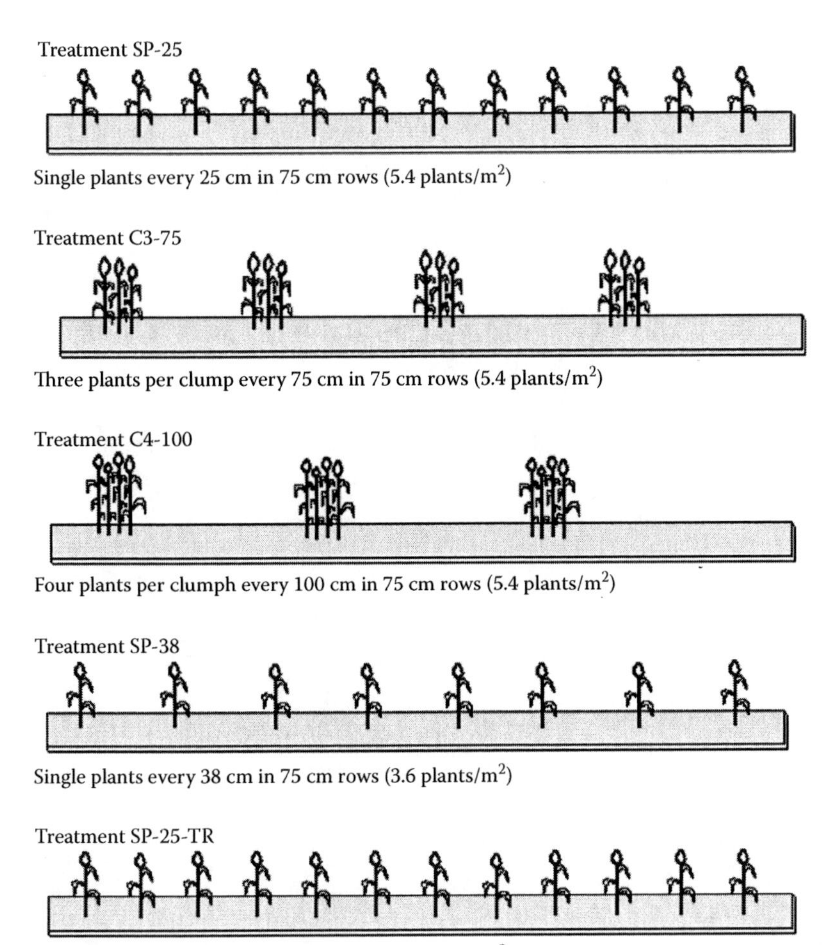

FIGURE 14.6 A schematic showing plant geometries for growing grain sorghum under dryland conditions at Bushland, TX, and Tribune, KS. (From Bandaru, V., Stewart, B.A., Baumhardt, R.L., Ambati, S., Robinson, C.A., and Schlegel, A., *Agron. J.*, 98, 1109–1120, 2006.)

in three different geometries—two clumps of three plants with 50 cm between the clumps, six plants spaced 16.7 cm apart, and six plants spaced 8.4 cm apart. The center of the box was halfway between the clumps, and between the thirdplant and the fourthplant for the other two geometries. For the maize plants, Angira grew three plants in each box, and the three geometries were as follows: a clump with three plants, three plants 17 cm apart, and three plants 33 cm apart with the middle plant in all geometries located in the center of the box. Water levels were maintained at the same levels as for the maize study. All boxes were covered with lids that fit around the plants but were in two parts secured by bungee straps that could be removed for adding water, but they essentially eliminated evaporation, so the total water use was considered transpiration. Figure 14.7 shows all results from the two studies as a relationship between water transpired and aboveground biomass produced. Maize and grain sorghum are both C_4 crops, and as discussed earlier, Sinclair and Weiss (2010) stated that the transpiration ratio was constant for C_4 crops and was 220 g water for each gram of plant growth. The data presented in Figure 14.7 show that maize and grain sorghum had essentially the same ratios, and the ratios were almost the same regardless of whether the plants were grown under adequate water or limited water conditions. The values shown in Figure 14.7 were only the total biomass. There were large differences in the grain, and the maize plants produced

TABLE 14.3

Mean Values of Measurements for Grain Sorghum as Affected by Five Planting Geometries in 75 cm Rows in Experiments Located on the Upper (Upper), Middle (Middle), and Bench (Bench) Positions of a No-Tilled Bench-Terraced Field at Bushland, TX, in 2004[a]

Planting Geometry[b]	Tillers per Plant 28 DAP[c] (no. per plant)	Biomass 42 DAP (kg/ha)	Leaf Area Index Values 42 DAP	Panicles per m² (no. per m²)	% Tillers with Panicles[d]	Grain (kg/ha)	Harvest Index Values	Aboveground Biomass (kg/ha)
Upper								
SP-25	2.2a[e]	2716a	1.30a	9.2a	33	2270b	0.29c	6866ab
C3-75	0.5c	1831c	1.10b	8.3b	106	2891a	0.38a	6673b
C4-100	0.5c	1622d	0.86c	7.7b	78	3011a	0.40a	6603b
SP-38	2.6b	2234b	1.29a	9.2a	61	2742a	0.33b	7288a
SP-25-TR	Removed	1550d	0.85c	5.4c	NA	2645ab	0.40a	5800c
Middle								
SP-25	2.3a	2924a	1.40a	10.3a	41	2690a	0.32a	7374ab
C3-75	0.7c	1897c	1.11c	9.7b	84	3338ab	0.40a	7320ab
C4-100	0.6c	1754d	0.90d	8.4b	103	3479a	0.42a	7266b
SP-38	2.7a	2352b	1.37b	10.3a	70	2954bc	0.33b	7852a
SP-25-TR	Removed	1634c	0.88d	5.4c	NA	2904c	0.39a	6561c
Bench								
SP-25	2.3a	3047a	1.45a	11.4a	50	4812a	0.41b	10295a
C3-75	0.9c	2088c	1.20c	9.1b	75	4968a	0.46a	9474b
C4-100	0.8c	1880d	0.93d	8.7b	84	4807a	0.46a	9167b
SP-38	2.8b	2478b	1.42b	11.5a	80	5070a	0.42b	10588a
SP-25-TR	Removed	1786d	0.92d	5.4c	NA	4222b	0.46a	8051c

Source: Bandaru, V., Stewart, B.A., Baumhardt, R.L., Ambati, S., Robinson, C.A., and Schlegel, A., *Agron. J.*, 98, 1109–1120, 2006.

[a] Separate but identical experiments were conducted on three positions that had different amounts of stored water at the time of seeding and different amounts of runoff or run-on during the cropping season.

[b] Planting geometries were SP-25 (plants ever 25 cm), C_3-75 (clumps of three plants every 75 cm), C_4-100 (clumps of four plants every 100 cm), SP-38 (plants every 38 cm), and SP-25-TR (plants every 25 cm with tillers removed by hand) in 75 cm rows.

[c] Days after planting.

[d] Percentages were obtained by dividing the number of plants plus the number of tillers; values about 100 are due to an experimental error.

[e] Means in columns for a position on the benched terrace followed by the same letter are not significantly different according to a protected LSD mean separation ($P < 0.5$ level); each position represents a separate experiment and cannot be compared statistically.

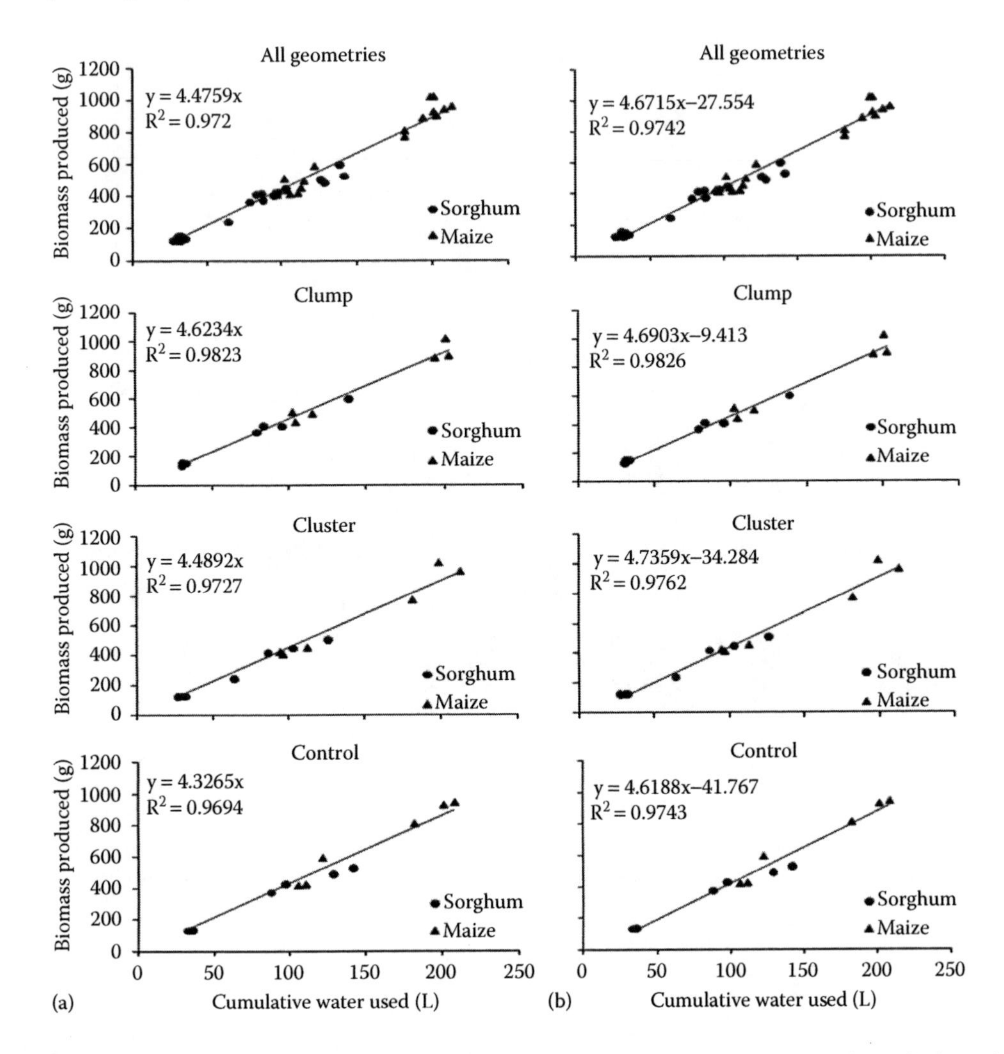

FIGURE 14.7 Grams aboveground biomass (y axis) as a function of liters water transpired (x axis) for maize and sorghum for all geometries and for different geometries; when line is statistically forced (b) and not forced (a) through the origin. (Figure based on the unpublished data of Reddy 2010 and Angira 2010.)

very little or no grain at the low water levels. The biomass, however, was directly proportional to water use as hypothesized earlier by Sinclair and Weiss (2010) in Figure 14.3.

The plants shown in Figure 14.8 under field conditions are in similar configurations to the maize plants grown in the greenhouse study presented above. The microclimate for the plants growing in clumps shown in the bottom half of the figure is likely to be more favorable during much of the time than that for the equally spaced plants shown in the top half, which should increase the transpiration efficiency.

14.3.2 Plant Spacing Effect on the Number of Tillers

Under dryland conditions, the most widely adopted strategy for growing grain crops such as maize and grain sorghum is to reduce the plant density. In irrigated areas and favorable precipitation regions, 80,000–90,000 plants/ha for maize and 200,000–250,000 plants/ha for grain sorghum are common. In dryland areas, the densities are often reduced by as much as 75%. Tiller numbers

FIGURE 14.8 Maize plants growing in a farmer's field in clumps (b) compared to equally spaced plants (a) at populations of 30,000 plants/ha.

for plants, particularly for grain sorghum, increase with decreasing plant densities, and this often negates most if not all of the anticipated benefit. The purpose of decreasing plant population is to decrease the amount of stored soil water used for vegetative growth so that there is more of it remaining for use during the grain-filling period. The tillers use water and nutrients (Figure 14.9), but under water-limiting conditions, they often produce little or no grain. Gerik and Neely (1987), Bandaru et al. (2006), Kapanigowda et al. (2010), and Krishnareddy et al. (2010) have clearly shown that maize and grain sorghum plants produce significantly more tillers at low plant densities than at high densities. Although tillers can contribute to grain yields under some conditions, they are more likely to decrease yields under severe water-stressed conditions because the tillers often senesce before they produce the number of leaves required before a panicle is formed. The tiller data shown in Table 14.3 for grain sorghum show that two to three tillers were common for each plant when plants were spaced 25–38 cm apart. Plants in clumps produced on average less than one tiller per plant. This greatly affected the harvest index values and was believed to be the primary reason for the higher grain yields produced by the plants grown in clumps. However, a more favorable microclimate in the clumps may have also contributed to the yield increase.

14.3.3 Effect of Plant Geometry on Soil Water Extraction

Aside from reducing plant populations, different spacing between rows and skip-row configurations have been widely used to enhance soil water contents later in the season. The most common spacing between rows of crops such as maize and grain sorghum is 75 cm, but 100 cm spacing between rows is often used in dryland areas. Configurations such as plant 1-skip 1, plant 2-skip 1, and plant 2-skip 2 are sometimes used, and this results in wide spacing between some rows (Lyon et al. 2009; Routley et al. 2003). Routley et al. (2003) found more consistent yields of grain sorghum in Australia when every third row or two rows of every four rows were left blank compared with uniformly spaced 1 m rows when yields were 2500 kg/ha or less. Yields were generally less for skip-row configurations at

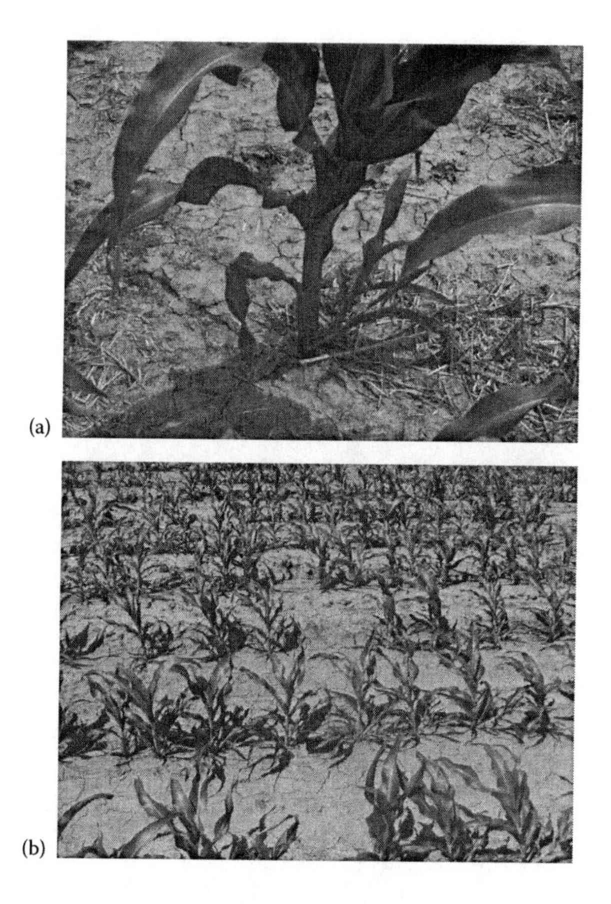

FIGURE 14.9 Maize (a) and grain sorghum (b) plants in farmer fields showing the formation of tillers when low plant populations are used.

higher yield levels. Lyon et al. (2009) grew maize at a number of locations for multiple years where annual precipitation amounts varied widely and compared yields for every 75 cm row planted to those of skip-row configurations of plant 2-skip 2, plant 2-skip1, and plant 1-skip 1. In trials where skip-row planting resulted in increased grain yields, the mean yield for the every-row treatment was 2800 kg/ha. In those trials where skip-row planting resulted in lower yields than the standard planting treatment, the mean yield was 8500 kg/ha.

Balancing soil water extraction with crop management is challenging. In general, stored soil water is more fully utilized by either closer row spacing or higher plant populations. The data in Table 14.4 clearly show that either increasing the plant density or decreasing the width between rows resulted in increased soil water extraction by grain sorghum, and this was particularly true between times of emergence and heading. The challenge is to try and save some of the stored water for the reproductive and grain-filling growth stages, and the data show that this can be done by wide rows or low population, but then much of the potential gain was lost because the water was not fully extracted. When most of the soil water is depleted during vegetative growth, the harvest index can be significantly reduced, resulting in low grain yields (Table 14.4). Finding the right balance is difficult because the amounts and distribution of precipitation in dryland regions are significantly different for every growing season.

Bandaru et al. (2006) showed that growing grain sorghum in clumps as illustrated in Figure 14.6 resulted in significantly less use of stored soil water early in the season as compared to equally spaced plants in rows. The differences at the end of the growing season were much smaller, but the equally spaced plants still used slightly higher amounts of soil water than the clump treatments (Table 14.5).

TABLE 14.4
Effect of Plant Population and Row Spacing on Dry Matter Yield, Harvest Index, and Deep (>0.9 m) and Shallow (<0.9 m) Soil Water Extraction by Grain Sorghum at Bushland, TX, in 1983

	Total Dry Matter (Mg/ha)	Harvest Index	Soil Water Extraction								
			Emergence to Heading (mm)			Heading to Maturity (mm)			Total Season (mm)		
			≤0.9 m	>0.9 m	Total	≤0.9 m	>0.9 m	Total	≤0.9 m	>0.9 m	Total
38[a]	5.3	0.38	75	11	86	30	19	49	105	30	135
76L	5.7	0.45	59	9	68	44	11	55	103	20	123
38M	6.4	0.39	85	15	100	27	21	48	112	36	148
76M	6.1	0.34	67	14	81	46	6	52	113	20	133
38H	6.2	0.25	98	17	115	28	14	42	126	31	157
76H	6.0	0.28	73	7	80	37	17	54	110	24	134

Source: Stewart, B.A. and Steiner, J.L., In: *Dryland Agriculture: Strategies for Sustainability*, eds. R.P. Singh, J.F. Parr, and B.A. Stewart, Advances Soil Science, vol. 13, pp. 151–173, 1990.

[a] L, M, H = low, medium, and high plant populations corresponding to 6.9, 13.1, and 18.6 plants/m^2, respectively; 38 and 76 represent 0.38- and 0.76 m spacing between rows, respectively.

TABLE 14.5
Total Soil Water (mm) in 180 cm Soil Profiles of Selected Treatments[a] at Various Times at Bushland, TX, in 2004

Position on Bench Terrace	29 July	18 August	14 September	4 October
Upper				
SP-25 (13 cm from plant)	474b[b]	429c	408c	487b
C4-100 (13 cm from clump)	532a	535bc	421bc	500ab
C4-100 (38 cm from clump)	530a	486a	453a	501ab
SP-25-TR (13 cm from plant)	634a	478ab	446ab	511ab
Middle				
SP-25 (13 cm from plant)	498b	447b	422b	499a
C4-100 (13 cm from clump)	558a	497a	453a	492a
C4-100 (38 cm from clump)	541a	506a	465a	493a
SP-25-TR (13 cm from plant)	552a	518a	480a	486a
Bench				
SP-25 (13 cm from plant)	548b	518b	481a	498a
C4-100 (13 cm from clump)	559a	545ab	492a	513ab
C4-100 (38 cm from clump)	541a	506ab	165a	493b
SP-25-TR (13 cm from plant)	552a	518a	480a	486a

Source: Bandaru, V., Stewart, B.A., Baumhardt, R.L., Ambati, S., Robinson, C.A., and Schlegel, A., *Agron. J.*, 98, 1109–1120, 2006.

[a] Treatments were SP-25 (plants every 25 cm), C4-100 (clumps of four plants every 100 cm), and SP-25-TR (plants every 25 cm with tillers removed by hand) in 75 cm rows.

[b] Numbers followed by the same letter within a column for a particular position are not significantly different according to a protected LSD mean separation ($P < 0.05$ level).

14.4 IMPLEMENTING COMPLEMENTARY TECHNOLOGIES IN DRYLAND AREAS

Successful crop production in dryland areas requires careful management, and every practice must be assessed in terms of how it affects precipitation capture, retention, and use because water is always the most limiting factor. This chapter focuses on the use of precipitation and stored soil water during the growing season to maximize the grain production of grain sorghum and maize. A strategy that includes multiple technologies is essential, and it must be fully understood that results from year to year will vary because the amount and timing of growing season precipitation are highly variable, and this affects technologies in different manners. However, a good understanding of basic principles of water management can greatly improve the sustainable crop production.

Grain yield is determined by multiplying the harvest index value times the amount of aboveground biomass, and the aboveground biomass for a given crop is directly proportional to the amount of transpiration that is affected by the vapor pressure deficit. While these principles apply to all soils and climates, the results are vastly different. Crop water use during a growing season is generally considered ET, which is the sum of the water evaporated from the soil surface and that transpired through the leaves between the date of planting and harvest. Dryland areas have less seasonal precipitation but require more transpiration to produce a unit of dry matter, and evaporation potential is also higher for water loss from the soil. It was discussed earlier that a crop can require almost two times more water in an arid climate to produce a unit of dry matter compared to a humid region. Therefore, practices and priorities for practices can differ greatly for different regions.

The first priority for growing a crop in a dryland region is to maximize water use. For dryland conditions, this is the sum of growing season precipitation and the amount of stored soil water used by the crop. Runoff should be minimized, and it is highly important to utilize as much of the stored soil water as feasible because it can greatly increase the yield, and in some areas such as the U.S. Great Plains, an entire growing season is sacrificed to increase the amount of water stored in the soil at the time of planting because growing season precipitation alone is not sufficient. Therefore, it is important to utilize this water.

The second priority is to utilize practices that proportion as much of the ET to T as feasible. This is difficult in dryland areas because not only is there less ET, but the evaporation potential is greater. The best strategies for decreasing the evaporation are to increase canopy cover to increase shading and to use mulch. Again, both of these strategies are difficult in dryland areas because reduced plant populations provide less ground cover, and there are often not enough crop residues to provide good ground cover even if no tillage is used. In many dryland regions, the situation is even worsened by the removal of crop residues for animal feed or household fuel.

The third priority is to utilize practices that can possibly increase transpiration efficiency but not reduce transpiration. Biomass production is tightly linked to transpiration, and the biochemistry of photosynthesis has not been improved genetically (Blum 2009). The effective use of water and not water-use efficiency should be emphasized for improving crop production under drought stress, and the effective use of water implies maximal soil moisture capture for transpiration, which also involves reduced nonstomatal transpiration and minimal water loss by soil evaporation (Blum 2009). Transpiration efficiency is dependent on the crop species, but environmental conditions determine the vapor pressure deficit that controls the amount of water required to produce a unit of dry matter. The transpiration environment is determined by air temperature, radiation, humidity, and wind speed. However, the environment surrounding a plant leaf may be considerably different than the environment in general. Growing plants in clumps may offer some potential for improving the microclimate, which will in turn increase the transpiration efficiency. It is conceivable that some of the benefit obtained by growing crops in skip-row configurations where plant populations are kept constant is that the space between plants is reduced by 50%, which may result in a more favorable microclimate. The closer spacing of plants also reduces the formation of tillers that can utilize water and nutrients but contribute little or no grain under water-stressed conditions. Lyon et al. (2009) recommended for maize growers in the U.S. central Great Plains with moderate riskaversion and likely yield levels of about 6 Mg or less to plant one row and skip one row but keep the plant population the same as if every row was planted. As discussed earlier, they attributed the increased yield to having more stored soil water available for use during the grain-filling period, but it is conceivable that the closer spacing of plants also resulted in a more favorable transpiration environment. Another potential benefit of using clump geometry or a skip-row configuration is that fertilizer use efficiency may be enhanced. Nutrient uptake could possibly be higher if a fertilizer was applied only to the soil adjacent to the clumps or only in the planted rows in skip-row configurations.

The harvest index is also of great importance and is affected by all of the priorities discussed. The harvest index is genetically controlled; it is maximum only when water is adequate to minimize stress. Under dryland conditions, water is always limited, so the harvest index values seldom approach the maximum values. Prihar and Stewart (1990) estimated the genetic harvest index for grain sorghum and maize to be approximately 0.55 and 0.60, respectively. The harvest index values can decline significantly when water is severely limited during the grain-filling period. Under extreme conditions, the harvest index can be zero, particularly for maize.

14.5 CONCLUSION

The importance of dryland agriculture production continues to increase because of the increasing world population, increasing prosperity that results in diet changes that demand more grain, decreasing rate of irrigated land expansion, urban sprawl taking agricultural land out of production, and other factors. During the last few decades, the most significant change in dryland agriculture has been to reduce tillage. In many cases, tillage has been essentially eliminated, and this has resulted in an increase in the amount of the scarce precipitation that is used for ET that has increased yields. While the basic principles are generally well understood, the implementation of these is difficult because the amount of growing season precipitation in dryland regions is always insufficient and the timing of the scarce precipitation is highly variable, and this affects different management practices in different ways. Thus, a practice that increases yield 1 year can actually decrease yield in another year because of the timing of rainfall events. This chapter has focused on manipulating plant geometries to increase water-use efficiency and yield. It is well established that the amount of water required to produce a unit of dry matter is highly dependent on the environment, which is primarily controlled by temperature, radiation, humidity, and wind speed, and these characteristics are widely measured and reported. However, it is the vapor pressure deficit at the leaf surface rather than the field environment that controls the transpiration ratio. While these are similar, there is evidence that the microclimate of the plants can be affected by practices such as row spacing, clump planting, skip-row configuration, spacing between plants, and other plant manipulation schemes. Additional studies are needed to determine the feasibility and reliability of such technologies for increasing yields in dryland areas.

ABBREVIATIONS

ET Evapotranspiration
FAO Food and Agriculture Organization
VPD Vapor pressure inside and outside a leaf

REFERENCES

Angira, B. 2010. Transpiration efficiency and yield parameters of maize in different plant geometries. MS thesis, West Texas A&M University, Canyon, TX.

Bandaru, V., B.A. Stewart, R.L. Baumhardt, S. Ambati, C.A. Robinson, and A. Schlegel. 2006. Growing dryland grain sorghum in clumps to reduce vegetative growth and increase yield. *Agron. J.* 98:1109–1120.

Blum, A. 2009. Effective use of water (EUW) and not water-use efficiency (WUE) is the target of crop yield improvement under drought stress. *Field Crops Res.* 112:119–123.

Bowden, L. 1979. Development of present dryland farming systems. In: *Agricultue in Semi-Arid Environments*, eds. A.E. Hall, G.H. Cannell, and H.W. Lawton, pp. 45–72. Springer-Verlag, Berlin.

Brown, W.L. 1985. New technology related to water policy—Plants. In: *Water and Water Policy in World Food Supplies*, ed. W.R. Jordan, pp. 37–41. Texas A&M University Press, College Station, TX.

Craufurd, P.Q., D.J. Flower, and J.M. Peacock. 1993. Effect of heat and drought stress on sorghum (*Sorghum bicolor*). I. Panicle development and leaf appearance. *Exp. Agric.* 29:61–67.

Dregne, H.E. 1976. *Soils of the Arid Regions*. Elsevier, Amsterdam.

Food and Agriculture Organization. 1993. *Key Aspects of Strategies for the Development of Drylands*. FAO, Rome, Italy.

Food and Agriculture Organization. 2000. *Land Resource Potential and Constraints at Regional and Country Levels*. World Soil Resources Report 90. FAO, Rome, Italy.

Gerik, T.J. and C.L. Neely. 1987. Plant density effects on main culm and tiller development of grain sorghum. *Crop Sci.* 27:1225–1230.

Hargreaves, M.W.M. 1957. *Dry Farming in the Northern Great Plains: 1900–1925*. Harvard University Press, Cambridge, MA.

Kapanigowda, M., B.A. Stewart, T.A. Howell, H. Kadasrivenkata, and R.L. Baumhardt. 2010. Growing maize in clumps as a strategy for marginal climatic conditions. *Field Crops Res.* 118:115–125.

Krishnareddy, S.R., B.A. Stewart, et al. 2010. Grain sorghum tiller production in clump and uniform planting geometries. *J. Crop Improv.* 24:1–11.

Loomis, R.S. 1983. Crop manipulations for efficient use of water: An overview. In: *Limitations to Efficient Water Use in Crop Production*, eds. H.M. Taylor, W.R. Gardner, and T.R. Sinclair, pp. 345–374. ASA, CSSA, SSSA, Madison, WI.

Lyon, D.J., A.D. Pavlista, G.W. Hergert, et al. 2009. Skip-row planting patterns stabilize corn grain yields in the Central Great Plains. Plant Management Network. Available at http://www.plantmanagementnetwork.org/pub/cm/research/2009/skip/(verified May 12, 2011).

Middleton, N.J. and D.S.G. Thomas. 1992. *World Atlas of Desertification*. Edward Arnold/UNEP, Seven Oaks, TN.

Noin, D. and J.I. Clark. 1997. Population and environment in arid regions of the world. In: *Population and Environment in Arid Regions*, eds. J. Clark and D. Noin, MAB/UNESCO, Vol. 19. The Parthenon Publishing Group, New York.

Peterson, G.A., D.G. Westfall, and N.C. Hansen. 2012. Enhancing precipitation use efficiency in the worlds's dryland agroecosystems. In eds. R. Lal, and B.A. Stewart, Chapter 16.

Prihar, S.S. and B.A. Stewart. 1990. Using upper-bound slope through origin to estimate genetic harvest index. *Agron. J.* 82:1160–1165.

Reddy, B.K. 2010. Impact of planting geometries on transpiration efficiency in grain sorghum. M.S. Thesis, West Texas A&M University, Canyon, TX.

Routley, R., I. Broad, G. McLean, J. Whish, et al. 2003. The effect of row configuration on yield reliability in grain sorghum: I. Yield, water use efficiency and soil water extraction. Proceeding Eleventh Australian Agronomy Conference, Geelong, VIC, January 2003.

Sinclair, T.R. 2009a. Taking measure of biofuel limits. *Am. Sci.* 97:400–407.

Sinclair, T.R. 2009b. Taking measure of biofuel limits. Available at http://climatesanity.wordpress.com/2009/09/24/taking-measure-of-biofuel-limits (verified April 20, 2011).

Sinclair, T.R. and A. Weiss. 2010. *Principles of Ecology in Plant Production*, 2nd edn. CAB International, Cambridge, MA.

Stewart, B.A. 1988. Dryland farming: The North American experience. In: *Challenges in Dryland Agriculture, A Global Perspective*, eds. P.W. Unger et al., pp. 54–59, Proceedings of International Conference on Dryland Farming, Amarillo/Bushland, TX, August 15–19, 1988. Texas Agric. Exp. Stn, College Station, TX.

Stewart, B.A. and E. Burnett. 1987. Water conservation technology in rainfed and dryland agriculture. In: *Water and Water Policy in World Food Supplies*, pp. 355–359, Proceedings of the Conference, College Station, TX, May 26–30, 1985. Texas A&M University, College Station, TX.

Stewart, B.A. and J.L. Steiner. 1990. Water-use efficiency. In: *Dryland Agriculture: Strategies for Sustainability*, eds. R.P. Singh, J.F. Parr, and B.A. Stewart. Advances Soil Science, vol. 13, pp. 151–173.

Stewart, B.A., P. Koohafkan, and R. Ramamoorthy. 2006. Dryland agriculture defined and its importance to the world. In: *Dryland Agriculture*, eds. G.A. Peterson, P.W. Unger, and W.A. Payne. Agronomy Monograph No. 23, 2nd edn, pp. 1–26. ASA, CSSA, SSSA, Madison, WI.

Unger, P.W., W.A. Payne, and G.A. Peterson. 2006. Water conservation and efficient use. In: *Dryland Agriculture*, eds. G.A. Peterson, P.W. Unger, and W.A. Payne. Agronomy Monograph No. 23, 2nd edn, pp. 39–85. ASA, CSSA, SSSA, Madison, WI.

15 Mulch Tillage for Conserving Soil Water

Paul W. Unger, R. Louis Baumhardt, and Francisco J. Arriaga

CONTENTS

15.1 INTRODUCTION

Mulching, that is, the practice of maintaining organic or inorganic materials on or applying them to the soil surface, is an ancient practice (Jacks et al. 1955; Lal 2006). Jacks et al. (1955), for example, stated that the practice of applying mulches to soil is possibly as old as agriculture itself. They mentioned that the ancient Romans placed stones on soils to conserve water and the Chinese used pebbles from the streambed for the same purpose. It is, indeed, well known that mulches provide water and soil conservation benefits. They maintain or improve the soil resource base and provide conditions for favorable plant growth and, from an agricultural viewpoint, satisfactory or enhanced crop productivity. In addition to providing for water and soil conservation, properly managed mulches provide soil temperature moderation; soil structure improvement; soil nutrient effects; and soil salinity, crop quality, and weed control (Unger 1995). The combination of these benefits, therefore, is highly important for achieving adequate food production for the ever-increasing human population.

A wide variety of materials are used as mulches, which may be organic or inorganic in nature (Unger 1995; Lal 2006). Mulching materials include crop residues, plant leaves and clippings, tree bark, manure, paper, plastic films, petroleum products, gravel, coal, and other materials. Crop residues are the main mulching materials under widespread agricultural conditions, but plastic mulches are also widely used in some countries and for some specialty crops.

The importance of conserving water for agricultural purposes has been recognized for centuries. Bennett (1939) cited numerous practices from ancient times that had water conservation benefits, even though they were used for other water-related purposes. In modern times, water conservation has become highly important because of the increasing competition for fresh water among nations, geographical regions, and segments of society that include agricultural, urban, industrial, and recreational users (Unger and Howell 1999).

Water for agriculture is derived from precipitation and also from streams, reservoirs, or aquifers where crops are irrigated. Precipitation may be limited or erratic and supplies in streams, reservoirs, or aquifers may be limited or declining. Water conservation for agriculture, therefore, is important because of the limited supplies, increasing competition for available water supplies, and the need for increased food, fiber, and fuel production for the ever-increasing world population. For this report, we emphasize mainly the use of tillage to manage crop residues and other organic materials (e.g., cover crops) as a mulch to conserve soil water.

15.2 TILLAGE METHODS/SYSTEMS

Mulch tillage is tillage or soil preparation that retains plant residues or other materials to cover the soil surface. Operationally, it is any full-width tillage and planting combination that results in at least 30% of the soil surface covered with crop residues (SSSA 2001). Because of the wide array of crops grown, amounts of residue produced, residues used for other purposes, soils involved, climatic conditions, and factors such as weed and insect problems, it is critical that appropriate tillage methods or systems be used to meet the required 30% residue cover. A 30% residue cover generally results in a 50% reduction in soil erosion (McCarthy et al. 1993; Hofman 1997), but the actual amount required to attain such reduction varies depending on the soil type, slope length and steepness, soil drainage, weather conditions, and presence of other conservation practices such as contour tillage and crop rotation (Hofman 1997; Jasa, nd). A 30% residue crop is also a requirement for compliance with federal farm bills (Myers, nd). Other terms for mulch tillage are mulch farming, trash farming, stubble-mulch tillage (SMT), and plowless farming (SSSA 2001).

Early emphasis in the United States for tillage to achieve soil water conservation was to deeply loosen the soil by plowing (Unger et al. 2010). Such practice was largely a carryover that early settlers had used in their home country, mainly Europe, and it virtually eliminated all surface residues. Elimination of surface residues unfortunately resulted in soils being highly susceptible to erosion, both by wind and water. It also resulted in surface sealing during rainstorms, which required additional tillage to loosen the surface crust to again achieve satisfactory water infiltration.

When the benefits of retaining residues on soils to conserve soil and water were first recognized, few tools or methods were available to produce crops under surface residue conditions. This changed in the 1930s when a farmer in Georgia (United States), J. Mack Gowder, used an implement having a 10-cm wide chisel point to loosen the soil and retain plant residues on the surface. By using that implement, which he called a "bull tongue scooter," he tried to mimic the surface cover conditions observed in a forest on his steeply sloping land. This method of tillage became known as "stubble-mulch farming," with that designation attributed to Dr. H.H. Bennett, the author of *Soil Conservation* published in 1939 (Barnett 1987).

Stubble-mulch farming (or tillage) quickly became a recommended practice to conserve soil when the value of keeping residues on the surface was recognized. The goal was to maintain the soil covered with residues as long as possible to reduce runoff and erosion. According to Barnett (1987), SMT was the forerunner of no-tillage (NT), which is "a procedure whereby a crop is planted directly into the soil with no primary or secondary tillage since harvest of the previous crop; usually a special planter is necessary to prepare a narrow, shallow seedbed immediately surrounding the seed being planted" (SSSA 2001).

SMT was widely promoted to help control erosion by wind in the U.S. Great Plains during the major drought of the 1930s. That drought, along with the clean tillage practices being used for crop production at that time, resulted in severe land devastation and hardships for people living in the region (Figure 15.1). Clean tillage involves plowing and cultivation that incorporates all plant residues and prevents growth of all vegetation (e.g., weeds) during the growing season, except for the crop being grown (SSSA 2001). The primary goal for using SMT at that time was to produce a cloddy soil surface and to retain any plant residues on the surface that may have been available. Other goals for using SMT are to perform the tillage at a shallow depth to control weeds; retain plant residues on the surface to decrease runoff, thereby reducing the potential for erosion by water and providing more time for water infiltration; reduce soil water losses due to evaporation; and improve soil water conservation (Allen and Fenster 1986).

The water conservation benefits of crop residues retained on the soil surface were recognized in the 1930s and early 1940s (Duley and Russel 1939; Borst and Woodburn 1942; Barnett 1987). Since then, extensive research dealing with soil water conservation as affected by SMT and other tillage methods that retain crop residues on the soil surface has been conducted at numerous locations, especially in drier regions. Water conservation research, however, has also been conducted at more humid regions and even under irrigated conditions because the need for soil water conservation is widespread.

The amount of residue produced by crops varies greatly, based on the crops grown and conditions under which they are grown. For example, the amount for winter wheat (*Triticum aestivum* L.) residue under dryland (nonirrigated) conditions ranged from 1.3 to 6.5 Mg/ha in studies in the southern U.S. Great Plains (Unger 1994a, 1996). With irrigation, wheat residue yields averaged 9.9 Mg/ha in a study on southern Great Plains (Unger 1994b). Similar differences in residue production occur also for other crops. Differences among crops, however, vary widely with regard to the amount of surface cover provided by a given weight of residue. For example, surface cover was 100% with 8 Mg/ha of wheat (hollow stem) residues, 90% with that amount of grain sorghum [*Sorghum bicolor* (L.) Moench] (pithy stalk) residues, and 37% with that amount of cotton (*Gossypium hirsutum* L.) (woody stalk) residues (Unger and Parker 1976).

FIGURE 15.1 Devastated land and buried machinery and car in a farmstead in South Dakota during the drought in the 1930s in the U.S. Great Plains. (1936 USDA Photo by Sloan, Image # 00DI0971.)

Because of the large differences in residue production by a given crop or different crops, it is highly important to use an appropriate tillage method for a given crop or cropping condition to maintain adequate surface residues to meet the requirements of mulch tillage. Approximate amounts of crop residues remaining on the soil surface after one pass with different tillage and planting implements are given in Table 15.1. Essentially all residues are retained on the surface when NT is used, which involves only minimal soil disturbance to place the seed in the soil. Under some conditions, in-row subsoiling is used in conjunction with NT, which results in greater soil disturbance, but residue retention would be similar to that shown in Table 15.1 for "machines that fracture soil."

When large amounts of residues are present, one pass with any of the implements mentioned in Table 15.1, except the moldboard or disk plow, may be satisfactory. If additional tillage is required to control weeds or for optimal seed placement, disk implements generally should be avoided. With small residue amounts present, implement selection becomes highly important for retaining adequate residues on the surface. In the case where surface residue amounts are extremely low, all soil disturbances by tillage should be avoided, except as needed to roughen the surface to help control erosion by wind or water. Provided the surface cover requirement is met, all tillage methods mentioned in Table 15.1, except moldboard or disk plowing, qualify as mulch tillage methods. Those methods also meet the surface residue requirement of conservation tillage, which is "operationally, a tillage or tillage and planting combination which leaves a 30% or greater cover of crop residues on the surface" (SSSA 2001). Other terms that denote less soil disturbance and possibly less residue incorporation with soil are minimum tillage and reduced tillage. Minimum tillage is "the minimum

TABLE 15.1
Percentages of Surface Residue Cover Remaining after One Pass with Various Implements

Tillage and Plating Implements	Residue Cover Remaining (%)
Moldboard and disk plows	0–10
Machines that fracture soil (paraplow; ripper, to 30–35 cm depth)	70–90
Chisel Plows	
Straight points	60–80
Twisted points	50–70
Sweeps and field cultivators (including stubble-mulch plows)	60–90
One-way disk	55–80
Tandem or Offset Disk	
25 cm or greater blade spacing	25–50
18–25 cm blade spacing	30–70
Drills and planters	60–95
Natural Weathering	
Overwinter following summer harvest of small grain	70–90
Overwinter following fall harvest of summer crop	80–95

Source: Adapted from Fenster, C.R., Woodruff, N.P., Chepil, W.S., and Siddoway, F.H., *Agron. J.*, 57, 52–55, 1965; Anderson, D.T. In *Conservation Tillage in the Great Plains* (Proc. Workshop, Lincoln, NE, 1968), pp. 83–91. Great Plains Agric. Counc. Publ. 32, 1968; Hill, P.R., Eck, K.J., and Wilcox, J.R., *Agronomy Guide AY-280*, Purdue Univ., West Lafayette, IN, 1994; Bradford, J.M. and Peterson, G.A. In *Handbook of Soil Science*, ed.-in-chief M.E. Sumner, pp. G-247–G-270, CRC Press, Boca Raton, FL, 1999.

use of primary and/or secondary tillage necessary for meeting crop production requirements under the existing soil and climatic conditions, usually resulting in fewer tillage operations than for conventional tillage" (SSSA 2001). Reduced tillage is "a tillage system in which the total number of tillage operations preparatory for seed planting is reduced from that normally used on that particular field or soil" (SSSA 2001). These tillage methods do not require surface residue retention, but may qualify as mulch tillage or conservation tillage if adequate residues are retained on the surface.

NT is essentially the ultimate, that is, the most effective mulch or conservation tillage method, provided residue amounts are sufficient to cover at least 30% of the soil surface. Because primary and secondary tillage operations are avoided, use of NT results in the maximum retention of surface residues and, therefore, is closely related to "conservation agriculture" (CA).

CA "is a concept for resource-saving agricultural crop production that strives to achieve acceptable profits together with high and sustained production levels while concurrently conserving the environment" (FAO 2008). The main objective of CA is economical, ecological, and socially sustainable crop production while achieving soil regeneration or reversing soil degradation. The system requires maintaining a permanent soil cover, minimum soil disturbance, and use of crop rotations.

The main principles of CA are minimum soil disturbance (avoiding tillage operations), maintaining permanent soil surface cover with organic materials, and using rotations involving more than two crops (FAO 2008). At its maximum condition, NT is very similar to CA, but by definition, NT does not meet the requirements for CA. For example, soil loosening with a subsoiler is possible with NT, but such disturbance is not appropriate with CA. Also, a permanent cover of the surface is required for CA, but not for NT. It is recognized, however, that a high level of surface cover by crop residues is essential for NT if soil and water conservation and soil C sequestration are priorities (Unger and Blanco-Canqui 2012). Finally, CA requires use of rotations involving more than two crop species, whereas one crop can be grown continually when using NT. Fortunately, both systems can provide for excellent erosion control and improve water conservation, crop production, and environmental conditions; therefore, either system may have its place in a given locale.

15.3 SOIL CONDITION AND CLIMATE EFFECTS

Agricultural soils vary widely in texture, structure, depth, surface slope, profile characteristics, salinity level, and potential for erosion by water or wind. In addition, agricultural soils occur in regions varying widely with respect to the precipitation level and prevailing temperatures. These conditions strongly influence which type of mulch tillage is best suited for a particular locale and also the potential for conserving (storing) soil water.

Some of the above conditions influence such factors as runoff and soil water-holding capacity, drainage from the profile, and settling. Mulch tillage generally is not appropriate for use on poorly drained soils, especially when large amounts of residues are present. On such soils, water is removed so slowly that the soil remains wet for an extended period and the water table is commonly at or near the surface during a considerable part of the year. Also, surface residue retention by use of mulch tillage may aggravate the poorly drained condition by reducing runoff and retaining water that would to some extent be removed by evaporation. If a properly installed surface or subsurface drainage system is used on these soils to improve productivity, some type of mulch tillage may be appropriate and even water conservation may be of importance at times.

Soil compaction may occur on almost any soil due to farm implement traffic across the surface. Compaction due to farm implement traffic is most likely to occur when soils are wet, which allows soil particles to be more easily rearranged, and when the soil organic matter content is low. A dense, slowly permeable soil layer, termed a plow pan, may develop at the depth of tillage by running the wheels of one side of the tractor in the dead furrow for steering purposes while performing the tillage operation (SSSA 2001). Such plow pan development would occur where moldboard plowing is used. Repeated tillage at the same depth with some implements also results in the formation of plow pans. Hardsetting is a characteristic of the horizons of some soils, usually the cultivated seedbeds that contain unstable aggregates (Greene 2006). Hardsetting results from soil strength development due to matric suction acting within interparticle and interaggregate spaces when the soil is still wet and from temporary cementation of dry soil by poorly ordered silica and aluminosilicates (Bresson and Moran 1995). Soil settling because of heavy rainfall and subsequent soil drying may cause hardsetting of some soils (Karunatilake and van Es 2002). While hardsetting soil horizons tend to be relatively soft while moist, they become unusually hard when dry (Franzmeier et al. 1996).

Compaction, plow pans, and hardsetting of soils may greatly hinder water conservation by reducing water infiltration and, therefore, crop growth and production, unless corrected by use of an appropriate tillage method. For hardsetting soils, NT generally has not been satisfactory (Mead and Chan 1985; Touchton et al. 1989). Generally, favorable crop growth and yields, however, are obtained when such soils are loosened before or at planting with chisel plows or other types of subsoiling implements (Touchton et al. 1989; Abu-Hamdeh 2003; Raper 2006). When adequately loosened, various types of mulch tillage implements, including NT, can be successfully used. Disrupting the dense soil layers also enhances the potential for achieving greater soil water storage.

As previously mentioned, mulch tillage was widely promoted for controlling erosion by wind during the major drought of the 1930s and it has remained an important practice for helping to control erosion by wind. Mulch tillage is also highly effective for controlling erosion by water. For both types of erosion, the level of control achieved increases with increases in the amount of surface cover provided by the residues (Figure 15.2). About 70% cover essentially eliminates erosion by wind and 80% cover essentially eliminates erosion by water. Greatest residue retention is achieved by using implements that cause least disturbance of the surface, namely, soil-fracturing tools, chisels, and sweep implements (Table 15.1). Of course, use of NT generally results in the greatest surface residue retention. Surface residue retention to control erosion is important for conserving the soil resource base and for achieving increased soil water conservation by reducing runoff, improving water infiltration, and reducing wind speeds at the soil surface.

Soil texture strongly influences the amount of water that can be stored in a soil (Figure 15.3) and, to some extent, the rate at which water infiltrates a soil. Sands have the least water-holding capacity, but water infiltration generally is rapid. Water-holding capacity increases with soil clay content and water infiltration is slow on some high-clay-content soils. With a well-developed structure, however, water infiltration may be rapid for soils having high clay content. By contrast, water infiltration usually is low for a soil having a high salt content. Soil texture alone strongly influences a soil's water-holding capacity, but soil texture in combination with soil structure, salt content, and organic matter content has a major impact regarding whether optimum water storage will be achieved for a given

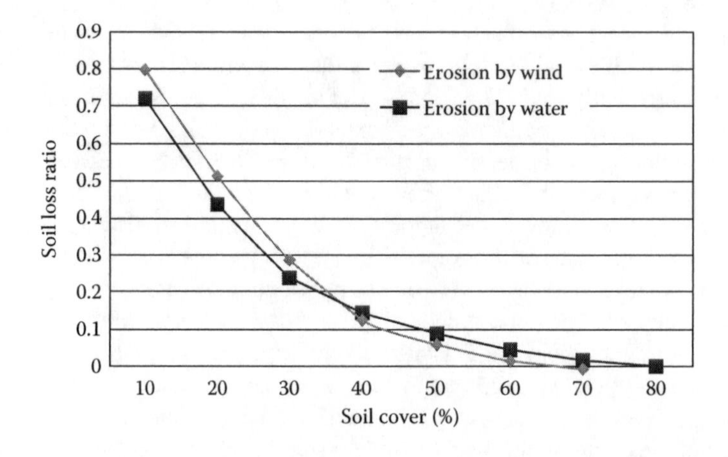

FIGURE 15.2 Relationship between soil loss (movement) ratio (loss with cover divided by loss from bare soil) and percentage of surface covered with residues. (Redrawn from Papendick, R.I., Parr, J.F., and Meyer, R.E., *Adv. Soil Sci.*, 13, 253–272, 1990.)

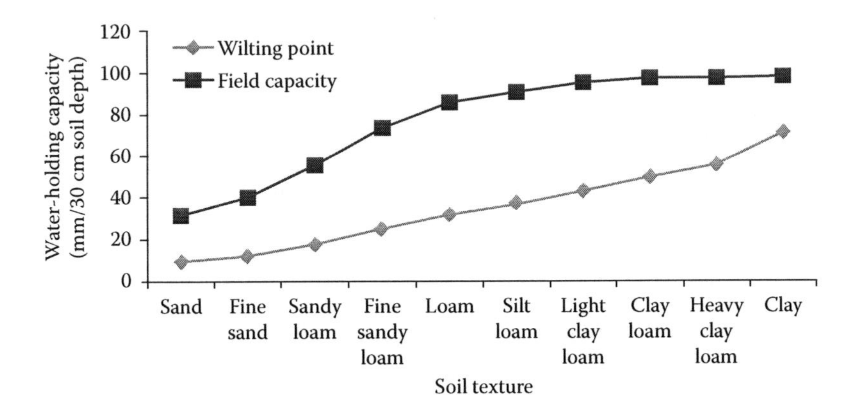

FIGURE 15.3 Typical water-holding characteristics of different-textured soils. The space between field capacity and wilting point values for a given texture represent the amount of water available to plants for that texture. (Redrawn from The Yearbook of Agriculture. In *Water, Yearbook of Agriculture*, ed. A. Stefferud, p. 120, U.S. Dept. of Agric. U.S. Gov. Print. Off., Washington, DC, 1955.)

soil. Regardless of soil structure development, mulch tillage with adequate residue retention can improve the potential for achieving optimum water storage on almost any soil. Use of a cover crop that provides a mulch may alleviate potential salt problems, thereby improving soil water conditions and enhancing crop yields (Mascagni et al. 2002).

15.4 INFILTRATION

The infiltration process represents the critical first step governing conservation of precipitation or irrigation as soil water because it is through infiltration that water enters the soil. Infiltration of water is driven as a result of the soil hydraulic gradient as limited by the conductive properties of both the soil surface and profile. The infiltration process itself is generally independent of climate, but can reflect consequences of cropping systems that incorporate production and management of mulches. Mulches from either crop or synthetic sources directly affect the infiltration process during irrigation or rainfall by intercepting water drop impact that often forms a thin flow-limiting soil surface layer known as a structural seal or crust.

Structural soil seals or crusts form as impacting drops of rain or irrigation water fracture aggregates and detach soil particles that, together with colloids, subsequently, are carried into and through the bulk soil matrix. These soil particles and aggregates transported by the infiltrating water precipitate and occlude the near-surface conducting pores of the bulk soil, thereby resulting in a "washed-in layer" (McIntyre 1958). Gradually, infiltration is reduced enough to cause surface ponding that further accelerates aggregate disruption through slaking (Moore 1981) and dispersion, depending on the electrolytes in the soil or water (Agassi et al. 1981). Fine soil material is deposited at the soil surface to form a thin "skin" layer (~0.1 mm) over any remaining conducting pores, which completes soil crust formation.

The conductive properties of a developing crust have been related, as a first approximation, to the cumulative rainstorm impact energy calculated as the product of rain intensity and the drop size–dependent kinetic energy rate or density (Baumhardt et al. 1990). In that paper, the cumulative rainstorm kinetic energy was linked to the physical changes in the surface soil matrix during crust formation that decreased hydraulic conductance through the crust, K_{cs}/L_c (i.e., crust hydraulic conductivity divided by crust thickness) (Figure 15.4). Crust formation was delayed during initial wetting of the surface aggregates, which absorbed drop impact with no corresponding change in the soil conductance from that of the bulk soil. Continued drop impact on the wetted surface soil fractured aggregates and released primary particles into the infiltrating water to form the crust and cause

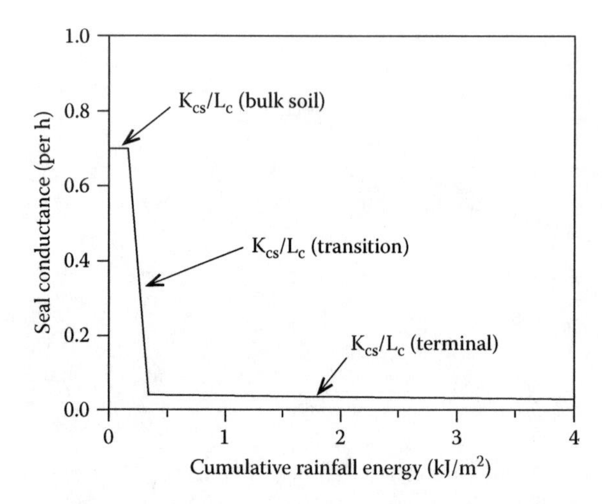

FIGURE 15.4 Example seal or crust conductance, calculated as the ratio of the crust saturated hydraulic conductivity, K_{cs}, divided by crust thickness, L_c, and plotted as a function of cumulative rainfall energy. Initially, crust conductance equals the bulk soil; however, surface wetting and the raindrop impact cause a rapid decrease in crust conductance that transitions to a terminal value. (From Baumhardt, R.L. and Schwartz, R.C. In *Encyclopedia of Soils in the Environment*, ed. D. Hillel, Vol. 1, pp. 347–356, Elsevier, Oxford, UK, 2004.)

a rapid decrease in the transitioning crust conductance to a final or terminal conductance value. Another implication is that in the absence of aggregate slaking or dispersion, an energy-absorbing barrier such as a synthetic material or crop residue mulch will prevent or delay crust formation and the surface soil conductive properties will remain unchanged from those of the underlying bulk soil.

Infiltration is limited to the rainfall intensity or irrigation application rate until the combined effect of the soil hydraulic gradient and conductive properties act to limit the water entry capacity, resulting in runoff and potential for soil erosion. Where soil surface conductive properties are depressed by water drop impact, increasing storm energy due to increased drop size or partial drop interception will govern the rate of crust formation. That is, increasing drop impact accelerates crust formation and, consequently, depresses the infiltration rate at corresponding times during the rainstorm (Figure 15.5). In this example, the transition from infiltration limited to the rainstorm intensity decreased from >40 min with no drop impact to 15 or 20 min for normal 0.0275 or partial 0.0114 kJ/m^2/mm storm energies, respectively. Likewise, the final infiltration rate at 120 min decreased from 17 mm/h to <10 mm/h and resulted in cumulative infiltration of 65 mm with no drop impact compared with 30–35 mm for the normal and partial storm energies, respectively.

15.4.1 Surface Protection by Residues

Except for obstruction of small surface microrelief channels or furrows that retard surface drainage and result in potentially greater opportunity time for infiltration (Musick et al. 1977), the principal benefit of a residue mulch for increasing infiltration is attributed to interception of irrigation or raindrops that delays or reduces soil crust formation. Using an overhead sprinkling device to apply water to nine soils with textures that varied from clay loam to sandy loam, Duley (1939) reported "no striking variability" for either total intake or the intake rate among soils in one of the earliest studies on this topic. Duley also observed that drop interception with straw mulch was almost as effective as a burlap energy barrier to prevent sealing and maintain infiltration rate compared with the corresponding bare soil. Like Duley (1939), others including McIntyre (1958), Morin and Benyamini (1977), and Baumhardt and Lascano (1996) observed that intercepting drop impact with an energy-absorbing barrier, like crop residue, reduces the formation of a crust.

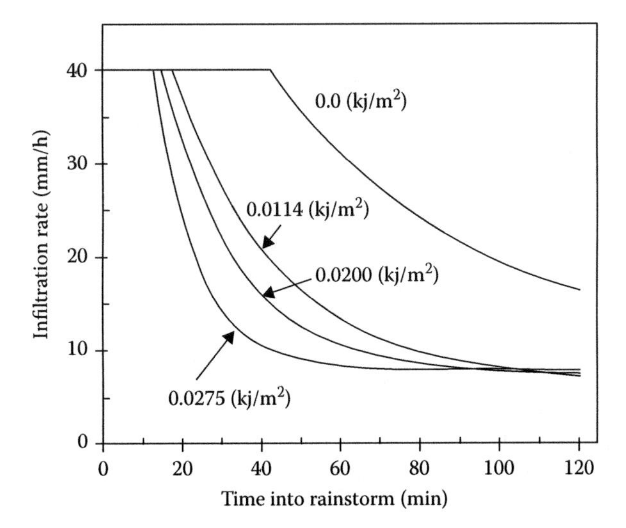

FIGURE 15.5 Infiltration rate (mm/h) is plotted with time during rainstorm for intercepted raindrop impact (0.0 kJ/m²/mm) compared with storms having progressively higher kinetic energy densities (0.0114–0.0275/m²/mm) (larger drop sizes). As rainstorm energy increased, crust formation accelerated and decreased infiltration more rapidly. Intercepting drop impact delayed any reduction in infiltration rate. (From Baumhardt, R.L. and Schwartz, R.C. In *Encyclopedia of Soils in the Environment*, ed. D. Hillel, Vol. 1, pp. 347–356, Elsevier, Oxford, UK, 2004.)

The efficacy of mulch tillage to control crust formation and determine infiltration under field conditions and, consequently, storm water runoff and erosion processes, will vary spatially depending on the soil properties and with the degree of residue cover. In a review article, Baumhardt and Schwartz (2004) indicated that tillage practices retaining sufficient residue to intercept drop impact on 40%–60% of the soil surface would be necessary to maintain high infiltration rates. Ruan et al. (2001) evaluated the effects of percent residue cover and geometry on infiltration of water from rainstorms varying in intensity from 25 to 250 mm/h into clay loams and loamy sands using a numerical model. While the cover required to maintain higher infiltration varied with both rainstorm intensity and soil properties, they observed that 40%–80% cover is needed to maintain infiltration. Mulch tillage is often managed to retain residue cover at this level; however, residue production by crops in semiarid regions often is limited. Furthermore, residue from a crop such as cotton often provides insufficient cover to protect the soil and reduce crust/seal formation (Baumhardt et al. 1993). As a result, crust formation progresses rapidly and infiltration of subsequent storms may be depressed by NT compared with SMT (Jones et al. 1994; Baumhardt and Jones 2002).

15.4.2 Improved Soil Conditions

Residue-conserving mulch tillage typically increases the near-surface soil organic carbon (SOC) compared with conventional tillage. This is because conventional tillage incorporates residue and soil microorganisms rapidly convert residue C into CO_2 (Lal et al. 2004), which is lost to the atmosphere. Conventional tillage and residue removal can accelerate soil erosion through earlier runoff initiation and greater sediment entrainment, which often results in nutrient and collateral SOC losses (Blanco-Canqui et al. 2009). A related benefit of using mulch tillage or a more intensive cropping system that increases SOC is the greater degree of water stable aggregation of surface soil (Liebig et al. 2006) and an increased mean weight diameter of dry aggregates (Pikul et al. 2006), both of which indicate reduced susceptibility to soil erosion.

Increased SOC that improves aggregate stability can support greater infiltration by delaying crust formation and by improving subsurface soil hydraulic properties. For example, the surface saturated hydraulic conductivity (K_s) of a silt loam in Ohio increased with NT that had mulches of 8 or 16 Mg/ha; however, this increased K_s and related infiltration were attributed to earthworm activity (Blanco-Canqui and Lal 2007a). By contrast, infiltration into semiarid Great Plains soils increased immediately following tillage and subsequently declined over time (Pikul et al. 2006), which was consistent with observations by Jones et al. (1994) and Baumhardt and Jones (2002). The related soil K_s was less for mulch tillage (NT) than for conventional tillage, in part, because of greater soil density, or reduced porosity, with NT (Evett et al. 1999).

Where residue production is sufficient to cover at least 30% of the soil surface, a conservation tillage system such as NT or mulch tillage typically benefited overall soil tilth by increasing SOC, which, in turn, improved soil aggregation. Related hydraulic factors such as K_s at the soil surface and with depth may likewise be improved with mulch tillage. Where residue production may not be sufficient to qualify as a mulch tillage system, the soil physical properties such as hydraulic conductivity and bulk density tended to favor conventionally tilled soil.

15.5 EVAPORATION

Controlling the evaporation process may represent the greatest challenge to soil water conservation for agricultural production systems in arid and semiarid regions. Evaporation is often partitioned into the loss of water from the soil and from crop (transpiration). Under fixed drying conditions, evaporation from bare soil has been represented as a three-stage process (Lemon 1956) that is initially limited by the amount of energy delivered to a wet surface (Figure 15.6). With continued drying, water flow to the surface is limited by the soil hydraulic properties and evaporation decreases as a function of the square root of time (Gardner 1959) or as "falling rate" Stage 2 evaporation (Ritchie 1972). The rate of evaporation during Stage 3 is nearly constant and very low because the soil has dried. Residue management efforts have been generally directed toward modifying Stage 1 and Stage 2 evaporation that maintained a progressive reduction in evaporation as residue cover increased, but not during Stage 3 evaporation (Bond and Willis 1969).

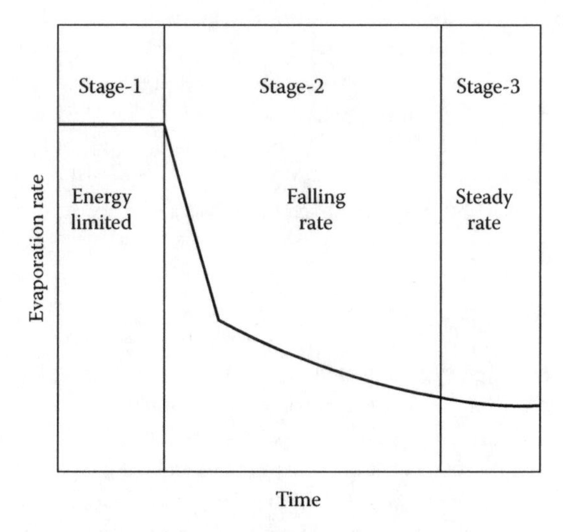

FIGURE 15.6 Conceptualized water evaporation rate plotted with time to illustrate a three-stage process for bare soil. Stage 1 evaporation is limited by energy delivered at the soil surface, while a "falling rate" of evaporation occurs during Stage 2, and evaporation ends with a low nearly steady rate during Stage 3.

Mulch tillage utilizes residue-retaining management practices to decrease evaporation and, consequently, increase storage of precipitation as soil water for subsequent crop use. Early experiments often led to the conclusion that reduced crusting, which resulted in increased infiltration, was a primary benefit of mulch tillage (Duley 1939). Further investigation during a 4-month period by Duley and Russel (1939) identified 54% rainfall conservation in straw-covered plots compared with 20% for bare plowed or disk-tilled soil and 28% for basin tillage that prevented any runoff. In that study, they concluded that evaporative water loss from bare soil was almost equivalent to the water conserved by preventing runoff using basin tillage.

In an effort to characterize crop residue effects on soil water conservation, Russel (1939) compared evaporation from soil cylinders under natural conditions that were shaded, shaded and wind-sheltered, or straw-mulched with evaporation from the cylinders exposed to normal wind and sun. Evaporation from the "densely shaded" soil decreased 36% because of reduced net radiation at the soil surface (R_{NS}). Sheltering the soil from wind further reduced evaporation by 17% as a result of greater aerodynamic boundary layer resistance. By contrast, straw mulching decreased evaporation by 73% from that of soil exposed to normal wind and sun because mulches reflect and intercept radiation, shelter against wind, and modify the vapor diffusivity at the soil surface. Surface energy balance factors governing evaporation that were indirectly compared experimentally by Russel (1939) have since been further investigated and described in a more recent review article by Horton et al. (1994). The surface energy balance can be revealing in quantifying crop residue effects on evaporation from soil and crops.

15.5.1 Energy at the Surface (Solar)

The soil surface energy balance equates the R_{NS} to various heat flux components including the soil heat flux (G) as described by Fourier's law of heat conduction, sensible heat flux into the air (H_{SOIL}), and latent heat flux (LE_{SOIL}) or evaporation according to Equation 15.1:

$$R_{NS} = G + H_{SOIL} + LE_{SOIL}, \tag{15.1}$$

assuming that energy flux into heat storage is negligible (Horton et al. 1994). For a bare soil system, the R_{NS} is the sum of incoming irradiance, both longwave (L_i) and global shortwave (S_g) adjusted for the albedo (α) reflectance correction, minus the reflected longwave radiation (L_o) according to Equation 15.2:

$$R_{NS} = (1-\alpha)S_g + L_i - L_o \tag{15.2}$$

The H_{SOIL} is a function of the volumetric heat capacity of air, the difference in soil surface and air temperatures, and the resistance to heat flux. Similarly, LE_{SOIL} is a function of the latent heat of vaporization of water, the difference in absolute humidity between the soil and air, and the resistance to LE_{SOIL}.

In mulch tillage systems, the residue cover acts directly at the soil surface to modify albedo, temperature, absolute humidity, and the resistance to sensible heat flux as well as LE_{SOIL} (Lascano and Baumhardt 1996). By increasing the reflection of incoming global irradiance and decreasing the surface emissivity, surface residues essentially displace radiation absorption upward from the soil surface. Residues are also less conductive to absorbed energy than mineral soils and resist surface heat exchange. Crop residue also affects both latent and sensible heat fluxes by increasing aerodynamic roughness, thus expanding the aerodynamic boundary layer, and by modifying soil surface temperature, water content, and related gradients. For example, total aerodynamic resistance increased by 15% after increasing the wheat residue height from 0.4 to 0.6 m, which also intercepted approximately 12% more global irradiance (Baumhardt et al. 2002).

Partitioning evapotranspiration into evaporation from the soil and transpiration from the crop canopy establishes the interacting effects of mulch tillage residues and a growing crop on water use and yield. Lascano and Baumhardt (1996) conducted one of the earliest efforts to calculate evapotranspiration from a crop grown on a residue-covered soil by using an energy balance procedure that estimated net irradiance, sensible heat exchange, and evaporation from the soil and canopy surfaces. The key feature of their approach was radiative energy transfer for distributing incoming irradiance to the cotton crop canopy, wheat residues at the soil surface, and any bare soil. During the early part of the cotton-growing season, wheat residues intercepted and reduced the net irradiance at the soil surface compared with the much higher net irradiance for conventional bare soil, which resulted in greater soil water evaporation. Because of a fuller canopy and high leaf area index during the later part of the growing season, the canopy net radiation was greater, which diminished the differences in evaporation or sensible heat flux at the soil surface, regardless of residue cover. Nevertheless, the combined seasonal soil and cotton canopy evaporation was approximately 330 mm over a 100-day period for both bare and residue-covered soils as a result of the common global irradiance. Estimated crop evaporation (transpiration) increased from 164 mm for bare soil to 223 mm for residue-covered soil. The increased crop transpiration resulted in a net cotton yield increase from 613 kg/ha for bare soil to 830 kg/ha with residue that when divided by the corresponding transpiration amount averaged a constant conversion of 0.38 Mg lint/m^3 water. The corresponding water-use efficiency values for the combined 330 mm of soil and crop evapotranspiration were approximately 0.19 and 0.26 Mg/m^3 for the bare and residue-covered soils, respectively (Lascano et al. 1994).

15.5.2 Temperature Effects

Mulch tillage effects on soil temperature are a result of the complex balance between the net irradiance that decreases with the greater albedo for soil that is untilled (Schwartz et al. 2010) or residue covered (Lascano and Baumhardt 1996) and the energy losses due to sensible and LE_{SOIL} at the soil surface. That is, residue reduces net radiation, resulting in cooler soil temperatures, and mulch tillage residue tended to dampen diurnal soil temperature fluctuation compared with bare soil possibly due to greater heat capacity of increased soil water. Unger (1978) reported progressively cooler soil temperature as straw mulches increased from 0 to 12 Mg/ha during sorghum-growing season. He also noted a reversal in soil temperatures resulting in warmer mulch-covered soil when bare soil temperature decreased below 0°C. This was similar to later observations during wheat growth on the North China Plain by Chen et al. (2007).

15.6 COVER CROPS

Cover crops are used to increase the amount of crop residue left on the soil surface. Typically, cover crops are planted in the autumn, after harvest of the cash crop, and grown until a few weeks before planting a crop the following spring. This practice helps reduce erosion by water and wind during the winter months, particularly with crops such as cotton and potatoes (*Solanum tuberosum*) that leave little residue on the field. Cover crops have also been used as green or living mulches to provide nitrogen to the cash crop (Pedersen et al. 2009; Ochsner et al. 2010). In this case, a legume is planted and grown until bloom to maximize the amount of nitrogen available to the following crop. At that time, the legume cover crop is terminated chemically or mechanically with the following crop planted very shortly afterward to optimize nitrogen use. An alternative is to plant the cash crop into the living legume cover (i.e., living mulch) and either terminate the cover crop in the entire field or just a strip over the cash crop row. Living mulches present some challenges because cover crops compete for resources, mainly water, with the cash crop.

Grass species such as oats (*Avena sativa*), rye (*Secale cereale*), ryegrass (*Lolium* genus), and wheat also are popular choices as cover crops. These can be planted with a drill or broadcast over the soil surface, but the seeding rate should be increased when broadcasting to obtain a good stand.

When managed properly, grass cover crops can produce a significant amount of biomass, with production of 4 Mg/ha or more dry matter having been reported for different cover species (Daniel et al. 1999; Bowen et al. 2000; Locke et al. 2005; Sainju et al. 2005). Management of such biomass can be challenging and some modifications to planting and tillage equipment might be needed. A NT planter equipped with row cleaners is usually required when planting into cover crop residue, either under NT or strip-till conditions. Planting of the cash crop can be done with the cover crop residue left standing or after it is flattened with a roller to form a dense mat of mulch over the soil surface (Kornecki et al. 2009). Either way, it is recommended to terminate the cover crop 2–3 weeks before planting the cash crop to allow the soil profile to be recharged with water and avoid competition (Reeves 1994).

Benefits of a high residue cover crop are many, including weed suppression, increased water availability, reduced erosion, nutrient recycling, and eventual improvement of soil quality from the addition of organic matter. Water availability is of special interest to this discussion, so we will discuss it in some detail. Gains in plant available water due to the use of cover crops is commonly attributed to increases in soil organic matter, which in turn improves soil aggregation, water infiltration, and soil water retention. Increasing organic matter content in the soil can take a considerable amount of time and varies with soil type and climate. However, some gains in water availability from cover crop use can be attributed to the mulch effect of the cover biomass, which reduces soil water evaporation and enhances infiltration by a reduction in soil surface crusting and by increased water flow into root channels. In central Alabama, an increase in soil water content was observed in the first year with the use of rye or wheat as a cover crop when compared with a no-cover control (Figure 15.7). This difference in soil water content was equivalent to approximately 25 mm of additional water in the top 50 cm of the soil profile. Although these cover crops produced a large amount of biomass (~4 Mg/ha), this increase in soil water content might not be typical and can be affected by other factors such as soil type, cash crop, cover crop management, and weather conditions.

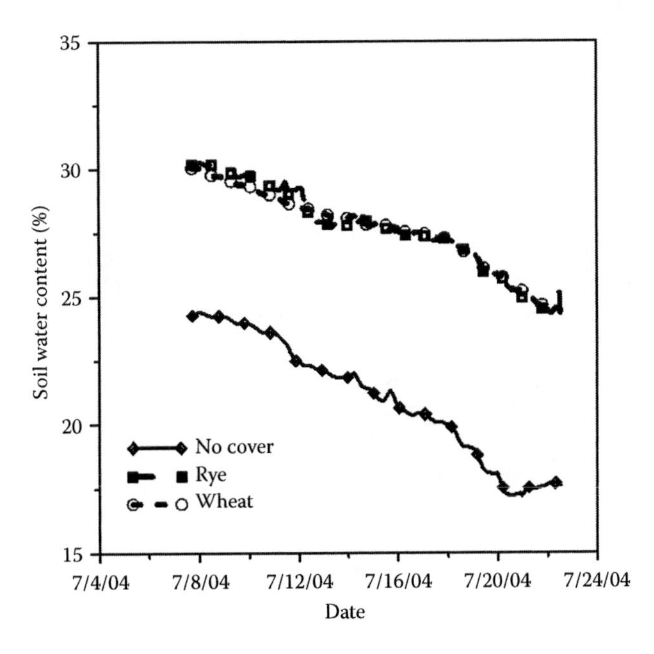

FIGURE 15.7 Volumetric soil water content during the first year of a study comparing rye and wheat as cover crops to a no-cover control on cotton near Prattville, in central Alabama. (Unpublished data of Arriaga, F.J. and Balkcom, K.S., USDA-Agricultural Research Service, National Soil Dynamics Laboratory, Auburn, AL.)

The amount of biomass produced is important because it affects the magnitude of the impacts the cover crop will have on the following crop. As the amount of biomass produced by the cover crop increases, benefits associated with the use of a cover crop seem to also increase. A cone penetrometer is a device consisting of a metal rod with a cone-shaped tip that is pushed into the ground to determine the relative compaction level of soil layers, known as penetration resistance or cone index. Cone index data collected from a study conducted on a loamy sand showed a decrease in cone index values in the soil profile as biomass levels increased (Figure 15.8). Penetration resistance decreased with increasing cover crop biomass amounts, which was probably caused by increased cover crop root growth and activity. In this case, differences in soil water were minimal because the penetrometer readings were collected when the soil water contents

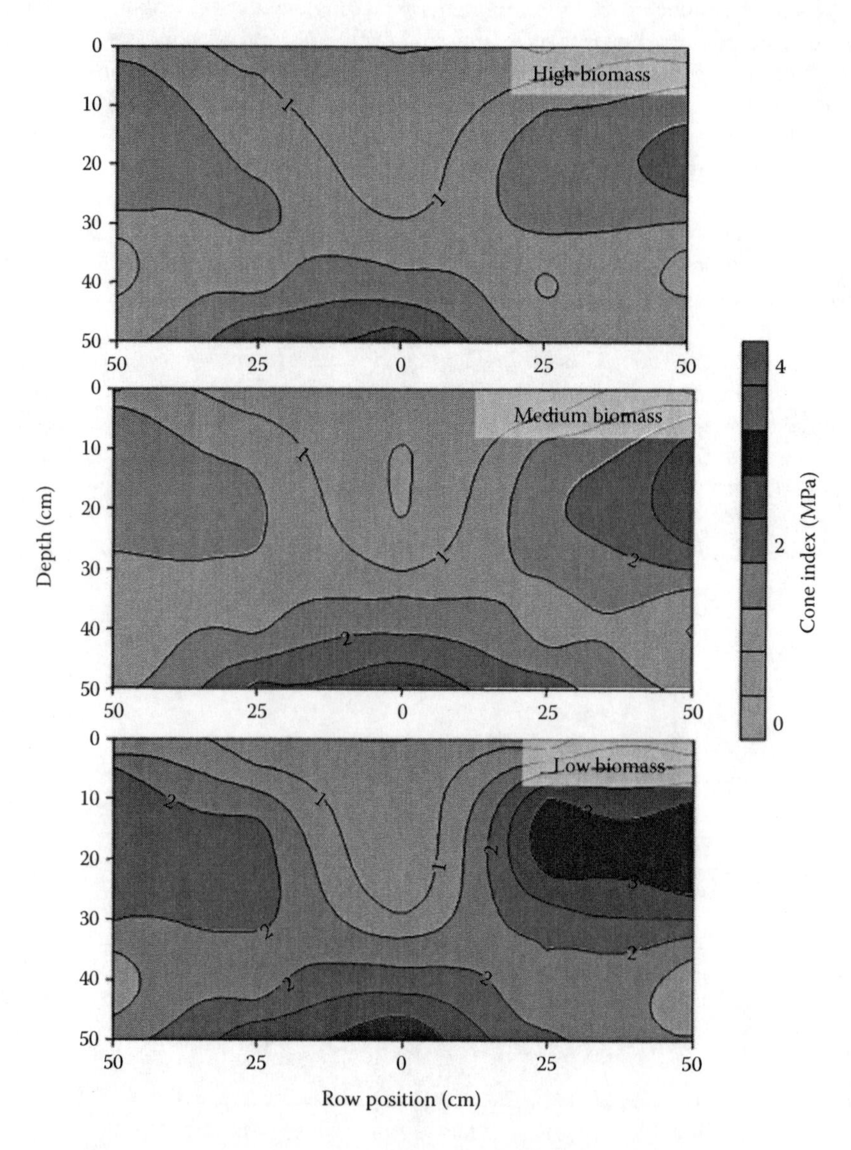

FIGURE 15.8 Penetration resistance for a coastal plain soil with three levels of rye biomass (high = 5.3 Mg/ha; medium = 2.6 Mg/ha; low = 0.2 Mg/ha) during the 2005 spring season before cash crop planting. Note that the areas with low cone index values in the center of the contour plots are associated with an in-row tillage operation and the higher values on the right with the trafficked row. (Unpublished data of Arriaga, F.J. and Price, A.J., USDA-Agricultural Research Service, National Soil Dynamics Laboratory, Auburn, AL.)

were at or near field capacity. Other researchers have concluded that a well-established and managed cover crop producing a large amount of biomass can be as effective as a strip-till operation (Raper et al. 2000, 2005). Cotton yield under NT conditions with rye as a cover crop was statistically similar to those of strip-till operations with a cover crop (i.e., there was no benefit from the tillage operation).

Recent interest on bioenergy production from cellulosic materials has created concerns over the long-term sustainability of agricultural production systems and environmental impacts. Cover crops can potentially be used to offset soil carbon imbalances that can be caused by the harvest of plant residues for bioenergy purposes. Harvest of plant residues would typically take place, for example, after corn (*Zea mays* L.) grain harvest. A winter cover crop could be planted after stover harvest. This cover crop would protect the soil from erosion due to snowmelt or rainfall during winter and early spring. Organic matter in the soil would be improved during the decomposition of the cover crop residue. Adding a cover crop to a crop management system essentially integrates a crop rotation within the same year, which is similar to the use of cover crops after cash crops such as potatoes and peanuts (*Arachis hypogaea*) that leave little residue after harvest.

Integrating cover crops into crop management rotations needs to be done carefully considering other factors such as precipitation amount and reliability in the region involved. Including cover crops in a cropping system can improve water conservation when mulch tillage is used to retain most of the residues on the soil surface, thereby improving water infiltration and reducing soil water evaporation. Use of cover crops is well adapted to humid regions where precipitation generally is adequate so that competition for soil water with crop plants is minimized. By contrast, in regions where precipitation is less abundant and reliable, competition for water generally increases water stress in crop plants when cover crops are involved, which results in lower yields. Although some favorable results with cover crops have been obtained in less humid regions (Folorunso et al. 1992; Reinbott et al. 2004), the use of cover crops generally is not advisable under dryland (nonirrigated) conditions in semiarid and arid regions where precipitation usually is limited and often highly erratic (Unger and Vigil 1998; Baumhardt and Lascano 1999).

15.7 CROP ROTATION IMPACTS

Use of crop rotations has long been an important management practice. A crop rotation is a planned sequence of crops grown in succession on the same field, as contrasted to growing one crop continually or several crops in a variable sequence (SSSA 2001). With respect to mulch tillage, use of crop rotations is especially important when the crops being grown differ widely in the amount of residues they produce. Under such conditions through careful management, residues from a crop that produces a large amount or provides for a high degree of surface cover may provide adequate surface cover throughout the growth period and even during the fallow period after harvest of the succeeding crop. This is illustrated in Figure 15.9 for a winter wheat–fallow–grain sorghum–fallow (WSF) rotation that results in two crops in 3 years. In the upper photo, some wheat residues remained on the surface after harvest of grain sorghum that was planted about 11 months after the wheat was harvested. In the lower photo, some sorghum residues remained after harvest of wheat that was planted about 11 months after the sorghum was harvested. In both cases, the combined surface residues provided adequate surface cover to meet the 30% residue cover requirement of mulch tillage. These results were for a dryland (nonirrigated) study in the U.S. southern Great Plains (Unger 1994a). In Australia, Freebairn et al. (2006) reported that carryover residues from a previous cereal crop can provide soil protection throughout the fallow period and crop phase for subsequent low residue producing crops such as sunflower (*Helianthus annuus* L.), canola (*Brassica* spp.), chickpea (*Cicer arietinum*), and cotton. Through use of an appropriate mulch tillage method, high amounts of residues from a crop such as corn can be managed to provide benefits for a low-residue-producing crop such as soybean (*Glycine max* L.) (USDA-SCS 1999) when such crops are grown in a rotation.

FIGURE 15.9 Residues of wheat and sorghum on the surface in a winter wheat–grain sorghum–fallow cropping system under dryland (nonirrigated) no-tillage conditions. Upper photo: standing sorghum (most recent crop) stalks with stubble of the previous wheat crop lying on the surface. Lower photo: standing wheat (most recent crop) stubble with stalks of the previous sorghum crop lying on the surface.

In addition to providing adequate residues to meet the surface cover requirements of mulch tillage, use of crop rotations also provides for insect, plant disease, and weed control (Peel 1998; Skillman 2001; Carr 2006; Freebairn et al. 2006; Rehman 2007). Crop rotations provide insect control because some insects feed only on specific plant species. By including totally different plant species in the rotation, the preferred species is no longer available, thus causing the particular insect to either leave or die (Poole, nd). Likewise, plant disease control is achieved because a given disease usually is prevalent only on a specific plant type. By limiting or controlling insect and disease problems, adequate plant growth is possible, thereby potentially providing adequate residues to be managed by mulch tillage to achieve water conservation. (Impacts of mulch tillage on weeds are discussed later.)

In addition to providing insect, disease, and weed control benefits, other benefits from using crop rotations include improved soil fertility, tilth, and aggregate stability; reduced soil erosion; improved soil water conditions; reduced allelopathic and phytotoxic effects; and possibly greater economic returns (Peel 1998; Carr 2006). These benefits generally are enhanced by using mulch tillage, thereby leading to potentially improved crop growth and production, which in turn can lead to improved conditions for the next crop in the rotation.

15.8 WEED CONTROL WHEN USING MULCH TILLAGE

Weeds have long been an important agricultural problem. Weeds compete with crop plants for nutrients, water, light, and space, and, if not effectively controlled, may greatly diminish crop yields. In extreme cases, crops may not produce a harvestable yield when weeds are not controlled. Effective weed control, therefore, is highly important for achieving satisfactory crop production and such control is possible through the use of tillage and herbicides. Modern cropping systems use cultivars that have been modified through genetic engineering to allow for over-the-top application of some herbicides. Each of these methods has a place where mulch tillage is used.

15.8.1 TILLAGE

Some type of tillage has been considered essential through the years to control weeds (Triplett and Dick 2008), with clean tillage being the prime method until mulch tillage was introduced in the 1930s and 1940s. Unfortunately, clean tillage had to be repeated whenever weed control again became necessary. Also, by incorporating all residues, surface soil crusting, high potential for erosion, and limited potential for conserving water were common where clean tillage was used.

While clean tillage, by definition, involves incorporating all surface residues, some retention of surface residues is implied by the term "conventional tillage." Conventional tillage involves the primary and secondary tillage operations normally used to prepare a seedbed and to cultivate a crop in a given geographical area. It usually results in <30% of the surface covered by residues after completing the tillage operations (SSSA 2001). All residues may be incorporated by using conventional tillage when, for example, moldboard plowing is used, and, therefore, it would be equivalent to clean tillage. By contrast, conventional tillage could also be similar to mulch tillage if the operations being used result in >30% surface cover by residues.

Mulch (conservation) tillage was used on 37% of the cropland in the United States in 1998, with a continuing upward trend indicated. Although use of mulch tillage is increasing with about 23% of the total cropland in the United States being planted under NT conditions in 2004 (Triplett and Dick 2008), conventional tillage continues to be widely used in the United States (Walters and Jasa, nd). By contrast, greater percentage adoption of mulch tillage than in the United States has occurred in some South American countries (Argentina, Bolivia, Brazil, Paraguay, and Uruguay) where NT is used on 69% or more of the cropland. High percentages of NT adoption have also occurred in some regions in Canada and Australia, with substantial adoption also occurring in some other countries (Derpsch and Friedrich 2009).

Clean or conventional tillage methods generally control annual and biennial weeds effectively because the entire plant is destroyed. Clean tillage, however, only suppresses perennial weeds that may grow back from vegetative parts. Some winter annual weeds are difficult to control with clean tillage because they become established along with the winter crop (e.g., winter wheat) (Kettler et al. 2000). Some summer weeds also are difficult to control with clean tillage (Wortman and Jasa 2009). Effective weed control can be achieved by using an appropriate mulch tillage method in many cases, but control also is difficult with mulch tillage for problem weeds as noted earlier for clean tillage. While the primary goal is to control the weeds, the long-term goal should be retaining adequate residues on the soil surface to achieve the benefits of using mulch tillage. Because all methods incorporate some residues (Table 15.1), tillage method selection is highly important, especially if additional tillage may be needed for further weed control. The method selected will be influenced by weed type and growth stage, surface residue level, soil conditions, and climatic region, among other factors.

For effective control, the tillage method should result in optimum separation of the weed roots from the soil or complete destruction of the weed. Soil fracturing and chisel implements (Table 15.1) are not satisfactory for controlling weeds. Generally, good control can be achieved with sweep-type implements (field cultivators and stubble-mulch plows), one-way disks, and disk harrows. If the

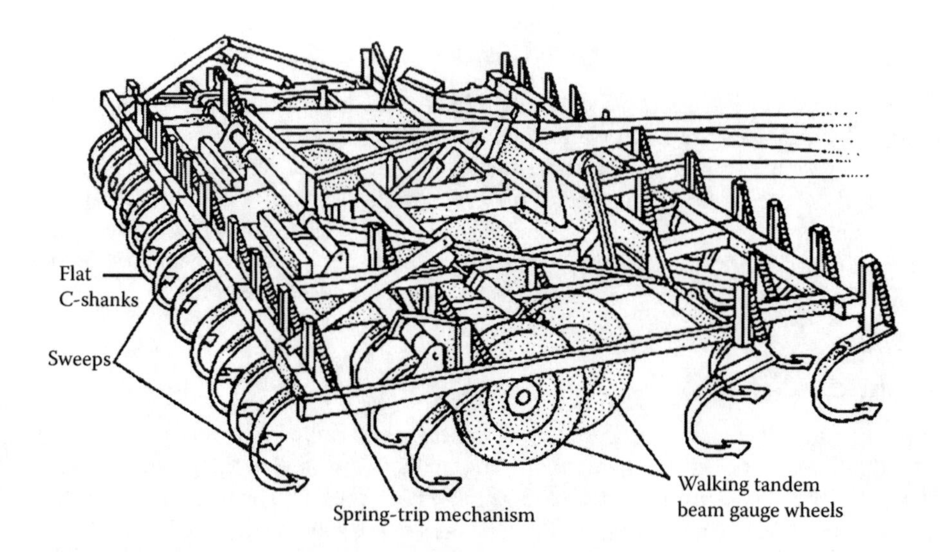

FIGURE 15.10 Schematic view of a field cultivator. (From SARE [Sustainable Agriculture Research and Education]. nd,a. Steel in the field: A farmer's guide to weed management tools [Dryland crops field cultivator (with sweeps)]. http://www.sare.org/publications/steel/pg88.htm [accessed June 10, 2010].)

goal of tillage in addition to controlling the weeds is to retain adequate surface residues, the residue amount present will strongly influence whether it is appropriate to use sweep-type implements or whether some type of disk implement can be used. Disk implements generally are highly effective for controlling weeds, but also incorporate a relatively high percentage of residues each time they are used (Table 15.1). With high amounts of residues initially present, one or possibly two operations with a disk implement may be satisfactory for controlling weeds and still retain adequate residues on the surface.

Field cultivators (Figure 15.10) and stubble-mulch plows (Figure 15.11) are mulch tillage implements used to till or prepare a soil in such a way that plant residues and other materials remain on the soil surface. These implements have V-shaped sweeps or straight blades that undercut the soil surface at a depth ranging from about 5 to 15 cm (Duley and Mathews 1947). Sweep widths vary widely among different plows, depending on the manufacturer or particular model of a plow. Likewise, widths of sweeps or blades on stubble-mulch plows also vary widely. Some stubble-mulch plows have blades that are 2 m or more wide. By undercutting the soil surface, these implements loosen and break up the soil if it has the proper water content. Such tillage is highly effective for controlling weeds, again if the soil water content is proper. Improved weed control is achieved when field cultivators or stubble-mulch plows are equipped with rod weeders (Figure 15.12). Rod weeders have a rotating subsurface rod that pulls and uproots weeds, thereby depositing them on the soil surface for exposure to the elements. Where a heavy infestation of weeds is present, the net effect of rod weeding can be to create mulch comprised of weed residues (SARE, nd,c).

To control weeds, tillage with a field cultivator or stubble-mulch plow generally is most effective where soils are relatively dry as frequently is the case in a region such as the U.S. Great Plains. SMT, however, was also found to be effective in a more humid region (New York) as early as the 1950s (Free 1953).

15.8.2 Herbicides

Herbicidal weed control began in the 1800s with copper sulfate being used first in 1821 (Reinhardt and Ganzel 2007) and iron sulfate being used to control broadleaf weeds first in 1896 (Tvedten 2001). The first synthetic organic chemical (2-methyl-4, 6-dinitrophenol) was introduced in 1932

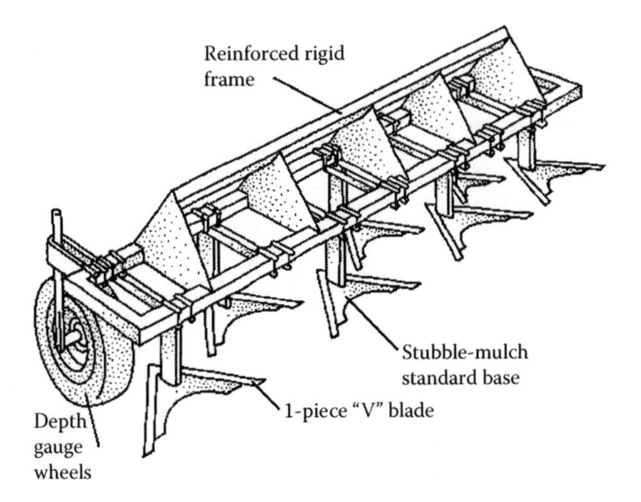

FIGURE 15.11 Schematic view of a stubble-mulch plow. (From SARE [Sustainable Agriculture Research and Education]. nd,b. Steel in the field: A farmer's guide to weed management tools [Dryland crops stubble mulch blade plow]. http://www.sare.org/publications/steel/pg90.htm [accessed June 10, 2010].)

(Reinhardt and Ganzel 2007). A new era of herbicidal weed control began in 1942 when 2,4-D [(2,4-dichlorophenoxy) acetic acid] was developed. Numerous herbicides now are available with applications possible before planting or after establishing a crop without damage to the crop. A sound understanding is essential regarding which herbicides can be used to avoid damage to the current crop or subsequent crops in a rotation.

Effective weed control during a crop's growing season is essential for reducing or eliminating direct competition between weeds and crops for soil water. Weed control in the interval between crops is also highly important to achieve soil water storage in preparation for the next crop. By using herbicides, less frequent or elimination of tillage, as with NT, is possible. By reducing or eliminating tillage to control weeds, exposure of moist soil to the atmosphere is limited, thereby reducing evaporative soil water losses. Also, more residues are retained on the surface when tillage is reduced or eliminated, thereby increasing the potential for improved soil water conservation as a result of reduced runoff and improved water infiltration.

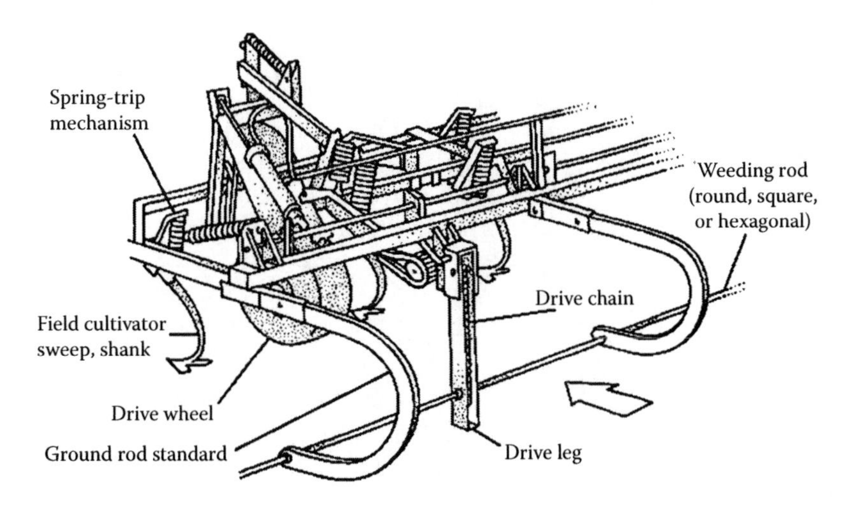

FIGURE 15.12 Schematic view of a field cultivator with an attached rod weeder. (From SARE [Sustainable Agriculture Research and Education]. nd,c. Steel in the field: A farmer's guide to weed management tools [Dryland crops rod weeder]. http://www.sare.org/publications/images/pg_92.gif [accessed June 9, 2010].)

Effective weed control is essential, especially during the crop's growth period, for successful crop production. With respect to water conservation, weed control in the interval between crops is important also, but where, for example, a dryland crop produces inadequate residues to provide erosion control and/or achieve satisfactory water conservation, allowing weed growth early in the interval between crops may be an option (Bennett 1939; Schillinger and Young 2000). Satisfactory soil water conservation is still possible under such conditions, provided the weeds are terminated in a timely manner and before they produce seed, but still allowing the weed residues to remain on the surface to reduce runoff, improve water infiltration, and reduce evaporation. Use of such practice, termed "delayed SMT," resulted in soil water contents when winter wheat was planted similar to those resulting from repeated use of SMT to control weeds throughout the fallow period (Johnson and Davis 1972).

As with tillage, there may be problems with herbicidal weed control, with undoubtedly the greatest problem being the development of herbicide resistance in some weeds. Herbicide resistance refers to the inherited ability of a weed or crop biotype to survive a herbicide application to which the original population was susceptible (Gunsolus 2008). Resistance of weeds to triazine herbicides was found in 1968 and 120 weed biotypes were resistant to those herbicides by 1991. Resistance to 15 other herbicide families also was found throughout the world by 1991, with additional cases of resistance found since that time (Gunsolus 2008). Where herbicide-resistant weeds are a problem, NT will not be appropriate because it is based on the use of herbicides for controlling weeds and alternative herbicides are not available for controlling the herbicide-resistant weeds. Under such conditions, SMT generally would be more appropriate. In some cases, NT fields are occasionally stubble-mulch plowed, which helps control such problem weeds and increases crop yields as compared with those in undisturbed NT fields (Kettler et al. 2000). Use of crop rotations that expands the diversity of herbicide mode-of-action can be crucial for reducing problems with herbicide-resistant weeds.

15.8.3 Herbicide-Resistant Crop Production Involving Genetic Engineering

With respect to weed control, genetic engineering involves selecting herbicide-tolerant genes that naturally occur in a given crop cultivar or cell culture and incorporating them through traditional breeding processes into crop varieties or hybrids (Penn State 2010). Through genetic engineering, it is now possible to use highly effective, quick-acting herbicides to control weeds in some actively growing crops that are resistant to those herbicides. Development of crops resistant to some herbicides has greatly expanded the opportunity to control problem weeds in some crops. For example, because of genetic engineering, glyphosate [N-(phosphonomethyl) glycine] can be applied to control weeds in glyphosate-resistant cultivars of soybean, corn, and canola (Padgette et al. 1995; Moll 1997; Rasche and Gadsby 1997). Growing of crops altered by genetic engineering is being rapidly adopted by producers in many countries worldwide (Duke 1999; McHughen 2006). Growing herbicide-resistant crops allows producers to more effectively use mulch tillage practices such as NT and reduced tillage (Duke 1999), thereby increasing the opportunity for improving soil water conservation when more residues are retained on the soil surface. When herbicide-resistant crops are grown, volunteer plants of the crop may cause problems because they also would be resistant to the herbicide, thus possibly requiring some tillage to control those plants.

15.9 IMPACT OF RESIDUE REMOVAL FOR BIOFUEL PRODUCTION

Crop residues retained on the soil surface, which are important for conserving soil water, also are highly important for controlling soil erosion by wind and water (Figure 15.2). Maintaining crop residues on the surface by using mulch tillage, therefore, is highly important for maintaining the soil resource base for sustainable crop production and possibly enhanced production for an ever-increasing world population. As discussed in previous sections, water and soil conservation generally increases with increases in residue retention on the surface. Removing residues for biofuel production, at least from some soils, therefore, could negate their beneficial effect on sustaining or improving conditions for sustained or improved crop production.

The use of crop residues as feedstock for cellulosic ethanol production has received much interest in the United States and some other countries, with some commercial plants being built for its production in the United States (USDOE 2007). Corn stover is widely considered the main cellulosic feedstock for biofuel production (Graham et al. 2007), but residues of wheat and sorghum are also feedstock for biofuel production (Sarath et al. 2008). Production of ethanol from renewable sources is important and should be pursued, but the practice of removing crop residues as biofuel feedstock must be carefully examined relative to its impacts on soil conditions, including water conservation, crop production, and environmental quality.

The removal of crop residues for any purpose can negatively and positively impact water and soil conservation, with negative impacts prevailing under long-term conditions (Wilhelm et al. 2007). Indiscriminate residue removal is not advisable in semiarid regions where precipitation is low and variable, thus resulting in low residue production. In more humid regions or with irrigation, it may be possible to remove some residues without adversely affecting soil conditions and water conservation, provided some residues are retained and managed through use of appropriate tillage methods.

The removal of crop residues influences crop production because it often reduces the amount of water available to plants. It may also affect production due to abrupt fluctuations in soil surface temperature and loss of plant nutrients. In Nebraska, complete removal of corn stover from a silty clay loam in a 4-year study involving NT reduced corn grain and biomass yields by about 23% (Wilhelm et al. 1986). The amount of crop residues on the surface may explain 95% of the variability in grain and biomass yields (Wilhelm et al. 1986; Blanco-Canqui and Lal 2007b).

Impacts of residue removal on corn production are site-specific and depend on factors such as soil type, topography, tillage method, climate, and duration of stover management. Crop production can be more adversely affected by residue removal from sloping, erosion-prone, and well-drained soils than from flat and clayey soils (Linden et al. 2000).

In some soils or ecosystems, it may be feasible to remove a portion of the crop residue for biofuel production in the short term without increasing the potential for erosion by water and wind, hindering soil water conservation, or reducing crop yields. Indeed, removing some residues may improve seed germination, facilitate planting, increase N mineralization, and reduce pest infestations. Lower soil temperatures under surface residues often results in slower seed germination, especially in colder regions as, for example, the northern United States and Canada. Corn emergence in Canada was as much as 30% lower in mulched plots than in unmulched plots (Dam et al. 2005). On three soils in Ohio, corn stover removal from long-term NT soils enhanced seed germination. Without stover removal, emergence was delayed by up to 3 days as compared with that on soils where all stover was removed (Blanco-Canqui et al. 2006). In both studies, the delayed emergence did not reduce crop yields. Plants in plots with low residue cover often grow taller during the first few weeks, but the height differences often diminish rapidly with time.

On wet and cold mulched soils, the combination of reduced germination, proliferation of weeds and pests, and nutrient immobilization may lower crop yields. In southwestern Wisconsin, corn yield decreased on two silt loams when stover cover was doubled (Swan et al. 1994). Yields of continual corn decreased during the last 4 years of a 13-year study when 2, 4, 8, or 16 Mg/ha of stover mulch was applied to a silty clay loam in Iowa (Morachan et al. 1972).

The impacts of removing residues on soil water conservation, other soil conditions, and crop production are site-specific and information regarding maximum permissible removal rates is limited. Involved are such factors as tillage methods and cropping systems being used, soil characteristics, and climatic zones. Based on a computer model, mainly for the U.S. Corn Belt region, 20%–50% of the residue produced may be removed without adversely affecting the residues needed to control erosion (Graham et al. 2007). Those estimates, however, did not deal with water conservation, which undoubtedly would require different amounts of surface residues to achieve optimum water conservation.

Use of current computer models along with tools such as remote sensing and geographic information system (GIS) is a promising approach to better understand the impacts of removing residues on soil water conservation on different soils and geographical areas (Green et al. 2003).

15.10 RESEARCH NEEDS

The benefits of retaining crop residues on the soil surface, as with mulch tillage, for conserving soil water are widely recognized. Therefore, it is highly important that research involving crop residues be continued to develop practices for further enhancing soil water conservation, thereby improving crop production to help meet the food, fiber, and fuel needs for an ever-expanding world population. The need for improved soil water conservation is also important because the supply of fresh water, which is used by agriculture, is limited, and there is increasing competition for such water among agricultural, urban, industrial, and recreational users. Agriculture must do its part to conserve and efficiently use its share of the available water.

With regard to using water efficiently, research is needed to develop crops that have improved drought tolerance, are of greater commercial or economic value, and have improved resistance to insects, plant diseases, and climatic extremes (e.g., unseasonably low or high temperatures).

Another issue regarding soil water conservation as affected by mulch tillage pertains to the current interest in using crop residues to produce biofuel. Soil water conservation increases with increases in the amount of crop residues retained on the soil surface under many conditions. Considerable research has been conducted regarding the surface residue amount needed to control soil erosion, but such information generally is not available regarding soil water conservation. Therefore, it is highly important to conduct water conservation research under a wide range of conditions (crop residue levels, soils, cropping systems, climate, etc.) so that effective recommendations can be made for achieving optimum water conservation. With good results from a wide range of conditions, it should be possible to develop a model that could assist consultants, agency personnel, and/or producers to readily determine the amount of residues needed under prevailing conditions to achieve the desirable level of soil water required for obtaining a given level of crop production.

15.11 SUMMARY

Mulching, that is, maintaining organic or inorganic materials on or applying them to the soil surface, is an ancient practice. Through the years, however, clean tillage that incorporated crop residues and also controlled weeds became the norm. In fact, frequent and deep tillage was promoted also for conserving soil water. Such tillage conserved water, but resulted in soil aggregate breakdown, surface sealing, and excessive runoff, all contributing to serious soil erosion by water. Clean tillage also contributed to the disastrous soil erosion by wind during the major drought in the U.S. Great Plains in the 1930s. SMT, which undercuts the soil surface and leaves crop residues on the surface, was developed to help control erosion. It was soon found that retaining residues on the surface also provided for conserving soil water, and extensive research involving various types of mulch tillage subsequently has been conducted at numerous locations throughout the world. Soil water conservation generally increases with increases in the amount of residues retained on the soil surface.

Mulch tillage is possible with a variety of implements, but careful implement selection is essential to retain the optimum amount of residues on the surface to achieve soil water conservation and also to achieve effective weed control. Weed control under mulch tillage conditions (as well as with other tillage methods) received a major boost with the development of herbicides, beginning in the 1940s. Improved herbicides have been developed through the years and it is now possible to achieve complete weed control with herbicides and produce crops by the ultimate mulch tillage method, namely, NT, under many conditions. Through use of NT, most crop residues are retained on the soil surface, thereby providing the greatest opportunity for conserving soil water and subsequently achieving favorable crop production.

ABBREVIATIONS

CA	Conservation agriculture
G	Soil heat flux
H_{SOIL}	Sensible heat flux into the air
K_s	Surface saturated hydraulic conductivity
LE_{SOIL}	Latent heat flux
NT	No-tillage
R_{NS}	Reduced net radiation at the soil surface
SMT	Stubble-mulch tillage
SOC	Soil organic carbon
WSF	Winter wheat–fallow–grain sorghum–fallow

REFERENCES

Abu-Hamdeh, N. 2003. Compaction and subsoiling effects on corn growth and soil bulk density. *Soil Sci. Soc. Am. J.* 67:1213–1219.

Agassi, M., I. Shainberg, and J. Morin. 1981. Effects of electrolyte concentration and soil sodicity on infiltration rate and crust formation. *Soil Sci. Soc. Am. J.* 45:848–851.

Allen, R.R. and C.R. Fenster. 1986. Stubble-mulch equipment for soil and water conservation in the Great Plains. *J. Soil Water Conservat.* 41:11–16.

Anderson, D.T. 1968. Field equipment needs in conservation tillage. In *Conservation Tillage in the Great Plains* (Proc. Workshop, Lincoln, NE, 1968), pp. 83–91. Great Plains Agric. Counc. Publ. 32.

Barnett, A.P. 1987. Fifty years of progress in soil and water conservation at the Southern Piedmont Conservation Research Center (SPCRC), Watkinsville, Georgia, 1937–1987 (Online). ftp://anyone:spc@128.192.164.106/schomberg (verified May 1, 2007).

Baumhardt, R.L. and O.R. Jones. 2002. Residue management and paratillage effects on some soil properties and rain infiltration. *Soil Till. Res.* 65:19–27.

Baumhardt, R.L. and R.J. Lascano. 1996. Rain infiltration as affected by wheat residue amount and distribution in ridged tillage. *Soil Sci. Soc. Am. J.* 60:1908–1913.

Baumhardt, R.L. and R.J. Lascano. 1999. Water budget and yield of dryland cotton intercropped with terminated winter wheat. *Agron. J.* 91:922–927.

Baumhardt, R.L. and R.C. Schwartz. 2004. CRUSTS/structural. In *Encyclopedia of Soils in the Environment*, ed. D. Hillel, Vol. 1, pp. 347–356. Oxford, UK: Elsevier.

Baumhardt, R.L., M.J.M. Römkens, F.D. Whisler, and J.-Y. Parlange. 1990. Modeling infiltration into a sealing soil. *Water Resour. Res.* 26:2497–2505.

Baumhardt, R.L., J.W. Keeling, and C.W. Wendt. 1993. Tillage and residue effects on infiltration into soils cropped to cotton. *Agron. J.* 85:379–383.

Baumhardt, R.L., R.C. Schwartz, and R.W. Todd. 2002. Effects of taller wheat residue after stripper header harvest on wind run, irradiant energy interception, and evaporation. In *Making Conservation Tillage Conventional: Building a Future on 25 Years of Research*, ed. E. van Santen, pp. 386–391. Proc. 25th Annual Southern Conservation Tillage Conference for Sustainable Agriculture, June 24–26, 2002, Auburn, AL. Spec. Rpt. No. 1. Auburn AL: Alabama Agric. Expt. Stn. and Auburn Univ.

Bennett, H.H. 1939. *Soil Conservation*. New York: McGraw-Hill Book Co.

Blanco-Canqui, H. and R. Lal. 2007a. Impacts of long term wheat straw management on soil hydraulic properties under no tillage. *Soil Sci. Soc. Am. J.* 71:1166–1173.

Blanco-Canqui, H. and R. Lal. 2007b. Soil and crop response to harvesting corn residues for biofuel production. *Geoderma* 141:355–362.

Blanco-Canqui, H., R. Lal, W.M. Post, R.C. Izaurralde, and L.B. Owens. 2006. Changes in long-term no-till corn growth and yield under different rates of stover mulch. *Agron. J.* 98:1128–1136.

Blanco-Canqui, H., R.J. Stephenson, N.O. Nelson, and D.R. Presley. 2009. Wheat and sorghum residue removal for expanded uses increases sediment and nutrient loss in runoff. *J. Environ. Qual.* 38:2365–2372.

Bond, J.J. and W.O. Willis. 1969. Soil water evaporation: Surface residue rate and placement effects. *Soil Sci. Soc. Am. J.* 33:445–448.

Borst, H.L. and R. Woodburn. 1942. The effect of mulching and methods of cultivation on runoff and erosion from Muskingum silt loam. *Agric. Eng.* 23:19–24.

Bowen, G., C. Shirley, and C. Cramer. 2000. *Managing Cover Crops Profitably*, 2nd edn. Sustainable Agriculture Network Handbooks Series, Book 3. Beltsville, MD: National Agricultural Library.

Bradford, J.M. and G.A. Peterson. 1999. Conservation tillage. In *Handbook of Soil Science*, ed.-in-chief M.E. Sumner, pp. G-247–G-270. Boca Raton, FL: CRC Press.

Bresson, L.M. and C.J. Moran. 1995. Structural change induced by wetting and drying in seedbeds of a hardsetting soil with contrasting aggregate size distribution. *Eur. J. Soil Sci.* 46:205–214.

Carr, P.M. 2006. Crop rotation benefits of annual forages. *2006 Annual Report, Agronomy Section.* Dickenson, ND: Dickenson Res. Ext. Ctr.

Chen, S.Y., X.Y. Yang, D. Pei, H.Y. Sun, and S.L. Chen. 2007. Effects of straw mulching on soil temperature, evaporation and yield of winter wheat: Field experiments on the North China Plain. *Ann. Appl. Biol.* 150:261–268.

Dam, R.F., B.B. Mehdi, M.S.E. Burgess, C.A. Madramootoo, G.R. Mehuys, and I.R. Callum. 2005. Soil bulk density and crop yield under eleven consecutive years of corn with different tillage and residue practices in a sandy loam soil in central Canada. *Soil Till. Res.* 84:41–53.

Daniel, J.B., A.O. Abaye, M.M. Alley, C.W. Adcock, and J.C. Maitland. 1999. Winter annual cover crops in a Virginia no-till cotton production system: I. Biomass production, ground cover, and nitrogen assimilation. *J. Cotton Sci.* 3:74–83.

Derpsch, R. and T. Friedrich. 2009. Global overview of conservation tillage adoption. In *Proceedings, Lead Papers*, pp. 429–438. 4th World Congress on Conservation Agriculture, February 4–7, 2009, New Delhi, India.

Duke, S.O. 1999. Weed management: Implications of herbicide resistant crops. Paper presented at the Workshop on Ecological Effects of Pest Resistance Genes in Managed Ecosystems, Bethesda, MD.

Duley, F.L. 1939. Surface factors affecting the rate of intake of water by soil. *Soil Sci. Soc. Am. Proc.* 4:60–64.

Duley, F.L. and O.R. Mathews. 1947. Ways to till the soil. In *Science in Farming, USDA Yearbook of Agriculture 1943–1947*, ed. A. Stefferud, pp. 518–526. Washington, DC: U.S. Gov. Print. Off.

Duley, F.L. and J.C. Russel. 1939. The use of crop residues for soil and moisture conservation. *Agron. J.* 31:703–709.

Evett, S.R., F.H. Peters, O.R. Jones, and P.W. Unger. 1999. Soil hydraulic conductivity and retention curves from tension infiltrometer and laboratory data. In *Proc. Int. Workshop Characterization and Measurement of the Hydraulic Properties of Unsaturated Porous Media*, ed. M.Th. van Genuchten, F.J. Leij, and L. Wu, pp. 541–551. Riverside, CA: Univ. of California.

FAO (Food and Agriculture Organization of the United Nations). 2008. Conservation agriculture. http://www.fao.org/ag/ca/la.html (accessed July 24, 2009).

Fenster, C.R., N.P. Woodruff, W.S. Chepil, and F.H. Siddoway. 1965. Performance of tillage implements in a stubble mulch system: III. Effects of tillage sequences on residues, soil cloddiness, weed control, and wheat yield. *Agron. J.* 57:52–55.

Folorunso, O.A., D.E. Rolston, T. Prichard, and D.T. Louie. 1992. Soil surface strength and infiltration rate as affected by winter cover crops. *Soil Technol.* 5:189–197.

Franzmeier, D.P., C.J. Chartres, and J.T. Wood. 1996. Hardsetting soils in Southeast Australia: Landscape and profile processes. *Soil Sci. Soc. Am. J.* 60:1178–1187.

Free, G.R. 1953. Stubble-mulch tillage in New York. *Soil Sci. Soc. Am. Proc.* 17:165–170.

Freebairn, D.M., P.S. Cornish, W.K. Anderson, S.R. Walker, J.B. Robinson, and A.R. Beswick. 2006. Management systems in climate regions of the world—Australia. In *Dryland Agriculture*, 2nd edn. Agronomy Monograph 23, eds. G.A. Peterson, P.W. Unger, and W.A. Payne, pp. 837–878. Madison, WI: Am. Soc. Agron., Crop Sci. Soc. Am., and Soil Sci. Soc. Am.

Gardner, W.R. 1959. Solutions of the flow equation for the drying of soils and other porous media. *Soil Sci. Soc. Am. Proc.* 23:183–187.

Graham, R.L., R. Nelson, J. Sheehan, R.D. Perlack, and L.L. Wright. 2007. Current and potential U.S. corn stover supplies. *Agron. J.* 99:1–11.

Green, T.R., L.R. Ahuja, and J.G. Benjamin. 2003. Advances and challenges in predicting agricultural management effects on soil hydraulic properties. *Geoderma* 116:3–27.

Greene, R.S.B. 2006. Hard setting soils. In *Encyclopedia of Soil Science*, ed. R. Lal, pp. 804–806. Boca Raton, FL: Taylor & Francis.

Gunsolus, J.L. 2008. Herbicide resistant weeds. Univ. of Minnesota Extension. http://www.extension.umn.edu/distribution/cropsystems/dc6077.html (accessed June 11, 2010).

Hill, P.R., K.J. Eck, and J.R. Wilcox. 1994. Managing crop residue with farm equipment. *Agronomy Guide AY-280.* West Lafayette, IN: Purdue Univ.

Hofman, V. 1997. Residue management for erosion control. Publ. DS-22-97. Fargo, ND: North Dakota State Univ. Ext. Service.

Horton, R., G.J. Kluitenberg, and K.L. Bristow. 1994. Surface crop residue effects on the soil surface energy balance. In *Managing Agricultural Residues*, ed. P.W. Unger, pp. 143–162. Boca Raton, FL: Lewis Publishers.

Jacks, G.V., W.D. Brind, and R. Smith. 1955. Mulching. *Tech. Commun. No. 49.* Commonwealth Bureau of Soil Sci. (England).

Jasa, P.J. nd. Conservation tillage. Univ. of Nebraska Publication. http://agecon.okstate.edu/isct/labranza/jasa/tillagesys.doc (accessed September 25, 2010).

Johnson, W.C. and R.G. Davis. 1972. Research on stubble-mulch farming of winter wheat. USDA Conserv. Res. Rpt. 16. Washington, DC: U.S. Gov. Print. Off.

Jones, O.R., V.L. Hauser, and T.W. Popham. 1994. No-tillage effects on infiltration, runoff, and water conservation on dryland. *Trans. ASAE* 37:473–479.

Karunatilake, U.P. and H.M. van Es. 2002. Rainfall and tillage effects on soil structure after alfalfa conversion to maize on a clay loam soil in New York. *Soil Till. Res.* 67:135–146.

Kettler, T.A., D.J. Lyon, J.W. Doran, W.L. Powers, and W.W. Stroup. 2000. Soil quality assessment after weed-control tillage in a no-till wheat–fallow cropping system. *Soil Sci. Soc. Am. J.* 64:339–346.

Kornecki, T.S., A.J. Price, R.L. Raper, and F.J. Arriaga. 2009. New roller crimper concepts for mechanical termination of cover crops in conservation agriculture. *Renew. Agric. Food Syst.* 24:165–173.

Lal, R. 2006. Mulch farming. In *Encyclopedia of Soil Science*, ed. R. Lal, pp. 1103–1110. Boca Raton, FL: Taylor & Francis.

Lal, R., M. Griffin, J. Apt, L. Lave, and M.G. Morgan. 2004. Managing soil carbon. *Science* 304:393.

Lascano, R.J. and R.L. Baumhardt. 1996. Effects of crop residue on soil and plant water evaporation in a dryland cotton system. *Theor. Appl. Climatol.* 54:69–84.

Lascano, R.J., R.L. Baumhardt, S.K. Hicks, and, J.L. Heilman. 1994. Soil and crop evaporation from cotton under strip tillage: Measurement and simulation. *Agron. J.* 86:987–994.

Lemon, E.R. 1956. The potentialities for decreasing soil moisture evaporation loss. *Soil Sci. Soc. Am. Proc.* 20:120–125.

Liebig, M., L. Carpenter-Boggs, J.M.F. Johnson, S. Wright, and N. Barbour. 2006. Cropping systems effects on soil biological characteristics in the Great Plains. *Renew. Agric. Food Syst.* 21:36–48.

Linden, D.R., C.E. Clapp, and R.H. Dowdy. 2000. Long-term corn grain and stover yields as a function of tillage and residue removal in east central Minnesota. *Soil Till. Res.* 56:167–174.

Locke, M.A., R.M. Zablotowicz, P.J. Bauer, R.W. Steinriede, and L.A. Gaston. 2005. Conservation cotton production in the southern United States: Herbicide dissipation in soil and cover crops. *Weed Sci.* 53:717–727.

Mascagni Jr, H.J., L.A. Gaston, and B. Guillory. 2002. Influence of irrigation and rye cover crop on corn yield performance and soil properties. In *Making Conservation Tillage Conventional: Building a Future on 25 Years of Research*, ed. E. van Santen, pp. 397–400. Proc. 25th Annual Southern Conservation Tillage Conference for Sustainable Agriculture, June 24–26, 2002, Auburn, AL. Spec. Rpt. No. 1. Auburn, AL: Alabama Agric. Expt. Stn. and Auburn Univ.

McCarthy, J.R., D.L. Pfost, and H.D. Currence. 1993. Conservation tillage and residue management to reduce soil erosion. http://extension.missouri.edu/publications/DisplayPub.aspx?P=G1650 (accessed September 24, 2010).

McHughen, A. 2006. Genetic engineering and testing methodologies. Publ. 8190, Agric. Biotech. in California Series. Davis, CA: Univ. of California.

McIntyre, D.S. 1958. Permeability measurements of soil crusts formed by raindrop impact. *Soil Sci.* 85:185–189.

Mead, J.A. and K.Y. Chan. 1985. The effect of deep tillage and fallow on yield of wheat on a hard setting soil. In *Crop and Pasture Production—Science and Practice*, ed. J.J. Yates, Proc. 3rd Australian Agronomy Conf., Univ. of Tasmania, Hobert, Tasmania.

Moll, S. 1997. Commercial experience and benefits from glyphosate tolerant crops. In *Proc. 1997 Brighton Crop Prot. Conf. –Weeds*, November 17–20, 1997, vol 3, pp. 931–940. Brighton, UK: British Crop Prot. Counc. (now BCPC).

Moore, I. D. 1981. Effect of surface sealing on infiltration. *Trans. ASAE* 24:1546–1553.

Morachan, Y.B., W.C. Moldenhauer, and W.E. Larson. 1972. Effects of increasing amounts of organic residues on continuous corn: I. Yields and soil physical properties. *Agron. J.* 64:199–203.

Morin, J. and Y. Benyamini. 1977. Rainfall infiltration into bare soil. *Water Resour. Res.* 13:813–817.

Musick, J.T., A.F. Wiese, and R.R. Allen. 1977. Management of bed-furrow irrigated soil with limited and no tillage systems. *Trans. ASAE* 20:666–672.

Myers, D.K. nd. Harvesting corn residue. Publ. AGF-003-92. Columbus, OH: Ohio State Univ. Ext.

Ochsner, T.E., K.A. Albrecht, T.W. Schumacher, J.M. Baker, and R.J. Berkevich. 2010. Water balance and nitrate leaching under corn in kura clover living mulch. *Agron. J.* 102:1169–1178.

Padgette, S.R., K.H. Kolacz, X. Delannay, D.B. Re, B.J. LaVallee, C.N. Tinius, et al. 1995. Development, identification, and characterization of glyphosate-tolerant soybean line. *Crop Sci.* 35:1451–1461.

Papendick, R.I., J.F. Parr, and R.E. Meyer. 1990. Managing crop residues to optimize crop/livestock production systems for dryland agriculture. *Adv. Soil Sci.* 13:253–272.

Pedersen, P., E.J. Bures, and K.A. Albrecht. 2009. Soybean production in a kura clover living mulch system. *Agron. J.* 101:653–656.

Peel, M.D. 1998. Crop rotations for improved productivity. EB-48 (Revised). http://www.ag.ndsu.edu/pubs/plantsci/crops/eb48-1.htm (accessed June 7, 2010).

Penn State. 2010. Pest management: Herbicide-resistant crops. *Agronomy Guide 2009–2010*. University Park, PA: The Pennsylvania State Univ.

Pikul, J.L., R.C. Schwartz, J.G. Benjamin, R.L. Baumhardt, and S. Merrill. 2006. Cropping system influences on soil physical properties in the Great Plains. *Renew. Agric. Food Syst.* 21:15–25.

Poole, T.E. nd. Crop rotation. *Fact Sheet 784*. College Park, MD: Maryland Cooperative Extension, Univ. of Maryland.

Raper, R.L. 2006. In-row subsoiling southeastern soils to reduce compaction and improve crop yields. In *Improving Conservation Technologies to Compete for Global Resources and Markets*, eds. R.C. Schwartz, R.L. Baumhardt, and J.M. Bell, pp. 85–94. Proc. 28th Annual Southern Conservation Tillage Conference for Sustainable Agriculture, June 26–28, 2006, Amarillo, TX. Rpt. No. 06-1. Bushland, TX: USDA-ARS Conservation and Production Research Laboratory.

Raper, R.L., D.W. Reeves, C.H. Burmester, and E.B. Schwab. 2000. Tillage depth, tillage timing, and cover crop effects on cotton yield, soil strength, and tillage energy requirements. *Appl. Eng. Agric.* 16:379–385.

Raper, R.L., E.B. Schwab, K.S. Balkcom, C.H. Burmester, and D.W. Reeves. 2005. Effect of annual, biennial, and triennial in-row subsoiling on soil compaction and cotton yield in southeastern U.S. silt loam soils. *Appl. Eng. Agric.* 21:337–343.

Rasche, E. and M. Gadsby. 1997. Glufosinate ammonium tolerant crops: International commercial developments and experiences. In *Proc. 1997 Brighton Crop Prot. Conf.—Weeds*, November 17–20, 1997, vol 3, pp. 941–946. Brighton, UK: Bristish Crop Prot. Counc. (now BCPC).

Reeves, D.W. 1994. Cover crops and rotations. In *Advances in Soil Science: Crops Residue Management*, eds. J.L. Hatfield and B.A. Stewart, pp. 125–172. Boca Raton, FL: Lewis Publishers.

Rehman, A. 2007. Zero tillage technology for rice and wheat crops. DAWN Group of Newspapers. http://www.dawn.com/2007/09/24/ebr6.htm (accessed June 3, 2010).

Reinbott, T.M., S.P. Conley, and D.G. Blevins. 2004. No-tillage corn and grain sorghum response to cover crop and nitrogen fertilization. *Agron. J.* 96:1158–1163.

Reinhardt, C. and B. Ganzel. 2007. Farming in the 1930s (Online). http://www.livinghistoryfarm.org/farminginthe30s/pests_06.html (verified June 14, 2007).

Ritchie, J.T. 1972. Model for predicting evaporation from a row crop with incomplete cover. *Water Resour. Res.* 8:1204–1213.

Ruan, H., L.R. Ahuja, T.R. Green, and J.G. Benjamin. 2001. Residue cover and surface sealing effects on infiltration: Numerical simulations for field applications. *Soil Sci. Soc. Am. J.* 65:853–861.

Russel, J.C. 1939. The effect of surface cover on soil moisture losses by evaporation. *Soil Sci. Soc. Am. Proc.* 4:65–70.

Sainju, U.M., W.F. Whitehead, and B.P. Singh. 2005. Carbon accumulation in cotton, sorghum, and underlying soil as influenced by tillage, cover crop, and nitrogen fertilization. *Plant Soil* 273:219–234.

Sarath, G., R.B. Mitchell, S.E. Sattler, D. Funnell, J.F. Pedersen, R.A. Graybosch, and K.P. Vogel. 2008. Opportunities and roadblocks in utilizing forages and small grains for liquid fuels. *J. Ind. Microbiol. Biotechnol.* 35:343–354.

SARE (Sustainable Agriculture Research and Education). nd,a. Steel in the field: A farmer's guide to weed management tools (Dryland crops field cultivator (with sweeps)). http://www.sare.org/publications/steel/pg88.htm (accessed June 10, 2010).

SARE (Sustainable Agriculture Research and Education). nd,b. Steel in the field: A farmer's guide to weed management tools (Dryland crops stubble mulch blade plow). http://www.sare.org/publications/steel/pg90.htm (accessed June 10, 2010).

SARE (Sustainable Agriculture Research and Education). nd,c. Steel in the field: A farmer's guide to weed management tools (Dryland crops rod weeder). http://www.sare.org/publications/images/pg_92.gif (accessed June 9, 2010).

Schillinger, W.F. and F.L. Young. 2000. Soil water use and growth of Russian thistle after wheat harvest. *Agron. J.* 92:167–172.

Schwartz, R.C., R.L. Baumhardt, and S.R. Evett. 2010. Tillage effects on soil water redistribution and bare soil evaporation throughout a season. *Soil Till. Res.* 110: 221–229.

Skillman, L. 2001. Crop rotation important part of farming. Univ. of Kentucky, College of Agric. http://www.ca.uky.edu/agc/news/2001/Jan/croprotate.htm (accessed June 7, 2010).

SSSA (Soil Science Society of America). 2001. *Glossary of Soil Science Terms, 2001.* Madison, WI: Soil Sci. Soc. of America.

Swan, J.B., R.L. Higgs, T.B. Bailey, N.C. Wollenhaupt, W.H. Paulson, and A.E. Peterson. 1994. Surface residue and in-row treatment effects on long-term no-tillage continuous corn. *Agron. J.* 86:711–718.

The Yearbook of Agriculture. 1955. Typical water-holding characteristics of different-textured soils. In *Water, Yearbook of Agriculture*, ed. A. Stefferud, p. 120. U.S. Dept. of Agric. Washington, DC: U.S. Gov. Print. Off.

Touchton, J.T., D.W. Reeves, and R.R. Sharpe. 1989. Cotton yields as affected by previous crop tillage and subsoiling for cotton. In *Conservation Farming: Preserving Our Heritage*, ed. I.D. Teare, pp. 23–25. Proc. 1989 Southern Conservation Tillage Conference on Sustainable Agriculture, July 12–13, 1989, Tallahassee, FL. Gainesville, FL: Univ. of Florida.

Triplett Jr, G.B. and W.A. Dick. 2008. No-tillage crop production: A revolution in agriculture. *Agron. J.* 100:153–165.

Tvedten, S. 2001. History of pest management—History of the development of organophosphate poisons (Online). http://www.safe2use.com/ca-ipm/01-04-27.htm (verified June 14, 2007).

Unger, P.W. 1978. Straw mulch effects on soil temperatures and sorghum germination and growth. *Agron. J.* 70:858–864.

Unger, P.W. 1994a. Tillage effects on dryland wheat and sorghum production in the southern Great Plains. *Agron. J.* 86:310–314.

Unger, P.W. 1994b. Residue management for winter wheat and grain sorghum production with limited irrigation. *Soil Sci. Soc. Am. J.* 58:537–542.

Unger, P.W. 1995. Role of mulches in dryland agriculture. In *Production and Improvement of Crops for Drylands*, ed. U.S. Gupta, pp. 241–270. New Delhi, Bombay, Calcutta: Oxford & IBH Publ. Co.

Unger, P.W. 1996. Dryland no-tillage winter wheat response to planter type. *J. Prod. Agric.* 9:256–260.

Unger, P.W. and H. Blanco-Canqui. 2012. Conservation tillage. In *Handbook of Soil Sciences: Resource Management and Environmental Impacts,* Second Edition, eds. P.M. Huang, Y. Li, and M.E. Sumner, pp. 25.1–25.31. Boca Raton: CRC Press.

Unger, P.W. and T.A. Howell. 1999. Agricultural water conservation—A global perspective. In *Water Use in Crop Production*, ed. M.B. Kirkham, pp. 1–36. New York: The Haworth Press.

Unger, P.W. and J.J. Parker. 1976. Evaporation reduction from soil with wheat, sorghum, and cotton residues. *Soil Sci. Soc. Am. J.* 40:938–942.

Unger, P.W. and M.F. Vigil. 1998. Cover crop effects on soil water relationships. *J. Soil Water Conservat.* 53:200–207.

Unger, P.W., M.B. Kirkham, and D.C. Nielsen. 2010. Water conservation for agriculture. In *Soil and Water Conservation Advances in the United States*, eds. T.M. Zobeck and W.F. Schillinger, pp. 1–45. SSSA Spec. Publ. 60. Madison, WI: Soil Science Society of America.

USDA-SCS (USDA-Soil Conservation Service; now National Resource Conservation Service). 1999. Intergrated crop management: Use crop residues for soil conservation. http://www.ipm.iastate.edu/ipm/icm/node/1319/print (accessed June 2, 2010).

USDOE (U.S. Department of Energy). 2007. DOE selects six cellulosic ethanol plants for up to $385 million in federal funding. Washington, DC: DOE. http://www.energy.gov/news/4827.htm (accessed January 31, 2009).

Walters, D. and P. Jasa. nd. Conservation tillage in the United States: An overview. Inst. of Agric. and Natural Resources, Univ. of Nebraska, Lincoln. http://agecon.okstate.edu/isct/labranza/walters/conservation.doc (accessed June 9, 2010).

Wilhelm, W.W., J.W. Doran, and J.F. Power. 1986. Corn and soybean yield response to crop residue management under no-tillage production systems. *Agron. J.* 78:184–189.

Wilhelm, W.W., J.M.F. Johnson, K.L. Douglas, and D.T. Lightle. 2007. Corn stover to sustain soil organic carbon further constrains biomass supply. *Agron. J.* 99:1665–1667.

Wortman, C.S. and P.J. Jasa. 2009. Choosing the right tillage system for row crop production. *NebGuide.* Lincoln, NE: Inst. of Agric. and Natural Resources, Univ. of Nebraska. http://www.ianpubs.unl.edu/epublic/pages/publicationD.jsp?publicationId=114 (accessed June 3, 2010).

16 Enhancing Precipitation-Use Efficiency in the World's Dryland Agroecosystems

G.A. Peterson, D.G. Westfall, and N.C. Hansen

CONTENTS

16.1 INTRODUCTION

Improving precipitation-use efficiency (PUE) in the world's dryland agroecosystems is more critical now than ever because the world's dependence on food produced in dryland areas continues to increase. Unger et al. (2006) expressed this global challenge clearly and succinctly:

> During the next several decades, dryland agriculture will play an increasingly important role in our efforts to maintain global food security. This is due to two relatively recent developments. The first is that, until about 1960, most increases in the world's food supply resulted from increasing the amount of land under production. Since then, most of the increasing demand for food in the world has been met by increasing yields. Additional lands still remain that could be brought into production, but as Evans (1998) pointed out, they tend to be unproductive, environmentally sensitive, remote, or otherwise unsuitable for agriculture. Indeed, many have argued that one of the most important reasons for continued yield increase is the need to protect environmentally sensitive land, including wildlife habitat.

Furthermore, in many developed regions, including the United States, existing agricultural lands are gradually being lost due to such processes as erosion, salinization, urbanization (or "suburbanization"), and contamination.

The second, and perhaps more alarming development, is that the world's supply of fresh water for irrigation is limited and increasingly the object of competition. Irrigated agricultural land, which constitutes less than one fifth of the world's arable land, has been the largest source of global yield increase for wheat (*Triticum aestivum* L.), rice (*Oryza sativa* L.), and other staple crops. Although there is now relatively little additional land left that is suitable for irrigation, a greater constraint in many regions is the availability of water for irrigation (Rothfeder 2001). Also, many large aquifers in the world already have been depleted to the point that remaining amounts of water are insufficient or pumping costs too great, for farmers to economically produce low value, bulk commodity crops. The world's most populous countries, China and India, are both depleting aquifers at alarming rates in order to feed their burgeoning populations. How, one must ask, are they to feed those populations when irrigation cannot keep pace with water demand?

To meet this challenge, dryland cropping systems in developed and developing countries alike must use precipitation as efficiently as possible for food production. To realize increased efficiency requires an understanding of how crop production is related to such determining factors as precipitation and evaporative demand, water capture, water retention, and crop management.

What are "dryland agroecosystems?" They are often confused with "rainfed agroecosystems," an all-inclusive term, which is any agroecosystem that does not have irrigation. Dryland agroecosystems are a subset of rainfed agroecosystems, but where lack of precipitation limits crop and/or pasture production in part(s) of the year (Stewart et al. 2006). Dryland agroecosystems are found in climatic regions classified as semiarid and arid in many parts of the world, regions where annual evaporation potential E exceeds annual precipitation (P). These regions are further characterized by sporadic and highly unpredictable precipitation events accompanied by temperature extremes that create the potential for plant water stress during the crop cycle.

For example, in the North American Great Plains, potential E exceeds P during most months of the year, which significantly affects PUE by crop plants. About 75% of the annual precipitation in the North American Great Plains is received from April through September and is accompanied by high temperatures (Figure 16.1) and low relative humidity. Note that open pan evaporation tracks closely with the average air temperature (Figure 16.2), and thus the potential for evapotranspiration (ET) is high (Peterson and Westfall 2004).

It is important to realize that large differences exist among dryland agroecosystems of the world. The extremes can be illustrated by contrasting the North American Great Plains' temperate climate with two types of Mediterranean climate (Morocco and Oregon, U.S.) as shown in Figures 16.3 and 16.4 (Peterson and Westfall 2004). The Morocco and Oregon agroecosystems receive most of their precipitation in the coolest months of the year, which means they have smaller evaporative losses during the time they receive their precipitation relative to the Great Plains' environment.

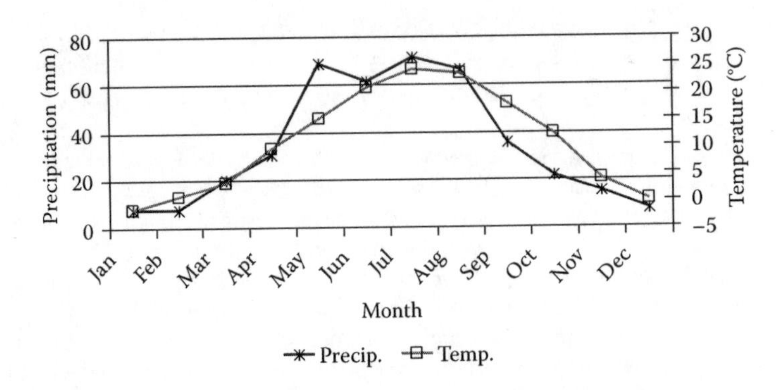

FIGURE 16.1 Long-term (1960–1990) precipitation and temperature distributions in eastern Colorado.

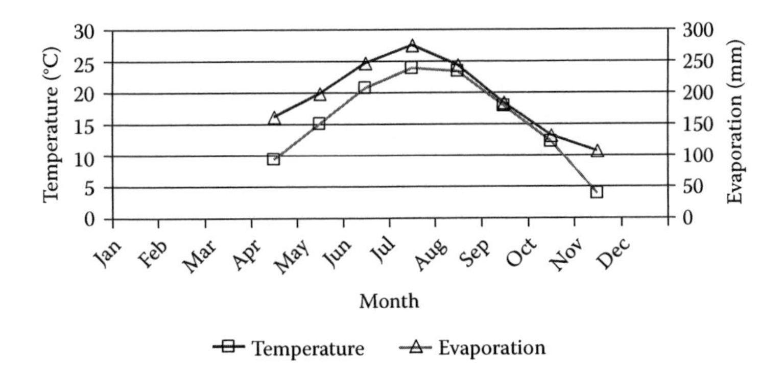

FIGURE 16.2 Long-term (1960–1990) temperature and open pan evaporation distributions for the summer growing season in eastern Colorado.

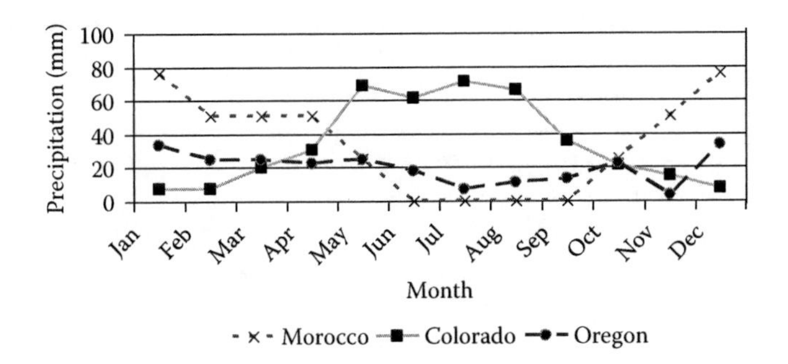

FIGURE 16.3 Long-term precipitation distributions in western Morocco (1934–2002), eastern Colorado (1960–1990), and eastern Oregon (1940–2002).

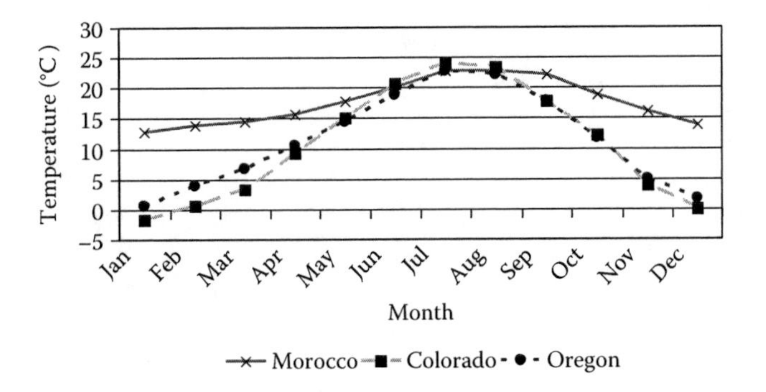

FIGURE 16.4 Typical temperature distributions in western Morocco (1934–2002), eastern Colorado (1960–1990), and eastern Oregon (1940–2002).

Also consider that air temperatures during the winter precipitation period in the Moroccan situation are 8°C–10°C warmer than in the Oregon situation. Water capture and retention issues differ greatly, even though both locations have Mediterranean climates. The contrast is heightened if one considers that snow management becomes an important issue in the Oregon environment, but not in Morocco. Precipitation distribution in relation to evaporation potential and other growing-season conditions dictate crop choices, crop sequences, and effectiveness of water conservation practices in any given climate.

Water availability is essential for plant establishment and successful production in all agroecosystems, especially in dryland systems. Dryland agroecosystems often have precipitation events that are short in length, high in intensity, and occur sporadically. Sporadic and intense precipitation events result in lower and less stable plant yields with more risk to the producers. Since precipitation in dryland agroecosystems is less reliable relative to rainfed agroecosystems, the risk of crop failure is higher. Successful crop production in dryland agroecosystems depends heavily on storing adequate soil water to sustain the crop until the next precipitation event. Thus, the focal point of the soil management practices in dryland agroecosystems is water conservation.

16.2 TILLAGE AND RESIDUE MANAGEMENT IN DRYLAND AGROECOSYSTEMS

Efficient precipitation management involves (1) maximization of precipitation capture in the soil, (2) minimization of stored soil water evaporation, and (3) maximization of plant water-use efficiency (WUE). Maximizing WUE involves proper variety selection, soil fertility, planting date, and a host of other management factors. Maximizing WUE of the plant is important to the ultimate system productivity, but it is of little consequence if one fails to capture the precipitation and retain it in the soil. Soil tillage and crop selection are the two primary tools available to manage precipitation storage in the soil. Reduced tillage or the complete avoidance thereof is the most effective approach to minimize evaporative soil water loss. Tillage choices, including type and timing, affect the amount of crop residue cover maintained on a soil surface and the soil pore size exposed to the atmosphere, which in turn control water capture and water retention.

The notion that maximizing water capture and minimizing evaporation is critical in dryland agroecosystems is not new, but only recently has the technology been developed to significantly change how the precipitation can be managed in dryland agroecosystems. Shaw (1911) and Widtsoe (1920), early in the twentieth century, recognized the principles of precipitation capture and retention, but the commonly accepted means of managing water in their era was tillage. Shaw (1911) stated: "The dominant idea in dry farming is in a sense two-fold. It seeks to secure to the greatest extent practicable the conservation and also the accumulation of moisture in the soil. To accomplish this end, the soil is stirred deeply, whether by the aid of the plow alone or by following the plow with the subsoiler, or by using some other implement, as the deep tilling machine. The ground is compressed subsequent to plowing, and a dust mulch is maintained upon the surface. The increase of organic matter in the soil is also sought." Widtsoe (1920) believed that water retention via dust mulching was the most important issue in dry farming, followed by its efficient use, but he could not effectively address water capture with the technology available to him. Shaw (1911) believed that water capture and retention were increased by more and deeper plowing, thinking it would increase storage capacity of the soil. Neither of these early researchers had any concept of crop residue retention and soil protection; tillage, as they recommended, often led to extensive soil erosion because of bare soil surfaces. Modern research has clearly demonstrated that intensive tillage has many adverse effects on the soil, including losses in soil organic matter (SOM). Careful management of crop residue via reduced and no-till (NT) technology allows us to more effectively and simultaneously address water capture and retention, as well as soil conservation (Unger et al. 2006). Soil management with reduced and NT systems has resulted in more sustainable dryland agroecosystem production.

Recognizing the relationships between climate and tillage is critical for efficient precipitation management. This chapter discusses the principles of managing dryland production scenarios from a systems perspective. Before specific climate and tillage interactions are explored, however, it is important to review the principles that govern precipitation capture and retention in the soil reservoir. Once these are understood, the reader should be able to address management issues in a wide range of dryland environments.

16.3 PRECIPITATION MANAGEMENT PRINCIPLES

16.3.1 Precipitation Capture in the Soil

The first step in water conservation in dryland agroecosystems is capturing the incident precipitation, whether it is from rain or snow. System sustainability depends on water capture maximization within the economic constraints of the system (Unger et al. 2006).

At first glance, water infiltration into soil appears to be a relatively simple process with water entering and simply displacing the soil air. It is more complex, however, because it involves both saturated and unsaturated water flows. Unsaturated flow is driven primarily by the attraction of water to dry solid surfaces (adsorption) and the surface tension of water held between the solids (capillarity). Together, adsorption and capillarity produce the matric potential energy state of water. So when the precipitation intensity is below the saturated flow rate of the soil surface layer, the water intake is governed by the unsaturated flow. As soon as the water application rate exceeds the soil's unsaturated intake rate, saturated flow is the dominant process (Unger et al. 2006).

Water entry rate under saturated conditions, the so-called infiltration rate, is controlled by surface soil porosity, soil water content, and soil profile permeability. Water capture is complex because the maximum infiltration rate occurs at the beginning of a rainfall event and decreases rapidly as water fills the surface pore space. The dry soil has a large storage capacity and a large potential energy gradient at the wetting front relative to the same soil in a uniformly moist condition. As the wetting front advances, the gradient and infiltration rate decline.

Water infiltration rates for a given landscape also can differ for reasons that are not subject to management, like soil texture. Surface soil macroporosity, a desirable soil property for rapid water infiltration, is highly governed by soil texture. Fine-textured soils generally have less macropore space and, consequently, lower infiltration rates than coarse-textured soils. Surface soil aggregation also regulates macropore space; soils of the same texture, but with different degrees of aggregation, can differ greatly in the amount of macropore space. Fortunately, the degree of soil aggregation can be altered by soil management practices. For example, increasing the SOM content and decreasing the tillage intensity can improve soil aggregation and in turn improve water infiltration rate. Management practices that improve soil structural stability, for example, those that increase SOM content, can help improve the infiltration rate. By contrast, soils with weak structure can quickly lose their ability to absorb water as the surface aggregates disintegrate and surface pore spaces become smaller. This can occur upon wetting and from raindrop impact.

Management of soil cover through tillage is a primary factor in improving water capture in dryland agroecosystems because of the soil aggregate protection it provides. The advent of herbicidal weed control created new scenarios for soil cover management, and there is the possibility of retaining much of the previous crop's residue on the soil surface.

Soil cover is defined as the sum of canopy cover and crop residue cover. Soil cover is highly dynamic and can range from 0% to 100%, all within a growing cycle of a crop, depending on the cropping and tillage system being used (Unger et al. 2006). For example, at planting, the soil cover consists only of the previous crop residue component. As the new crop grows, the cover becomes increasingly dominated by the plant canopy. Meanwhile the residue component is in decline as microbial decomposition and physical deterioration occur. When the crop matures and the canopy dies, residues once again become the primary soil cover. Crop cultivation during the growth period also influences the total cover at a given point in the cycle.

Residue cover or crop canopy cover over a soil protect soil aggregates from raindrop impact energy. Surface soil structure that has no cover is easily damaged by raindrops because the raindrop energy causes soil particles to "slake" from aggregates, thus leading to their destruction; thus, the surface soil macroporosity is degraded. Soil water content affects how easily the aggregates slake; wetter soil slakes more readily. The ultimate result of soil aggregate slaking is a crusted soil surface after the rainfall event, which has little macroporosity and hence a very low water infiltration rate.

When crusting occurs, tillage is needed to break the crust, creating macroporosity and thus allowing water from the next precipitation event to infiltrate. However, tillage increases the evaporation rate by exposing moist soil to the dry atmosphere, thereby causing a net water loss. Thus, there is a negative feedback loop where tillage creates a soil surface prone to crusting, which requires additional tillage for water capture. The frequent tillage gradually degrades SOM and soil structure, making the soil surface even more vulnerable to crusting. It is best to avoid crust formation, thereby eliminating the need for tillage. Crusting can be minimized by protecting the soil surface with residues and/or crop canopy cover.

A plant canopy can intercept up to 45% of the raindrops (Troeh et al. 1991), which means that the plant leaves are absorbing the drop energy and the water drips to the soil surface with greatly reduced impact, resulting in less soil aggregate damage; the surface soil pores remain open. As the leaf area index increases during the growing season, the ground cover increases, which further decreases the raindrop energy impact on the soil surface. Unger et al. (2006) stated that "Benefits in water capture resulting from canopy development are greatest in areas with summer precipitation; for example, corn (*Zea mays* L.) or grain sorghum (*Sorghum bicolor* (L.) Moench) production cycles in the Great Plains of North America occur during a period when 75% of the annual precipitation is received. By contrast, dryland areas that receive primarily low-intensity winter precipitation, like the Pacific Northwest in the United States, do not have active canopy development during the period when most of the precipitation is received. Early establishment of fall-sown crops to obtain partial soil cover also is recognized as an important deterrent to soil detachment and runoff during winter months."

Crop residues can protect soil aggregates from raindrop impact and destruction just as crop canopies do. Crop residues play a highly beneficial role in dryland agriculture by protecting the soil against erosion by wind and water and decreasing the soil water evaporation rate. They also have competing economic roles such as providing animal fodder and/or as a fuel source. Obviously, the latter roles are not independent of the others.

When available in sufficient amounts, residues can physically block water runoff and slow the evaporation rate after a rain event, thus allowing water to move into the profile before being lost by evaporation. The water and energy balances of residue-covered soils constitute complex processes and the amount of water conserved by the residues varies with specific circumstances (Papendick and Campbell 1988).

Tillage practices greatly alter the amount of residue cover on a soil surface. Tillage is practiced for reasons ranging from weed control to seedbed preparation and often these operations influence soil water capture by their disruption of soil crusts and alterations in surface aggregate size distribution. Tillage operations are never independent of crop residue cover because even the most minimal soil disturbance tends to incorporate some residues, which decreases the cover on the soil surface (Unger et al. 2006). Most often tillage creates large open macropores at the soil surface that greatly increase the initial water infiltration rate. If a soil has little cover, the aggregate sizes decrease during rainfall events and the infiltration rate decreases in proportion to the rainfall intensity and duration due to macropore destruction. Tillage management choices can alter water capture, but they are not independent of the tillage effects on the residue cover. Ideally, a large amount of soil macroporosity, with at least 50% soil cover, should result in a high infiltration rate that is sustainable throughout a rainfall event. Jones et al. (1994) illustrated this principle; they reported higher infiltration rates and lesser amounts of runoff with stubble-mulch tillage than with NT management. Reduced tillage minimizes crust formation, preventing the need for additional tillage. The increased organic matter at the soil surface improves aggregation and further improves infiltration. This positive feedback loop is ultimately more sustainable that a system dependent upon frequent tillage.

Soil structure deteriorates with increasing tillage intensity and/or years of cultivation. Tillage has negative effects on soil aggregates for two main reasons: (1) physical grinding that reduces aggregate size and (2) increased SOM oxidation that occurs because of macroaggregate destruction and subsequent increased exposure of organic compounds to soil organisms. As aggregate size distributions shift to smaller diameters, microporosity increases at the expense of macroporosity, which eventually results in decreased water infiltration rates. Implementation of reduced and NT systems has the potential to reverse the effects of excessive tillage over the long term.

16.3.2 Retention of Stored Soil Water

After precipitation has been captured in the soil, it must be retained for subsequent use by a crop. Successful retention involves reducing the losses due to evaporation and transpiration by weeds. Some intercepted quantities of precipitation evaporate from the canopy and residue cover before they can be stored in soil; but even after water is stored in the soil, it is very susceptible to evaporative loss. Evaporation occurs both before a crop is planted and during a crop's growing season. Losses before planting are especially critical because they reduce the amount of water available for the ensuing crop and may also affect crop establishment.

Soil water evaporation rate occurs in three stages (Hillel 1998; Lemon 1956). First-stage evaporation depends on the net effect of environmental conditions (wind speed, temperature, relative humidity, and radiant energy). The evaporation rate decreases rapidly during the second stage when the water content in the soil decreases; at this point, the evaporation rate depends mainly on the soil conditions that control water flow to the surface. During the third stage, when water is moving to the surface as vapor, the evaporation rate is low and controlled mainly by the adsorptive forces at the solid–liquid interface (Hillel 1998; Lemon 1956). The greatest potentials for decreasing evaporation lie in the first two stages of evaporative water loss from a soil system.

The soil water evaporation process is a highly complex process because it involves water flow in the soil in response to water potential differences, soil temperature gradients, and atmospheric conditions. Evaporation is greatest when a soil is wet and the air is dry (low humidity or vapor pressure). The soil water potential changes constantly in response to the decreases in water content due to evaporation, use by plants, or deep percolation and increases due to precipitation. As a soil dries at the surface, water must flow to the surface to replenish the loss by evaporation. With continued evaporation, the flow distance increases, which results in increasingly slower flow rates to the surface as liquid or vapor, resulting in lower rates of evaporation. Eventually, water flow is only in the vapor phase, which results in even lower evaporation rates. These constantly changing water potential conditions result in constant changes in the water flow rate to the surface. The water potential of air also changes constantly due to climatic changes. The evaporation cycle restarts each time water is added to the soil, by precipitation (Unger et al. 2006).

Many practices have been evaluated regarding their effect on soil water evaporation. Effective practices to reduce evaporation form a barrier to prevent vapor movement from the soil, negate the energy available for evaporation, minimize the vapor pressure gradients at the soil–atmosphere interface, or disrupt water flow within the soil. Mulch cover on the soil surface is the most effective and practical method of reducing evaporative losses from the soil. A mulch is "any material such as straw, leaves, plastic film, loose soil, etc., that is spread or formed upon the surface of the soil to protect the soil and/or plant roots from the effects of raindrops, soil crusting, freezing, evaporation, etc." (SSSA 1997). This chapter limits discussion to the effects of straw (crop residues) and loose soil on soil water evaporation.

Crop residues are plant materials (straw, stover, stalks, leaves, cobs, etc.) remaining after harvest of a crop for its grains, lint, etc. In many areas of the world, these materials have little economic value and remain in the fields after harvesting the crops (Unger et al. 2006). However, straw, stover, etc., are used in some cultures as feed for animals, fuel, manufacturing, or shelter (Parra and Escobar 1985; Powell and Unger 1998), which produces a competitive environment for the materials. There are numerous competitive uses for residues, but the discussion will be confined to their value for controlling evaporation.

Crop residue characteristics that affect evaporation are their orientation (standing, flat, or matted), which affects the thickness and porosity of the layer; layer uniformity; reflectivity, which affects the radiant energy balance at the surface; and the aerodynamic roughness resulting from the residues (Van Doren and Allmaras 1978).

Smika (1983) measured soil water losses that occurred during a 35-day period without precipitation. Soil water loss was 23 mm from bare soil, 20 mm with flat wheat residues, 19 mm with 75% flat and 25% standing residues, and 15 mm with 50% flat and 50% standing residues on the surface. Nielsen et al. (1997) showed that the potential evaporation decreased as the residue height increased. Height was especially important when stem populations were <215 stems per square meter, and the effect decreased with increasing stem populations. Smika (1976) measured soil water contents 1 day after a 13.5 mm rain and again at 34 days without additional rain. Initial soil water contents were similar to a depth of 15 cm where conventional-, minimum-, or no-tillage treatments were imposed after harvesting winter wheat. The treatments resulted in surface residue amounts of 1.2, 2.2, and 2.7 Mg/ha, respectively. At 34 days, the soil water content was <0.1 m^3/m^3 to depths of 12 cm with conventional tillage, but only to a depth of 9 cm with minimum tillage. By contrast, under NT, the soil had only dried to a 5 cm depth. Total water remaining in the soil was greatest under NT management, where the surface residue amount was also the greatest.

The data reported in the foregoing studies clearly show that crop residue mulch retained on the soil surface can reduce soil water evaporation and thereby conserve water for crop use. The evaporation reductions in these cases result primarily from reduced turbulent transfer of water vapor to the atmosphere.

A major limitation to using residues as a mulch in dryland agroecosystems is that sometimes the cropping systems do not produce enough residues and resultant soil cover to have a significant effect on evaporation. In some cases, virtually no residues remain on the surface because they are removed for fodder or fuel. Under such conditions, other means of controlling evaporation have been investigated. These generally involved reducing capillary water flow to the soil surface, which can be achieved by tillage at a shallow depth. This practice is known as dust mulching (also soil mulching) and is a form of clean tillage (Unger et al. 2006). Dust mulching can reduce evaporation, but it is most effective where a distinct rainy season is followed by a distinct dry season or where water moves to the surface from deeper soil layers or a water table (Papendick et al. 1973; Papendick and Miller 1977).

However, dust mulching leaves soils vulnerable to erosive forces. When considering overall soil management of precipitation capture, water retention, and soil erosion control, it is best to keep cover on the soil and not rely on dust mulch.

Another major issue in retaining stored soil water is minimizing water use by weeds. Soil water use by weeds must be avoided or minimized to obtain optimum soil water storage at crop planting time. Weeds present before planting decrease the soil water supply for later use by the crop and those present during the growing season compete directly with the crops for the available water supply. Weed control can be achieved by tillage alone, herbicides alone, or a combination of the two methods. Crop rotations also are a management tool for reducing weed pressure (Wiese 1983). Regardless of the control method used, timely weed control is essential because weeds may use as much as 5 mm of soil water each day (Wicks and Smika 1973). When tillage is used, a balance is needed between water use by developing weeds and that lost due to exposing moist soil to the atmosphere. Because water loss after each tillage operation may amount to 5–8 mm (Good and Smika 1978), the decision about when to till for weed control is not an easy one, and it must be made by the individual manager for a given situation. Herbicidal weed control is the most water-efficient approach because no additional water is lost from the soil by tilling and the crop residue is not disturbed.

Hand weeding is commonly practiced by small-scale farmers in many countries, such as those in sub-Saharan Africa (Twomlow et al. 1997). As with tillage, repeated weeding by hand may be needed. In Zimbabwe, for example, WUE and grain yields were greater when weeding for corn was at 2, 4, and 6 weeks after emergence than with a single weeding at 2 weeks after emergence. The soil was driest and the yields were lowest when the weeds were not controlled.

Cover crops can be grown to protect the soil against erosion, especially in places where residue production is low and adequate water for such crop is available. Although such crops are not considered to be weeds, their effect on soil water conservation is essentially the same as that of weeds. Cover crops may have beneficial effects such as soil erosion reduction and improved water infiltration, but they also use water. In most cases, growing cover crops in dryland agroecosystems decreases the total water available to the subsequent crop, and thus this practice is usually not recommended. There are specific climatic situations where the use of cover crops may be of value and the reader can learn more about those cases in Unger et al. (2006).

16.4 APPLICATION OF WATER CAPTURE AND RETENTION PRINCIPLES IN A SYSTEMS CONTEXT

To gain a perspective of how to best manage precipitation capture and to maximize PUE, the west central Great Plains of North America is herein used as a test case. Dryland agriculture in the west central Great Plains was developed around wheat production in a wheat–fallow agroecosystem. The wheat–summer fallow system was developed and adopted to decrease the risk of crop failure (Peterson et al. 1996). The soil water stored during the summer fallow period increases the probability of a successful wheat crop. Spring wheat is grown in the northern Great Plains and is planted in late spring (March) with a late summer harvest (August). The fallow period varies in length from 18 to 21 months, depending on the exact planting and harvest dates. The actual time the wheat plants are growing in the field is only 3–6 months out of the 24-month cycle for spring wheat and about 10 months for winter wheat. Weed control during the fallow period in both the spring and winter wheat regions is critical because maximum water storage can only occur if the fields are weed-free. Prior to the advent of herbicides, the only feasible weed control was tillage, which usually left the soil surface barren of residue cover. Following the 1930s Dust Bowl era, summer fallow became a way of life for the farmers in the west central Great Plains. Higher crop prices during and after World War II and much improved tractor power systems and implements facilitated tillage for weed control during the fallow period (Greb 1979). Haas et al. (1974) estimated that there are more than 6.1 million hectares (Mha) of summer-fallowed land in the U.S. Great Plains alone, and since the Canadian provinces also use the summer fallow technique, the total area was even larger. In 1979, Greb chronicled the progress in soil water storage efficiency in winter wheat–fallow systems from the early 1900s through to 1977 and then projected progress through 1990 (Table 16.1). Over time, changes in fallow tillage systems have improved water storage, fallow efficiency (% of fallow precipitation stored as soil water), winter wheat yield, and PUE. As tillage type changed, the number of tillage operations per fallow period decreased, and the amount of crop residue remaining on the soil surface increased. PUE doubled from 1916 to 1975, increasing from 1.22 to 2.78 kg of wheat/ha mm of precipitation. This was largely due to improved fallow-period soil water storage efficiencies, which increased from 19% to 33% over the same time period. Greb (1979) predicted that fallow efficiency would increase to 40% by 1990, resulting in a PUE of 3.25 kg/ha mm. The mechanisms that allowed the improvements in fallow efficiency and PUE are a complex array of interacting factors that were reviewed earlier in the chapter. In summary they include (1) maintained water infiltration rates because the residue cover absorbs raindrop impact energy, thus protecting the soil aggregates; (2) decreased first-stage evaporation rates due to cooler soil temperatures under the residue; (3) reduced opportunity for stimulated evaporation due to fewer tillage events; (4) decreased wind speed at the soil surface because of residue protection; (5) improved fertilization practices; (6) increased opportunity for weed control; (7) improved semidwarf wheat cultivars; and (8) increased timely fallow operations because of more tractor horsepower and better equipment (Peterson et al. 1996). No single factor has changed the system, but all have worked in concert to create a net positive outcome.

TABLE 16.1
Progress in Fallow Systems and Winter Wheat Yields, U.S. Central Great Plains Research Station, Akron, Colorado

Years	Changes in Fallow Systems	Number of Tillages	Fallow Water Storage[a] (mm)	Fallow Efficiency[c] (%)	Winter Wheat Yield (kg/ha)	Precipitation Use Efficiency (kg/ha mm)
1916–1930	Maximum tillage; plow harrow (dust mulch)	7–10	102	19	1070	1.22
1931–1945	Conventional tillage; shallow disk, rod weeder	5–7	112	24	1160	1.43
1946–1960	Improved conventional tillage; begin stubble mulch 1957	5–7	137	27	1730	2.06
1961–1975	Stubble mulch; begin minimum tillage with herbicides (1969)	2–4	157	33	2160	2.78
1976–1990	Projected estimate; minimum tillage; begin no-till 1983	0–2	183[b]	40	2690	3.25[c]

Source: Adapted from Greb, B.W. Reducing drought effects on croplands in the west-central Great Plains. USDA Information Bulletin No. 420, 31 pp. U.S. Govt. Printing Office, Washington, DC, 20402, 1979.

[a] Based on 14-month fallow, mid-July to second mid-September.

[b] Assuming 2-year precipitation per crop in a wheat–fallow system.

[c] Fallow efficiency = soil water stored during fallow/precipitation received during fallow period.

16.4.1 Modern Soil Water Storage Potential in Fallow

As is observable in Table 16.1, Greb (1979) predicted increased fallow storage efficiency based on improved residue management and more economical NT methods. Yet, it has not been possible to improve upon the 35% storage efficiency that Greb achieved in the early 1970s. Fallow efficiency reports from the 1980s and 1990s are generally less than 40%, regardless of the climatic zone where the data were collected (Table 16.2). The range of efficiencies reported by McGee et al. (1997) in Colorado under NT conditions, 17%–28%, are disturbingly low. These data represent a wide range of climate and soil combinations, and it appears it will be difficult to improve upon fallow efficiency with the current fallow technology. Fallow storage efficiency was equally low in the northern climates of Canada and North Dakota, despite their lower evaporation potentials. Their much longer fallow period, 21 months for spring wheat–fallow systems compared with 14 months for winter wheat–fallow systems, also contributes to the low efficiencies in the region with lower evaporation.

Greb et al. (1967) and Unger (1978) demonstrated that surface residue greatly increases fallow water storage, but that residue amounts in excess of 6 mg/ha are required to achieve fallow efficiencies greater than 35%–40%. Unfortunately, the residue amounts in the west central Great Plains at wheat harvest, the maximum residue accumulation point in the system cycle, are commonly in the 2.2–5.6 mg/ha range. With favorable precipitation they can reach 7 mg/ha, and on soils that receive runoff water from surrounding hills, the residue levels may even reach 10 mg/ha; but the latter cases are rare. Since the residue levels in the west central Great Plains area usually do not reach the 6 mg/ha threshold, it is highly likely that the projected fallow efficiency of 40% (Greb et al. 1967) will not be achieved.

TABLE 16.2
Modern Soil Water Storage Efficiencies of No-Till and Reduced Till Summer Fallow Systems over a Range of Environments in the Great Plains of the United States

Water Storage Efficiency (%)			
Range	Range	State or Province	Reference
18[a]	—	Saskatchewan, Canada	Campbell et al. (1987)
31[a]	(26–36)	North Dakota	Deibert et al. (1986)
37[a]	32–42	Montana	Tanaka (1989)
49	—	Colorado	Smika (1990)
22	17–28	Colorado	McGee et al. (1997)
25	10–37[b]	Kansas	Schlegel (1990)
30	25–35	Kansas	C.A. Norwood[c]
10	—	Texas	Jones and Johnson (1993)

Source: Peterson, G.A., Schlegel, A.J., Tanaka, D.L., and Jones, O.R., *J. Prod. Agric.*, 9, 180–186, 1996.

[a] Spring wheat (21-month fallow), all other data for winter wheat (14-month fallow).

[b] Reduced tillage.

[c] Personal communication, Southwest Kansas Research Center, Kansas State University, Garden City, Kansas.

16.4.2 Management and Water Storage Interactions

When the residue amounts are too small to allow further reductions in evaporation rates and totals, other approaches must be found for increasing PUE. The challenges are (1) to increase water capture, (2) to decrease the water storage time in the soil, and (3) to choose plant species and rotations that use water more efficiently.

Water capture and storage efficiency are usually greatest when the soil surface is dry and in a receptive condition for rainfall. At wheat harvest time in the Great Plains (July), soils often are at zero plant available water content and can absorb water rapidly. At this point in the crop cycle, there is maximum residue cover on the soil; thus, the soil conditions are very receptive to water infiltration.

Smika and Wicks (1968) reported that substituting herbicidal weed control for tillage greatly improved water storage during the early portion of the fallow period in a winter wheat system. Conventional plow tillage treatments stored no water in the early fallow period, while minimum tillage treatments stored 12% of the precipitation and NT increased the water storage efficiency to 24%. By spring of the following year, which is only 8 months into the 14-month fallow period, the plow tillage had stored only 16% of the precipitation (56 mm of water), while the minimum till and NT systems had stored 40% (140 mm of water) and 60% (210 mm of water), respectively.

When water is stored early in the fallow season, the storage becomes less efficient in the latter part of the fallow period. For spring wheat–fallow systems, Haas and Willis (1962) also found that little or no soil water was stored in summer fallow after July 1. These data point to the possibility of terminating the fallow period, in NT and reduced tillage systems, before July, which will permit the planting of a summer crop to use the water via transpiration rather than lose it by evaporation.

Management techniques that foster early water capture and retention after wheat harvest and during the winter and spring periods usually result in moist surface soils that are near field capacity by May. When a rain event occurs in the summer period, as shown in Figure 16.1, the greatly reduced infiltration rates allow water to remain on the NT soil surface for longer periods compared with tilled conditions. The high temperatures in the late fallow period from July to September, in concert with high vapor pressure deficits in the air, accelerate evaporation and keep precipitation capture to a minimum. On sloping land, runoff is increased by the compact soil surface and water capture is decreased even more than on level land (Jones et al. 1994).

In conventional winter wheat–fallow systems, where weeds are not controlled after harvest and/or tillage is used for weed control, the soil water contents in spring are much lower and the water storage potential during the May–September period is greater than in NT systems. Since these soils are tilled multiple times for weed control during the fallow period, they have more macroporosity at the surface and water infiltration is not impeded. However, any water stored in the tillage layer is rapidly lost by evaporation because the fallow weed control tillage hastens evaporation by exposing moist soil during the hottest period of the year.

Reports by Black and Power (1965), Deibert et al. (1986), and Norwood (1994) all substantiate the inefficiency of water storage in NT wheat–fallow systems during the latter portion of the summer fallow period, whether it is a winter or spring wheat production system. In early NT research, Black and Power (1965) working with a spring wheat–summer fallow system found that the fallow storage efficiency from harvest to the following May was 66%, from May to September was 9%, and from September to seeding the next spring was 19%. Deibert et al. (1986) working with spring wheat–summer fallow systems reported early fallow storage efficiencies of 56%–59% (90–125 mm of water stored), but efficiencies of only 26%–36% after a full 21-month fallow period (112–117 mm of water stored). Norwood (1994) in Kansas reported a storage efficiency of 46% for the 11-month period from winter wheat harvest to spring sorghum planting (175 mm of water stored). Norwood's wheat–summer fallow system, for the entire 14-month fallow period, only had an efficiency of 23% (137 mm of water stored). There was a 38 mm water loss during the late fallow period. A long fallow period appears to decrease the amount of stored water under most dryland production systems.

Most data indicate that there can be as much or more stored water in NT-managed soils in the spring after wheat harvest, as there will be if fallow is continued until fall wheat planting in September. It appears that intensifying the cropping pattern, by shortening the summer fallow period and using the precipitation nearer to the time it is received, would increase the overall system PUE and ultimately increase the soil productivity via the increased annual amounts of residue added to the soil.

16.4.3 Testing the Hypothesis

Peterson et al. (1993) established a long-term agroecosystem project in 1985 to test the hypothesis that it was possible to decrease the length of the fallow period by using NT techniques and that more intensive cropping systems could be successful. Specifically, they were looking for ways to reduce the amount of summer fallow time and to reverse the soil degradation that has occurred because of the tilled wheat–fallow cropping system.

The experiment had three variables: (1) climate regime; (2) soils; and (3) management systems (Peterson et al. 1993). The climate variable was based on three locations with varying levels of potential evapotranspiration; the soil variable was achieved using a soil catena with three distinct soils at each site; their management system variable was increasingly intensive cropping systems ranging from wheat–summer fallow to continuous cropping. The system responses were assessed via total aboveground plant productivity, WUE, changes in the soil chemical, physical, microbiological properties, and economic evaluations. All soil samples, dry matter yields, soil water measurements, etc., were collected from the benchmark areas within each experimental unit, and the details are reported in Peterson et al. (1993). It was possible to decrease the length of the fallow period by using NT techniques, and the more intensive cropping systems could be successfully used (Peterson and Westfall 2004). In fact, cropping systems were successfully intensified across the entire climate gradient included in their experiment; annualized grain yields increased by more than 75% relative to the yield of the wheat–fallow system. The largest step gain in annualized yield was achieved with the addition of maize or sorghum to the system (two crops in a 3-year system). Increasing cropping intensity to the 4-year system (three crops in 4 years) only resulted in small yield increases relative to the 3-year system. Net income for farmers increased from 25% to 40% relative to the traditional wheat–fallow system.

They also tested a continuous cropping system that included annual forage crops. Yields of total aboveground biomass were used to compare the continuous cropping system with the systems containing only grain crops. Continuous cropping produced annualized total biomass yields superior to all systems containing fallow. Avoiding a fallow year maximized total aboveground biomass production. However, the monetary value of the forage crop in their continuous systems was so low that the continuous cropping system was not as profitable as the 3- or 4-year grain crop systems that included a summer fallow period (Peterson and Westfall 2004).

They were able to determine that systems with fewer summer fallow periods increased WUE; the 3-year system had a 27% increase in grain WUE relative to wheat–summer fallow, and the 4-year system had a WUE 37% greater than that of wheat–summer fallow. Crop use of stored soil water that was usually lost by evaporation in a wheat–summer fallow system was responsible for the increased grain and biomass yields in the more intensive rotations, which resulted in the increased WUE.

Peterson et al. (1993) also had hypothesized that reducing the amount of summer fallow time would help reverse the soil degradation, especially the weak soil structure that has resulted from many years of tillage and less amounts of crop residue return in the typical wheat–fallow cropping system. Residue return is particularly low in a wheat–fallow system because a crop is produced only once in every 2 years. Weak, unstable soil structural units reduce the receptivity of the soil surface to water infiltration. Soils in dryland farming areas are usually low in SOM because they were formed in semiarid environments where only small amounts of vegetative carbon were returned to the soil for potential storage. Despite this condition, these soils often had good aggregate strength when first placed under cultivation. In the Great Plains of North America in particular, the native grass vegetation, with its highly developed fibrous surface root system, provided a well-structured soil that allowed pioneer farmers to do a good job of water capture within the limits of the technology of their era. Since all of the cropping systems relied on frequent tillage events, the soil organic C (SOC) depleted rapidly and the aggregate size and strength diminished quickly. Surface crusting began to occur after the rainfall events and the farmers' only alternative was more tillage to create macropores to improve the water capture potential of the next precipitation event. Thus, both water capture and retention capabilities have declined with time. Peterson and Westfall (2004) reported that NT management practices, coupled with cropping intensification, did begin to reverse the negative effects of past management. Properties of the immediate surface soil layer (2.5 cm), such as bulk density, porosity, and macroaggregation, affect the pore space and pore size and thus are useful indicators of changes in water capture potential.

Cropping system intensification under NT management decreased the bulk density of the surface soil layer (Shaver et al. 2002). For example, continuous cropping decreased the soil bulk density from 1.32 g/cm^3 in the wheat–summer fallow system to 1.22 g/cm^3. This reduction in the bulk density resulted in an increase of 0.04 m^3/m^3 in the total porosity, meaning that there is 4% more space to infiltrate water from a rainfall event. Furthermore, there was an absolute increase of 5% in effective pore space, which means that even if the surface soil is at field capacity water content, there is 5% more space to accommodate water from a rainfall event. The improvements in surface soil porosity were related to an increase in the proportion of macroaggregates relative to microaggregates (Shaver et al. 2002).

In the Great Plains environment, rapid water intake during and after a rainfall event is critical because water that ponds on the soil surface, for even a short time, is rapidly evaporated. Improvements in macroaggregation and effective porosity thus increase the opportunity for more efficient water capture.

The causal agent for the improvement in the physical properties was the addition of more crop residue biomass and less soil stirring (tillage) in the more intensive cropping systems relative to the wheat–fallow system (Shaver et al. 2003). Coupled with the lack of soil disturbance in an NT environment, the additional residue carbon promoted aggregation and increased aggregate stability. SOC levels increased after only 12 years of intensively cropped NT management (Sherrod et al. 2003). Each step of increased cropping intensity tended to increase the surface SOC at all soil depths, but the increases were most significant in the surface 0–2.5 and 2.5–5 cm soil layers. Continuous cropping, with no summer fallow period, increased the organic C content of the surface 0–2.5 cm of soil by 39% relative to wheat–fallow. The SOC increases were closely associated with the changes in physical properties reported by Shaver et al. (2003). Furthermore, the increases in soil C were directly linked to increased crop residue biomass returned to the soil over the 12-year life of the experiment (Shaver et al. 2003).

Peterson and Westfall (2004) concluded from their long-term agroecosystem experiment (Peterson et al. 1993) that NT technology has greatly altered our ability to manage precipitation in dryland systems. Specifically, it improved the potential for precipitation capture and for soil water retention. In turn, the improved precipitation capture and the retention permitted increased the cropping intensity, which has proven to be both agronomically and economically sound in the west central Great Plains of North America. Furthermore, these management strategies have provided positive feedbacks to the soil system that should improve long-term productivity. Positive feedbacks included increased SOC levels that improved surface soil structure, which in turn has increased the water infiltration rates of the soils. Increased water capture because of greater infiltration has the potential to increase PUE by providing more water for plant production and additional crop residue return to the soil. Greater season-long cover over the soil, whether it be crop canopy or crop residue, and higher water infiltration rates all have major impacts in decreasing soil erosion by wind and water.

16.4.4 Principles and Inferences

Farahani et al. (1998) provided an insightful dissection of the fallow period in an NT winter wheat–summer fallow system common in the west central Great Plains. It has proven to be very instructive in terms of how we might best maximize the PUE for a particular agroecosystem. Using soil water storage data from their long-term experiment in the Great Plains of North America, they divided the 14.5-month summer fallow period into three stages (Figure 16.5). Stage I is the first 2.5-month period after wheat harvest until mid-September; Stage II is the next 7.5-month over winter period

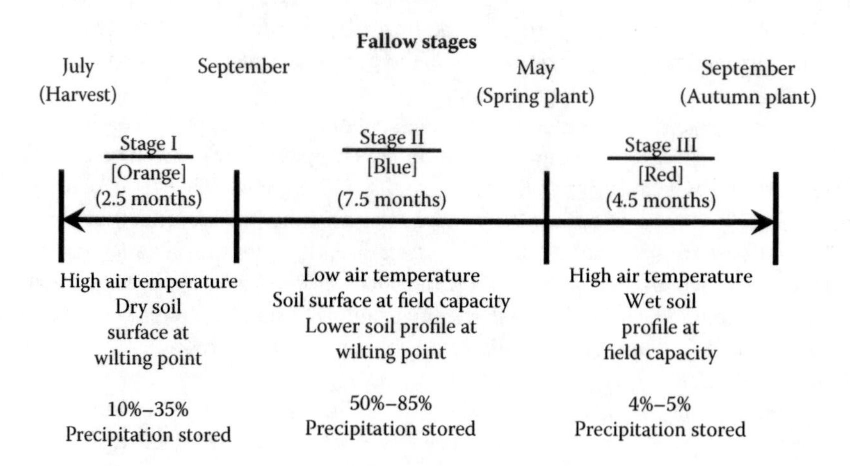

FIGURE 16.5 Fallow stages and their characteristics for dryland cropping systems in eastern Colorado.

from fall to early May; and Stage III is the 4.5-month late fallow period from spring until wheat planting in mid-September. Their partitioning strategy was not arbitrary; each stage coincided with a combination of soil water and air temperature conditions. They coded Stages I, II, and III as "orange," "blue," and "red" zones to depict the water storage potential of each stage. Note that the various crop and noncrop phases of 3- and 4-year systems also fit well with these periods.

Water storage efficiencies are greatest (50%–85%) in the "blue" zone because air temperatures are lowest and precipitation is received as snow and/or low-intensity storms (Figure 16.5). Storage efficiency in the "orange" zone is much lower (10%–35%) because air temperatures are much higher during this stage. However, surface soil conditions are dry (near wilting point), and thus some water infiltrates before it is lost by evaporation. Rainstorm intensity governs whether one is at the low or high end of the range. The "red" zone has high air temperatures (similar to the "orange" zone), but surface soil and profile water contents are near field capacity, evaporation is high, and no water is stored during this time period. These fallow period stages can be used to determine what has changed as cropping systems are intensified and can give us insight into what might be feasible for future improvements.

Their analysis of the cropping systems revealed that switching from a 1 crop in 2-year system, such as wheat–summer fallow, to a 2 crops in 3-year or 3 crops in 4-year system did not appreciably change the proportion of time in total fallow (Table 16.3). In fact, the proportion of time in fallow (no crop in the field) actually increased as the cropping was intensified, while the proportion of time in crop decreased. Note, however, that the proportion of time in Stage I fallow decreased from 10% of the total system time in a 2-year system to 5% in a 4-year system with summer crops. Furthermore, the proportion of time in Stage III fallow, the worst water storage period, decreased from 19% to 11% for the same comparison.

Since precipitation storage efficiency in Stage I was only 10%–35% and was essentially 0% in Stage III, having a smaller proportion of the fallow time in these stages and more in Stage II, which is a highly efficient storage time, benefits the overall system. These shifts in fallow timing partially explain the advantage of the more intensive systems. An analysis of when precipitation is received relative to the fallow stages and the cropping season completes the explanation.

Farahani et al. (1998) demonstrated that during fallow Stages I and II one can expect to save 10%–35% and 50%–85% of the precipitation, respectively, which is in stark contrast to the 0% storage efficiency expected in Stage III. Data in Table 16.4 show that cropping intensification did not appreciably alter the proportion of the total agroecosystem precipitation received during fallow Stages I and II, which are the best water storage periods. However, intensification decreased the proportion of the precipitation occurring during fallow Stage III from 34% in a 2-year wheat–summer fallow system to 17% for a 4-year system such as wheat–maize–millet–fallow. Since no water can usually be stored in Stage III, this is a gain for the system. Furthermore, it dramatically increased the proportion of the precipitation received during the time when a summer crop could be grown.

TABLE 16.3
System Analysis: Time in Crop and Time in Various Fallow Stages for Three Cropping Systems in the Great Plains

System	Total Time in Crop	Total Time in Fallow	Time in Stage I	Time in Stage II	Time in Stage III
			Total Rotation Time (%)		
2-year (WF)	40	60	10	31	19
3-year (WMF)	39	61	7	42	12
4-year (WMPF)	38	62	5	46	11

Note: WF = wheat–fallow; WMF = wheat–maize–fallow; WMPF = wheat–maize–proso millet–fallow.

TABLE 16.4
System Analysis: Proportion of Precipitation Received in the Various Fallow Stages and during the Cropping Season for Three Cropping Systems

System	Precipitation in Stages I and II	Precipitation in Stage III	Precipitation in Cropping Season
	Total Precipitation for a Given System (%)		
2-year (WF)	34	34	32
3-year (WMF)	35	23	42
4-year (WMPF)	36	17	47

Note: WF = wheat–fallow; WMF = wheat–maize–fallow; WMPF = wheat–maize–proso millet–fallow.

In the Great Plains, this period of the year is especially favorable for summer crops such as maize, sorghum, sunflower, and millet.

Switching from a 2-year (wheat–summer fallow) to a 4-year system that includes summer crops such as maize or sorghum increased the proportion of the precipitation that fell while crops were present in the field from 32% in a 2-year system to 47% for a 4-year system. Receiving precipitation during the cropping period resulted in improved system PUE because the crop canopy and underlying residues left by the NT system absorb the raindrop impact, keep the soil surface cooler, which decreases evaporation, and the overall result is a net increase in soil water storage. Runoff also is diminished and the opportunity time for water infiltration is increased. An additional factor that contributes to improved water conservation during the cropping season is that the plants continuously exhaust the available water from the surface soil layers, which improves the infiltration rate because a dry surface soil is more receptive to water than a wet surface soil.

16.4.5 Application of Analysis to Other Climate Situations

Peterson and Westfall (2004) demonstrated that these principles of fallow stages can be applied to other agroecosystems. In the Great Plains situation analyzed by Farahani et al. (1998), most of the precipitation falls during the warmest time of the year, which is within the growing season for several well-adapted plant species. Analysis of a wheat–fallow agroecosystem in a Mediterranean climate in the eastern part of the U.S. state of Oregon, with an annual precipitation of approximately 400 mm like the Great Plains situation, revealed that the proportion of time in each fallow stage was the same as for wheat–summer fallow in the Great Plains because the fall planting and summer harvest dates of the winter wheat crop were essentially identical in both environments (Table 16.5). In a warmer Mediterranean climate near Settat, Morocco (annual precipitation approximately 400 mm), however, there were substantial differences in the proportion of time in each fallow stage and the time in crop. Wheat was planted later and harvested earlier in the Moroccan situation, which decreased the proportion of time in crop to 29% in contrast to 40% in the two cooler environments and increased the proportion of time in Stages I and III fallow. In the Great Plains environment, such a shift would have a negative effect on soil water storage because of the low precipitation storage efficiency possible in Stages I and III, but this was not the case in the warmer Moroccan environment as will be seen later.

Much larger contrasts appeared when the three environments were compared on the basis of precipitation distribution. In the Great Plains, only 32% of the precipitation fell while a wheat crop was in the field, which contrasted sharply with 46% for the Oregon environment and 50% for the Moroccan environment (Table 16.5). One of the major water savings in the Great Plains

TABLE 16.5
System Analysis: Proportion of Time in Fallow Stages and Proportion of Precipitation Received in a Wheat–Fallow System in Colorado and Oregon, U.S., and Morocco

Fallow and Precipitation Distributions	Climate	Stage I	Stage II	Stage III	Crop
			Rotation Time (%)		
Proportion of system[a] time in crop or fallow	Colorado	10	31	19	40
	Oregon	10	31	19	40
	Morocco	15	31	25	29
			Precipitation (%)		
Proportion of system[a] precipitation in each stage	Colorado	21	13	34	32
	Oregon	5	37	12	46
	Morocco	0	47	3	50

[a] System = 24 months for complete crop–fallow cycle.

environment was avoiding Stage III fallow because water storage efficiency was 0% in that stage. However, in Oregon, only 12% of the precipitation fell during Stage III, and in Morocco, only 3%, making avoidance of Stage III fallow much less critical in the two Mediterranean environments (Table 16.5). Furthermore, in Oregon and Morocco, 37% and 47% of the precipitation was received in Stage II fallow, respectively, and because of the cool temperatures, this was an excellent period in which to store water. In the Great Plains environment, which also has the potential for maximum storage in Stage II, there is much less precipitation to be saved. Note, however, that in the Oregon situation, extra attention to snow and snow melt retention would be required. In Morocco, there are no frozen soils and snow melt runoff problems, resulting in an even better period for soil water storage. Based on this analysis, it is plain that cropping intensification with warm-season plants such as maize or sorghum would have little chance for success in either the Oregon or Moroccan climate. Wheat and other cool season species obviously fit the Mediterranean environment very well relative to the Great Plains.

Peterson and Westfall (2004) suggested that scientists could use this approach to analyze precipitation and temperature distributions in relation to various crop species and thus identify the best intervention points for improved management. If fallow is used in a cropping system, one should make every attempt to avoid the periods when precipitation storage is likely to be grossly inefficient. The greatest PUE will be achieved if noncrop periods can be decreased in length and if crops are grown during the times when the precipitation is being received. The following are example applications of the analysis to climates with widely varying combinations of temperature and precipitation distributions.

16.4.5.1 Lanzhou, China (Lat. = 36°3′N; Long. = 103°47′E) (Annual precipitation = 328 mm)

The proportion of time in Stages I, II, and III and in crop production for this location, 13%, 33%, 16%, and 38%, respectively, were very similar to the Great Plains situation (Figure 16.6 and Table 16.6). The highest rainfall months coincide with the warmest temperatures and thus present an ideal growing season for warm-season plants. Although winter wheat is a common crop in the Lanzhou area, the precipitation and temperature distributions favor the production of summer crops such as

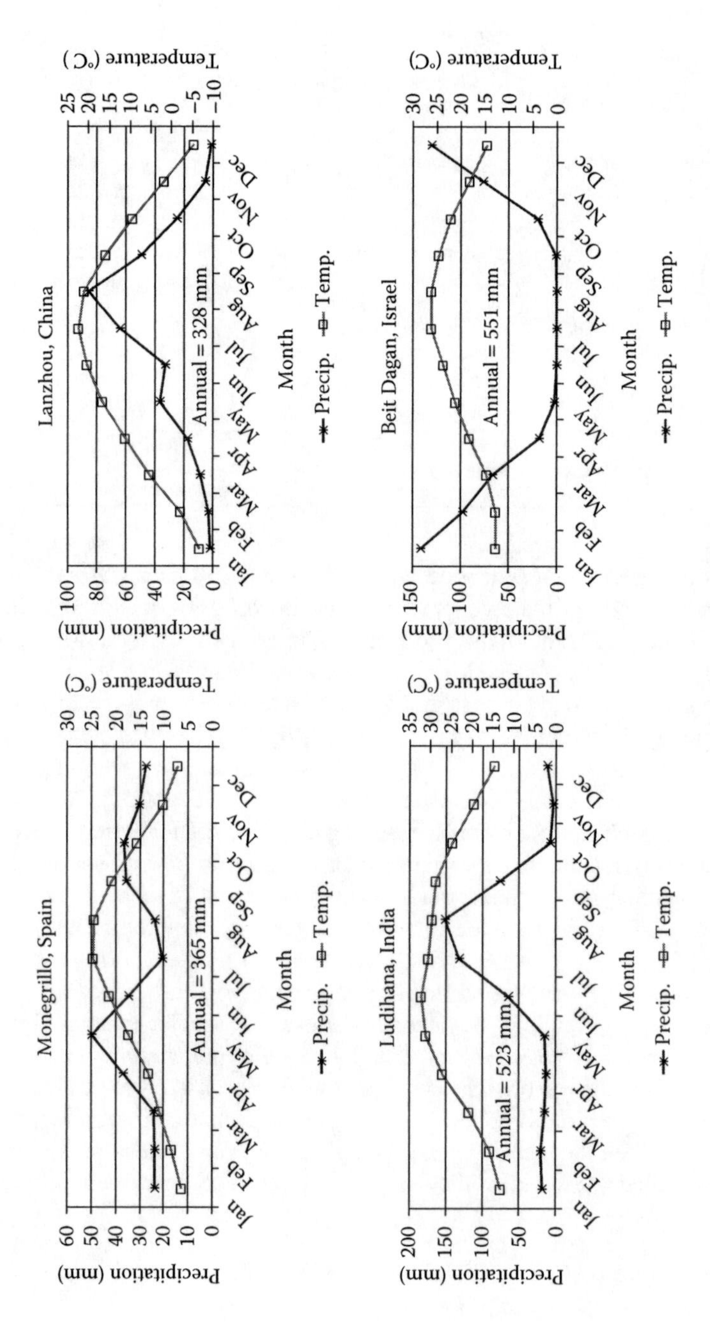

FIGURE 16.6 Long-term precipitation and temperature distributions for Lanzhou, China; Monegrillo, Spain; Ludhiana, India; and Beit Dagan, Israel.

TABLE 16.6
System Analysis: Proportion of Time in Fallow Stages and Proportion of Precipitation Received in a Wheat–Fallow System for the United States, Morocco, China, Spain, India, and Israel

Fallow and Precipitation Distributions	Climate	Stage I	Stage II	Stage III	Crop
			Rotation Time (%)		
Proportion of system[a] time in crop or fallow	Eastern Colorado	10	31	19	40
	Eastern Oregon	10	31	19	40
	Settat, Morocco	15	31	25	29
	Lanzhou, China	13	33	16	38
	Monegrillo, Spain	16	25	25	34
	Ludhiana, India	29	21	25	25
	Beit Dagan, Israel	16	25	25	34
			Precipitation (%)		
Proportion of system[a] precipitation in each stage	Eastern Colorado	21	13	34	32
	Eastern Oregon	5	37	12	46
	Settat, Morocco	0	47	3	50
	Lanzhou, China	30	15	35	20
	Monegrillo, Spain	16	23	27	34
	Ludhiana, India	44	7	42	7
	Beit Dagan, Israel	2	48	2	48

[a] System = 24 months for complete crop–fallow cycle.

maize, sorghum, sunflower, and millet, particularly if NT management is used. Interestingly, the August and September precipitation at Lanzhou tends to be greater than that of the Great Plains, which may permit the successful production of soybeans. Dryland soybean production has been unsuccessful in the west central Great Plains environment because of lack of adequate August precipitation.

Effective snow management will be necessary in this environment to maximize PUE since the winter precipitation is in the form of snow. Maintaining a standing residue cover during the winter months will provide good opportunity to capture snow and retain runoff, which can ideally be accomplished in an NT management system.

Based on this analysis, cropping intensification with warm season plants such as maize or sorghum should be good production possibilities in Lanzhou. Wheat and other cool-season species could be grown effectively in rotation with warm-season plants, and it is highly likely that no summer fallow is necessary at all.

16.4.5.2 Monegrillo, Spain (Lat. = 41°38′N; Long. = 00°25′W) (Annual precipitation = 365 mm)

The proportion of time in Stages I (16%) and III (25%) for this location are comparable to the Moroccan situation (Figure 16.6). The locations differ in that at Monegrillo, there is less time in Stage II (favorable water storage period) than in Morocco, but Monegrillo has more time in crop production. These conditions would seem to favor the production of cool-season crops such as wheat and barley at the Monegrillo site (Table 16.6). More significant, however, is that the precipitation distribution at Monegrillo differs from that of Morocco. During the fallow stages and the cropping period, the precipitation distribution is more similar to the Great Plains than to Morocco (Table 16.6). Monegrillo has less rainfall to store during the favorable storage in Stage II, more rainfall during the unfavorable storage in Stage III, and less rainfall during the crop period, compared to Morocco.

Based on this analysis, the Monegrillo environment is less favorable for cool-season crops such as barley and wheat than the Moroccan situation. Because summer temperatures are not as extreme and because there is some summer rainfall in Monegrillo relative to the Moroccan situation, summer crops such as sorghum and sunflower should be good possibilities for the Monegrillo site.

16.4.5.3 Ludhiana, India (Lat. = 30°54′N; Long. = 75°51′E) (Annual precipitation = 523 mm)

The distribution of fallow stages and crop production time at Ludhiana is more uniform relative to all other locations that have been compared to this point; Stage I = 29%, Stage II = 21%, Stage III = 25%, and the crop period = 25% (Figure 16.6 and Table 16.6). However, the precipitation distributions relative to the fallow stages and to the crop production period are totally nonuniform; Stage I = 44%, Stage II = 7%, Stage III = 42%, and the crop period = 7%. Since the summer temperatures in Ludhiana also are extremely high, the potential for soil water storage during either Stage I or III would be nearly zero. This situation, coupled with the small amount of the annual precipitation (7%) received during the favorable water storage Stage II, indicates that summer fallowing would be a very ineffective practice, even with the best NT practices. In this situation, production of adapted annual crops during the coolest times of the year would be the only feasible solution under nonirrigated conditions.

16.4.5.4 Beit Dagan, Israel (Lat. = 32°0′N; Long. = 34°49′E) (Annual precipitation = 551 mm)

The proportion of the crop rotation in the fallow stages and cropping period for Beit Dagan are identical to Monegrillo, Spain (Figure 16.6 and Table 16.6), which suggests that summer fallowing would be a marginal practice because over 40% of the time is either in Stage I or III fallow, which are the least favorable stages for soil water storage. However, the precipitation distribution pattern at Beit Dagan differs widely from that of Monegrillo, Spain; 48% of the precipitation at Beit Dagan occurs during the favorable soil water storage stage; and at Monegrillo, only 23% occurs during Stage II. Furthermore, at Beit Dagan, 48% of the precipitation occurs during the crop period, and at Monegrillo, only 34% occurs during this same period. Therefore, summer fallowing would be a better option at Beit Dagan than at Monegrillo. Both locations have high summer temperatures and thus favor the production of crops such as barley and wheat during the cooler periods of the year.

16.5 SUMMARY

Improving the PUE of dryland cropping systems in the semiarid regions of the world obviously is critical to adequate food production for many nations. We have found that adoption of NT management has permitted cropping system intensification in our environment, which has increased the system yield and profit. Furthermore, these new systems are very beneficial to the environment, since they decrease soil erosion by both wind and water. Understanding the constraints and opportunities that exist in a particular climatic situation is necessary to maximize the PUE. Our analysis uses precipitation, amount and distribution, and seasonal temperatures to estimate what type of cropping systems may be feasible.

In this chapter, we applied the analysis to four widely varying climate regimes. Each case was discussed, assuming the use of an NT management system, so as to maximize precipitation capture and retention in the soil. Each analysis resulted in a recommendation regarding the value of summer fallow and a recommendation regarding the production of either cool or warm season crops. To provide more specific recommendations, it would be necessary to know specific information about soils, crop markets, and infrastructure. It is our hope that the reader will perform similar analyses of their climate and production system and identify the potential ways for improvement.

REFERENCES

Black, A. and J.F. Power. 1965. Effect of chemical and mechanical fallow methods on moisture storage, wheat yields, and soil erodibility. *Soil Sci. Soc. Am. Proc.* 29: 465–468.

Campbell, C.A., R.P. Zentner, and H. Steppuhn. 1987. Effect of crop rotations and fertilizers on moisture conserved and moisture use by spring wheat in southwestern Saskatchewan. *Can. J. Soil Sci.* 67: 457–472.

Deibert, E.J., E. French, and B. Hoag. 1986. Water storage and use by spring wheat under conventional tillage and no-till in continuous and alternate crop-fallow systems in the northern Great Plains. *J. Soil Water Conserv.* 41: 53–58.

Evans, L.T. 1998. *Feeding the 10 Billion—Plants and Population Growth.* Cambridge University Press, Cambridge, UK.

Farahani, H.J., G.A. Peterson, and D.G. Westfall. 1998. Dryland cropping intensification: A fundamental solution to efficient use of precipitation. *Adv. Agron.* 64: 197–223.

Good, L.G., and D.E. Smika. 1978. Chemical fallow for soil and water conservation in the Great Plains. *J. Soil Water Conserv.* 33: 89–90.

Greb, B. W. 1979. Reducing drought effects on croplands in the west-central Great Plains. USDA Information Bulletin No. 420, 31 pp. U.S. Govt. Printing Office, Washington, DC, 20402.

Greb, B.W., D.E. Smika, and A.L. Black. 1967. Effects of straw-mulch rates on soil water storage during summer fallow in the Great Plains. *Soil Sci. Soc. Am. Proc.* 31: 556–559.

Haas, H.J. and W.O. Willis. 1962. Moisture storage and use by dryland spring wheat cropping systems. *Soil Sci. Soc. Am. Proc.* 26: 506–509.

Haas, H.J., W.O. Willis, and J.J. Bond. 1974. Introduction. In *Summer Fallow in the Western United States.* USDA Conservation Research Report No. 17, 1–12. U.S. Govt. Printing Office, Washington, DC, 20402.

Hillel, D. 1998. *Environmental Soil Physics.* Academic Press, San Diego, CA.

Jones, O.R. and G.L. Johnson. 1993. Cropping and tillage systems for dryland grain production. USDA-ARS Rep. no. 93-10, Conserv. and Prod. Res. Lab, Bushland, TX.

Jones, O.R., S.J. Smith, and L.M. Southwick. 1994. Tillage system effects on water conservation and runoff water quality—Southern High Plains dryland. In Proceedings of *Great Plains Residue Management Conference*, pp. 67–76, Amarillo, TX, August, 15–17, 1994. Great Plains Agricultural Council Bulletin No. 54.

Lemon, E.R. 1956. The potentialities for decreasing soil moisture evaporation loss. *Soil Sci. Soc. Am. Proc.* 20: 120–125.

McGee, E.A., G.A. Peterson, and D.G. Westfall. 1997. Water storage efficiency in no-till dryland cropping systems. *J. Soil Water Conserv.* 52: 131–136.

Nielsen, D.C., R.M. Aiken, and G.S. McMaster. 1997. Optimum wheat stubble height to reduce erosion and evaporation. Conservation Tillage Fact Sheet No. 4-97. USDA-ARS, USDA-NRCS, and Colorado Conservation Tillage Association, Akron, CO.

Norwood, C.A. 1994. Profile water distribution and grain yield as affected by cropping system and tillage. *Agron. J.* 86: 558–563.

Papendick, R.I. and G.S. Campbell. 1988. Water conservation in dryland farming. In *Challenges in Dryland Agriculture, A Global Perspective,* eds. P.W. Unger, T.V. Sneed, W.R. Jordan, and R. Jensen, pp. 119–127. Procedings of International Conference on Dryland Farming, Amarillo/Bushland, TX, August 15–19, 1988. Texas Agricultural Experiment Station, College Station, TX.

Papendick, R.I. and D.E. Miller. 1977. Conservation tillage in the Pacific Northwest. *J. Soil Water Conserv.* 32: 49–56.

Papendick, R.I., M.J. Lindstrom, and V.L. Cochran. 1973. Soil mulch effects on seedbed temperature and water during fallow in eastern Washington. *Soil Sci. Soc. Am. Proc.* 37: 307–314.

Parra, R. and A. Escobar. 1985. Use of fibrous agricultural residues in ruminant feeding in Latin America. In *Better Utilization of Crop Residues and By-Products in Animal* Feeding, eds. T.R. Preston, V.L. Kossila, J. Goodwin, and S.B. Reed, pp. 81–98. Proceedings of FAO/ILCA Expert Consultation, Addis Ababa, Ethiopia, March 1984.

Peterson, G.A., A.J. Schlegel, D.L. Tanaka, and O.R. Jones. 1996. Precipitation use efficiency as affected by cropping and tillage systems. *J. Prod. Agric.* 9: 180–186.

Peterson, G.A. and D.G. Westfall. 2004. Managing precipitation use in sustainable agroecosystems. *Ann. App. Biol.* 144: 127–138.

Peterson, G.A., D.G. Westfall, and C.V. Cole. 1993. Agroecosystem approach to soil and crop management research. *Soil Sci. Soc. Am. J.* 57: 1354–1360.

Powell, J.M. and P.W. Unger. 1998. Alternatives to crop residues for sustaining agricultural productivity and natural resource conservation. *J. Sustainable Agric.* 11: 59–84.

Rothfeder, J. 2001. *Every Drop for Sale.* Penguin Putnam Inc., New York, NY.

Schlegel, A.J. 1990. Effect of cropping system and tillage practices on grain yield and soil water accumulation and use. pp. 1–12. In eds. J.S. Hickman and E. Schofield. *Conservation Tillage Research.* Kansas Agric. Exp. Stn. Rep. of Progress 598.

Shaver, T.M., G.A. Peterson, L.R. Ahuja, D.G. Westfall, L.A. Sherrod, and G. Dunn. 2002. Surface soil properties after twelve years of dryland no-till management. *Soil Sci. Soc. Am. J.* 66: 1296–1303.

Shaver, T.M., G.A. Peterson, L.A. Sherrod, L.R. Ahuja. 2003. Cropping intensification in dryland systems improves soil physical properties: Regression relations. *Geoderma* 113: 149–164.

Shaw, T. 1911. *Dry Land Farming.* The Pioneer Co., St. Paul, MN.

Sherrod, L.A., G.A. Peterson, D.G. Westfall, and L.R. Ahuja. 2003. Cropping intensity enhances soil organic carbon and nitrogen in a no-till agroecosystem. *Soil Sci. Soc. Am. J.* 67: 1533–1543.

Smika, D.E. 1976. Seed zone water conditions with reduced tillage in the semiarid central Great Plains. Proceedings of 7th Conference on International Soil Tillage Research Organization, Upsalla, Sweden.

Smika, D.E. 1983. Soil water changes as related to position of straw mulch on the soil surface. *Soil Sci. Soc. Am. J.* 47: 988–991.

Smika, D.E. 1990. Fallow management practices for wheat production in the Central Great Plains. *Agron. J.* 82: 319–323.

Smika, D.E. and G.A. Wicks. 1968. Soil water during fallow in the Central Great Plains as influenced by tillage and herbicide treatments. *Soil Sci. Soc. Am. Proc.* 32: 591–595.

SSSA (Soil Science Society of America). 1997. *Glossary of Soil Science Terms, 1996.* SSSA, Madison, WI.

Stewart, B.A., P. Koohafkan, and K. Ramamoorthy. 2006. Dryland agriculture defined and its importance in the world. In *Dryland Agriculture*, eds. G.A. Peterson, P.W. Unger, and W.A. Payne, 2nd edn, pp. 1–26. Agronomy Monograph No. 23. ASA-CSSA-SSSA, Madison, WI.

Tanaka, D.L. 1989. Spring wheat plant parameters as affected by fallow methods in the northern Great Plains. *Soil Sci. Soc. Am. J.* 53: 1506–1511.

Troeh, F.R., J.A. Hobbs, and R.L. Donahue. 1991. *Soil and Water Conservation.* 2nd edn. Prentice Hall, Englewood Cliffs, NJ.

Twomlow, S., R. Riches, and S. Mabasa. 1997. Weeding—Its contribution to soil water conservation in semi-arid maize production. In *Proceedings of 1997 Brighton Crop Protection Conference—Weeds*, pp. 185–190. Brighton, UK.

Unger, P.W. 1978. Straw-mulch rate effect on soil water storage and sorghum yield. *Soil Sci. Soc. Am. J.* 42: 486–491.

Unger, P.W., W.A. Payne, and G.A. Peterson. 2006. Water conservation and efficient use. In *Dryland Agriculture.* eds. G.A. Peterson, P.W. Unger, and W.A. Payne, 2nd edn, pp. 39–85. Agronomy Monograph No. 23. ASA-CSSA-SSSA, Madison, WI.

Van Doren Jr, D.M. and R.R. Allmaras. 1978. Effect of residue management practices on the soil physical environment, microclimate, and plant growth. In *Crop Residue Management Systems*, ed. W.R. Oschwald, pp. 49–83. ASA Spec. Publ. 31, ASA, CSSA, and SSSA, Madison, WI.

Wicks, G.A. and D.E. Smika. 1973. Chemical fallow in a winter wheat-fallow rotation. *J. Weed Sci. Soc. Am.* 21: 97–102.

Widtsoe, J.A. 1920. *Dry-Farming.* The MacMillan Co., New York.

Wiese, A.F. 1983. Weed control. In *Dryland Agriculture*, eds. H.E. Dregne and W.O. Willis, pp. 463–488. Agronomy Monograph No. 23. ASA, CSSA, and SSSA, Madison, WI.

17 Historical and Present Usage of *Shatian* Gravel Mulch for Crop Production in Arid and Semiarid Regions of Northwestern China

Liang Wei-li, Gao Wang-sheng, Xu Qiang, and Huang Gao-bao

CONTENTS

Work the father to death, enrich the son, and impoverish the grandson.

—Lament of *Shatian*

17.1 INTRODUCTION

In some marginal areas, theoretically not suitable for crop production in terms of quantity of annual precipitation in northwestern China, farmers have survived for many generations and are trying to be better off by practicing a unique dryland farming system. The system involves growing crops on gravel-, pebble-, and/or sand-mulched land. Such a gravel-based cropland, along with the relevant farming system, is called *shatian* in Chinese.

While it is difficult to provide an exact terminological translation of *shatian*, it is easier to comprehend its significant effect on preserving rainwater in the soil. In a dry spring while soil moisture content in *shatian* is 13%–14%, sufficient for seed germination and seedling emergence, soil moisture content in the nonmulched field is about half (7%), which is not adequate for sowing a crop. Realizing the benefits of *shatian*, a satisfactory crop yield can be obtained even in an arid climate of 200–300 mm annual precipitation (Chen et al. 2008). This chapter specifically focuses on the description, origin, distribution, field operation, effects (on erosion, soil moisture, fertility, temperature, salinity, crop growth and development, yield, and quality), new advances, challenges, and research needs of *shatian*.

17.2 HISTORY AND DISTRIBUTION

17.2.1 ORIGIN OF *SHATIAN*

Shatian originated in Lanzhou of Gansu province of China, but there is no consensus on the exact year of its development and use. Some researchers claim that it was developed during the middle of the Ming dynasty dating back to 400–500 years ago (Li and Zhang 1982). By contrast, others believe that it was evolved during the Qing dynasty about 200–300 years ago (Xin 1993). Yet, some

BOX 17.1 AN OLD FARMER AND THE INVENTION OF *SHATIAN*

During the Kangxi period (~1662–1722 AD) of the Qing dynasty, there was a severe and perpetual drought prevailing in the Lanzhou region of Gansu province. Crops failed and people were suffering from extreme starvation. One day, an old farmer who lived in Qin-wang-chuan (the king of Qin's flat area) went to his farm to see the crops. Sad and depressed at the state of the crops, he walked away to answer nature's call. During this detour he came across a vigorously growing and healthy wheat (*Triticum aestivum*) plant. Surprised, he carefully examined the plant and its surroundings. The farmer observed that the wheat plant was growing at the opening of a burrow made by a rodent. Thus, the base of the plant was covered with gravel, pebbles, and sand. Upon removing the gravel and sand, the farmer noticed that the soil underneath was moist. Upon returning to the farm, the old farmer mulched his crop fields also with pebbles and gravel. His family became food-secure thereafter.

others argue that *shatian* was developed some 2000 years ago. Despite the controversy about the specific period of its origin, there is a general consensus regarding a popular story about the invention of *shatian* (see Box 17.1). It is thus hypothesized that *shatian* was invented much earlier but became popular only 200–300 years ago.

Farmers in Lanzhou are well experienced in crop production using *shatian*. They have developed special implements and field management techniques for this unique farming system for producing a wide range of crops (Figure 17.1). Spring wheat (*Triticum aestivum*), foxtail millet (*Setaria italica*), common millet (*Panicum miliaceum*), potato (*Solanum tuberosum*), and garlic (*Allium sativum*) are grown on rainfed *shatian*. By contrast, melons (*Cucumis dudaim*), vegetables, and fruits are grown in suburbs on irrigated *shatian*. Lanzhou owes its fame as the city of melons and fruits to the adoption of *shatian* (Huo et al. 2000). After the 1970s, cash crops have gradually replaced the food grains and are successfully grown on *shatian* (Yang et al. 2007).

17.2.2 Classification

Shatian is classified into different types depending on the type of mulch, with and without irrigation, age, and so on. Some commonly used *shatian* types are discussed below.

17.2.2.1 Composition of Mulching Material

Shatians are classified into two types depending on the nature of mulching material: *clear shatian* and *mixed shatian*. The clear *shatian* is mulched with gravel or pebbles, and little amount of soil is mixed in the gravel. Gravel or pebbles used for mulching are usually mined from the subsoil of nearby hills. The mixed *shatian* is mulched with gravel or pebbles with some soil mixed in. Mulching material for mixed *shatian* is obtained either from the hills or from a riverbed. In terms of crop yields, quality and agronomic productivity of mixed *shatian* are not as good as those of the clear *shatian*.

17.2.2.2 With or without Irrigation

Nonirrigated or rainfed *shatian* can have a life span of 40–60 years or even longer because there is no soil/silt brought in along with the irrigation water. *Irrigated shatian* plots are usually developed for growing cash crops, and are widely distributed in the vicinity of cities and towns. The mulch layer of irrigated *shatian* is relatively thinner than that of the rainfed *shatian*. Because silt and sediments are brought in with irrigation water, and frequent soil disturbance is needed for an intensive sequential cropping system, life span of irrigated *shatian* is usually 3–5 years—much shorter than that of rainfed *shatian*.

17.2.2.3 "Age" of *Shatian*

Depending on the age (duration since development), a *shatian* may be classified as new, mid-life, old, or duplex. The "age" is also indicative of the condition for its use for crop production.

New *shatian* are those developed within 15–20 years for rainfed *shatian* and 2 years for irrigated *shatian*. Mid-life *shatian* is developed within 20–30 years for rainfed *shatian* and in the 3rd year for irrigated *shatian*. Old *shatian* is developed beyond 30 years for rainfed *shatian* and after 4 years for irrigated *shatian*. Duplex *shatian* fields are those re-mulched with a new layer of gravel or pebbles on top of the previously mulched layer, to extend the life span of the *shatian* field for another 10–25 years after the previous one loses its functions.

17.2.2.4 Types of Mulching Materials

Different types of *shatian* are mulched with pebbles (round or semipolished gravel or *luan-shi shatian* in Chinese), flat gravel (flat rock fragment or *pian-shi shatian* in Chinese), and sands (*mian shatian* in Chinese. "*Mian*" means "soft" in Chinese). In the context of their effectiveness for water

FIGURE 17.1 *Shatian* and crops: (a) gravel *shatian*; (b) a profile of the mulch layer of a gravel *shatian*; (c) spring wheat grown on *shatian*; (d) a *shatian* watermelon field; (e) pebble *shatian*; (f) chili pepper grown on *shatian*; (g) sesame grown on *shatian*; (h) millet grown on *shatian*; and (i) linseed grown on *shatian*.

conservation, sand *shatian* is not as good as that mulched with gravel or pebbles. Further, sand *shatian* has a much shorter life span than those mulched with gravel or pebbles.

This chapter focuses only on gravel- or pebble-mulched *shatian*.

17.2.3 Distribution of *Shatian*

Dress in furs in the early morning, silk at noon, and sit aside a stove and eat watermelon.

—Local lyrics

Major distribution areas of *shatian* are located in the Lanzhou district of Gansu Province, and in the counties of Zhongwei, Haiyuan, Xingren, and Zhongning of Ningxia Hui Autonomous Region, with a specific concentration in Gaolan and Baiyin Counties of Gansu and in the Xiangshan area of Zhongwei County of Ningxia (Figure 17.2). There are also scattered distributions of *shatian* in Xinjiang, Qinghai, Shaanxi, and Shanxi provinces or autonomous regions. The total area of *shatian* is estimated to be 0.17 million ha (Mha), of which 0.067 Mha is in Gansu and 0.068 Mha in Ningxia (Lu 2007; Xu et al. 2009a).

Principal soils in the major distribution areas of *shatian* are developed on the loess parent material, and are prone to erosion. *Shatian* fields are built on the summit and side slopes of hills with an elevation ranging between 1400 and 2500 meters above sea level (m.a.s.l). The climate in these areas is primarily arid or semiarid, characterized by inland continental and temperate monsoons, with cold dry winter, dry spring, and hot summer. With an annual precipitation of 180–400 mm, with a high variation between years and among seasons, more than 60% of it occurs between July and September. The mean annual pan evaporation of the region ranges between 1500 and 2200 mm (Yang et al. 2005).

Predominant ecological characteristics of the regions where *shatian* is practiced include dry, windy, stormy, highly variable precipitation, light soil, and large diurnal temperature variations. These characteristics are reflected in the lyrics "Dress in furs in the early morning, silk at noon, and sit aside a stove and eat watermelon." Box 17.2 gives two examples of the environmental feature of these areas.

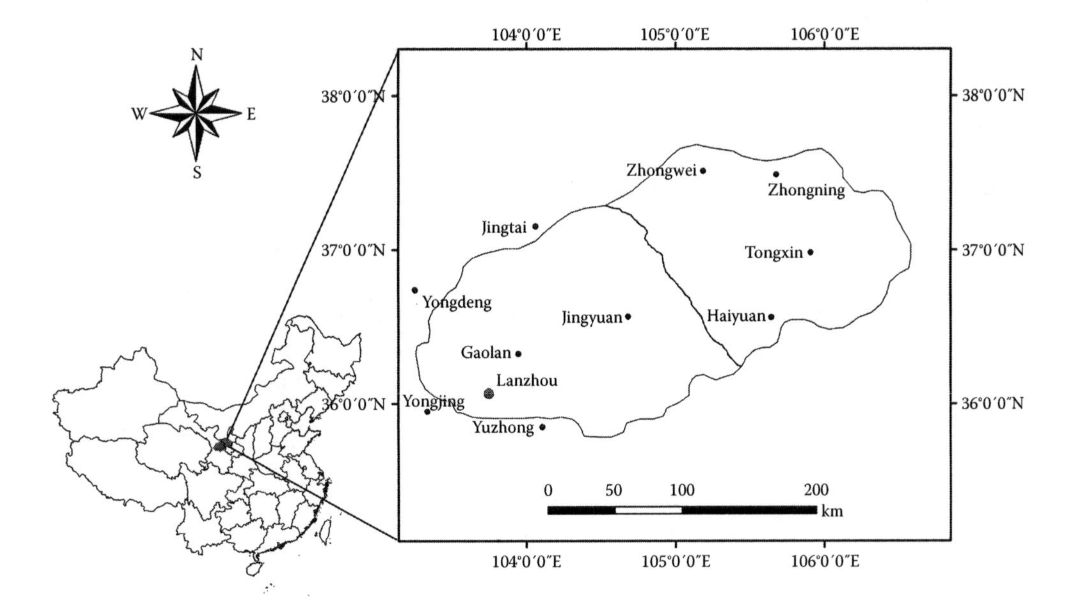

FIGURE 17.2 Major distribution area of gravel- or pebble-mulched *shatian*.

BOX 17.2 ENVIRONMENTAL FACTORS OF TWO TYPICAL *SHATIAN* DISTRIBUTION AREAS

Zhonghe, a township of Gaolan County, Lanzhou, Gansu Province, is located 1800 m.a.s.l. Soils are prone to erosion by water and wind. Soils contain 1.0%–1.1% organic matter (OM), 70–90 mg/kg total nitrogen (TN), 198–210 mg/kg total potassium (TP), 70–100 mg/kg available phosphorus (AP), and 120–160 mg/kg available potassium (AK), contained in the top 20 cm layer. The entire region is covered by sierozems. The soil bulk density of the 0–15 layer is 1.20 g/cm^3. The average annual precipitation is 263 mm and ranges between 154.9 and 392.4 mm. As much as 70% of the precipitation is received between June and September. The mean annual temperature is 7.1°C, the average temperature in the coldest month of January is −9.1°C and in the hottest month of July is 20.7°C. The accumulative thermal period above 0°C is 3324.5 degree days and above 10°C is 2798.3 degree days. There are 2768.1 h of sunshine. The average number of annual days of sandstorm is 2.6 with a maximum frequency of 9 (Yang 2004).

The Xiangshan area of Zhongwei County, Ningxia Hui Autonomous Region, is located on the extension of the Qilian Mountains. It has an elevation of 1500–2361 m.a.s.l. The original native vegetation in the area is arid desert steppe. The sierozem soils contain 0.3%–1.0% OM in the top 20-cm layer. The predominant *shatian* in this area is rainfed. The average annual temperature is 6.8°C. Accumulative thermal period above 0°C is 2332 degree days. Diurnal temperature variation ranges from 12°C to 16°C. With 2963.1 sunshine hours, the annual solar radiation load is 567.1 kJ/cm^2. Average annual precipitation is 247.4 mm, while the annual pan evaporation is 2172.3 mm. There are 146 frost-free days in a year. With strong wind and light soil, this region is among the major dust sources of Ningxia. Traditional crops of the region are millets, linseed (*Linum usitatissimum*), and wheat. The grain yield ranges between 1.5 and 2.0 tons per hectare for millets, between 1.8 and 2.3 tons per hectare for wheat, and between 2.23 and 2.7 tons per hectare for linseed. Crops fail completely in years of severe drought (Lu 2007; Xu et al. 2009b; Wang et al. 2010).

17.3 FIELD OPERATIONS

17.3.1 ESTABLISHING *SHATIAN*

Procedures of building *shatian* (Song 1994) include land selection, land preparation, basal application fertilizers, and application of mulch comprising laying of gravel or pebbles. Principal steps involved in establishing a *shatian* are the following:

Step 1: Land selection: Land with satisfactory soil fertility, flat or with a slope $<15°$ is suitable for building *shatian*.

Step 2: Land preparation: Plow the chosen land once or twice, to the depth of 30–40 cm, level soil surface with a drag or a harrow, and then compact with a stone roller. In recent years, many *shatian* plots have been established without plowing the land before laying gravel or pebbles.

Step 3: Applying basal fertilizers: Broadcast fertilizers on the soil surface, and mix it in the soil by plowing or harrowing.

Step 4: Laying gravel or pebbles: This is preferably done during the winter when the soil is frozen and there are no other farm activities. The trafficability of heavily loaded vehicles is good on frozen soils, and the impact of vehicular traffic on soils is less. Further, labor is easily available during winter to uniformly apply the gravel and pebbles, which is an important criterion that must be met in establishing *shatian*.

Thickness of mulch: There are no formal standard criteria regarding the suitable thickness of mulch in building *shatian*. A wide range of thicknesses proposed include 10–15 cm (Song 1994), 5–10 cm (Yang et al. 1995), and 10–13 cm (Yang et al. 2005). Thickness of mulch is important because it determines the cost of building a *shatian* plot. The thickness also has a significant effect on the functionality of *shatian*. The commonly used thickness of mulch in Zhongwei of Ningxia is 7–15 cm. In general, a thin layer of mulch is suitable only for irrigated *shatian* because of the lesser dependence on conserving soil water (Lu 2007). A mulch layer thicker than 15 cm is good for conserving soil water, but it has three drawbacks: (i) increase in cost of establishing *shatian*, (ii) increase in labor hours for planting crops, and (iii) reduced penetration of rainfall water into soil beneath the mulch layer for rains of low amount. Lu (2007) recommended that the ideal thickness of mulch be 12–13 cm for rainfed *shatian*.

17.3.2 Tillage of *Shatian*

Traditionally, the mulch layer of *shatian* is usually chiseled after the harvesting of the previous crop and before onset of rains. Chiseling is done by using the boot of an animal-drafted seed drill to enhance transport of the soil particles in the mulch to be washed down to the soil beneath. The repeated tillage operations are performed perpendicular to the previous tillage and to the same depth limited within the mulch layer. The field is harrowed just prior to the rainfall season (Yang et al. 1995). Two objectives of chiseling the mulch layer are (i) to remove soil particles in the mulch layer to the soil underneath and (ii) to break the crust in the mulch layer (Lu 2007). Nowadays, the tillage can be done with a tractor-drafted chisel plow (see Figure 17.3).

17.3.3 Sowing Crops

High population density crops are sown with a seed drill while low population density crops (i.e., melons) are planted by hand. The procedure of planting watermelon is opening the mulch layer to shape a hole (see Figure 17.4) → soften the soil underneath with a hand shovel → open the soil 3–4 cm deep → put one or two germinated seeds into the opening → cover the seeds with soil → surround the hole with four gravel or pebbles of about 10 cm diameter → put another pebble on top to cover the seeded hole → uncover the hole after the seedling emerges. More than 60% of watermelons planted in *shatian* in the Zhongwei County of Ningxia follow this procedure (Lu 2007).

FIGURE 17.3 A farmer is tilling his *shatian* with self-made chisel-like implement.

FIGURE 17.4 Planting watermelon on *shatian*: (a) making a hole and sowing by hand and (b) the sowing hole.

17.3.4 Applying Fertilizers on *Shatian*

The traditional approach of applying fertilizers on *shatian* is to remove the mulch layer in rows or open the mulch layer, soften the soil, apply fertilizers, and then level and compact the soil, and replace the mulch layer (Yang et al. 1995).

17.3.5 Restoration of Degraded *Shatian*

Shatian fields do not last forever, and have a fixed life span. A popular saying about *shatian* is "Work the father to death, enrich the son, and impoverish the grandson" vividly reflects the tremendous workload in establishing *shatian*, high productivity and profitability of new and mid-life *shatian*, and decline in productivity after the normal life span.

There are several factors influencing the life span of *shatian*. Important among these are the following:

1. Initial soil fertility: Since traditionally most rainfed *shatian* fields are not fertilized after establishment, the soil fertility and productivity decline with age. The difference in initial soil fertility could result in a difference in life span of *shatian* by 10–30 years.
2. Soil content in the mulch material used for establishing *shatian*: The lower the soil content in the mulching material, the longer is the life span.
3. Field management practices: All field management practices (i.e., tillage, fertilization, and sowing crops) should be implemented in a way that would minimally mix soil particles into the mulch layer. However, some mixing of soil particles into the mulch layer is unavoidable. Therefore, the function of *shatian* is weakened with cropping. *Shatian* loses its functions when soil particle content in the mulch layer exceeds 66% (Du 1993).
4. Sediment content in irrigation water: The reason that irrigated *shatian* has a shorter life span than the nonirrigated one is the silt content of the irrigation water. It is estimated that irrigation with water from the Yellow River brings 30 m^3 sediment per hectare into the field per year (0.3 cm/year).
5. "Dying out" of the underneath soil: The mellow top soil becomes less "active" because of a continuous cover as subsoil under the mulch layer for a long time without an adequate exposure to sunshine and air. A continuous gravel cover adversely impacts soil fertility and productivity of *shatian*.

A principal method of restoring a degraded *shatian* field is to remove the used mulch material, leave the land under fallow to restore fertility, or grow some low-nutrient demanding crops for 1 or 2 years, and then remulch the field with new material. In some cases, when the new mulching material is not easily available, the previous mulch material can be reused after sieving or washing off the soil particles in it. Because of higher soil particle content, which is equivalent to that of mid-life *shatian*, reusing the old material does not conserve soil water as well as using the new material (Wu and Feng 2006).

Field studies have shown that soil particle content in the mulch layer of *shatian* decreases within 1 year under fallow. Meanwhile, available soil nutrients and soil water storage are increased compared to the cropped *shatian*. Thus, it is possible to reclaim and sustain productivity of *shatian* by some means such as fallowing (Wu et al. 2008).

17.4 EFFECTS OF GRAVEL OR PEBBLE MULCHING

17.4.1 Conserving Soil and Water

The most important effect of *shatian* is conserving soil water. The gravel or pebble mulch reduces water runoff, increases infiltration into soil, and decreases evaporation (E). The mulch protects the soil surface against wind and sunshine, disrupts capillary rise of soil moisture rising to the surface, and reduces soil E. These processes increase soil moisture storage in *shatian* (see Figure 17.5).

Yang (2004) summarized the data of past research and concluded that soil moisture content in *shatian* is significantly higher than in unmulched control. The data also showed higher moisture content in subsoil than in the surface layers. Therefore, rainwater in *shatian* infiltrates to a deeper soil depth than in a nonmulched field. This theme needs more systematic and in-depth study and analyses. Yang also concluded from the evidence of a higher increase in soil moisture in *shatian* than the increase in nonmulched field following a heavy rainfall (31.3 mm in 1 day) that *shatian* could accept more rainwater. While increase in infiltration of rainfall reduces runoff, the mulch also protects the soil against direct raindrop impact and significantly reduces soil and water erosion (Zhao et al. 2009). Li (2003) observed in a field experiment that among the 91 rainfall events, 18 produced a total runoff of 48.4 mm from the control plots, while only 6 produced 3.4 mm runoff from the *shatian* plots. *Shatian* significantly reduced runoff by increasing infiltration. Similar to reduction in water erosion, wind erosion in *shatian* is also significantly reduced (Wang et al. 2003; Xie et al. 2003). For example, Zheng (2004) observed that dust collected in *shatian* at 24 cm above the land surface was only 20% of that collected in a nonmulched field. The results of a research experiment in Zhongwei of Ningxia show that, compared to the nonmulched field, soil moisture in the top 20-cm layer during the period from April to June was 32.5% higher, average soil temperature in the 0–20 cm soil layer was 0.96°C higher, and the dust collected within 2 m height was 43.9%

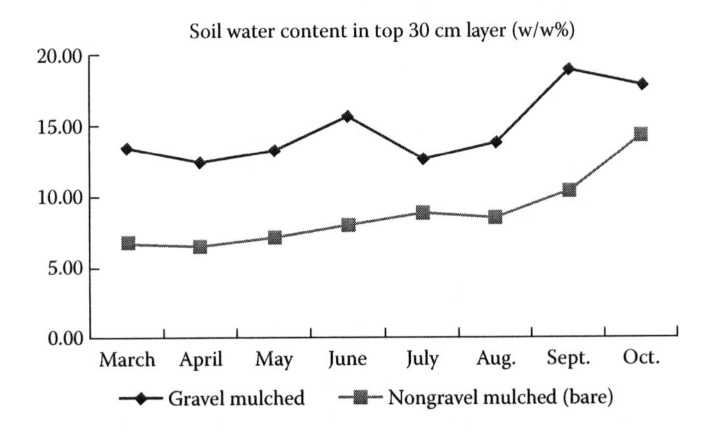

FIGURE 17.5 Comparison of soil moisture content in 0–30 cm layer in a *shatian* and nonmulched field. (Graph was drawn based on data calculated from the report of a gravel mulching survey team of Gansu Province, mimeographed unpublished document, 1964.)

less in *shatian* than in the unmulched field (Xu et al. 2009a). The result of a wind tunnel experiment (Li 2003) reveals that wind erosion of soil decreases constantly with the increase of pebble coverage at wind velocities of 10, 18, and 26 m/s, and the relation between wind erosion rate and pebble coverage rate follows a negative exponential function. Li's data from a field experiment (2003) implemented on *shatian* mulched with 10-cm-thick washed pebbles 5.0 cm in diameter at different surface coverages show a positive exponential correlation between surface pebble coverage and dust quantity trapped in *shatian* during the period from December 10, 2000 through May 10, 2001, the most windy period during a year.

Shatian not only conserves rainfall water in the soil during the growing season, it also preserves rainwater that falls during the postcropping season in the autumn and stores it in the soil for the crop grown in the following spring. Gao (1984) measured soil moisture content in *shatian* and nonmulched fields before and after the rainy season. The results showed that soil moisture content in *shatian* measured immediately after the rainfall season was approximately 20% higher, and that measured in the next spring before the rainy season began was 100% higher than that in the control. Research reports also show that during a year with severe drought, with a meager rainfall of hardly 10–15 mm, crops could not be sown in nonmulched fields but were successfully grown in a *shatian* field. Crops sown in *shatian* had a good stand, and yielded approximately 750 kg/ha compared with complete failure in the control (Song 1994; Yang 2004). Experimental data showed that soil water E in a period of 60 days in *shatian* was only 18.3% of that in a nonmulched field (Lü and Chen 1955).

Soil moisture content to 80-cm depth in *shatian* of different "ages" (ranging from 1 to 17 years) is higher than those in the adjacent nonmulched fields, although its effectiveness declines as *shatian* becomes "older" (Xu et al. 2009b). Soil moisture content in the 80 cm profile of *shatian* of 17 years age was 49.6% less than that under the newly established one, showing an average decline of 3.1% per year. After being cropped for more than 20 years and abandoned for 2 years, soil moisture content in the 80 cm profile of *shatian* was still significantly higher than that in the adjacent nonmulched field. Xu and colleagues also observed an interesting phenomenon that soil moisture content increased again in the subsoil layers of the profile after *shatian* was abandoned (fallowed) for 2 years. Wang et al. (2010) also reported that *shatian* of 1, 3, 5, 15, and 25 years of age stored 68.5%, 60.3%, 50.1%, 44.4%, and 40.4%, respectively, more soil moisture than a nonmulched field.

17.4.2 Increase in Soil and Air Temperatures

Gravel and pebbles have a lower heat capacity than soil, and thus warm up quicker under sunshine during the daytime than unmulched soil. Temperature of the soil beneath the mulch layer rises thereafter. At night, because of a higher moisture content and hence higher heat capacity, temperature of soil in *shatian* decreases slowly. Thus, soil temperature in *shatian* is higher than it is in nonmulched field during day and night. The difference in soil temperature between the two field types can be 1–2°C in spring and 3–4°C in summer (Gao 1984). Besides, because of a higher albedo and radiation reflection from gravel or pebbles, temperature of the near-ground air in *shatian* is higher than that in a nonmulched field. Lü and Chen (1955) reported that soil temperature in *shatian* was significantly higher than that in a nonmulched field in early spring and throughout the cropping season. The difference in soil temperature between the two field types was 1.5°C at night. Increase in soil temperature is most significant within the seedbed (0–8 cm soil depth). Diurnal variation in soil temperature in *shatian* is smaller than that in a nonmulched field. Lü and colleagues also observed that newly built *shatian* had a 2–3°C higher soil temperature than a nonmulched field and the date of soil being frozen was 20 days later and the date of thaw was 15 days earlier, which is extremely favorable to growing tropical crops (Lü and Chen 1955). Because of the favorable effects of *shatian* on soil moisture and temperature regimes, cotton has been successfully cultivated since the Jiaqing period (1796–1820 AD) of the Qing Dynasty in Lanzhou (Zhao 2009). Without *shatian*, the region is not favorable at all in terms of temperature and rainfall during growing cotton (Li and Zhang

1982). In Jingyuan County of Gansu Province, crops grown in *shatian* mature 10–20 days earlier than those grown in nonmulched fields (Song 1994). Research by Xu et al. (2009a) shows that soil temperature in *shatian* is significantly higher during the night and early in the morning, but is lower in 0–15 cm depth in the afternoon than in a nonmulched field. Li (2003) has also reported that soil temperature of *shatian* at 10 cm depth is 0.5°C–4.5°C higher based on hourly or daily measurements, air temperature at 20 cm height above the *shatian* surface is approximately 0.5°C higher than in the control, and daily variance of soil temperature is smaller in *shatian* than in the bare control. However, Li has observed consistently higher soil temperature in *shatian* during the day and night, which is different to Xu's observation. This is a topic of further in-depth and systematic study and analysis.

17.4.3 Effects on Plant Growth, Crop Yield, and Produce Quality

Wheat grown in *shatian* has a bigger root system, 20%–30% larger leaf area, and higher photosynthesis and transpiration (T) rates than in the unmulched control. Crops in *shatian* emerge and mature earlier. Spring wheat emerges 3–5 days earlier and matures 10 days earlier, tomato (*Lycopersicon esculentum*) matures 17 days earlier, Chinese cabbage (*Brassica rapa pekinensis*) matures 14 days earlier, and spring cabbage (*Brassica oleracea* L. var. *capitata* L.) matures 47 days earlier (Xie et al. 2003) than in the control field. In addition, the presence of the gravel or pebble mulch layer significantly reduces the incidence of insects, diseases, and weeds (Zhao et al. 2009).

In normal years, cereals grown in an unmulched field yield around 750 kg/ha and no more than 1125 kg/ha. By contrast, cereals grown in *shatian* can yield 1500–2250 kg/ha with a maximum of 3000 kg/ha. A crop grown in *shatian* seldom fails even in drought years (Song 1994). Since the adoption of *shatian* changes the microenvironment of the crop grown, it has turned large land areas with problems of inadequate water, low soil temperature, and/or high salinity into relatively highly productive and reliable arable lands, especially those lands in high-risk areas. Increase in yield of some crops grown in *shatian* can be as much as 50%–80% in cotton, 100%–300% in watermelon, and 40% in tropical vegetables (Yang 2004).

Crops grown on *shatian* not only yield more but also have better quality. Lanzhou's fame in high-quality products—wheat flour in the old brand of "He-shang-tou" (monk's head), honeydew (*Cucumis melo*), and rainfed watermelon—is entirely due to the contribution of *shatian* (Yang et al. 2004). Wheat on *shatian* has 10%–30% higher gluten and 21% higher protein contents than that in the control (Yang 2004). The sucrose content of honeydew grown on *shatian* is usually higher than 14%, and watermelon produced on *shatian* is much bigger in size, easier to transport, and can be preserved for a longer time period (can be consumed during the spring festival). Results from a field experiment show that yield, soluble carbohydrate content (SCC), and water-use efficiency (WUE) of watermelon are significantly higher for *shatian* than for the control treatments (Xie et al. 2006). Furthermore, since gravel and pebbles contain selenium, there is a higher selenium content of this health-beneficial microelement in *shatian* produce (Lu 2007).

Productivity of *shatian* declines with duration. Watermelons grown on *shatian* plots of 5–8, 10, 15, and 20 years of age yield 33.3%, 38.9%, 55.6%, and 71.1% less compared to those grown on 13-year-old *shatian* (Wang et al. 2010).

17.4.4 Effects on Soil Salinity

Greatly increased water infiltration into the soil and decreased evaporation from the soil in *shatian* reduce risks of soil salinity in the top layers. Data from field experiments show that soil salinity in the top 20 cm decreased sharply from 0.74 mg/g to 0.29 mg/g after 1 year of gravel mulching, continued to decrease slowly in the following 2 years, and was maintained at 19–22 mg/g in the succeeding years. Salinity in the top 20 cm soil in 25-year-old *shatian* increased slightly to 0.27 mg/g. Soil salinity had the same trend in 20–40 cm depth, with a slightly higher salt content (Wang et al. 2010). Xu et al. (2009b) reported a continuous and more significant salinity decline—0.21 mg/g in the first year and 0.05 mg/g in the seventeenth year after the *shatian*. Similar results have been reported by Li (2003) and Yang (2004).

17.4.5 Effect on Soil Fertility

Shatian under rainfed conditions is traditionally not fertilized. Xu et al. (2009b) compared rainfed nonfertilized watermelon *shatian* fields of 1–17 years in age with that abandoned for 2 years after more than 20 years of continuous cropping. Their results showed that after 17 years of continuous cropping, soil particle content in the mulched layer increased from 9.24% (the first year) to 36.15%; moisture content in 0–20 cm soil decreased from 25.3% to 13.7%; AP declined from 2.42 mg/kg to 2.26 mg/kg; AK declined from 172 mg/kg to 109 mg/kg; and OM, TN, and available nitrogen (AN) in the 0–20 cm soil layer increased in the first 4–5 years and declined continuously thereafter. The concentration of TN in 0–20 cm soil layer of the fifth year of *shatian* was 0.32 mg/g, 53.7% higher than in the 1st year of *shatian*; and AN in 0–20 cm soil layer of the 4th year of *shatian* was 15.1 mg/kg, which was 53.7% higher than that in the 1st year of *shatian*. While the AN concentration in the soil layer declined to 8.3 mg/kg in the 17th year of *shatian*, the soil OM content in the 17th year of *shatian* was 15.0% lower than that in the first year of *shatian*. Soil OM in *shatian* that was abandoned for 2 years after more than 20 years of cropping was 3.8% higher, AP was 2.4% higher, and AK was 31.2% higher than those in the 17-year *shatian*. These results support the conclusion that fallowing can restore soil fertility of *shatian* (Xu et al. 2009b).

Wang et al. (2010) reported a continuous decline of all soil nutrients in *shatian* being cropped from 1 to 25 years.

In recent years, even rainfed *shatian* fields are being fertilized, but soil nutrient dynamics under this new system have not yet been studied.

17.5 NEW DEVELOPMENTS IN PRODUCTION PRACTICES AND ADVANCES IN RESEARCH

New developments in production practices on *shatian* mainly consist of change in major crops, incorporation of plastic film with mulching, supplemental irrigation using harvested rainwater, intercropping and mechanization, and so on. Recent research efforts have been focused on high value crop production, mainly on watermelon and honeydew.

17.5.1 Change of Major Crops on *Shatian*

Risks of food insecurity prior to 1978 were the deciding factor, and *shatian* was mainly used for production of food grains. In recent years, however, *shatian* has been used for growing high-value cash crops, mainly watermelon and honeydew, although there are still small areas of spring wheat, Chinese date (jujube) (*Ziziphus zizyphus*), millets, potato, garlic, chili pepper (*Capsicum annuum*), soybean (*Glycine max*), and so on. Consequently, recent research efforts are aimed at change of the focus on increasing productivity, quality, and profitability of major cash crops (Yang 2004; Lu 2007).

17.5.2 Incorporation of Plastic Film

Plastic films have been incorporated in *shatian* in order to conserve and use the very precious water in a more efficient way and to further increase crop productivity and profitability. Plastic films are used in conjunction with supplemental irrigation using harvested rainwater. Several techniques have been developed for incorporating plastic in *shatian*, that is, single-film mulching, double-film mulching, triple-film mulching, and greenhousing (Wu and Zheng 1995; Yang et al. 2005, 2007). Patterns of the incorporation include *shatian* + ground plastic filming, *shatian* + plastic cloche, *shatian* + ground plastic filming + plastic cloche, and *shatian* + ground plastic filming + plastic cloche + plastic greenhouse (Ma et al. 2010). Recent studies have shown strong and positive effects of plastic film mulching in *shatian* in terms of quicker emergence and earlier maturity (Jia et al. 1998), preserving more soil moisture to ensure crop growth and development and yield improvements during the late phenological stages (Zhang and Zheng 2006; Xie et al. 2006). These positive effects are especially important during seasons of severe drought, and noticeable on seedling stand, early maturity, and yield (Yang et al. 2005). Early maturity is important to avoid early frost. By using plastic films, soil temperature at 10 cm depth can be increased by 4–6°C, soil moisture content in 0–5 cm layer by 15.23% in late April, and crop canopy receives more reflected radiation (Zhang and Zheng 2006; Lu 2007).

Many practical patterns of incorporation of plastic film have been developed by *shatian* farmers, which include the following:

- Covering the planting holes with plastic films: this is done by mulching the planting hole and its adjacent surrounding field surface, by making a mini plastic cloche above the hole, or by covering the hole with a transparent plastic bowl or cup (Jia et al. 1998; Lu 2007).
- Mulching the planted rows with plastic film on top of gravel or pebbles (Figure 17.6). Usually transparent films are used, but colored films are also used in places where aphids are a serious problem and black films are used where weeds are problematic (Yang 2004; Lu 2007). A *shatian* with plastic film mulch achieves a 52.5% higher watermelon yield than nonfilmed *shatian* (Jia et al. 1998). This is the most widely used pattern of *shatian* + film incorporation for its relatively lower cost, satisfying performance in conserving soil moisture, and ease of implementation by machines.

The effectiveness of different techniques of applying plastic film in terms of conserving soil moisture and increasing temperature are in the order of planting row mulching > planting hole mulching > planting hole covering by bow or cup (Zhang and Zheng 2006; Lu 2007). Different practices include the following:

- Cloche the planting row: immediately after crops (melons or any other cash crops) are seeded or transplanted, a transparent plastic film cloche is built over the planting row. The film is removed after flowering or fruit setting. With this technology, melons mature 25 days earlier than those on noncloched *shatian* (Yang 2004; Lu 2007).

(a)

(b)

FIGURE 17.6 Incorporation of *shatian* with plastic film: (a) laying films and (b) young watermelon seedlings grown on *shatian* under film mulch.

Since the late 1990s, some more expensive and more complicated incorporation technologies have been developed (Yang 2004). Important among these are:

- Double use of plastic film: *Shatian* is mulched and then cloched with plastic film.
- Greenhousing: A plastic greenhouse built on *shatian* further shortens the growing period of crops. Watermelon and honeydew are grown on *shatian* in greenhouses equipped with modern ecological technologies of pest control, organic fertilizer application, and light-reflecting film to produce green food melons (Jia et al. 1998). Greenhouses built on *shatian* are widely seen in the Ping Chuan zone, Baiyin County of Lanzhou, Gansu Province.
- *Shatian* + ground plastic filming + plastic cloche + plastic greenhouse: The highest obtained yield of honeydew grown with this technology was 7.4 ton/ha in Shichuan township, Gaolan County of Lanzhou (Jia et al. 1998).

17.5.3 Mechanization

- A gravel/pebble *mulching machine* has been developed by Ningxia University (Figure 17.7a). The machine can mulch 0.33–0.47 ha with 10–15-cm-thick gravel/pebbles per hour.
- A mulch layer *softening and fertilization machine* has been developed by Jianfeng Machinery Factory of Zhongwei County (Figure 17.7b). The machine can also do weeding and sow wheat, buckwheat (*Fagopyrum esculentum*), sunflower (*Helianthus annuus*), and so on. It can loosen and fertilize 0.53–0.67 ha/hour, applying fertilizers up to the rate of 1200–2250 kg/ha.
- A planting–*hole-making machine* has also been developed by Jianfeng Machinery Factory. The machine can dig planting holes to a depth of 10–15 cm with an efficiency of 0.4–0.53 ha/hour.

FIGURE 17.7 Machinery for *shatian* operations: (a) gravel/pebble-laying machine; (b) a *shatian* driller, which can also be used for softening the mulch layer, applying fertilizers, and sowing narrow row spacing crops (i.e., cereals); (c) gravel/pebble-sieving machine at work, and (d) farmers planting watermelon using water supplied by a water tank for supplemental irrigation.

- A prototype of a *gravel/pebble-sieving machine* (Figure 17.7c) to separate gravel/pebbles from sands/soil particles has been developed by Ningxia University and the Ningxia Association of Senior Scientific Workers.
- Several kinds of *supplemental irrigation machines* have been developed. Those machines can supply 1.6–6.4 m^3 water per hour (Figure 17.7d).

17.5.4 Supplemental Irrigation

In conditions of less rainfall, supplemental irrigation can greatly increase yield, profitability, and WUE of crop production on *shatian* (Zhang 1998; Wang et al. 2003; Tian et al. 2003). Supplemental irrigation by using the harvested rainwater at critical stages of crop growth increased watermelon yield by nearly three times (Tian et al. 2003). However, melons grown on rainfed *shatian* sell better than irrigated melons because of higher sucrose content and longer shelf life (can be preserved for consumption during the spring festival). Soluble solid content (SSC) is positively related to sucrose content, and thus is an easy-to-measure indicator of produce quality of watermelon. Research data show a close relation of SSC to soil water content which is significantly affected by rainfall, mulching, and supplemental irrigation. There are no significant differences in SSC among irrigated treatments below 68 mm under *shatian* conditions. However, SSC in the nonirrigated control is significantly higher than that with 68 mm irrigation (Wang et al. 2004). Applying irrigation water <45 mm has no significant effect on the sucrose content of watermelon, and that up to 67.5 mm can decrease the sucrose content by 4.4% or more (Wang et al. 2003; Xie et al. 2003). Thus, attention must be paid to the trade-offs between quality and yield.

17.5.5 Intercropping of Melons with Other Crops on *Shatian*

Melons can be intercropped with peanut (*Arachis hypogaea*) (Wu and Lin 1999), chili pepper, Chinese date, soybean, sesame, and so on. Intercropping increases land productivity, increases profitability, and decreases the rate of degradation of *shatian*. Intercropping of melons with perennial crops (i.e., jujube) has the potential of using an aging *shatian* and restoring it (Wang and Zhang 2006; Lu 2007).

17.5.6 Research Advances in Supplemental Irrigation on *Shatian* Incorporated with Film Mulching

Major recent advances in *shatian* technology include supplemental irrigation and its incorporation with plastic films. Xu et al. (2009b) studied soil moisture dynamics for a 0–80 cm profile with three mulching practices and a nonmulched watermelon field under rainfed and supplemental irrigated conditions. The results showed that average moisture content in the 0–80 cm soil profile for the entire growing season of all three mulched fields was much higher than that in the nonmulched field. Soil moisture content in the 0–80 cm profile in the fields mulched with both gravel and plastic film, with gravel only, and with films only was 44.2%, 34.0%, and 23.2% higher than those in the nonmulched field, respectively, under supplemental irrigation; and 71.9%, 73.7%, and 21.2% higher, respectively, under rainfed conditions (see Table 17.1). The effectiveness of mulching in conserving soil moisture is more significant under rainfed conditions. The small amount of supplemental irrigation (three applications of 22.5 mm total by watering only the planting hole) significantly increased the crop yield (Table 17.2).

Without considering the labor involved, profitability of the four treatments would be in the order of gravel + plastic film > gravel > plastic film > nonmulched under supplemental irrigated condition and gravel + plastic film > gravel > nonmulched > plastic film in rainfed conditions. Plastic film mulching made a small profit under irrigated conditions but could not recover the cost under rainfed conditions.

TABLE 17.1
Comparison of Soil Water Content in 0–80 cm Profile of Different Mulching Treatments and Nonmulched Watermelon Field (w/w%)

	Irrigated				Rainfed			
	G + P	G	P	N	G + P	G	P	N
Young seedling	21.6A	18.5B	17.7C	16.2D	17.3Aa	16.6Ab	14.3Bc	13.3Bd
Stem elongation	19.4A	18.1B	17.1C	16.8D	17.6A	15.6B	12.2C	10.7D
Flowering	19.5A	17.5B	16.6C	15.3D	18.1A	15.6B	12.1C	9.3D
Fruit setting	18.7Aa	16.7Bb	16.8Bb	12.3Cc	15.7A	17.0B	11.2C	8.7D
Fruit enlarging	17.1A	15.1B	14.2C	10.3D	14.6A	17.0B	10.2C	8.4D
Harvesting	18.4A	20.6B	15.5C	8.5D	16.5A	18.9B	10.1C	7.6D

Source: Xu, Q. et al. A study on soil moisture dynamics in dryland under different mulch materials. Unsubmitted/unpublished paper 2010.

Note: The experiment is randomized block designed.

G + P = gravel + plastic film mulched; G = gravel mulched; P = plastic film mulched; N = nonmulched.

Different letters in uppercase indicate statistically significant difference of 99% probability; different letters in lowercase indicate statistically significant difference of 95% probability.

Differences between irrigated and rainfed treatments at all growing stages are statistically significant at 99% probability.

Evaporation can be reduced significantly when the gravel surface in the watermelon field is mulched with plastic film (Xie et al. 2006). Obviously, gravel or pebble mulch is effective in increasing water infiltration into the soil but is not as good as the plastic film in reducing soil water evaporation. In comparison, a plastic film is effective in reducing soil water evaporation but also decreases the water infiltration into the soil. Therefore, plastic mulching has positive effects only under satisfactory soil moisture or irrigated conditions. In general, *shatian* is better adapted to the arid and semiarid regions than is the plastic film technology. However, a combination of the traditional and the new technology improves crop performance and increases the profit margin.

TABLE 17.2
Effects of Different Mulching Practices on Yield and WUE of Watermelon

	Irrigated				Rainfed			
	G + P	G	P	N	G + P	G	P	N
Yield (ton/ha)	24.80A	19.16B	7.41C	2.40D	20.38A	16.05B	4.21C	0.68D
Relative yield (%)	3628	2803	1084	351	2981	2349	616	100
ET (mm)	90.8a	91.0a	90.5a	91.1a	76.9A	81.1B	84.3C	87.8D
WUE (kg/ha/mm)	24.14A	18.65B	7.21C	2.33D	19.83A	15.62B	4.10C	0.67D

Source: Xu, Q. et al. A study on soil moisture dynamics in dryland under different mulch materials. Unsubmitted/unpublished paper 2010.

Note: The experiment is randomized block designed.

G + P = gravel + plastic film mulched; G = gravel mulched; P = plastic film mulched; N = nonmulched.

Different letters in uppercase indicate statistically significant difference of 99% probability; different letters in lowercase indicate statistically significant difference of 95% probability.

Differences between irrigated and rainfed treatments in ET and WUE are statistically significant at 99% probability, while yield difference between irrigated and rainfed treatments is statistically significant at 95% probability.

17.6 SOME CHALLENGES AND RESEARCH NEEDS

Major shortcomings of *shatian* include increased labor demand (extremely laborious), declining soil fertility, and mixing of soil particles with gravel or pebbles. Increasing efficiencies of *shatian* in high crop production for the entire life span of *shatian* and on a sustainable basis are among the principal objectives of future research.

- *Technical standards*: Thickness of gravel/pebble mulch and size of gravel/pebble affect cost of establishing *shatian* and the subsequent field operations. There is an ongoing debate on suitable size of gravel or pebble and gravel/sand or pebble/sand ratio to be used as the mulching material. Some researchers believe that the smaller the size, the worse the mulching material in conserving soil moisture because of lesser infiltration, stronger capillary function, and hence greater evaporation (Luo 1991). By contrast, others (Guan and Feng 2009) believe that a bigger size of mulching material conserves less soil moisture because of increased porosity and hence higher losses by evaporation. Lü et al. (1958) built *shatian* mulched only with gravel >5 mm in diameter (those <5 mm were removed by sieving). Results of a 3-year experiment showed that there was no difference in soil temperature between *shatian* mulched with pure gravel and that with mixture of gravel and sand; pure gravel mulch was more effective in conserving soil moisture; and yield of spring wheat grown on the pure gravel-mulched *shatian* (clear *shatian*) was higher than that grown on traditional *shatian* mulched with the mixture of gravel and sand. They concluded that this kind of *shatian* has a longer life span and is easier to perform mechanized operations on (Lü et al. 1958). Xie et al. (2006) reported that evaporation and E/ET ratio of *shatian* increased linearly with the increase in gravel size. Mulching with 2–5 mm diameter sand and gravel resulted in significantly less evapotranspiration (ET) than mulching with 5–20 mm and 20–60 mm diameter sand and gravel. They reported that gravel size had no significant effect on yield, but SCC of watermelon was significantly higher and WUE was significantly lower on *shatian* mulched with 20–60 mm diameter gravel than that on *shatian* mulched with 2–5 mm diameter gravel. They concluded that mulching with smaller-sized gravel improved WUE because of decrease of soil moisture evaporation, and the increased SCC with larger-sized gravel was because of an increase in diurnal soil temperature variance. These experiments indicate the need for further in-depth and systematic research on the suitable thickness of mulch, type and size of gravel/pebble, and their effects on soil moisture, temperature, and fertility.
- *Fertilization*: Difficulty in applying fertilizer has been one of the major shortcomings of *shatian* for a long time. Traditionally, fertilizers are not applied in most rainfed *shatian* fields, which results in decline of soil fertility over time. This problem must be addressed for the sustainable development of *shatian* (Chen et al. 2008). Fertilizer formulations and doses suitable for application on diverse crops and relevant easy-to-implement technology of fertilizer application in *shatian* must be given a high priority in future research (Tian et al. 2003; Wang et al. 2005).
- *Crop water demand in* shatian *and suitable areas for* shatian *development*: Under rainfed conditions, a minimum rainfall is needed for crop growth and yield. For example, watermelons grown on *shatian* need a minimum of 120 mm of rainfall during the growing season. In 2005, there was only 60 mm of annual precipitation in Zhongwei County of Ningxia, which resulted in a complete failure of melons in the county (Lu 2007). The water requirement of different crops is not known, especially the thresholds at different growth stages and in relation to soil moisture storage and rainfall patterns are not clearly understood. Yet, this is key important information for developing new *shatian*. In this context, long-term experiments are needed under different climatic conditions.

- *Crop rotation for disease control*: Since *shatian* is generally used for growing a limited number of crops, it is difficult to follow crop rotations. Lack of crop rotation results in the buildup of pathogens and soilborne diseases, which reduce crop yields. For example, watermelon is grown continuously without a crop rotation. Year after year of growing the same crop on the same piece of land can cause serious breakout of diseases with drastic yield losses (Liu and Zhou 2000). Research has shown that after 5–6 years of continuous cropping, watermelon should either be rotated with another crop or the land should lie fallow for 1–2 years (Luo 1991; Xu et al. 2009b). Currently, farmers either do not plant the melon in the same hole as was planted the previous year or grow low-yielding wheat. Long-term experiments on crop–soil–pathogen–disease dynamics are urgently needed for designing optimal crop rotation schedules.
- *Utilization of degraded shatian*: Degradation of *shatian* is inevitable, although its life span can vary greatly depending on the quality of mulching materials and field management practices. Restoration of degraded *shatian* to extend its useful life span is a possible alternative to building new *shatian*. Yet, little research has been done on this topic, except a limited research on the practice of intercropping watermelon with jujube in an old *shatian* field (Lu 2007).

17.7 SUMMARY

Research data and historical experiences indicate that *shatian* is indispensible because of its effectiveness in soil and water conservation. In the context of the present scenario in which economic status, market, means of production, public concerns for resources and environment, and so on have drastically changed, crop production on *shatian* has evolved into an entirely new farming system compared to the traditional system.

Some researchers argue that development of *shatian* should be restricted because it may result in unregulated mining of gravel and pebbles, which may damage the environment, destroy the vegetation, and degrade the land. However, it is not merely a question of the unregulated mining of gravel/pebbles. The principal issue is of a better administration and management (Yang 2004). There is also a concern that abandonment of the degraded *shatian* by the farmers would lead to the creation of an anthropogenic desert. Fortunately, there have been some practical solutions to address this concern. Important among these are new advances in *shatian* renewing technology and utilization of degraded *shatian*, and there are more and better solutions on the horizon. If people have the skills and innovations to build *shatian*, they must also be able to avoid the risks of creating an "artificial Gobi."

Those who benefit from *shatian* are now expecting even more from it—to thrive by making full use of this miracle farming system. However, using *shatian* only for economic reasons would soon "degrade" it, as has been the case with the traditional *shatian*. Thus, sustainable use of *shatian* implies economic, ecological, and social aspects. Evolution of such a system requires in-depth scientific research.

ACKNOWLEDGMENT

The authors wish to sincerely thank Dr. Stewart and Dr. Lal for inviting them to write this chapter, for their patience, and especially for their time spent on revising the manuscript. Most of the photos used in this chapter were taken by Kang Jian-hong and Wu Hong-liang from Professor Xu Qiang's research team at Ningxia University. Other photos were provided by Professor Huang Gao-bao's research team and Professor Chai Shou-xi of Gansu Agricultural University. The authors are most thankful for their very helpful contribution. Our thanks also go to Dr. Li Ling-ling of Gansu Agricultural University, who prepared the figure of the major distribution areas of *shatian* and helped in referencing.

ABBREVIATIONS

AK	Available potassium
AN	Available nitrogen
AP	Available phosphorus
E	Evaporation
ET	Evapotranspiration
OM	Organic matter
SCC	Soluble carbohydrate content
SSC	Soluble solid content
T	Transpiration
TN	Total nitrogen
WUE	Water-use efficiency

REFERENCES

Chen, N.L., D.H. Liu, X.W. Wang, et al. 2008. Research and development of *shatian* production in Gansu province. *China Melons and Vegetables* (2): 29–31.

Du, Y.Z. 1993. Role of *shatian* in soil and water conservation in dry areas. *China Soil and Water Conservation* (4): 35–39.

Gao, B.S. 1984. *Shatian* in Gansu. *China Soil and Water Conservation* (1): 10–12.

Guan, H.J. and H. Feng. 2009. Effects of mulching thickness and gravel size on soil evaporation. *Journal of Irrigation and Drainage* (8): 41–44.

Huo, G.Y., G.Z. Zhou, Z.Z. Zhang, et al. 2000. *History of Agriculture in Lanzhou*. Lanzhou: Lanzhou University Press.

Jia, D.Y., X.L. Zeng, Y.Y. Zhang, et al. 1998. Experiment on plastic filming production of seed watermelon in rain-fed *shatian*. *China Watermelon and Muskmelon* (1): 20–21.

Li, F.G. and B. Zhang. 1982. Study on *shatian* in middle Gansu. *China Agricultural History* (1): 36–42.

Li, X.Y. 2003. Gravel–sand mulch for soil and water conservation in the semiarid loess region of northwest China (in English). *Catena* 52: 105–127.

Liu, D.X. and G.H. Zhou. 2000. *Technique Book of Watermelon Production*. Beijing: China Agricultural Press.

Lu, C.C. 2007. *Watermelon Production on shatian in Xiangshan of Zhongwei*. Beijing: China Economics Press.

Lü, Z.S. and B.Y. Chen. 1955. Study on *shatian* in Gansu. *Journal of Agriculture* 6(3): 65–69.

Lü, Z.S., B.Y. Chen, and C.R. Tian. 1958. A methods of improving *shatian* in Gansu. *Journal of Soil Science* 6(1): 65–69.

Luo, H.X. 1991. Function of *shatian* in Baiyin on drought preventing and its tillage. *Agricultural Research in the Arid Areas* (1): 37–44.

Ma, Z.M., S.P. Du, and L. Xue. 2010. Current status, problems and solutions of gravel mulched melon production. *China Melons and Vegetables* 23(3): 60–63.

Song, W.F. 1994. *Shatian* in Gansu. *Hydrological Technology in Gansu* (2): 56–58.

Tian, Y., X.L. Li, F.M. Li, et al. 2003. Effect of supplemental irrigation with harvesting rainwater on yield of watermelon and soil moisture in sand-filed. *China Desert* 23(4): 459–463.

Wang, F., Y.H. Li, T.C. Zhao, et al. 2005. Considerations on the sustainable development of gravel mulched watermelon and honeydew production. *Ningxia Agriculture and Forestry Technology* (5): 60–61.

Wang, Y.H. and X.F. Zhang. 2006. Experimental study on adaptation of red date to *shatian* in Zhongwei County. *Ningxia Agriculture and Forestry Technology* 4: 20.

Wang, Y.J., Z.K. Xie, Z.S. Zhang, et al. 2003. Effect of rainwater harvesting for supplementary irrigationon watermelon in gravel and plastic mulched field in Gansu. *Journal of Desert Research* 23(3): 300–305.

Wang, Y.J., Z.K. Xie, F.M. Li, et al. 2004. The effect of supplemental irrigation on watermelon (*Citrullus lanatus*) production in gravel and sand mulched fields in the Loess Plateau of northwest China (in English). *Agricultural Water Management* 69: 29–41.

Wang, Z.J., Q. Jiang, J.L. He, et al. 2010. Characteristic analysis of soil fertility of gravel-mulched land around Xiangshan mountain area in Ningxia. *Journal of Soil and Water Conservation* 24(2): 201–204.

Wu, D.K. and X.H. Feng. 2006. *Questions and Answers on Watermelon Production Technology*. Yinchuan: Ningxia People's Press.

Wu, H.L., J.H. Liang, and Q. Xu. 2008. Study on biodiversity on abandoned *shatian*. In *Advances in Farming Systems Research in China*, ed. W.S. Gao, pp. 339–343, Shenyang: Liaoning Science and Technology Press.

Wu, J.Y. and X.R. Zheng. 1995. Production technology of high quality honeydew on *shatian* in greenhouse. *China Water Melon and Musk Melon* (4): 18–19.

Wu, Y.A. and S.M. Lin. 1999. Technology of intercropping honeydew with peanut and autumn vegetables. *China Water Melon and Musk Melon* (1): 22–24.

Xie, Z.K., Y.J. Wang, S.H. Chen, et al. 2003. Effect of supplemental irrigation with harvested rainwater on watermelon (*Citrullus lanatus*) production in gravel-and-plastic mulched fields in the Loess Plateau of Northwest China. *Acta Ecologica Sinica* 23: 2033–2039.

Xie, Z.K., Y.J. Wang, W.L. Jiang, and X.H. Wei. 2006. Evaporation and evapotranspiration in a watermelon field mulched with gravel of different sizes in northwest China (in English). *Agricultural Water Management* 81: 173–184.

Xin, X.X. 1993. On the invention and origination of *shatian* in Gansu. *Gansu Agricultural Science and Technology* (5): 5–7.

Xu, Q., L. Qiang, H.L. Wu, et al. 2009a. Study on sandy field ecosystem effection. *Journal of Ningxia University* (Natural Science Edition) 30(2): 180–182.

Xu, Q., H.L. Wu, L. Qiang, et al. 2009b. Study on evolution characteristics of sandy field in arid region. *Agricultural Research in the Arid Areas* 27(1): 37–41.

Yang, G.Q., J.Q. Yang, and M.X. Zhang. 1995. Role and benefits of gravel mulched land on sustainable agricultural development in arid mountainous areas. *China Soil and Water Conservation* (5): 31–33.

Yang, J.Q. 2004. Building *shatian* is an effective way for being better-off in arid mountainous areas. *Agriculture in Gansu* (12): 62.

Yang, L.N., X.F. Ma, X.Z. Liu, et al. 2005. Meteorological effects of different mulching methods on gravel mulched watermelon field. *Ningxia Agriculture and Forestry Science and Technology* (5): 323–325.

Yang, L.S. 2004. Effects of gravel mulching and other mulching methods on temperature and moist and on growth and development of honeydew. MS thesis of Northwestern University of Agricultural and Forestry Science and Technology.

Yang, L.S., F.Q. Guo, Z.F. Ma, et al. 2004. *History of Melons and Fruits in Lanzhou*. Lanzhou: Lanzhou University Press.

Yang, L.S., Z.Y. Xi, L. Li, et al. 2007. Application and development of *shatian* in Lanzhou. *China Melons and Vegetables* (3): 32–33.

Yang, X.Q., Z.C. Wei, X.S. Liu, et al. 2005. *China Melons and Vegetables* (6): 39–40.

Zhang, G.H. 1998. Effect of gravel and plastic film mulching on soil water in apple orchard. *Gansu Agriculture Science and Technology* (4): 31–32.

Zhang, Y.L. and Y.F. Zheng. 2006. A preliminary study on temperature increasing effect of different mulching methods in gravel mulched watermelon field. *China Agricultural Meteorology* 27(4): 323–325.

Zhao, X.G. 2009. A study on the rise and decline of cotton planting in Lanzhou district in the Qing Dynasty. *Ancient and Contemporary Agriculture* (2): 72–76.

Zhao, Y., C.J. Li, J.H. Kang, et al. 2009. Study on development and application of *shatian* in Ningxia. *Journal of Agricultural Science* 30(2): 35–38.

Zheng, H.P. 2004. On the comprehensive effects of conservation tillage practices and evaluation on its ecological and economical benefits. Ph.D. thesis of Gansu Agricultural University.

18 Improving Wheat Yield and Water-Use Efficiency under Semiarid Environment

U.S. Southern Great Plains and China's Loess Plateau

Qingwu Xue, Wenzhao Liu, and B.A. Stewart

CONTENTS

18.1 INTRODUCTION

Wheat (*Triticum aestivum* L.) is grown in a wide range of environments around the world and has the broadest adaptation of all cereal crop species (Briggle and Curtis 1987). Although the majority of wheat is produced in the range of 25–50° latitude, wheat production has expanded to the lower latitude (15°) as a cool season crop and to the higher latitude (60°N) as a warm season crop (Musick and Porter 1990). In addition, wheat is the number one food grain consumed directly by humans and provides more nourishment for people than any other food source. Wheat grains are the most important source of carbohydrates in the countries of temperate zone (Briggle and Curtis 1987). Therefore, wheat production plays a critical role in the world economy and food security.

Winter wheat is a major crop grown in the U.S. central and southern Great Plains (SGP). From 2008 to 2010, the wheat production from Colorado, Kansas, Oklahoma, and Texas was 27%–32% of the U.S. total production, ranging from 16.5 to 19.5 million tons (NASS 2010). In the U.S. SGP (the area includes the Texas High Plains, the Oklahoma Panhandle, parts of eastern New Mexico, southwestern Kansas, and southeastern Colorado), winter wheat is widely grown under dryland (rainfed), full-irrigation, and deficit-irrigation production systems and produced for both grain and winter cattle forage (Musick and Dusek 1980; Musick et al. 1994; Howell et al. 1995). The area has a semiarid climate with annual precipitation ranging from 380 mm in the southwest to 580 mm in the northeast and averaging about 480 mm. Growing season precipitation for wheat production averages about 250 mm (Musick et al. 1994). The seasonal evapotranspiration (ET) for winter wheat growth ranges from 700 to 950 mm when irrigation is used to prevent plants from water stress (Musick and Porter 1990; Musick et al. 1994; Howell et al. 1995, 2007). Therefore, the seasonal precipitation for winter wheat can only meet one-third of the ET required for maximum grain yield. As a result, wheat yield and water-use efficiency (WUE) are primarily limited by soil water deficits from late spring to early summer (Musick et al. 1994; Howell et al. 1997). Under dryland conditions, wheat production is largely determined by the amount and effective use of soil water storage and seasonal precipitation (Jones and Popham 1997). However, dryland wheat yields are generally much lower than irrigated wheat in the area (Musick et al. 1994).

In the U.S. SGP, irrigation application provides a means to maintain high yields and productivity in wheat. The irrigation water is mainly from the Ogallala Aquifer (Figure 18.1), and development of irrigation in this region significantly increased during the 1950s. The Ogallala is essentially a closed system with minimal recharge capacity, and the dramatic increase in water extraction for crop irrigation resulted in a significant decline in the water table; some areas have experienced up to 50% reduction in predevelopment saturated thickness. Irrigated land area has decreased from a peak of 2.4 million ha in 1974 to 1.9 million ha in 2000 (Colaizzi et al. 2008).

Similar to many other agricultural regions in the United States, the southern High Plains faces considerable challenges related to the domestic and global economy, climate change, and societal concerns over the impacts of agriculture on environments. Currently, several key issues hinder an adequate supply of wheat in the future. First, the growing world population continuously requires more production of food, forage, and fiber, particularly the major food crops such as wheat and corn (*Zea mays* L.). Second, in recent years, there is an increasing demand for

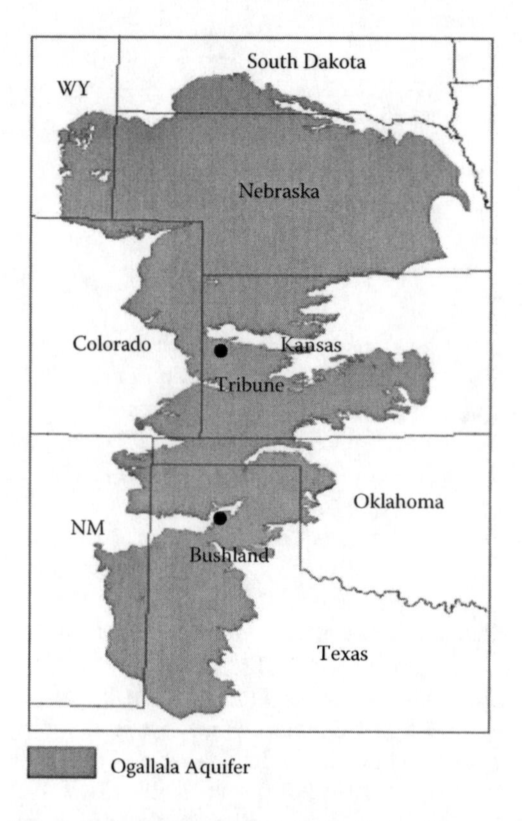

FIGURE 18.1 The map of Ogallala Aquifer in the U.S. Great Plains. The two locations (Bushland, Texas, and Tribune, Kansas) for long-term wheat experiments are shown on the map.

biofuel production in the United States, adding another pressure to current arable land areas. The increasing demands for energy independence and development of bioenergy crops in the United States mandate to consistently increase crop yields (Biomass Research and Development Board 2009). Third, the declining water table in the Ogallala Aquifer and increasing pumping costs will inevitably reduce irrigation (Musick et al. 1994; Stone and Schlegel 2006; Colaizzi et al. 2008). Fourth, the possibility for increasing frequency and severity of drought stress as well as other abiotic and biotic stresses under changing climate will likely reduce crop yields more frequently. The wheat yield and economic losses from the historic 2011 drought in Texas resulted in $5.2 billion agricultural losses and the losses from wheat alone were $243 million. The last two drought years in 2006 and 2009 resulted in $4.1 and $3.6 billion economic losses in Texas agriculture, respectively (Fannin 2011).

The Loess Plateau (LSP) of China is located in the middle reaches of the Yellow River, with a total area of over 600,000 km^2 and a population of 82 million (Figure 18.2). The region also has a semiarid climate with average annual precipitation in the range of 300–600 mm (Kang et al. 2002; He et al. 2003; Fan et al. 2005a). Precipitation distribution is very uneven, with more than 60% of precipitation falling in three summer months (July, August, and September). Wheat is the major crop in the region, and winter wheat is grown mainly in the southern part and spring wheat in the northwestern part of the LSP (Wang et al. 2009). The majority of crops including wheat are grown under dryland, which accounts for 90% of the total cropland area (Fan et al. 2005a). In the southern part of the LSP, continuous wheat (CW) is the most common cropping system, in which wheat is generally harvested in June and planted in late September or early October after a short

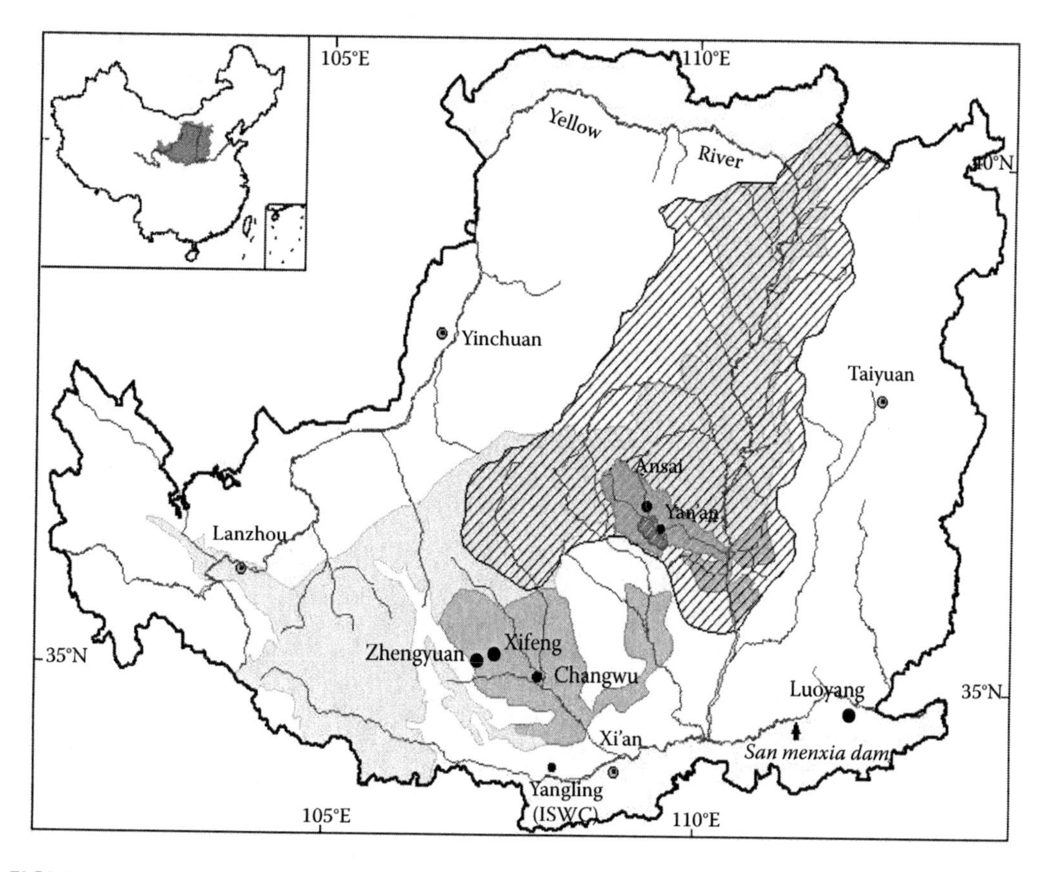

FIGURE 18.2 The map of China's Loess Plateau with an insert of the national map. The locations of long-term field experiment sites in winter wheat are shown on the map.

3 month fallow period. Wheat yields have increased significantly in the last 30 years in the LSP. For example, Huang et al. (2003) reported that average wheat yield has increased from 1.7 Mg/ha in the middle of the 1980s to 2.9 Mg/ha in the middle of the 1990s, as a result of increasing fertilizer applications. Recent reports showed that dryland winter wheat yields can be as high as 6 Mg/ha in research plots (Wang et al. 2011).

Like any other dryland farming area in the world, wheat production in the LSP is highly related to soil water storage and growing season precipitation for high yield (Li 1983; Huang et al. 2003; Liu et al. 2010). Therefore, drought stress is the most common factor limiting wheat yield. Although irrigation is possible in some areas along the rivers (Yellow River and its branches) or at surface water resources (e.g., dams and reservoirs), the available water resources in the region have been declining due to increasing water consumption for crop production and urban residential and industrial uses, hydraulic project constructions, and more frequent occurrences of drought (He et al. 2003). Therefore, the future wheat production in the LSP will likely depend on dryland production. Currently, the wheat production in the region also faces other challenges such as interests shifting from grain crops to cash crops (e.g., vegetables and produces), increasing nonfarming job opportunities, raising costs of fertilizers and farm machinery, environmental quality degradation, and possibly more unpredictable drought and heat stresses due to climate change (Fang et al. 2010).

Since water is the most important factor affecting crop production in the semiarid areas, the development of crop management practices to conserve water, optimize water use, and improve crop WUE becomes critical, particularly under a changing climate condition. Although increasing yield is an ultimate goal to any crop production system, maximizing WUE is of particular importance under water-limited conditions since lower WUE sometimes indicates a poor management and inefficient use of available soil water (ASW) (Musick et al. 1994; Passioura and Angus 2010). In this chapter, we review the research progress and identify the differences and similarities for improving wheat yield and WUE in the U.S. SGP and LSP of China. The development of

TABLE 18.1
Data Sets Used for Analysis of Yield, Evapotranspiration (ET), Water-Use Efficiency (WUE), Biomass and Harvest Index (HI), and Their Relationships in the U.S. Southern Great Plains and Loess Plateau of China

Region	Site	Location	Treatments	Years	References
U.S. Southern Great Plains	Bushland, Texas	35°11′ N 102° W	Irrigation and N rates	1981–1999	Eck (1988); Howell et al. (1995); Xu et al. (1998); Schneider and Howell (2001); Xue et al. (2006)
	Bushland, Texas	35°11′ N 102° W	Tillage and rotation	1984–1993	Jones and Popham (1997)
	Tribune, Kansas	38°28′ N 101° 45′W	Tillage and rotation	1973–2004	Stone and Schlegel (2006)
Loess Plateau of China	Changwu, Shaanxi	35°11′ N 107° 40′E	Fertilization and irrigation	1986–2009	Kang et al. (2002); Huang et al. (2004); Liu et al. (2007); Wang et al. (2011)
	Xifeng, Gansu	35°40′ N 107° 51′E	Crop rotation	1988–1991	Li et al. (2000, 2002)
	Zhengyuan, Gansu	35°30′ N 107° 28′E	Irrigation and mulching	1997–2003	Fan et al. (2005a)
	Luoyang, Henan	34°30′ N 113° E	Tillage and rotation	1999–2005	Su et al. (2007)

management strategies is the major focus, although breeding has played an important role for yield improvement. First, we examine the wheat yield—ET, WUE, and their relationships under a wide range of management practices. Then, we identify the current knowledge gaps and research priorities for improving yield and WUE to face future climate change and social economic challenges. In the entire chapter, we define WUE as a ratio of grain yield and seasonal ET, with a unit of kg/m^3. Another unit of WUE widely used in the literature is kg/ha/mm, which can be converted to kg/m^3 by multiplying by 0.1. The data sets for analysis in this chapter were mostly from long-term field experiments in both regions (Table 18.1).

18.2 YIELD DETERMINATIONS UNDER WATER-LIMITED CONDITIONS

For wheat, Passioura (1977) introduced a framework for identifying the important components for grain yield under water-limited conditions. Grain yield is determined by three components:

$$\text{Yield} = \text{ET} \times \text{WUE}_{\text{bm}} \times \text{HI}, \tag{18.1}$$

where ET is the seasonal evapotranspiration, WUE_{bm} is the water-use efficiency for biomass production, and HI is the harvest index, that is, the fraction of biomass partitioning to grains (Passioura 1977). Since these three components are likely to be largely independent of each other, an improvement in any one of them should result in an increase in yield as well as WUE. This framework has been proved to be very useful for identifying management strategies under water-limited conditions (Richards et al. 2002). Since the product of ET and WUE_{bm} is biomass at maturity (BM), Equation 18.1 can be rewritten as (Equation 18.2)

$$\text{Yield} = \text{BM} \times \text{HI}. \tag{18.2}$$

Improving biomass production, HI, or both will lead to higher yield under water-limited conditions (Blum 2009). In the SGP, Howell (1990b) showed that wheat improvement over the years was mainly contributed by increased biomass. The potential HI was relatively stable in the semi-dwarf cultivars. However, HI is generally low due to water stress during reproductive stages under semiarid environment (Schneider and Howell 2001; Xue et al. 2006).

18.2.1 YIELD, ET, AND WUE RELATIONSHIPS

In the U.S. SGP, the long-term dryland wheat yields ranged from 0 to 5 Mg/ha and seasonal ET ranged from 200 to 600 mm (Musick et al. 1994; Jones and Popham 1997; Xue et al. 2003, 2006; Stone and Schlegel 2006). In Bushland, Texas, the dryland yield was mostly 1–2 Mg/ha in long-term tillage and rotation studies (Jones and Popham 1997). However, dryland yields in irrigation studies were frequently over 3 Mg/ha (Musick et al. 1994; Xue et al. 2006). In Tribune, Kansas, the average dryland yield (3 Mg/ha) was generally higher than those in Bushland, Texas (1–2 Mg/ha) because Tribune is located north of Bushland where the evaporative demand is lower and the distribution of precipitation is more favorable during the reproduction period. Nevertheless, there were years with zero yields at both locations as a result of severe drought during the growing season (Musick et al. 1994; Stone and Schlegel 2006). The irrigated wheat yields ranged from 3.0 to 7.7 Mg/ha and ET from 400 to over 900 mm, depending on irrigation timing and frequency (Musick et al. 1994; Howell et al. 1995; Xue et al. 2003, 2006; Schneider and Howell 2001). Wheat yields under full irrigation were in the range of 5.3–7.7 Mg/ha and required about 700–950 mm ET (Howell et al. 1995; Schneider and Howell 2001; AgriPartners 2007). Under dryland conditions, WUE ranged from 0 to 0.8 kg/m^3, with an average of 0.4 kg/m^3. The WUE for irrigated wheat was higher than for dryland wheat and ranged from 0.5 to 1.2 kg/m^3 (Musick et al. 1994; Xue et al. 2006).

In the LSP of China, there exists a wide range of dryland wheat yields, ranging from 1.2 to 6.0 Mg/ha. The seasonal ET generally ranges from 200 to 500 mm. In the data collected in this paper, none of the data sets showed zero yields. The irrigated wheat yield in the region ranges from 3 to 7 Mg/ha but ET is mostly in the range of 300–500 mm, similar to the range of dryland wheat. Huang et al. (2004) showed a wider range of ET but the maximum ET was only about 650 mm. The wheat WUE in the LSP is generally higher than that in the U.S. SGP. For both dryland and irrigated wheat plots, WUE mostly ranges from 0.8 to 1.6 kg/m^3.

The wheat yield–ET relationship has been reported in different studies in the two regions (Musick et al. 1994; Schneider and Howell 2001; Kang et al. 2002; Huang et al. 2004; Stone and Schlegel 2006). Musick et al. (1994) summarized the yield–ET relationship based on long-term data in dryland and irrigated plots from 1958 to 1992 at Bushland, Texas. They showed a linear relationship between grain yield and seasonal ET pooling dryland and irrigated data together. The linear regression between yield and ET resulted in a slope of 1.22 kg grain yield per cubic meter of seasonal ET (kg/m^3) and a threshold of 206 mm ET (Musick et al. 1994). The threshold ET is defined as the minimum amount of ET required before any grain is produced (Musick et al. 1994; Kang et al. 2002). We analyzed the yield–ET relationship again by using the data from Table 18.1 and the results are shown in Figure 18.3. Similarly, there was a significant linear relationship between yield and ET, and the regression resulted in a slope of 1.06 kg/m^3 and a threshold of 175 mm ET ($Y = 0.0106X - 1.7393$, $R^2 = 0.75$, $P < 0.001$). The new data showed a slightly lower slope but a lower ET threshold as compared to those of Musick et al. (1994). Stone and Schlegel (2006) summarized dryland wheat data from 1974 to 2004 and showed a linear relationship between wheat yield and ET ($Y = 0.01X - 1.838$, $R^2 = 0.64$, $P < 0.0001$). The slope was 1.0 kg/m^3 and the ET threshold was 183 mm, which were close to the results from Bushland in Figure 18.3. The WUE based on the above linear regression analysis was about 1.0 kg/m^3 in the SGP region.

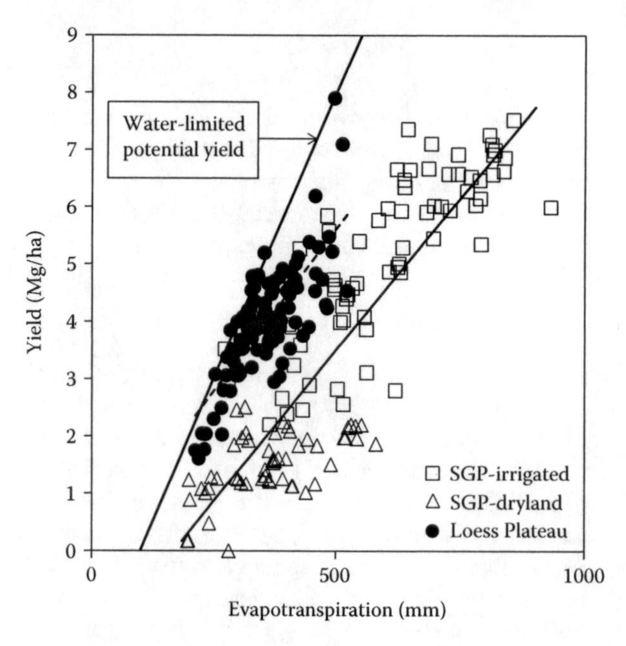

FIGURE 18.3 The linear relationship between yield and evapotranspiration (ET) in the U.S. southern Great Plains (SGP) (Bushland, Texas) and the Loess Plateau of China (LSP) based on the published data from experiments listed in Table 18.1. The lower solid line is the regression line for the SGP ($Y = 0.0106X - 1.7393$, $R^2 = 0.75$, $P < 0.001$), the dashed line is the regression line for the LSP ($Y = 0.0114X - 0.0645$, $R^2 = 0.60$, $P < 0.001$), and the upper solid line is the water-limited potential yield as a function of ET with a slope of 2.0 kg/m^3 and a threshold ET of 100 mm. The slope of 2.0 kg/m^3 represents an upper limit of WUE in wheat under water-limited conditions. (From Passioura, J.B. and Angus, J.F., *Adv. Agron.*, 106, 37–75, 2010.)

In the LSP, the yield–ET relationship was also a linear function ($Y = 0.0114X - 0.0645$, $R^2 = 0.60$, $P < 0.001$), pooling all data sets from Table 18.1. The linear regression of yield and ET in the LSP had a greater slope (1.14 kg/m^3) than that of the U.S. SGP (1.0 kg/m^3). Wheat yields are mostly higher in the LSP than in the SGP at any given ET level (Figure 18.3). Therefore, wheat in the LSP generally had higher WUE than in the U.S. SGP. The estimate of threshold ET for the LSP based on the linear regression in Figure 18.3 is unrealistically low (6 mm). The threshold ET may be estimated based on individual data sets. Kang et al. (2002) found a quadratic function between yield and ET and estimated the threshold ET of 152 mm. Huang et al. (2004), however, showed a linear regression equation between yield and ET ($Y = 0.0112X - 1.1358$, $R^2 = 0.66$, $P < 0.001$) and the threshold ET was 102 mm. Fan et al. (2005a) reported another linear relationship between yield and ET ($Y = 0.0115X - 0.696$, $R^2 = 0.55$, $P < 0.001$) and the estimated threshold ET was 61 mm. Obviously, the threshold ET for grain production was lower in LSP (61–152 mm) than in the U.S. SGP (175–206 mm).

The threshold ET is associated with the seasonal water loss through soil evaporation (Es). However, the threshold ET is not true Es and the accurate seasonal Es can be calculated in different methods (Howell 1990a). Estimate of Es based on linear regression of biomass and ET varied from 80 to 120 mm, with an average of 100 mm (Angus and van Herwaarden 2001). The Es generally counted 30% of total ET in some early studies in wheat (Angus et al. 1983; Howell 1990a). The threshold ET represented about 20% of maximum ET at Bushland, Texas, and still about 30% of maximum ET at Tribune, Kansas (Figure 18.3; Stone and Schlegel 2006). The threshold ET in the LSP was generally less than that in the SGP, and represented 12% and 16% of ET in the data sets from Fan et al. (2005a) and Huang et al. (2004), respectively. However, the threshold ET was 30% of ET based on Kang et al. (2002). It should be mentioned that threshold ET is lower in the areas with lowering seasonal ET (Musick et al. 1994). In the SGP, the maximum ET can be as high as 950 mm (Howell et al. 1995). However, the maximum ET in the LSP is generally less than 650 mm (Huang et al. 2004). The lower percentage of threshold ET (12%) from Fan et al. (2005a) was because of the use of plastic film mulching in their study, which can significantly reduce soil evaporation (Fan et al. 2005a).

The analysis of yield–ET relationship provides an important tool to identify wheat potential yield and yield gap between actual yield and attainable yield under water-limited conditions (Howell 1990a; Musick et al. 1994; Angus and van Herwaarden 2001). Angus and van Herwaarden (2001) set an upper limit of WUE in wheat under water-limited conditions based on the work of French and Schultz (1984). The upper solid line in Figure 18.3 is the water-limited potential yield as a function of ET with a slope of 2.0 kg/m^3 and a threshold ET of 100 mm. The slope of 2.0 kg/m^3 represents an upper limit of WUE in wheat under water-limited conditions (Angus and van Herwaarden 2001; Passioura and Angus 2010). In the LSP, there were some data showing that the wheat yield achieved to upper limit level. However, most of the yield data fall under the upper limit. In the case of the SGP, almost none of the data sets could reach the upper limit of yield potential. Figure 18.3 clearly showed that there is a large gap between potential and actual yields under water-limited conditions in both regions, and the gap was larger in the U.S. SGP. For each region, environmental conditions and management practices may both be responsible for the yield gaps. However, management practices can be more important to increase yield potential since environmental conditions are hard to control.

The relationship between wheat WUE and seasonal ET is shown in Figure 18.4. In the LSP, there was no clear relationship between WUE and ET. At any ET level (200–500 mm), WUE varied from as low as 0.7 kg/m^3 to as high as 1.5 kg/m^3. Although there was a positive relationship between WUE and ET in the SGP by pooling dryland and irrigated data together, the WUE also varied largely at the same ET level. The lack of significant relationship between WUE and ET indicated that WUE can be achieved under a wide range of ET. The variability in the amount and distribution of seasonal precipitation could be a major source of variation in ET and WUE (Musick et al. 1994). In the U.S. SGP, timing and frequency of limited irrigation might also contribute to part of

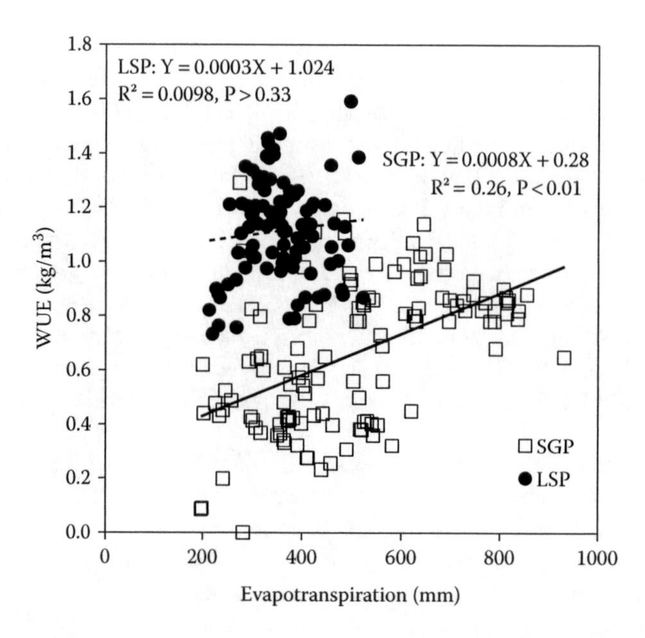

FIGURE 18.4 The linear relationship between water-use efficiency (WUE) and evapotranspiration (ET) in the U.S. southern Great Plains (SGP) (Bushland, Texas) and the Loess Plateau of China (LSP) based on the published data from experiments listed in Table 18.1.

the variation in WUE and ET. For example, at a similar ET level, irrigation between booting and anthesis stages could result in higher yield than irrigation from tillering to jointing (Musick et al. 1994). In the LSP, fertilization may significantly contribute to the variation in WUE given the same level of ET. Wang et al. (2011) showed that the average WUE in treatments fertilized with N, P, and manure was about 1.4 kg/m^3. However, WUE in treatments without fertilization or fertilized only with N averaged about 0.5 kg/m^3. All the treatments had about the same ET of 300 mm.

The relationship between WUE and yield is shown in Figure 18.5. In the LSP, WUE increased linearly as yield increased when pooling all the data from Table 18.1. In the region, a yield increase in 1 Mg/ha would lead to in an increase of 0.12 kg/m^3 in WUE. From individual data sets, Huang et al. (2004) showed a quadratic relationship between WUE and yield, and WUE did not increase further when yield was over 5 Mg/ha. In the U.S. SGP, the WUE–yield relationship was a quadratic function when the full range of yield was considered, which was similar to Huang et al. (2004). The WUE increased linearly when yield increased up to 4–5 Mg/ha. When yield increased further, WUE maximized and even tended to decrease. However, the initial slope between WUE and yield relationship was higher in the SGP than that in the LSP, with an increase of 0.28 kg/m^3 for every 1 Mg/ha of yield increase. Figure 18.5 indicated that higher WUE generally can be achieved with higher yields. However, a curvilinear relationship between WUE and yield showed that WUE might not be the highest when yield was in the high range. This is particularly true in the SGP. The ET demand could be as high as 12 mm/day and was frequently over 60 mm/week in irrigated wheat due to high winds and associated high vapor pressure deficit (Howell et al. 1995). As a result, it is difficult to achieve a very high WUE with full irrigated wheat in the SGP.

18.2.2 Biomass and Harvest Index

The relationship between aboveground biomass and ET in the SGP (based on Eck 1988; Schneider and Howell 2001; Xue et al. 2006) and the LSP (based on Kang et al. 2002) is shown in Figure 18.6a. In both regions, biomass increased linearly as ET increased. The linear regression of biomass and ET resulted in the same slope in the two regions (3.43 kg/m^3). However, biomass was always

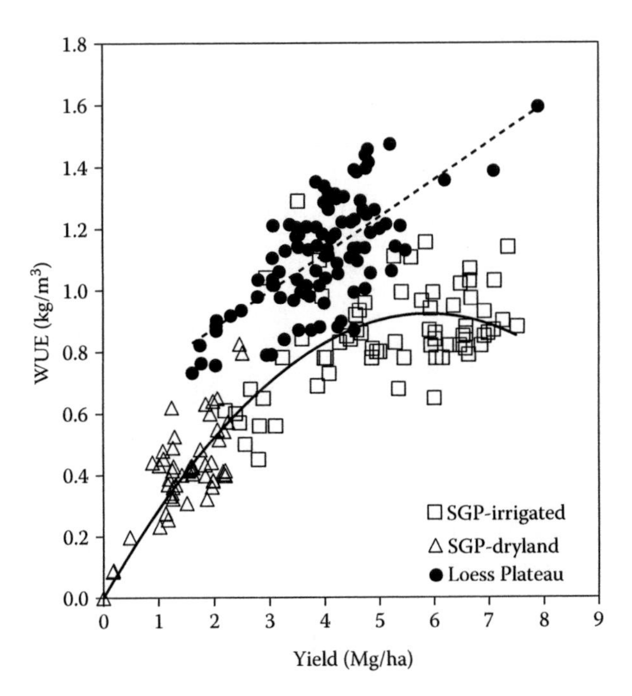

FIGURE 18.5 The relationship between water-use efficiency (WUE) and yield in the U.S. southern Great Plains (SGP) (Bushland, Texas) and the Loess Plateau of China (LSP) based on the published data from experiments listed in Table 18.1. (Regression equations: SGP: $Y = -0.0267X^2 + 0.3138X$, $R^2 = 0.785$, $P < 0.001$; LSP: $Y = 0.1206X + 0.6346$, $R^2 = 0.4746$, $P < 0.001$).

higher in the LSP than that in the SGP at a given ET level, indicating that wheat has a higher WUE_{bm} in the LSP environment (Figure 18.6a). However, the biomass difference at a given ET between two regions was much smaller than the yield difference (Figure 18.3). This suggests that the WUE_{bm} in the two regions is close but the WUE is very different. In the SGP, there was a wide range of biomass in different irrigation treatments (Eck 1988; Schneider and Howell 2001; Xue et al. 2006).

In both regions, grain yield is a linear function of biomass (Figure 18.6b). However, the slope of linear regression of yield and biomass was smaller in the SGP (0.23) than in the LSP (0.35), indicating that wheat in the LSP has higher HI than in the SGP. Kang et al. (2002) showed that the HI ranged from 0.25 to 0.45 in the LSP. Liu et al. (2007) reported a very high range of HI (0.41–0.62) in an irrigated study. In the SGP, Schneider and Howell (2001) showed a low HI range in irrigated wheat (0.21–0.32). Although Eck (1988) reported some high HI values (0.36–0.50), the HI is about 0.35 based on long-term wheat yield and biomass data (Howell 1990b). Nevertheless, increased WUE is largely contributed by increased HI under water-limited conditions as shown in Figure 18.7. The improved WUE as a result of higher HI was also reported in the Northern China Plain (Zhang et al. 1998) and Mediterranean environment (Oweis et al.2000).

18.3 MANAGEMENT PRACTICES FOR IMPROVING YIELD AND WUE

The above analysis of yield, ET, WUE, biomass, HI, and their relationships indicated that there were large variations in yield and WUE at different ET levels in both regions. As such, development of management practices is important to improve yield and WUE under water-limited conditions. Generally, there are two ways to improve crop performance: breeding and management practice. Improving wheat yield and WUE through breeding has been a major focus in semiarid environments (Richards et al. 2002, 2010; Reynolds et al. 2005). In the SGP, changes of grain yield under full irrigation over the years can reflect the genetic improvement of yield potential. In Bushland,

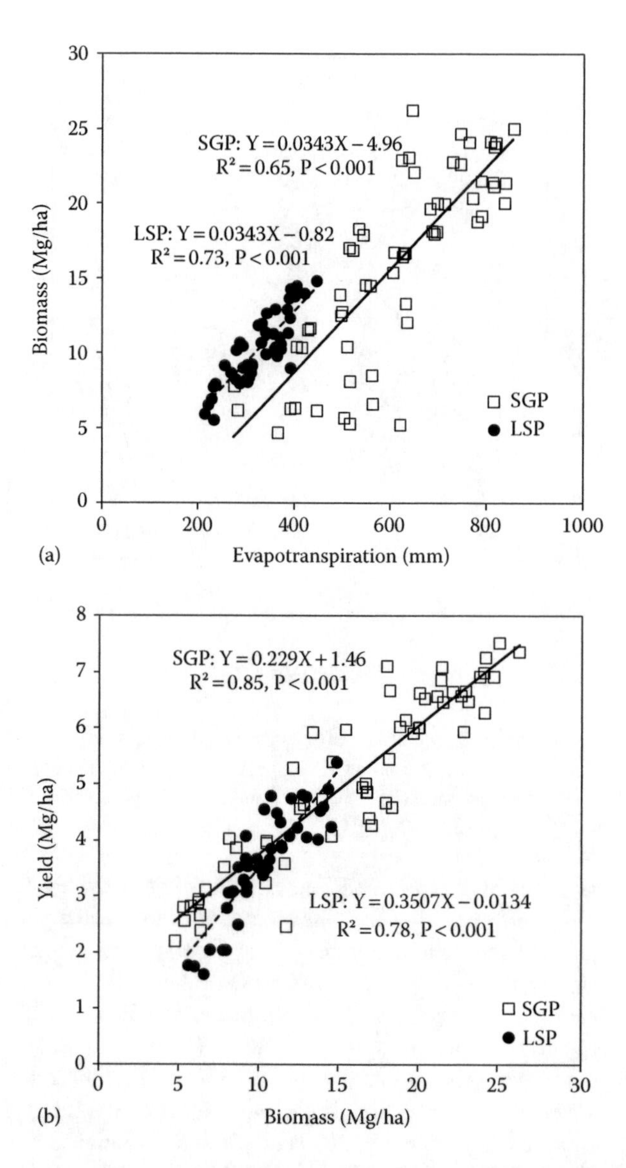

FIGURE 18.6 The relationship between (a) biomass and ET and (b) yield and biomass (SGP: Southern Great Plains; LSP: Loess Plateau). Data for SGP are from Eck, H.V., *Agron. J.*, 80, 902–908, 1988; Schneider, A.D. and Howell, T.A., *Trans. ASAE*, 44, 1617–1623, 2001; and Xue, Q., Zhu, Z., Musick, J.T., Stewart, B.A., and Dusek, D.A., *J. Plant Physiol.*, 163, 154–164, 2006. Data for LSP are from Kang, S., Zhang, L., Liang, Y., Hu, X., Cai, H., and Gu, B., *Agric. Water Manag.*, 55, 203–216, 2002.

Texas, the yield at full irrigation increased from 5.5 Mg/ha in the 1970s to 7.1 Mg/ha in the 1990s (Musick and Dusek 1980; Musick et al. 1994; Schneider and Howell 2001; Xue et al. 2006). The direct comparisons of cultivars released indifferent years showed that yield and WUE of new cultivars (released in the 2000s) increased over 20% as compared to those of an old cultivar (released in the 1970s) (Xue et al. 2010). Wheat yield increased significantly through breeding in China (Rudd 2009; Wang et al. 2009; Fang et al. 2010). In LSP, Zhang (1998) demonstrated that cultivars developed for irrigated conditions generally had higher WUE than dryland cultivars. In addition to yield improvement, WUE has also been increased through breeding. Zhang et al. (2010) demonstrated that wheat WUE increased from 1.0–1.2 kg/m³ for cultivars from the early 1970s to 1.4–1.5 kg/m³ for recently released cultivars in the Northern China Plain.

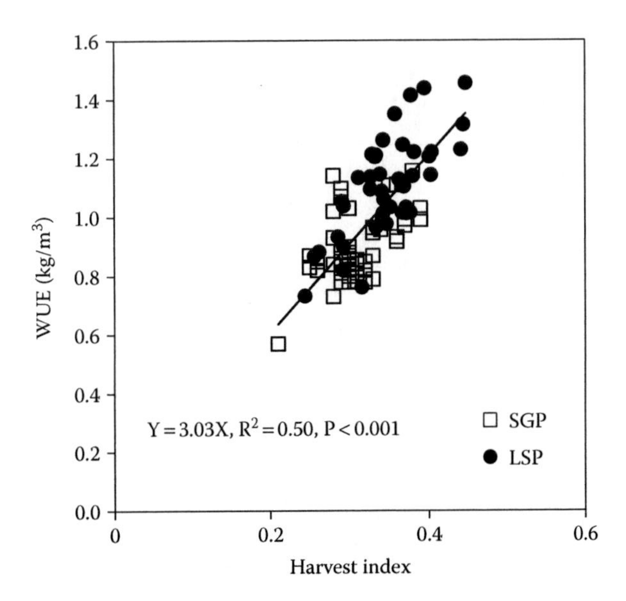

FIGURE 18.7 The relationship between water-use efficiency (WUE) and harvest index (HI). Data are from Xu, M., Xue, Q., and Musick, J.T., In *Ecophysiological Basis of Dryland Farming*, The Chinese Academic Press, Beijing (in Chinese), 1998; Kang, S., Zhang, L., Liang, Y., Hu, X., Cai, H., and Gu, B., *Agric. Water Manag.*, 55, 203–216, 2002; Schneider, A.D. and Howell, T.A., *Trans. ASAE*, 44, 1617–1623, 2001; and Xue, Q., Zhu, Z., Musick, J.T., Stewart, B.A., and Dusek, D.A., *J. Plant Physiol.*, 163, 154–164, 2006.

Management practices are as important as breeding to improve crop yield, water use, and WUE under water-limited conditions (Passioura and Angus 2010; Richards et al. 2010). Optimizing major management practices is important to maximize crop yields under water-limited conditions. Improved crop management is responsible for a large portion of increased productivity under water-limited conditions (Anderson 2010).

18.3.1 Soil Water Conservation

Under semiarid environmental conditions, dryland wheat production is determined by both soil water storage and precipitation during the growing season since precipitation can meet only part of the seasonal ET requirement. Soil water storage at planting has long been a major focus in dryland wheat production (Musick et al. 1994; Stone and Schlegel 2006). In the SGP, Musick et al. (1994) analyzed the relationship between wheat yield and ASW at planting (1.8-m profile) based on 34 years of dryland wheat data in Bushland, Texas. There was a linear relationship between yield and ASW at planting ($Y = 0.0157X - 0.94$, $R^2 = 0.34$, $P < 0.001$). The linear regression resulted in a yield response of 1.57 kg/m³ to soil water storage in a 1.8-m profile. In other words, increasing 1 mm of ASW at planting would lead to 15.7 kg/ha yield increase (Musick et al. 1994). Stone and Schlegel (2006) showed another linear relationship between ASW at emergence using 30-year dryland wheat data in Tribune, Kansas ($Y = 0.0098X + 0.828$, $R^2 = 0.32$, $P < 0.0001$) and resulted in a yield response of 0.98 kg/m³ to soil water storage at emergence in a 1.8-m profile. In the LSP, soil water storage at planting generally accounted for 35.4% of the seasonal ET, but it can be as high as 52.6% in the years with low growing season precipitation (Li and Su 1991). Fan et al. (2005a) also found a linear relationship between wheat yield and ASW at planting in a 2-m profile based on a 7-year field study ($Y = 0.0172X + 2.422$, $R^2 = 0.38$, $P < 0.001$). The linear regression resulted in a slope of 1.72 kg/m³. The linear relationships between yield and ASW at planting or emergence emphasize the importance of preseason soil water storage and conservation of precipitation for dryland wheat production in both regions. For the linear regressions from Musick et al. (1994) and

Fan et al. (2005a), the slope of the regression between yield and ASW at planting was much higher than that of the regression between yield and ET (1.57 vs. 1.22 kg/m^3 for the SGP, 1.72 vs. 1.15 kg/m^3 for the LSP). This suggests that ASW at planting is important for wheat plants to effectively use precipitation during the growing season (Fan et al. 2005a). The lower R^2 of regression between yield and ASW at planting or emergence (<0.40) indicated that wheat yield is affected by other factors. In the SGP, seasonal precipitation (from October to June) accounted for the majority of dryland wheat yield variation (55%) (Musick et al. 1994). Huang et al. (2004) showed that precipitation during the late growing season (May and June) was important for dryland yield in the LSP.

Soil water storage is determined by the amount of precipitation as well as the efficiency of precipitation storage during the fallow period. In the SGP, management practices for increasing soil water storage have been focused on crop rotation, tillage systems, and residue management (Musick et al. 1994; Jones and Popham 1997; Stone and Schlegel 2006; Stewart et al. 2010). Dryland wheat may be grown under different rotation systems such as CW, wheat–fallow (WF), and wheat–sorghum (*Sorghum bicolor* L.) – fallow (WSF). The tillage systems include conventional tillage (CT) and no-till (NT) with wheat residue mulch. Jones and Popham (1997) investigated the effects of rotation and tillage systems on dryland wheat and sorghum production and WUE in a 10-year study (1984–1993) at Bushland, Texas. Among the different rotation systems, ASW at planting was always lower in the CW system than in the WF and WSF systems. The 10-year average ASW amounts at planting were 156, 212, and 205 mm for CW, WF, and WSF, respectively. For the tillage systems, NT significantly increased ASW at planting (199 mm) as compared to stubble mulch (183 mm). The precipitation storage efficiency (PSE) was generally low in Bushland, Texas, for wheat. Among the various treatments, WSF generally had higher PSE (about 17%) than WF (11%) and CW under stubble mulch (14%). However, NT with wheat residue mulch in CW resulted in a higher PSE (20%) than WSF, indicating that NT is an important practice for improving PSE in the Great Plains environment (Jones and Popham 1997). The above-mentioned long-term study clearly demonstrates that the WSF system under NT provided significant benefits for improving wheat yield and WUE, as a result of increasing ASW at planting and PSE. On average, the annual grain production was 1.8 Mg/ha for WSF (wheat and sorghum) but only 1.0 Mg/ha for CW (wheat only) (Jones and Popham 1997). Norwood (1994) compared conventional and NT tillage under WF and WSF systems. Although they did not find differences in wheat WUE between CT and NT, less evaporation and runoff in the NT system promoted water moving to deeper soil profile in the WF and WSF systems. Stone and Schlegel (2006) sorted the wheat yield and WUE data by tillage and showed that NT significantly increased dryland wheat yield and WUE. The wheat yield response to ET was greater with NT (1.38 kg/m^3) than with CT (0.86 kg/m^3). Stone and Schlegel (2006) also summarized the additional water increase in the soil profile during the fallow period as a result of NT, compared with CT in the Great Plains. The soil water increase due to NT ranged from 15 mm in the SGP to 87 mm in the central Great Plains, depending on rotation systems.

In the LSP, winter wheat is predominately grown under the CW system under CT. The fallow period is only for 3 months (July–September). Since most of the precipitation (>60%) falls during the wheat fallow period, improving PSE becomes critical for enhancing ASW at planting (Shangguan et al. 2002; Jin et al. 2007; Wang et al. 2011). Shangguan et al. (2002) showed that the PSE in the LSP was about 35%–40%. Jin et al. (2007) and Wang et al. (2011) showed a wide range of PSE (12%–43%), depending on year, tillage system, and fertilization. The conventional deep tillage has long been used to improve PSE (Fang et al. 2010). For example, Wang et al. (2002) showed that about 90% precipitation can be stored in soil by deep tillage. However, deep tillage may affect soil properties without residue mulching (Fang et al. 2010). Jin et al. (2007) investigated the effect of five tillage systems on dryland wheat yield, PSE, and precipitation use efficiency (PUE) in a 5 year field study (2001–2005). The tillage systems included CT, subsoiling with mulch (SS), NT with mulch (NT), reduced tillage (RT), and two crops per year (TC, winter wheat and peanut). Among the tillage systems, SS resulted in highest PSE, PUE, and crop yield. Although the NT system was slightly less effective than SS, NT had a significant effect on water conservation and increased PSE by 12%

as compared to CT (Jin et al. 2007). In contrast to the SGP, the NT practice is generally not popular in the LSP. However, the NT may provide a long-term benefit and continuous NT practice would result in a higher yield in the long run (Jin et al. 2007).

Crop residue mulch can provide an effective way to increase PSE and WUE. Jin et al. (2007) found that the highest PSE (41.6%) was obtained under NT with wheat straw mulch. Deng et al. (2006) indicated that using crop residue mulching can reduce soil evaporation by 36% in winter wheat. Wheat straw mulching can be easily implemented and has been increasingly adopted by producers in the LSP (Deng et al. 2006). Plastic film mulching is another technique to conserve soil water. Fan et al. (2005a) reported PSE, yield, and WUE in a 7-year field study using plastic film mulching. They showed that plastic film mulching resulted in a very high PSE (70%) as compared to without plastic film mulching (35%). Meanwhile, the ASW at planting (2-m profile) increased from 62 to 123 mm as a result of plastic film mulching. However, plastic film mulching is not widely used in field crops. Instead, it is applied more often in some high-value cash crops such as vegetable and oil crops. In addition, plastic film mulching may cause environmental problems such as difficulty of removal and disposal (Gao et al. 2009; Fang et al. 2010). Fertilization is an important practice to increase PSE in the LSP. Wang et al. (2011) reported PSE, yield, and WUE in a long-term fertilization study. They found that wheat plots fertilized with N (120 kgN/ha) and P (60 kgP_2O_5/ha) significantly increased PSE from 28% to 34% as compared to those without fertilization. As a result, fertilization significantly increased wheat yield and WUE (Wang et al. 2011).

18.3.2 Seeding Factors

Management practices related to seeding and wheat stand establishment include seeding date, rate, depth, row spacing, and quality (seed size, protein content, etc.). Among the seeding factors, seeding date and rate have long been studied since they are easily controlled by producers in most cropping systems (Paulsen 1987). Selection of optimum seeding date and rate can be important for wheat yield and WUE (Musick and Porter 1990; Chen et al. 1991; Winter and Musick 1993; Shangguan 1998). In the U.S. SGP, there is a wide range of seeding date for winter wheat (August–November). The wide range in seeding date in the region is because of dryland and irrigated cropping systems either with or without grazing, and highly variable weather conditions. For grain production, early October is normally considered as optimum seeding date. However, the seeding date may be delayed to November due to dry weather conditions (Winter and Musick 1993). Musick and Dusek (1980) investigated the seeding date and irrigation effects on wheat phenological development and yield. In general, early (September 15) and normal (October 12) seeding dates resulted in a higher yield than a later seeding date (November 7). Early to normal seeding significantly increased biomass production as compared to late seeding. Seeding date affected soil water use and root growth pattern. Winter and Musick (1993) showed that wheat plants seeded in August had low grain yield since the plants used most of the soil water in the fall and experienced severe water stress in the spring. Wheat seeded in October used as much soil water as August seeding by anthesis and produced high grain yield. By contrast, wheat seeded in November had a very shallow root system, did not use much soil water, and produced low grain yield (Figure 18.8). In the U.S. Great Plains, the most widely recommended seeding rate for wheat is 67 kg/ha or 200 seeds/m^2 (Paulsen 1987). However, seeding rate varied with cropping systems (dryland or irrigated) and seeding date (early or late) (Paulsen 1987; Musick and Porter 1990; Musick et al. 1994). Seeding rate ranged from 60 to 100 kg/ha (176–300 seeds/m^2) in irrigated wheat and from 35 to 50 kg/ha (100–150 seeds/m^2) in dryland wheat (Musick and Porter 1990; Musick et al. 1994). Seeding date and rate can interact to affect wheat grain yield (Dahlke et al. 1993; Spink et al. 2000). Higher seeding rate is normally required if the seeding date is delayed from optimum date because any delay in seeding normally reduces plant growth and tiller production (Dahlke et al. 1993; Spink et al. 2000). In general, the seeding rate should be increased by 10% per week of delay in seeding after the optimum seeding date (Paulsen 1987).

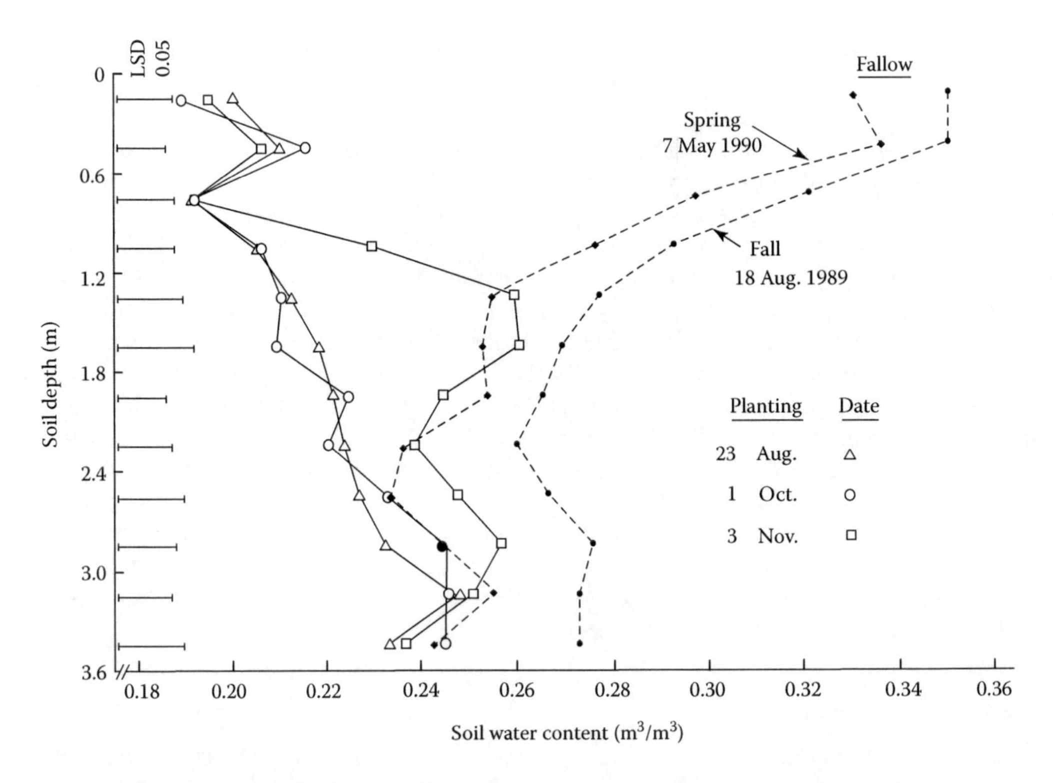

FIGURE 18.8 Soil water content near anthesis (May 7, 1990) for wheat at three planting dates (solid lines) and a fallow check plot on two dates (dashed lines). (From Winter, S.R. and Musick, J.T., *Agron. J.*, 85, 912–916, 1993.)

In the LSP, the range of seeding date for winter wheat is generally narrow because seeding normally occurs at the end of the rainy season. However, seeding date is still an important factor affecting yield and WUE (Chen et al. 1991; Shangguan 1998; Zhang 1998). Chen et al. (1991) reported that the optimum seeding date was in the middle of September (14–18). Either early or late seeding for just 1 day from optimum seeding date would reduce yield by about 150 kg/ha. The seeding date also interacts with N fertilization. For example, plants in early seeding were generally more responsive to N fertilizer than those in late seeding (Chen et al. 1991). The seeding rate in the LSP ranged from 150 to 200 kg/ha (Chen et al. 1991; Li et al. 2000; Huang et al. 2004) for both dryland and irrigated wheat. The seeding rate for dryland wheat is much higher than that in U.S. Great Plains.

In the SGP, early studies used wide row spacing (25–34 cm; Musick et al. 1994), while new studies generally used narrow row spacing (15 cm; Xue et al. 2010). When grown under conditions of relatively high yield potentials, narrow row spacing increased yields (Musick and Porter 1990). Seed quality can significantly affect wheat yield and WUE. Xue and Stougaard (2006) observed that spring wheat grown from large seeds generally had higher yield and WUE than those grown from small seeds. Growing large seeds significantly reduced soil evaporation (Xue and Stougaard 2006).

18.3.3 Fertilization

Plant growth and development are dependent on soil nutrients supply. Improving nutrient status can significantly increase wheat yield and WUE. In both the U.S. SGP and the LSP of China, nitrogen (N) and phosphorus (P) are two major fertilizers applied for wheat production. The soils in both regions generally have sufficient potassium (K) (Musick and Porter 1990; Wang et al. 2011). Nitrogen is often the most limiting nutrient for wheat production and represents one of the highest input costs in agricultural systems (Thomason et al. 2002). Due to the semiarid environmental conditions, N fertilization had a significant interaction with soil water levels. In the SGP, Eck (1988) investigated the interaction of N and irrigation in wheat. The yield and WUE response to N fertilizer rates were largely related to irrigation frequency and water stress. For full irrigation treatment, yield and WUE increased as N rate increased up to 140 kg/ha. For limited irrigation treatments, 70 kg/ha was sufficient to obtain maximum yield and WUE. For dryland treatment, N had little effect on yield and WUE since yield was mainly limited by water stress. Nevertheless, wheat yield (2–3 Mg/ha) and WUE (0.45–0.65 kg/m^3) were low without N application under irrigated conditions. Application of N fertilizer increased yield to 5–6 Mg/ha and WUE to 1.03 kg/m^3. Among the yield components, application of N considerably increased spikes per square meter and seeds per spike but did not affect seed weight (Eck 1988).

In the LSP, application of N and P fertilizers has significantly increased wheat yield and WUE in the last three decades (Huang et al. 2003; Wang et al. 2011). Huang et al. (2003) and Wang et al. (2011) reported a long-term fertilization study in dryland winter wheat (1984–2009). Although yield and WUE varied from year to year due to precipitation variation, wheat with N and P fertilization always had higher yield and WUE. Without fertilization, yield was generally less than 2 Mg/ha and WUE was between 0.4 and 0.5 kg/m^3. Application of N (120 kg/ha) and P (60 kg/ha P_2O_5) increased the yield up to 4–5 Mg/ha and WUE to 1.4 kg/m^3 (Figure 18.9). Adding manure to N and P fertilizers also resulted in a higher yield and WUE (Wang et al. 2011). The main reason for increased yield and WUE by fertilization is that it promoted root growth and enhanced water use from soil water

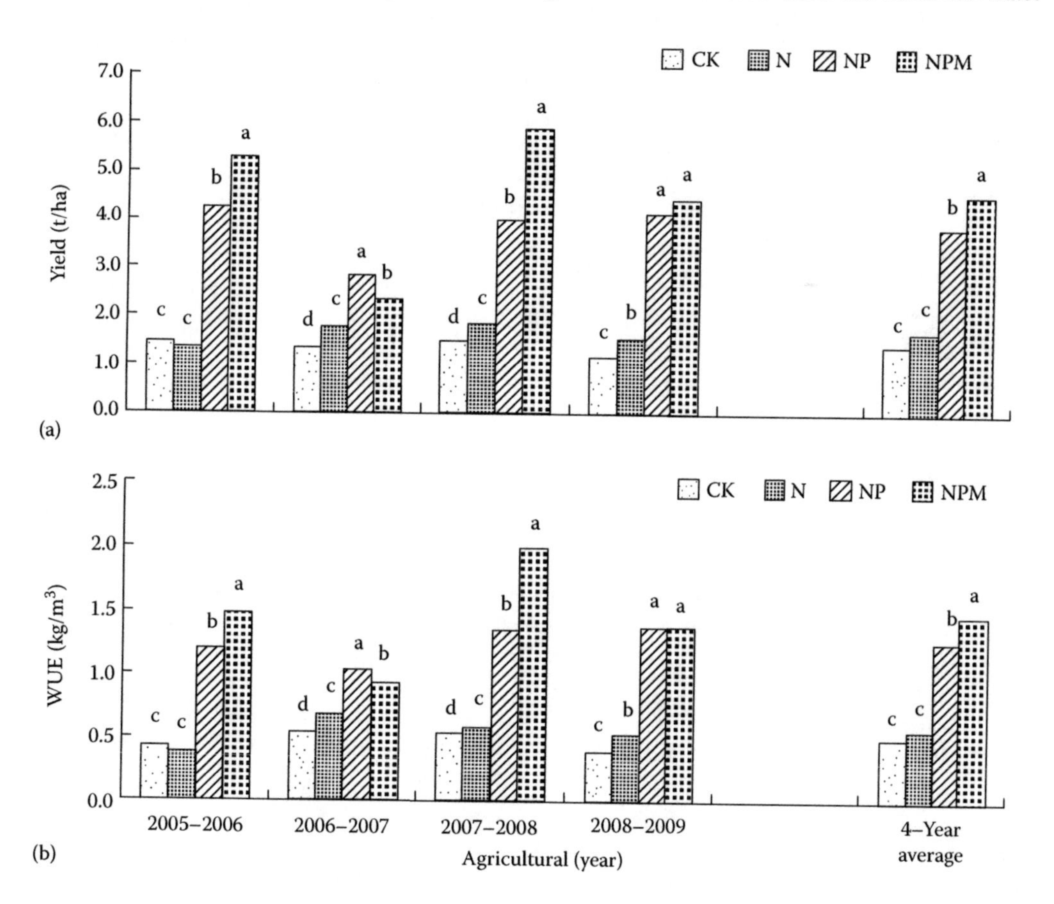

FIGURE 18.9 (a) Winter wheat yield and (b) water-use efficiency (WUE) in different fertilization treatments in a long-term fertilization study. (From Wang, J., Liu, W., and Dang, T., *Plant Soil*, 2011. DOI:10.1007/s11104-011-0764-4.) (CK: without fertilization; N: 120 kg/ha N; NP: 120 kg/ha N, 60 kg/ha P_2O_5; NPM: 120 kg/ha N, 60 kg/ha P_2O_5, and 76 Mg/ha manure.)

storage (Li and Su 1991). As a result, years of increasing fertilization combined with other yield-enhancing management practices have caused gradual depletion of soil water in the layer of 2–3 m profiles due to intensive plant water use (Li 2001; Huang et al. 2003). There is an increasing concern about the effect of current fertilization practices on sustainable wheat production and WUE in the region (Liu et al. 2010). On a regional scale, soil water conservation practices and intensive farming have induced soil water depletion, decreased surface runoff, and reduced movement of water into the soil profile (He et al. 2003). Although the amount of soil water storage was reduced, fertilization with N and P or with N, P, and manure significantly improved PSE and reduced soil water loss during the fallow period. Such improvement of PSE may offset some negative effects of low ASW at planting (Wang et al. 2011). Since manure application had positive effect on soil properties and wheat yield, manure should be included in fertilization practice for long-term sustainable wheat production (Wang et al. 2011). In addition, application of straw mulch may further improve wheat yield and WUE when combined with fertilization, particularly at high N rates (Gao et al. 2009).

Similar to the SGP, N and soil water had a significant interaction for wheat growth and yield in the LSP. In particular, plant response to fertilization was related to ASW at planting. Chen et al. (1991) showed that the application of fertilizers did not increase the yield when ASW at plant was lower than 330 mm in a 3-m profile. As ASW at planting increased from 330 mm, fertilizer application significantly increased the wheat yield. Xue and Chen (1990) showed that leaf photosynthetic rate and stomatal conductance in high N rates were more sensitive to water stress than those in low N rates. As such, wheat with higher N fertilizer rates may reduce yield under severe water stress. In a study in Australia, van Herwaarden et al. (1998) demonstrated that water deficit during grain filling reduced assimilation and consequently grain yield, and crops with high N rates suffered greater yield reduction than unfertilized crops. Since soil N availability is affected by many factors such as soil type, tillage system, crop residue, and rotation, the responses of wheat yield and WUE varied greatly (Zhang 1998). In addition, different cultivars also had different responses to N fertilization (Chen et al. 1991). Excessive N fertilizer application is not uncommon in LSP. For example, in some irrigated areas, N fertilizer rates averaged 450 kg/ha per year, that is, 225 kg/ha per crop (either wheat or corn) (Gao et al. 2009). High N fertilizer rates not only increase the risk to contaminate underground water but also may result in low N use efficiency (NUE) (Foulkes et al. 2009; Gao et al. 2009). Therefore, one future challenge for wheat production in the LSP is to improve NUE.

18.3.4 Deficit Irrigation

In the U.S. SGP, with the declining irrigation water supplies from Ogallala Aquifer and increasing energy costs, application of less irrigation water than the plants require for high yield will be the primary practice in the future for irrigated wheat production. Toward this end, deficit irrigation has been studied and practiced for over three decades and shown to be a viable management practice for improving yield and WUE (Eck 1988; Musick et al. 1994; Schneider and Howell 2001; Xue et al. 2003, 2006). Deficit irrigation is defined as the application of less water than is required for full ET and maximum yield, resulting in conservation of limited irrigation water and an increase in WUE (English 1990; Musick et al. 1994). Compared to dryland wheat, deficit irrigation significantly increased grain yield and WUE. In a field study in Bushland, Texas, deficit irrigation of 100 mm at booting stage increased the yield by 46% and WUE by 23% as compared to dryland treatment. Compared to full irrigation of 400 mm, a deficit irrigation of 220 mm at jointing and anthesis achieved 84% of the yield at full irrigation and resulted in 45% irrigation water savings (Table 18.2). In another study at the same location, Schneider and Howell (2001) showed that irrigation application of 50% ET requirement resulted in 86%–95% yield of full irrigation (100% ET). Deficit irrigation is an optimizing strategy under which crops are deliberately allowed to sustain some degree of water deficit and yield reduction (English 1990). Maintaining high ASW at planting is important for the successful practice of deficit irrigation for the efficient use of both precipitation

TABLE 18.2
Seasonal Evapotranspiration (ET), Biomass, Grain Yield, Yield Components, Harvest Index (HI), and Water-Use Efficiency in Yield (WUE) and in Biomass (WUE_{bm}) in Winter Wheat as Influenced by Irrigation

TRT	Total Irrigation (mm)	Seasonal ET (mm)	Biomass (Mg/ha)	Yield (Mg/ha)	Spikes (m^2)	Seeds per Spike	Seed Weight (mg)	HI	WUE (kg/m^3)	WUE_{bm} (kg/m^3)
T-1	0	414e	10.4f	3.2f	674e	16.6d	29.2c	0.31c	0.782d	2.51b
T-2	100	498d	12.8de	4.6e	797abc	20.4bc	28.8c	0.36ab	0.933b	2.57ab
T-3	100	494d	14.0cd	4.7e	772bcd	21.3b	29.3c	0.34bc	0.958ab	2.84a
T-4	140	496d	12.6e	4.6e	730cde	18.5cd	33.8a	0.36ab	0.918bc	2.54b
T-5	140	427e	11.6ef	3.6f	688de	16.6d	31.8b	0.31c	0.842cd	2.72ab
T-6	220	604b	15.5b	6.0c	801abc	22.2b	33.7a	0.39a	0.992ab	2.57ab
T-7	220	547c	14.6bc	5.4d	756cde	23.0b	31.5b	0.37ab	0.993ab	2.69ab
T-8	300	686a	18.2a	6.7b	864a	26.9a	28.4c	0.37ab	0.973ab	2.65ab
T-9	400	691a	18.0a	7.1a	857ab	26.5a	31.8b	0.39a	1.030a	2.62ab

Source: Xue, Q., Zhu, Z., Musick, J.T., Stewart, B.A., and Dusek, D.A., *J. Plant Physiol.*, 163, 154–164, 2006.
In each column, different letters represented the significant difference at level of 0.05 based on the LSD test.
T-1: dryland; T-2–T-5: 1 irrigation at jointing (100 mm), booting (100 mm), anthesis (140 mm), and mid-grain filling (140 mm), respectively; T-6: 2 irrigations at jointing (100 mm) and anthesis (120 mm); T-7: 2 irrigations at booting (100 mm) and grain filling (120 mm); T-8: 3 irrigations at jointing (100 mm), booting (100 mm), and anthesis (100 mm); T-9: 4 irrigations at jointing (100 mm), booting (100 mm), anthesis (100 mm), and grain filling (100 mm); seasonal precipitation: 254 mm.

and irrigation water (Musick et al. 1994). Xue et al. (2003) demonstrated that high ASW at planting resulted in deep root system and could delay the irrigation until anthesis if only one irrigation was allowed. When only a limited amount of irrigation is available, irrigation must be applied at critical stages for wheat yield determinations. Critical growth stages for irrigating winter wheat generally occur from early spring growth (floral initiation) to early grain development (Musick et al. 1994; Schneider and Howell 2001). Xue et al. (2006) examined the physiological mechanisms that contributed to increased yield and WUE under deficit irrigation. The increased WUE was largely related to an improved HI in different irrigation treatments (Figure 18.7). The HI is determined during grain filling by both current photosynthesis and remobilization of preanthesis carbon reserve from stems. For the maintenance of current photosynthesis to meet the carbohydrate supply, higher photosynthesis rate and longer green leaf area duration are advantageous under drought conditions. When the photosynthesis during grain filling is reduced by drought or heat stresses, the remobilization of carbon reserves can be important to grain filling. The contribution of remobilized carbon reserves to grain yield in wheat varied from 5% to 90%, depending on environmental conditions (Foulkes et al. 2002; Asseng and van Herwaarden 2003; Xue et al. 2006). The increased HI under appropriate deficit irrigation was due to both increased current photosynthesis and the remobilization of preanthesis carbon reserves (Xue et al. 2006).

Although winter wheat is grown mainly under dryland conditions in the LSP, irrigation is still possible in some areas with limited water resources. A small amount of irrigation at critical stages can significantly increase yield and WUE as compared to dryland. Fan et al. (2005a) reported a 3-year study with different amounts of supplementary irrigation (12–48 mm) at various stages (tillering, jointing, booting, and heading). Although yield varied due to precipitation, supplementary irrigation always increased yield as compared to dryland. The supplementary irrigation significantly increased WUE. For example, 36 mm irrigation at jointing stage resulted in a WUE of 2.17 kg/m^3 as compared to a WUE of 1.07 kg/m^3 in dryland treatment (Fan et al. 2005a). Based on a long-term field study, Huang et al. (2004) demonstrated that three irrigations (87.5 mm) at jointing, booting, and anthesis resulted in high yield and WUE as compared to dryland. However, one irrigation (87.5 mm) at booting stage still resulted in 78% yield of full irrigation treatment (four irrigations) and 75% of water savings (Huang et al. 2004). Kang et al. (2002) demonstrated that increased yield and WUE under limited irrigation conditions were largely related to improved HI, which is similar to findings from Xue et al. (2006) and Zhang et al. (1998). In the Northern China Plain, Zhang et al. (1998) showed that one irrigation at booting stage produced comparable yield and resulted in 24%–30% WUE increase as compared to four-irrigation treatment. The increased WUE was contributed by deep root system and higher HI.

In the SGP, irrigation technology has changed significantly in the last four decades, from furrow irrigation in the early years (1950s–1970s) to the current central pivot sprinkler systems (Musick et al. 1990; Colaizzi et al. 2008). The subsurface drip irrigation (SDI) system has also been used in the region in recent years. Over the years, irrigation efficiency has improved significantly (Howell 2001; Colaizzi et al. 2008). However, with the declining trend of well capacities in many areas, the future challenge will be to efficiently use reduced amount of irrigation in central pivot systems. SDI systems may have some advantages over central pivot for delivering small amounts of irrigation water. In the LSP, the traditional irrigation system is furrow flood irrigation. Although some new irrigation technologies (e.g., water delivered with low pressure pipes, and sprinkler and drip irrigation) have been tested and proved to be more efficient than furrow irrigation, these technologies have not been widely advocated or adopted (Deng et al. 2006).

18.4 FUTURE PERSPECTIVES

Although progress has been made to improve wheat yield and WUE in the last few decades in the U.S. SGP and China's LSP, the current WUE level is still below the generally accepted upper limit (2.0 kg/m^3) in most cases, particularly in the SGP. As such, developing better management strategies is still a challenge for agricultural scientists. In both regions, ASW at planting is extremely important to increasing wheat yield and WUE. For dryland wheat production in the SGP, management practices (e.g., NT, residue mulching, and rotation) resulted in significant gain in soil water storage during the fallow period. Future research must address how to efficiently use limited growing season precipitation (Stewart et al. 2010). Soil water is critical for yield during late developmental stages (e.g., from booting to anthesis) for grain production. Therefore, one important challenge is to save soil water from early season precipitation for use during the critical grain production period (Stewart et al. 2010). Using relatively lower seeding rate and optimum seeding date may provide some benefits to manipulate plant water use. However, highly variable seasonal precipitation makes it difficult to draw a general conclusion. The amount, intensity, and timeliness of precipitation during the growing season are often so erratic that successful practices for one year often fail in subsequent years.

Since the drought stress is inevitable during the wheat-growing season, improving drought resistance through breeding will be an important part of overall crop improvement. Currently, breeders are not well equipped to make selections in drought resistance because information is lacking on (i) the extent of variation for drought resistance within elite germplasm; (ii) key physiological and morphological traits that contribute to drought resistance, WUE, and yield determination; and (iii) genetic factors that control these traits, and empirical and molecular tools for selection of superior germplasm in breeding programs (Ober 2008). A better understanding of crop response to drought stress and identification of plant traits will lead to the development of improved germplasm and cultivars in the region. Improving transpiration efficiency (TE, or WUE_{bm}) has been successful for increasing yield and WUE in Australia (Richards et al. 2002, 2010). In the SGP, genetic variability of TE has been found in sorghum (Balota et al. 2008; Xin et al. 2008, 2009). However, the genetic variation in TE among wheat genotypes has not been well understood and needs to be investigated in future research.

The declining irrigation water continues to challenge the irrigated wheat production in the SGP. Although several management factors (improved cultivars, fertilization, pest control, etc.) contributed to wheat yield and WUE improvements, irrigation played a vital role in increasing wheat yield and WUE in the region. Irrigated wheat yields can be two to four times higher than dryland yields (Figure 18.3; Howell 2001). Therefore, irrigation will be an important management practice for a long time. Since irrigation water is becoming limited, deficit irrigation will be the primary practice in the future for irrigated wheat production. Toward this end, the adoption of drought-resistant cultivars is also important for implementing the deficit-irrigation practice.

In the LSP, the future wheat production will likely be dependent on dryland production because of the very limited irrigation resources. Since the majority of precipitation falls in the wheat fallow period, improving PSE becomes extremely important. Currently, the PSE level is generally low (<40%) and varies from year to year (Shangguan et al. 2002; Jin et al. 2007; Wang et al. 2011). Therefore, the future research must address how to increase PSE. Traditionally in the region, wheat straw has been harvested by farmers for various purposes and the field is deep-plowed after harvest for improving precipitation infiltration (Shangguan et al. 2002). Recent studies indicated that leaving the straw as a mulch combined with NT significantly increased PSE (Jin et al. 2007). Proper fertilization also increased PSE (Wang et al. 2011). Currently, the cropping and rotation systems in the LSP are not diversified and CW is still a dominant cropping system. Using diversified cropping and rotation systems may provide opportunities to increase PSE, PUE, and total crop yield in the future. For example, Li et al. (2002) demonstrated that winter wheat–legume–summer crop was a viable rotation system to increase overall PUE. The future improvement of PSE and PUE may be dependent on the integration of various management practices such as straw mulching, RT, and fertilization, and adopting more diversified cropping systems. The N and P fertilization has played and will continue to play an important role in increasing wheat yield and WUE in the region. There are two major challenges for maintaining long-term benefits of fertilization. First, long-term application of chemical fertilizers and unbalanced fertilization may result in low soil fertility and soil organic carbon (SOC) (Fan et al. 2005b). Combined chemical fertilizers and organic materials (either crop residues or manure) will not only result in sustainable crop productivity but also increase SOC, which provides numerous environmental benefits. Second, excessive application of N fertilizer is a common practice in the LSP. As such, future research is needed to address NUE under various N fertilizer rates.

ABBREVIATIONS

ASW	Available soil water
BM	Biomass at maturity
CT	Conventional tillage
CW	Continuous wheat
ET	Evapotranspiration
HI	Harvest index
LSP	Loess Plateau
NASS	National Agricultural Statistical Service
NT	No-till
PSE	Precipitation storage efficiency
PUE	Precipitation use efficiency
RT	Reduced tillage
SGP	Southern Great Plains
SS	Subsoiling with mulch
TC	Two crops per year
TE	Transpiration efficiency

WF Wheat–fallow
WSF Wheat–sorghum–fallow
WUE Water-use efficiency in grain yield
WUE_{bm} Water use efficiency in biomass.

REFERENCES

AgriPartners. 2007. Irrigation and cropping demonstrations. Texas AgriLife Research and Extension Center at Amarillo. http://amarillo.tamu.edu/amarillo-center-programs/agripartners/

Anderson, W. K. 2010. Closing the gap between actual and potential yield of rainfed wheat. The impacts of environment, management and cultivar. *Field Crops Res.* 116: 14–22.

Angus, J. F. and van Herwaarden, A. F. 2001. Increasing water use and water use efficiency in dryland wheat. *Agron. J.* 93: 290–298.

Angus, J. F., Hasegawa, S., Hsaio, T. C., Liboon, S. P., and Zandstra, H. G. 1983. The water balance of post-monsoonal dryland crops. *J. Agric. Sci.* 101: 699–710.

Asseng, S. and van Herwaarden, A. F. 2003. Analysis of the benefits to yield from assimilates stored prior to grain filling in a range of environments. *Plant Soil* 256: 217–229.

Balota, M., Payne, W. A., Rooney, W., and Rosenow, D. 2008. Gas exchange and transpiration ratio in sorghum. *Crop Sci.* 48: 2361–2371.

Biomass Research and Development Board. 2009. Increasing feedstock production for biofuels: Economic drivers, environmental implications and the role of research. Available at: www.brdisolutions.com/Site20%Docs/Increasing%20Feedstock_revised.pdf.

Blum, A. 2009. Effective use of water (EUW) and not water-use efficiency (WUE) is the target of crop yield improvement under drought stress. *Field Crops Res.* 112: 119–123.

Briggle, R. W. and Curtis, B. C. 1987. Wheat worldwide. In *Wheat and Wheat Improvement*, ed. E. G. Hey, 2nd edn, pp. 1–32. ASA-CSSA-SSSA, Madison, WI.

Chen, P. Y., Li, Y., and Chen, J. 1991. The study of high yielding cultivation models in winter wheat in Loess Plateau. In *A Comprehensive Research on High Efficient Eco-Economic System in Wangdonggou Watershed of Changwu County*, ed. Y. Li, pp. 88–109. Science and Technology Document Press, Beijing (in Chinese).

Colaizzi, P. D., Gowda, P. H., Marek, T. H., and Porter, D. O. 2008. Irrigation in the Texas High Plains: A brief history and potential reductions in demand. *Irrig. Drain.* 58:DOI:10.1002/ird.418.

Dahlke, B. J., Oplinger, E. S., Gaska, J. M., and Martinka, M. J. 1993. Influence of planting date and seeding rate on winter wheat grain yield and yield components. *J. Prod. Agric.* 6: 408–414.

Deng, X. P., Shan, L., Zhang, H. P., and Turner, N. C. 2006. Improving agricultural water use efficiency in and semiarid areas of China. *Agric. Water Manag.* 80: 23–40.

Eck, H. V. 1988. Winter wheat response to nitrogen and irrigation. *Agron. J.* 80: 902–908.

English, M. J. 1990. Deficit irrigation: Observations in the Columbia Basin. *J. Irrig. Drain. Eng.* 116: 413–426.

Fan, T., Stewart, B. A., Payne, W. A., Wang, Y., Song, S., Luo, J., and Robinson, C. A. 2005a. Supplemental irrigation and water-yield relationships for plasticulture crops in the Loess Plateau of China. *Agron. J.* 97: 177–188.

Fan, T., Stewart, B. A., Wang, Y., Luo, J., and Zhou, G. Y. 2005b. Long-term fertilization effects on grain yield, water-use efficiency and soil fertility in the dryland of Loess Plateau in China. *Agric. Ecosyst. Environ.* 106: 313–329.

Fang, Q. X., Ma, L., Green, T. R., Yu, Q., Wang, T. D., and Ahuja, L. R. 2010. Water resources and water use efficiency in the North China Plain: Current status and agronomic management options. *Agric. Water Manag.* 97: 1102–1116.

Fannin, B. 2011. Texas agricultural drought losses reach record $5.2 billion. Texas AgriLife News, August 17, 2011. http://agrilife.org/today/2011/08/17/texas-agricultural-drought-losses-reach-record-5-2-billion/.

Foulkes, M. J., Scott, R. K., and Sylvester-Bradley, R. 2002. The ability of wheat cultivars to withstand drought in U.K. conditions: Formation of grain yield. *J. Agric. Sci.* 138: 153–169.

Foulkes, M. J., Hawkesford, M. J., Barraclough, P. B., Holdsworth, M. J., Kerr, S., Kightley, S., and Shewry, P. R. 2009. Identifying traits to improve the nitrogen economy of wheat: Recent advances and future prospects. *Field Crops Res.* 114: 329–342.

French, R. J. and Schultz, J. E. 1984. Water use efficiency of wheat in a Mediterranean type environment. I. The relation between yield, water use and climate. *Aust. J. Agric. Res.* 35: 743–764.

Gao, Y., Li, Y., Zhang, J., Liu, W., Dang, Z., Cao, W., and Qiang, Q. 2009. Effects of mulch, N fertilizer, and plant density on wheat yield, wheat nitrogen uptake, and residual soil nitrate in a dryland area of China. *Nutr. Cycl. Agroecosyst.* 85: 109–121.

He, X. B., Li, Z. B., Hao, M. D., Tang, K. L., and Zheng, F. L. 2003. Down-scale analysis for water scarcity in response to soil-water conservation on Loess Plateau of China. *Agric. Ecosyst. Environ.* 94: 355–361.

Howell, T. A. 1990a. Relationships between crop production and transpiration, evapotranspiration, and irrigation. In *Irrigation of Agricultural Crops.* Agronomy Monograph 30, eds. B. A. Stewart and D. R. Nielson, pp. 391–434. ASA-CSSA-SSSA, Madison, WI.

Howell, T. A. 1990b. Grain, dry matter yield relationships for winter wheat and grain sorghum—Southern High Plains. *Agron. J.* 82: 914–918.

Howell, T. A. 2001. Enhancing water use efficiency in irrigated agriculture. *Agron. J.* 93: 281–289.

Howell, T. A., Steiner, J. L., Schneider, A. D., and Evett, S. R. 1995. Evapotranspiration of irrigated winter wheat: Southern High Plains. *Trans. ASAE* 38: 745–759.

Howell, T. A., Steiner, J. L., Schneider, A. D., Evett, S. R., and Tolk, J. A. 1997. Seasonal and maximum daily evapotranspiration of irrigated winter wheat, sorghum, and corn southern High Plains. *Trans. ASAE* 40: 623–634.

Howell, T. A., Tolk, J. A., Evett, S. R., Copeland, K. S., and Dusek, D. A. 2007. Evapotranspiration of deficit irrigated sorghum and winter wheat. *Proceedings USCID 4th International Conference*, pp. 223–239. Sacramento, California, USA.

Huang, M. B., Dang, T. H., Gallichand, J., and Goulet, M. 2003. Effect of increased fertilizer applications to wheat crop on soil–water depletion in the loess plateau, China. *Agric. Water Manag.* 58: 267–278.

Huang, M. B., Gallichand, J., and Zhong, L. 2004. Water-yield relationships and optimal water management for winter wheat in the Loess Plateau of China. *Irrig. Sci.* 23: 47–54.

Jin, K., Cornelis, W. M., Schiettecatte, W., et al. 2007. Effects of different management practices on the soil–water balance and crop yield for improved dryland farming in the Chinese Loess Plateau. *Soil Tillage Res.* 96: 131–144.

Jones, O. R. and Popham, T. W. 1997. Cropping and tillage systems for dryland grain production in the southern High Plains. *Agron. J.* 89: 222–232.

Kang, S., Zhang, L., Liang, Y., Hu, X., Cai, H., and Gu, B. 2002. Effects of limited irrigation on yield and water use efficiency of winter wheat in the Loess Plateau of China. *Agric. Water Manag.* 55: 203–216.

Li, Y. S. 1983. The properties of water cycle in soil and their effect on water cycle for land in the Loess Plateau. *Acta Ecol. Sin.* 3: 91–101 (in Chinese with English abstract).

Li, Y. S. 2001. Fluctuation of yield on high-yield field and desiccation of the soil on dryland. *ActaPedol. Sin.* 38: 353–356 (in Chinese with English abstract).

Li, Y. S. and Su, S. M. 1991. Dryland crop water productivity potential and water-fertility-yield relationships in Webei plateau, Shaanxi. In *A Comprehensive Research on High Efficient Eco-Economic System in Wangdonggou Watershed of Changwu County*, ed. Y. Lipp, pp. 115–125. Science and Technology Document Press, Beijing (in Chinese).

Li, F. R., Zhao, S. L., and Geballe, G. T. 2000. Water use patterns and agronomic performance for some cropping systems with and without fallow crops in a semi-arid environment of northwest China. *Agric. Ecosyst. Environ.* 79: 129–142.

Li, F. R., Gao, C.-Y., Zhao, H. L., and Li, X. Y. 2002. Soil conservation effectiveness and energy efficiency of alternative rotations and continuous wheat cropping in the Loess Plateau of northwest China. *Agric. Ecosyst. Environ.* 91: 101–111.

Liu, L., Xu, B. C., and Li, F. M. 2007. Effects of limited irrigation on yield and water use efficiency of two sequence-replaced winter wheat in Loess Plateau, China. *Afr. J. Biotechnol.* 6:1493–1497.

Liu, W. Z., Zhang, X. C., Dang, T. H., Ouyang, Z., Li, Z., Wang, J., Wang, R., and Gao, C. Q. 2010. Soil water dynamics and deep soil recharge in a record wet year in the southern Loess Plateau of China. *Agric. Water Manag.* 97: 1133–1138.

Musick, J. T. and Dusek, D. A. 1980. Planting date and water deficit effects on development and yield of irrigated winter wheat. *Agron. J.* 72: 45–52.

Musick, J. T. and Porter, K. B. 1990. Wheat. In *Irrigation of Agricultural Crops.* Agronomy Monograph 30, eds. B. A. Stewart and D. R. Nielson, pp. 597–638. ASA-CSSA-SSSA, Madison, WI.

Musick, J. T., Pringle, F. B., Harman, W. L., and Stewart, B. A. 1990. Long-term irrigation trends—Texas High Plains. *Appl. Eng. Agric.* 6: 717–724.

Musick, J. T., Jones, O. R., Stewart, B. A., and Dusek, D. A. 1994. Water–yield relationships for irrigated and dryland wheat in the U.S. southern plains. *Agron. J.* 86: 980–986.

NASS. 2010. Small Grains 2010 Summary, September 2010. http://usda.mannlib.cornell.edu/usda/nass/SmalGraiSu//2010s/2010/SmalGraiSu-09-30-2010_new_format.pdf.

Norwood, C. 1994. Profile water distribution and grain yield as affected by cropping system and tillage. *Agron. J.* 86: 558–563.

Ober, E. S. 2008. Breeding for improving drought tolerance and water use efficiency. HGCA Conference—Arable Cropping in Changing Climate, January 2008. http://www.hgca.com/publications/documents/S2_3_E_Ober_5.pdf.

Oweis, T., Zhang, H., and Pala, M. 2000. Water use efficiency of rainfed and irrigation bread wheat in a Mediterranean environment. *Agron. J.* 92: 231–238.

Passioura, J. B. 1977. Grain yield, harvest index and water use of wheat. *J. Aust. Inst. Agric. Sci.* 43: 117–120.

Passioura, J. B. and Angus, J. F. 2010. Improving productivity of crops in water-limited environments. *Adv. Agron.* 106: 37–75.

Paulsen, G. 1987. Wheat stand establishment. In *Wheat and Wheat Improvement.* Agronomy Monograph 13, ed. E. G. Heyne, 2nd edn, pp. 384–389. ASA-CSSA-SSSA, Madison, WI.

Reynolds, M. P., Mujeeb-Kazi, A., and Sawkins, M. 2005. Prospects for utilising plant-adaptive mechanisms to improve wheat and other crops in drought- and salinity-prone environments. *Ann. Appl. Biol.* 146: 239–259.

Richards, R. A., Rebetzke, G. J., Condon, A. G., and van Herwaarden, A. F. 2002. Breeding opportunities for increasing the efficiency of water use and crop yield in temperate cereals. *Crop Sci.* 42: 111–121.

Richards, R. A., Rebetzke, G. J., Watt, M., Condon, A. G., Spielmeyer, W., and Dolferus, R. 2010. Breeding for improved water productivity in temperate cereals: Phenotyping, quantitative traits loci, markers and the selection environment. *Funct. Plant Biol.* 37: 85–97.

Rudd, J. C. 2009. Success in wheat improvement. In *Wheat: Science and Trade*, ed. B. F. Carver, pp. 387–395. Wiley-Blackwell, Ames, IA.

Schneider, A. D. and Howell, T. A. 2001. Scheduling deficit irrigation with data from an evapotranspiration network. *Trans. ASAE* 44: 1617–1623.

Shangguan, Z. P. 1998. Crop community physiology and high yielding cultivation. In *Ecophysiological Basis of Dryland Farming*, eds. L. Shan and P. Chen, pp. 357–369. The Chinese Academic Press, Beijing (in Chinese).

Shangguan, Z. P., Shao, M. A., Lei, T. W., and Fan, T. L. 2002. Runoff water management technologies fro dryland agriculture on the Loess Plateau of China. *Int. J. Sustain. Dev. World Ecol.* 9: 341–350.

Spink, J. H., Semere, T., Sparkes, D. L., Whaley, J. M., Foulkes, M. J., Clare, R. W., and Scott, R. K. 2000. Effect of sowing date on the optimum planting density of winter wheat. *Ann. Appl. Biol.* 137: 179–188.

Stewart, B. A., Baumhardt, R. L., and Evett, S. R. 2010. Major advances of soil and water conservation in the U.S. southern Great Plains. In *Soil and Water Conservation Advances in the United States, Special Publication 60*, eds. T. M. Zoebeck and W. F. Schillinger, pp. 103–129. Soil Science Society of America, Madison, WI.

Stone, L. R. and Schlegel, A. J. 2006. Yield–water supply relationships of grain sorghum and winter wheat. *Agron. J.* 98: 1359–1366.

Su, Z., Zhang, J., Wu, W., et al. 2007. Effects of conservation tillage practices on winter wheat water-use efficiency and crop yield on the Loess Plateau, China. *Agric. Water Manag.* 87: 307–314.

Thomason, W. E., Raun, W. R., Johnson, G. V., Freeman, K. W., Wynn, K. J., and Mullen, R. W. 2002. Production system techniques to increase nitrogen use efficiency in winter wheat. *J. Plant Nutr.* 25: 2261–2283.

van Herwaarden, A. F., Farquhar, G. D., Angus, J. F., Richards, R. A., and Howe, G. N. 1998. 'Haying-off', the negative grain yield response of dryland wheat to nitrogen fertiliser. I. Biomass, grain yield, and water use. *Aust. J. Agric. Res.* 49: 1067–1081.

Wang, H. X., Liu, C. M., and Zhang, L. 2002. Water-saving agriculture in China: An overview. *Adv. Agron.* 75: 135–171.

Wang, F. Z., Sayre, K., Li, S., Si, J., Feng, B., and Kong, L. 2009. Wheat cropping systems and technologies in China. *Field Crops Res.* 111: 181–188.

Wang, J., Liu, W., and Dang, T. 2011. Responses of soil water balance and precipitation storage efficiency to increased fertilizer application in winter wheat. *Plant Soil*, DOI:10.1007/s11104-011-0764-4.

Winter, S. R. and Musick, J. T. 1993. Wheat planting date effects on soil water extraction and grain yield. *Agron. J.* 85: 912–916.

Xin, Z., Franks, C., Payton, P., and Burke, J. J. 2008. A simple method to determine transpiration efficiency in sorghum. *Field Crops Res.* 107: 180–183.

Xin, Z., Aiken, R., and Burke, J. J. 2009. Genetic diversity of transpiration efficiency in sorghum. *Field Crops Res.* 111: 74–80.

Xu, M., Xue, Q., and Musick, J. T. 1998. The study of yield and water relations in winter wheat under limited irrigation. In *Ecophysiological Basis of Dryland Farming*, eds. L. Shan and P. Chen, pp. 174–188. The Chinese Academic Press, Beijing (in Chinese).

Xue, Q. and Chen, P. 1990. Effects of nitrogen nutrition on water status and photosynthesis in wheat under soil drought. *ActaPhytophysiol. Sin.*16: 49–56 (in Chinese with English abstract).

Xue, Q. and Stougaard, R. N. 2006. Spring wheat seeding rate and seed size affect water-use efficiency under wild oat competition. Proceedings of 2nd International Symposium on Soil Erosion and Dryland Farming, Yangling, Shaanxi, China.

Xue, Q., Zhu, Z., Musick, J. T., Stewart, B. A., and Dusek, D. A. 2003. Root growth and water uptake in winter wheat under deficit irrigation. *Plant Soil* 257: 151–161.

Xue, Q., Zhu, Z., Musick, J. T., Stewart, B. A., and Dusek, D. A. 2006. Physiological mechanisms contributing to the increased water-use efficiency in winter wheat under deficit irrigation. *J. Plant Physiol.* 163: 154–164.

Xue, Q., Rudd, J., Lu, H., Bean, B., Colaizzi, P., Mahan, J., Jessup, K. E., Devkota, R., and Payton, P. 2010. Physiological basis for improving yield and water use efficiency of wheat in the southern High Plains. ASA-CSSA-SSSA, 2010 International Annual Meeting, Long Beach, CA.

Zhang, S. Q. 1998. Effects of inorganic nutrients on crop growth, development and water use efficiency under drought conditions. In *Ecophysiological Basis of Dryland Farming*, eds. L. Shan and P. Chen, pp. 233–246. The Chinese Academic Press, Beijing (in Chinese).

Zhang, J. H., Sui, X. Z., Li, B., Su, B. L., Li, J. M., and Zhou, D. X. 1998. An improved water-use efficiency for winter wheat grown under reduced irrigation. *Field Crops Res.* 59: 91–98.

Zhang, X., Chen, S., Sun, H., Wang, Y., and Shao, L. 2010. Water use efficiency and associated traits in winter wheat cultivars in the North China Plain. *Agric. Water Manag.* 97: 1117–1125.

Section V

Policy and Economics

19 Sustainable Land and Water Management Policies

Claudia Ringler and Ephraim Nkonya

CONTENTS

19.1 INTRODUCTION

A critical challenge for agriculture over the next half century is to increase food production within the constraints of scarce land and water resources and overused soils—conditions that are growing in severity in many parts of the world. While difficult to measure, land (including soils) and water constraints may easily reduce the food production potential by 5%–10% today and 20% by 2025. For example, Rosegrant et al. (2002) estimated—for the group of developing countries—the loss of cereal production potential as a result of growing water scarcity over time to 2025, and in 2008, they expanded their estimates to 2050 (Figure 19.1). While in 1995 about 5% of developing-country grain production potential was lost as a result of water scarcity, by 2025 this share is expected to increase to 11% of potential production and by 2050 to 14% of global cereal production. At the same time, land degradation remains one of the major challenges to food production. A recent study shows that 24% of the global land area is degraded (Bai et al. 2008). The largest degraded areas are concentrated in Africa, south of the equator, which account for 13% of total degraded area (Bai et al. 2008). The same study also showed a positive correlation between land degradation and infant mortality, suggesting a positive relationship between poverty and land degradation. Estimates of the cost of land degradation vary widely. Requier-Desjardins (2006) suggests that land degradation in Africa leads to a loss of 3%–9% of agricultural gross domestic product (AGGDP), a loss that underscores the importance of land degradation in developing countries. To reduce growing water

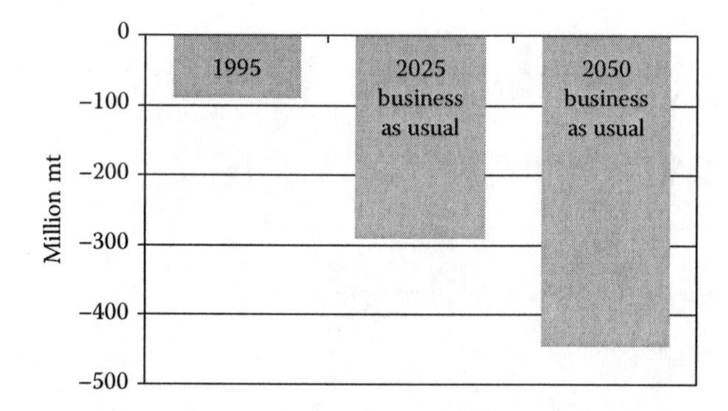

FIGURE 19.1 Loss of grain production potential due to water scarcity in developing countries. Loss of production potential refers to the grain quantity that could not be produced as a result of lack of water availability. (From Rosegrant, M.W., Cai, X., and Cline, S.A., World water and food to 2025: Dealing with scarcity. Joint publication. Washington, DC, and Colombo, Sri Lanka: International Food Policy Research Institute and International Water Management Institute, 2002; IFPRI IMPACT simulations (2008).)

and land shortages is partly a technical problem, but much more so a challenge of better policies, rights, and institutions.

Appropriate policies, rights, and institutions are key ingredients for sustainable land and water management (SLWM). They can provide incentives to users for appropriate use based on the scarcity value of the resource and for investing in land and water resource conservation. However, land and water management policies have been slow to adjust to the rapid pace of economic change of the last few decades and are generally inadequate to address growing land and water challenges. As a consequence, in many parts of the world, the resources are unnecessarily degrading, and particularly so in the group of developing countries. One way to enhance policy outcomes for both water and land could be through the development of policies that take account of the growing scarcity of both resources and their interrelationships in a holistic fashion, at least in cases of important synergies (Binswanger-Mkhize et al. 2011).

The focus of this chapter will be on policies and institutions that directly impact land and water outcomes. However, we acknowledge that there are a host of policies and institutions in the agriculture and nonagriculture sectors with potentially large positive and negative impacts on land and water resources. An assessment of the entire scope of global, regional, and national policies that can affect land and water outcomes is, however, outside the scope of this chapter.

19.2 CHALLENGES TO SUSTAINABLE LAND AND WATER MANAGEMENT

A key challenge to SLWM remains the continued need to increase food supplies—for a population estimated at 9 billion by 2050, up from 6 billion in 2000. Growth of incomes and urbanization are leading to shifts in consumer demand toward food products that are more water-intensive. On the supply side, investments in land and water are expected to increase only slowly and at growing costs, leading to continued degradation of the land and water base. Distorted macroeconomic and trade policies will continue to favor food production and natural resource extraction in some areas without comparative advantage. Moreover, agricultural input and output price subsidies and other distorted incentives in many countries further degrade land and water resources, contributing to groundwater depletion and poor fertilizer application practices in much of South Asia. Moreover, overgrazing of fragile pastures and grasslands is common in many parts of the world.

On the demand side, over the last several decades, competition from industrial, household, environmental, and aquaculture water uses has significantly increased challenges for water resource management for agriculture. In addition to adverse impacts on agricultural economies, industrial and household uses often return poor-quality water to the environment and irrigation, particularly in developing countries, with potential adverse impacts on the productivity, health, and environment of farmers, crops, livestock, and fisheries (Meinzen-Dick and Ringler 2008; Rosegrant and Ringler 1998).

On the land side, expanding cities continue to absorb small, but highly productive lands. Indonesia, for example, has documented encroachment of housing, industrial estates, and other economic infrastructure of around 1 million hectares on agriculture land between 1983 and 1993 on Java alone (BPS, Agriculture Census 1993). Importantly, biofuel production using sugarcane or corn and other feedstocks has diverted significant areas of land away from food production and increasingly threatens sustainable land, soil, and water use in addition to food prices and thus affordability for the poor. Higher energy prices have also increased the costs of intensive agricultural production—which had helped reducing deforestation.

Thus, after decades of being considered of little or no interest to both policymakers and investors, the food price crisis of 2005–2008 brought the need for investments in sustainable land and water resources back to the forefront; but the subsequent financial crisis and the ensuing (short-term) drop in food prices have somewhat dampened renewed interest.

19.2.1 Growing Scarcity of Land and Water Resources

While irrigation water supply is expected to increase slowly, nonirrigation demands are expected to double out to 2050, putting pressure on supplies available for irrigation, particularly in the group of developing countries where changes in population, economic growth, and urbanization are fastest. Some nonirrigation and nonagricultural land-use demands will come at a direct cost to agriculture, particularly in countries with high economic growth coupled with limited natural resources, such as China, and countries with extreme water scarcity, such as the Middle East and North Africa.

Irrigated harvested area—taking multiple cropping and estimates of informal irrigation into account—is expected to increase from 421 million hectares in 2000 to 473 million hectares by 2050 at 0.23% per year. As has been stated earlier, water scarcity alone has already reduced the cereal production potential in the mid-1990s and is expected to account for a larger loss of potential in the future, even without taking into account adverse impacts from climate change and further land degradation (Rosegrant et al. 2002; IFPRI IMPACT Simulations 2008). Another measure of growing water shortages for agriculture is the irrigation water supply reliability index (IWSR), which measures the availability of water relative to full water demand for irrigation. This index is projected to decline from 0.71 globally in 2000 to 0.66 by 2050 (IFPRI IMPACT Simulations 2008), with steeper declines in water-scarce basins. Furthermore, irrigators are hurt not only on average but because water availability becomes more susceptible to downside risk in low rainfall years. In much of the world, the problem will be compounded by increasing intra-annual variability in rainfall and significant increases in the number and severity of droughts (Meehl et al. 2007).

Agricultural harvested area is estimated to grow only at 0.39% annually from 2000 to 2025 and 0.07% per year during 2025–2050; productivity growth in much of current rainfed agriculture could be significantly higher if the challenges of turning low-potential areas into high-potential areas could be overcome in much of sub-Saharan Africa, parts of Latin America, and parts of the former Soviet Union. These challenges include poor infrastructure, poor access to markets, and lack of incentives for small farmers and investors alike. Of course such conversion will need to take into account environmental and carbon sequestration needs and opportunities. Per capita land area has been falling rapidly in sub-Saharan Africa but also in Latin America and the Caribbean and South Asia. Per capita arable area declined from 0.5 ha per capita to 0.25 ha per capita from 1970 to 2005 (Figure 19.2). Per capita arable land area in the East Asia and Pacific region was smallest throughout but it maintained a stable trend over the 30-year period. Decreasing land area leads to conversion of forest lands to agriculture and expansion of crop production into more fragile areas, which in turn leads to more severe land degradation.

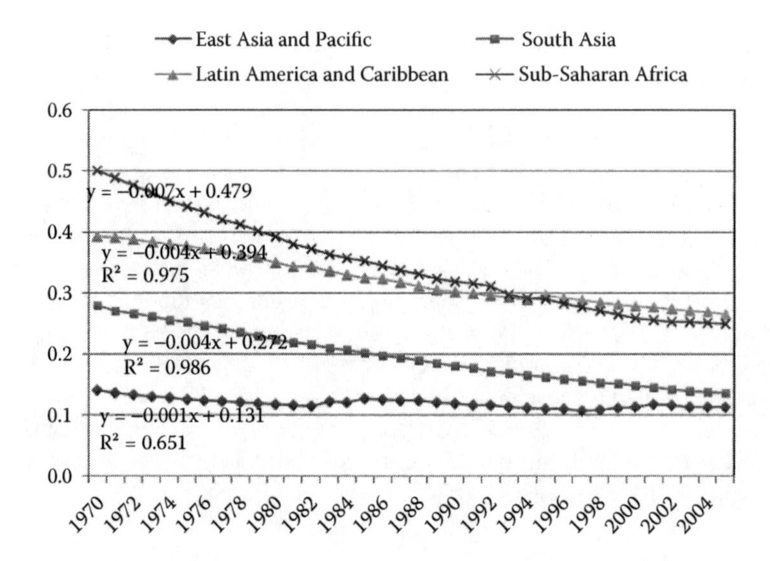

FIGURE 19.2 Trend of per capita arable land area in developing countries. (From FAOSTAT [FAO Statistical Databases], 2010. Accessed online at http://faostat.fao.org/.)

In conclusion, growing scarcities of water and land are projected to increasingly constrain food production growth, causing adverse impacts on food security and human well-being goals and were an underlying factor of the food price spikes during 2005–2007.

19.2.2 Increasing Competition for Nonfood Uses (Including Biofuels)

The production of biofuels affects land and water resources in two ways: directly through land reallocation from food to nonfood crops and water withdrawals for irrigation and the industrial processes of feedstock conversion; and indirectly by increasing water loss through evapotranspiration that would otherwise be available as runoff and groundwater recharge (Berndes et al. 2003). Biofuel production can also affect water quality by increasing nutrient loads in rivers and lakes and by accelerating soil erosion. Even though globally the amount of water withdrawn for the production of biofuels is modest, local water scarcity problems may worsen due to the irrigation of feedstocks or when rainfed feedstocks have higher evapotranspiration requirements than previous land uses. It is unlikely that investors in large-scale biofuel plantations would rely on rainfall only to support their investments (Rosegrant et al. 2008). In countries with little land and water available for biofuel expansion, the use of water for biofuel production is likely to affect existing water allocation both across sectors as well as within agriculture and can involve serious trade-offs between energy, environment, food security, and livelihood protection (McCornick et al. 2008; Muller et al. 2008). Rosegrant et al. (2008) show that continued rapid expansion of biofuel production will have significant impacts on the food sector. Under a "biofuel expansion" scenario, 2020 world prices are 26% higher for maize, 18% higher for oilseeds, 12% higher for sugar, 11% for cassava, and 8% for wheat compared with the 2020 prices in the baseline scenario, presenting a clear "food-versus-fuel" trade-off. These impacts include substantial price increases for food commodities, reductions in the availability of calories, and increased levels of malnourishment at the regional level, particularly in sub-Saharan Africa. Careful land and water-use planning focusing on rainfed and marginal land and water-using feedstocks, such as sweet sorghum and Jatropha, could help mitigate some adverse impacts (McCornick et al. 2008).

Moreover, second-generation biofuels will still require water and land resources and, importantly, access to rural infrastructure that is generally poorly developed in countries with potential for area expansion, such as parts of sub-Saharan Africa and Latin America. To ensure sustainable development of biofuels without undue impact on land and water resources will require sound water and land policy and right systems that reveal the full set of costs and benefits of biofuel developments; these are described in detail in the following sections.

There have also been significant efforts to develop pharma plants using maize and tobacco particularly since the 1990s; but so far transgenic pharma crops have not gone forward as a result of regulations, the cost of development, and commercialization, as well as possible market and litigation risks in the United States. Producing pharma crops in open fields requires a full segregation and aiming for zero percent tolerance for comingling, which may be tricky, particularly for crops like maize (Guillaume Gruere, personal communication).

19.2.3 Degradation of the Water and Land Base

Major land and water degradation processes include the degradation of water and soil in irrigated areas, the depletion of groundwater, the degradation of water-related ecosystems, increasing water pollution, desertification processes, the growing overgrazing of pasture lands, and the depletion of agricultural soils in rainfed and dryland areas, especially in sub-Saharan Africa. Issues are more severe in sub-Saharan Africa because of a lack of rural infrastructure and availability and access to other agricultural inputs. In much of the rest of the world, degradation of irrigation infrastructure investments is proceeding apace, increasing the need for modernization and rehabilitation. Similarly, agrochemical pollution, overgrazing, soil nutrient depletion, and degraded soils require significant investments for sustainable soil and land management.

Poor irrigation practices accompanied by inadequate drainage have often damaged soils through oversaturation and salt buildup. A notable example is the Aral Sea Basin in Central Asia, which shrank drastically as a result of excessive withdrawals for irrigation. It is estimated that on a global scale there are about 20–30 million hectares of irrigated land severely affected by salinity. The global annual losses due to salinity are estimated to be US$12 billion (Pitman and Läuchli 2004). An additional 60–80 million hectares are affected to some extent by waterlogging and salinity (FAO 1996). In arid and semiarid areas, pumping groundwater at unsustainable rates has contributed to significant lowering of groundwater tables, particularly in South Asia and in parts of China, but also in parts of the United States, and to saltwater intrusion in some coastal areas (Vorosmarty et al. 2005).

19.2.4 Growing Pollution of Water and Land

Growing water pollution is a further constraint to SLWM. Poor water quality increasingly constrains agricultural and economic development in regions that experience water scarcity and are plagued by poor wastewater treatment, particularly in densely populated Asia. Water pollution reduces agricultural production and threatens fish and other aquatic life and human health. Agrochemical pollution as a result of poor nutrient management practices is a growing concern in much of East and Southeast Asia, parts of the former Soviet Union, and some plantations in Central and Latin America. Subsidies for agrochemicals, poor extension services, and lack of water quality monitoring and enforcement of existing standards are the cause for much of today's agrochemical pollution. China clearly leads in this type of pollution, consuming more than 30% of the world's nitrogen fertilizer, which is applied to only 7% of the world's agricultural area (FAOSTAT 2010).

Biological and industrial nitrogen fixation converts about 120 million tons of nitrogen annually (Rockström et al. 2009) but about two-thirds of the fixed nitrogen finds its way to waterways (Corcoran et al. 2010). Nitrogen runoff causes eutrophication in lakes and pollution of coastal reefs at river mouths, both of which have severe impacts on fish stock and human health. Similarly, about 10 million tons of phosphorus is deposited in water globally (Corcoran et al. 2010). Improved fertilizer use efficiency could reduce this pollution and save both costs and pollution. Unfortunately nutrient use efficiency technologies have not been widely promoted in developing countries. They include optimizing the timing of fertilizer application, split application, soil testing, balanced nutrient application, and use of crop varieties with a strong response to fertilizer application (Roy et al. 2002).

Freshwater biodiversity and associated fisheries are on a decline in almost all developing countries with negative impacts on protein availability for the poor. Rapid economic growth in Asia is increasing political pressures to remedy these pollution efforts, and the induced policy and institutional changes will eventually lead to reductions in pollution, as they have in most of the developed world (Lomborg 2007).

19.2.5 Climate Change Impacts on Land and Water for Agriculture

The principal water-related climate changes include changes in the volume, intensity, and variability of precipitation and higher crop water evapotranspiration needs as a result of higher temperature. Increases in precipitation are mainly expected in high latitudes while decreases are expected in subtropical and lower latitude regions (Bates et al. 2008). Furthermore, rising temperatures will increase the rate of snow cap and glacier melt, affecting agricultural production in river basins fed by mountain ranges. Sea-level rise due to the thermal expansion of seawater and the melting of continental glaciers will lead to inundation of low-lying coastal areas, with significant adverse effects, including salinization of coastal agricultural lands, damage to infrastructure, and tidal incursions into coastal rivers and aquifers (Kundzewicz et al. 2007).

Analyses of multiple climate change scenarios indicate that climate change will likely have a slight to moderate negative effect on crop yields (Parry et al. 2004; Cline 2007; Nelson et al. 2009), and crop irrigation requirements would increase (Frederick and Major 1997; Doll 2002; Fischer et al. 2006), as would overall water stress in many areas dependent on irrigation (Arnell 1999; Fischer et al. 2006). Nelson et al. (2009) estimated that irrigation and water-use efficiency will account for about 42% of the estimated US$7.3 billion annual investment required to reduce childhood malnutrition to a level without climate change.

19.3 POLICIES, INSTITUTIONS, AND INVESTMENTS FOR SUSTAINABLE LAND AND WATER MANAGEMENT

There is not one "optimal" institutional arrangement for water or land; rather, it is critical to understand the potential contributions, facilitating conditions, and limitations of each (Merrey et al. 2007; Meinzen-Dick 2007).

19.3.1 The Need for Demand Management for Water

In the past, increasing the supply of water through new water development has been a common strategy to manage water resources. However, in maturing water economies, which are characterized both by increasing scarcity of water (Randall 1981) and by increasing transfers of water both in scale and amount, managing the demand for water becomes more important. The task of demand management is to generate both physical savings of water and economic savings by increasing the output per unit of evaporative loss of water, by reducing water pollution, and by reducing nonbeneficial water uses. Bhatia et al. (1995) differentiate four types of policy instruments for demand management: (1) enabling conditions, which are actions to change the institutional and legal environment in which water is supplied and used, such as policies of development of institutions, water rights and collective action mechanisms, and water user associations, but also the privatization of utilities; (2) market-based incentives, which directly influence the behavior of water users by providing incentives to conserve water use, including water pricing, water markets, as well as effluent charges and other taxes and subsidies; (3) nonmarket instruments, such as quotas, licenses, and pollution controls; and (4) direct interventions, such as investments in efficiency-enhancing water infrastructure, or conservation programs. The specific set of water policies that will be used varies from location to location depending on the status of economic development, the level of water scarcity, historic development, and institutional capability. In most situations, a mix of all four types of policy instruments is applied. Despite the wide range of water policies, implementation remains generally rather complex, given the variety of sources, ranging from precipitation to groundwater, and various surface water bodies, the fluidity of the resource, the many claimants on its uses, and the distinction between consumptive and nonconsumptive uses.

19.3.2 Complexities in Water Management

Moreover, water policies are implemented at different scales, ranging from the local level to the district, national, and regional levels up to the global level. While most statutory-based water policies are generated at the national level, increased decentralization processes have moved the actual implementation and applications to lower levels of authority, in particular, the province or district level—providing both new opportunities and new challenges (e.g., in Indonesia). At the same time, some water and related policies have moved up to higher levels, such as global climate policy, which is being discussed by the United Nations Framework Convention on Climate Change (UNFCCC) and assessed by international working groups, such as the Intergovernmental Panel on Climate Change (IPCC). Moreover, water policies can also be implemented at the basin boundary or subcatchment level, which tends to dissect various administrative scales. Furthermore, some water policies follow customary use rights, generally those on a small scale, while others are based on statutory laws and regulations. Thus, multiple legal and normative frameworks coexist, and the dynamics between statutory and customary water policies are fluid and in constant motion (see also the literature on legal pluralism, e.g., Bruns and Meinzen-Dick 2000).

As a further complexity, in many countries, including both developed and those in development, water policies are developed and implemented by different agencies or ministries, including those focusing on the environment, agriculture, public health, construction, energy, fisheries, and water proper. In addition to fragmentation across agencies, the source itself is often fragmented across different agencies with surface water sources managed separately and by agencies that differ from groundwater sources.

19.3.3 Water and Land Rights as the Foundation for Sustainable Management

The key basis of sustainable water policy is water rights. Although some system of water rights is found to operate in virtually any setting where water is scarce, systems that are not firmly grounded in formal or statutory law are likely to be more vulnerable to expropriation. On the other hand, if well-defined rights are established, the water user can benefit from investing in water-saving technology. Growing water scarcity and the possibility to exclude other users has led the trend to the establishment of private property rights to water. Trends have been similar for land, where private property rights are most common (Bruns and Meinzen-Dick 2000, 2005; Rosegrant and Binswanger 1994; Binswanger-Mkhize et al. 2011).

Improvement in land tenure systems would enhance the enforcement of agricultural land preservation and promote sustainability objectives. Rural populations under weak land preservation regulations have failed to invest in long-term land improvements on existing agricultural land and have also, in many countries, abandoned land in favor of migration to forest and other marginal lands. Policies related to land tenure and resources access are of great relevance for the sustainable management and use of natural resources in all countries of the world where the majority of the population relies heavily on land to provide income, employment, and livelihoods. In recent years, there is a growing recognition of the centrality of the land tenure in sustainable development processes as witnessed by new land policies and increasing local and international demands for relevant and sound land tenure laws and regulations (Rosegrant et al. 2009).

When property rights are difficult to define or enforce, as for example for common pool resources, such as small reservoirs or communal forest lands, collective action is needed to achieve sustainable water management (Ostrom 1990). While scarcity itself and access to markets may drive the emergence of collective action and/or property rights, appropriate institutions are needed to enable and administer property rights and to support collective action. If property rights to water or land have not been established by statutory means or if customary rights are not recognized by government authorities, local water users might lose out when biofuel plantations are established through government sales of concessions.

While private property rights are increasingly important, the need for enhanced state capacity for water and land management is increasing as well as a result of growing challenges of water scarcity and competition. For water, state capacity is increasingly needed to coordinate water users across systems and across sectors, bridging between irrigation, municipal, industrial, and environmental uses and users and between water quantity and quality. For land, state capacity is needed to register land titles and enforce land rights (Binswanger-Mkhize et al. 2011).

19.3.4 Economic Incentives for Land and Water Management

Economic incentives for water management include prices, taxes, subsidies, quotas, and ownership/rights. These incentive measures, when implemented appropriately, can affect the decision-making process and motivate water users to conserve and use water efficiently in irrigation and other uses. Economic incentives play out differently under differing enabling institutions. The most important among these enabling institutions are water rights for farmers and other water users. Other institutions that influence economic incentives include the rule of law and good governance, the relative focus on public systems versus private development, the role of decentralization in the form of farmer management of irrigation systems, and the existence or not of river basin organizations.

Water pricing is the most common economic instrument used. In a review of the World Bank irrigation and drainage portfolio covering 68 projects, water pricing was most common (52 out of 68 projects) (Dinar 2001). However, prices are generally far below full capital cost recovery in both the developing and the industrialized countries (OECD 1999; Dinar and Subramanian 1998; Barker and Rosegrant 2007). At the zero or low levels of current irrigation water prices in many countries, irrigation water use is highly price inelastic; and prices high enough to induce significant changes in water allocation (or recover capital costs) would severely reduce farm income (Ringler 2005; Perry 2001; Löfgren 1996). However, water pricing policies can improve efficiency and sustainability when combined with appropriate supporting policies (Rosegrant et al. 1995; Dinar and Mody 2004; Gardner 1983).

For many developing countries facing high water scarcity and rapid urban-economic growth, paying farmers for using less water can be an attractive alternative that supports rural-to-urban water transfers while providing compensation to long-term irrigation water users. If adequately compensated, irrigators can then invest in on-farm advanced irrigation technologies and/or switch to less-water-consuming crops (Rosegrant et al. 2005).

Water markets can also provide important economic incentives. They empower water users, by requiring user consent to any reallocation of water and by compensating users for water transferred. Thus, marketable rights induce water users to consider the full opportunity cost of water, including its value in alternative uses, thus providing incentives to economize on the use of water and gain additional income through the sale of saved water. Finally, a properly managed system of tradable water rights provides incentives for water users to internalize the external costs imposed by their water use, reducing the pressure to degrade resources (Rosegrant and Binswanger 1994).

19.3.5 Sustainable Land Management

Land and water used for agricultural production have complex and inseparable linkages suggesting benefits from joint management (Bossio et al. 2007; Binswanger-Mkhize et al. 2011). With the advent of climate change, this linkage has become even more imperative, especially in dry and humid areas.

For land resources, studies have shown that integrated land management practices that strategically integrate organic and inorganic fertility management practices lead to higher and sustainable yields (Vanlauwe et al. 2010; Tittonell et al. 2008) and were found to be more profitable than practices that use either one of the two alone in Africa (Doraiswamy et al. 2007; Tschakert 2004; Sauer et al. 2007; Nkonya et al. 2009). Box 19.1 provides an example from Nigeria, showing high returns of joint fertilizer and manure management but low adoption.

Despite the high returns to sustainable land management (SLM) practices, their adoption rates have remained low due to the their high labor intensity, limited extension services promoting integrated soil fertility management practices, and poor rural services supporting efficient marketing (Sauer et al. 2007; Pender 2009; Nkonya et al. 2009). The results suggest the need to use multipronged approaches to address the current land degradation and the low adoption of SLM practices. Consistent with Boserup (1965), for example, a long-term study by Tiffen et al. (1994) showed "more people, less erosion" in Machakos, Kenya, an area with better market access to market. Similarly, a recent study also showed a negative correlation between population density and land degradation (Bai et al. 2008). These results suggest that in areas with high market access and other favorable agricultural policies, farmers use SLM practices successfully (Boyd and Slaymaker 2000; Mazzucato and Niemeijer 2001).

BOX 19.1 BENEFIT–COST ANALYSIS OF NIGERIA LAND MANAGEMENT PRACTICES

A study was done in Nigeria to determine the costs and benefits of a combination of land management practices including manure, fertilizer application, and incorporation of crop residues. Using 30-year crop simulation data and household survey data, the study showed that a combination of crop residues (100%), manure (5 tons/ha), and fertilizer (80 kgN/ha) gave higher returns than those which use one of the three treatments alone (Nkonya et al. 2009). However, household survey data showed that the adoption of the most profitable land management practice (combination of fertilizer, manure, and crop residues) was only 8% compared to 18% for manure alone and 31% for fertilizer alone (Figure 19.3).

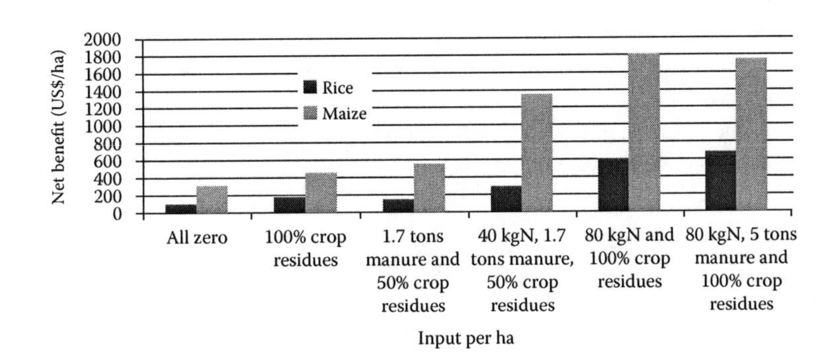

FIGURE 19.3 Returns to integrated soil fertility management practices for maize and rice in Nigeria.

These favorable findings are not universal, however. Other studies have shown more severe land degradation in densely populated areas (e.g., Cleaver and Schreiber 1994). In such cases, market failure and other unfavorable policies have been the major challenges for adopting SLM practices.

19.3.6 Increasing Crop Yields

Increasing crop yields, for example, through closing the yield gap between developed and developing regions and between rainfed and irrigated crops can save significant water resources and help conserve ecosystems and remaining forest areas in the developing world. Breeding strategies in the past have focused on crop yield enhancement. More recently, the focus has been on increasing stress tolerance, particularly drought resistance, allowing crops to survive in warmer, water-scarcer environments. Other important water- and land-saving research strategies include increasing the nitrogen use efficiency of fertilizers and range from zero-till systems that conserve soil moisture and help sequester carbon to laser-leveling systems for optimal agricultural input use. To support the development and adoption of new technologies requires not only profitability at the farm level but also a supportive national agricultural research and extension system, intellectual property rights protection, and the existence of efficient agricultural input and output markets (Rosegrant et al. 2009).

19.4 CONCLUSION

Growing scarcity of water and land is projected to progressively constrain food production growth, slowing progress toward food security and goals of human well-being. In the absence of institutional, policy, and investment reform, water and land for food production will increasingly conflict with other uses and users in many parts of the world. Negative impacts will be even more pronounced under climate change, especially in developing countries. Even under moderate climate change scenarios, impacts are projected to be negative for dryland areas in Africa, Asia, and the Mediterranean. Increasing water scarcity for agriculture not only limits crop area expansion but also slows irrigated cereal yield growth in developing countries. Similarly, increasing land diversion for nonfood crops, and urban and industrial encroachment, as well as continued loss to degradation will slow agricultural productivity growth.

Joint land and water management has been attempted in pastoral systems and in watershed management, but the time frame for implementation is generally very long and there are few successful cases. Addressing land and water degradation can benefit from a watershed approach, including problem identification, land-use planning, and institutions for coordinating between upstream and downstream areas. Integrated land and water management through watershed rehabilitation has been successful in China's Loess Plateau area, for example (World Bank 2003, 2007).

Payment for Environmental Services (PES) has increased in importance as a mechanism for the conservation of land and water resources and has most often been used in watersheds. PES typically connects upstream land users, often farmers, with downstream water users. Upstream land users may be paid to maintain current forest areas or plant additional trees, for not grazing on sloping lands, or any other land activities that could affect water quality and quantity for downstream cities, reservoirs, industries, or tourism areas (Binswanger-Mkhize et al. 2011).

Similarly, pasture management requires both access to water for the animals wherever they migrate to and access to pasture lands. The location of water sources influences pastoralists' choices of grazing areas, opens up new pasture areas, and thus improves animals' nutritional status. Consequently, water rights are the key to control and utilization of arid and semiarid areas. However, many pastoralists have formal rights neither over land nor over water (Binswanger-Mkhize et al. 2011).

The most promising avenue for addressing land and water scarcity is the reform of policies, rights, and institutions for SLWM to provide incentives to users for appropriate use. Reform efforts must start from the recognition that there are no one-size-fits-all solutions to policies, rights, and institutions, and no ideal solutions that apply everywhere, but that these have to be adapted to a country's or region's situation. However, some system of water and land rights should be at the start of any water and land management reform process.

ABBREVIATIONS

AGGDP	Agricultural gross domestic product
IPCC	Intergovernmental Panel on Climate Change
PES	Payment for Environmental Services
IWSR	Irrigation water supply reliability index
SLM	Sustainable land management
SLWM	Sustainable land and water management
UNFCCC	United Nations Framework Convention on Climate Change

REFERENCES

Arnell, N.W. 1999. Climate change and global water resources. *Global Environmental Change* 9: S31–S49.

Bai, Z.G., D.L. Dent, L. Olsson, and M.E. Schaepman. 2008. Global assessment of land degradation and improvement 1: Identification by remote sensing. Report 2008/01, FAO/ISRIC–Rome/Wageningen.

Barker, R. and M.W. Rosegrant. 2007. Establishing efficient use of water resources in Asia. In *Reasserting the Rural Development Agenda: Lessons Learned and Emerging Challenges in Asia*, eds. A. Balisacan and F. Nobuhiko. Laguna, Philippines: Institute of the South East Asian Studies and South East Asia Regional Center for Graduate Study and Research in Agriculture (SEARCA) College.

Bates, B.C., Z.W. Kundzewicz, S. Wu, and J.P. Palutikof (eds). 2008. Climate change and water. Technical Paper VI of the Intergovernmental Panel on Climate Change. Geneva: IPCC Secretariat.

Berndes, G., M. Hoogwijk, and R. van den Broek. 2003. The contribution of biomass in the future global energy supply: A review of 17 studies. *Biomass and Bioenergy* 25: 1–28.

Bhatia, R., R. Cestti, and J. Winpenny. 1995. Water conservation and reallocation: Best practice cases in improving economic efficiency and environmental quality. A World Bank–Overseas Development Institute joint study. Washington, DC: World Bank.

Binswanger-Mkhize, H.P., R. Meinzen-Dick, and C. Ringler. 2011. *Policies, Rights, and Institutions for Sustainable Management of Land and Water Resources.* SOLAW Background Thematic Report - TR09. Accessed online at http://www.fao.org/fileadmin/templates/solaw/files/thematic_reports/TR_09_web.pdf

Boserup, E. 1965. *The Conditions of Agricultural Growth.* New York: Aldine Publishing Company.

Bossio, D., W. Critchley, K. Geheb, G. van Lynden, and B. Mati. 2007. Conserving land—Protecting water. In *Water for Food, Water for Life*, ed. D. Molden, pp. 551–584. London and Colombo, Sri Lanka: Earthscan and International Water Management Institute.

Boyd, C. and T. Slaymaker. 2000. Re-examining the 'more people less erosion' hypothesis: Special case or wider trend? *Natural Resource Perspectives* 63: 1–6.

BPS (Badan Pusat Statistik). 2003. Agricultural survey on land use for Indonesia. Various years of publication, 1980–2000. Jakarta: BPS.

Bruns, B.R. and R.S. Meinzen-Dick (eds). 2000. *Negotiating Water Rights.* London and Delhi, India: Intermediate Technology Publications and Vistaar.

Bruns, B.R. and R.S. Meinzen-Dick. 2005. Frameworks for water rights: An overview of institutional options. In *Water Rights Reform Lessons for Institutional Design*, eds. B.R. Bruns, C. Ringler, and R.S. Meinzen-Dick. Washington, DC: IFPRI.

Cleaver, K.M. and G.A. Schreiber. 1994. Reversing the spiral: The population, agriculture, and environment nexus in sub-Saharan Africa. Washington, DC: World Bank.

Cline, W.R. 2007. Global warming and agriculture: Impact estimates by country. Washington, DC: Center for Global Development.

Corcoran, E., C. Nellemann, E. Baker, R. Bos, D. Osborn, and H. Savelli (eds). 2010. Sick water? The central role of wastewater management in sustainable development. A rapid response assessment. United Nations Environment Programme, UN-HABITAT, GRID-Arendal. www.grida.no.

Dinar, A. 2001. Review of active bank irrigation and drainage portfolio: Use of economic incentives. In eds. F.J. Gonzalez and S.M.A. Salman, *World Bank Technical Paper* No. 524. Washington, DC: World Bank.

Dinar, A. and J. Mody. 2004. Irrigation water management policies: Allocation and pricing principles and implementation experience. *Natural Resource Forum* 28: 112–122.

Dinar, A. and A. Subramanian. 1998. Policy implications from water pricing experiences in various countries. *Water Policy* 1: 239–250.

Doll, P. 2002. Impact of climate change and variability on irrigation requirements: A global perspective. *Climatic Change* 54: 269–293.

Doraiswamy, P.C., G.W. McCarty, E.R. Hunt Jr., R.S. Yost, M. Doumbia, and A.J. Franzluebbers. 2007. Modeling soil carbon sequestration in agricultural lands of Mali. *Agricultural Systems* 94: 63–74.

FAO (Food and Agriculture Organization of the United Nations). 1996. Food production: The critical role of water. Rome: World Food Summit.

FAOSTAT (FAO Statistical Databases). 2010. Accessed online at http://faostat.fao.org/.

Fischer, G., F.N. Tubiello, H. van Velthuizen, and D. Wiberg. 2006. Climate change impacts on irrigation water requirements: Global and regional effects of mitigation, 1990–2080. Tech Forecasting Soc., Ch. 74.

Frederick, K.D. and D.C. Major. 1997. Climate change and water resources. *Climatic Change* 37: 7–23.

Gardner, B.D. 1983. Water pricing and rent seeking in California agriculture. In *Water Rights Scarce Resources Allocation, Bureaucracy, and the Environment*, T.L. Anderson ed. Pacific Studies in Public Policy, Cambridge, MA: Ballinger Publishing Company.

IFPRI IMPACT simulations. 2008. Mimeo. Washington, DC: IFPRI.

Kundzewicz, Z.W., L.J. Mata, N.W. Arnell, P. Doll, P. Kabat, B. Jimenez, K.A. Miller, T. Oki, Z. Sen, and I.A. Shiklomanov. 2007. Freshwater resources and their management. *Climate Change 2007: Impacts, Adaptation, and Vulnerability*, eds. M.L. Parry, O.F. Canziani, J.P. Palutikof, P.J. van der Linden, and C.E. Hanson, pp. 173–210. Contribution of Working Group II to the Fourth Assessment Report of the Intergovernmental Panel on Climate Change. Cambridge, UK: Cambridge University Press.

Löfgren, H., 1996. The cost of managing with less: Cutting water subsidies and supplies in Egypt's agriculture. TMD Discussion Paper 7. Washington, DC: International Food Policy Research Institute.

Lomborg, B. 2007. *Cool It. The Skeptical Environmentalist's Guide to Global Warming*, 253 pp. New York: Knopf.

Mazzucato, V. and D. Niemeijer. 2001. Overestimating land degradation, underestimating farmers in the Sahel. Drylands Issue Paper No. 101. London, UK: International Institute for Environment and Development.

McCornick, P.G., S.B. Awulachew, and M. Abebe. 2008. Water–food–energy–environment synergies and tradeoffs: Major issues and case studies. *Water Policy* 10(1): 23–36.

Meehl, G.A., T.F. Stocker, W.D. Collins, P. Friedlingstein, A.T. Gaye, et al. 2007. Global climate projections. In *Climate Change 2007: The Physical Science Basis*, eds. S. Solomon, D. Qin, M. Manning, Z. Chen, M. Marquis, et al., pp. 747–845. Contribution of Working Group I to the Fourth Assessment Report of the Intergovernmental Panel on Climate Change. Cambridge, UK: Cambridge Univ. Press.

Meinzen-Dick, R. 2007. Beyond panaceas in irrigation institutions. *Proceedings of the National Academy of Sciences of the United States of America* 104: 15200–15205.

Meinzen-Dick, R. and C. Ringler. 2008. Water reallocation: Drivers, challenges, threats, and solutions for the poor. *Journal of Human Development* 9(1): 47–64. DOI: 10.1080/14649880701811393.

Merrey, D.J., R.S. Meinzen-Dick, P.P. Mollinga, and E. Karar. 2007. Policy and institutional reform processes for sustainable agricultural water management: The art of the possible. In *Water for Food, Water for Life*, ed. D. Molden, pp. 193–232. London: Earthscan.

Muller, A., J. Schmidhuber, J. Hoogeveen, and P. Steduto. 2008. Some insights in the effect of growing bioenergy demand on global food security and natural resources. *Water Policy* 10(1): 83–94.

Nelson, G., M. Rosegrant, J. Koo, et al. 2009. Climate change. Impact on agriculture and costs of adaptation. IFPRI Food Policy Report.

Nkonya, E., J. Koo, H. Xie, and P.S. Traore. 2009. SLM advisory services: Key institutional, financing, and economic elements for scaling up sustainable land management in Nigeria. IFPRI mimeo.

OECD. 1999. *Water Subsidies and Environment*. Paris: OECD.

Ostrom, E. 1990. *Governing the Commons*. The Evolution of Institutions for Collective Action. Cambridge, UK: Cambridge University Press.

Parry, M.L., C. Rosenzweig, A. Iglesias, M. Livermore, and G. Fischer, 2004. Effects of climate change on global food production under SRES emissions and socio-economic scenarios. *Global Environmental Change* 14: 53–67.

Pender, J. 2009. Food crisis and land. The world food crisis, land degradation, and sustainable land management: Linkages, opportunities and constraints. Pretoria, South Africa: A TerraAfrica partnership publication.

Perry, C.J. 2001. Charging for irrigation water: The issues and options, with a case study from Iran. Research Report 52. Colombo, Sri Lanka: International Water Management Institute.

Pitman M. and A. Läuchli. 2004. Global impact of salinity and agricultural ecosystems. In *Salinity: Environment–Plant–Molecule*, eds. M. Läuchli and U. Luttee, pp. 3–20. Netherlands: Springer.

Randall, A. 1981. Property entitlements and pricing policies for a maturing water economy. *Australian Journal of Agricultural Economics*, 25: 195–212.

Requier-Desjardins, M. 2006. The economic costs of desertification: A first survey of some cases in Africa. *International Sustainable Development* 9(2): 199–209.

Ringler, C. 2005. The role of economic incentives for the optimal allocation and use of water resources—case study of the Dong Nai River Basin in Vietnam. In *Water and Sustainable Development*, ed. P.M. Schmitz, pp. 61–92. Frankfurt: Peter Lang GmbH.

Rockström, J., W. Steffen, K. Noone, et al. 2009. Planetary boundaries: Exploring the safe operating space for humanity. *Ecology and Society* 14(2): 32, [online] URL: http://www.ecologyandsociety. org/vol14/iss2/art32/.

Rosegrant, M.W. and H.P. Binswanger. 1994. Markets in tradable water rights: Potential for efficiency gains in developing country water resource allocation. *World Development* 22(11): 1613–1625.

Rosegrant, M.W. and C. Ringler. 1998. Impact on food security and rural development of transferring water out of agriculture. *Water Policy* 1(6): 567–586.

Rosegrant, M., R.G. Schleyer, and S. Yadav. 1995. Water policy for efficient agricultural diversification: Market based approaches. *Food Policy* 20(3): 203–223.

Rosegrant, M.W., X. Cai, and S.A. Cline. 2002. World water and food to 2025: Dealing with scarcity. Joint publication. Washington, DC, and Colombo, Sri Lanka: International Food Policy Research Institute and International Water Management Institute.

Rosegrant, M.W., C. Ringler, and C. Rodgers. 2005. The water brokerage mechanism—Efficient solution for the irrigation sector. Proceedings of XII World Water Congress Water for sustainable development—Towards innovative solutions, November 22–25, 2005. New Delhi: India.

Rosegrant, M.W., T. Zhu, S. Msangi, and T. Sulser. 2008. Global scenarios for biofuels: Impacts and implications. *Review of Agricultural Economics* 30(3): 495–505.

Rosegrant, M.W., M. Fernandez, A. Sinha, et al. 2009. Looking into the future for agriculture and AKST. In *International Assessment of Agricultural Knowledge, Science and Technology for Development (IAASTD): Global Report*, eds. B.D. McIntyre, H.R. Herren, J. Wakhungu, and R.T. Watson. Washington, DC: Island Press.

Roy, R., R. Misra, and A. Montanez. 2002. Decreasing reliance on mineral nitrogen—Yet more food. *Ambio* 31(2):177–183.

Sauer, J., H. Tchale, and P. Wobst. 2007. Alternative soil fertility management options in Malawi: An economic analysis. *Journal of Sustainable Agriculture* 29(3): 29–53.

Tiffen, M., M. Mortimore, and F. Gichuki. 1994. *More People—Less Erosion: Environmental Recovery in Kenya.* London: Wiley and Sons.

Tittonell, P., M. Corbeels, M. van Wijk, B. Vanlauwe, and K. Giller. 2008. Combining organic and mineral fertilizers for integrated soil fertility management in smallholder farming systems of Kenya: Explorations using the crop-soil model FIELD. *Agronomy Journal* 100(5): 1511–1526.

Tschakert, P. 2004. The costs of soil carbon sequestration: An economic analysis for small-scale farming systems in Senegal. *Agricultural Systems* 81: 227–253.

Vanlauwe, B., A. Bationo, J. Chianu, et al. 2010. Integrated soil fertility management Operational definition and consequences for implementation and dissemination. *Outlook on Agriculture* 39(1): 17–24.

Vorosmarty, C., C. Leveque, and C. Revenga. 2005. Fresh water. In *Ecosystems and Human Well-Being: Current State and Trends.* Findings of the Condition and Trends Working Group. Millennium Ecosystem Assessment. Island Press.

World Bank. 2003. Implementation completion report for Loess Plateau watershed rehabilitation project. Report No. 25701. Washington, DC.

World Bank. 2007. Project performance assessment report for second Loess Plateau watershed rehabilitation project and Xiaolangdi multipurpose project I & II and Tarim Basin II project. Report No. 41122. Washington, DC.

Section VI

Tools of Watershed Management

20 Watershed Management for Erosion and Sedimentation Control Case Study

Goodwin Creek, Panola County, MS

Seth M. Dabney, F. Doug Shields, Ronald L. Bingner, Roger A. Kuhnle, and James R. Rigby

CONTENTS

20.1 INTRODUCTION

Erosion and sedimentation are major global problems, with specific issues associated with loss of agricultural production, adverse impacts on infrastructure, and environmental degradation (Walling and Webb 1996). Erosion problems are particularly severe in hilly areas of the southeastern coastal plain of the United States, where annual watershed sediment yield is three to six times the national average for watersheds of similar size (Shields et al. 1995).

The Goodwin Creek watershed, located within this region of elevated sediment yield, provides an instructive case study of the dynamism of land use, where hydrologic, geomorphologic, and water quality characteristics can change over time; demonstrates how piecemeal application of accepted best management practices can have unanticipated long-term consequences; and illustrates the challenges inherent in attempting to assess the impact of conservation practices at the watershed scale (Bingner et al. 2007).

The Goodwin Creek watershed is located in north central Mississippi (89.8659 W, 34.2559 N). It is a tributary to Long Creek, and subsequently to the Yocona, Tallahatchie, Yazoo, and Mississippi

Rivers. At its confluence with Long Creek, Goodwin is a fourth-order stream. At the watershed outlet (Station #1), Goodwin has a drainage area of 21.3 km^2. It receives an average of about 1400 mm rainfall annually that is more or less evenly distributed over the months but often falls with higher intensity during the summer.

Goodwin Creek is located in the "bluff-hills" region, an area of thick loessal deposits located adjacent to and within about 35–70 km to the east of the Mississippi River alluvial plain ("The Delta"). Pleistocene loess overlies older coastal plains sand, clay, and gravel deposits (Grissinger et al. 1981; Grissinger and Murphey 1983). The loess thins to the east and is thickest (up to ~30 m) adjacent to the Delta. The Goodwin Creek watershed lies 15–25 km east of the bluff line (Delta/bluff-hills boundary), so prior to European settlement, the upland soils may have had loessal caps 2–3 m thick.

20.2 EARLY EUROPEAN SETTLEMENT

European settlers acquired control of the last 2,543,055 ha of Chickasaw Nation tribal lands, which included the Goodwin Creek watershed, in 1832 through terms of the Treaty of Pontotoc. Thereafter, rapid development involved land clearing and conversion to agriculture. The rolling or gently undulating upland loessal table lands were among the first lands cleared of their native oak/hickory forests because they were recognized to be productive, well drained, and easy to work.

One of the preeminent soil scientists of the nineteenth century, Eugene W. Hilgard, held faculty position at the University of Mississippi from 1855 through 1873. He also served as the state geologist of Mississippi. His writings describe and document the geology, soils, and early agricultural development of the region. According to Hilgard (1860, p. 293), one farmer characterized the loessal soil as "good enough as long as it will stay." However, when organic matter became depleted, productivity was lost because of low fertility and soil sealing that caused excessive runoff and erosion. Faced with depleted soils, the settlers moved on and cleared new lands, while "turning out" or fallowing older fields. Short periods of fallow improved fertility (Hilgard 1860, p. 206) but, Hilgard (1860, p. 295) warned, "It is highly important … to prevent the washes from penetrating the loam into the underlying sand or hardpan. Should the sand be loose, the moment the water reaches it, an undermining process will begin, which will cause the land to waste with greatly increased rapidity. Should it, on the contrary be an impervious hardpan, as is very frequently the case, the increased mass and velocity of the water will rapidly widen its channel, casting away the sides of the gully." The civil war greatly increased the amount of abandoned land, and the effects of uncontrolled gully growth during that period are recorded in valley sediment deposits of the region (Happ 1937).

According to the 1880 census, 34% of Panola County was classified as tilled lands, and of this, 45% was planted with cotton (*Gossypium* sp.) (Hilgard 1884). The cropland figure did not include the abandoned lands that grew at a rate and fashion alarming to Hilgard (1884, p. 238): "… unfortunately there has been a great deal of almost irretrievable damage done to these lands by allowing them to be washed and finally gullied by the rains, the water ultimately cutting into the underlying sand, and thereafter undermining the soil stratum and converting the hill lands into unavailable sand-hills, while the valleys also have been filled up with a mixture of sand and soil, the former usually predominating, rendering them almost as unavailable for cultivation as the hills … every year the evil increases in a geometrical ratio, and if unchecked must result in the serious and permanent injury to the agricultural interests of one of the fairest and naturally most favored portions of the state."

Nearly all the virgin timber in the region had been cut by 1910. The uncontrolled growth of upland gullies in the bluff-hills had proceeded. State geologist E.N. Lowe (1910) described the region as "the badlands of Mississippi." "The timber has been very largely removed, and much of the surface lies out in old fields." "The forests must be cleared and the lands must be cultivated for man's sustenance, and as long as he gives intelligent care to the soil washing need not result. It is only when he relinquishes it after clearing and cultivating it that it goes to destruction." Erosion problems were exacerbated by the common practice of the day of burning fallowed fields in the fall to improve visibility for hunting.

The period between 1910 and 1930 saw erosion of the upland areas at its worst. Most of the land had once been cropped and much was now abandoned and subjected to uncontrolled gully expansion. Forest cover was at a minimum. Steep-sided gullies grew until they approached the hill crests. Sandy sediments choked the stream channels, creating alluvial fans where they intersected floodplains. These sediment deposits diverted runoff flows onto the floodplains causing flooding from even minor storms of 2–3 cm (Happ 1937). Sandy splay deposits covered the more fertile natural soils and reduced productivity of the floodplain soils. Floodplain aggradation with "post-settlement alluvium" was frequently 2–3 m thick (Happ et al. 1940), and sometimes more than 8 m thick. The thickest deposits were located near the head of permanent stream flow, with 65% of the total deposition occurring within the upper 10 km of some Mississippi creek valleys (Happ 1975).

The original stream system had evolved in equilibrium with the runoff and sediment loads associated with intensive rainfall on mature forests growing on silty soils. The channels did not have the transport capacity to transport the sands washed from the upland gullies. In addition to creating flooding damage, sedimentation also eliminated navigation of the major rivers in the region. For example, state geologist Lowe (1922; as cited by Bennett (1939)) stated: "The Tallahatchie was formerly a navigable stream. Even as late as 1900 a small steamer drawing 4 ft of water plied the Tallahatchie from Batesville downstream. Now the stream is choked with sand bars, and can easily be waded at almost any place ..." The severity of historical erosion within Goodwin Creek watershed was evidenced in the soil descriptions in the Panola County, Mississippi, soil survey (USDA-SCS 1963): only 2% of the land area was described as a nominal series, 11% was characterized as eroded, 40% as severely eroded, 26% as gullied, and 21% as floodplains.

20.3 DRAINAGE DISTRICTS

Watson et al. (1997) described the watershed development activities during the period from 1910 to 1939 as the "Drainage District Period." The first drainage law in Mississippi was enacted in 1886, and such laws became common in subsequent years. These laws empowered local districts to organize, issue bonds to be repaid with local taxes, and execute channel improvements. Most drainage works involved straightening the channels or digging laterals through the floodplains in order to increase conveyance and sediment transport capacity. The activity was related to cost–benefit calculations; the interest in drainage increased with cotton prices (Watson et al. 1997). The drainage district encompassing Goodwin Creek was called the Long Creek Drainage District No. 3 and was organized in 1922 (Olsen and Dumm 1941). This drainage district constructed about 10 km of ditches paid for with $19,000 worth of bonds. Construction of the "Smith Creek Lateral," which is the lowest 800 m of Goodwin Creek, was part of this activity. Gauging Station #1 of the Goodwin Creek Experimental Watershed is actually located within the constructed lateral at a point about 130 m south of the original channel's nearest approach (Figure 20.1).

Typically, several districts planned and constructed drainage works along a stream without coordination with each other. Often, channel activities of an upstream portion of a stream by one drainage district sent too much discharge and sediment to an unchannelized downstream reach of a different district (Watson et al. 1997). The equally ineffective approach of treating the channel system without first controlling upland management was documented by Happ (1937) for the case of the Wells Drainage District in adjacent Lafayette County that spent $30,000 to excavate bottom land drainage laterals in 1920 without treating the eroding upland gullies whose sand had clogged the natural stream channels. Benefits amounting to $54,000 were anticipated beyond the taxes totaling $55,000 to be collected to retire bonds by 1943. However, within 5 years of construction, sedimentation had caused the canals to cease to function properly.

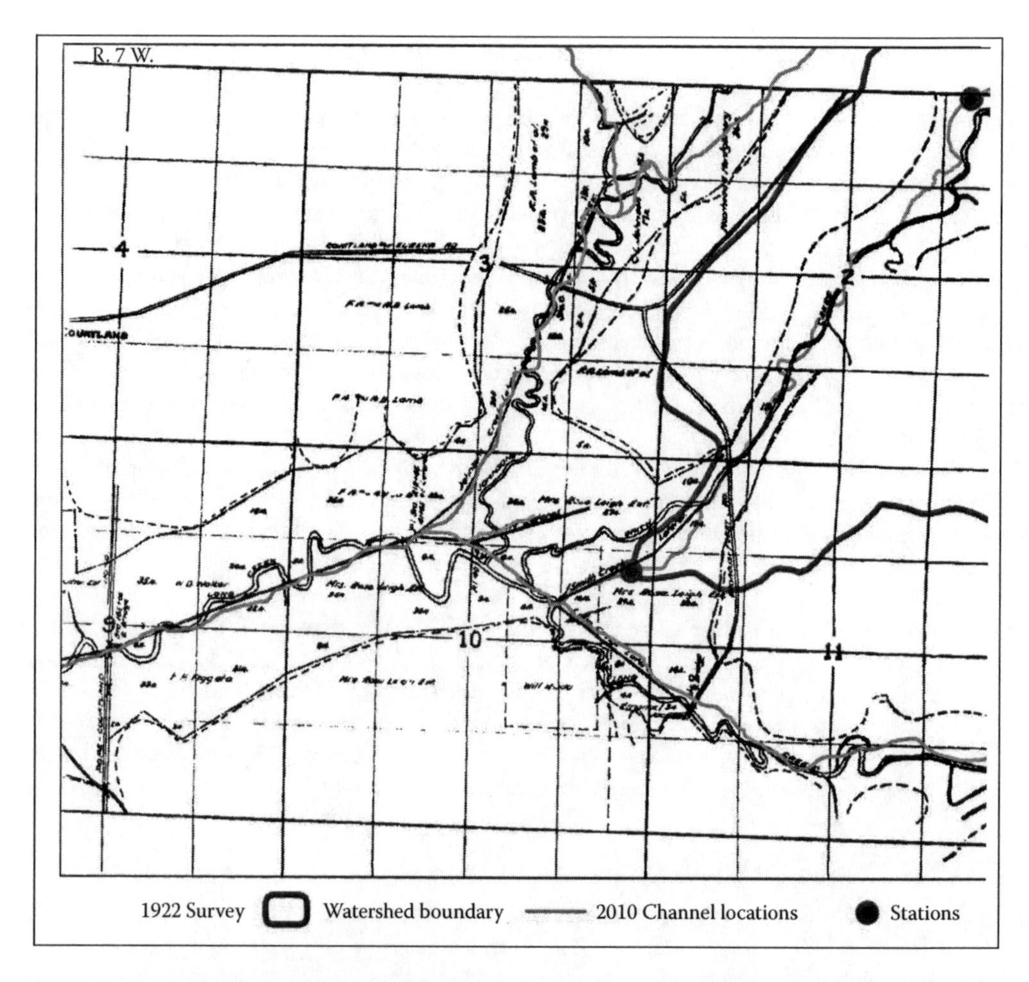

FIGURE 20.1 The Long Creek Drainage District No. 3 constructed 10 km (6.25 miles) of ditches and issued $19,000 in bonds in 1922. The "Smith Creek Lateral" is now the lowest 800 m of Goodwin Creek. The gray line shows that the affected channels in 2010 remain considerably straighter, and therefore shorter and steeper, than before the drainage district project. Also shown are the locations of gauging stations #1 and #2 and the Goodwin Creek watershed boundary. Note that gauging station #1 is located on the constructed lateral at a point about 130 m south of the original channel.

20.4 FEDERAL INVOLVEMENT

20.4.1 Flood Control Dams

Following a devastating flood of the Mississippi River in 1927 (Barry 1997), the Federal Flood Control Act was passed in 1928 that established the U.S. Waterways Experiment Station at Vicksburg and empowered the Secretary of Agriculture to conduct research related to flood control such as reforestation. In 1936, legislation was passed mandating that floods on the Yazoo River were to be controlled with channel improvements, levee construction, and the construction of flood control reservoirs on the Coldwater, Tallahatchie, Yocona, and Yalobusha Rivers (Figure 20.2). One of these, the Enid Dam on the Yocona River, was constructed about 11 km upstream of the confluence with Long Creek with the Yocona. As discussed later, the closure of the dam in 1953 effectively lowered the base level of tributaries to the Yocona downstream of the dam.

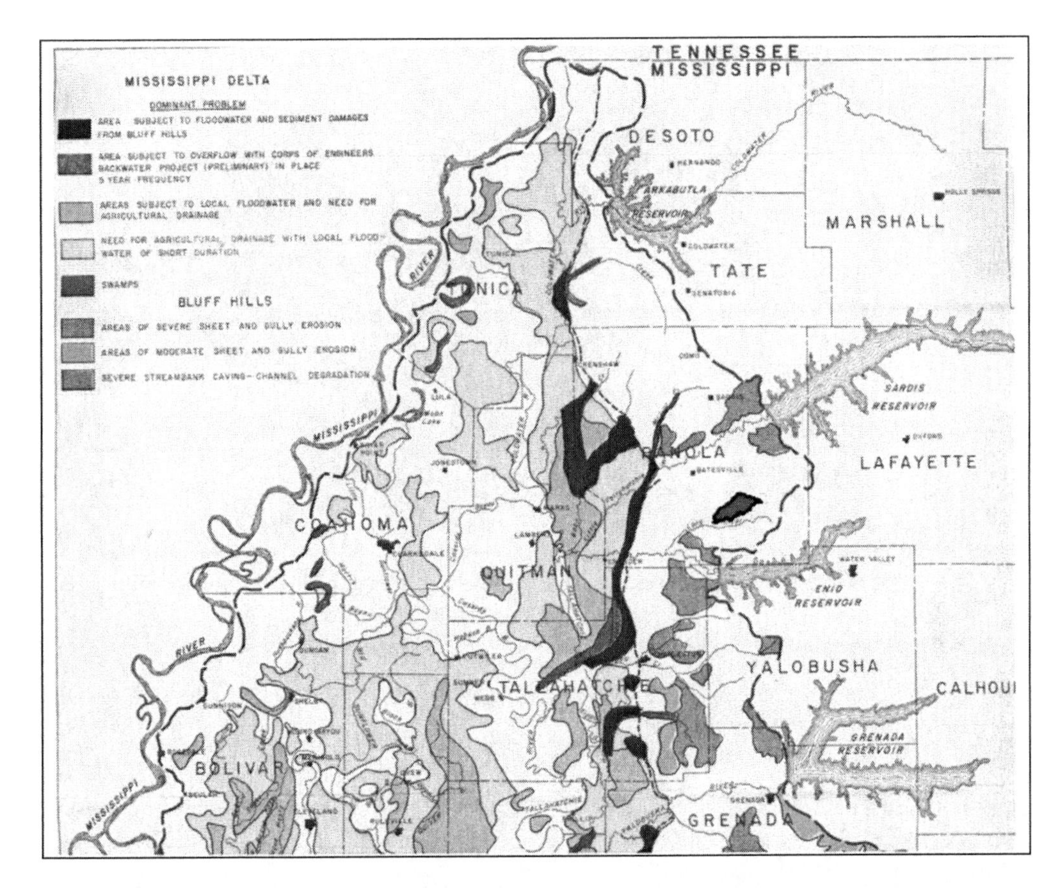

FIGURE 20.2 Following the construction of four major flood control reservoirs and the successful stabilization of upland areas by the YLT project, floodplains became an increasing source of sediment. Problems of streambank erosion were first recognized, as illustrated in this figure adapted from USDA-SCS (1961), in areas along the loess bluff line bordering the Mississippi Delta. In the following decades, headcut migration and streambank erosion became the dominant sources of sediment leaving upland watersheds in the region. Note the location of Goodwin Creek Experimental Watershed, which is a tributary to Long Creek and then to the Yocona River below the Enid Reservoir.

20.4.2 Upland Erosion Control and the YLT Project

Federal funds began to be expended to directly provide technical assistance to reduce the upland soil erosion on private lands following the creation of The Civilian Conservation Corps and the Soil Erosion Service in 1933, which was renamed the Soil Conservation Service (SCS) in 1935. In 1944, congress passed The Flood Control Act, PL 534, which established 11 watersheds throughout the United States, including the Yazoo and Little Tallahatchie River (YLT) watersheds in Mississippi, and authorized the Secretary of War, The Chief of the Corps of Engineers, and the Secretary of Agriculture to carry out works of improvement on them. The YLT Flood Prevention Project was officially launched in 1947 and lasted until 1985. On upland areas, USDA-SCS and USDA-FS worked together to reclaim gullies and return lands to constructive uses. Technology was developed and transferred and cost sharing was provided. Where land owners were interested, the worst gullied lands were afforested with loblolly pine (*Pinus taeda*) (Figure 20.3), which research showed provided the best and most rapid cover of pine straw mulch. Between 1948 and 1982, 339,000 ha of land was established in pine (Williston 1988; Duffy and Ursic 1991), while numerous farmland best management practices (BMPs) such as contour farming, terraces, grassed waterways, farm ponds, and grade control structures were applied on farmlands by SCS. The effectiveness of these upland

FIGURE 20.3 The YLT project reforested the most badly gullied land. In the case illustrated, strips of weeping lovegrass (*Eragrostis curvula*) were established to stabilize the actively eroding sandy gully sidewalls, and in the following winter, loblolly pine (*Pinus taeda*) seedlings were transplanted. Brush dams were constructed to stabilize thalweg areas before tree planting. (From Williston, H.L., The Yazoo–Little Tallahatchie Flood Prevention Project. A history of the Forest Service's role. U.S. Forest Service. Southern Forest Experiment Station. Forestry Report R8-FR 8, pp. 63, 1988; Duffy, P.D. and Ursic, S.J. *Land Use Policy*, 8, 196–205, 1991.)

conservation practices in increasing the amount of land covered with vegetation is evident in the time series of aerial photographs (1940, 1944, 1953, 1957, 1963, 1968, and 1976) of Goodwin Creek presented by Whitten and Patrick (1981).

20.4.3 DEC and CEAP Eras

As the amount of sediment generated from upland areas was reduced, a new problem of channel bank and bed erosion developed (Figure 20.2). Severe problems were first reported at the western boundary of the bluff-hills (USDA-SCS 1961). However, these problems soon propagated upstream and became widespread throughout the YLT watersheds, consistent with the conceptual models of incised channel evolution (e.g., Simon 1989). Happ (1975) concluded that in Yazoo River tributaries that had been dredged, or trenched by erosion headward from the dredged channels, channel and floodplain erosion exceeded the overbank deposition so that the valleys had become the sources rather than the repositories of sediment. Channel incision along the mainstem of Goodwin Creek produced steep, continually failing banks 4–5 m high and elevated sediment yield to ~1000 t/km^2 (Figure 20.4).

The growing recognition of economic losses associated with streambank erosion caused Congress to pass the River and Harbors Act of 1968 and the Streambank Erosion Control Demonstration Act of 1974, PL 93-251. Based on the 1974 legislation, the U.S. Army Corps of Engineers (COE) entered into a data collection and evaluation program with the USDA-ARS National Sedimentation Laboratory to test the concepts and designs of bed and bank stabilization techniques and to collect data needed for models evaluation over an indefinite period of time. The Goodwin Creek Experimental Watershed was ultimately selected and established in 1977. Two of the reasons why Goodwin Creek was selected included the following: (1) that it was less highly gullied than nearby watersheds and thus supported a wider range of land uses and (2) that it contained both straightened and natural stream reaches.

Prior to 1981, the COE constructed 14 supercritical grade control structures to stabilize the channel and serve as gauging stations operated by the Agricultural Research Service (ARS) (Figure 20.5).

FIGURE 20.4 Concave bank of Katherine Leigh's bendway, Goodwin Creek, Mississippi, in 1986.

Land use, stream flow, water quality, and aquatic habitat data collection within the Goodwin Creek watershed became part of the ARS research program in 1980. Resulting data are available at http://www.ars.usda.gov/Main/docs.htm?docid=5120. In 1984, Congress authorized the Demonstration Erosion Control Project (DEC) to demonstrate the effectiveness of tools developed under PL 93-251 within the designated subwatersheds of the Yazoo River Basin. DEC funding for construction and monitoring continued until the late 1990s. Demonstration, evaluation, monitoring, and modeling research has continued in Goodwin Creek since 1980 and was intensified when Goodwin Creek was included as an ARS benchmark watershed in the cropland Conservation Effects Assessment Program (CEAP) in 2003. The goal of this assessment was to provide a better understanding of the role agricultural conservation practices and programs play in achieving the nation's environmental objectives—clean air and water, healthy soils, and functioning habitat for wildlife (http://www.ars.usda.gov/Research/docs.htm?docid=18645).

FIGURE 20.5 Goodwin Creek supercritical flow flume (Station #2) that serves as a grade control structure and flow and sediment discharge monitoring station.

20.4.4 Channel Erosion Control

Several attempts were made to stabilize the channel of Goodwin Creek and rehabilitate stream corridor habitats (Table 20.1). Construction of the aforementioned 14 supercritical flow flumes was completed prior to 1981 at a cost of $1.9 million (costs not adjusted for inflation; U.S. Army Corps of Engineers 1981). Flumes were designed to serve as grade control structures to prevent further channel incision and as monitoring stations to allow the measurement of water discharge and collection of sediment samples. The placement of these grade control structures arrested downcutting, and an average of 0.6 m of sediment deposition was reported for lower reaches in early 1989 (Neill and Johnson 1989). However, consistent with Stage IV/V of the incised channel evolution model (Simon 1989), lateral erosion and migration continued (Little et al. 1982; Figure 20.5). To address this erosion, stone stabilization structures were placed on the selected portions of the bank between gauging stations 1 and 2 in 1990–1991. Stabilization structures were longitudinal stone toe (6 ton/m) or short stone spurs ("groins"), which were generally placed on the outside of bends. These stabilization structures were effective in arresting lateral channel erosion, and stabilized banks were soon covered with deposited sediments and colonizing vegetation. However, bank failure following channel incision had widened the channel several fold, and the bank protection measures did not project far enough into the backflow channel to create scour pools that would benefit fish habitat. Fish habitat was generally degraded in Goodwin Creek by shallow depths, flashy hydrology, shifting substrate, and a lack of large wood and associated features (e.g., cover and scour pools) (Shields et al. 1994, 1998a). Reaches near the mouth had bank heights of 4–5 m, widths of 20–70 m, and thalweg sinuosity of only 1.12. Only 32% of the bank line was dominated by woody vegetation, and estimated canopy over the baseflow channel averaged 14%. Mean water depth was only 0.20 m, and large wood density was an order of magnitude smaller than for a nearby lightly impacted stream.

These deficiencies were partially addressed by extending 17 of the groins to create low weirs and by adding 11 spur-type extensions to the stone toe. In addition, 1729 dormant willow (*Salix nigra*) posts were planted adjacent to the modified stone structures. Few of the willows survived due to infertile soils (Grissinger and Bowie 1984; Pezeshki and Shields 2006) and competition from the exotic vine, *Pueraria lobata* (kudzu). The combined effects of grade control, groins, and spur dikes reduced bank and bed erosion and essentially stopped the thalweg migration. This is illustrated based on the analysis of aerial photographs from 1937, 1957, 1979, 1996, and 2006 at a location

TABLE 20.1
Channel Erosion Control and Habitat Restoration Projects, Goodwin Creek, Mississippi

Construction Date	Measures Emplaced	Length of Channel Treated (m)	Cost (Dollars Not Adjusted for Inflation)	Reference
1978–1980	14 grade control flumes	Systemic	$1,911,448	U.S. Army Corps of Engineers (1981)
1980	Toe protection and upper bank planting	160	$86,297	Bowie (1982, 1995), Shields et al. (1995)
1990–1991	Stone toe and groins	3000	$350,000 (est.)	n/a
1993	Stone weirs, spurs, and willow post plantings	1200	$50,000 (est.)	Shields et al. (1998a, 2007)
1993–1994	Woody and herbaceous plantings	200	n/a	Snider (1996)
2007	Toe protection, bendway weirs, upper bank plantings	100	$33,000	Simon et al. (2008)

defined as Katherine Leigh's bendway (Figure 20.6). Channel meandering was so destructive that some floodplain lands were taken out of production.

Physical aquatic habitat improvement was also almost immediate. Mean pool habitat increased from 32% to 78% of the water area. Biological response was more muted than the physical response, but fish collections in the 3 years following habitat rehabilitation produced fewer, larger fish than in the 2 prior years. Species composition shifted toward game and sportfish species that prefer pool habitats. The reach treated with spur-type extensions experienced a stronger biological response (Figure 20.7; Shields et al. 1998b). Ecological improvements persisted for at least 10 years (Shields et al. 2007).

More local efforts to control bank erosion using combinations of vegetation and structure and to accelerate the revegetation of the overwidened incised channel were attempted in 1980, 1995, and 2007. In 1980, the concave bank of a single bendway was treated with bank shaping; toe protection with stone, chain link, and pilings; and soil amendment and planting with native willow and six herbaceous species (Figure 20.8; Bowie 1982, 1995). The opposite convex bank was planted, but no toe protection was placed there. Thirteen years after construction, banks were stable and supported a lush mix of three of the planted species and several native invader species (Shields et al. 1995). In 1993–1994, test plots on a large point bar comprised of sand and gravel and selected portions of the opposite convex bank were planted with 7 herbaceous species and 17 woody species (Snider 1996). Plant materials included rooted seedlings, rhizomes, unrooted cuttings (stakes), and fascines (bundles of stems). Survival rates were generally low due to extremely harsh conditions imposed

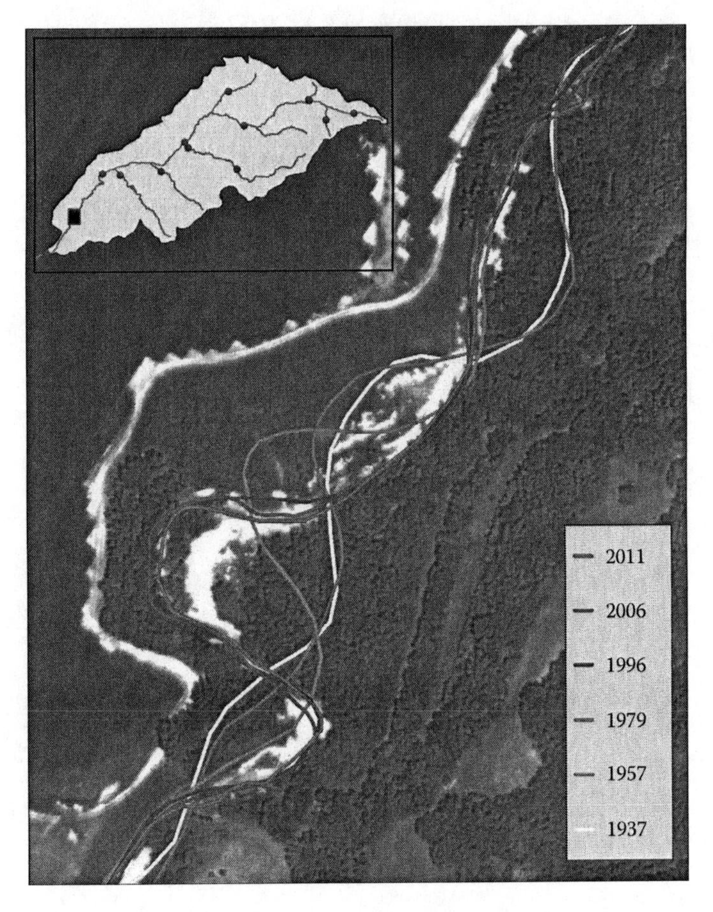

FIGURE 20.6 Migration of Goodwin Creek's Katherine Leigh bendway between Stations 1 and 2 from 1937 to 2011. Note that thalweg migration was greatly reduced after 1996.

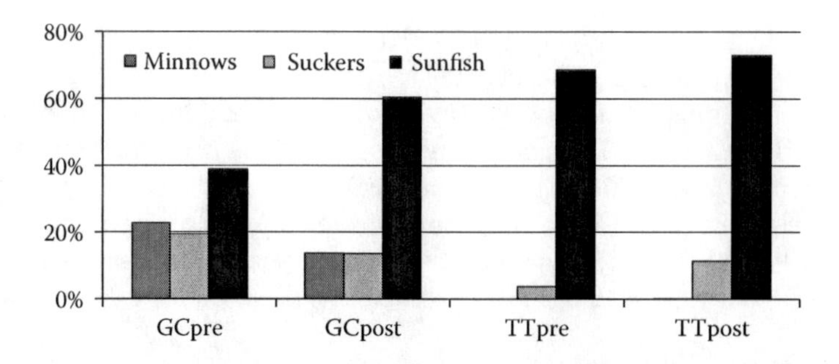

FIGURE 20.7 Impact of habitat rehabilitation on the distribution of fish biomass. GCpre = 1991–1992, prior to the construction of weirs, GCpost = 1993–1995, following weir construction. TTpre and TTpost are concurrent samples from a lightly degraded reference stream nearby. (After Shields, F.D., Jr, Knight, S.S., and Cooper, C.M., *Hydrobiologia*, 382, 63–86, 1998.)

by erosion, deposition, erosive flows, and drought. Shrubby willow species and giant reed (*Arundo donax*) performed well (Snider 1996), but the use of the latter is discouraged as it is an invasive exotic (Dudley 2000). In 2007, a 100-m steep, eroding convex bank was treated using stone toe, spur-type structures ("bendway weirs"), and upper bank plantings (Simon et al. 2008). The protection of an existing road necessitated a steeper bank slope (1V:1H) than stability criteria dictated for rapid drawdown conditions. The design involved geotechnical modeling of bank stability; the use of riparian vegetation to increase soil shear strength via root reinforcement to allow the use of the steep slope was a key aspect of the implemented design. As of this writing, the project has remained free from failure for 5 years.

FIGURE 20.8 Vegetative streambank stabilization studies in Goodwin Creek proved successful, though expensive (From Bowie, A.J. *Transactions of the ASAE*, 25, 1601–1606, 1982; Bowie, A.J. Use of vegetation to stabilize eroding streambanks. U.S. Department of Agriculture, Conservation Research Report. 43, 24, 1995.) and stable reaches have survived without maintenance for more than 20 years. Prior to construction, banks averaged 3.2 m in height and channel bottom widths ranged from 9 to 12 m. The bed gradient is 0.0049, and the catchment area above the study reach is approximately 14 km^2.

20.4.5 Land Use and Sediment Load Changes

Fine sediment (less than 0.062 mm in diameter) decreased between the years of 1982 and 1991 on Goodwin Creek (Figure 20.9; Kuhnle et al. 1996). The mean fine sediment concentration decreased from approximately 3000–1000 parts per million by weight (ppmw). The analysis of fine sediment concentration versus stream discharge showed a positive relationship between these variables, although the rating curve varied between rising and falling limbs of the hydrograph. During the period 1993 through 2003, researchers halted the sediment analysis, relying instead on the rating curve to calculate the sediment yield. Sediment sampling was reinitiated in 2003, and concentrations were found to be similar to those in 1993.

The decrease observed in the sediment concentration between the early 1980s and the mid-1990s corresponded to a reduction in the fraction of the land that was in cropland (Kuhnle et al. 2008; Figure 20.10), the previously described efforts at channel bank stabilization, and continued construction of ponds (Figure 20.11). Another structural practice that was commonly applied is "drop pipes," which allow surface runoff from fields to be conveyed to the level of incised streams without causing gully erosion (Figure 20.12). Peak runoff rates are also damped (Dabney et al. 2006). Sediment concentrations were higher during the third of the year from April through July, when average rainfall erosivity is higher and cropland is disturbed by tillage (Figure 20.9). Rating curve studies showed that stream fine sediment concentrations were higher at high flow rates, particularly during the rising limb of the hydrograph, so the observed lower concentrations and runoff rates may be due to either climatic or land management changes. Increased areas of grassland and forestland and more and larger ponds could reduce the peak runoff rates for equivalent rainfall inputs. Kuhnle et al. (2008) reported that the change in the percentage of cultivated land was coincident with a significant change in the runoff to rainfall relations for the April to July period.

20.4.6 Determination of Fine Sediment Sources

Sediment sources in the watershed have been determined using the naturally occurring radionuclides of ^{7}Be and $^{210}Pb_{xs}$. These radionuclides are produced continuously in the atmosphere and are

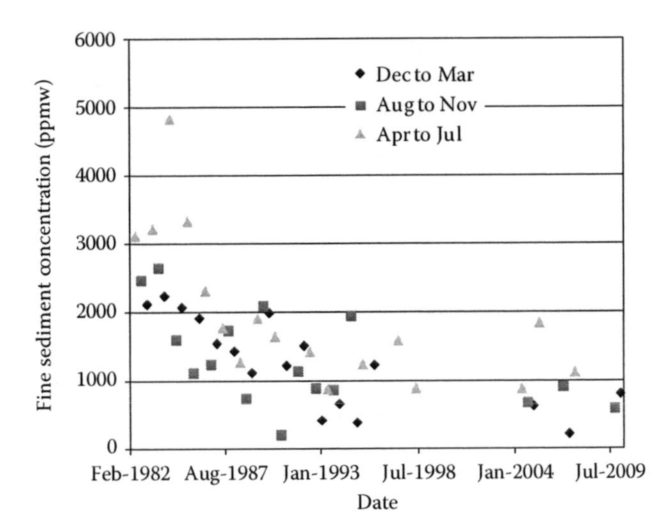

FIGURE 20.9 Coincident with the decreasing fraction of cropland in the watershed (Figure 20.10), the increase in the number of ponds (Figure 20.11), and the stabilization of streambank reaches, the flow-weighted average concentration of sediment <62 μm diameter declined. A significant seasonal effect—the higher spring/summer sediment corresponds to both tillage periods and periods—higher rainfall intensity are noted.

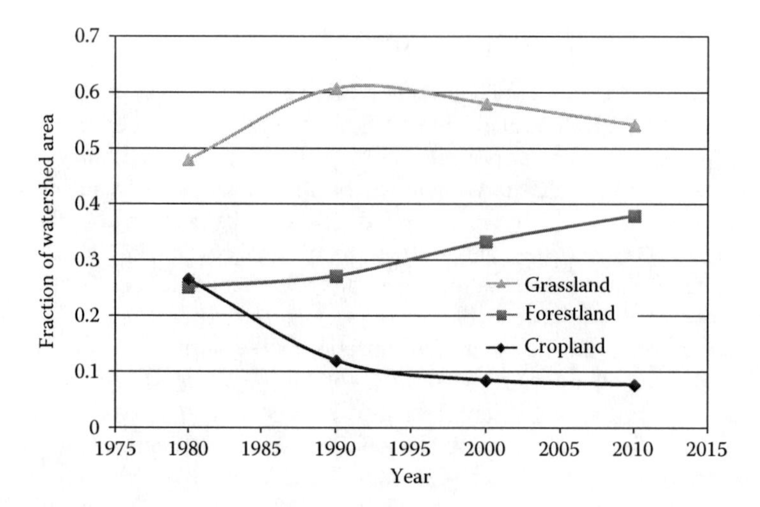

FIGURE 20.10 The fraction of the watershed area in cropland decreased substantially between 1980 and 1990, coinciding with conservation provisions of the 1985 Food Security Act and the creation of the Conservation Reserve Program. Subsequent to 1990, there has been a gradual increase in forested area and a gradual decline in cropland and grassland areas.

deposited on the land surface predominantly during rainfall events. The adsorption of radionuclides has been found to be restricted to the finer silt- and clay-sized particles, and thus source information is only available for these sizes using this technique. The unique signatures of the ratios of ^{7}Be to $^{210}Pb_{xs}$ on surface soils and channel banks yielded an indicator that was compared to the radionuclide ratio of sediment transported in the channels. It was found that eroded surface soils make up the dominant portion of fine sediment in the channels during the first parts of runoff events, while later in a runoff event, sediment from bank sources predominates. Source information integrated over entire runoff events has revealed that 78% of fine sediment originated from channel sources on Goodwin Creek (Wilson et al. 2008). These findings are consistent with the estimates of erosion of

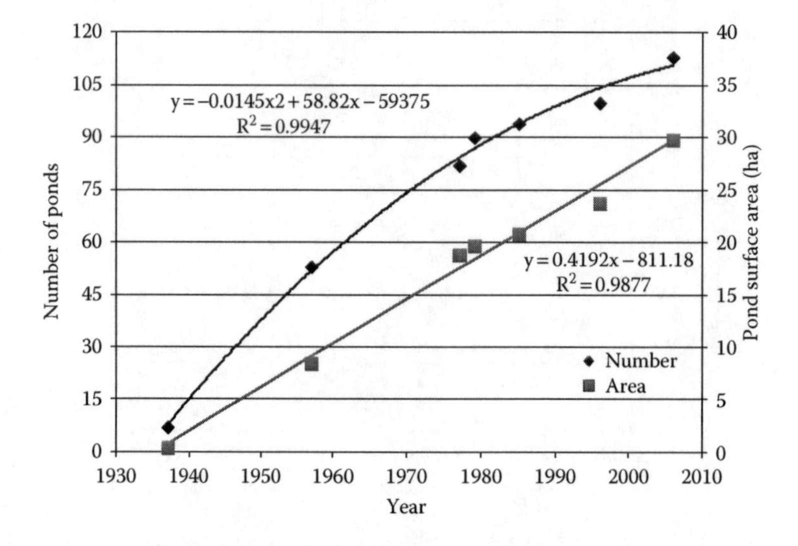

FIGURE 20.11 Based on the analysis of aerial photos, the number of ponds increases by about 2.3/year during the first 20 years, by 1.5/year during the next 30 years, and by about 0.9/year during the last 20 years. There has been a linear increase in the net pond surface area, which comprised 1.4% of the watershed area in 2010.

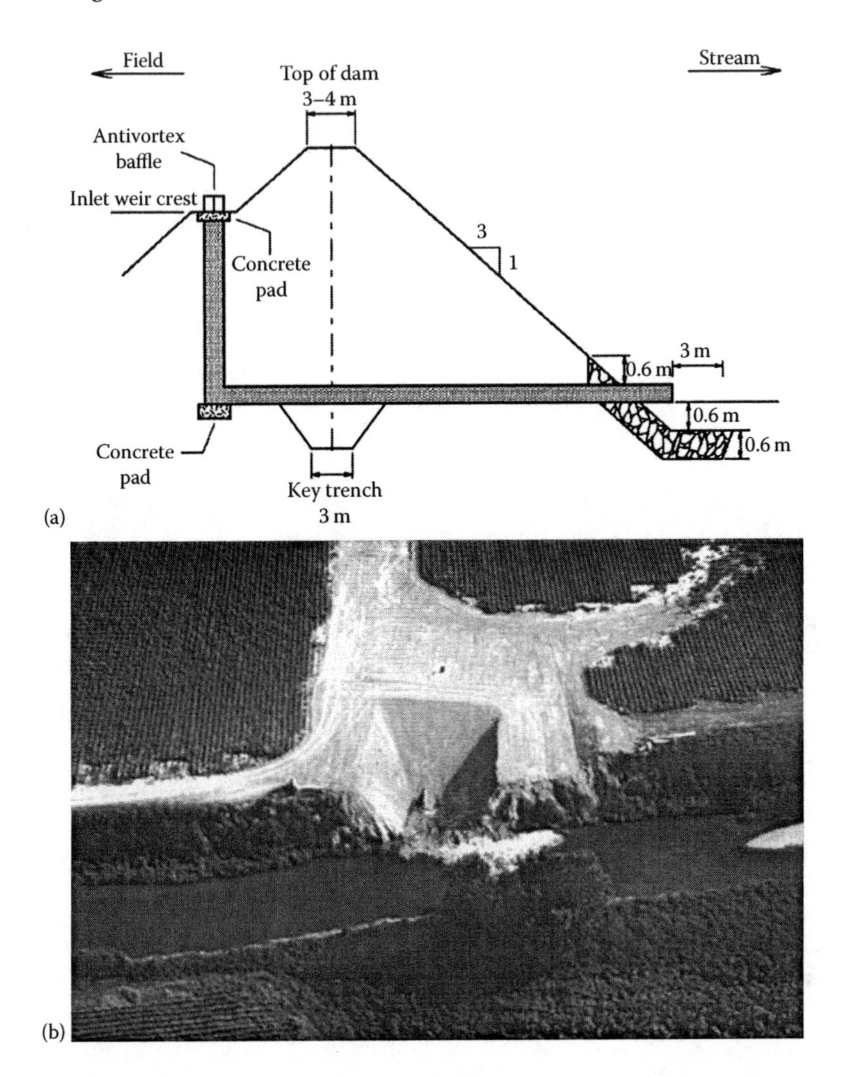

FIGURE 20.12 (a) A schematic of drop pipe structure including earthen embankment and (b) oblique air photo of a recently completed drop pipe viewed from the downstream (stream channel) side of the embankment.

channel beds and banks accounting for 75%–85% of the sediment transported through the channel system based on repetitive channel cross-section surveying (Grissinger et al. 1991) and watershed-scale simulation modeling (Kuhnle et al. 1996).

20.5 DISCUSSION AND IMPLICATIONS

While historically upland gully erosion was a serious sediment source in the region, at the current time, the erosion of incised channels is the predominant source of sediment transported from Goodwin Creek watershed. Efforts to stabilize the channel system during the last 30 years have produced mixed results (Watson et al. 1997). The main reason for the instability of the channel system is an imbalance between sediment transport capacity and sediment supply. Conservation practices applied to upland areas since the 1930s greatly reduced the upland sediment yield to channels. The reduction in runoff yield has been smaller, partly because of the persistence of topographical channel (gully) networks within the forested and pastured uplands.

Piecemeal drainage district efforts to alleviate flooding by straightening the stream system increased the stream gradients both directly (activities within the watershed) and indirectly by lowering the downstream base level (activities in the Delta). The construction of Enid Dam on the Yocona River also lowered the base level in the tributaries below the dam in two ways: (1) by retarding floods, the Enid Dam lowers the water levels in the lower Yocona during periods of intense runoff and (2) the discharge of sediment-free water from the dam resulted in scouring and lowering of the Yocona bed, thus steepening the gradient of tributaries (such as Long Creek) that join the Yocona downstream of the dam. Over time, the lower base level has propagated upstream in the form of headcuts.

The relatively sediment-free runoff flowing in steeper channels has unsatisfied sediment transport capacity, energy available to erode stream banks and beds. As channel erosion proceeds, the deeper and wider channels have the capacity to contain a greater percentage of peak flows, further concentrating the power of the flows within the channel system and reducing the connection of the streams to their floodplains. Practices to stabilize this type of inherently unstable channel system are difficult and expensive to implement (Table 20.1), and efficacy is uncertain (Shields 2008).

Land use has changed dramatically within Goodwin Creek watershed during the last 200 years. In 1800, the land was largely forested, and downstream rivers were reported to be deep and clear. By 1900, most of the land had been cleared, upland erosion was excessive, sediments were clogging channels, and streams were flooding bottomlands frequently. Today, cropland again represents less than 10% of the land area, and that mostly on the floodplains. Erosion of the channel beds and banks now accounts for 75%–85% of the sediment transported through the channel system.

In trying to assess the effect of field-scale Natural Resources Conservation Service (NRCS) conservation practices applied to cropland on water quality at the watershed scale, which is a main focus of CEAP, it is useful to reflect on the magnitude of upland BMPs that have already been applied in the Goodwin Creek watershed. The area has been part of at least one federal conservation demonstration program for most of the past 60 years. Ninety percent of what was once cropland has been converted to a conserving use (trees or grass cropland was 34% of Panola County in 1880; Hilgard 1884), with most other land either producing forages for draft animals or remaining idle; forests would soon be virtually eradicated. By contrast, during 2002, cropland had been reduced to 20% and reforested areas covered over 30% of the land (CTIC 2002). Although the percentage of cropland planted to cotton was about the same (45%) in 1880 and 2001, 95% of the cotton in Panola County in 2002 was planted no-till (CTIC 2002). Thus, although most of the land has received some conservation treatment, water quality in Goodwin Creek nevertheless remains impaired by sediment with stream banks and gullies being the predominant sources.

Over 50 years ago, King et al. (1956, p. 19) stated: "One outstanding need is some means of evaluating the benefits derived from land treatments … the comparative benefits derived from various individual treatments plus the overall cumulative effects of all treatments combined on a watershed basis." That statement was made relative to the joint USDA-FS and USDA-ARS research associated with the YLT program, but it could equally well apply to the objectives of CEAP today.

Goodwin Creek watershed is characterized with high rainfall erosivity, highly erodible soils, and (currently) relatively steep (0.004) and seriously incised channels. Total watershed sediment yield is among the highest in the nation (14.5 t/ha/year at station #1; Kuhnle et al. 2008). Suspended sediment concentrations regularly reach up to thousands of milligrams per liter during storm runoff events. High suspended sediment loadings have been shown to impact aquatic biota in streams and rivers in this region. The long historical record combined with 30 years of hydrological measurements provides a contrasting sequence of watershed management strategies during which uplands have been severely degraded and floodplains have acted first as sinks and now as sources of sediment. These experiences illustrate the possibilities, pitfalls, and limitations of using best management practices and watershed management to control erosion and sedimentation. Channelization as a management practice has been especially deleterious, creating a cascade of legacy impacts. Despite the almost irreversible environmental and economic damage caused by channel straightening, this management practice continues to be widely employed (Shields et al. 2011). The location and geography of Goodwin Creek watershed caused watershed degradation and adjustment processes to proceed more rapidly and to a greater extent than in other parts of the country, but the processes observed and lessons learned are believed to be broadly applicable.

20.6 SUMMARY AND CONCLUSION

The history of Goodwin Creek since European settlement can be read as a series of attempts to mitigate the unintended consequences of prior management. With the establishment of European control of lands drained by Goodwin Creek in 1832, the Goodwin Creek watershed underwent a phase of rapid land clearing and development resulting in the loss of virtually all forests in the watershed by 1910 in favor of pasture, cropland, and abandoned old fields. In the absence of forested area, the upland loessal soils suffered extreme erosion, thereby choking the streams with sediment. Because of the excess sediment supply to streams, a significant fraction of mobilized upland soils remained in the watershed forming new "post-settlement" deposits in channels and floodplains, in some cases contributing as much as 8 m of aggradation at the base of the upland slopes. This aggradation aggravated local flooding, which in turn motivated the establishment of institutional Drainage Districts to organize and manage engineering flood control projects. Drainage Districts set to the work of flood control primarily by straightening stream channels and constructing ditches, which increased the local stream gradients, draining the land more rapidly without addressing the excessive upland sediment supply, effectively only pushing the problem downstream within the larger, heavily agricultural, Yazoo River Basin. With the advent of federal involvement, regional flood control dams were constructed to manage flooding in the larger Yazoo River and Mississippi River Basins. The closure of Enid Dam in 1953 buffered the Yazoo Basin from flooding while also lowering the base level in the Goodwin watershed, adding further energy to an increasingly unstable system. Concurrent with the construction of Enid Dam, the Yazoo–Little Tallahatchie Flood Prevention Project initiated the reclamation of upland areas from widespread gullying, thereby reducing the sediment supply to channels in Goodwin Creek even as Enid Dam effectively steepened the channel gradients. With the supply of sediment from upland slopes greatly reduced by reclamation efforts, the banks and channels of Goodwin Creek became the primary source of sediment to dissipate the excess stream energy of the twice-steepened channels. While channel erosion control within the lower reaches of the watershed has been successful, even now large portions of the watershed are characterized by channel instabilities, which serve as the predominant source of degrading sediment.

From this sequence, at least four lessons can be gleaned. The first is the absolute necessity of considering the "downstream" effects of proposed management. In this sense, "downstream" may denote a spatial relationship within the watershed or even a temporal relationship as adjustment is a time-dependent process. In Goodwin Creek, the straightening of channel reaches "solved" the flooding problem locally by pushing the problem downstream. Similarly, the reclamation of the uplands from gullying reduced the sediment supply but did not consider the effect of a reduced sediment supply on the stability of stream channels.

The second lesson is an obvious corollary to the first. The management of erosion and sedimentation must always have in mind the dynamic balance of sediment and water. This point is highlighted in watersheds such as Goodwin Creek where few, if any, geologic constraints control erosion. As mentioned already, best management practices succeeded in stabilizing and reclaiming gullies on the uplands without equally decreasing the amount or energy of runoff in the streams. Kuhnle et al. (1996) summarized the effect of changing land use in the Goodwin Creek watershed in this way: "The major benefit in this watershed from a shift of a highly erodible land use to a nonerodible land use was not in the reduction of the amount of sediment from the upland areas but in the runoff leaving the upland areas and then its subsequent effect to reduce channel erosion and sediment transport." Thus, while erosion and sedimentation are the more visible vestiges of a problem, runoff remains the underlying supply of energy for channel erosion.

A third potential lesson from Goodwin Creek's history concerns the impact of institutions on watersheds. Institutional involvement in Goodwin Creek occurred primarily through the work of the Drainage Districts and subsequently by the flood control efforts of the federal government. The channelization of Goodwin Creek under the direction of the Drainage District reduced local flooding while pushing sedimentation and flooding problems downstream in the watershed. Clearly, local institutions have nonlocal effects. At the regional scale, the construction of flood control reservoirs produced downstream effects (base-level lowering), which then propagated upstream through the tributary watersheds such as Goodwin Creek. Thus, large-scale institutions and projects will directly affect the stability of smaller-scale systems within the region and may create upstream-propagating disturbances such that no portion of a large watershed is fully decoupled from nonlocal practices. An important lesson to take away, therefore, is that increases in the scale of management are not necessarily equivalent to increasing the necessary coordination of management within a basin or watershed.

Lastly, Goodwin Creek demonstrates that the effects of past management decisions cannot simply be undone. While reforestation of the watershed led to the stabilization of the uplands, it also destabilized the channels themselves because of the very flood control efforts aimed at mitigating the unintended effects of upland erosion. Furthermore, the behavior of reforested land can be expected to differ substantially from the same land when it was covered by virgin forests due to the degradation of surface soils and the presence of remnant gully channels that increase connectedness and reduce times of concentration. In a tightly coupled system, if management is always only aimed at mitigation, we risk always chasing our own unintended consequences in circles through the logic of watershed adjustment. This irreversible aspect of watershed adjustment demands management of the system based on its present state and functioning rather than on how it functioned when the previous management practices were put in place and led to unintended consequences (which have almost certainly changed the state of the system).

In sum, the message presented by the history of management efforts in Goodwin Creek is the necessity of *coordinated* watershed management. The many efforts in Goodwin Creek demonstrate the development of technical expertise and effectiveness in addressing local issues of upland erosion and channel stability. The more general challenge is to coordinate this expertise throughout the physically coupled system so that one fix does not make another problem worse. Evidence has been presented that Goodwin Creek channels have not yet stabilized to current channel gradients, runoff rates, and sediment loads. The continued construction of private ponds within the watershed may be an effective means of reducing runoff rates in the near term. However, only time, perhaps a century or more, will allow the soil depths and infiltration capacities of degraded areas now covered by forestlands to gradually improve, thereby reducing hydrograph peaks and increasing base flow, and for the stream channel system to gradually evolve to a stable lower-gradient state.

REFERENCES

Barry, J.M. 1997. *Rising Tide: The Great Mississippi Flood of 1927 and How It Changed America*. Simon & Schuster, New York, NY, 528 pp.

Bennett, H.H. 1939. *Soil Conservation*. McGraw-Hill Book Co., New York. 993 pp.

Bingner, R.L., R.A. Kuhnle, and C.V. Alonso. 2007. Goodwin Creek experimental watershed: A historical perspective. In *Environmental and Water Resources: Milestones in Engineering History*, ed. Rogers, J.R. ASCE, Reston, VA, pp. 113–117.

Bowie, A.J. 1982. Investigations of vegetation for stabilizing eroding streambanks. *Transactions of the ASAE* 25(6): 1601–1606.

Bowie, A.J. 1995. Use of vegetation to stabilize eroding streambanks. U.S. Department of Agriculture, Conservation Research Report. 43, 24 pp.

CTIC. 2002. National Crop Residue Management Survey. Conservation Technology Information Center, West Lafayette, IN. http://www.ctic.purdue.edu/CTIC/CRM.html (accessed April 1, 2005).

Dabney, S.M., M.T. Moore, and M.A. Locke. 2006. Integrated management of in-field, edge-of-field, and after-field buffers. *Journal of the American Water Resources Association*, 42(1): 15–24.

Dudley, T.L. 2000. Noxious wildland weeds of California: *Arundo donax*. In: *Invasive Plants of California's wildlands*, eds. C. Bossard, J. Randall, and M. Hoshovsky.

Duffy, P.D. and S.J. Ursic. 1991. Land rehabilitation success in the Yazoo Basin, USA. *Land Use Policy* 8(3): 196–205.

Grissinger, E.H. and A.J. Bowie. 1984. Material and site controls of stream bank vegetation. *Transactions of the ASCE* 27(6): 1829–1835.

Grissinger, E.H. and J.B. Murphey. 1983. Present channel stability and late Quaternary valley deposits in northern Mississippi. Special Pub. #6. International Association of Sedimentologists, pp. 241–250.

Grissinger, E.H., A.J. Bowie, and J.B. Murphey. 1991. Goodwin Creek instability and sediment yield. In *Proceedings of the Fifth Federal Interagency Sedimentation Conference*, Vol. 2, PS-32–PS-39. U.S. Gov. Print. Off., Washington, DC.

Grissinger, E.H., J.B. Murphey, and W.C. Little. 1981. Problems with Eocene stratigraphy in Panola County, northern Mississippi. *Southeastern Geology*, 22(1): 19–29.

Happ, S.C. 1937. Fertile valleys laid waste by upland erosion. *Soil Conservation* 2(9): 194–198.

Happ, S.C. 1975. Valley sedimentation as a factor in sediment-yield determinations. In *Present and Prospective Technology of Predicting Sediment Yields and Sources, Proceedings of the Sediment-Yield Workshop*. USDA Sedimentation Laboratory, Oxford, MS, November 28–30, 1972. ARS-S-40. pp. 57–60.

Happ, S.C., G. Rittenhouse, and G.C. Dobson. 1940. Some principles of accelerated stream and valley sedimentation. Technical Bulletin Number 695. USDA, Washington, DC.

Hilgard, E.W. 1860. Report on the geology and agriculture of the state of Mississippi. E. Barksdale, State Printer, Jackson, MS, 391 pp.

Hilgard, E.W. 1884. Report on the cotton production of the state of Mississippi, with a discussion of the general agricultural features of the state. U.S. Bur. of the Census, 10th Census U.S., 1880, pp. 197–366.

King, D.B., S.J. Ursic, and J.L. Smith. 1956. Project analysis: Watershed management research in the mid-South. Southern Forest Experiment Station, USDA-FS, Tallahatchie Research Center, Oxford, MS, 53 pp.

Kuhnle, R.A., R.L. Bingner, G.R. Foster, and E.H. Grissinger. 1996. Effect of land use changes on sediment transport in Goodwin Creek. *Water Resources Research* 32(10): 3189–3196.

Kuhnle, R.A., R.L. Bingner, C.V. Alonso, C.G. Wilson, and A. Simon.2008. Conservation practice effects on sediment load in the Goodwin Creek experimental watershed. *Journal of Soil and Water Conservation* 63(6): 496–503.

Little, W.C., C.R. Thorne, and J.B. Murphey.1982. Mass bank failure analysis of selected Yazoo Basin Streams. *Transactions of the ASAE* 25(5): 1321–1328.

Lowe, E.N. 1910. Our waste lands, a preliminary study of erosion in Mississippi. Mississippi Geological Survey. Brandon Publishing, Nashville, TN, 23 pp.

Neill, C.R. and J.P. Johnson. 1989. *Long Creek Watershed Field Investigation and Geomorphic Analysis*. Northwest Hydraulic Consultants, Inc., Kent, WA.

Olsen, J.T. and L.D. Dumm. 1941. Summary report of organized drainage districts in Mississippi. Mississippi Board of Development and USDA-SCS, a summary report of Work Projects Administration Projects, O.P. No. 665-62-3-130 and O.P. No. 65-1-62-60. Investigation of organized drainage districts. Jackson, MS, 31 pp.

Pezeshki, S.R. and F.D. Shields Jr. 2006. Black willow cutting survival in streambank plantings, southeastern United States. *Journal of the American Water Resources Association* 42(1): 191–200.

Shields Jr, F.D. 2008. Effects of a regional channel stabilization project on suspended sediment yield. *Journal of Soil and Water Conservation* 63(2): 59–69.

Shields Jr, F.D., S.S. Knight, and C.M. Cooper. 1994. Effects of channel incision on base flow stream habitats and fishes. *Environmental Management* 18(1): 43–57.

Shields Jr, F.D., A.J. Bowie, and C.M. Cooper. 1995. Control of streambank erosion due to bed degradation with vegetation and structure. *Water Resources Bulletin*. 31(3): 475–489.

Shields Jr, F.D., S.S. Knight, and C.M. Cooper. 1998a. Rehabilitation of aquatic habitats in warmwater streams damaged by channel incision in Mississippi. *Hydrobiologia* 382: 63–86.

Shields Jr, F.D., S.S. Knight, and C.M. Cooper. 1998b. Addition of spurs to stone toe protection for warmwater fish habitat rehabilitation. *Journal of the American Water Resources Association* 34(6): 1427–1436.

Shields Jr, F.D., S.S. Knight, and C.M. Cooper. 2007. Can warmwater streams be rehabilitated using watershed-scale standard erosion control measures alone? *Environmental Management* 40: 62–79. DOI 10.1007/s00267-006-0191-0.

Shields Jr, F.D., E.J. Langendoen, R.E. Thomas, and A. Simon. 2011. Cyclical fluvial response caused by rechannelization. *Proceedings, World Environmental and Water Resources Congress 2011*. American Society of Civil Engineers, Reston, VA. (CD-ROM).

Simon, A. 1989. A model of channel response in disturbed alluvial channels. *Earth Surface Processes and Landforms* 14(1): 11–26.

Simon, A., D. Derrick, C.V. Alonso, and N.L. Bankhead. 2008. Application of a deterministic bank stability model to design a reach-scale restoration project. *Environmental and Water Resources Institute World Congress Proceedings*. Honolulu, HI, May 2008. 10 p. (CD-ROM).

Snider, J.A. 1996. Biotechnical erosion control. Technical note 12(2), USDA-NRCS Jamie L. Whitten Plant Materials Center, Coffeeville, MS, 8 pp.

U.S. Army Corps of Engineers. 1981. Final Report to Congress–The Streambank Erosion Control Evaluation and Demonstration Act of 1974, Section 32, Public Law 93-251, Appendix F—Yazoo Basin Projects. U.S. Army Corps of Engineers, Vicksburg, MS.

USDA-SCS. 1961. Report on survey and investigation of and water resources, Yazoo-Mississippi River Basin. Part I: General report and summary analysis of land use and water resources. USDA-SCS in cooperation with Mississippi Board of Water Commissioners, Jackson, MS. 106 p.

USDA-SCS. 1963. Soil Survey of Panola County, Mississippi, Series 1960, No. 10, 63 pp., 123 maps.

Walling, D.E. and B.W. Webb (eds). 1996. Erosion and sediment yield: Global and regional perspectives. International Association for Hydrological Sciences, Wallingford, UK.

Watson, C.C., N.K. Raphelt, and D.S. Biedenharn. 1997. Historical background of erosion problems in the Yazoo Basin. In *Proceedings of the Conference on Management of Landscapes Disturbed by Channel Incision*, eds. S.Y. Wang, E.J. Langendoen, and F.D. Shields Jr, pp. 115–119, May 20–22, 1997, Oxford, MS, University of of Mississippi.

Whitten, C.B. and D.M. Patrick. 1981. Engineering geology and geomorphology of streambank erosion (report 2: Yazoo River Basin uplands, Mississsippi), Tech Report GL-79-7 (report 2 of a series). Geotechnical Laboratory, U.S. Army Engineer Waterways Experiment Station, Vicksburg, MS.

Williston, H.L. 1988. The Yazoo–Little Tallahatchie Flood Prevention Project. A history of the Forest Service's role. U.S. Forest Service. Southern Forest Experiment Station. Forestry Report R8-FR 8, February 1988, 63 pp.

Wilson, C.G., R.A. Kuhnle, D.D. Bosch, J.L. Steiner, P.J. Starks, M.D. Tomer, G.V. Wilson. 2008. Quantifying relative contributions from sediment sources in conservation effects assessment project watersheds. *Journal of Soil and Water Conservation* 63(6): 523–532.

Section VII

Research and Development Priorities

21 Toward Enhancing Storage of Soil Water and Agronomic Productivity

Rattan Lal

CONTENTS

21.1 INTRODUCTION

The renewable freshwater resources (RFWR) are limited (Table 21.1) and are determined by the global hydrological cycle (Oki and Kanae 2006). With the increase in population from 7 billion in 2011 to 9.2 billion in 2050, there will be a greater competition for water use by agriculture than for industrial and urban uses. There will also be additional water use for improving the environment (i.e., C sequestration) for nature conservancy (i.e., restoration of wetlands, and wildlife habitat) and for recreational purposes. Limiting and decreasing water resources must be judiciously managed for both people and nature (Postel and Richter 2003). The availability of freshwater resources will also be confounded by the projected climate change. The hydrological cycle is likely to be intensified by climate warming (Huntington 2010), with possible adverse effects on the availability of fresh water for competing uses. Thus, sustainable management of water resources, using the strategies of integrated water management (IWM) (Bouwer 2002) and ecohydrological approach (Falkenmark and Rockström 2005) among others, is essential to meeting the growing demands. While agriculture currently uses about 70% of the total water withdrawal (Molden 2007), decline in its share of the scarce resource is inevitable because of numerous competing but essential uses. Estimates of calories produced per cubic meter of water range from 1000 to 7000 for corn, 1260 to 3360 for legumes, 500 to 2000 for rice, and 60 to 210 for beef (Molden et al. 2007). Therefore, understanding the hydrological processes in agricultural and related ecosystems is essential to improving the use efficiency by decreasing the losses. The themes of Chapters 1–20 presented in this volume address the scientific processes, technological options, and policy interventions to enhance effective water use. This chapter is focused on addressing the research and development priorities in improving soil water storage for enhancing and sustaining agronomic productivity with the goal to advance the global food security, while improving the environment.

TABLE 21.1
Global Reserves of Water

Reservoir	Storage (10^3 km^3)
Sea	1,338,000
Glaciers and snow	24,064
Permafrost	300
Lakes	175
Wetlands	17
Soils	17
Water vapor	13
• Over ocean	(10)
• Over land	(3)
River	2
Biological	1
Total	*24,589*

Source: Recalculated from Oki, T. and Kanae, S., *Science*, 313, 1068–1072, 2006.

21.2 FACTORS AFFECTING SOIL WATER STORAGE

There is a large water footprint for agricultural production (Khan and Hanjra 2009), and its magnitude differs widely among ecoregions, land uses, and soil types. Soil water, its storage, and its efficient use are critical to reducing the footprint of agriculture. Global soil water storage is estimated

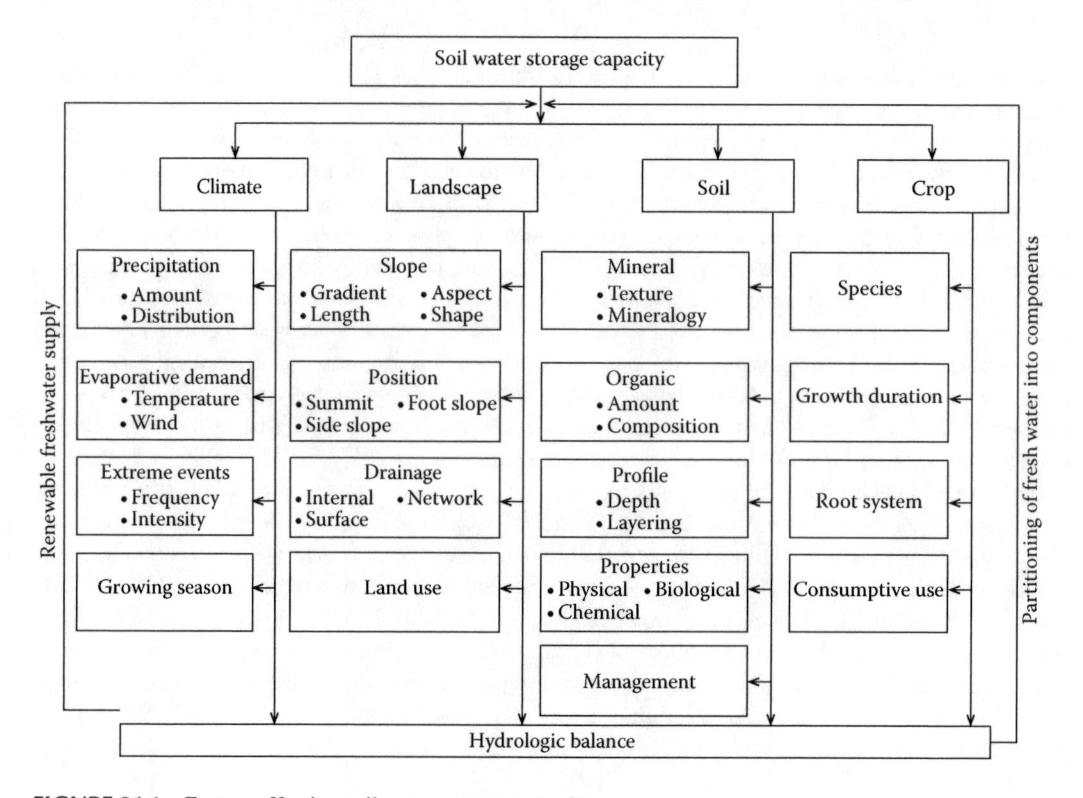

FIGURE 21.1 Factors affecting soil water storage capacity.

TABLE 21.2
Factors Affecting Capacity and Use Efficiency of Soil Water

Soil Properties	Soil Management	Crop Management	Water Management
Texture	Tillage methods, nutrients	Root system	Drainage
Structure	Mulching, conservation agriculture, cover crops, biomass	Crop rotations, farming system	Quality of irrigation water
Soil organic matter	Compost, manure, INM, mulching, conservation tillage	Complex crop rotations, agroforestry, agropastoral systems	Water conservation, irrigation, wastewater use
Compaction	Guided traffic, chiseling, soil fauna	Deep-rooted crops	Runoff control, infiltration management
Low water retention	Mulching (plastic, gravel, crop residues), conservation tillage	Crops and cropping systems, time of sowing	Water harvesting microirrigation, wastewater use
Low soil fertility	Integrated nutrient management, balanced nutrient application	Legume-based rotations, grazing management	Soil water conservation, irrigation with treated gray water
High salt concentration	Conservation tillage, mulching, balanced nutrient application, composting	Tolerant crop species, rotation, use of saltbush, halophytes	Microirrigation using marginal and saline water for irrigation

at 17×10^3 km^3 (Oki and Kanae 2006). Among three types of drought (i.e., meteorological, hydrological, and agronomic), agronomic drought depends on the soil's available water capacity (AWC), and meteorological drought is caused by climatic factors. While building resilience in drought-prone agriculture is important (Falkenmark and Rockström 2008), it can be realized through increasing soil water storage and enhancing its water productivity (WP) and use efficiency. The soil water storage of 17×10^3 km^3 (Table 21.1) is the source of water for all terrestrial life. It is augmented by storage in glaciers/snow, lakes, permafrost, wetlands, rivers, and atmospheric water vapor. Yet, it is crucial to all life. There is a wide range of factors that affect capacity and use efficiency of soil water (Figure 21.1 and Table 21.2). Principal among soil factors are profile/solum depth, texture, mineralogy, structure, soil organic matter (SOM) content, AWC, salt concentration, and nutrient reserves and availability. While the solum thickness and the effective rooting depth are determined by natural forces, other factors (i.e., soil structure, fertility, compaction, salt concentration, and SOM content) can be managed to enhance AWC and its use efficiency. There are several soil, crop, and water management options (Table 21.2). Important among these are those that enhance soil and water conservation (i.e., conservation tillage, mulch farming, contour hedges), soil fertility, and nutrient reserves (i.e., integrated nutrient management [INM] involving manuring, composting, biological N fixation, mycorrhizae), and supplemental irrigation (see Section 21.4). The goal is to restore degraded soils and desertified ecosystems to improve water infiltration and AWC. Similar to soils, crops can be managed to enhance plant water uptake. The choice of an appropriate crop, along with cropping sequence and crop combinations, is important to moderate growth duration and consumptive water use. The latter is also influenced by the nature of root systems and soil properties. Factors (Figure 21.1) and management options (Table 21.2) can be managed to influence the hydrologic balance at soilscape, landscape, and watershed scale to improve water availability for agroecosystems. Management also depends on economic and social factors, which comprise the human dimensions of natural resource management.

21.3 CLIMATE CHANGE AND SOIL WATER REGIME

While the anthropogenic increase in the atmospheric concentration of greenhouse gases (i.e., CO_2, CH_4, and N_2O) is undisputable, there is much uncertainty about the magnitude of changes in climate, soil properties, and hydrological parameters. It is probable that the projected climate change and increase in temperature and evaporation may adversely affect the supply of RFWR through alterations in a wide range of ecosystem attributes and processes (Figure 21.2). Notable among these are alterations in microclimatic and mesoclimatic factors (i.e., precipitation amount and distribution, evaporation, run-on/runoff and groundwater recharge, and the components of hydrologic balance). There are widespread reports of decrease in rainfall throughout the Mediterranean region, Sahel, southern Africa, the Aral Sea Basin, and Australia, since the early twentieth century (Ragab and Prudhomme 2002). A decline in precipitation, in conjunction with an increase in temperature and evaporation, would severely jeopardize the availability of freshwater resources. The adverse effects on soil water storage will be exacerbated by a decline in soil hydraulic properties and processes (i.e., moisture retention characteristics, infiltration capacity, and rooting depth) and changes in land use and plant species. It is widely argued that climate change may intensify the hydrologic cycle (Huntington 2010) and increase evaporation, evapotranspiration, atmospheric water pool, and (in some cases) total precipitation. However, global trends in soil water storage are uncertain because of the effects of other confounding factors (Figure 21.2). In addition to soil factors, increase in frequency and intensity/duration of extreme events (i.e., drought, runoff, erosion, and floods) may also affect the soil water storage (Huntington 2010). These factors may adversely affect the water resources in the Near East (region covering 29 countries and an area of 18.9×10^6 km^2, from Mauritania/Morocco in the west to Pakistan in the east), which is about 14% of the world's total area (Ragab and Prudhomme 2002). Although the rainfall over the Sahara may increase, it is of little agronomic consequence. The average annual precipitation may also decrease in the Thar Desert (India, Pakistan, and Afghanistan).

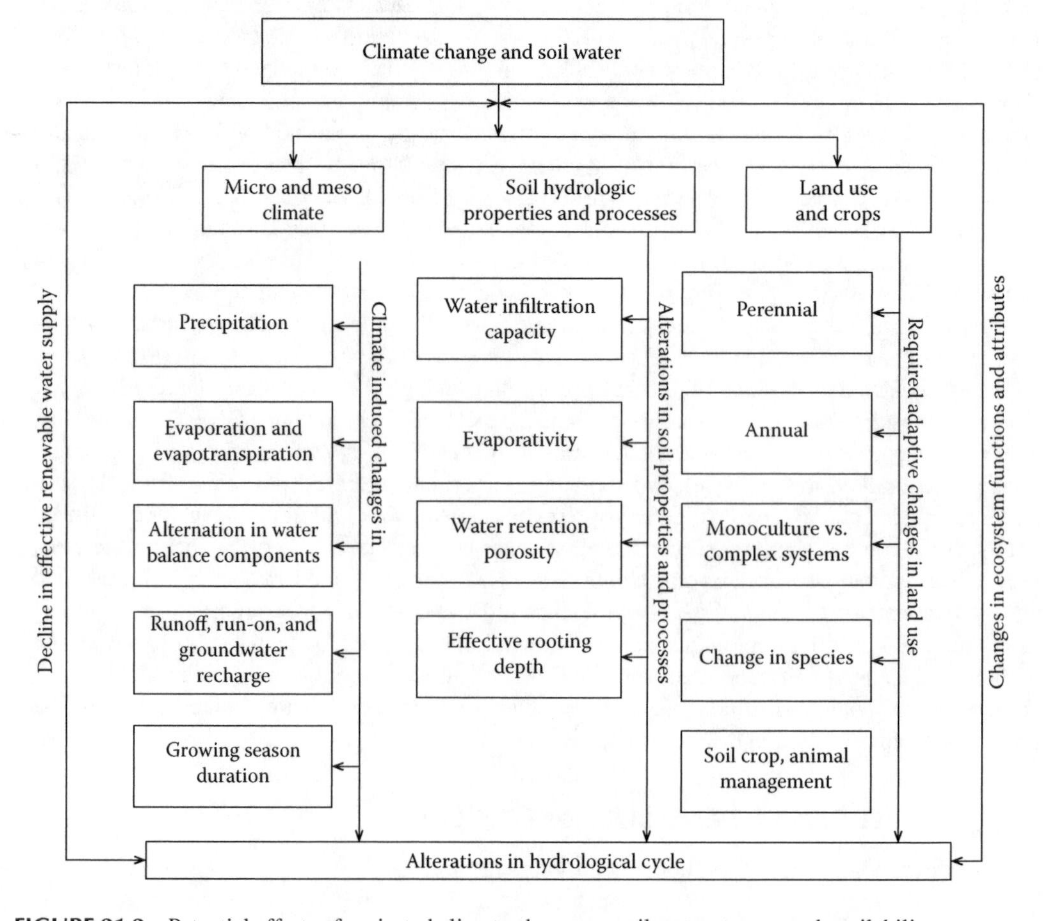

FIGURE 21.2 Potential effects of projected climate change on soil water storage and availability.

21.4 AGRONOMIC PRODUCTIVITY

Implications of climate change on soil water storage and agronomic productivity (Rosenzweig and Hillel 1998) are not clearly understood. A major uncertainty is due to the confounding interaction with CO_2 enrichment or the CO_2 fertilization effect (Polley 2002), which itself depends on the availability of soil moisture and plant nutrients. Soil- and site-specific research is needed for the measurement and modeling of the agronomic effects. The role of soil/pedospheric processes on climate-induced changes in agronomic productivity cannot be overemphasized. For example, changes in SOM content, because of increase in temperature-related decomposition, on risks of accelerated soil erosion, can affect soil quality and use efficiency of water and nutrients. Decline in SOM content to below the threshold level may have severe agronomic, ecologic, and economic implications. While increasing the terrestrial C pool (soil and biota) is considered a win-win option, there are notable trade-offs between C and water. Afforestation and establishing plantations, as well as options for increasing the C pool of degraded and desertified ecosystems, may decrease the stream flow and also exacerbate risks of soil salinization and acidification (Jackson et al. 2005).

The agronomic data, based on well-designed field experiments to assess the impact of climate change on crop/tree/animal performance, are scanty. The scarcity of data is specifically imminent in climate-induced changes in soil properties and plant AWC. The latter is affected by a range of interacting soil properties and processes (Figure 21.3). Rather than a sole determinant, effects of soil properties and processes are highly interactive both within and outside the pedosphere. While assessment of the changes within the pedosphere is a challenge in itself, it is even more difficult for those caused by interaction with the environment (Figure 21.4). Specific interactive processes that are a high researchable priority include the CO_2 fertilization effect (in relation to water and nutrients supply), biological weathering by bioturbation processes, transport of dissolved and suspended materials within and

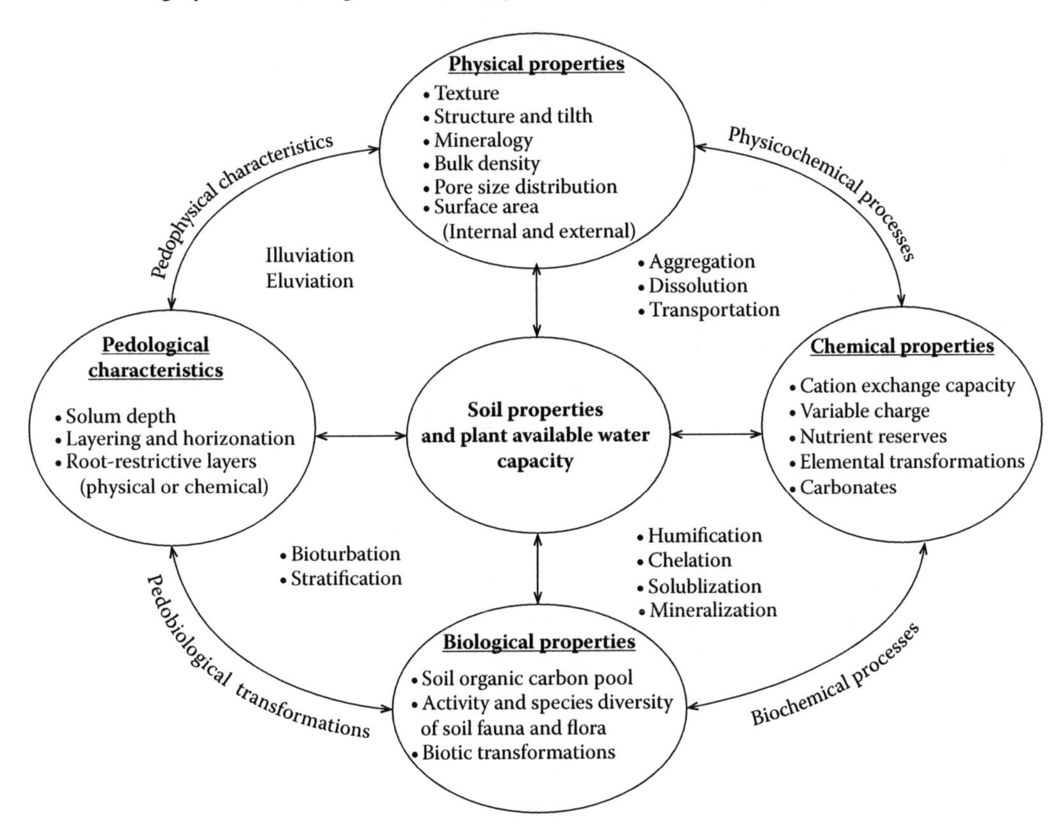

FIGURE 21.3 Potential effects of climate change on soil properties and plant available water capacity.

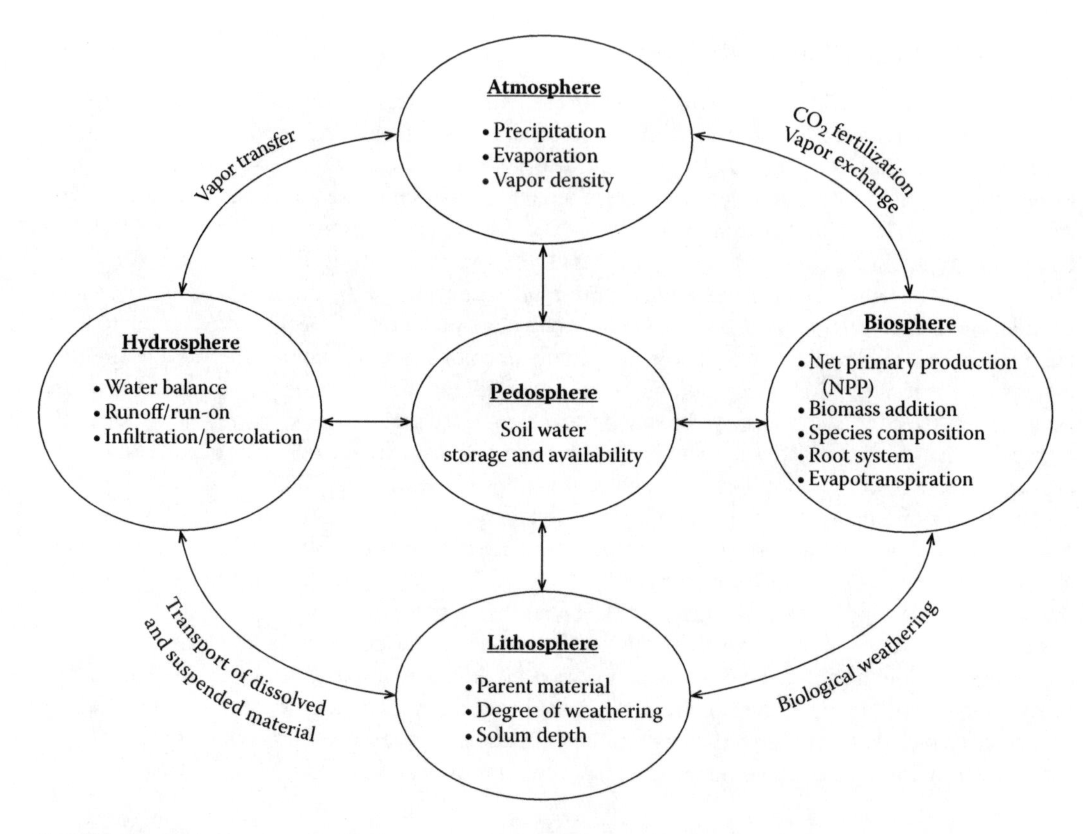

FIGURE 21.4 Interactive effects of pedosphere with the environment in relation to soil water storage and availability.

outside of the pedosphere, and vapor exchange between the atmosphere on one hand and biosphere and hydrosphere on the other (Figure 21.4). Soil- and ecoregion-specific data are also needed concerning the interaction between the pedosphere and the atmosphere for gaseous exchange (CO_2, CH_4, and N_2O), biosphere for nutrient/elemental cycling (N, P, and S), lithosphere for material transport, and hydrosphere for leaching (illuviation and eluviation) and erosion (Figure 21.4).

It is the complexities outlined in Figures 21.3 and 21.4 that create numerous uncertainties with regard to the impact of climate change on agronomic productivity. The effects on agronomic productivity, discussed by Rosenzweig and Hillel (1998) and Ragab and Prudhomme (2002), may differ among temperate and tropical regions (Table 21.3). The adverse effects on agronomic production,

TABLE 21.3
Potential Effects of Climate Change on Agronomic Productivity

Parameter	Temperate/Boreal	Tropics/Subtropics
Growing season duration	Increase	Decrease
CO_2 fertilization effect	Positive	Negative on grain yield (positive on biomass)
Pest incidence	Increase	Increase
Water deficit/drought	Uncertain	Exacerbate
Net radiation	Uncertain	Decrease
Soil water storage	Uncertain	Decrease
Agronomic yield	Increase	Decrease

more severe in the tropics than in temperate climate, may aggravate the food insecurity, which is already a serious problem in the tropics.

21.5 ENHANCING USE EFFICIENCY AND PRODUCTIVITY IN RAINFED AND IRRIGATED AGRICULTURE

Irrigated agriculture, with an arable land area of <20%, produces 40% of the world's agronomic output comprising cereals, legumes, roots, and tubers. Therefore, sustainable management of irrigated agriculture is a high priority, especially in consideration of the fact that less fresh water may be available for maintaining the existing area and bringing new area under irrigation. The proposed 30% increase in irrigated area by 2030 (Kijne 2001) can only be implemented by diverting surface (blue water) and municipal (gray water) water to soil water (green water). With rapidly increasing and competing demands for blue water, there will be an increasing use of brackish water (refer to Chapter 11) and municipal wastewater (Jueschke et al. 2008). However, the use of municipal water for agriculture would necessitate pretreatment to minimize health risks (WHO 2006). Such facilities for treating wastewater are not available in developing countries of the tropics and subtropics, where the risks of health hazards are already serious.

The WP under traditional flood irrigation is low. Considerable savings in water can be realized by adopting modern irrigation systems including drip subirrigation, other microirrigation techniques, and partial root-drying strategies (Lindbolm and Nordell 2006). Subsurface irrigation by condensation of humid air (condensation irrigation or CI) is also an option, especially for hot/arid regions. The CI is a system that combines a solar desalination system with subirrigation (Lindbolm and Nordell 2006). Solar stills are used to evaporate brackish water, and the humid air is then carried into a network of underground pipes to condense it into liquid water at cooler temperature. The condensed water and humid air irrigates the root zone of a crop. As much as 3 kg (3L) of H_2O can be condensed per meter of the cooling pipe (Lindbolm and Nordell 2006).

Rice paddy, grown under flooded conditions in a puddled soil, is a water-intensive production system. The rice–wheat system practiced in the Indo-Gangetic Plains of South Asia (Chapter 6) has severely depleted the groundwater reserves (Rodell et al. 2009; Kerr 2009). Severe depletion of groundwater is also reported in China (Chapter 12), the Middle East (Chapter 7), and the U.S. Great Plains (Chapters 4 and 5). With adequate weed control and seeding equipment, specific varieties of aerobic rice can be grown under upland conditions, similar to the cultivation of wheat. There are other water management options for rice, especially suited to conditions of water scarcity (Bouman et al. 2007).

Water harvesting and recycling for supplemental irrigation have been practiced for millennia, especially in arid regions of Asia. Water harvesting, used in conjunction with innovated irrigation technologies discussed earlier, is important in enhancing WP. A runoff-based agroforestry system can also improve water-use efficiency (WUE) in arid climates. The strategy is to combine deep-rooted perennials with shallow-rooted annuals (Droppelman et al. 2000).

Using plant hydraulic lift by perennials to benefit annuals grown in arid climates is another option. Roots of some desert trees (i.e., shepherd's tree or *Boscia albitrunca*) can extract water from 68 m below the surface (Canadell et al. 1996), because plant roots can carry water from deep soil layers (Clothier and Green 1997). Termites and other soil fauna (harvester ants) also carry water from subsoil to their nests on the surface. The process by which water is transported upward through root systems from the subsoil (moist layer) to the surface soil (dry layer) is termed hydraulic lift (Liste and White 2008). Water transport by the hydraulic lift of perennials with a deep root system can be used to cultivate annuals grown in association (as an agroforestry system) to produce some agronomic yields even under harsh climates where complete crop failure is inevitable (Espeleta et al. 2004; Oliveira et al. 2005). The process can be used in ecoregions that experience frequent surface droughts (agronomic drought) but have water reserves deep in the subsoil (i.e., Mediterranean climates, arid and cool temperate regions, seasonally dry tropical and subtropical environments, and the Mojave Desert). The magnitude and pattern of hydraulic lift may differ among perennials and companion annuals and also among soil types and the depth of the moist layer.

21.6 RESEARCH AND DEVELOPMENT PRIORITIES

While the general principles of soil, water, and crop management may be known, site-specific research is essential to identify practices that can adapt to changing climate. Some researchable priorities (Table 21.4) have been discussed in the previous sections. The strategy is to validate and fine-tune the practices for sustainable management of soil (i.e., conservation tillage, mulch farming, nutrient requirement, AWC, and SOM management), crop (i.e., choice of species and varieties; time, configuration, and methods of sowing; consumptive water use; CO_2 fertilization effect; hydraulic lift; and growing season duration), and microclimate (i.e., temperature, precipitation, extreme event, hydraulic balance, and radiation budget). Adaptation is possible when basic information on soil, crops, and microclimate and mesoclimate is known. This information is needed at soilscape, landscape, and watershed (agricultural watersheds and river basin) scales.

Adoption of recommended management practices (RMPs) for saving water and enhancing WP can be promoted through payments to land managers for ecosystem services. Similar to soil C credits, green water credits is a mechanism for payments to land managers for adopting water-saving practices (Dent 2006). Incentivizing the land managers would save scarce water resources and minimize losses and wasteful use. Policies must also be identified and implemented so that groundwater and surface waters are not prone to the "tragedy of the commons." Undervaluing scarce water resources (surface waters and groundwater) can lead to their abuse (i.e., free electricity for pumping the groundwater for irrigation in Punjab, India). Land managers must be charged for the electricity and water used, and the cost must be eventually paid by the consumer. There is no such thing as free soil or free water. Channels of communications need to be established between researchers on the one hand and policy makers and land managers on the other. The lack of communication, due to weak institutions and inadequate extension/outreach, is a serious issue, which must be addressed.

TABLE 21.4
Site-Specific Researchable Priorities in Soil Water and Agronomic Productivity

Climate/Hydrology	Soil	Plant
Changes in temperature	Physical and hydrological properties	Choice of crops/plants, and rotational/spatial configuration
Alterations in hydrologic/energy balance	Potential and actual soil water storage capacities	Growing season duration, onset and end
Frequency of extreme events	Periods of soil water deficit	Nutrient–water interaction in relation to NPP
Probability of rainless periods (>10 days) during the growing season	Soil water regime in the subsoil	CO_2 fertilization effect
Predicting onset and cessation of rains during the growing season	Practices to reduce soil evaporation and runoff	Potential of the hydraulic lift
Runoff, infiltration, percolation relationships for extreme events	Techniques to enhance water productivity (aerobic rice, microirrigation, condensation irrigation)	Crop evaporative demand and consumptive use at different phenological stages

21.7 CONCLUSION

All terrestrial life depends directly or indirectly on soil water. The finite reserve, which comprises only 0.07% of renewable fresh water (17×10^3 km^3 out of a total of $24{,}589 \times 10^3$ km^3), may be prone to climate-induced alterations in the global hydrological cycle. The soil water reserve (green water) can be supplemented by the diversion of blue water (lakes and river) and recycling of adequately treated municipal (gray) water. Yet, soil water storage capacity and its efficient use depend on a wide range of site-specific factors related to climate, soil, land use, and cropping/farming systems. The potential effects of climate change on soil water storage and agronomic productivity also differ among ecoregions (temperate vs. tropical climate) and soil types. Therefore, it is important to identify site/soil-specific RMPs. Research data are needed with regard to the magnitude of projected climate change (temperature, precipitation, extreme events, growing season duration, and hydrologic balance), soil properties (physical and hydrologic properties, soil water storage capacity, periods of soil water deficit, and practices to enhance WP under rainfed and irrigated conditions), and plants (species, rotation sequences, seeding configuration, time of sowing, hydraulic lift, consumptive water use, nutrient–water interaction, and CO_2 fertilization effect). Policy interventions are needed to minimize excessive use of water, promote adoption of RMPs through payments for green water credit, and create a mechanism to charge a just price to the water user to avoid it being used as a common resource. Channels of communication must be developed between researchers, land managers, and policy makers for creating public awareness about the scarcity of this precious resource and the need for its sustainable use.

ABBREVIATIONS

AWC	Available water capacity
INM	Integrated nutrient management
IWM	Integrated water management
RFWR	Renewable freshwater resources
RMPs	Recommended management practices
WP	Water productivity
WUE	Water-use efficiency

REFERENCES

Bouman, B.A.M., R.M. Lampayan and T.P. Toung. 2007. *Water Management in Irrigated Rice.* Lost Banos, Philippines: IRRI.

Bouwer, H. 2002. Integrated water management for the 21st century: Problems and solutions. *J. Irrig. Drain. Eng.* 128(4): 193–201.

Canadell, J., R.B. Jackson, J.R. Ehleringer, et al. 1996. Maximum rooting depth of vegetation types at the global scale. *Oecologia* 108: 583–595.

Clothier, B.E. and S.R. Green. 1997. Roots: The big movers of water and chemicals in soil. *Soil Sci.* 162: 534–543.

Condon, A.G., R.A. Richards, G.J. Robetzke et al. 2002. Improving intrinsic water-use efficiency and crop yield. *Crop Sci.* 42: 122–131.

Dent, D. 2006. *Green Water Credits: Basin Identification.* Holland: ISRIC.

Droppelman, K.L., J. Lehmann, J.E. Ephrath, et al. 2000. Water use efficiency and uptake patterns in a runoff agroforestry system in an arid environment. *Agroforestry Systems* 49: 223–242.

Espeleta, J.F., J.B. West and L.A. Donovan. 2004. Species specific patterns of hydraulic lift in co-occurring adult trees and grasses in a sandhill community. *Oecologia* 138: 341–349.

Falkenmark, M. and J. Rockström. 2005. *Balancing Water for Human and Nature.* London: Earthscan.

Falkenmark, M. and J. Rockström. 2008. Building resilience to drought in desertification-prone savannas in sub-Saharan Africa: The water perspective. *Nat. Res. Forum* 32: 93–102.

Huntington, T.G. 2010. Climate-warming induced intensification of the hydrologic cycle: An assessment of the published record and potential impacts on agriculture. *Adv. Agron.* 109: 1–53.

Jackson, R.B., E.G. Jobbagy, R. Avissar, et al. 2005. Trading water for carbon with biological C sequestration. *Science* 310: 1944–1947.

Jueschke, E., B. Marschner, J. Tarchitzky, et al. 2008. Effects of treated wastewater irrigation on the dissolved and soil organic carbon in Israeli soils. *Water Sci. Tech.* 57.5: 728–733.

Kerr, R.A. 2009. North India's groundwater is going, going, going. *Science* 325: 798.

Khan, S. and M.A. Hanjra. 2009. Foot prints of water and energy inputs in food production-global perspectives. *Food Policy* 34: 130–140.

Kijne, J.W. 2001. Preserving the (water) harvest: Effective water use in agriculture. *Water Sci. Tech.* 43: 133–139.

Lindbolm, J. and B. Nordell. 2006. Subsurface irrigation by condensation. In *Sustainable Irrigation Management Technologies and Polices*, eds. G. Lorenzini and C.A. Brebbia, p. 390, South Hampton: WIT Press.

Liste, H.-H. and J.C. White. 2008. Plant hydraulics lift of soil water-implications for crop production and land restoration. *Plant Sci.* 31: 1–7.

Molden, D. 2007. *Water for Food, Water for Life.* Colombo, Sri Lanka/London: IWMI/Earthscan.

Molden, D., T.Y. Oweis, P. Steduto, et al. 2007. Pathways for increasing agricultural productivity. In *Comprehensive Assessment of Water Management for Agriculture: Water for Food, Water for Life*, ed. D. Molden, pp. 278–310, Colombo, Sri Lanka/London: IWMI/Earthscan.

Oki, T. and S. Kanae. 2006. Global hydrological cycles and world water resources. *Science* 313: 1068–1072.

Oliveira, R.A., T.E. Dawson, S.S.O. Burdess, et al. 2005. Hydraulic redistribution in three Amazonian trees. *Oecolgia* 145: 354–363.

Polley, H.W. 2002. Implications of atmospheric and climate change for crop yield and water use efficiency. *Crop Sci.* 42: 131–140.

Postel, S. and B. Richter. 2003. *Rivers for Life: Managing Water for People and Nature.* Washington, DC: Island Press.

Ragab, R. and C. Prudhomme. 2002. Climate change and water resources management in arid and semi-arid regions: Prospective and challenges for the 21st century. *Biosyst. Eng.* 81: 3–34.

Rodell, M., V. Isabella, and J.S. Famiglietti. 2009. Satellite-based estimated of groundwater depletion in India. *Nature* 460: 999–1002.

Rosenzweig, C. and D. Hillel. 1998. *Climate Change and the Global Harvest: Potential Impact on the Greenhouse Effect on Agriculture.* Oxford: Oxford University Press.

WHO. 2006. WHO guidelines for safe use of wastewater, excreta and grey water. Geneva, Switzerland: WHO/UNEP/FAO.

Index

D

G

H

I

J

L

M

N

O

P

R

S

T

U

V

W

Y